INTERNATIONAL AGENCY FOR RESEARCH
INTERNATIONAL PROGRAMME ON CHEMICAL SAFETY (WHO)
COMMISSION OF THE EUROPEAN COMMUNITIES

LONG-TERM AND SHORT-TERM ASSAYS FOR CARCINOGENS: A CRITICAL APPRAISAL

Reports of an ad-hoc Working Group which met in Lyon, 2-6 December 1985

EDITORS

R. MONTESANO, H. BARTSCH, H. VAINIO,
J. WILBOURN & H. YAMASAKI

MAIN RAPPORTEURS:

R.A. GRIESEMER & S. VENITT

IARC SCIENTIFIC PUBLICATIONS NO. 83

LYON 1986

The International Agency for Research on Cancer (IARC) was established in 1965 by the World Health Assembly, as in independently financed organization within the framework of the World Health Organization. The headquarters of the Agency are at Lyon, France.

The Agency conducts a programme of research concentrating particularly on the epidemiology of cancer and the study of potential carcinogens in the human environment. Its field studies are supplemented by biological and chemical research carried out in the Agency's laboratories in Lyon and, through collaborative research agreements, in national research institutions in many countries. The Agency also conducts a programme for the education and training of personnel for cancer research.

The publications of the Agency are intended to contribute to the dissemination of authoritative information on different aspects of cancer research.

Distributed for the International Agency for Research on Cancer
by Oxford University Press, Walton Street, Oxford OX2 6DP, UK

London   New York   Toronto
Delhi   Bombay   Calcutta   Madras   Karachi
Kuala Lumpur   Singapore   Hong Kong   Tokyo
Nairobi   Dar es Salaam   Cape Town
Melbourne   Auckland

Oxford is a trade mark of Oxford University Press

Distributed in the USA
by Oxford University Press, New York

ISBN 92 832 1183 9
ISSN 0300 5085

© International Agency for Research on Cancer 1986
150 cours Albert Thomas, 69372 Lyon Cedex 08, France

The authors alone are responsible for the views expressed in the signed articles in this publication. All rights reserved. No part of this publication may be reproduced, stored in a retrieval system, or transmitted, in any form or by any means, electronic, mechanical, photocopying, recording, or otherwise, without the prior permission of Oxford University Press

PRINTED IN SWITZERLAND

CONTENTS

Foreword 1
List of participants 3
Introduction 7

Bioassays for carcinogenicity in animals

    Report 1   Long-term assays for carcinogenicity in animals 13
    Report 2   Early preneoplastic lesions 85
    Report 3   Assays for initiating and promoting activities 103

Short-term assays to predict carcinogenicity

    Report 4   DNA damage and repair 129
    Report 5   Short-term assays using bacteria 143
    Report 6   The host-mediated assay 163
    Report 7   Assays for genetic changes in mammalian cells 167
    Report 8   Assays for germ-cell mutations in mammals 245
    Report 9   Mammalian cell transformation in culture 267
    Report 10 In-vitro assays that may be predictive of tumour-
       promoting agents 287
    Report 11 Assays for genetic changes in fungi 303
    Report 12 Assays for genetic activity in *Drosophila melanogaster* 351
    Report 13 Assays for aneuploidy in *Drosophila melanogaster* 395
    Report 14 Short-term assays for the analysis of body fluids and
       excreta 409
    Report 15 Metabolic activation 439
    Report 16 Statistical analysis of data from in-vitro assays of
       mutagenesis 457

Report 17   Testing of complex chemical mixtures 483

Use of structure-activity relationships in predicting carcinogenesis
    H.S. Rosenkranz, M.R. Frierson & G. Klopman 497

Genetic toxicology at the crossroads: a personal overview of the deployment of
   short-term tests
    B. Bridges 519

Deployment of short-term assays for the detection of carcinogens: genetic and
   molecular considerations
    C. Ramel 529

Subject index 553

FOREWORD

In 1980, the IARC published as Supplement 2 to the *IARC Monographs* a series of critical reports on the use of long-term and short-term assays for the detection of chemical carcinogens. A major objective was to encourage the scientific community to meet certain basic requirements when carrying out and reporting the results of these often expensive and time-consuming studies. The availability of results from adequately conducted and critically analysed experiments is essential to strengthening measures for the primary prevention of cancer. The present reports result from the deliberations of experts in chemical carcinogenesis and mutagenesis who worked together to bring up to date the previous publication, taking into consideration progress made since 1980.

I would like to acknowledge with gratitude the collaboration and financial support of the International Programme on Chemical Safety (WHO - ILO - UNEP) and of the Commission of European Communities. Particular thanks goes also to to the Chairmen and Vice-chairmen of the meeting, Dr P.N. Magee, Dr G. Della Porta, Dr M.L. Mendelsohn, Dr N.P. Napalkov and Dr M. Hofnung, and to the main rapporteurs, Dr R.A. Griesemer and Dr S. Venitt.

L. Tomatis, MD
Director
IARC

LIST OF PARTICIPANTS

Anders, F., Genetic Institute of the Justus-Liebig University, Heinrich-Buff-Ring 58-62, 6300 Giessen, Federal Republic of Germany

Bannasch, P., Deutsches Krebsforschungszentrum, Abteilung für Cytopathologie, 69 Heidelberg 1, Federal Republic of Germany

Barrett, J.C., Laboratory of Pulmonary Pathobiology, Department of Health and Human Services, National Institute of Environmental Health Sciences, PO Box 12233, Research Triangle Park, NC 27709, USA

Bartsch, H., Unit of Environmental Carcinogens and Host Factors, International Agency for Research on Cancer

Becker, R., Unit of Mechanisms of Carcinogenesis, International Agency for Research on Cancer

Becking, G., World Health Organization, International Programme on Chemical Safety, Interregional Research Unit, PO Box 12233, MD A206, Research Triangle Park, NC 27709, USA

Breslow, N., Department of Biostatistics SC-32, School of Public Health and Community Medicine, University of Washington, Seattle, WA 98195, USA

Cabral, J.R.P., Unit of Mechanisms of Carcinogenesis, International Agency for Research on Cancer

Carere, A., Unit of Mutagenesis and Carcinogenesis, Istituto Superiore di Sanita, Viale Regina Elena 299, 00161 Rome, Italy

Chu, E.H.Y., Department of Biological Sciences, 223 Herrin Laboratories, Stanford University, Stanford, CA 94305, USA

Connors, T.A.[1], MRC Toxicology Unit, Medical Research Council Laboratories, Woodmansterne Road, Carshalton, Surrey SM5 4EF, UK

De Carli, L., Dipartimento di Genetica e Microbiologia, Universita di Pavia, Via San Epifanio 14, 27100 Pavia, Italy

Della Porta, G., Istituto Nazionale per lo Studio e la Cura dei Tumori, Via Venezian 1, 20133 Milano, Italy

Ehling, U.H., Institut für Genetik, Gesellschaft für Strahlen- und Umweltforschung MBH, Ingolstadter Landstrasse 1, Neuherberg, 8042 Oberschleissheim, Federal Republic of Germany

Evans, H.J., MRC Clinical and Population Cytogenetics Unit, Western General Hospital, Crewe Road, Edinburgh EH4 2XU, UK

---

[1]Unable to attend

LIST OF PARTICIPANTS

Feron, V.J., Department of Biological Toxicology, TNO-CIVO Toxicology and Nutrition Institute, Utrechtsweg 80, Postbus 360, 3700 AJ Zeist, The Netherlands

Fuchs, R.P.P., Institut de Biologie Moléculaire et Cellulaire, 15 rue Descartes, 67084 Strasbourg Cedex, France

Griesemer, R.A., Biology Division, Oak Ridge National Laboratory, PO Box Y, Oak Ridge, TN 37831, USA

Hayashi, M., Unit of Biostatistics and Field Studies, International Agency for Research on Cancer

Heseltine, E., Editorial and Publications Services, International Agency for Research on Cancer

Henschler, D., Institut für Pharmakologie und Toxikologie der Universität Wurzburg, Versbacherstrasse 9, 8700 Wurzburg, Federal Republic of Germany

Hofnung, M., Institut Pasteur, 28 rue du Dr Roux, 75724 Paris Cedex 15, France

Ito, N., First Department of Pathology, Nagoya City University Medical School, 1 Kawasumi, Mizuho-cho, Mizuho-ku, Nagoya 467, Japan

Kaldor, J., Unit of Biostatistics and Field Studies, International Agency for Research on Cancer

Kakunaga, T., Department of Oncogene Research, Research Institute for Microbial Diseases, Osaka University, 3-1 Yamadaoka, Suita, Osaka 565, Japan

Kroes, R., National Institute of Public, Health and Environmental Hygiene, PO Box 1, 3720 BA Bilthoven, The Netherlands

Kuroki, T., Department of Cancer Cell Research, The Institute of Medical Science, The University of Tokyo, 4-6-1 Shirokanedai, Minato-ku, Tokyo 108, Japan

Lambert, B., Department of Clinical Genetics, Karolinska Sjukhuset, 104 01 Stockholm 60, Sweden

Magee, P.N. (*Chairman*), Fels Research Institute, Temple University School of Medicine, Philadelphia, PA 19140, USA

Malaveille, C., Unit of Environmental Carcinogens and Host Factors, International Agency for Research on Cancer

Matsushima, T., Department of Molecular Oncology, Institute of Medical Science, University of Tokyo, 4-6-1 Shirokanedai, Minato-ku, Tokyo 108, Japan

McKnight, B., Unit of Biostatistics and Field Studies, International Agency for Research on Cancer

Mendelsohn, M.L., Biomedical and Environmental, Sciences Division, Lawrence Livermore National Laboratory, University of California, POB 5507, Livermore, CA 94550, USA

LIST OF PARTICIPANTS

Mercier, M., International Programme for Chemical Safety, WHO, 1211 Geneva 27, Switzerland

Mohr, U.[1], Institut für experimentelle Pathologie, Medizinische Hochschule Hannover, Karl-Wiechert-Allee 9, Postfach 610180, 3000 Hannover, Federal Republic of Germany

Montesano, R., Unit of Mechanisms of Carcinogenesis, International Agency for Research on Cancer

Morgenroth, V.H.M., Chemicals Division Environment Directorate, Organization for Economic Cooperation and Development, 2 rue Andre Pascal, 75775 Paris Cedex 16, France

Moustacchi, E., Section de Biologie, Institut Curie, 26, rue d'Ulm, 7523, Paris Cedex 05, France

Napalkov, N. P. (*Vice-chairman*), Petrov Research Institute of Oncology, 68 Leningradskaya St, Pesochny 2, 188646 Leningrad, USSR

Nesnow, S., Carcinogenesis and Metabolism Branch, MD-68, US Environmental Protection Agency, Health Effects Research Laboratory, Research Triangle Park, NC 27711, USA

Neubert, D., Institut für Toxikologie und Embryonalpharmakologie, Freie Universität Berlin, Garystrasse 5, 1000 Berlin (West) 33, Federal Republic of Germany

Ott, H.[1], Environment Research Programme, Commission of the European Communities, Rue de la Loi 200, 1049 Brussels, Belgium

Partensky, C., Unit of Carcinogen Identification and Evaluation, International Agency for Research on Cancer

Pegg, A.E., Department of Physiology, The Milton S. Hershey Medical Center, The Pennsylvania State University, PO Box 850, Hershey, PA 17033, USA

Rajewsky, M.F., Institut für Zellbiologie (Tumorforschung), Universitat Essen, Hufelandstrasse 55, 4300 Essen 1, Federal Republic of Germany

Ramel, C., Wallenberg Laboratory, Stockholms Universitet, 10691 Stockholm, Sweden

Rao, G.N., Laboratory Animal Management, Chemical Pathology Branch, Toxicology Research and Testing Program, National Institute of Environmental Health Sciences, PO Box 12233, Research Triangle Park, NC 27709, USA

Roberfroid, M., Departement de Biochimie et de Biologie Cellulaire, Faculté de Medecine, Université Catholique de Louvain, UCL 7369, Avenue E. Mounier 73, 1200 Brussels, Belgium

---

[1]Unable to attend

Rosenkranz, H.S., Case Western Reserve University, Centre for the Environmental Health Sciences, School of Medicine, Cleveland, OH 44106, USA

Slaga, T.[1], University of Texas System Cancer Center, Science Park, Research Division, PO Box 389, Smithville, TX 78957, USA

Thilly, W.G., Department of Applied Biological Sciences, Massachusetts Institute of Technology, E18-666, Cambridge, MA 02139, USA

Tomatis, L., Director, International Agency for Research on Cancer

Trosko, J.E., Department of Pediatrics and Human Development, Michigan Sate University, College of Human Medicine, East Lansing, MI 8824, USA

Turusov, V.S.[1], All-Union Cancer Research Center, The USSR Academy of Medical Sciences, Kashirskoye shosse 6, Moscow 115478, USSR

van der Venne, M.T., Health and Safety Directorate, Directorate General of Social Affairs and Education, Commission of the European Communities, Jean Monnet Building C4/83, L2920 Luxembourg

Vasiliev, J.M., All-Union Cancer Research Center, The USSR Academy of Medical Sciences, Kashirskoye shosse 24, 115478 Moscow, USSR

Venitt, S., The Institute of Cancer Research, Royal Cancer Hospital, F Block, Clifton Avenue, Sutton, Surrey SM2 5PX, UK

Wahrendorf, J., Unit of Biostatistics and Field Studies, International Agency for Research on Cancer

Wilbourn, J.D., Unit of Carcinogen Identification and Evaluation, International Agency for Research on Cancer

Williams, G.M., Naylor Dana Institute for Disease Prevention, American Health Foundation, Department of Pathology, Dana Road, Valhalla, NY 10595, USA

Würgler, F.E., Institute of Toxicology, Swiss Federal Institute of Technology and University of Zurich, Schorenstrasse 16, 8603 Schwerzenbach bei Zurich, Switzerland

Vainio, H., Unit of Carcinogen Identification and Evaluation, International Agency for Research on Cancer

Yamasaki, H., Unit of Mechanisms of Carcinogenesis, International Agency for Research on Cancer

---

[1] Unable to attend

INTRODUCTION

Clinical observations and epidemiological studies, as well as experimental studies, have contributed to the identification of specific factors that are causally associated with cancer in humans. Such findings have resulted from the observation of rare cancers by alert physicians or of an overwhelming difference in the incidence of a given tumour between populations with a particular habit, such as tobacco smoking, or with a given occupation, like dyestuff workers exposed to aromatic amines; in parallel, evidence of carcinogenicity in experimental animals has often preceded findings in humans (see Tomatis, 1977; Doll & Peto, 1981; Tomatis, 1986). Although important contributions have thus been made to the identification of cancer-causing agents in humans, the specific risk factors or exposures responsible for some major types of cancer (e.g., of the stomach, colon/rectum, oesophagus, breast, cervix and prostate) are unknown or ill-defined. Even for well-known risk factors like tobacco smoking and chewing, the specific agents that, alone or as a result of their interaction, are responsible for their carcinogenicity have not been well characterized or understood. Some dietary habits have been associated with various cancers in man (National Research Council, 1982), but the role of specific components of the diet in the induction of cancer is less well defined.

Elucidation of these issues would represent a major contribution to the primary prevention of cancer. It is difficult to foresee how this could be achieved by epidemiological studies alone without a substantial contribution of laboratory findings based on progress in our understanding of the mechanisms of carcinogenesis and of the results of assays for potential carcinogens (see Tomatis *et al.*, 1982; Weinstein, 1985; Montesano *et al.*, 1986).

There is also a deficiency of epidemiological investigations to follow up experimental evidence for the carcinogenicity of certain environmental chemicals. Data from the *IARC Monographs on the Evaluation of the Carcinogenic Risk of Chemicals to Humans*, which include evaluations of over 700 chemicals and types of exposures, indicate that, apart from the 30 chemicals, mixtures of chemicals and nine industrial exposures that are carcinogenic to humans, for over 100 chemicals there is *sufficient evidence* of carcinogenicity to experimental animals but no epidemiological data (Vainio *et al.*, 1985).

It is even more difficult to see how epidemiological studies alone could reliably assess the role in carcinogenesis of naturally occurring chemicals (Ames, 1983), of carcinogens formed in the human body (Bartsch & Montesano, 1984), and of dietary components (Armstrong & Mann, 1985; Sugimura & Sato, 1983), or the relative contributions of agents acting at different stages in the etiopathogenesis of human cancer (IARC, 1983). One of the aims of this publication is to assess how the long-term and short-term testing of chemicals for carcinogenicity could facilitate this task.

This volume contains a series of reports on the use of such assays for the detection of chemical carcinogens. The reports were prepared in the light of present understanding of the carcinogenesis process and of the scientific basis of the various endpoints used in these assays. Considerable progress has been made since the previous publication in 1980 (IARC, 1980) in understanding the value and limitations of these assays.

In the reports on long-term carcinogenicity assays (Reports 1, 2 and 3), particular attention was given to (i) the role of pharmacokinetic data in the design and evaluation of such assays, (ii) the role of early preneoplastic lesions in carcinogenesis, and (iii) the relative contributions that carcinogens may make to the various stages of the carcinogenesis process (see also Reports 9 and 10).

The main developments in short-term testing in recent years are a better understanding of the endpoints measured, better validation of the capacity of these short-term assays to detect carcinogens, and characterization of new assays using prokaryotic and eukaryotic cells. New strains of bacteria have been developed that are sensitive to mitomycin C, X-rays and oxidative mutagens (Reports 4 and 5). Very sensitive immunological and physicochemical methods permit determination of the presence of DNA damage in human tissues, providing a marker of individual exposure (Report 4). In the reports on assays using mammalian cells, the discussion is devoted in particular to the detection *in vivo* and *in vitro* of genetic changes (gene mutation, chromosomal translocation and deletion, aneuploidy and gene amplification), which have been shown consistently to occur in tumour cells (Reports 7 and 8). Recent studies (see Yunis, 1983) show the high specificity of these changes at the molecular and chromosomal levels, providing a rationale for the use of these assays in testing for potential carcinogens.

While the use of assays for genetic changes in bacteria and mammalian cells to detect potentially carcinogenic chemicals is based mainly on the theoretical (and empirical) correlation between mutagenesis and carcinogenesis, use of tests for neoplastic transformation of cultured cells is based on the evidence that the endpoints measured have a direct bearing on the natural history of cancer development *in vivo* (Report 9). In addition, this assay appears to be one of the most promising for identifying at which stage of the carcinogenic process a chemical carcinogen might act (Report 10).

In the 1980s, there were high expectations that the use of short-term assays might be a possible alternative to long-term carcinogenicity testing. In the light of the complexity of the carcinogenesis process, this was a simplistic approach, but it did have some justification from the public health aspect. It is now evident that the correlation between mutagenic and carcinogenic effects is far from perfect (see Mendelsohn, 1986). There is no doubt, however, that the intelligent use of short-term assays (i) plays an essential complementary role in the planning of long-term carcinogenicity testing and in the evaluation of the results; (ii) contributes to increasing the sensitivity and specificity of epidemiological studies (see Perera & Weinstein, 1982); (iii) assists in understanding the mechanisms of carcinogenesis/mutagenesis when combined with biochemical and genetic studies; (iv) contributes to the identification of mutagenic and/or carcinogenic compounds in complex mixtures (Friesen *et al.*, 1985); and (v) is useful in ascertaining the absence of possible adverse effects in the development of new drugs, pesticides, food additives and industrial chemicals.

In view of the complexity of these issues, it was felt that it would not be profitable to attempt to standardize a schema for the deployment of short-term tests in different situations, as the optimal battery of tests would depend on the particular investigation. Two chapters on this topic are presented on pp. 519 and 529 of this publication, which represent the personal view of the authors. This issue was also addressed at the previous IARC meeting (IARC, 1980), and many of the points discussed at that time are still valid today. Another chapter is included (p. 497), which discusses the use of structure-activity relationships in predicting carcinogenicity.

It is reassuring that the present understanding of the molecular biology of carcinogenesis is highly consistent with what has been known up to now about the biochemistry and biology of cancer induced by chemical carcinogens, which serves as the scientific basis for many of the assays described in this publication. A more unified concept of chemical and viral carcinogenesis is appearing (see Temin & Miller, 1984; Barbacid, 1986) which may result in the near future in a further improvement in the value of these tests for the primary prevention of cancer.

The Editors

REFERENCES

Ames, B.N. (1983) Dietary carcinogens and anticarcinogens. Oxygen radicals and degenerative diseases. *Science*, *221*, 1256-1264

Armstrong, B. & Mann, J. (1985) *Diet*. In: Vessey, M.P. & Gray, M., eds, *Cancer Risk and Prevention*, Oxford, Oxford University Press, pp.68-98

Barbacid, M. (1986) Oncogenes and human cancer: Cause or consequence? *Carcinogenesis*, 7, 1037-1042

Bartsch, H. & Montesano, R. (1984) Commentary: Relevance of nitrosamines to human cancer. *Carcinogenesis*, 5, 1381-1393

Doll, R. & Peto, R. (1981) *The Causes of Cancer*, Oxford, Oxford Medical Publications

Friesen, M., O'Neill, I.K., Malaveille, C., Garren, L., Hautefeuille, A., Cabral, J.R.P., Galendo, D., Lasne, C., Sala, M., Chouroulinkov, I., Mohr, U., Turusov, V., Day, N.E. & Bartsch, H. (1985) Characterization and identification of 6 mutagens in opium pyrolysates implicated in oesophageal cancer in Iran. *Mutat. Res.*, 150, 177-191

IARC (1980) *IARC Monographs on the Evaluation of the Carcinogenic Risk of Chemicals to Humans*, Suppl. 2, *Long-term and Short-term Screening Assays for Carcinogens: A Critical Appraisal*, Lyon

IARC (1983) *Approaches to Classifying Chemical Carcinogens According to Mechanism of Action (IARC Internal Technical Report No. 83/001)*, Lyon

Mendelsohn, M.L. (1986) Can chemical carcinogenicity be predicted by short-term tests? *Ann. N.Y. Acad. Sci.* (in press)

Montesano, R., Parkin, D. & Tomatis, L. (1986) *Environmental causes of cancer in man*. In: Waring, M.J. & Ponder, B.A.J., eds, *Cancer Biology and Medicine*, Vol. 1, *Biology of Carcinogenesis*, Lancaster, MTP Press Ltd (in press)

National Research Council (1982) *Diet, Nutrition and Cancer*, Washington DC, National Academy Press

Perera, F.P. & Weinstein, I.B. (1982) Molecular epidemiology and carcinogen-DNA adduct detection: New approaches to studies of human cancer causation. *J. chron. Dis.*, 35, 581-600

Sugimura, T. & Sato, S. (1983) Mutagens-carcinogens in foods. *Cancer Res.*, 43, 2415s-2421s

Temin, H.M. & Miller, C.K. (1984) Insertion of oncogenes into retrovirus vectors. *Cancer Surv.*, 3, 229-246

Tomatis, L. (1977) *The value of long-term testing for the implementation of primary prevention*. In: Hiatt, H.H., Watson, J.D. & Winsten, J.A., eds, *Origins of Human Cancer (Cold Spring Harbor Cell Proliferation Series Vol. 4)*, Cold Spring Harbor, NY, CSH Press, pp. 1339-1357

Tomatis, L. (1986) The contribution of the IARC Monographs programme to the identification of cancer risk factors. *Ann. N.Y. Acad. Sci.* (in press)

Tomatis, L., Breslow, N.E. & Bartsch, H. (1982) *Experimental studies in the assessment of human risk*. In: Schottenfeld, D. & Fraumeni, J.F., eds, *Cancer Epidemiology and Prevention*, Philadelphia, W.B. Saunders, pp. 44-73

Vainio, H., Hemminki, K. & Wilbourn, H. (1985) Data on the carcinogenicity of chemicals in the IARC Monographs programme. *Carcinogenesis*, 6, 1653-1665

Weinstein, I.B., (1985) *Chemical carcinogenesis*. In: Calabresi, P., Schein, P.S., & Rosenberg, S.A., eds, *Medical Oncology*, New York, Macmillan, pp. 63-82

Yunis, J.J. (1983) Chromosomal basis of human neoplasia. *Science*, 221, 227-236

BIOASSAYS FOR CARCINOGENICITY IN ANIMALS

Chairman: G. Della Porta

Vice-chairman: N.P. Napalkov

Rapporteurs: P. Bannasch & R.A. Griesemer

REPORT 1

LONG-TERM ASSAYS FOR CARCINOGENICITY IN ANIMALS

Prepared by:

P. Bannasch, R. A. Griesemer (Rapporteurs),
F. Anders, R. Becker, J.R. Cabral, G. Della Porta (Chairman),
V.J. Feron, D. Henschler, N. Ito, R. Kroes, P.N. Magee, B. McKnight,
U. Mohr, R. Montesano, N.P. Napalkov, S. Nesnow, A.E. Pegg,
G.N. Rao, V.S. Turusov, J. Wahrendorf and J. Wilbourn

CONTENTS

1. General introduction 17
2. Design of experiments 18
    2.1   Introduction 18
    2.2   Personnel responsibilities 18
    2.3   Test substances 21
    2.4   Animal species, strain and sex 23
    2.5   Route of administration 26
    2.6   Pharmacokinetics 30
    2.7   Selection of doses 34
    2.8   Inception, duration of exposure and observation period 36
    2.9   Number of animals 36
    2.10 Methods of randomization 38
    2.11 Observations 39
    2.12 Data acquisition, processing and storage 41
    2.13 Animal husbandry 41
    2.14 Diet 42
    2.15 Safety measures 45
    2.16 Written working protocol 46
3. Conduct of experiments 46
    3.1   Introduction 46
    3.2   Technical staff and their responsibilities 48
    3.3   Receipt and quarantine of test animals 48
    3.4   Selection of caging/bedding 50
    3.5   Allotment to treatment groups 52
    3.6   Health monitoring 54
    3.7   Diagnostic laboratory procedures 58
    3.8   Differentiating disease and toxic response 59
    3.9   Reducing losses due to autolysis 62
    3.10 Observation of tumours at necropsy 63
    3.11 Histopathology 63
    3.12 Data collection 64
    3.13 Waste disposal 64

4. Analysis and reporting 65
    4.1   Validation of data 65
    4.2   Adequacy of the experiment 65
    4.3   Chemically related effects on growth, morbidity and survival 65
    4.4   Categories of tumour 66
    4.5   Statistical analyses 67
    4.6   Reporting the results 71

5. Other assays for carcinogenicity in animals 71
    5.1   Introduction 71
    5.2   Strain A mouse lung tumour bioassay 71
    5.3   Breast cancer induction in female Sprague-Dawley rats 72
    5.4   Subcutaneous injection 73

6. References 74

1. General introduction

The essence of long-term testing is to observe test animals for a major portion of their lifespan for the development of neoplastic lesions after or during exposure to various doses of a test substance by an appropriate route. Such an assay requires careful planning and documentation of the experimental design, good animal care, leading to long survival, a high standard of pathology and unbiased statistical analysis. These requirements are well known and have not changed in principle during the last two decades.

A steadily increasing demand for testing of chemicals for carcinogenicity has led to the involvement in such studies of numerous laboratories and investigators in many countries. The establishment of internationally acceptable basic requirements for carrying out tests for the purpose of assessing the possible carcinogenicity of environmental chemicals and for reporting the results has been useful.

The main stages of long-term assays for carcinogenicity of chemicals include design, conduct, analysis and reporting. These are described below, not in terms of rigid rules, but rather as guidelines that meet the basic requirements for carcinogenicity testing. Extra observations, or different experimental methods, might be necessary for the assay of chemicals for effects other than carcinogenicity.

In this report — a revision of Supplement 2 to the *IARC Monographs on the Evaluation of the Carcinogenic Risk of Chemicals to Humans* (IARC, 1980) — special attention has been given mainly to newer information on pharmacokinetics and statistics. The discussions on statistical methods in the original report (Peto *et al.*, 1980) stimulated the preparation of a more comprehensive text on the use of statistics in carcinogenicity bioassays (Gart *et al.*, 1986), the contents of which are highlighted in this report. Also included are brief descriptions of three further, limited animal bioassays that may be useful for screening chemicals for carcinogenic activity.

No new concept for the testing of chemicals for carcinogenic activity in chronic experiments is suggested here, but some recommendations of other national and international groups are given, since they may be useful for obtaining internationally comparable and reliable results. The intent is not to supersede previous monographs or recommendations of national cancer centres and such international organizations as the World Health Organization, the International Union Against Cancer and the Organisation for Economic Co-operation and Development, but rather to present for general consideration various suggestions based in part on the experience of IARC in attempting to interpret the results of long-term carcinogenicity tests carried out in many countries, in the hope of achieving more uniform quality.

Interested readers might also consult: Shubik and Sice (1956), Weisburger and Weisburger (1967), Arcos *et al.* (1968), Boyland (1968), Berenblum (1969), WHO (1969), Magee (1970), US Food and Drug Administration (1971), Golberg (1973), Health and Welfare Canada (1973), Roe and Tucker (1973), Sontag *et al.* (1976), Munro (1977), Page (1977a,b), Committee on Carcinogenicity of Chemicals in Food, Consumer Products and the Environment (1978), Dayan and Brinblecombe (1978), US Food and Drug Administration (1978), WHO (1978a), Interagency Regulatory Liaison Group (1979), Organisation for Economic Co-operation and Development (1981), US Food and Drug Administration (1982), US Environmental Protection Agency (1983), Office of Science and Technology Policy (1984), US National Toxicology Program (1984), Milman and Weisburger (1985) and Turusov and Parfenov (1986).

Testing of chemicals in experimental animals should take into account principles of good laboratory practice (Organisation for Economic Co-operation and Development, 1982). Principles for the humane care of animals used in biomedical research have been described by the Council for International Organizations of Medical Friends (1985).

2. Design of experiments

2.1 *Introduction*

A long-term carcinogenicity study is a time-consuming and costly undertaking which requires adequate physical facilities and qualified and experienced personnel of various disciplines, in order to attain high standards of animal husbandry, pathology and data acquisition and processing. A detailed, written plan should be produced by close cooperation among chemists, toxicologists, laboratory animal specialists, pathologists and statisticians in the light of all available information on the nature and proposed uses of the substance to be tested.

Although many aspects of experimental design cannot be altered once the experiment is in progress, plans should be flexible enough to permit necessary changes during the experiment. Changes should be made only under the conditions that: (i) they are made only after consultation between all the responsible investigators, including the statistician (and also perhaps between the investigators and the sponsor or relevant regulatory agency); (ii) the reason for the change is compelling; and (iii) the change is well explained and adequately documented.

2.2 *Personnel responsibilities*

The planning of a long-term carcinogenicity study involves laboratory animal science, pathology and biostatistics. Depending on the bioassay, it may also involve any of the following: analytical chemistry, biochemistry, clinical chemistry, epidemiology, haematology, pharmacokinetics, microbiology, nutrition and computer programming. In addition, it is useful if individuals not closely associated with the conduct or design of the experiment periodically ascertain that the protocol is being followed, the quality of the work is satisfactory, and the animals are humanely cared for.

A single individual should be the study director, bearing ultimate responsibility for the development of an adequate protocol and its implementation. It may also be useful to appoint a deputy study director.

The roles of the toxicologist, pathologist and biostatistician in the design of the experimental protocol are generally well recognized, and the responsibilities of these individuals in the conduct of the study should be clearly defined. The execution, interpretation and reporting of a study are greatly facilitated by integration of the efforts of the toxicologist, the pathologist and the biostatistician throughout the study. The programme should be monitored by key staff; and, in particular, the toxicologist and pathologist should jointly establish an animal health monitoring programme to minimize the loss of animal and tissue samples for histological evaluation.

A laboratory animal veterinarian may supervise the general conditions for animal housing, nutrition and treatment. He or she should be accompanied by a toxicologist or veterinary clinician, who should observe animals for clinical signs of disease and contribute to recommending dose levels for each test phase.

The pathologist conducting and supervising gross necropsies should, if possible, also interpret the tissue sections for the same study. Experience in laboratory animal pathology, particularly in tumour and toxicological pathology, is essential. The selection of tissue samples by the pathologist for histopathological examination is a critical step in the bioassay.

A clinical laboratory scientist working in the field of laboratory animal haematology, clinical chemistry and clinical pathology would be helpful for the conduct of those aspects of the bioassays and the interpretation of results.

An analytical chemist is essential for the multidisciplinary approach required for the conduct of toxicological studies. The participation of a chemist is needed for supervision of, and/or consultation with, personnel performing stability analysis, dosage preparation, dosage analysis, chemical analysis and the monitoring of inhalation exposure concentrations. Additionally, the monitoring of diets for homogeneity of mixing and the contents of various nutrients and contaminants, such as heavy metals, pesticides, oestrogenic materials, mycotoxins and nitrosamines, is the responsibility of the analytical chemist. The animals' drinking-water may contain toxic chemicals, chlorinated compounds, calcium carbonate (National Academy of Sciences/- National Research Council, 1974) or metals that might affect experimental results; therefore, it might be appropriate to analyse the water periodically for contaminants.

The advice of a nutritionist is required when test chemicals are incorporated in the diet and, more importantly, when using a semisynthetic diet or when formulating a special diet (McLean, 1973; Newberne, 1974; Almeida *et al.*, 1978; Institute of Laboratory Resources, 1978).

A microbiologist can contribute to a chronic study in several areas. For example, it may be prudent to test the drinking-water for bacterial content (Fox et al., 1979), to evaluate the bacterial and viral profiles of newly received animals and to monitor water-bottles, cages and feed containers after washing to ascertain whether the washing machines are properly cleaning and sanitizing the equipment.

A health and safety officer who has the authority to bring unsafe conditions to the attention of higher management and the study director should have experience in occupational health and safety hazard control and industrial hygiene as well as in local, state and federal environmental protection.

Adequately trained individuals should be responsible for monitoring each study to assure that the facilities, equipment, personnel, methods, practices, records and controls conform with legal and administrative requirements. The members of this 'quality assurance unit' should demonstrate an organizational independence from the personnel engaged in the direction and conduct of the study.

Biometrical advice should be available for all studies. A biostatistician should be involved in a carcinogenicity study from its inception until the last parameter has been statistically evaluated, and must ensure that no unnecessary bias or confounding factor is introduced into the study or into the interpretation of results.

Computer-based data acquisition and retrieval systems are used to monitor many parameters of an animal experiment (Munro et al., 1972; Kalbach et al., 1977). Computer-processed data concerning body weight, feed and water consumption, clinical observations and gross and histopathological evaluations can also be a useful component of a health monitoring programme and can be programmed to signal outlying data as they are collected, for reverification.

For pulmonary inhalation studies, individuals experienced in generating dusts and monitoring test atmospheres are required.

Optimal conditions for storing wet and dry specimens, raw data, records and reports must be ensured. Therefore, archival and associated personnel should be under the direction of the study director. Histological tissue blocks, slides and remaining test material should be stored and registered.

Moreover, the technical personnel should be well trained in the standards of animal experimental studies. It is desirable that administrative and technical training courses enable personnel to keep up to date. In particular, personnel handling radioactive materials must be trained in concert with applicable legal or administrative requirements. Expertise and experience in the safe handling and use of radioactive materials in scientific methods are therefore required.

The responsibilities of each of the individuals involved in the conduct of the experiment should be clearly defined and documented.

2.3 Test substances

Before initiating a long-term carcinogenicity study, the investigator(s) should obtain all available, relevant information on the characterization, storage and handling of the test material as well as data on control parameters such as:

(*a*)   synonyms and trade names;

(*b*)   structural and molecular formulae and molecular weight;

(*c*)   method of analysis, including chemical and physical properties of the pure substance, e.g.,

description,

boiling-point,

melting-point,

density,

refractive index,

spectroscopy data,

solubility in water and organic solvents,

volatility,

reactivity,

conversion factor (ppm in air/mg per m^3),

photochemical properties,

extent of ionization,

particle size, shape and density (important in inhalation studies with aerosols), and

stability: the stability of the test compound should be examined under conditions of temperature and pH similar to those under which it is to be administered (e.g., in the diet, in a vehicle or in inhaled atmospheres). The frequency with which treatment mixtures are prepared depends on the stability of the test agent under the conditions of storage and administration. For agents to be tested by inhalation exposure, it is particularly important to avoid changes in the test material, such as oxidation or polymerization.

(*d*)   adequate specifications of commercial products, including knowledge of the percentage of additives such as stabilizers and emulsifiers;

(*e*)   impurities and methods of manufacture or route of synthesis: the purity of each test substance should be determined prior to its testing for carcinogenicity. Additives and contaminants should be identified and quantified using analytical techniques.

(f) sources and batch number(s): it is usually preferable to use only one manufactured lot of the test chemical for the entire study, otherwise the purity of each lot must be ascertained separately.

(g) date(s) when received;

(h) methods of storage: a proper method of long-term storage of a sufficient quantity of the test material should be determined prior to the start of the experiment. The stability of the material under defined conditions of storage determines the frequency at which either fresh batches of the material should be prepared or purifications should be carried out.

(i) probable daily exposure level for humans: if possible, the investigators should be informed about the usual range of dietary doses (for foods or food additives), residue or contamination levels (for pesticides), dosage schedules (for drugs), and concentrations in the atmosphere (for occupational chemicals and air pollutants).

(j) biochemical information: data on absorption and pharmacokinetics, including metabolism, accumulation and excretion, should be available prior to the start of the study.

Recent experience has indicated that impurities and manufacturing by-products may be responsible (or may subsequently be suspected of having been responsible) for some of the effects of a test substance. The manufacturing procedures and chemical synthesis processes should be examined carefully with a view to detecting secondary reactions that could lead to the production of impurities (Munro, 1977). Factors such as variations in purity between batches, packaging, handling and processing into food or drug preparations and reactions with other chemical can influence the toxicity of the substance (WHO, 1978b). The further testing of any impurity would depend upon the amount and nature of the impurity. If mixtures of materials with carriers or vehicles are used, the particular influence of the carrier on the characteristics of the materials must be considered.

One of the earliest and most difficult decisions that must be taken during the planning phase of a long-term study is the selection of the source of the material to be studied. Several aspects of this problem have been discussed (Health and Welfare Canada, 1973; WHO, 1978b). One of these is the testing of 'technical grade' *versus* 'highly purified' material: the use of highly purified test agents for establishing the mode of action or metabolism of suspected carcinogens is a generally accepted practice. Since humans may be exposed to impure compounds, however, priority should be given, if possible to testing the material to which humans are exposed (WHO, 1969; Health and Welfare Canada, 1973). Another, more comprehensive approach is the simultaneous and comparative testing of both the technical grade and the highly purified material. In addition, separate testing of important impurities might be considered. Since this is costly and may be of doubtful relevance, the use of technical-grade material is to be preferred in general. If a positive effect is seen, a step-by-step approach may be necessary (possible second step: pure material; possible third step: impurities). But there may be circumstances that necessitate the immediate initiation of an extensive programme of various long-term studies.

2.4 *Animal species, strain and sex*

(*a*) *Species*

The ideal animal species for carcinogenicity studies would be one in which the biological response to the test substance is identical to that in humans. Since no such animal species is known, it is generally recommended that a material be tested in at least two species. A particular species may be indicated by metabolic or pharmacokinetic data, by susceptibility to a particular class of carcinogens, or by the proposed route of administration. The main criteria that could be applied to the selection process for test animals in a carcinogenicity study can be listed as follows:

availability,

economy,

sensitivity to carcinogens,

stable response, and

similarity to humans with regard to metabolic, toxicological and pathological responses.

The species most frequently used in carcinogenicity studies are rats, mice and Syrian golden hamsters. Rats and mice have been used more extensively than hamsters, but the latter are particularly suitable for studies on respiratory and urinary-tract carcinogenesis (Nettesheim, 1972; Shubik, 1972; Adams *et al.*, 1978; Mohr, 1979). The wide use of these three rodent species, which are still the species of choice for carcinogenicity testing, is not based on biochemical, physiological or anatomical similarities between these animals and humans, but rather on practical considerations, such as relatively short lifespan, small size and availability. Another major consideration is the existence of considerable biological knowledge on these species, including their physiology, anatomy, genetics, husbandry, nutrition, spontaneous diseases, spontaneous tumour incidences and susceptibilities to tumour induction.

Primates and dogs are seldom used in carcinogenicity bioassays because of their relatively long lifespans, genetic heterogeneity, limited availability and higher maintenance costs. There are few indications that the species of nonhuman primates and the breeds of dogs commonly used in biomedical research are any closer to humans in their metabolic capabilities than rodents. Their routine use is not recommended. Small fish have also been proposed for carcinogenicity testing (see Hoover, 1984 and Report 3).

(b) *Strain*

Strains should be well characterized, free from interfering congenital defects and as far as possible free from spontaneous diseases such as amyloidosis. Ideally, preference in strain selection should be given to strains with a consistent low incidence of spontaneous tumours (1-5%) and diseases and a high and specific susceptibility to human carcinogens. In practice, the chief determinant of which strain workers in a particular laboratory use is often their experience and background knowledge of their colony of such animals over the years. For, although we know that the choice of strain may be important in a particular bioassay, we have in general insufficient knowledge to permit reliable prediction in advance of which strain to prefer. Indeed, there is not even general agreement as to whether inbred, F_1 hybrid or outbred animals are usually to be preferred. Reasons for the selection of outbred animals are:

(i) the strains with which many laboratories have experience are for largely historical reasons outbred strains; and

(ii) it has generally been considered that outbred or random-bred strains are more resistant to disease than inbred animals (although with progressively better husbandry, inbred animals now survive quite well, and F_1 hybrids survive excellently).

However, neither of these considerations is overriding and the purely scientific considerations are perhaps more important. The problem is that within a single outbred strain there can easily be enormous differences in the incidence of cancer resulting from a standard treatment. For example, in a mere five generations of selective breeding for and against skin tumorigenesis, Boutwell (1964) produced two substrains of mice that differed by two orders of magnitude in their papilloma response to skin initiation plus promotion. Similar degrees of variation of intrinsic susceptibility to standard insult are strongly suggested by other data as well.

It is difficult to feel comfortable when organizing bioassays in which such enormously important variables are uncontrolled (Festing, 1975), and so some of the strains routinely used in bioassay programmes are now inbred or F_1 hybrid. At present, Fischer 344 rats, Sprague-Dawley rats, B6C3F_1 mice, ICR Swiss albino mice, CD-1 mice and outbred hamsters are most often used in the USA and Japan, whereas Wistar rats and NMRI mice are often used in some countries in Europe. The disadvantage of studying genetically identical animals is, of course, that if the test agent is a carcinogen of only moderate strength one may inadvertently study an F_1 or inbred strain that is resistant to its effects.

A compromise has been advocated by Haseman and Hoel (1979) whereby if several different F_1 or inbred strains could be found that would survive well under standard laboratory conditions, each treatment group could initially contain exactly the same proportions of each of these genetically different strains. The advantages would be, firstly, that the effects of genetic heterogeneity would be controlled (necessarily, each group would initially be genetically identical to every other treatment group) and, secondly, that the risk that the treated group is comprised entirely of animals resistant to the test carcinogen is reduced. The practical disadvantages of such designs are, firstly, that it would be difficult to find a suitable set of robust F_1 hybrids, and, secondly, that separate 90-day toxicity studies might have to be done to see whether one maximum tolerated dose would apply to each separate strain.

If appropriate statistical methods are used (Mantel & Ciminera, 1979), then, especially if in some of the strains the majority of treated animals get tumours, it is likely that such designs would be more sensitive than those of studies with conventional outbred strains, and there would be less risk of obtaining false-negative results than in studies with conventional inbred strains. They should perhaps, therefore, be pursued more actively than heretofore. Until more F_1 hybrids have been developed, and experience with such unconventional genetically stratified designs has been accumulated, most investigators will continue the ad-hoc procedure of using the strain with which they have most experience.

Whatever strain of animals is selected, the incidences of naturally occurring cancers at various body sites will be characteristic of that strain. The influence of high or low background tumour incidences on the outcome of carcinogenesis bioassays is uncertain as yet. Some investigators believe that body sites at which tumours commonly occur are sensitive sites at which to detect carcinogens. Others believe that induction of additional tumours at commonly occurring sites is weaker evidence of carcinogenicity than induction of tumours at sites where they rarely occur naturally. In any case, background variability in tumour incidences must be taken into account in evaluating carcinogenic effects (see section 4.5).

In general, the following guidelines should be considered before the beginning of an experiment (Society for Laboratory Animal Science, 1985; Working Committee for the Biological Characterization of Laboratory Animals/GV-SOLAS, 1985):

(i) breed or designation of stock or strain must be outlined in the experimental protocol, preferably using international nomenclature;

(ii) the genetic status (strain or stock, hybrid or mutant), if not obvious from stock/strain designation, must be defined separately;

(iii) the age, weight and sex of the animals must be determined at the start of the experiment;

(iv) the definition of the microbiological status should include the specification of pathogens or groups of pathogens from which the animals are free (It is not required that the conventional microbiological flora be specified. In the special case of studies with specific microorganism-associated gnotobiotic animals, all microorganisms present must be specified.); and

(v) methods of delivery and transport as well as the quarantine or acclimatization periods of the animals must be specified.

(c) *Sex*

Sex differences in response to carcinogens have been well documented. Therefore, equal numbers of animals of each sex should usually be included in a carcinogenicity study.

2.5 *Route of administration*

There is general agreement that the test compound should be administered by the predominant route of human exposure, unless pharmacokinetic studies show that other routes are equivalent. The need to achieve adequate dose levels for an appropriate study has to be taken into account. The three main routes of human exposure are oral intake, skin absorption and inhalation.

(a) *Oral exposure*

When the route of exposure is *per os*, the test substance can be administered (i) mixed in the diet, (ii) in the drinking-water, or (iii) by gastric intubation. The advantages and disadvantages of each of these methods of administration have been described (Page, 1977b; Committee on Carcinogenicity of Chemicals in Food, Consumer Products and the Environment, 1978).

Incorporation in the diet or drinking-water is appropriate for most substances ingested by humans, and generally most closely simulates the pattern of human exposure. Sometimes oral gavage, which has the advantage that the doses administered are known, is undesirable, because the high concentration of material administered may result in local injury to the upper digestive tract. Moreover, administration by gavage is time-consuming and may lead to a loss of animals by faulty intubation. However, if this route is unavoidable, then the test compound should be administered daily on seven days a week. The ultimate choice most often largely depends on the characteristics of the test material, such as volatility, corrosiveness, solubility in water, stability in the diet, palatability, etc. A given dosage (mg/kg body weight) may produce very different profiles of absorption, blood level and elimination, dependent upon whether it is administered by gavage as compared to in the diet or drinking-water. Therefore, the kinetics of uptake, distribution, conversion and elimination are very important if administration by gavage is anticipated (see section 2.6).

If the test material is given mixed in the diet or in the drinking-water, exposure is fairly continuous. The use of this route, however, imposes a strict surveillance of possible contamination of the laboratory environment and possible exposure of personnel. If the substance is administered by gavage, the usual frequency is once a day for five or preferably seven days a week. A frequency of two or three times a week is also employed, though this is less desirable, unless supported by pharmacokinetic data.

(b) Cutaneous exposure

Skin exposure in carcinogenicity testing may be undertaken for two reasons: (i) to simulate a main route of human exposure, e.g., for cosmetics, or (ii) as a model system for induction of skin tumours (see also Report 3).

The kinetics of absorption and metabolism and the potential for local or systemic effects should be considered in determining whether skin application is appropriate. Although the chief human exposure may be *via* the skin, if animal absorption is much greater by inhalation or ingestion, these routes are nevertheless preferred. It should, however, be realized that in this case the dosage may be far higher in relation to human exposure. Moreover, one must determine whether the metabolism of the test substance is comparable by each route of administration. When testing by skin application is undertaken to simulate human exposure, the basic experimental design will be the same as for oral exposure. The hair should be clipped to allow maximum contact of the chemical with the skin and to facilitate its penetration. The preparation of the test area as well as the cleaning procedures for the skin must also be determined. When necessary to reduce local irritation, alternative sites of application may be used.

(c) Inhalation exposure

In humans, the respiratory tract is an important route of entry for many substances. Therefore, inhalation exposure has become more and more important in carcinogenicity testing, despite its higher cost.

One difficulty in inhalation exposures of rodents is that particulate materials have substantially different deposition patterns than in humans because rodents have small-diameter airways, highly turbinated nasal passages and variations in bronchial branching patterns. In addition, the pattern of exposure of obligatory nose breathers, such as rats and mice, is likely to be considerably different from that of humans.

Properly designed exposure chambers and special equipment for generating, sampling and monitoring the test atmospheres are needed for inhalation studies using vapours. Other specialized equipment is needed for the generation and characterization of aerosols (WHO, 1978c). Aerosol exposures are somewhat unsatisfactory because contamination of the animals' fur also results in oral intake of the test material during grooming. Another limitation of aerosol exposure is that animals may filter the atmosphere by hiding their noses in their own or in each others' fur. Individual housing reduces the extent of such filtration, and satisfactory equipment for head-nose-only exposure exists. However, such studies remain expensive and time-consuming.

In those studies in which dose-response data are desired, the kinetics of retention and lung clearance of particles should be investigated. Particles found deep in the lung in chronic experiments are a mixture of recently inhaled particles and of particles that were inhaled early in the study and cleared slowly. Clearance measurements with labelled particles can distinguish between the two types of clearance and will indicate when clearance is in equilibrium or overloaded (Vostal et al., 1982; Snipes et al., 1983; Muhle et al., 1986). For additional information see the general guidelines for the performance of inhalation studies presented by the Organisation for Economic Co-operation and Development (1981).

A daily exposure of six hours after equilibration of chamber concentrations is recommended for five days a week to simulate workplace (intermittent exposure) exposures, or 22-24 hours of exposure per day on seven days a week (continuous exposure, with about an hour for feeding the animals and maintaining the chambers) to simulate environmental exposures. A major difference between intermittent and continuous exposure is that, with the former, there is a 17-18-hour period in which animals may recover from the effects of each daily exposure, and an even longer recovery period during weekends. The choice of intermittent or continuous exposure depends on the objectives of the study and on the human experience that is to be simulated. However, certain technical difficulties must be considered. For example, the advantages of continuous exposure for simulating environmental conditions may be offset by the necessity of watering and feeding during exposure (WHO, 1978c) and by the need for greater quantities of test materials.

Intake air to chambers should be filtered through absolute filters, and exhaust air must be treated to remove the test substances. The animals should be tested in inhalation chambers designed to sustain a dynamic flow of sufficient fresh air to assure removal of waste gases such as ammonia and an evenly distributed atmosphere. There should be a back-up ventilation system in the event of the failure of the primary system supplying conditioned air to the chambers. For all test materials, a slight negative pressure inside the chambers is generally maintained to prevent leakage of the test substance into the surrounding area. Animals should be housed individually or in limited numbers in wire mesh cages within the chambers.

The following measurements should be taken to avoid major fluctuations in the air concentrations or major discrepancies in the operation of the chambers:

(i) air flow: the rate of air flow through the chamber should be monitored continuously;

(ii) temperature and humidity: it is suggested that, for rodents, the temperature be maintained at 23 (± 2)°C and the humidity within the chamber at 30-70% even when water is used to suspend the test substance in the atmosphere. There should be an alarm system for warning of excess temperature and humidity fluctuations; and

(iii) chamber concentrations: wherever possible, continuous monitoring and recording of chamber concentrations is recommended.

The particle size distribution should be determined by the mass median aerodynamic diameter and the geometric standard deviation of the distribution. The aerosol particles should be of respirable size for the test animal used. Samples of the chamber atmospheres should be taken at the breathing level of the animals. The air sample should be representative of the distribution of the particles to which the animals are exposed and should account for, on the basis of gravimetric and particle size distribution, all of the suspended aerosol, even when much of the aerosol is not respirable. Size analyses should be carried out frequently during development of the generating system to ensure the stability of the aerosol, and thereafter during the exposure as often as necessary to determine adequately the consistency of the particle distribution to which the animals are exposed.

The method of generating the test atmosphere must be reproducible so that the average chamber concentration does not vary by more than $\pm 10\%$.

(d) *Intratracheal instillation*

Intratracheal instillation should be considered as a possible alternative to inhalation exposure for aerosols. This technique has been widely used in experimental respiratory-tract carcinogenesis (Saffiotti *et al.*, 1968; Nettesheim, 1972). Since humans are more likely to inhale through the mouth than are rodents, thereby bypassing filtration by the nasal passages, intratracheal instillation might imitate more closely human inhalation of particulate material (e.g., tobacco smoke particles) than inhalation experiments.

Intratracheal instillations are generally given once, or at most twice, a week, to avoid mechanical damage.

(e) *Other routes*

Repeated *subcutaneous injection* is often regarded as inappropriate, because this route is generally different from the one by which humans are exposed. However, in certain situations, use of this route has advantages, for example, when a compound is poorly absorbed after oral administration or when one wishes to minimize contamination of the laboratory. This topic is discussed further in section 5.4.

Intraperitoneal, *intramuscular* and *intravenous injection*s are not widely used for testing the potential carcinogenicity of compounds, because these routes do not resemble probable conditions of exposure in humans. However, they can be considered for testing materials that are not readily absorbed from the digestive tract by rodents but may be absorbed by man.

If drugs are used parenterally in humans, parenteral administration might also be used in animals; but, if the metabolism of the parenterally administered drug is the same as by oral administration, the latter might nevertheless be preferred. Daily injections of drugs by these routes for months may be difficult.

2.6 Pharmacokinetics

Pharmacokinetics is the study of the dynamics of the fate of chemicals in the body. Relatively recent reviews of this field have been published by Gibaldi and Perrier (1982) and Dayton and Sanders (1983).

Metabolic and pharmacokinetic studies have now become a common feature in toxicological evaluations, and many articles indicate the importance of pharmacokinetic considerations in toxicology (reviewed by Gehring et al., 1976; Food Safety Council, 1978; Young & Holson, 1978; Jollow, 1980; Smyth & Hottendorf, 1980; O'Flaherty, 1981). It has been argued that, whenever possible, pharmacokinetic considerations should be taken into account in the design of long-term carcinogenicity studies and in their interpretation, particularly when extrapolating results obtained in such animal tests to human situations involving much lower exposures but much larger numbers of individuals (Reitz et al., 1980; Andersen et al., 1984; Grice et al., 1984a).

Knowledge of pharmacokinetics can indicate the extent to which the level of a compound or its active metabolite (or even DNA adducts) in the organism is proportional to the dose administered. A variety of factors can lead to non-linear dose-response relationships: these include saturation of uptake, metabolic activation, metabolic detoxification, depletion of a protective molecule such as glutathione, or saturation of a repair protein such as O^6-methylguanine-DNA methyltransferase. Other possibilities include the greater uptake of a substance at doses sufficient to damage a protective barrier, the saturation of binding by a protective extracellular protein, the saturation of an excretion process, and non-linear uptake due to the need for mechanical and/or biochemical dispersal of a material administered in a solid form.

If extrapolations are made from results obtained with doses in this non-linear region, potential effects could be seriously overestimated (if detoxification or repair was saturated) or underestimated (if activation or uptake was saturated). An important parameter to be obtained from pharmacokinetic measurements, therefore, is the relationship between dose and the bioavailability of the compound. In some circumstances, the even more valuable determination of the relationship between dose and the level of DNA adducts or other biological events may be approachable; but, in most cases, the limit of information that can be obtained readily would be the effect of dose and time on the plasma level of the compound.

Pharmacokinetics may also assist in selecting the mode of administration to be employed in a long-term carcinogenicity test. For example, an oral bolus given by intubation may not lead to the same bioavailability of a compound in the plasma as dosing with the compound in the food or drinking-water, which spreads the administration over a longer period of time. Similarly, uptake following exposure by inhalation may lead to widely different exposure of certain tissues and peak blood levels than would be produced by the same dose if given orally. Some means of administration are much more convenient experimentally than others, and an understanding of pharmacokinetics can be used to determine whether the route chosen is likely to be comparable to the most probable environmental exposure.

(a) Type of pharmacokinetic information needed

Suggestions as to what data on pharmacokinetics are desirable prior to the start of a long-term study have been given in several recent references (Gehring *et al.*, 1976; Smyth & Hottendorf, 1980; Andersen *et al.*, 1984; Grice *et al.*, 1984a). In particular, the paper by Andersen *et al.* (1984) and the preceding pages in the same document (US National Toxicology Program, 1984) give a detailed consideration of the type of information necessary, the experimental work needed to provide it and suggestions for its use. However, it should be recognized that it is not possible to provide a single approach to obtaining the necessary background data for the wide range of compounds that might be subjected to testing. In fact, in some cases it may not be possible to obtain these data at all.

First, a means must be found to quantify the amount of the compound in samples of tissues and physiological fluids obtained from experimental animals. Many compounds can be determined with a high degree of sensitivity by standard analytical chemistry techniques including separation by gas chromatography or high-performance liquid chromatography and quantification by fluorescence or absorption, thermoenergy detection or mass spectroscopy, after derivitization as necessary. When these methods cannot be employed, another approach is to use a radioactive form of the test compound. Radioactively labelled forms of the compound are also particularly useful for studies of metabolism.

The analytical method can then be used to measure the time course of elimination of the compound from the blood or other tissues after administration of a range of doses by the route to be used in the assay and by the intravenous route. The doses should be selected to cover a range which encompasses the highest dose tolerated and the lowest dose for which the analytical methodology is adequate. One or more doses should be in the range of probable human exposures.

The results can be plotted as graphs of plasma concentration against time. The area under these curves should then be expressed as a function of the administered dose; this can be used to establish whether non-linearity in bioavailability with dose occurs. Such information can then be used in the selection of appropriate doses for the experiments. Comparison of the plasma concentrations and the elimination rates between the selected route of administration and the intravenous route can indicate whether uptake mechanisms, saturable protective barriers to uptake or first-pass clearance influence the results by the chosen method of administration.

If a radioactive form of the compound is available, the major routes and form of elimination can be determined as a function of dose. If the metabolites can be identified structurally, this too may aid in determining whether metabolism is responsible for activation or detoxification. The principle reason for such studies, however, is to obtain evidence of lack of linearity with administered dose. Complete characterization is not essential as long as it is clear that the analytical separation is adequate.

If the techniques are available, DNA adducts can be measured as a function of dose. Such measurements can also be used to give an indication of the linearity or lack of linearity of damage to the genetic material. This may be possible in only a few cases at present, but these measurements have the advantage that saturable or inducible DNA repair mechanisms are also taken into consideration. Studies with both single and multiple doses could also be designed to provide information on the effects of the compound on cell division and regenerative cell replication as a function of dose.

Finally, a limited study should be carried out in which the same measurements of blood levels, clearance and metabolites are measured in animals that have been treated with multiple doses for two to four weeks. Such experimental results can then be compared with the single dose studies and should yield information on the possible induction or repression of activation and detoxification mechanisms. A further advantage of such studies is that they can be used to confirm calculations of the bioaccumulation of the compound, which are made on the basis of the pharmacokinetics of single doses.

These studies with multiple-dosed animals represent a major undertaking. The number of animals to be used and the time period covered, or even if they should be done at all, must be decided on a case-by-case basis. Such measurements are particularly valuable in the case of those compounds believed to induce metabolic enzymes or DNA repair pathways, and for compounds that deplete protective molecules such as glutathione, which must then be regenerated.

These studies should be carried out principally in animals of one sex, but a limited investigation to confirm or refute the applicability of the information obtained to animals of the other sex is also needed. In addition to possible variations with sex, pharmacokinetic parameters may also differ with age or diet (Grice *et al.*, 1984b), but the extent to which experimental measurements can be made to examine these possibilities is obviously limited.

(b) Difficulties in using pharmacokinetic measurements

(i) If the sole objective is to determine whether a compound is carcinogenic under some circumstances, the normal procedure of administering the compound at the maximally tolerated dose over a lifetime to two or more animal species should be appropriate even without regard to pharmacokinetics. It is unlikely that this approach will totally obscure any possible carcinogenic effect, although it is quite conceivable that such doses may not lead to a maximal effect. The effect may not be greater or may even be less than that with lower doses because (1) uptake into the relevant tissues and/or activation to the reactive intermediate may be saturated: if this is the case, lower doses may have the same effect as higher ones; and (2) the high doses may induce detoxification pathways and repair mechanisms, which could lead to a situation in which a greater effect is produced by lower doses. Thus, some knowledge of the pharmacokinetics would aid interpretation of the results.

(ii) For some compounds of major importance, it may be essential to carry out bioassays even though the methodology for determination of even the most basic pharmacokinetic parameters is not available, and the experiments cannot be designed in the light of pharmacokinetic knowledge. However, if an effect is observed, a further effort to define such parameters may be worthwhile.

(iii) The present state of knowledge of the mechanism of action of chemical carcinogens may be insufficient to use the pharmacokinetic data in a predictive manner. Although it has been argued, for example that there is a linear relationship between DNA adduct formation and tumour response (Hoel *et al.*, 1983), this proposition is not yet sufficiently well documented to be generally accepted. Even for well-known carcinogens, little may be known concerning the actual lesions in DNA that are responsible for the initiation of tumours. It is also clear that other factors, which are poorly understood, influence further development through promotion or progression of initiated cells to form the final tumour; these factors cannot be taken into account in the kinds of pharmacokinetic modelling that are feasible at the present time. Therefore, it would be unwise to ignore a positive result in a long-term animal screening test on the basis of a difference in pharmacokinetics between humans and the test species.

(iv) Many of the factors that must be taken into account in the derivation of pharmacokinetic parameters may change during the course of a long-term feeding study. Thus, the metabolism of a foreign compound and rates of cell replication or DNA repair may be inducible and may change as a function of age. Although it may be possible to study these changes, the level of detail obtainable in a preliminary investigation of pharmacokinetics is not sufficient to ensure that these changes can be allowed for in the course of a continuous exposure study, which may take as long as two or more years.

(v) The responses of human individuals to drugs and toxic agents may differ widely due to genetic variations, different exposures to other agents, dietary factors which modify metabolism, or protective factors. The results of a controlled study with laboratory animals kept under standard conditions are likely to be in much closer agreement with each other. When the results of a carcinogenicity test are extrapolated to provide an estimate of potential risk to humans, it is clearly advantageous to know whether the pharmacokinetic parameters are similar in the animal test species and in humans. However, there may be sufficient interindividual variation in the human population that a significant proportion of them will differ greatly from the animal model. In this case, the most prudent approach is to assume the 'worse-case' scenario (i.e., that humans are at least as sensitive as the animal species tested and that the effect will be proportional to the dose). However, once again, it should be kept in mind that saturation of activation mechanisms may lead to an underestimation of risk from high-dose experiments.

(vi) Information on comparable pharmacokinetic parameters in humans may be entirely lacking, making comparisons with experimental animals impossible.

2.7 Selection of doses

(a) Number of dose levels

The selection of doses and the number of dose levels becomes a difficult task if the bioassay is to fulfill more than one objective. If the purpose of the long-term study is to discover whether or not the test substance is carcinogenic, one dose level may be sufficient, although two dose levels plus control are often used to allow for misjudgements in selecting the higher dose level. Moreover, if both dose levels produce an observable tumour response in a dose-related manner, the weight of the evidence for carcinogenicity is increased. If the bioassay is intended also to provide dose-response data for cancer risk assessment (and possibly other indicators of chronic toxicity as well), three or more dose levels are typically employed. The more dose levels, the greater the information obtained about the shape of the dose-response curve, as long as all but one of the dose levels are in the range at which effects are observed. When testing much lower dose levels, with which tumour responses are expected to be low, it may be necessary to adjust the group sizes to retain the sensitivity of the test (Gart et al., 1986). A sequential or staged approach, in which the nature of the tumour response is determined before the dose-response study is performed, has the advantage that a more precise and cost-effective dose-response study can be designed but the disadvantage that several more years may be required to obtain the results (IARC, 1980; Grice et al., 1984a; US National Toxicology Program, 1984).

(b) High dose

Selection of the high test dose is one of the most important elements in a long-term bioassay. It should be sufficiently high that the capacity of the test substance to produce cancer is detected in a relatively small number of experimental animals. If it is too high, excess mortality or morbidity might diminish the probability of detecting carcinogenicity; if too low, doubt may arise about whether the substance has been tested adequately.

Two general approaches are used to select the high dose level — an observational approach, in which predictions about chronic toxicity are based on previous subchronic toxicity studies, and a pharmacokinetic approach, in which the optimal dose level and rate to produce maximal and steady tissue concentrations are approximated. When available, both types of information are used in selecting the dosing regimen. Knowledge of the chronic toxicity and metabolic activation and detoxification of structurally related compounds may help to predict the highest dose that can be tolerated.

Definition of the high dose (or the 'estimated maximum tolerated dose' as it is sometimes called) as given in the previous IARC guidelines (IARC, 1980) is still applicable. The high dose in chronic toxicity testing is one that is expected on the basis of subchronic studies to produce some toxicity when administered for the duration of the test period. It is estimated separately for animals of each sex of each species used. The high dose should not, however, induce toxic manifestations that are predicted to reduce the lifespan of the animals materially except as the result of neoplastic development or a 10% or greater retardation of body weight gain as compared with control animals. The recommendation regarding retardation of body weight gain is empirical, however, and if exceeded in practice does not necessarily invalidate the bioassay. In two-generation studies, the high dose should not be significantly detrimental to conception rates, fetal or neonatal survival, or postnatal development. In this respect, it may be necessary to use different dosing schedules for the dams, neonates, juveniles and the F_0 and F_1 adults. (See also section 3.5.)

The above criteria are recommended for selecting the high dose, even if the metabolic pathways at the high dose differ from those that operate at lower doses. In one series, more than two-thirds of the carcinogenic effects detected in feeding studies would have been missed had the high doses been reduced to one-half (Haseman, 1985). However, pharmacokinetic data, preferably established in repeated dose experiments, should be taken into consideration in selecting lower doses for study.

These criteria for high dose selection are inapplicable in the complete absence of toxicity. In such instances, the high dose in dietary studies may be set arbitrarily at about 5% of the diet under the assumption that that dose level will not interfere significantly with nutrition. At that level or higher, however, preliminary nutritional studies may be needed, whether or not the test substance is a nutrient, to demonstrate that the nutritional status of the animals is adequate at the high dose level selected. For inhalation studies with particulates, the rate of clearance should be taken into account in selecting the concentrations to be used.

(c) *Lower doses*

Spacing of dose levels and the corresponding allocations of animals to dose groups are intended to optimize the experimental design in respect to (i) its sensitivity to detect a carcinogenic effect and (ii) its precision in estimating effects at dose levels below the experimental range. The selection of lower dose levels is based on pharmacokinetic data, subchronic toxicity studies and anticipated human exposure. The steepness of the dose-response curve in subchronic studies aids in spacing lower dose levels. Although test animals are usually distributed in equal numbers among the dosed and control groups, theoretical models have suggested different allocations, such as sample size ratios of 1:2:2:1 for equally spaced doses (Krewski *et al.*, 1982) and 1:1:2:1 for dose levels in the ratio of 0:1/4:1/2:1 (Portier & Hoel, 1983).

2.8 *Inception, duration of exposure and observation period*

It is generally recommended that exposure to the test substance be started no later than a few weeks after weaning and be continued for the major portion of the animals' lifespan. It should be kept in mind, however, that because of the duration of long-term studies, the treatment might potentiate or be obscured by age-associated diseases. For reasons of expense, and because statistical treatment of the results is less straightforward, prenatal and two-generation studies are not recommended for routine screening for carcinogenicity (US Food and Drug Administration, 1971; Committee on Carcinogenicity of Chemicals in Food, Consumer Products and the Environment, 1978; WHO, 1978a).

For carcinogenicity testing, some investigators have advocated a duration of 24 months for rats and 18 months for mice and hamsters (Peck, 1974; WHO, 1978a; Grice *et al.*, 1984b), while others prefer a test period extending over the entire lifespan of the animals (Committee on Carcinogenicity of Food, Consumer Products and the Environment, 1978). For economic reasons and confounding factors such as changes in background tumour incidences with age, the latter procedure is not very suitable for carcinogenicity screening (Food Safety Council, 1978; WHO, 1978a) and should therefore be avoided. Although generally still acceptable, the former procedure seems too rigid and perhaps even inappropriate when the cumulative mortality at the time of sacrifice is still very low, e.g., less than 10%. Today, an acceptable duration would be of the order of 24 months for rats and mice, and 18 to 20 months for hamsters, unless there are appreciable losses from compound-related fatal tumours.

Preferably, the animals should be treated throughout the entire observation period. However, some investigators prefer to discontinue dosing towards the very end of their experiment and to leave the old animals under observation for the remaining few weeks of the study. If this is to be done, it is recommended that treatment continue at least to week 104 of age for rats and mice and week 80 for hamsters.

2.9 *Number of animals*

To allow meaningful biological and statistical evaluation of the results, a sufficient number of animals should be used in each group. Fifty males and 50 females in each test group and concurrent control group is generally accepted as a minimum adequate number and as a reasonable compromise between cost and avoidance of false negative results in routine carcinogenicity studies with rodents. A moderate increase in group size will produce only a small decrease in the likelihood of obtaining false-negative results (Peck, 1974; Page, 1977a), whereas the increase in work-load is considerable.

Interim sacrifices are sometimes desirable to assess late toxic effects other than cancer, but these require that extra animals be included in the experiment; otherwise the sensitivity of the cancer test may be reduced (even though the statistical methods that are recommended can incorporate into the main analysis whatever data are available from animals intended for, or subjected to, interim sacrifice, especially if those killed at any one time are drawn equally from all treatment groups). It is advisable that animals be chosen randomly for interim sacrifice in the initial randomization to form specific satellite groups.

Control groups

A control group of untreated animals should always be included in a carcinogenicity test. An adequate control group is one that is separated at random from the exposed group and is thereafter handled identically to the treated groups in every respect of diet, husbandry and observation, except, of course, exposure to the test material(s). Consequently, when a test compound is administered in a vehicle, the control animals should also receive that vehicle.

When cross-contamination is not a problem, the controls should be housed in the same room as the treated animals, but if cross-contamination cannot be avoided, each test group should preferably be housed in a separate room with, as far as can be ensured, identical physical conditions.

The various aspects of including a 'positive control' group in carcinogenicity studies have been discussed extensively (Hoffmann & Wynder, 1974; Weisburger, 1974; Page, 1977b; Committee on Carcinogenicity of Chemicals in Food, Consumer Products and the Environment, 1978). In routine carcinogenicity testing, positive controls need not be used.

Accumulated background information about the colony of animals to be used is of great importance, and in any laboratory in which a series of tests is being conducted on particular strains of animal it should be ensured that information is accumulated on cancer incidence among control groups. The data should ideally be tabulated by sex, strain, site and age and time of occurrence, distinguishing between tumours that directly or indirectly caused the death of their host and tumours that were found incidentally in animals that died (or were killed) for unrelated reasons (see section 4.4).

2.10 *Methods of randomization*

(a) *Unstratified allocation*

There is no reasonable doubt that some form of randomization procedure should be used to allocate animals to different groups. Doubt does exist, however, about:

(i) whether animals should be 'stratified' with respect to litter before allocation, and whether exactly one animal from each litter be allocated to each treatment group (Mantel & Ciminera, 1979); or

(ii) whether animals should be subdivided into a few strata with respect to body weight (e.g., very light, light, average, heavy, very heavy), with equal numbers from each stratum allocated to each group; or

(iii) whether animals should likewise be stratified with respect to birth date.

Experimentalists, fearing perhaps that these variables might be of some relevance, often prefer to control them; but statisticians, realizing that these effects (and the other, much more important, unmeasured constitutional effects) are controlled by randomization, frequently prefer not to. If randomization is stratified, then, to be strictly logical, a stratified statistical analysis should also be undertaken, which is more complex than it would otherwise be.

Perhaps the best compromise is to allocate animals purely at random and merely to record for each those characteristics (birth date, weight, or litter) that might have been used for stratification. During the eventual statistical analysis, the true relevance of these features to cancer incidence may be assessed; and, if (improbably) they do turn out to matter, then a 'retrospectively stratified' analysis can easily be done in which only 'alike' animals are compared with each other in order to reach an overall assessment of carcinogenicity. The logic is identical with that governing unstratified allocation and possible retrospective stratification in standard clinical trial methodology, as discussed, for example, by Peto *et al.* (1976, 1977).

(b) *Animal placement*

Two questions must be posed. First, should animals be caged individually? This costs more, but it facilitates husbandry and the avoidance of cannibalism. It also eliminates positive or negative correlations between the status of animals in one cage, e.g., those due to common infections or to the establishment of a social hierarchy which affects feeding, fighting and hormonal status. While individual housing of animals has been shown to produce stress (Hatch *et al.*, 1965; Sigg *et al.*, 1966), the aggressive behaviour often observed when animals are group-housed is likewise a stressful situation. When individually housed animals are attended to daily and are handled frequently during routine examinations, the stress of isolation is much reduced (Fox *et al.*, 1979).

The second question is whether cages containing treated and experimental animals should be intermingled, and, if so, whether this should be done merely by assigning whole rows of cage or racks of cages to particular treatment groups (perhaps moving those groups around systematically); or whether it is advisable in a four-group experiment to divide the whole room into 'blocks' of four cages, with one cage per treatment group in each block. Such arrangements would clearly be undesirable in studies involving volatile agents where cross-contamination might occur, and for such studies separate rooms (or at least effective barriers to cross-contamination) are probably more important than randomized position. In general, systematic differences between different groups in the height, light or ventilation of cages should be avoided.

2.11 Observations

(a) Clinical signs and mortality

Physical examination of each animal should be conducted by experienced personnel at least once a week and preferably daily. In the high-dose and control groups, ophthalmological examination should be carried out. If changes in the eye are detected, all animals should be studied in the same way.

Clinical signs and mortality should be recorded for all animals. Special attention must be paid to tumour development: the time of detectable onset, location, dimensions, appearance and progression of each grossly visible or palpable tumour should be recorded.

(b) Body weight and food consumption

Body weights of all animals should be recorded individually once a week during the first few months of the test period and at least once every four weeks thereafter.

Food intake should be determined for at least 20 animals/sex per group during the first few months of the study and then during periods of a few weeks at intervals. It is also desirable to monitor food consumption in inhalation studies and in other experiments in which the test compound is not administered in the diet.

(c) Haematology and clinical chemistry

Procedures such as repeated sampling of blood and urine are not a necessary part of a routine cancer test and, in addition, may affect the animals, thereby hampering interpretation of the results. If, therefore, routine haematological and clinical chemical examinations are deemed desirable in a carcinogenicity bioassay, satellite groups of animals should be included in the study so that, whether or not those animals are affected, the main experiment will not be.

(d) Pathology

Several recent papers contain detailed recommendations concerning necropsy facilities and procedures for obtaining optimum results from macroscopic and microscopic pathological examinations (Arnold *et al.*, 1977; Page, 1977a; Committee on Carcinogenicity of Chemicals in Foods, Consumer Products and the Environment, 1978; Food Safety Council, 1978; WHO, 1978b; Committee of the Health Council, 1979). Therefore, only a few of the major aspects to be considered when designing a long-term bioassay are mentioned here:

(i) Gross pathology should be performed on all animals by a trained laboratory animal pathologist or by experienced technicians under the guidance and direct supervision of the pathologist. The technicians must be able to recognize and properly describe gross lesions.

(ii) A check list should be available to ensure that all organs are inspected and sampled. Lesions observed must be recorded.

(iii) Personnel should be available during weekends and holidays to perform autopsies on animals found dead or killed *in extremis*.

(iv) A wide range of organs and tissues (listed by WHO, 1978d) from all animals should be preserved in an adequate fixative, such as 4% aqueous neutral phosphate-buffered formaldehyde solution.

(v) The urinary bladder should be inflated with fixative. All other hollow organs should be opened, either before or after fixation.

(vi) In inhalation studies, and preferably in all studies, the lungs should be fixed by intratracheal infusion of the fixative at constant pressure (e.g., 10 cm water pressure). Special attention should be paid to the processing of the nose (decalcification; at least four transverse sections taken at standardized sites) and of the larynx, trachea, main bronchi and pulmonary lobes (at least three longitudinal sections of each lobe).

(vii) Clinical signs as well as the results of chemical and haematological examinations should be available to the pathologist prior to macroscopic and microscopic examination.

(viii) A terminal blood smear should be prepared from all animals for diagnosis of leukaemia.

(ix) Detailed histological examinations should be conducted on the control and the highest effective dose groups. If no significant difference is observed between these groups, if clinical signs and laboratory findings do not indicate a specific disease entity and if no significant difference is observed in relevant gross lesions between control animals and those at the lower dose levels, then histological examination could be restricted to the lesions observed at gross necropsy in animals at the lower dose levels and to all sites where a significant lesion was observed in animals at the high dose.

2.12 Data acquisition, processing and storage

An essential part of the design of a long-term bioassay is an adequate system for collecting, processing, reporting, collating and storing the large amount of data produced during or as a result of the experiment. The person responsible for the study should issue instructions to the animal house and the various laboratories (e.g., haematology, pathology) on defined dates. This is most easily organized if there is a computerized data storage and retrieval system. In the long run, such an automatic system seems to be an inevitable necessity for each testing faculty, because of the ever-increasing complexity of chronic experiments.

Microfiches or originals of all raw data, log books, protocols, records, and interim and final reports pertaining to a carcinogenicity study, together with all slides and paraffin blocks, should be kept for a period (commonly ten years) after completion of the final report for possible external review. The storage period for wet tissues is to be left to the judgement of the pathologist, but should not be shorter than 12 months after completion of the final report. Applicable legal and administrative requirements should be adhered to when necessary.

2.13 *Animal husbandry*

Detailed procedures for animal husbandry can be found in the literature (Sontag *et al.*, 1976; Page, 1977b), in animal welfare publications and in codes of good laboratory practices (US Food and Drug Administration, 1978; Canadian Council on Animal Care, 1980, 1984; Institute of Laboratory Animal Resources, 1985). Therefore, it is enough to mention here a few main principles that are particularly relevant to long-term bioassays using rodents.

Factors involved in good laboratory practice, such as housing conditions, animal care facilities, intercurrent diseases, impurities in diet, air, water and bedding, cannibalism and autolysis, can significantly influence the outcomes of animal experiments. Animals should be kept in well-ventilated rooms with controlled lighting, temperature and humidity. Animals from outside sources should not be placed on test without a period of quarantine and acclimatization (see also Sontag *et al.*, 1976; Page, 1977b; Committee on Carcinogenicity of Chemicals in Food, Consumer Products and the Environment, 1978). Access to animal quarters should be restricted to individuals essential to the carcinogenicity study.

The control of chronic respiratory disease and other infections (e.g., parasites in the intestines and urinary bladder) is facilitated if rats and mice are bred and maintained under conditions free from pathogenic organisms. Maintenance of disease-free animals under conventional, clean conditions is perhaps even preferable to very strict specific pathogen-free conditions, since a full barrier system may be unnecessarily costly and considerably hampers biotechnical work.

Cages, racks and other equipment must be constructed of materials that permit easy and frequent cleaning. Cages to be used in inhalation chambers should be made of wire mesh (stainless-steel). The use of disinfectants and pesticides should be avoided in general, but if their use is necessary, care should be taken to ensure that equipment is adequately rinsed afterwards. In particular, in the case of inhalation studies, the use of formalin-containing mixtures should be avoided.

Animals should be identified individually at the onset of the study using a standard method of identification, such as ear clipping. Contaminated waste, faeces, dead bodies and radioactive waste should be eliminated according to applicable legal or administrative requirements.

2.14 *Diet*

Diet is an important environmental variable in carcinogenicity testing. The ingredients, nutrient composition and contaminant concentrations of the diet may influence the physiology and the health of the animals and thus, possibly, the animals' responses to chemicals. Therefore, the quality of the basal diet for long-term studies should be defined and standardized (Sontag *et al.*, 1976; Committee on Animal Nutrition, 1978).

(*a*) *Types of diet*

Diets for laboratory animals can be classified as 'natural-ingredient' diets and 'purified' diets (Institute of Laboratory Animal Resources, 1978). The natural-ingredient diets contain predominantly plant and animal products and can be further classified as 'open-formula' in which the composition is known and 'closed-formula' diets in which it is not (American Institute of Nutrition, 1977). There are advantages and disadvantages to each of these types of diet (Institute of Laboratory Animal Resources, 1978). If natural-ingredient diets are used, it is preferable to use the 'open-formula' diets to minimize the variability in response that can be introduced when the ingredients or their proportions are altered unknowingly (American Institute of Nutrition, 1977). Purified semisynthetic or synthetic diets are not generally used as yet in carcinogenesis bioassays because of uncertainty about their effects on the development and growth of tumours.

(*b*) *Nutrients*

The quantitative nutrient requirements for most species of laboratory animals may differ by strain or stock and by the stage of the life-cycle (growth, reproduction and maintenance) (Committee on Animal Nutrition, 1978). The nutrient requirements of laboratory animals, except for growth and in reproductive stages, have not been well established. Nutrients in the diet may interact with test chemicals and influence the toxic and carcinogenic responses of laboratory animals (Rogers *et al.*, 1980; Conner & Newberne, 1984).

Dietary protein concentrations within the range required to support growth do not consistently decrease or increase carcinogenic responses to chemicals (Newberne, 1974). Severe protein deficiency may retard tumour growth and depress hepatic drug metabolizing enzymes (McLean & McLean, 1966). Animals on high-protein diets are less sensitive to the toxic effects of some pesticides; low dietary protein is associated with higher toxicity of cholinesterase inhibitors and chlorinated hydrocarbons (Shakman, 1974). High levels of protein may markedly increase the severity of spontaneous kidney lesions in rats and hamsters (Newberne, 1978; Feldman *et al.*, 1982).

The protein requirement for maintenance of adult animals is much lower than for growth, reproduction and lactation (Committee on Animal Nutrition, 1978). Most of the diets recommended for long-term studies (American Institute of Nutrition, 1977) are formulated for growth and reproduction and may contain an excess of some nutrients, such as protein, for maintenance. However, the use of one diet during the growth phase and another (e.g., low-protein diet) for maintenance after the animals are five to six months of age in the long-term study would require changing the diet during the course of a study. Advantages and disadvantages of changing the composition of the diet during the course of long-term carcinogenicity studies of unknown chemicals are yet to be determined and practical logistic problems in implementing such changes in the conduct of chronic studies should be considered.

Dietary fat levels have not consistently increased or decreased tumour development due to chemicals (Haseman *et al.*, 1985). The ingestion of high-fat diets, however, is associated with a high incidence of mammary tumours in laboratory animals (Rogers, 1983). The essential lipotropic methyl-donating nutrients (methionine, choline, folic acid and vitamin B_{12}) in rodents may influence the toxicity and carcinogenicity of chemicals, but not in a consistent pattern (Rogers & Newberne, 1980). A fat content of approximately 5%, containing mostly polyunsaturated fatty acids, appears to be adequate for growth and maintenance. High-fat content may increase caloric intake, obesity and mammary tumour incidence, and would appear to have no consistent advantage in evaluating unknown chemicals for carcinogenic activity. Fat contents of diets may be increased in retrospective studies, however, to elucidate the mechanisms of toxic and carcinogenic responses observed with some chemicals. Low-fat content (e.g., $<1\%$) may decrease the palatability of food and may not provide adequate levels of the essential fatty acids required for the growth and well-being of laboratory rodents.

Consistent effects on tumour induction or development have not been demonstrated within the range of vitamin and mineral concentrations sufficient to maintain growth (Newberne, 1974). However, marked deficiency of vitamins A, B_{12} and E and selenium and amounts several-fold higher than the normal requirements of these micronutrients could markedly influence tumour induction and growth in the presence of some chemicals.

(c) Contaminants

The natural-ingredient diets may contain chemical contaminants contributed by the ingredients used to formulate the diets. Among the contaminants of concern in long-term studies are N-nitrosamines (Edwards *et al.*, 1979; Walker, 1979), heavy metals, mycotoxins (Page, 1977a,b), pesticide residues, halogenated polycyclic hydrocarbons and oestrogens. The concentrations of these contaminants must be limited to achievable practical levels. This can be accomplished by establishing quality-control procedures in the selection of ingredients for these diets.

(d) Purified diets

Purified diets may be the most appropriate diets for animals in chemical carcinogenicity studies, mainly because the refined ingredients used in these diets can be well defined, can be reproduced year after year without much variation, and may be free of enzyme inducers, antioxidants, and other contaminants. However, standardized purified diets considered optimal for long-term studies on different species are not well established, and adequate historical data on growth, survival, incidences of neoplastic and non-neoplastic lesions in animals on these diets are not yet available. They are not recommended for routine bioassays.

(e) Storage of diets

Natural-ingredient diets may be stored in a cool, dry place or in an environment with controlled temperature and relative humidity that is similar to that of the experimental animal room. Under these conditions, all the nutrients of the diet will be adequate for more than 168 days (Fullerton *et al.*, 1982), and so this type of diet may be used for five to six months from the date of manufacture. Storage of natural-ingredient diets at 4°C may not be necessary. However, natural-ingredient diets to which vitamin C has been added may be used for only three months after manufacture due to the gradual loss of vitamin C activity. Use of radiation to eliminate microorganisms after packaging the diet may markedly increase the shelf-life of these natural-ingredient diets, but possible denaturation of protein may introduce another variable. If purified diets are selected, they should be stored at 4°C to reduce the loss of vitamins A and B_1 and to retard rancidity or peroxide formation (Fullerton *et al.*, 1982).

(f) Feeding of animals

The test animals should become accustomed to the type and texture of the diet prior to the start of exposure to the test chemical. Fresh feed must be supplied at least once a week, preferably twice a week, or as often as necessary to provide adequate diet for daily consumption. If the test chemical is incorporated into the basal diet, it should not markedly influence the palatability at the high dose selected for the long-term study. Decreased palatability will cause decreased caloric intake and may amount to diet restriction. Decreased caloric intake causes lower body-weight gain and may markedly decrease tumour incidence (Tannenbaum & Silverstone, 1957; Gilbert *et al.*, 1958; Ross & Bras, 1971). The nature and cause of a reduction in feed consumption or in body-weight gain should be established by a subchronic study.

The chemical-diet mixtures may be prepared once every two weeks or more often, depending upon the stability of the test chemical in the diet. In studies involving chemically inert substances, the chemical-diet mixture may be prepared at longer intervals. Regardless of the frequency of chemical-diet mixing, precautions must be taken to prevent microbial contamination during mixing, and the mixture should be stored at about 4°C in air-tight containers to retard microbial growth.

If the test material is a nutrient and can be metabolized to supply caloric intake (e.g., irradiated food product, industrially-treated protein), high levels of these materials may induce nonspecific changes related to caloric intake and other nutrients (Conner & Newberne, 1984). If it is necessary to study these materials at high levels in the diet (20-60%), the diet at each dose level may have to be reformulated to balance the nutrients and caloric value.

(g) Documentation

Long-term study protocols should specify the type of diet to be used, with ingredients and formulation. Each batch of diet used should be analysed for macronutrients to confirm the formulation, for selected labile micronutrients to ascertain adequate levels, and for contaminants that might be present in the selected ingredients. In addition, randomly-selected batches of diet should be analysed more extensively for micronutrients to confirm the adequacy of the diet for the strain and species on study. Such analyses may be done by the manufacturers or the users of the diet, and the results should be retained and included in the final report on each test material.

2.15 *Safety measures*

Each test compound should be regarded as potentially hazardous (carcinogenic) to humans. Thus, precautions should be taken to prevent inadvertent exposure of personnel and the environment (Sontag *et al.*, 1976; Page, 1977b; Sansone *et al.*, 1977).

Major elements of a safety and health plan for the handling of potential carcinogens and toxic substances include:

(*a*) precautions for transport and storage of the test material within the laboratory;

(*b*) protective clothing for all personnel who may be exposed to the test substance or to contaminated apparatus or animals;

(*c*) disposal of contaminated waste, including faeces and dead bodies;

(*d*) cleaning of contaminated cages, rooms, ventilation systems, mixers and other equipment;

(*e*) protection of maintenance workers and emergency personnel;

(*f*) reduced pressure in animal rooms;

(*g*) written and posted instructions for emergencies;

(*h*) reporting and recording of all accidents; and

(*i*) medical surveillance of all individuals involved in the operation for evidence of absorption of the test chemical.

After oral administration or skin application of a volatile test compound, inhalatory contact of biotechnicians with the test compound should be avoided as far as reasonably possible, for example, by housing the test animals in ventilated safety cabinets. Suppression of dust from diets may be done by the addition of small amounts of oil. Personnel may have to wear safety equipment such as masks. Inhalation chambers should be operated at slightly negative air pressure to prevent leakage (WHO, 1978a).

Information on the above and other aspects of the safe handling of carcinogens in the laboratory is given in an IARC publication (Montesano *et al.*, 1979). Methods for the decontamination and destruction of several classes of chemical carcinogen in laboratory wastes are given in another series of IARC publications (Castegnaro *et al.*, 1980, 1982, 1983a,b, 1984, 1985a,b, 1986).

2.16 *Written working protocol*

For each bioassay, a detailed protocol should be prepared, indicating what procedures are to be followed during each sacrifice, procedures for determining body weight and feed consumption, necropsy procedures, when interim sacrifices will be performed, etc. Another function of a working protocol is to delineate the responsibility of each individual associated with the bioassay. It is also used by each laboratory (e.g., chemical, clinical) that will participate in the bioassay programme to ensure that its capacity will not be exceeded.

3. Conduct of experiments

3.1 *Introduction*

Although many monographs and review articles deal with procedures to be used in the conduct of chronic studies (Barnes & Denz, 1954; Shubik & Sice, 1956; Boyland, 1958; Clayson, 1962; Della Porta, 1964; Weisburger & Weisburger, 1967; Arcos *et al.*, 1968; Loomis, 1968; Berenblum, 1969; WHO, 1969; Benitz, 1970; Magee, 1970; US Food & Drug Administration, 1971; Roe & Tucker, 1973; Peck, 1974; Dayan & Brinblecombe, 1978; Food Safety Council, 1978; WHO, 1978e), it is apparent that not all laboratories performing chronic studies are using optimal experimental methods. Rigid guidelines for experimental procedures to be used in chronic studies are inappropriate due to the continual updating of toxicological methodology. In addition, initiative and ingenuity are required of the toxicologist if the study is to provide the most useful and relevant information. This view is in keeping with those expressed by previous WHO scientific groups (WHO, 1969). However, certain well-established principles must be adhered to in the conduct of chronic studies if they are to provide reproducible results that will be acceptable internationally.

Inherent in the conduct of any chronic study is the requirement for an animal facility that provides adequate control of the environment within the animal room, allowing only minor changes in temperature and humidity. In close proximity to the animal rooms there must be adequately equipped laboratory facilities appropriate for the conduct of the clinical and post-mortem examinations that are integral to such studies. A detailed discussion of these items is found in the literature (Jonas, 1965; Weihe, 1971; Poiley, 1974; Institute of Laboratory Animal Resources, 1976).

The objective of any chronic study is to ascertain what effect repeated administration of a chemical will have upon tissues or organ systems in test animals of either sex. The attainment of this objective requires:

(*a*) a well-devised and explicit protocol, coupled with sufficient supervision to monitor daily activities, to ensure that all items of protocol and any changes therein are understood and are being followed. Any deviation from the protocol must be well documented as to the reason(s) for the deviation, its extent and its nature.

(*b*) technical staff who thoroughly understand their responsibilities and duties, as well as a management that recognizes the importance of the technical staff in the conduct of a carcinogenicity study and is supportive of the staff;

(*c*) a record-keeping system that is accurate, reliable, secure and complete (Cranmer *et al.*, 1976);

(*d*) a health monitoring programme that will ensure accurate diagnosis of disease or toxic states, with a minimal loss of tissue samples for histological examination;

(*e*) accurate and continuous identification of all specimens, comprising precise marking of date, study, animal and organ;

(*f*) a detailed record of gross findings and the corresponding histopathological examinations; and

(*g*) an archive for storing the test data, protocols and specimens to allow for possible reevaluation in the light of future studies.

3.2 *Technical staff and their responsibilities*

Any toxicological study is dependent upon the competence and diligence of the technical staff working with the animals, on their aptitude for handling animals and their willingness to care for them. There is a growing recognition that technical staff should receive appropriate formal education and training so they are better able to assist in the conduct of toxicological studies. This need is being met in some countries by various technical institutions and associations that provide certificate programmes. Any certification or degree programme should be supplemented by continuing in-house training.

3.3 *Receipt and quarantine of test animals*

The acquisition of healthy animals from a reliable producer is the initial step towards obtaining accurate and reproducible experimental results. Many commercial breeding laboratories now monitor their animals for microbial and viral status as part of a continuing programme to upgrade the quality of their animals. Although such animals tend to cost more, when total costs are considered, disease-free animals are cheaper to use than diseased animals. It is desirable to examine these animals again in the research laboratory to ensure and document their health status before and during the experiments.

It is essential that the experimental animals be treated according to principles and practices that ensure humane care.

(a) *Disease surveillance programme*

It may be appropriate for institutions that undertake chronic studies on a routine basis to establish an animal disease surveillance programme. The animals to be used exclusively for the disease surveillance programme can be placed on test and killed at various times during the experimental period to ascertain through laboratory studies if latent disease is present (Institute of Laboratory Animal Resources, 1976).

(b) *Quarantine procedures*

Newly received animals are quarantined to prevent the introduction of unwanted pathogens and to provide a period for the animals to adjust to their new environment and recover from the stress of the shipment. Although it has been customary to quarantine animals in rooms that will not be used for the actual study, many commercial breeders supply animals of sufficient quality that they may be introduced directly into the experimental room upon receipt. The use of shipping boxes with polyester-fibre filters and the use of transit times often of less than 24 hours have minimized potential contact with infectious organisms and have reduced transit stress. Apart from latent viral infection, the outbreak of disease within a few days of the arrival of animals is generally attributable to situations inherent in the supplier's animal colony. However, in-transit stress could activate a latent disease condition that may not have been readily apparent in the supplier's colony.

To minimize the introduction of unwanted organisms into an animal house, appropriate procedures must be established. Such a programme starts with a standard operating procedure for the disposition of all incoming animals. The technical staff must be familier with this procedure and know who to contact when problems arise. The following husbandry practices are beneficial in this regard: each room is sanitized before new batches of animals are housed therein; specific procedures concerning the traffic flow of personnel, supplies and equipment are established; and frequent damp mopping of floors and sponging of walls and places where dust might accumulate is undertaken to reduce formation of aerosols.

Several physical conditions and management procedures that help to minimize the spread of disease are the following:

(i) Each room should have 100% exhausted air with individual temperature and humidity controls. The potential problems associated with recirculated air (Institute of Laboratory Animal Resources, 1976; Sansone & Fox, 1977; Sansone *et al.*, 1977; Sansone & Losikoff, 1978) and faulty environmental controls (Fox, 1977) are well documented.

(ii) Monitoring devices should be affixed to the washing equipment to ensure that the water temperature is appropriate for sanitation.

(iii) Preferably, water bottles should not be refilled between washings, i.e., a clean, freshly-filled water bottle should be used.

(iv) The diet should be kept in special rooms under dry conditions to avoid contamination or contact with unauthorized persons.

(v) Eating, drinking, smoking or the application of cosmetics should not be allowed in the animal rooms.

(vi) Personal clothing should be covered by appropriate garments, or, preferably, personal clothing should be removed and only suitable garments be worn in the animal rooms. Footwear should consist of safety shoes with steel toes and skid-resistant soles and heels. Neither the footwear nor outer garments should be removed from the animal quarters except when the outer garments require laundering. Only in-house laundries should be used to launder garments soiled with hazardous test chemicals.

(vii) Animals in solid-bottom cages should be transferred at least once a week to clean, sanitized cages; the papers in collection trays under suspended cages should be changed at least once a day and the cages sanitized at least twice a week. Testing of chemicals that produce diuresis or diarrhoea will require alteration of these schedules.

(viii) Preventive maintenance should be performed on the ventilation system, scales and other mechanical equipment at frequent, prescribed intervals.

3.4 *Selection of caging/bedding*

Selection of the correct caging system for chronic studies is critical, because not only does the cage microenvironment directly influence the biological response of the animal to the test chemical (Fox, 1977) but there have been instances in mouse studies in which the incidences of skin (Fare, 1965) and mammary tumours (Andervont, 1944) have been affected by the caging system. The type of caging selected should minimize the spread of communicable disease and the formation of the test chemical into an aerosol (since this may pose a potential health hazard for laboratory personnel and could inadvertently expose the control animals to the test chemical).

The ideal number of animals per cage, i.e., that which is least stressful to the animals, permits their ready scrutiny, minimizes cannibalism, allows for accurate estimates of feed consumption, and which is economically acceptable, has yet to be established for chronic studies.

(a) *Laminar flow caging*

This system provides unidirectional air flow after passage through a high-efficiency particular aerosol (HEPA) filter, which removes particles as small as 0.3 μm from the air. This system also reduces variability in the microenvironment and may enable animals to maintain homeostasis, thus more easily minimizing variation in basic physiological parameters (Fox *et al.*, 1979). However, HEPA filters do not protect the animal from volatile chemicals within the animal room, and air movement across the tops of the cages may help to distribute the vapours of a noxious test chemical within the animal room. Laminar air-flow systems with a negative air-flow movement across the animal cage are also available. In these systems, the air passes through HEPA filters prior to recirculation within the animal room or exhausting to the outside.

(b) *Solid-bottom caging*

This type of caging is fabricated from such materials as stainless-steel, polycarbonate or polypropylene. The cages can be fitted with filter tops, which decrease the likelihood of exposure to aerosolized microorganisms but increase the levels of humidity, ammonia, carbon dioxide and temperature in the cage (Serrano, 1971; Briel *et al.*, 1972). Temperature and gas concentrations within the cage are affected by several other factors as well (Simmons *et al.*, 1968). Such conditions may affect experimental results, since ammonia levels in excess of 25 ppm increase the severity of some disease states in rats (Broderson *et al.*, 1976).

This type of caging affords rodents a greater opportunity for coprophagy, which is normal and necessary for them, whereby the test chemical and its in-vivo metabolites are reingested.

(c) Wire-bottom caging

The stainless-steel, wire-mesh cage offers the convenience of durability and individual housing while conserving space and providing ready visibility of animals. However, in these cages the animal is not protected from microbial agents, nor does the cage minimize contamination of the general environment from test chemicals being administered to the animals (Sansone & Fox, 1977). Other disadvantages of this caging system are that the wire-mesh floors can cause decubital ulcers on the plantar surface of rodents' feet during long-term studies; it is also associated with a low incidence of fractures, dislocation of jaws and injured teeth. Mice housed in this type of caging may require a higher ambient temperature than those housed in solid-bottom cages (Murakami & Kinoshita, 1978).

However, in inhalation studies, wire-mesh cages must be used during exposure to ensure a quick and uniform distribution of the material in the atmosphere.

(d) Bedding

The ideal bedding material should be absorbent, dust-free, sterilizable and not contaminated with pesticides or other materials that might produce physiological changes. While there are brands of commercial bedding that have these properties, cedarwood or white-pine bedding may produce elevated levels of some hepatic microsomal enzymes and should be avoided (Wade *et al.*, 1968). Apparently, a mixture of beech, birch and maple does not create this problem (Vesell *et al.*, 1976).

(e) Additional considerations

Standard procedures have dictated that control and treated animals be housed within the same animal room, and often in the same racks, to avoid 'rack and room variability'. However, recent studies have shown that control animals may inadvertently be exposed to the test chemical under some experimental conditions (Sansone & Fox, 1977; Sansone *et al.*, 1977; Sansone & Losikoff, 1978). Therefore, it is not inconceivable that exposure of control animals to the test chemical could sufficiently distort the incidence of toxicological manifestations so as to result in erroneous experimental findings. Some investigators are now splitting or duplicating the control group for each chronic study — keeping one half in the same room as the tested animals and the second half in a room containing only control animals. Obviously, if such a practice is followed, the two rooms should be environmentally identical. Ideally, the only significant difference should be the fact that one group is at risk of exposure to the test chemical and the other is not, and, if this can be achieved by controlling the circulation of air and material in one room, the control group should share the same room as the treated group. In the case of inhalation studies, the control group is housed in a separate inhalation chamber.

3.5 Allotment to treatment groups

(a) Conventional studies

Due to the manner in which animal suppliers cage, select and ship test animals, proper randomization is essential in order to ensure an unbiased allocation of experimental animals to various treatment groups. A completely randomized design, which is simple and certainly adequate for most purposes, calls for the random assignment of animals to groups without regard for possible blocking factors such as litter status or body weight. Results from such a study may be assessed using statistical tests which provide a means of determining whether any observed difference among treatments is larger than would be expected if the experimental variation were due to randomization alone.

Alternatively, inter-litter variation may be controlled through the use of a litter-matched design, in which pups within litters culled to equal size are allocated randomly among various treatment groups. Since treatment comparisons are then between littermates, this design may be expected to be more efficient than the completely randomized design in the presence of appreciable litter effects, although special statistical techniques are required for the analysis of time-to-tumour data in order for the litter-matched design to remain advantageous (Mantel & Ciminera, 1979). Even if litter effects are negligible, the inefficiency of the litter-matched design relative to the completely randomized design may be minimal in a large-scale study (Scheffe, 1959; Fleiss, 1973).

Another allocation procedure that might be considered involves the random assignment of entire litters (culled to equal size) to individual treatment groups. This design may be expected to be less efficient than the completely randomized design, since any treatment comparison will be entirely between animals from different litters. In addition, the effective number of experimental units may be less than in the completely randomized design, since each litter, rather than each individual pup, would represent one unit. Even if litters of one size were used in order to maintain similarity with the completely randomized design in terms of the number of experimental units, the variation among units in the former case might be expected to exceed that in the latter case due to the larger cross-section of litters involved.

Since intentional caloric restriction has been shown to decrease tumour incidence both in rats (Ross & Bras, 1971) and mice (Gilbert *et al.*, 1958), differences in body weight may also be taken into account when assigning animals to treatment groups. It has been suggested that the available experimental animals be divided into several homogeneous weight groups prior to assigning individuals to treatment groups (Sontag *et al.*, 1976). As is the case with the litter-matched design, however, this block structure should, in principle, at least be taken into account in the analysis of the results obtained.

(b) *Two-generation studies*

Incorporation of transplacental and/or neonatal exposure into carcinogenicity testing has also been discussed (Grice *et al.*, 1981). Some aspects include:

(i) *Selection of second-generation animals*

One procedure that may be used for this purpose (Grice, 1978) involves culling litters to a maximal size of eight at four or five days of age in order to prevent natural selection of more vigorous offspring and to balance the burden placed on the dams. At weaning, an equal number of pups of each sex should be randomly selected from each litter to continue on test in the second generation. The selection of one pup of each sex per litter at weaning will ensure maximal statistical sensitivity of the study, although this will require an increased number of dams in the parent generation. While it may be argued that animals with physical abnormalities might not survive a chronic study and should therefore be excluded from the selection process, it is likely that the vast majority of anomalies which are not fatal at weaning are in fact compatible with longevity.

Another concern in the selection of second-generation animals is the relative magnitude of inter- and intra-litter variation in transplacental exposure. A recent study, in which pregnant rats received a single dose of ^{14}C-saccharin *via* the jugular vein on the 20th day of gestation, demonstrated that the inter-litter variation exceeded the intra-litter variation by a factor of about three (Munro & Willes, 1978; Ruddick *et al.*, 1978). While there have been some investigations of whether body weight, prenatal mortality and the incidence of spontaneous congenital malformations are affected by uterine location (Kalter, 1975), no data are presently available regarding whether uterine location affects subsequent toxicological responses. The work of Ruddick *et al.* (1978) also showed greater inter- than intra-litter variation. Therefore, with the random allocation of the dams and sires in the parent generation, efficient and valid statistical inferences may be drawn from the F_1 generation if one male and one female are selected randomly from a litter.

(ii) *Dose*

A number of factors are to be taken into account when deciding what doses to administer to the dam and how these are reflected in the level to which the neonate is exposed. The dose that juvenile and F_1 adults receive must also be chosen carefully. For example, how is the time or duration of dosing prior to breeding established? Is the test compound fetotoxic, and, if so, does this limit the amount of compound that can be administered to the dam? Should the dosage levels for juvenile and F_1 adults have some relationship to human exposure levels? Such questions constitute a major challenge in use of this experimental protocol.

Other considerations regarding dose selection, which are applicable to both one- and two-generation studies, are whether the compound should be administered as a constant percentage of the diet or whether dietary levels should be changed weekly so that a constant weight per body weight is received by the test species. This is a particularly important point when young animals are started on test, since growing animals consume three to four times the amount of diet per kg of body weight as do adults. Consequently, the interpretation of experimental results regarding a possible no-effect level may be compromised due to the greater amount of test chemical consumed per kg body weight during the initial weeks of the chronic study.

(iii) *Other concerns*

The major potential for differences in effect between standard chronic exposure studies and in-utero protocols relates to the possibility that the test chemical is affected metabolically by the placenta and mammary glands. The chemical may also affect a developing tissue or organ from its primordial state to the stage of development reached at weaning, but the effect may not be apparent until several months later. Such effects would not be observed with any of the standard reproductive or teratogenic protocols.

Teratologists have been concerned with the relationship between the species tested and humans, due to differences in placental structure. These concerns will not be reiterated here since it is assumed that teratological studies would have been undertaken with a compound prior to initiation of chronic studies.

3.6 *Health monitoring*

Monitoring the health of animals is an essential component of chronic studies and requires attentive personnel, who, because of their training, experience and familiarity with the animals, are able to detect subtle changes in the animals' behaviour that may signal a toxic response, a disease process or infestation with parasites.

The importance of having well-trained personnel conduct thorough clinical examinations cannot be overstressed. One of the major faults with some previous chronic studies has been the high loss of tissues due to cannibalism and autolysis when animals died during the test. To compensate for such losses, the practice was to include more animals on test. However, it is not inconceivable that if the animals that died on test had been available for histological examination, a more precise appraisal of the toxicological properties of the test chemical would have been possible. Likewise, the animals that survive may be the most resistant to the toxicological effects of the chemical.

It is important that the individuals who are responsible for executing a health monitoring programme be familiar, too, with clinical signs that are indicative of tumour development and be able to differentiate these processes from inflammatory swellings, abscesses, cysts and other masses. A health monitoring programme will also help to define more precisely the latent period for tumours and to determine the appropriate time for euthanasia, thereby minimizing tissue losses due to autolysis. The achievement of these objectives depends on the training of the technical staff and their ability to select and use appropriate clinical and laboratory procedures (Arnold *et al.*, 1977, 1978).

(a) *General observations*

Several clinical observations provide a good indication of an animal's general health status. If the animal's physiological habits appear normal, i.e., it is eating, drinking, defecating, urinating and moving about in a normal fashion and its coat is sleek and the mucuous membranes are a normal colour, then it is likely that death is not imminent. This is a very basic tenet, and all personnel should be attuned to these features of the animal's general health status. The experience, training and responsibility of the technical staff concerned with monitoring animal health has been discussed recently (Arnold *et al.*, 1977), and only a summary is presented here.

(b) *Routine daily monitoring*

It is preferable that the individuals responsible for the day-to-day health monitoring and care of the animals be assigned to the study for its duration. This allows them to gain familiarity with the test animals, so that subtle behavioural changes are readily noticed, and the animals are not unduly stressed by the handling techniques of several persons.

Each animal is monitored twice a day — first thing in the morning and last thing in the afternoon. This does not mean that the technician merely glances casually into the cage to see if the animal is still breathing. On each inspection, the technician opens every cage to assess the general health status of the animal, as described above. If these features are not normal, a more extensive examination is required and could include visual examination for:

(i) *Behavioural status*: observing the animal's response to the disturbance of the examination — normal activity, inactivity, hyperactivity. If the animal does not respond in the expected manner, it should be picked up and examined more closely.

(ii) *Respiratory signs*: nasal discharge, its nature, texture, colour; respiratory rate and depth; dyspnoea; abnormal breathing.

(iii) *Skin/fur*: the fur or hair should be sleek and close to the body. Commonly observed signs that are indicative of altered health status include dry or wet fur, alopecia, loss of hair gloss, and erect, unkempt or unclean hair.

(iv) *Eyes and mucous membranes*: disease signs include chromodacryorrhoea, discolouration, abnormal secretion or discharge, and lack of normal colouration. Pale mucous membranes suggest anaemia and, possibly, the need for a haematological evaluation.

(v) *Bleeding*: from various body orifices or surfaces. Attempts should be made to establish the site and nature of the bleeding.

(vi) *Excretory products*: amount, colour, presence of blood, consistency and shape of faecal pellets. The use of wire-mesh cages facilitates such observations, but with solid-bottom cages the animal must be removed from the cage and possibly induced to defecate and urinate.

(vii) *Food and water consumption*: in the case of decreased or increased food and water intake, the body weight should be measured and clinical tests performed.

(viii) *Muscle tone*: when the animal is handled, its general response and muscle tone should be determined.

(ix) *Tumour development*: the first occurrence of palpable and/or visible tumours and their localization should be documented. The growth behaviour of the tumours should be described by recording daily their size and appearance. Animals bearing tumours must be observed more attentively than others.

Changes in the health status of individual animals are recorded, using a list of common clinical signs in technical and lay terms, in the individual animal's clinical file. It is important that complete and accurate clinical files be maintained, since the pathologist requires access to them when performing autopsies and interpreting histological slides.

(c) Routine weekly observations

It is common practice to record and statistically analyse data on feed consumption and body weight during a chronic study. However, if the test chemical is administered in the drinking-water, the amount of water/test chemical consumed must also be recorded on a routine basis. Most institutions monitor body weight and feed consumption on a weekly basis for the first few weeks and then switch to a two-weekly or monthly schedule for the remainder of the study. Since animals must be handled to determine their body weight, it is a convenient time to palpate and examine each animal in more detail than during the daily visual examination. As the animals get older and the diseases associated with ageing start to become apparent (from about 18 months in rats and 14 months in mice), it may be appropriate to increase the frequency of determinations of body weight and feed consumption, possibly returning to weekly determinations. If the test chemical induces tumours prior to this time, the reinstatement of weekly examinations is well advised. As the animals become older, more emphasis must be placed upon weekly and detailed examinations, in order to preclude unmonitored mortality.

All clinical examinations that include palpation are best conducted early in the morning to preclude unmonitored mortality losses due to internal haemorrhaging, as may occur with large liver tumours.

(d) Detailed clinical examinations

A detailed clinical examination is not conducted on a scheduled basis but is performed on sick animals by highly skilled individuals. It is designed to provide a diagnosis, which occasionally may require clinical laboratory confirmation, as to whether the sick animal is afflicted by a communicable disease or by toxic manifestations. It should include:

(i) visual examination for signs of disease processes, such as pale mucous membranes (anaemia), yellow discolouration of ears and/or mucous membranes (jaundice), external changes, nodules, abrasions, cuts, other lesions; oral cavity anomalies, for example, malocclusion of teeth and excessively long teeth;

(ii) palpation of the unanaesthetized animal for subcutaneous nodules, lymph node enlargement, and changes in abdominal viscera (Generalized enlargement of the superficial lymph nodes may suggest malignant lymphatic disease, while a concurrently enlarged spleen may provide further evidence of haematopoietic malignancy.);

(iii) assessment of gait and basic postural reflexes for detection of possible neurological deficit;

(iv) body temperature;

(v) visual assessment of excretory products (diuresis or diarrhoea), as well as possible urinalysis and testing of faeces for occult blood;

(vi) auscultation with a stethoscope for abnormalities of the heart and lung;

(vii) assessment of changes in body weight and food consumption;

(viii) sensitivity to pain (i.e., tail pinch) and other neurological tests; and

(ix) ophthalmoscopic examination of the eyes.

The findings of these observations are likewise recorded in the animal's individual clinical file.

If the tentative or presumptive diagnosis suggests a health disorder, the animal cage should be flagged or the animal put in a special rack in the same animal room, to ensure that it receives additional observation.

(e) Isolation of sick animals

A segregated rack within the test room may be utilized for the purpose of housing sick animals, or each cage containing a sick animal may be tagged, thus allowing immediate scrutiny of these animals. However, the isolation of sick animals is imperative if group housing is used. The animals placed in the segregated rack are monitored at least twice a day, and weight change and feed consumption of each animal are determined daily. The evaluation of these data and the results of a detailed clinical examination provide the basis for deciding whether the animal should be returned to its original cage, remain where it is, or be killed.

The isolation of sick animals during inhalation experiments is difficult, unless they are not exposed during their illness, which effectively removes them from the experiment.

Animals are usually placed in segregated racks for one or more of the following reasons:

(i) consistent weight loss, often concurrent with anorexia and adypsia;

(ii) overt and occult blood loss: moderate to severe haematuria, which may be indicative of a variety of pathological changes in the urinary tract; vaginal bleeding; malaena; cuts or abrasions;

(iii) apparent anaemia of unknown cause;

(iv) severe malocclusion;

(v) respiratory difficulties;

(vi) markedly enlarged liver, spleen or lymph nodes;

(vii) abdominal tumour or large mammary tumour(s);

(viii) cutaneous or subcutaneous tumours with erosions, ulcers and/or bleeding;

(ix) central nervous system abnormalities (most commonly, these are associated with pituitary tumours);

(x) urinary bladder distension;

(xi) failure to urinate or defecate;

(xii) ascites; and

(xiii) marked lethargy and weak muscle tone.

3.7 *Diagnostic laboratory procedures*

The degree of sophistication that can be employed while undertaking diagnostic procedures with animals involved in a chronic study is somewhat limited by factors inherent to small animals. However, following the detailed clinical examination described above, a technician should be able to diagnose most of the diseases commonly encountered with the strain of animals on test (Fox, 1977).

Several laboratory procedures can be used to assist the toxicologist and veterinarian in diagnosing or confirming disease conditions.

(*a*) *Biochemical analysis*

The advent of micromethods has made it possible to take blood samples for biochemical analysis without unduly stressing an animal by blood loss. While there is a wide variety of tests to choose from, each has its limitations, and the choice may often be made because a particular test is that with which there is familiarity or for which a correlation exists with a known, specific pathological change. Due to homeostatic mechanisms, many of the determinations of serum enzymes do not detect minor or subtle pathological changes or may reflect only transient changes in organ homeostasis that produce no lasting toxicological effect (Grice *et al.*, 1971; Korsrud *et al.*, 1972).

(b) Haematological tests

Although various haematological tests are often performed at intervals prescribed by the protocol, they may also be undertaken for specific diagnostic purposes. The basic haematological profile constitutes one of the important aspects of monitoring the clinical status of a laboratory animal. Changes observed in the blood and/or the blood-forming organs may be physiological or pathological, experimentally induced or spontaneous. Relatively simple tests may be useful tools for diagnosing haematological disorders or assessing the extent of haematotoxicity (Linman, 1966).

When one or more of the following conditions are observed, a haematological assessment of the animal's condition is often necessary:

(i) overt clinical anaemia;

(ii) conditions that could result in anaemia, such as haematuria, melaena, haemorrhagic tumour, vaginal bleeding; and

(iii) enlarged spleen, liver or lymph nodes, or physical trauma.

(c) Urinalysis

Urine examinations might be performed when problems in the urinary tract become evident from such clinical signs as anuria or polyuria, haematuria, palpable kidney abnormality, or suspected bladder tumour.

Overnight or 24-hour urinary collections provide information on urine volume and the urinary excretion of various substances (i.e., protein, calcium, sodium, phosphate, creatinine, chloride and potassium) but are considered unreliable for assessing parameters such as urinary pH, presence of microcalculi, malignant cells and osmolarity (Arnold *et al.*, 1977).

(d) Faecal examination

If either soft stools or diarrhoea is observed, a sample of faeces should be submitted to the laboratory for examination to aid in diagnosis of the condition.

3.8 *Differentiating disease and toxic response*

Many of the common diseases previously observed in laboratory animals have either been dramatically reduced or eliminated. The major problem remaining at the present time is the presence of latent oncogenic viruses (Canadian Association for

Laboratory Animal Science, 1986). Many of the diseases common to laboratory rodents have been reviewed elsewhere (Balazs, 1961; Fox, 1977; Burek, 1978; Holmes, 1984). The following is a very brief summary of the more common diseases, by organ system.

(a) Integument

The severity of dermal lesions may range from alopecia to severe dermatitis. Abnormal nutrition, hormonal disturbance, fighting, trauma from superficial wounds and even grooming may cause dermal lesions. Various infectious agents, which may be mycotic, bacterial or viral, can cause dermatitis, which may be exacerbated by physical or dietary factors (Fox, 1977).

Decubital ulcers on the plantar surface of the feet can result in substantial blood loss, leading to anaemia and enlargement of the regional lymph nodes. This problem is primarily associated with the maintenance of rodents on wire-mesh cages for prolonged periods.

The occurrence of ectoparasites in laboratory animals has declined in recent years; but several species of fleas, flies, lice, ticks and mites can produce such effects as restlessness, scratching, allergic responses, oedema, erythema, scar formation, anaemia and, in severe instances, an unthrifty appearance or death. Details of the diagnosis and control of these various ectoparasites have been discussed extensively by Flynn (1973). While some infestations with ectoparasites may produce ulcerative dermatitis, it should be ascertained that the lesion is not of bacterial origin (Fox *et al.*, 1977).

(b) Respiratory system

Pneumonia and upper respiratory disease are caused by a variety of bacteria, viruses and mycoplasma, and respiratory infections are refractory to treatment with antibiotics. However, barrier-maintained, caesarian-derived animals can be used to avoid many of the organisms that cause these diseases (Fox, 1977).

(c) Optic system

Disease processes in the eye can be indicative of metabolic disorders or systemic infections. Cataracts, which are more common in older rodents, may indicate the presence of diabetes. Conjunctivitis can result from several bacterial infections, and chromodacryorrhoea suggests the presence of sialodacryoadenitis virus (Fox, 1977).

Infections or tumours of the Harderian glands can produce exophthalmus or blepharitis in rodents (Heywood, 1973; Jacoby *et al.*, 1975; Tucker & Baker, 1967).

It should be recognized that attempts to puncture the orbital sinus to obtain blood may result in severe damage to the eye.

(d) Auditory system

Rodents kept under conventional conditions are susceptible to middle and inner ear infections caused by several microorganisms (Fox, 1977), although in some instances this disease appears to be age-related (Matheson et al., 1955). From a diagnostic viewpoint, middle-ear infection is asymptomatic and becomes evident only when there is involvement of the inner ear with accompanying signs of disequilibration.

(e) Urinary system

Infectious diseases of the urinary system are relatively infrequent. A degenerative condition of the kidney that is seen frequently in older rats of some strains, and to a lesser extent in older mice and hamsters, has been termed, variously, chronic nephropathy, chronic progressive nephrosis, glomerulosclerosis, glomerular hyalinosis and tubular atrophy (Anver & Cohen, 1979; Schmidt et al., 1983). Primary amyloidosis is a common degenerative renal disease observed in Syrian hamsters, in which amyloid deposition in the renal glomeruli begins in animals under one year of age and progresses with age (Casey et al., 1978). Possible clinical signs of degenerative renal disease include chronic wasting, pronounced proteinuria, ascites, diuresis and anuresis and uraemia.

(f) Mammary gland

While non-neoplastic lesions are relatively uncommon in rats, mammary tumours can be induced with relative ease, and considerable numbers of mammary neoplasms develop spontaneously in most rat strains (Greaves & Faccini, 1984). Spontaneous tumours of the breast are very rare in hamsters and have a variable frequency in different mouse strains, depending on differences in genetic susceptibility and the presence or absence of milk-transmitted 'mammary tumour virus' (Squartini, 1979).

(g) Digestive system

Malocclusion and overgrowth of incisors are encountered in several laboratory species, but they can be easily trimmed with bone-cutting forceps.

Most rat colonies are relatively free of gastrointestinal disease, while mice are affected by several latent viruses that can affect the gastrointestinal tract as well as the liver and spleen (Fox, 1977). In hamsters enteritis is one of the most common diseases (Schmidt et al., 1983). Some clinical signs that might be observed in conjunction with gastrointestinal disease include abdominal distension and sensitivity to handling and palpation.

(h) Cardiovascular system

Thrombosis of the heart chambers, particularly of the left atrium with concomitant congestion of the vascular system is frequently observed in Syrian hamsters (Pour et al., 1976). Aged rodents, mainly rats, quite commonly show mild focal myocardial degeneration, necrosis and fibrosis (Schmidt *et al.*, 1983; Greaves & Faccini, 1984). Intimal plaque formation, medial degeneration, calcification of vascular walls and arteritis have been reported as frequent, age-related vascular lesions in rodents (Ayers & Jones, 1978).

(i) Pituitary gland

Some strains of rats have a particularly high incidence of spontaneous pituitary tumours; these are age-related, and, in female Sprague-Dawley and Fischer 344 rats from some suppliers, they are seen in almost every animal over 18 months of age. Large pituitary neoplasms are sometimes clinically associated with signs of lethargy and progressive anorexia but may be clinically silent.

(j) Thyroid gland

Hyperplasia of the thyroid C cells, which may lead to adenomatous proliferation and malignant C-cell (medullary) carcinomas, is a common finding in ageing rats of certain strains, especially in Long Evans rats (Boorman & DeLellis, 1983). Such tumour-bearing animals have been reported to have elevated serum calcitonin levels (DeLellis *et al.*, 1979).

(k) Adrenal gland

While relatively infrequent in most rat and mice strains, neoplasms of the adrenal gland, particularly adenomas of the adrenal cortex are the most common spontaneous tumours in Syrian hamsters (Cardesa *et al.*, 1982). Hamster adrenal glands are also frequently involved in generalized amyloidosis with intense amyloid deposition and degeneration (Schmidt *et al.*, 1983).

(l) Other organs

Depending on the species and strain of animals, tumours in other organs such as Leydig-cell tumours of the testis, granulosa-cell tumours of the ovaries as well as lymphomas and leukaemias are very common in rodents.

3.9 Reducing losses due to autolysis

It is important to establish a set of criteria for determining when a test animal that is sick or moribund should be killed; these require a competent diagnostician. Numerous factors are to be considered. For example, if the presence of a potentially malignant tumour is suspected, the longer the animal remains on test, the more likely it is that the tumour will invade or metastasize and provide irrefutable evidence of malignancy. However, if the animal dies suddenly, the tissues may be lost due to cannibalism or autolysis. Similarly, if the animal suffers cachexia, due either to a disease or to an induced toxic state, the condition may obscure subtle changes associated with the effects of the test chemical. It is important, therefore, to kill the animal at the most appropriate time. Obviously, animals should not be killed prematurely simply to establish a low incidence of tissue loss in the experiment.

Clinical signs are not always sufficiently clear-cut to provide a tentative diagnosis, and there are no reliable diagnostic criteria for conditions such as mesenteric arteritis or periarteritis nodosa (Yang, 1965). Cases of sudden death are encountered in which there is no history of significant weight loss or decreased food consumption. At necropsy, the only gross pathological finding may be a slight enlargement of the liver or kidney. However, in such cases, microscopic examination quite often reveals interstitial pneumonitis or membranous glomerular nephritis.

3.10 *Observation of tumours at necropsy*

All grossly visible lesions should be recorded on the individual animal's necropsy sheet, including the location, size, number, shape, texture and consistency of tumours. The statistical methods that may be applied (section 4.4) might require information on whether a tumour found at necropsy was an incidental finding (i.e., the animal died of other causes) or whether it was fatal (i.e., contributed to the death of the animal). The context of observation of a tumour influences the choice of the statistical method. Determination of context of observation may be difficult, but one practicable compromise is to devise a four-point scale for each organ in which a tumour was observed, according to whether it was found:

(1) definitely in an incidental context,

(2) probably in an incidental context,

(3) probably in a fatal context, or

(4) definitely in a fatal context.

This classification should eventually be applied to both observations made at necropsy and histopathological evaluations of each animal with a tumour.

It is important that the sampling method used to select normal-appearing tissues and lesions for histopathological analysis be applied systematically to all animals in all groups.

3.11 *Histopathology*

For the histopathological evaluation of carcinogenesis bioassays, the quality of histological slide preparation is of critical importance, because the slides document morphological differences between treated and control animals directly.

Factors to be considered in pathological examination are the following:

(*a*) Gross tissue examination and collection should be conducted by qualified persons.

(b) Tissue samples should be fixed in appropriate sizes by methods that assure optimal preservation without loss of tissue due to autolysis.

(c) Further processing of tissue specimens, including sectioning and staining, should be carried out by trained technicians.

(d) An accurate record-keeping system must be chosen which documents exactly the origin of each tissue section for final diagnosis and reporting.

(e) The scope of selected normal and abnormal tissues to be examined microscopically must fulfill the conditions needed for statistical analysis.

(f) Histological examinations for carcinogenesis bioassays must be done by a qualified pathologist.

In certain cases, it might be necessary to use special staining procedures or ultrastructural examination.

The relevance of a bioassay also depends on the number of organs and the thoroughness with which they have been examined both grossly and microscopically. The fewer the tissue samples, the less the value of the study in providing evidence of a carcinogenic hazard.

3.12 *Data collection*

Many of the data recorded during a chronic study cannot be redetermined or verified even a few days after they were obtained. Consequently, it is imperative that all data collection be monitored closely, and, when an observation appears to be erroneous, it is best to verify it then. Once the data have been verified as being accurate, they must be stored, perhaps in duplicate, in a manner which allows for quick retrieval and analysis. Frequent, periodic reports are necessary during a chronic study to avoid missing any observation that was not expected on the basis of the results of subchronic studies and to preclude any unforeseen problems that might invalidate the study if not noticed in sufficient time.

3.13 *Waste disposal*

The problem of disposal of animal bedding, test diets and carcasses faces any institution in which animal studies are conducted. The procedure whereby such wastes are contained should be performed in such a manner as to minimize any possibility of contaminating other test animals, people or the environment. Before packaged wastes are removed from a facility they are usually collected and stored for short periods of time in one or more centrally located areas which are accessible to large vehicles. Such a collection area must be operated so as to minimize vermin infestations, odours and disease hazards without using volatile chemicals which could be spread throughout the facility *via* the ventilation system.

The ultimate disposal of such wastes in many countries is often dictated by local municipalities, which have either allowed such wastes to be used in sanitary landfills or required that they be incinerated, often at temperatures which are thought to reduce everything to carbon dioxide and water. However, in situations in which an institution is solely responsible for disposition of its wastes, they should be handled in such a manner as to protect the environment from any possible noxious agents contained therein.

4. Analysis and reporting

4.1 *Validation of data*

Before attempting to interpret the results of an experiment it is necessary to seek out errors in the data. Transcriptional errors may occur in collating many primary data elements: and, when machines are used to store data, systematic errors may occur.

Because experiments are performed to determine whether a dosed group is associated with a higher incidence of cancer than an undosed control group, it is important that the diagnoses of cancer be accurately and consistently recorded. The terms used to identify lesions should be the same for all groups of animals. When many pathologists have contributed to a study, it is necessary for one of them to review the diagnostic criteria and terminology employed by the others and to re-examine microscopically the possible target sites for tumour induction in doubtful cases to ensure uniformity and accuracy of diagnosis. Consideration should be given to appraisal of the performance of the pathologists, perhaps by review of a sample of the histopathological material by a disinterested pathologist. Review of a sample of coded tissue sections, of unknown relation to the experimental procedure, has also been proposed (Ribelin & McCoy, 1971; Weinberger, 1973; Fears & Schneiderman, 1974; Zbinden, 1975) and is used in some institutions as a quality control step.

4.2 *Adequacy of the experiment*

The first step in the analysis is to review generally the adequacy of the experiment, and to decide whether it is reportable. For example, do the results indicate that the doses used were in the proper range? Did sufficient numbers of animals survive long enough for the development of late-appearing tumours, or were there unforeseen events such as an outbreak of infectious disease that might invalidate the experiment?

4.3 *Chemically related effects on growth, morbidity and survival*

The administration of test chemicals may affect the growth and development of animals. When given in doses in the low toxic range, dose-related reductions in body weight may be observed, especially in the first months. It is desirable in a carcinogen screening test that one of the doses produce some clinical signs of toxicity, so that it can be inferred that the chemical was tested at, or near to, the highest possible level within the conditions of the test. This is particularly important for interpretation when the incidence of tumours in dosed animals is not increased. If administration of the chemical causes marked weight depression as compared with control animals, however, the dosed animals may have a lower rate of tumour formation, perhaps giving rise to false-negative or weakened positive results.

To the extent that clinical signs of toxicity can be quantified, those signs related to compound administration should be dose-related. This comparison may help check the effectiveness of the doses used. The first analysis to be performed in any long-term study should usually, therefore, be the construction of two simple data sets — one giving percent survival against age, and the other mean body weight of survivors against age.

4.4 *Categories of tumour*

One method sometimes used to evaluate the carcinogenicity of chemicals is to compare the incidences of total tumour-bearing animals in dosed and control groups. This endpoint is of little value in long-term animal experiments in which the majority of animals have one or more tumours; but it may be useful in shorter experiments, of perhaps six to 18 months, when the test compound has considerable carcinogenic activity or produces many types of tumour.

Since it appears that cancers arise independently in various parts of the body, it has become customary to treat each potential target site (e.g., brain, kidney, bladder) as a separate experiment for evaluation. It is not certain, however, that cancers arise independently in all experiments (Haseman, 1983); but, until new evidence is obtained, analysis of tumour incidences should continue to be based on the assumption that the formation of tumours at different target sites can be evaluated separately without serious error. One general exception is evaluation of types of tumours that may be multicentric in origin, including leukaemias and, possibly, tumours originating in blood vessels or nerves (e.g., haemangioendothelioma, neurofibrosarcoma).

Judgement is also required in grouping lesions for statistical analysis. The experienced pathologist uses his knowledge of histogenesis and pathogenesis to decide, for example, whether a carcinoma of the stratified epithelium of the upper larynx should be separated for evaluative purposes from a carcinoma in the respiratory epithelium of the lower larynx, not far away. In general, it is accepted practice to group together tumours of the same general histological type that arise in the same type of tissue. For example, all papillomas of the urothelium in the renal pelvis, ureter, bladder and urethra might be considered together, although the individual numbers should also be given. When judgements about groupings are difficult, it is desirable to evaluate them in both ways — divided and combined: e.g., should fibrosarcomas in all parts of the body be considered together when those in the subcutis are not uncommon in incidence and those in the spleen are rare? If each is statistically significantly different from its counterpart in control animals, interpretation is easier.

Benign and malignant tumours

Preneoplastic lesions have been described in a number of tissues, especially in the liver (Anon., 1976; Bannasch, 1984). These lesions appear prior to the development of certain tumour types in target tissues of the carcinogen (see Report 2 for details) and should be considered in the evaluation of the carcinogenic risk from chemicals in bioassays. When present along with the tumours, they add to the weight of evidence for carcinogenicity.

Sometimes, a tumour arises in an organ which seems to be benign for a long time but which ultimately progresses into frank malignancy. Indeed, it is frequently a matter of arbitrary definition (on which expert pathologists may disagree) as to how to designate those tumours that are on the borderline in the continuum between benign and malignant. Moreover, it seems that, although certain viruses are capable of producing strictly benign lesions with a negligible likelihood of progression (e.g., warts), few, if any, chemicals exist which produce only benign tumours (and no malignant tumour) in any species. Consequently, although in most experiments benign and malignant tumours are still reported separately, the practical difficulties that often arise in categorizing certain tumours as benign or malignant are no longer viewed as seriously as they once were. Chemical agents that markedly increase the incidence of benign tumours are now viewed as potential human hazards to almost the same extent as they would have been if the induced tumours had been malignant. Likewise, a chemical agent that produces a clear excess of some recognized precancerous lesion is also viewed with considerable suspicion. Finally, if a marginally significant excess of treated animals have tumours in a particular organ, but, in addition, there is a highly significant excess of preneoplastic lesions in that organ, then the experiment as a whole suggests that the test agent did indeed cause some of the tumours.

4.5 *Statistical analyses*

This section provides an overview of statistical issues with references for details of the appropriate methods. A comprehensive review is given in a recent IARC publication (Gart *et al.*, 1986).

(*a*) *Goals of statistical analysis*

Before statistical analysis of bioassay data can be considered, the endpoint must be chosen. It is this endpoint, or response, that is estimated, modelled or compared by the statistical techniques. In this discussion of statistical methods, the age-specific risk of developing a first tumour of the type of interest is considered to be the implicit endpoint.

This choice is made for two reasons. The age-specific incidence rate is important because the comparison of non-age-adjusted incidence measures can be biased by differences in mortality. Interest in the incidence of first tumours is expressed because the appearance of second tumours may be influenced by differences in survival due to the effects of the agent on the course of the first tumour rather than on carcinogenicity *per se*.

Statistical methods may be applied to (i) compare individual treated groups with an untreated control group, (ii) investigate whether the response increased with increasing dose, or (iii) fit particular dose-response models. When testing for carcinogenicity the presence of a positive trend (point ii above) plays a dominant role.

(b) Source of bias

A potential source of bias in long-term animal bioassays is a mortality rate among tumour-free animals that differs across treatment groups. If animals without tumours die earlier in one treatment group, then at each age there are fewer animals at risk of developing a tumour in that group; consequently, fewer animals in that group develop tumours over a lifetime than would have developed tumours if the mortality rate had not increased.

Excess mortality is most common in groups treated with extremely high doses, in which toxic effects of the agent being tested increase mortality in tumour-free and tumour-bearing animals alike. Consequently, in the statistical analysis of a long-term experiment, one would both investigate the mortality patterns and use methods for the tumour analysis that adjust for intercurrent mortality. The corresponding range of statistical techniques is outlined below and is described in detail by Gart *et al.* (1986).

(c) Description of the data

The appropriate hypothesis tests to be applied to data from a long-term carcinogenesis bioassay are described below, where it is clear that the choice of the test and model depends on the subjective judgement of the data analyst. Because such judgements are open to criticism, and because hypothesis tests and fitted models reduce considerably the information contained in the raw data, it is important to include as much purely descriptive information as possible in a report on the experimental results.

At a bare minimum, the description should include the following information for each treatment group:

    (i)   the number of animals at the beginning of the experiment;

    (ii)   the number of animals still alive at the appearance of the first tumour of the type of interest in any of the treatment groups;

    (iii)   the number of animals that died naturally with the tumour of interest and the number that died without the tumour of interest during the course of the experiment; and

(iv) the numbers of tumour-bearing and of non-tumour-bearing animals that were sacrificed at each of any serial sacrifice times and at the terminal sacrifice.

In addition, if the context of observation is recorded, the natural deaths with the tumour should be divided into those for which the tumour was observed in a fatal context and those for which it was observed in an incidental context, and the number of each should be reported.

If there is intercurrent mortality in any of the treatment groups, particularly if it occurs during the part of the lifespan when tumours occur, plots of life-table estimates of the probabilities of tumour-free survival over time, as described by Gart *et al.* (1986, Chapter 5), should be included for each group.

Finally, if space permits, tables or plots should be included that indicate for each group the time to appearance or detection of the first tumour of the type of interest in each animal.

(d) Unadjusted testing methods

When the mortality rates do not differ among the experimental groups, simple comparisons of the proportions of animals that develop at least one tumour of the type of interest can lead to unbiased comparison of tumour incidence in the various treatment groups. A broad discussion of the appropriateness of unadjusted methods is given by Gart *et al.* (1979, 1986).

Proportions in two groups can be compared by approximate or exact methods, and corresponding tests for trend can be applied (chi-square test for heterogeneity, Fisher-Irwin exact test, Cochran-Armitage test for trend). These are described in detail by Gart *et al.* (1986, Chapter 5).

(e) Mortality-adjusted testing methods

When the mortality rate is the same in all groups, all of the tests mentioned in the previous section yield unbiased comparisons of the tumour incidence in different treatment groups. However, the power of these tests to detect differences between control and treated groups may be limited and could be improved by age-adjusted methods. In addition and most importantly, age-adjusted methods can remove biases due to differential mortality. The particular type of mortality-adjusted method that is appropriate for the data of a given experiment depends on the biological nature of the tumour and the observational procedure. The various special cases are as follows:

(i) *Observable or rapidly lethal occult tumours.* These are tumours which are, in the former case, visible or palpable or, in the latter case, observed in a fatal context, and the onset time of which, or an approximation of it, can be used for statistical analysis.

(ii) *Non-lethal occult tumours.* These are tumours that do not affect the risk of death for an animal and are thus observed at necropsy in an incidental context.

(iii) *Occult tumours of identified lethality*. These tumours can occur in either an incidental or fatal context; a distinction between the two contexts is made for each tumour.

(iv) *Occult tumours of unidentified lethality*. These are tumours for which no distinction has been made between the incidental and fatal context of observation.

For observable or rapidly lethal tumours, the animals experiencing the event of interest among those surviving, and thus at risk to do so, form the basis for the statistical analysis. Life-table methods such as the log-rank test, the generalized Wilcoxon test or the proportional hazards model are appropriate (see Gart *et al.*, 1986, Chapters 5 and 6).

For non-lethal occult tumours, the animals with the tumour of interest among those dying during a certain period form the basis for the statistical analysis. Prevalence methods utilizing the Mantel-Haenszel approach or logistic regression are appropriate (see Gart *et al.*, 1986, Chapters 5 and 6).

For occult tumours of identified lethality, a combination of life-table and prevalence methods is appropriate. This analysis was proposed and outlined in the Annex to Supplement 2 (IARC, 1980) and is also discussed by Gart *et al.* (1986, Chapter 5).

For occult tumours of unidentified lethality, the probable lethality of the tumours must be assumed, and this judgement will determine the choice between life-table and prevalence methods. If such a judgement is difficult to make, analyses by both methods may be informative.

(f) Serial sacrifice studies

When an experiment includes serial sacrifices, analyses that do not require specification of the context of observation are available. These include fits of the log-linear and logistic models of Mitchell and Turnbull, as described by Gart *et al.* (1986), and the general non-parametric methods of Dewanji and Kalbfleisch (1985).

(g) Dose-response models

Fitting of dose-response models can in principle be integrated into the above-mentioned methods for testing for carcinogenicity. However, it is usually considered apart, and methods for fitting dose-response models to unadjusted proportions of tumour-bearing animals have received broad attention, in particular in respect to estimating certain measures of carcinogenic activity and evaluating risks at low dose levels. Additional animal-specific information, such as time to death or tumour, permits the use of more specific parametric models, such as the Weibull model, which has been widely used in carcinogenesis studies, and allows the estimation of mortality-adjusted measures of carcinogenic activity (for details, see Gart *et al.*, 1986, Chapter 6).

4.6 Reporting the results

In reporting the results of animal tests for carcinogenicity, inadequacies of the design and conduct of the experiment must be presented. The evidence for carcinogenicity should be fully described, along with the basis for the author's interpretations.

For reports of studies in which carcinogenicity is not demonstrated, it is important (for comparison with other studies) that the conditions of the experiment be described in reasonable detail.

It is unfortunate that no medium exists for the publication of all of the experimental data available from a carcinogenesis test. It is important, therefore, that, in addition to reporting the results, investigators make available data and pathological materials to interested scientists for review and full understanding.

5. Other assays for carcinogenicity in animals

5.1 Introduction

In addition to long-term bioassays for carcinogenicity in animals, several more limited animal bioassays have been developed that may be useful for screening chemicals for carcinogenic activity. These take advantage of model systems in which there is (*a*) a high background rate of cancer and thus an apparent high sensitivity to induction of cancer and/or (*b*) cancer formation can be detected early, by direct observation or palpation. The bioassays described briefly here have the advantages that the results are obtained relatively quickly (one-half to one year), and that the endpoint measured is actual tumour formation. Among the disadvantages are that the test material is applied to only one body site in one species, and that the assays may merely shorten the latent period of formation of commonly occurring tumours rather than increase the incidence of tumours. Accordingly, positive results in these assays may be strongly suggestive of carcinogenicity but are not conclusive by themselves, and negative results do not abrogate the need for long-term, full-fledged bioassays.

5.2 Strain A mouse lung tumour bioassay

In contrast with other strains of mice, strain A mice develop a high incidence of primary lung tumours, starting with a few nodules at three or four months of age, increasing steadily to a nearly 100% incidence of multiple lung tumours by 24 months of age. The assay, developed more than 40 years ago, is based on the earlier appearance of multiple tumours in the lung following treatment with carcinogens (reviewed by Stoner & Shimkin, 1982). In a typical assay, carcinogens are administered intraperitoneally at several dose levels to groups of 30 weanling strain A mice of each sex, three times per week for eight weeks (a total of 24 injections). The animals are killed 24 weeks after the start of treatment, the lungs are removed and fixed and the tumours are counted.

Urethane is usually used as a positive control. The sensitivity of the test can be enhanced even further by administration of small doses of butylated hydroxytoluene (Witschi & Kehrer, 1982).

More than 300 substances from a variety of chemical classes have been tested in this relatively short-term (six months) assay. In general, many chemicals shown to be carcinogenic in other assay systems give positive results in the strain A lung-tumour assay. Few false-positive results have been reported, but false-negative results are not uncommon, and this greatly reduces the utility of the assay. False-negative results are presumably related to the kinetics of the distribution of carcinogens to the target tissue and the alveolar epithelium, and to the metabolic capacity of the lung. In one review of 116 substances tested in both long-term bioassays in mice or rats and in the strain A mouse lung bioassay, 50 of 93 (54%) carcinogens were identified in the lung-tumour assay (Pereira & Stoner, 1985).

5.3 *Breast cancer induction in female Sprague-Dawley rats*

The unusual sensitivity of some strains of rats to the induction of mammary tumours by certain chemical carcinogens was recognized by Shay *et al.* (1949) and developed into an assay in female Sprague-Dawley rats by Huggins *et al.* (1959). The assay has so far been used primarily for polynuclear aromatic hydrocarbons, polycyclic nitro and amino derivatives, and certain coumarins. Its possible use for other classes of chemicals is not clear. The propensity of female Sprague-Dawley rats to develop mammary tumours, however, may limit the use of this animal for some kinds of carcinogenicity research.

In a typical bioassay, virgin female Sprague-Dawley rats are administered the test substance during the stage of maximum breast development — 40-70 days of age. A single administration of test substance is usually given around day 55, the peak of developmental and hormonal activity. Griswold *et al.* (1968) have shown that there are some advantages to multiple administrations. Lipid-soluble test materials are usually administered in an oily vehicle by gavage so that the administered doses are known. The animals are then examined periodically for the time to detectable tumours, tumour incidence and tumour multiplicity for six (Huggins *et al.*, 1959) or nine (Griswold *et al.*, 1968) months. About 3% of vehicle-control rats can be expected to develop mammary tumours in nine months. A positive control group administered the carcinogen 7,12-dimethylbenz[*a*]anthracene (usually included in each assay) develops the first tumours after about two months and essentially a 100% incidence of multiple mammary tumours, nearly all of which are malignant, after nine months of observation. A variety of mammary tumours are produced, including the fibroadenomas that commonly occur in rats but also adenomas, adenocarcinomas and sarcomas. Tumours may arise in other organs, too, but, for most substances tested, the incidence of tumours at other sites has been low within the nine-month experimental period. A positive response in this assay has usually been confirmed in other animal bioassays. A negative response does not prove lack of potential carcinogenicity.

5.4 *Subcutaneous injection*

Subcutaneous injection is a simple method for administering single or repeated, precisely known doses of test materials to a predetermined local site. Historically, the first evidence for the carcinogenicity of ethylene oxide and epichlorohydrin was obtained in subcutaneous tests in mice (IARC, 1976, 1985). This method is sensitive since high concentrations of test substance come into contact with local tissue and, depending on solubility and rates of metabolism, may stay in contact for prolonged periods. Developing sarcomas at injection sites can be detected by palpation and the time to tumour formation approximated. Moreover, the test material may be distributed from the local site throughout the body, giving rise to tumours at distant sites.

Evaluation of this assay is made difficult, however, by variations in protocols in which different species of animals are used (mice, rats, hamsters, rabbits) and the number of injections and the vehicle or solvent vary. Whether or not distant body sites were examined has not always been reported. However, two reviews of the testing of substances by subcutaneous injection, disregarding variations in experimental design, have been reported. Tomatis (1977) summarized the results from the first 11 volumes of the *IARC Monographs*. Of 102 substances that had been tested by both the subcutaneous and another route of administration, 69 gave positive results in both assays and 18 gave negative results in both assays. Nine substances gave positive results by the subcutaneous route only, and six gave negative results subcutaneously but positive results by another route. In a later review of substances evaluated in the first 26 volumes of the *IARC Monographs*, Theiss (1982) found that 112 substances gave positive results by subcutaneous injection and had been tested by other routes of administration. Thirty-six substances produced local tumours only, 31 produced both local and distant tumours, and 45 produced tumours at distant sites only. All but two of the 76 substances that produced distant tumours also produced tumours by other routes of administration, indicating the utility of this approach. Of the 36 substances that produced only local tumours, however, 15 (42%) were not tumorigenic by other routes of administration. While this latter result might be attributed to retention at the injection site or to low tumour potency of the substances, these and other possibilities have not been systematically examined as yet. Moreover, it has been well established that some solid substances and persistent foreign body reactions can elicit local sarcomas in rodents, as can injections of hypertonic or acidic solutions.

It can be concluded that the production of tumours at distant body sites after subcutaneous injections of test substances into rodents is indicative of carcinogenicity, but that the production of tumours at injection sites only indicates the need for further study.

6. References

Adams, R.A., DiPaolo, J.A. & Homburger, F. (1978) The Syrian hamster in toxicology and carcinogenesis research. *Cancer Res.*, *38*, 2642-2645

Almeida, W.F., de Mello, D. & Rodrigues-Puga, F. (1978) *Influence of nutritional status on the toxicity of food additives and pesticides.* In: Galli, C.L., Paoletti, R. & Vettorazzi, G., eds, *Chemical Toxicology of Food*, Amsterdam, Elsevier, pp. 169-184

American Institute of Nutrition (1977) Report of the American Institute of Nutrition Ad Hoc Committee on Standards for Nutritional Studies. *J. Nutr.*, *107*, 1340-1348

Andersen, M.E., Divincanzo, G.D., Gehring, P.J., Maronpot, R.R., Ramsey, J.C. & Reitz, R.H. (1984) *Position paper on pharmacokinetics.* In: *Report of the NTP ad hoc Panel on Chemical Carcinogenesis Testing and Evaluation*, Washington DC, US Department of Health and Human Services, pp. 150-164

Andervont, H.B. (1944) Influence of environment on mammary cancer in mice. *J. natl Cancer Inst.*, *4*, 579-581

Anon (1976) Symposium on early lesions and the development of epithelial cancer. *Cancer Res.*, *36*, 2475-2706

Anver, M.R. & Cohen, B.J. (1979) *Lesions associated with aging.* In: Baker, H.J., Lindsey, J.R. & Weisbroth, S.H., eds, *The Laboratory Rat*, New York, Academic Press, pp. 377-399

Arcos, J.C., Argus, M.F. & Wolf, G. (1968) *Chemical Induction of Cancer*, Vol. 1, New York, Academic Press

Arnold, D.L., Charbonneau, S.M., Zawidzka, Z.Z. & Grice, H.C. (1977) Monitoring animal health during chronic toxicity studies. *J. environ. Pathol. Toxicol.*, *1*, 227-239

Arnold, D.L., Fox, J.G., Thibert, P. & Grice, H.C. (1978) Toxicology studies. I. Support personnel. *Food Cosmet. Toxicol.*, *16*, 479-484

Ayers, K.M. & Jones, S.R. (1978) *The cardiovascular system.* In: Benirschke, K., Garner, F.M. & Jones, T.C., eds, *Pathology of Laboratory Animals*, Vol. 1, New York, Springer, pp. 1-69

Balazs, T. (1961) Common diseases of laboratory animals in Canada. *Can. vet. J.*, *2*, 179-182

Bannasch, P. (1984) Sequential cellular changes during chemical carcinogenesis. *J. Cancer Res. clin. Oncol.*, *108*, 11-22

Barnes, J.M. & Denz, F.A. (1954) Experimental methods used in determining chronic toxicity: a critical review. *Pharmacol. Rev.*, *6*, 191-242

Benitz, K.-F. (1970) *Measurement of chronic toxicity.* In: Paget, G.E., ed., *Methods in Toxicology*, Oxford, Blackwell, pp. 82-131

Berenblum, I., ed. (1969) *Carcinogenicity Testing (UICC Technical Report Series, No. 2)*, Geneva, International Union Against Cancer

Boorman, G.A. & DeLellis, R.A. (1983) *Medullary carcinoma, thyroid, rat.* In: Jones, T.C., Mohr, U. & Hunt, R.D., eds, *Monographs on Pathology of Laboratory Animals: Endocrine System*, Berlin (West), Springer, pp. 200-204

Boutwell, R.K. (1964) Some biological aspects of skin carcinogenesis. *Prog. exp. Tumor Res.*, *4*, 207-250

Boyland, E. (1958) The biological examination of carcinogenic substances. *Br. med. Bull.*, *14*, 93-98

Boyland, E. (1968) *Carcinogenicity.* In: Boyland, E. & Goulding, R., eds, *Modern Trends in Toxicology*, London, Butterworths, pp. 107-129

Briel, J.E., Kruckenberg, S.M. & Besch, E.L. (1972) *Observations on Ammonia Generation in Laboratory Animal Quarters*, Manhattan, KS, Institute for Environmental Research

Broderson, J.R., Lindsey, J.R. & Crawford, J.E. (1976) The role of environmental ammonia in respiratory mycoplasmosis of rats. *Am. J. Pathol., 85,* 115-130

Burek, J.D. (1978) *Pathology of Aging Rats,* Boca Raton, FL, CRC Press pp. 1-230

Canadian Association for Laboratory Animal Science (1986) *Panel discussion.* In: *Quality Research Needs Quality Animals,* Ottawa (in press)

Canadian Council on Animal Care (1980) *The Guide to the Care and Use of Experimental Animals,* Vol. 1, Ottawa

Canadian Council on Animal Care (1984) *The Guide to the Care and Use of Experimental Animals,* Vol. 2, Ottawa

Cardesa, A., Handler, A.H. & Kelman, A.D. (1982) *Tumours of the adrenal gland.* In: Turusov, V.S., ed., *Pathology of Tumours in Laboratory Animals*, Vol. III, *Tumours of the Hamster (IARC Scientific Publications No. 34)*, Lyon, International Agency for Research on Cancer, pp. 281-292

Casey, H.W., Ayers, K.M. & Robinson, F.R. (1978) *The urinary system.* In: Benirschke, K., Garner, F.M. & Jones, T.C., eds, *Pathology of Laboratory Animals*, Vol. 1, New York, Springer, pp. 116-173

Castegnaro, M., Hunt, D.C., Sansone, E.B., Schuller, P.L., Siriwardana, M.G., Telling, G.M., van Egmond, H.P. & Walker, E.A., eds (1980) *Laboratory Decontamination and Destruction of Aflatoxins B_1, B_2, G_1, G_2 in Laboratory Wastes (IARC Scientific Publications No. 37),* Lyon, International Agency for Research on Cancer

Castegnaro, M., Eisenbrand, G., Ellen, G., Keefer, L., Klein, D., Sansone, E.B., Spincer, D., Telling, G. & Webb, K., eds (1982) *Laboratory Decontamination and Destruction of Carcinogens in Laboratory Wastes: Some N-Nitrosamines (IARC Scientific Publications No. 43),* Lyon, International Agency for Research on Cancer

Castegnaro, M., Grimmer, G., Hutzinger, O., Karcher, W., Kunte, H., Lafontaine, M., Sansone, E.B., Telling, G. & Tucker, S.P., eds (1983a) *Laboratory Decontamination and Destruction of Carcinogens in Laboratory Wastes: Some Polycyclic Aromatic Hydrocarbons (IARC Scientific Publications No. 49),* Lyon, International Agency for Research on Cancer

Castegnaro, M., Ellen, G., Lafontaine, M., van der Plas, H.C., Sansone, E.B. & Tucker, S.P., eds (1983b) *Laboratory Decontamination and Destruction of Carcinogens in Laboratory Wastes: Some Hydrazines (IARC Scientific Publications No. 54),* Lyon, International Agency for Research on Cancer

Castegnaro, M., Benard, M., van Broekhoven, L.W., Fine, D., Massey, R., Sansone, E.B., Smith, P.L.R., Spiegelhalder, B., Stacchini, A., Telling, G. & Vallon, J.J., eds (1984) *Laboratory Decontamination and Destruction of Carcinogenis in Laboratory Wastes: Some N-Nitrosamides (IARC Scientific Publications No. 55),* Lyon, International Agency for Research on Cancer

Castegnaro, M., Alvarez, M., Iovu, M., Sansone, E.B., Telling, G.M. & Williams, D.T., eds (1985a) *Laboratory Decontamination and Destruction of Carcinogens in Laboratory Wastes: Some Haloethers (IARC Scientific Publications No. 61),* Lyon, International Agency for Research on Cancer

Castegnaro, M., Barek, J., Dennis, J. Ellen, G., Klibanov, M., Lafontaine, M., Mitchum, R., van Roosmalen, P., Sansone, E.B., Sternson, L.A. & Vahl, M., eds (1985b) *Laboratory Decontamination and Destruction of Carcinogens in Laboratory Wastes: Some Aromatic Amines and 4-Nitrobiphenyl (IARC Scientific Publications No. 64)*, Lyon, International Agency for Research on Cancer

Castegnaro, M., Adams, J., Armour, M.A., Barek, J., Benvenuto, J., Confalonieri, C., Goff, U., Luderman, S., Reed, D., Sansone, E.B. & Telling, G., eds (1986) *Laboratory Decontamination and Destruction of Carcinogens in Laboratory Wastes: Some Antineoplastic Agents (IARC Scientific Publications No. 73)*, Lyon, International Agency for Research on Cancer

Clayson, D.B. (1962) *Chemical Carcinogenesis*, London, J. & A. Churchill

Committee of the Health Council (1979) *The Evaluation of the Carcinogenicity of Chemical Substances*, Leidschendam, Ministry of Public Health & Environment

Committee on Animal Nutrition (1978) *Nutrient Requirements of Laboratory Animals*, No. 10, 3rd revised ed., Washington DC, National Academy of Sciences

Committee on Carcinogenicity of Chemicals in Food, Consumer Products and the Environment (1978) *Guidelines on Carcinogenicity Testing (Report on Health and Social Subjects 25)*, London, Her Majesty's Stationery Office

Conner, W.M. & Newberne, P.M. (1984) Drug-nutrient interactions and their implications for safety evaluations. *Fundam. appl. Toxicol.*, 4, S341-S356

Council for International Organizations of Medical Friends (1985) International guiding principles for biomedical research involving animals. *Int. Digest. Health Legislation*, 36, 508-512

Cranmer, M.F., Lawrence, L.R., Konvicka, A.J., Taylor, D.W. & Herrick, S.S. (1976) Research data integrity: a result of an integrated information system. *J. Toxicol. environ. Health*, 2, 285-299

Dayan, A.D. & Brinblecombe, L.W., eds (1978) *Carcinogenicity Testing: Principles and Problems*, Lancaster, UK, MTP Press

Dayton, P.G. & Sanders, J.E. (1983) Dose-dependent pharmacokinetics: emphasis on phase 1 metabolism. *Drug Metab. Rev.*, 14, 347-405

DeLellis, R.A., Nunnemacher, G., Bitman, W.R., Gagel, R.F., Tashjian, A.H., Jr, Blount, M. & Wolfe, H.J. (1979) C cell hyperplasia and medullary thyroid carcinoma in the rat: an immunohistochemical and ultrastructural analysis. *Lab. Invest.*, 40, 140-154

Della Porta, G. (1964) The study of chemical substances for possible carcinogenic action. *Proc. Eur. Soc. Study Drug Toxic.*, 3, 29-40

Dewanji, A. & Kalbfleisch, J.D. (1985) *Non-parametric methods for survival/sacrifice experiments*. In: *Proceedings of the Symposium on Long-term Animal Carcinogenicity Studies: A Statistical Perspective*, Washington DC, American Statistical Association, pp. 100-106

Edwards, G.S., Fox, J.G., Policastro, P., Goff, U., Wolf, M.H. & Fine, D.H. (1979) Volatile nitrosamine contamination of laboratory animal diets. *Cancer Res.*, 39, 1857-1858

Fare, G. (1965) The influence of number of mice in a box on experimental skin tumour production. *Br. J. Cancer*, 19, 871-877

Fears, T.R. & Schneiderman, M.A. (1974) Pathologic evaluation and the blind technique. *Science*, 183, 1144-1145

Feldman, D.B., McConnell, E.E. & Knapka, J.J. (1982) Growth, kidney disease, and longevity of Syrian hamsters (*Mesocricetus auratus*) fed varying levels of protein. *Lab. Anim. Sci.*, 32, 613-618

Festing, M.F.W. (1975) A case for using inbred strains of laboratory animals in evaluating the safety of drugs. *Food Cosmet. Toxicol., 13*, 369-375

Fleiss, J. (1973) *Statistical Methods for Rates and Proportions*, New York, Wiley p. 88

Flynn, R.J. (1973) *Parasites of Laboratory Animals*, Ames, IA, The Iowa State University Press

Food Safety Council (1978) Chronic toxicity testing. Proposed system for food safety assessment. *Food Cosmet. Toxicol., 16*, Suppl. 2, 97-108

Fox, J.G. (1977) Clinical assessment of laboratory rodents in long-term bioassay studies. *J. environ. Pathol. Toxicol., 1*, 199-226

Fox, J.G., Niemi, S.M., Murphy, J.C. & Quimby, F.W. (1977) Ulcerative dermatitis in the rat. *Lab. Anim. Sci., 27*, 671-678

Fox, J.G., Thibert, P., Arnold, D.L., Krewski, D.R. & Grice, H.C. (1979) Toxicology studies. Part II. The laboratory animal. *Food Cosmet. Toxicol., 17*, 661-676

Fullerton, F.R., Greenman, D.L. & Kendall, D.C. (1982) Effects of storage conditions on nutritional qualities of semipurified (AIN-76) and natural ingredient (NIH 07) diets. *J. Nutr., 112*, 567-573

Gart, J.J., Chu, K.C. & Tarone, R.E. (1979) Statistical issues in interpretation of chronic bioassay tests for carcinogenicity. *J. natl Cancer Inst., 62*, 957-974

Gart, J.J., Krewski, D., Lee, P.N., Tarone, R.E. & Wahrendorf, J., eds (1986) *Statistical Methods in Cancer Research*, Vol. III, *The Design and Analysis of Long-term Animal Experiments (IARC Scientific Publications No. 79)*, Lyon, International Agency for Research on Cancer (in press)

Gehring, P.J., Watanabe, P.G. & Blau, G.E. (1976) *Pharmacokinetic studies in evaluation of the toxicological and environmental hazard of chemicals*, In: Mehlman, M.A., Shapiro, R.E. & Blumenthal, H., eds, *New Concepts in Safety Evaluation*, New York, Wiley, pp. 195-270

Gibaldi, M. & Perrier, D. (1982) *Pharmacokinetics*, New York, Marcel Dekker, pp. 1-494

Gilbert, C., Gillman, J., Loustalot, P. & Lutz, W. (1958) The modifying influence of diet and physical environment on spontaneous tumor frequency in rats. *Br. J. Cancer, 12*, 565-593

Golberg, L., ed. (1973) *Carcinogenesis Testing of Chemicals*, Cleveland, OH, CRC Press

Greaves, P. & Faccini, J.M. (1984) *Rat Histopathology*, Amsterdam, Elsevier

Grice, H.C., Barth, M.L., Cornish, H.H., Foster, G.V. & Gray, R.H. (1971) Correlation between serum enzymes, isoenzyme patterns and histologically detectable organ damage. *Food Cosmet. Toxicol., 9*, 847-855

Grice, H.C., Munro, J.C., Krewski, D.R. & Blumenthal, H. (1981) In utero exposure in chronic toxicity/carcinogenicity studies. *Food Cosmet. Toxicol., 19*, 373-379

Grice, H.C., Arnold, D.L., Blumenthal, H., Emmerson, J.L. & Krewski, D. (1984a) *The selection of doses in chronic toxicity/carcinogenicity studies.* In: Grice, H.C., ed., *Current Issues in Toxicology*, Berlin (West), Springer, pp. 6-49

Grice, H.C., Arnold, D.L., Blumenthal, H., Emmerson, J.L. & Krewski, D. (1984b) *Age-associated (geriatric) pathology: its impact on long-term toxicity studies.* In: Grice, H.C., ed., *Current Issues in Toxicology*, Berlin (West), Springer, pp. 57-107

Griswold, D.P., Casey, A.E., Weisburger, E.K. & Weisburger, J.H. (1968) The carcinogenicity of multiple intragastric doses of aromatic and heterocyclic nitro or amino derivatives in young female Sprague Dawley rats. *Cancer Res., 28*, 924-933

Haseman, J.K. (1983) Patterns of tumour incidence in two-year cancer bioassay feeding studies in Fischer 344 rats. *Fundam. appl. Toxicol.*, *3*, 1-9

Haseman, J.K. (1985) Issues in carcinogenicity testing: dose selection. *Fundam. appl. Toxicol.*, *5*, 66-78

Haseman, J.K. & Hoel, D.G. (1979) Statistical design of toxicity assays: role of genetic structure of the test animal population. *J. Toxicol. environ. Health*, *5*, 89-101

Haseman, J.K., Huff, J.E., Rao, G.N., Arnold, J.E., Boorman, G.A. & McConnell, E.E. (1985) Neoplasms observed in untreated and corn oil gavage control groups of F344/N rats and (C57BL/6N × C3H/HeN) F1 mice. *J. natl Cancer Inst.*, *75*, 975-984

Hatch, A., Wiberg, G., Zawidzka, Z., Cann, M., Airth, J. & Grice, H. (1965) Isolation syndrome in the rat. *Toxicol. appl. Pharmacol.*, *7*, 737-745

Health & Welfare Canada (1973) *The Testing of Chemicals for Carcinogenicity, Mutagenicity and Teratogenicity*, Ottawa

Heywood, R. (1973) Some clinical observations on the eyes of Sprague-Dawley rats. *Lab. Anim.*, *7*, 19-27

Hoel, D.G., Kaplan, N.L. & Anderson, M.W. (1983) Implication of nonlinear kinetics on risk estimation in carcinogenesis. *Science*, *219*, 1032-1037

Hoffmann, D. & Wynder, E.L. (1974) *Positive controls in environmental respiratory carcinogenesis.* In: Golberg, L., ed., *Carcinogenesis Testing of Chemicals*, Cleveland, OH, CRC Press, pp. 35-39

Holmes, D.D. (1984) *Clinical Laboratory Animal Medicine*, Ames, IA, Iowa State University Press

Hoover, K.L., ed. (1984) *Use of Small Fish in Carcinogenicity Testing* (*Natl Cancer Institute Monograph No. 65*), Bethesda, MD, National Cancer Institute

Huggins, C., Briziarelli, G. & Sutton, H., Jr (1959) Rapid induction of mammary carcinoma in the rat and the influence of hormones on the tumours. *J. exp. Med.*, *109*, 25-41

IARC (1976) *IARC Monographs on the Evaluation of Carcinogenic Risk of Chemicals to Man*, Vol. 11, *Cadmium, Nickel, Some Epoxides, Miscellaneous Industrial Chemicals and General Considerations on Volatile Anaesthetics*, Lyon, pp. 131-140

IARC (1980) *IARC Monographs on the Evaluation of the Carcinogenic Risk of Chemicals to Humans*, Suppl. 2, *Long-term and Short-term Screening Assays for Carcinogens: A Critical Appraisal*, Lyon, pp. 23-83

IARC (1985) *IARC Monographs on the Evaluation of the Carcinogenic Risk of Chemicals to Humans*, Vol, 36, *Allyl Compounds, Aldehydes, Epoxides and Peroxides*, Lyon, pp. 189-226

Institute of Laboratory Animal Resources (1985) *Guide for the Care and Use of Laboratory Animals* (*US Department of Health and Human Services Publication No. (NIH)85-23*), Washington DC, National Research Council

Institute of Laboratory Animal Resources (1976) *Long Term Holding of Laboratory Rodents*, Washington DC, National Academy of Sciences

Institute of Laboratory Animal Resources (1978) *Control of Diets in Laboratory Animal Experimentation*, Washington DC, National Academy of Sciences

Interagency Regulatory Liaison Group (1979) *Scientific Bases for Identification of Potential Carcinogens and Estimation of Risks*, Washington DC

Jacoby, R.O., Bhatt, P.M. & Jonas, A.M. (1975) Pathogenesis of sialodacryoadenitis in gnotobiotic rats. *Vet. Pathol.*, *12*, 196-209

Jollow, D.J. (1980) Glutathione thresholds in reactive metabolite toxicity. *Arch. Toxicol., Suppl. 3*, 95-110

Jonas, A.M. (1965) Laboratory animal facilities. *J. Am. vet. med. Assoc., 146*, 600-606

Kalbach, H., McCallum, W.F., Konvicka, A.J. & Holland, M.T. (1977) Computer record keeping in a large rodent colony. *Lab. Anim. Sci., 27*, 660-666

Kalter, H. (1975) Prenatal epidemology of spontaneous cleft lip and palate, open eyelid and embryonic death in AlJ mice. *Teratology, 12*, 245-258

Korsrud, G.O., Grice, H.C. & McLaughlan, J.M. (1972) Sensitivity of several serum enzymes in detecting carbon tetrachloride-induced liver damage in rats. *Toxicol. appl. Pharmacol., 22*, 474-483

Krewski, D., Kovar, J. & Bickis, M. (1982) *Optimal experimental designs for low dose extrapolation.*, In: Dwived, T.W., ed., *Topics in Applied Statistics*, New York, Marcel Dekker

Linman, J.W. (1966) *Principles of Hematology*, New York, MacMillan

Loomis, T.A. (1968) *Essentials of Toxicology*, Philadelphia, Lea & Febiger

Magee, P.N. (1970) *Tests for carcinogenic potential*. In: Paget, G.E., ed., *Methods in Toxicology*, Oxford, Blackwell, pp. 158-196

Mantel, N. & Ciminera, J.L. (1979) Use of logrank scores in the analysis of litter-matched data on time to tumor appearance. *Cancer Res., 39*, 4308-4315

Matheson, B.H., Grice, H.C. & Cornell, M.R.E. (1955) Studies of middle ear disease in rats. I. Age of infection and infecting organisms. *Can. J. comp. Med., 19*, 91-97

McLean, A.E.M. (1973) *Diet and the chemical environment as modifiers of carcinogenesis*. In: Doll, R. & Vodopija, I., eds, *Host Environment Interactions in the Etiology of Cancer in Man (IARC Scientific Publications No. 7)*, Lyon, International Agency for Research on Cancer, pp. 223-230

McLean, A.E.M. & McLean, E.K. (1966) The effect of diet and 1,1,1-trichloro-2,2-bis(*p*-chlorophenyl)-ethane (DDT) on microsomal hydroxylating enzymes and on sensitivity of rats to carbon tetrachloride poisoning. *Biochem. J., 100*, 564-571

Milman, H.A. & Weisburger, E.K., eds (1985) *Handbook of Carcinogenicity Testing*, Park Ridge, NJ, Noyes Publications

Mohr, U. (1979) The Syrian golden hamster as a model in cancer research. *Prog. exp. Tumor Res., 24*, 245-252

Montesano, R., Bartsch, H., Boyland, E., Della Porta, G., Fishbein, L., Griesemer, R.A., Swan, A.B. & Tomatis, L., eds (1979) *Handling of Chemical Carcinogens in the Laboratory: Problems of Safety (IARC Scientific Publications No. 33)*, Lyon, International Agency for Research on Cancer

Muhle, H., Bellmann, B. & Heinrich, U. (1986) Overloading of lung clearance after chronic exposure of experimental animals to particles. *Ann. occup. Hyg.* (in press)

Munro, I.C. (1977) Considerations in chronic toxicity testing: the chemical, the dose, the design. *J. environ. Pathol. Toxicol., 1*, 183-197

Munro, I.C. & Willes, R.F. (1978) *Reproductive toxicity and the problem of in utero exposure*. In: Galli, C.L., Paoletti, R. & Vettorazzi, G., eds, *Chemical Toxicology of Food*, Amsterdam, Elsevier, pp. 133-145

Munro, I.C., Charbonneau, S.M. & Willes, R.F. (1972) An automated data acquisition and computer-based compilation system for application to toxicological studies in laboratory animals. *Lab. Anim. Sci.*, *22*, 753-756

Murakami, H. & Kinoshita, K. (1978) Temperature preference of adolescent mice. *Lab. Anim. Sci.*, *28*, 277-281

National Academy of Sciences/National Research Council (1974) *Report to FDA on the 'Safety of Saccharin and Sodium Saccharin in the Human Diet' (Publication 6238 137)*, Washington DC

Nettesheim, P. (1972) Respiratory carcinogenesis studies with the Syrian golden hamster. A review. *Prog. exp. Tumor Res.*, *16*, 185-200

Newberne, P.M. (1974) *Diets*. In: Golberg, L., ed., *Carcinogenesis Testing of Chemicals*, Cleveland, OH, CRC Press, pp. 17-21

Newberne, P.M. (1978) *Nutritional diseases*. In: Benirschke, K., Garner, F.M. & Jones, T.C., eds, *Pathology of Laboratory Animals*, Vol. II, New York, Springer, pp. 2153-2154

Office of Science and Technology Policy (1984) Chemical carcinogens; notice of review of the science and its associated principles, Part II. *Fed. Regist.*, *49* (100)

O'Flaherty, E.J. (1981) *Toxicants and Drugs: Kinetics and Dynamics*, New York, Wiley

Organisation for Economic Co-operation and Development (1981) *OECD Guidelines for Testing Chemicals: Carcinogenicity Studies*, No. 451, 452, 453, Paris

Organisation for Economic Co-operation and Development (1982) *Good Laboratory Practice in the Testing of Chemicals*, Paris

Page, N.P. (1977a) Chronic toxicity and carcinogenicity guidelines. *J. environ. Pathol. Toxicol.*, *1*, 161-182

Page, N.P. (1977b) *Concepts of a bioassay program in environmental carcinogenesis*. In: Kraybill, H. & Mehlman, H., eds, *Environmental Carcinogenesis*, Washington DC, Hemisphere, pp. 87-171

Peck, H.M. (1974) *Design of experiments to detect carcinogenic effects of drugs*. In: Golberg, L., ed., *Carcinogenesis Testing of Chemicals*, Cleveland, OH, CRC Press, pp. 1-15

Pereira, M.A. & Stoner, G.D. (1985) Comparison of rat liver foci assay and strain A mouse lung tumour assay to detect carcinogens: a review. *Fundam. appl. Toxicol.*, *5*, 688-699

Peto, R., Pike, M.C., Armitage, P., Breslow, N.E., Cox, D.R., Howard, S.V., Mantel, N., McPherson, K., Peto, J. & Smith, P.G. (1976) Design and analysis of randomized clinical trials requiring prolonged observation of each patient. *Br. J. Cancer*, *34*, 585-612

Peto, R., Pike, M.C., Armitage, P., Breslow, N.E., Cox, D.R., Howard, S.V., Mantel, N., McPherson, K., Peto, J. & Smith, P.G. (1977) Design and analysis of randomized and clinical trials which require observation of each patient. II. Analysis and examples. *Br. J. Cancer*, *35*, 1-39

Peto, R., Pike, M.C., Day, N.E., Gray, R.G., Lee, P.N., Parish, S., Peto, J., Richards, S. & Wahrendorf, J. (1980) *Guidelines for simple, sensitive significance tests for carcinogenic effects in long-term animal experiments*. In: *IARC Monographs on the Evaluation of the Carcinogenic Risk of Chemicals to Humans,*, Suppl. 2, *Long-term and Short-term Screening Assays for Carcinogens: A Critical Appraisal*, Lyon, pp. 311-426

Poiley, S.M. (1974) *Housing experiments — general considerations*. In: Melby, E.C., Jr & Altman, N.H., eds, *CRC Handbook of Laboratory Animal Science*, Vol. 1, Cleveland, OH, CRC Press, pp. 21-60

Portier, C. & Hoel, D. (1983) Optimal design of the chronic animal bioassay. *J. Toxicol. environ. Health*, *12*, 1-19

Pour, P., Mohr, U., Althoff, J., Cardesa, A. & Kmoch, N. (1976) Spontaneous and common diseases in two colonies of Syrian hamsters. IV. Vascular and lymphatic systems and lesions of other sites. *J. natl Cancer Inst.*, *56*, 963

Reitz, R.H, Quast, J.F., Schumann, A.M., Watanabe, P.G. & Gehring, P.J. (1980) Non-linear pharmacokinetic parameters need to be considered in high dose/low dose extrapolations. *Arch. Toxicol., Suppl. 3*, 79-94

Ribelin, W.E. & McCoy, J.R. (1971) *The Pathology of Laboratory Animals*, Springfield, IL, Charles C. Thomas, pp. 417-418

Roe, F.J.C. & Tucker, M.J. (1973) Recent developments in the design of carcinogenicity tests on laboratory animals. *Proc. Eur. Soc. Study Drug Toxic.*, *15*, 171-177

Rogers, A.E. (1983) Influence of dietary content of lipids and lipotropic nutrients on chemical carcinogeneses in rats. *Cancer Res.*, *43*, 24775-24843

Rogers, A.E. & Newberne, P.M. (1980) Lipotrope deficiency in experimental carcinogenesis. *Nutr. Cancer*, *2*, 104-112

Rogers, A.E., Lenhart, G. & Morrisson, G. (1980) Dietary components that influence chemical carcinogenesis in rats. *Cancer Res.*, *40*, 2802-2807

Ross, M.H. & Bras, G. (1971) Lasting influence of early caloric restriction on prevalence of neoplasms in the rat. *J. natl Cancer Inst.*, *47*, 1095-1113

Ruddick, J.A., Ashanullah, M., Craig, J. & Stavric, B. (1978) *Uptake and distribution of ^{14}C-saccharin in the rat fetus*. In: *Proceedings, Canadian Federation of Biological Societies, 21st Annual Meeting, London, Ontario*, Abstract 635, p. 159

Saffiotti, U., Cefis, F. & Kolb, L.H. (1968) A method for the experimental induction of bronchogenic carcinoma. *Cancer Res.*, *28*, 104-124

Sansone, E.B. & Fox, J.G. (1977) Potential chemical contamination in animal feeding studies: evaluation of wire and solid bottom caging systems and gelled feed. *Lab. Anim. Sci.*, *27*, 457-465

Sansone, E.B. & Losikoff, A.M. (1978) Contamination from feeding volatile test chemicals. *Toxicol. appl. Pharmacol.*, *46*, 703-708

Sansone, E.B., Losikoff, A.M. & Pendleton, R.A. (1977) Potential hazards from feeding test chemicals in carcinogen bioassay research. *Toxicol. appl. Pharmacol.*, *39*, 435-450

Scheffe, H. (1959) *The Analysis of Variance*, New York, Wiley, pp. 298-303

Schmidt, R.E., Eason, R.L., Hubbard, G.B., Young, J.T. & Eisenbrandt, D.L. (1983) *Pathology of Aging Syrian Hamsters*, Boca Raton, FL, CRC Press

Serrano, L.J. (1971) Carbon dioxide and ammonia in mouse cages. Effect of cage covers, population and activity. *Lab. Anim. Sci.*, *21*, 75-85

Shakman, R.A. (1974) Nutritional influences on the toxicity of environmental pollutants. *Arch. environ. Health*, *28*, 105-113

Shay, H., Aegerter, E.A., Gruenstein, M. & Komarov, S.A. (1949) Development of adenocarcinoma of the breast in the Wistar rat following the intragastric instillation of methylcholanthrene. *J. natl Cancer Inst.*, *10*, 255-266

Shubik, P. (1972) The use of the Syrian golden hamster in chronic toxicity testing. *Prog. exp. Tumor Res.*, *16*, 176-184

Shubik, P. & Sice, J. (1956) Chemical carcinogenesis as a chronic toxicity test: *a review. Cancer Res., 16*, 728-742

Sigg, E., Day, C. & Colombo, C. (1966) Endocrine factors in isolation-induced aggressiveness in rodents. *Endocrinology, 78*, 679-684

Simmons, M.L., Robie, D.M., Jones, J.B. & Serrano, L.J. (1968) Effect of a filter cover on temperature and humidity in a mouse cage. *Lab. Anim., 2*, 113-120

Smyth, R.D. & Hottendorf, G.H. (1980) Application of pharmacokinetics and biopharmaceutics in the design of toxicological studies. *Toxicol. appl. Pharmacol., 53*, 179-195

Snipes, M.B., Boecker, B.B. & McClellan, R.O. (1983) Retention of monodisperse or polydisperse aluminosilicate particles inhaled by dogs, rats and mice. *Toxicol. appl. Pharmacol., 69*, 345-362.

Society for Laboratory Animal Science (1985) Guidelines for data reporting on laboratory animals, conduct of animal experiments and analysis and reporting of results [in German]. *Z. Versuchstierk., 27*, 49-52

Sontag, J.M., Page, N.P. & Saffiotti, U. (1976) *Guidelines for Carcinogen Bioassay in Small Rodents (National Cancer Institute Carcinogenesis Technical Report Series No. 1; DHEW Publ. No. NIH 76-801)*, Washington DC, Department of Health, Education, and Welfare

Squartini, F. (1979) *Tumours of the mammary gland*. In: Turusov, V.S., ed., *Pathology of Tumours in Laboratory Animals*, Vol. II, *Tumours of the Mouse (IARC Scientific Publications No. 23)*, Lyon, International Agency for Research on Cancer, pp. 43-67

Stoner, G.D. & Shimkin, M.B. (1982) Strain A mouse lung tumor assay. *J. Am. Coll. Toxicol., 1*, 145-169

Tannenbaum, A. & Silverstone, H. (1957) *Nutrition and the genesis of tumors*. In: Raven, R.W., ed., *Cancer*, Vol. 1, London, Butterworths, pp. 306-334

Theiss, J.C. (1982) Utility of injection site tumorigenicity in assessing the carcinogenic risk of chemicals to man. *Regul. Toxicol. Pharmacol., 2*, 213-222

Tomatis, L. (1977) Comment on methodology and interpretation of results. *J. natl Cancer Inst., 59*, 1341

Tucker, M.J. & Baker, S.B. de C. (1967) *Diseases of specific pathogen-free mice*. In: Cotchin, E. & Roe, F.J.C., eds, *Pathology of Laboratory Rats and Mice*, Oxford, Blackwell, pp. 787-824

Turusov, V.S. & Parfenov, J.D. (1986) *Methods of Detection and Reglementation of Chemical Carcinogens* [in Russian], Moscow, Meditsina

US Environmental Protection Agency (1983) Toxic substances control; good laboratory practice standards. *Fed. Regist., 48* (230), Part III, pp. 53922-53944: Part IV, pp. 53946-53969

US Food and Drug Administration (1971) Advisory Committee on Protocols for Safety Evaluation: Panel on Carcinogenesis report on cancer testing in the safety evaluation of food additives and pesticides. *Toxicol. appl. Pharmacol., 20*, 419-438

US Food and Drug Administration (1978) Non-clinical laboratory studies; good laboratory practice regulations. *Fed. Regist., 43* (247), Part II, pp. 59886-60024

US Food and Drug Administration (1982) *Toxicological Principles for the Safety Assessment of Direct Food Additives and Color Additives Used in Food*, Washington DC, Bureau of Foods

US National Toxicology Program (1984) *Report of the NTP ad hoc Panel on Chemical Carcinogenesis Testing and Evaluation*, Washington DC, US Department of Health and Human Services

Vesell, E.S., Lang, C.M. & White, W.J. (1976) Environmental and genetic factors affecting the response of laboratory animals to drugs. *Fed. Proc., 35*, 1125-1132

Vostal, J.J., Schreck, R.M., Lee, P.S., Chan, T.L. & Soderholm, S.C. (1982) *Deposition and clearance of diesel particles from the lung.* In: Lewtas, J., ed., *Toxicological Effects of Emissions from Diesel Engines*, New York, Elsevier, pp. 143-159

Wade, A.E., Holl, J.E., Hilliard, C.C., Moulton, E. & Greene, F.E. (1968) Alteration of drug metabolism in rats and mice by an environment of cedarwood. *Pharmacology, 1*, 317-328

Walker, E.A., Castegnaro, M. & Griciute, L. (1979) *N*-Nitrosamines in the diet of experimental animals. *Cancer Lett., 6*, 175-178

Weihe, W.H. (1971) *The significance of the physical environment for the health and state of adaptation of laboratory animals.* In: *Defining the Laboratory Animals*, Washington DC, National Academy of Sciences, pp. 353-378

Weinberger, M.A. (1973) The blind technique. *Science, 181*, 219-220

Weisburger, J.H. (1974) *Inclusion of positive control compounds.* In: Golberg, L., ed., *Carcinogenesis Testing of Chemicals*, Cleveland, OH, CRC Press, pp. 29-35

Weisburger, J.H. & Weisburger, E.K. (1967) Tests for chemical carcinogens. *Meth. Cancer Res., 1*, 307-398

WHO (1969) *Principles for the Testing and Evaluation of Drugs for Carcinogenicity (WHO tech. Rep. Ser. No. 426)*, Geneva

WHO (1978a) *Carcinogenicity and mutagenicity.* In: *Principles and Methods for Evaluating the Toxicity of Chemicals*, Part 1, Geneva, pp. 236-272

WHO (1978b) *Factors influencing the design of toxicity studies.* In: *Principles and Methods for Evaluating the Toxicity of Chemicals*, Part 1, Geneva, pp. 62-94

WHO (1978c) *Inhalation exposure.* In: *Principles and Methods for Evaluating the Toxicity of Chemicals*, Part 1, Geneva, pp. 199-235

WHO (1978d) *Morphological studies.* In: *Principles and Methods for Evaluating the Toxicity of Chemicals*, Part 1, Geneva, pp. 178-198

WHO (1978e) *Principles and Methods for Evaluating the Toxicity of Chemicals (Environmental Health Criteria 6)*, Part 1, Geneva

Witschi, H. & Kehrer, J.P. (1982) Adenoma development in mouse lung following treatment with possible promoting agents. *J. Am. Coll. Toxicol., 1*, 171-184

Working Committee for the Biological Characterization of Laboratory Animals/GV-SOLAS (1985) Guidelines for specification of animals and husbandry methods when reporting the results of animal experiments. *Lab. Anim., 19*, 106-108

Yang, Y.H. (1965) Polyarteritis nodosa in laboratory rats. *Lab. Invest., 14*, 81-85

Young, J.F. & Holson, J.F. (1978) Utility of pharmacokinetics in designing toxicological protocols and improving interspecies extrapolation. *J. environ. Pathol. Toxicol., 2*, 169-186

Zbinden, G. (1975) *Progress in Toxicology*, Vol. 2, *Special Topics*, Berlin (West), Springer, pp. 14-15

REPORT 2

EARLY PRENEOPLASTIC LESIONS

Prepared by:
P. Bannasch, R.A. Griesemer (Rapporteurs),
F. Anders, R. Becker, J.R. Cabral, G. Della Porta (Chairman),
V.J. Feron, D. Henschler, N. Ito, R. Kroes, P.N. Magee,
R. Montesano, N.P. Napalkov, S. Nesnow, G.N. Rao and J. Wilbourn

1. Introduction

The appearance of cancer is usually, if not always, preceded by noncancerous cellular changes, which are considered preneoplastic if they are known to be part of the process of the development of cancer (Anon., 1976; Bannasch, 1984). The more advanced precancerous lesions (dysplasia, neoplastic nodules, some benign tumours) are well known to pathologists and have been regarded as ancillary information in the evaluation of carcinogenicity. Considered here are even earlier phenotypically altered cell populations that are thought now to represent preneoplastic changes and which may have counterparts in in-vitro cell transformation assays. The early lesions may give rise to benign or malignant tumours, but not all progress to cancer. The fate of individual early lesions cannot be predicted unequivocally, and thus, by themselves, they are not definitive indicators of carcinogenicity. The detection of a high incidence of such lesions may be useful, however, for deciding which chemicals to test more extensively and for indicating possible target organs.

The early preneoplastic lesions best characterized thus far are foci of cellular changes in rat liver and, to a lesser extent, in mouse liver. Putative preneoplastic lesions in several other organs are discussed briefly to illustrate developing research.

2. Hepatic preneoplasia

Hepatic preneoplasia has been studied extensively in rats and mice, by many workers using various experimental models (for recent reviews see Peraino *et al.*, 1983; Farber, 1984; Scherer, 1984; Ward, 1984; Bannasch *et al.*, 1985a; Moore & Kitagawa, 1986). In rats, preneoplastic hepatic foci have been considered endpoints of carcinogenicity testing by several authors (Ito *et al.*, 1980; Pereira, 1982; Williams, 1982; Pereira & Stoner, 1985). Many studies have indicated that, with time, foci may progress to expansively growing neoplastic nodules and frank carcinomas (Bannasch *et al.*, 1980;

Emmelot & Scherer, 1980; Pitot & Sirica, 1980; Williams, 1980; Moore *et al.*, 1983a). However, under certain experimental conditions, foci that are similar to preneoplastic lesions may regress to a large extent (Farber, 1984; Moore & Kitagawa, 1986).

2.1 *Phenotypic patterns*

A landmark for studying early stages of hepatocarcinogenesis was the finding that treatment of rats with nitrosamines produced microscopic focal liver lesions characterized by an excessive storage of glycogen (Bannasch, 1968), and usually, also a reduction in the activity of the microsomal enzyme, glucose-6-phosphatase (Friedrich-Freska *et al.*, 1969), as demonstrated by cytochemical methods. In a series of experiments, a sequence of cellular changes has been inferred, which leads from clear, acidophilic hepatocytes storing glycogen in excess to basophilic hepatoma cells poor in glycogen but rich in free or membrane-bound ribosomes (Bannasch, 1968). The preneoplastic clear, acidophilic glycogen-storage cells constitute foci which persist after withdrawal of the carcinogen and may progress through intermediate, mixed and basophilic cell foci to neoplastic nodules (adenomas) and hepatocellular carcinomas. This developmental sequence has been adopted for the classification of specific hepatocellular lesions in rats; 'foci of altered hepatocytes' were separated from 'neoplastic nodules' by two working groups (Squire & Levitt, 1975; Stewart *et al.*, 1980). Recently, a somewhat more sophisticated nomenclature has been proposed which takes into consideration a certain type of basophilic focus, the 'tigroid cell focus', which apparently differs in many respects from the basophilic foci described earlier (Bannasch *et al.*, 1985a). The size of the various types of foci may vary from a few cells to larger areas including one or even more liver lobules. Foci of altered hepatocytes that exhibit cytomorphological changes similar to those seen in rats have also been detected in other species, including mice and primates (for reviews, see Bannasch, 1983; Ward, 1984). There are some indications that the predictive value of the various types of foci may differ (Ward, 1984; Bannasch *et al.*, 1985a).

In addition to changes in the amount of glycogen, endoplasmic reticulum and ribosomes, and alterations in the activity of glucose-6-phosphatase, a large number of other cellular changes have been described as 'negative' and 'positive' markers for carcinogen-induced foci (for reviews, see Emmelot & Scherer, 1980; Pitot & Sirica, 1980; Williams, 1980; Peraino *et al.*, 1983; Farber, 1984; Scherer, 1984; Bannasch *et al.*, 1985a; Moore & Kitagawa, 1986). Examples of enzymes that frequently show decreased activity, like glucose-6-phosphatase, are membrane-bound adenosine triphosphatase (Schauer & Kunze, 1976), acid and alkaline nucleases (Taper *et al.*, 1971) and glycogen phosphorylase (Hacker *et al.*, 1982). Increased activities or content of γ-glutamyl transpeptidase (Kalengayi & Desmet, 1975), glucose-6-phosphate dehydrogenase (Hacker *et al.*, 1982), epoxide hydrolase (Enomoto *et al.*, 1981; Kuhlmann *et al.*, 1981), uridine diphosphate glucuronyltransferase (Fischer *et al.*, 1983), various isoenzymes of cytochrome P450 (Schulte-Hermann *et al.*, 1984; Buchmann *et al.*, 1985), glutathione (Deml & Oesterle, 1980) and glutathione transferases (Sato *et al.*, 1984; Buchmann *et al.*, 1985) have been found in many (albeit not all) foci. Other alterations described include resistance to experimental haemosiderosis (Williams *et al.*, 1976) and decreased lipid peroxidation (Benedetti *et al.*, 1984). The cytochemical changes appearing in focal liver lesions induced with chemicals in mice differ in some respects from those in rats, but they also show many similarities (Ward, 1984; Vesselinovitch *et al.*, 1985).

Observations in different strains and species indicate that the cytochemical pattern of the preoplastic foci may be rather heterogeneous and is apparently influenced by many factors, such as dose and duration of carcinogen treatment, localization of the focal lesions within the liver lobule, and the age, sex and strain of the animals (see Rabes *et al.*, 1972; Hirota & Williams, 1979; Pitot & Sirica, 1980; Deml *et al.*, 1981; Bannasch *et al.*, 1982; Peraino *et al.*, 1984; Moore & Kitagawa, 1986). Of particular interest is the finding that the activity of γ-glutamyl transpeptidase, which has been considered to be a reliable indicator of preneoplastic changes in the liver by a number of investigators (see Hanigan & Pitot, 1985), may be partly or totally lacking in certain types of preneoplastic foci in both rat and mouse liver (Butler *et al.*, 1981; Moore *et al.*, 1983a; Rao *et al.*, 1984; Bannasch *et al.*, 1985a; Vesselinovitch *et al.*, 1985). A periportal focal increase in the activity of this enzyme may also occur in rat liver with age (Kitagawa *et al.*, 1980) or after partial hepatectomy (Bone *et al.*, 1985).

In old, untreated rodents, various types of foci may develop 'spontaneously' (Burek, 1978; Ward, 1981, 1984). An exceptionally high incidence of spontaneous foci in certain rat strains suggests that genetic factors most probably also play an important role (Burek, 1978; Ward, 1981).

2.2 *Phenotypic instability*

In various experimental models it was found that most phenotypic 'markers' of hepatocarcinogenesis that have been described so far are not stable and may not necessarily be related to neoplastic transformation. Thus, under certain experimental conditions, foci (or nodules) have been observed which resemble in their phenotype the persistent lesions but may disappear after cessation of treatment.

It has been known for some time that clear-cell areas or foci induced by carcinogens may be partly reversible after withdrawal of the chemical (Bannasch, 1968; Newberne, 1976; Flaks & Basley, 1979; Hacker *et al.*, 1982). A partial reversibility of cytomorphological and cytochemical changes, including clear, acidophilic and basophilic features which are similar to those in persistent preneoplastic foci, has been observed (Moore *et al.*, 1983b) in focal lesions induced in rat liver by the short-term procedure of Solt and Farber (1976). A 'reversion', 'remodelling', 'neodifferentiation' or 'maturation' of carcinogen-induced focal liver lesions has also been reported by a number of other authors (Kitagawa, 1971; Ito *et al.*, 1976; Kitagawa, 1976; Williams & Watanabe, 1978; Ogawa *et al.*, 1979; Tatematsu *et al.*, 1983). Even if some observations seem to indicate that 'remodelled' lesions may reappear after long lag periods (Williams & Watanabe, 1978; Tatematsu *et al.*, 1983), use of the adjectives 'neoplastic' or 'preneoplastic' for such reversible nodules should be avoided. The terms 'hepatocyte nodule', proposed by Farber (1982), or 'proliferating hepatic nodule' appear to be much more appropriate. The only reliable way to distinguish between reversible and persistent focal lesions, the latter being considered preneoplastic, are 'stop' or reversibility experiments. In these experiments, the test compound is applied for a limited period, and the animals are investigated weeks or months after cessation of the chemical exposure (Bannasch *et al.*, 1982).

2.3 Development and dose dependence

Many authors have investigated the kinetics of cell proliferation in altered foci identified by enzymatic changes (Schauer & Kunze, 1968; Rabes *et al.*, 1972; Barbason *et al.*, 1977; Pugh & Goldfarb, 1978; Kuhlmann *et al.*, 1981; Schulte-Hermann *et al.*, 1981; Rabes, 1983). In all of these studies, a considerable increase in cell proliferation was found in foci as compared with the surrounding tissue or with the liver parenchyma of untreated controls. However, most of these studies did not investigate the different cell populations composing the foci. When these cell populations were investigated separately, it was found that cell proliferation was increased only slightly, if at all, in clear and acidophilic glycogen-storage foci (Bannasch *et al.*, 1982). A pronounced and steadily increasing cell proliferation was linked instead with the appearance of mixed and basophilic cell populations in foci, nodules and carcinomas.

The dose-dependence of the sequential cellular changes induced in rat liver by stop experiments has been investigated by morphometric methods (Moore *et al.*, 1982). In experiments with N-nitrosomorpholine, the majority of foci developed after withdrawal of the carcinogen. With all dose levels studied, a sequence was established leading from clear-cell, acidophilic-cell, glycogen-storage foci through mixed-cell foci and neoplastic nodules to hepatocellular carcinomas. The first appearance, the frequency and the kinetics of evolution of the different lesions investigated proved to depend on the dose of carcinogen administered. There was no indication of reversibility of the focal lesions under these experimental conditions. On the contrary, the foci became larger and acquired phenotypic markers closer to neoplasia without further action of the carcinogen. A progressive development has also been revealed for basophilic foci induced with a single injection of N-nitrosodiethylamine in infant mice (Goldfarb *et al.*, 1983; Vesselinovitch & Mihailovich, 1983).

Histological transitions between hepatic foci, adenomas (neoplastic nodules) and carcinomas have been described by many authors (for reviews, see Bannasch, 1968; Schauer & Kunze, 1976; Williams, 1980). While these observations indicate that the adenomas originate from the foci and may progress to carcinomas, it is probable that the latter can also develop directly from the foci without going through a nodular intermediate stage (Bannasch, 1976).

Not only these cytomorphological and morphometric results, but also observation of alterations in a number of functional cellular changes support the concept of a close relation among foci, nodules and carcinomas. Thus, a similar decrease or increase in the activity of many enzymes has been demonstrated by enzyme histochemical methods in these lesions (see Bannasch *et al.*, 1980; Emmelot & Scherer, 1980; Farber, 1980; Pitot & Sirica, 1980; Williams, 1980). Several authors have demonstrated in rats or mice that preneoplastic foci, neoplastic nodules and carcinomas share a common defect, in that they do not accumulate iron in a siderotic liver produced artificially (Williams *et al.*, 1976; Lipsky *et al.*, 1979).

Of particular interest are data concerning the dose-dependence of development of enzyme-altered foci, as published by several groups (see Emmelot & Scherer, 1980; Kunz *et al.*, 1982; Goldsworthy *et al.*, 1984; Goldsworthy & Pitot, 1985; Préat *et al.*, 1985). Although the methods used varied somewhat, all the results indicate that quantitative correlations exist between the size and number of foci and the dose and duration of carcinogen treatment. In this context, it should be mentioned, however, that a direct relationship between focal liver lesions, nodules and carcinomas has been questioned by some (O'Connor & Woodward, 1981). Large discrepancies between the number of foci appearing early during hepatocarcinogenesis and final tumour yield are indeed a common finding (Scherer & Emmelot, 1975; Emmelot & Scherer, 1980; Scherer, 1984; Kaufmann *et al.*, 1985).

2.4 *Usefulness in bioassays*

There appears to be general agreement that the evaluation of bioassays may be augmented by consideration of preneoplastic focal liver lesions. If a test compound produces significantly more foci of altered hepatocytes in treated animals than in concurrent untreated controls, this suggests a carcinogenic potential of the respective chemical. In long-term bioassays, it may be useful to have an additional group of animals for sequential killing and detection of foci of altered hepatocytes. Because of a possible reversion-linked phenotypic instability of the lesions, stop experiments, which allow a clear distinction between reversible and persistent preneoplastic foci, should be conducted whenever hepatic foci of disputed significance appear (Bannasch *et al.*, 1982). Although all hepatocarcinogens investigated so far induced some type of focal hepatic lesions prior to the development of hepatic tumours, it is not yet clear whether this is always the case. Hence, the absence of hepatic foci after administration of a chemical does not preclude a carcinogenic potential of the compound.

A comparison of the results obtained in a number of experimental models is hampered by both the remarkable variation in the experimental approaches in different laboratories and the frequent lack of a clear separation of the (preneoplastic) foci of altered hepatocytes from the (neoplastic) nodular lesions. Treatment of animals with the test compound for several weeks (Williams, 1982) and stop experiments with a longer observation period after withdrawal of the carcinogen (Bannasch et al., 1982) have been suggested. Extreme types of stop experiments are represented by models in which a single dose of the test compound is administered to newborn animals (Vesselinovitch et al., 1979; Vesselinovitch, 1980; Vesselinovitch & Michailovich, 1983) or to adult animals after partial hepatectomy (Scherer et al., 1972; Scherer & Emmelot, 1975). Frequently, the additional application of phenobarbital has been used to 'promote' the 'initiated' cells possibly induced by the test compound, either in young animals (Peraino et al., 1981; Cater et al., 1985) or in adult animals after partial hepatectomy (Pitot et al., 1978; Pitot & Sirica, 1980) or without partial hepatectomy (Williams, 1982). This approach has been adopted for the so-called 'rat liver foci bioassay' (Pereira, 1982; Pereira & Stoner, 1985). Although the additional application of phenobarbital renders the system more 'sensitive', this 'two-stage model' poses problems since some authors imply that the operational distinction between initiating and promoting effects allows mechanistic explanations but it is as yet unknown whether phenobarbital promotes initiated cells or acts together with the 'initiator' as a syncarcinogen (Rossi et al., 1977; Feldman et al., 1981; Goldsworthy et al., 1984). Nevertheless, the liver foci bioassay has been used during the past few years (Pereira et al., 1982; Williams, 1982; Laib et al., 1985a,b; Pereira & Stoner, 1985).

In a recent review, Pereira and Stoner (1985) compared the rat-liver foci assay with the strain A mouse-lung tumour assay to detect carcinogens. The authors stated that the rat-liver foci assay was sensitive to 69% of 54 compounds found to be carcinogenic in long-term bioassays in rats or mice and the strain A lung-tumour assay to 54% of 93 carcinogens. None of ten compounds found to be noncarcinogenic in long-term animal bioassays was active in the liver-foci assay.

More complex models have been developed on the basis of an experiment described by Teebor and Becker (1971), and additional considerations derive from other studies (Solt & Farber, 1976; Cayama et al., 1978; Ito et al., 1980; Tatematsu et al., 1980; Tsuda et al., 1980; de Gerlache et al., 1982). From a review taking into consideration 80 compounds tested, Parodi et al. (1983) concluded that the 'preneoplastic nodules' (focal and nodular lesions were not separated in these studies) induced by these complex procedures were more predictive than the Ames' test. There was a somewhat better predictivity of 'preneoplastic' nodules for liver tumours, as compared to tumours at other sites. The additional application of chemicals, some of which are potent carcinogens in these complex test systems, hampers the interpretation of the results.

3. Preneoplastic lesions in other tissues

Putative preneoplastic lesions have also been described in a number of other tissues (Anon., 1976; Bannasch, 1984), but as yet they have not been used systematically in the evaluation of carcinogenesis bioassays and will be summarized only briefly. A more detailed analysis of such lesions will be important not only for the histological diagnosis of tumour pre-stages but also perhaps for elucidation of the hitherto poorly understood organotropism and cytotropism of carcinogens. The following examples do not include the 'dysplastic' lesions widely regarded as preneoplastic changes in various tissues, such as the uterine cervix, skin and breast (for reviews, see Farber & Cameron, 1980; Carter, 1984), but concentrates, as for the liver (section 2), on early phenotypically altered cell populations that may represent preneoplastic lesions.

3.1 *Intrahepatic biliary system*

It has been known for a long time that hepatotropic carcinogens frequently induce a characteristic lesion called 'cholangiofibrosis' (or 'adenofibrosis'), in addition to the preneoplastic or neoplastic changes of the liver parenchyma discussed earlier (Schauer & Kunze, 1976; Stewart *et al.*, 1980; Bannasch *et al.*, 1985b). Cholangiofibrosis develops especially after short-term application of high doses of carcinogens or in long-term studies of dose levels that produce considerable necrotic change of the liver parenchyma. The pathogenesis and significance of cholangiofibrosis for liver neoplasia have been subjects of controversy (see Stewart, 1975; Farber & Cameron, 1980; Bannasch, 1983), but many findings suggest that the lesion is preneoplastic and may progress to cystic cholangioma or cholangiofibroma and eventually also to cholangiocarcinoma (Bannasch *et al.*, 1985b).

3.2 *Pancreas*

The morphogenesis of pancreatic carcinomas has been studied extensively in the past decade in rodents treated with various carcinogens (see Levitt *et al.*, 1977; Pour *et al.*, 1977; Flaks *et al.*, 1980a,b; Rao & Reddy, 1980; Takahashi *et al.*, 1980; Pour *et al.*, 1981; Takahashi *et al.*, 1981; Moore *et al.*, 1983c,d; Longnecker *et al.*, 1984, 1985). The apparent species specificity of the carcinogens used and the seeming disparity of the target cells in the different experimental models have generated considerable controversy as to the cells of origin of pancreatic carcinoma. In the azaserine-treated rat model, the earliest morphological alterations were described in acinar cells, leading to the development of nodules of atypical acinar cells and eventually to acinar-cell carcinoma (Longnecker *et al.*, 1984). The phenotypic alterations that characterize the cells in the atypical acinar-cell nodules include (i) reduced cytoplasmic basophilia, (ii) reduced zymogen content and increased cytoplasmic basophilia, and (iii) nuclear abnormalities such as enlargement and/or irregular shape. The early morphological changes in models in which hamsters are treated with *N*-nitrosobis(2-hydroxypropyl)amine or *N*-nitrosobis(2-oxypropyl)amine (Levitt *et al.*, 1977; Flaks *et al.*, 1980a,b; Moore *et al.*, 1983c,d), the guinea-pig *N*-methyl-*N*-nitrosourea model (Rao & Reddy, 1980) and the rat 7,12-dimethylbenz[*a*]anthracene model (Dissin *et al.*, 1975;

Bockman et al., 1978) have been described as duct-like structures, the origin of which is alternatively proposed as pre-existing centroacinar and ductular cells or 'dedifferentiated' acinar cells (Parsa et al., 1985). Other early lesions during pancreatic carcinogenesis in hamsters are transient changes in carbohydrate metabolism with an accumulation of mucus in goblet cells, occasional storage of glycogen and increased activity of glucose-6-phosphate dehydrogenase in regions of epithelial atypia (so-called 'dysplastic lesions') (Moore et al., 1983c,d).

3.3 Lung

Studies on preneoplasia in the lungs of humans and experimental animals have been reviewed by Gusterson (1984). Ultrastructural investigations comparing preneoplastic lesions in the bronchial tree of hamsters and humans have demonstrated close correlations between the preneoplastic lesions and the tumours produced in the two species (Becci et al., 1978a,b; Trump et al., 1978). When a mixture of benzo[a]pyrene and ferric oxide is instilled into the trachea of Syrian hamsters, a sequence of changes takes place, starting with an acute reaction within 24 hours to produce 'goblet cell hyperplasia' (Becci et al., 1978a). With continued application of the carcinogen, the epithelium becomes hyperplastic, forming five to ten cell layers. At this stage, the mucosa takes on the light-microsopic features of squamous metaplasia, but at the ultrastructural level the cells still contain mucus in addition to tonofilaments. However, metaplastic changes are not necessarily a manifestation of preneoplasia (Nettesheim, 1976) and the significance of the early lesions is uncertain. In other experiments with 7,12-dimethylbenz[a]anthracene in rat trachea, the early metaplastic lesions disappeared within a few months and were replaced by atrophic, atypical epithelial cells, followed by a later return to atypical metaplasia and by dysplasia before carcinoma development (Griesemer et al., 1977).

Sequential morphological alterations in a different bronchial cell type, amine precursor uptake and decarboxylase activity (APUD) cells, of hamsters treated with nitrosamines have been described during tumorigenesis by Reznik-Schüller (1977a,b). The same author claimed that, after three weeks' treatment with N-nitrosomorpholine, lamellar bodies, similar to those normally seen in Type II pneumocytes, developed in the non-ciliated Clara cells of the terminal bronchus. These cells are reported to increase in number and eventually develop into invasive tumours resembling alveolar carcinomas.

3.4 Kidney

Renal carcinogenesis has been studied mainly in rats treated with nitrosamines (Hard & Butler, 1971; Ito, 1973; Bannasch et al., 1974; Hard, 1976). 'Proliferative tubules' lined by an irregular epithelium (Magee & Barnes, 1962), which occasionally show clear cells (Ito et al., 1966; Hard & Butler, 1971; Dees et al., 1980), were the first early lesions described. Systematic studies on the morphogenesis of epithelial kidney tumours in the rat revealed that at least three different types of presumably preneoplastic tubular lesion can be distinguished, each of which appears to be the precursor of a cytologically characteristic tumour type (Bannasch et al., 1986). Thus, clear-cell tubules storing excess glycogen regularly precede the development of clear- and granular-cell tumours. Oncocytic (distal) tubules are pre-stages of renal oncocytomas, and chromophobic and/or basophilic (proximal) tubules, which frequently store acid mucopolysaccharides, seem to be closely related to the genesis of chromophobic and basophilic renal epitheliomas. Changes in the activities of a number of enzymes have recently been shown in oncocytic and basophilic tubules. It is noteworthy that all of these cytologically different types of tubular lesions and tumours frequently develop side by side in the same kidney. As a rule, the preneoplastic tubular changes appear multicentrically and bilaterally.

3.5 *Urinary bladder*

Morphologically, preneoplasia in the urothelium may be described as a slight or moderate qualitative abnormality in basal or intermediate cells without an appreciable increase in the number of cells, i.e., without significant hyperplasia in a flat lesion (Cohen et al., 1984). In rats, mice and dogs treated with chemical carcinogens, a focal loss in the activity of alkaline phosphatase has been demonstrated in non-hyperplastic and hyperplastic lesions of the bladder epithelium by a number of authors (Kunze & Schauer, 1971; Stiller & Rauscher, 1971; Highman et al., 1975; Ito, 1976; Hicks & Chowaniec, 1978; Radomski et al., 1978; Kunze, 1979). Increased activities of nonspecific esterases and of β-glucuronidase have also been described in hyperplastic bladder lesions of these species, but comparable enzymatic changes did not occur in bladder mucosa of hamsters and guinea-pigs treated with N-butyl-N-(4-hydroxybutyl)-nitrosamine (Ito, 1976). In rats, the foci deficient in alkaline phosphatase persisted throughout the preneoplastic phase, and a similar loss in enzyme activity was found in many, albeit not all, papillomas and transitional-cell carcinomas (Kunze, 1979). However, similar losses of alkaline phosphatase activity are seen in reversible regenerative hyperplasias of the bladder induced by various methods, such as intraperitoneal cyclophosphamide injection or ulceration of the bladder by freezing. Under these conditions, alkaline phosphatase activity returns to normal levels as the hyperplastic changes resolve. Studies of β-glucuronidase demonstrate that, as for alkaline phosphatase, the levels decrease in reversible regenerative hyperplasia as well as in transitional-cell carcinomas (Cohen et al., 1984). Vanderlaan et al. (1982) reported that early bladder lesions induced in rats with N-butyl-N-(4-hydroxybutyl)nitrosamine usually contained no γ-glutamyl transpeptidase but showed increased activity of NADH diaphorase.

4. References

Anon. (1976) Symposium on early lesions and the development of epithelial cancer. *Cancer Res.*, 36, 2475-2706

Bannasch, P. (1968) The cytoplasm of hepatocytes during carcinogenesis. Light and electron microscopic investigations of the nitrosomorpholine-intoxicated rat liver. *Recent Results Cancer Res.*, 19, 1-100

Bannasch, P. (1976) Cytology and cytogenesis of neoplastic (hyperplastic) nodules. *Cancer Res.*, 36, 2555-2562

Bannasch, P. (1983) *Strain and species differences in the susceptibility to liver tumor induction*. In: Turusov, V. & Montesano, R., eds, *Modulators of Experimental Carcinogenesis (IARC Scientific Publications No.51)*, Lyon, International Agency for Research on Cancer, pp. 9-38

Bannasch, P. (1984) Sequential cellular changes during chemical carcinogenesis. *J. Cancer Res. clin. Oncol.*, 108, 11-22

Bannasch, P., Schacht, U. & Storch, E. (1974) Morphogenesis and micromorphology of epithelial kidney tumours in nitrosomorpholine-treated rats [in German]. *Z. Krebsforsch.*, 81, 311-331

Bannasch, P., Mayer, D. & Hacker, H.J. (1980) Hepatocellular glycogenosis and hepatocarcinogenesis. *Biochim. biophys. Acta*, 605, 217-245

Bannasch, P., Moore, M.A., Klimek, F. & Zerban, H. (1982) Biological markers of preneoplastic foci and neoplastic nodules in rodent liver. *Toxicol. Pathol.*, 10, 19-34

Bannasch, P., Zerban, H. & Hacker, H.J. (1985a) *Foci of altered hepatocytes, rat*. In: Jones, T.C., Mohr, U. & Hunt, R.D., eds, *Monographs on Pathology of Laboratory Animals*, Berlin (West), Springer, pp. 10-30

Bannasch, P., Benner, U. & Zerban, H. (1985b) *Cholangiofibroma and cholangiocarcinoma, liver, rat*. In: Jones, T.C., Mohr, U. & Hunt, R.D., eds, *Monographs on Pathology of Laboratory Animals*, Berlin (West), Springer, pp. 52-65

Bannasch, P., Hacker, H.J., Tsuda, H. & Zerban, H. (1986) Aberrant carbohydrate metabolism and metamorphosis during renal carcinogenesis. *Adv. Enzyme Regul.*, 25 (in press)

Barbason, H., Fridman-Manduzo, A., Lelievre, P. & Betz, E.H. (1977) Variations of liver cell control during diethylnitrosamine carcinogenesis. *Eur. J. Cancer*, 13, 13-18

Becci, P.H., McDowell, E.M. & Trump, B.F. (1978a) The respiratory epithelium. IV. Histogenesis of epidermoid metaplasia and carcinoma in situ in the hamster. *J. natl Cancer Inst.*, 61, 577-586

Becci, P.H., McDowell, E.M. & Trump, B.F. (1978b) The respiratory epithelium. VI. Histogenesis of lung tumors induced by benzo[a]pyrene-ferric oxide in the hamster. *J. natl Cancer Inst.*, 61, 607-618

Benedetti, A., Malvadi, G., Fulceri, R. & Comporti, M. (1984) Loss of lipid peroxidation as a histochemical marker for preneoplastic hepatocellular foci of rats. *Cancer Res.*, 44, 5712-5717

Bockman, D.C., Black, O., Jr, Mills, L.R. & Webster, P.D. (1978) Origin of tubular complexes developing during induction of pancreatic adenocarcinoma by 7,12-dimethylbenz(a)anthracene. *Am. J. Pathol.*, 90, 645-658

Bone, S.N., III, Michalopoulos, G. & Jirtle, R.L. (1985) Ability of partial hepatectomy to induce γ-glutamyltranspeptidase in regenerated and transplanted hepatocytes of Fischer 344 and Wistar-Furth rats. *Cancer Res.*, 45, 1222-1228

Buchmann, A., Kuhlmann, W.D., Schwarz, M., Kunz, H.W., Wolf, C.R., Moll, E., Friedberg, T. & Oesch, F. (1985) Regulation and expression of four cytochrome P-450 isoenzymes, NADPH-cytochrome P-450 reductase, the glutathion transferase B and C and microsomal epoxide hydrolase in preoplastic and neoplastic lesions in rat liver. *Carcinogenesis, 6*, 513-521

Burek, J.D. (1978) *Pathology of Aging Rats. A Morphological and Experimental Study of the Age-associated Lesions in Aging BN/Bi, WAG/Rij and (WAGxBN) F_1 Rats*, West Palm Beach, FL, CRC Press, pp. 58-68

Butler, W.H., Hempsall, V. & Stewart, M.C. (1981) Histochemical studies on the early proliferative lesions induced in the rat liver by aflatoxin. *J. Pathol., 133*, 325-340

Carter, R.L., ed. (1984) *Precancerous States*, London, Oxford University Press

Cater, K.C., Gandolfi, A.J. & Sipes, I.G. (1985) Characterization of dimethylnitrosamine-induced focal and nodular lesions in the livers of newborn mice. *Toxicol. Pathol., 13*, 3-9

Cayama, E., Tsuda, H., Sarma, D.S.R. & Farber, E. (1978) Initiation of chemical carcinogenesis requires cell proliferation. *Nature, 275*, 60-62

Cohen, S.M., Greenfield, R.E., Jacobs, J.B. & Friedell, G.H. (1984) *Precancerous and noninvasive lesions of the urinary bladder.* In: Carter, R.L., ed., *Precancerous States*, London, Oxford University Press, pp. 278-303

Dees, J.H., Heatfield, B.H., Reuber, M.D. & Trump, B.F. (1980) Adenocarcinoma of the kidney. III. Histogenesis of renal adenocarcinomas induced in rats by *N*-(4'-fluoro-4-biphenylyl)acetamide. *J. natl Cancer Inst., 64*, 1537-1545

Deml, E. & Oesterle, D. (1980) Histochemical demonstration of enhanced glutathione content in enzyme-altered islands induced by carcinogens in rat liver. *Cancer Res., 40*, 490-491

Deml, E., Oesterle, D., Wolff, T. & Greim, H. (1981) Age-, sex-, and strain-dependent differences in the induction of enzyme-altered islands in rat liver by diethylnitrosamine. *J. Cancer Res. clin. Oncol., 100*, 125-134

Dissin, J., Mills, L.R., Mainz, D.L., Black, O., Jr & Webster, P.D. III. (1975) Experimental induction of pancreatic adenocarcinoma in rats. *J. natl Cancer Inst., 55*, 857-864

Emmelot, P. & Scherer, E. (1980) The first relevant cell stage in rat liver carcinogenesis. A quantitative approach. *Biochim. biophys. Acta, 605*, 247-304

Enomoto, K., Ying, T.S., Griffin, M.J. & Farber, E. (1981) Immunohistochemical study of epoxide hydrolase during experimental liver carcinogenesis. *Cancer Res., 41*, 3281-3287

Farber, E. (1980) The sequential analysis of liver cancer induction. *Biochim. biophys. Acta, 605*, 149-166

Farber, E. (1982) The biology of carcinogen-induced hepatocyte nodules and related liver lesions in the rat. *Toxicol. Pathol., 10*, 197-205

Farber, E. (1984) Pre-cancerous steps in carcinogenesis. Their physiological adaptive nature. *Biochim. biophys. Acta, 738*, 171-180

Farber, E. & Cameron, R. (1980) The sequential analysis of cancer development. *Adv. Cancer Res., 31*, 125-226

Feldman, D., Swarm, R.L. & Becker, J. (1981) Ultrastructural study of rat liver and liver neoplasms after long-term treatment with phenobarbital. *Cancer Res., 41*, 2151-2162

Fischer, G., Ullrich, D., Katz, N., Bock, W.K. & Schauer, A. (1983) Immunohistochemical and biochemical detection of uridine-diphosphate-glucuronyl-transferase (UDP-GT) activity in putative preneoplastic liver foci. *Virchows Arch. B. Cell Pathol., 42*, 193-200

Flaks, B. & Basley, W.A. (1979) Acute fine structural changes in rat hepatocytes induced by a single large dose of 2-acetylaminofluorene. *Virchows Arch. B. Cell Pathol., 29*, 309-320

Flaks, B., Moore, M.A. & Flaks, A. (1980a) Ultrastructural analysis of pancreatic carcinogenesis: morphological characterization of *N*-nitrosobis(2-hydroxypropyl)amine-induced neoplasms in the Syrian hamster. *Carcinogenesis, 1*, 423-438

Flaks, B., Moore, M.A. & Flaks, A. (1980b) Ultrastructural analysis of pancreatic carcinogenesis. III. Multifocal cystic lesions induced by *N*-nitrosobis(2-hydroxypropyl)amine in the hamster exocrine pancreas. *Carcinogenesis, 1*, 693-706

Friedrich-Freska, H., Papadopulu, G. & Gössner, W. (1969) Histochemical investigations on carcinogenesis in rat liver following administration of diethylnitrosamine for limited time periods [in German]. *Z. Krebsforsch., 72*, 240-253

de Gerlache, J., Lans, M., Taper, H., Préat, V. & Roberfroid, M. (1982) *Promotion of the chemically initiated hepatocarcinogenesis.* In: Sorsa, M. & Vainio, H., eds, *Mutagens in Our Environment,* New York, Alan R. Liss, pp. 35-46

Goldfarb, S., Pugh, T.D., Koen, H. & He, Y.-Z. (1983) Preneoplastic and neoplastic progression during hepatocarcinogenesis in mice injected with diethylnitrosamine in infancy. *Environ. Health Perspectives, 50*, 149-161

Goldsworthy, T.L. & Pitot, H.C. (1985) The quantitative analysis and stability of histochemical markers of altered hepatic foci in rat liver following initiation by diethylnitrosamine administration and promotion with phenobarbital. *Carcinogenesis, 6*, 1261-1269

Goldsworthy, T., Campbell, H.A. & Pitot, H.C. (1984) The natural history and dose-response characteristics of enzyme-altered foci in rat liver following phenobarbital and diethylnitrosamine administration. *Carcinogenesis, 5*, 67-71

Griesemer, R.A., Nettesheim, P., Martin, D.H. & Caton, J.E. (1977) Quantitative exposure of respiratory airway epithelium to 7,12-dimethylbenz[*a*]anthracene. *Cancer Res., 37*, 1266-1271

Gusterson, B.A. (1984) *Precancerous changes in the lungs and the potential of cells to have modulated phenotypes.* In: Carter, R.L., ed., *Precancerous States*, London, Oxford University Press, pp. 161-184

Hacker, H.J., Moore, M.A., Mayer, D. & Bannasch, P. (1982) Correlative histochemistry of some enzymes of carbohydrate metabolism in preneoplastic and neoplastic lesions in the rat liver. *Carcinogenesis, 3*, 1265-1272

Hanigan, M.H. & Pitot, H.C. (1985) Gamma-glutamyl transpeptidase — its role in hepatocarcinogenesis. *Carcinogenesis, 6*, 165-172

Hard, G.C. (1976) *Tumours of the kidney, renal pelvis and ureter.* In: Turusov, V.S., ed., *Pathology of Tumours in Laboratory Animals*, Vol. I, *Tumours of the Rat*, Part 2 (*IARC Scientific Publications No. 6*), Lyon, International Agency for Research on Cancer, pp. 73-101

Hard, G.C. & Butler, W.H. (1971) Morphogenesis of epithelial neoplasms induced in the rat kidney by dimethylnitrosamine. *Cancer Res., 31*, 1496-1505

Hicks, R.M. & Chowaniec, J. (1978) Experimental induction, histology, and ultrastructure of hyperplasia and neoplasia of the urinary bladder epithelium. *Int. Rev. exp. Pathol., 18*, 199-280

Highman, B., Frith, C.H. & Littlefield, N.A. (1975) Alkaline phosphatase activity in hyperplastic and neoplastic urinary bladder epithelium of mice fed 2-acetylaminofluorene. *J. natl Cancer Inst., 54*, 257-261

Hirota, N. & Williams, G.M. (1979) The sensitivity and heterogeneity of histochemical markers for altered foci involved in liver carcinogenesis. *Am. J. Pathol., 95*, 317-328

Ito, N. (1973) Experimental studies on tumors of the urinary system of rats induced by chemical carcinogens. *Acta pathol. Jpn., 23,* 87-109

Ito, N. (1976) Early changes caused by *N*-butyl-*N*-(4-hydroxybutyl) nitrosamine in the bladder epithelium of different animal species. *Cancer Res., 36,* 2528-2531

Ito, N., Johno, J., Marugami, M., Konishi, Y. & Hiasa, Y. (1966) Histopathological and autoradiographic studies on kidney tumors induced by *N*-nitrosodimethylamine in rat. *Gann, 57,* 595-604

Ito, N., Hananouchi, M., Sugihara, S., Shirai, T., Tsuda, H., Fukushima, S. & Nagasaki, H. (1976) Reversibility and irreversibility of liver tumors in mice induced by the α-isomer of 1,2,3,4,5,6-hexachloro-cyclohexane. *Cancer Res., 36,* 2227-2234

Ito, N., Tatematsu, M., Nakanishi, K., Hasegawa, R., Takano, T., Imaida, K. & Ogiso, T. (1980) The effects of various chemicals on the development of hyperplastic liver nodules in hepatectomized rats treated with *N*-nitrosodiethylamine or *N*-2-fluorenylacetamide. *Gann, 71,* 832-842

Kalengayi, M.M.R. & Desmet, V.J. (1975) Sequential histological and histochemical study of the rat liver during aflatoxin B_1-induced carcinogenesis. *Cancer Res., 35,* 2845-2852

Kaufmann, W.K., Mackenzie, S.A. & Kaufman, D.G. (1985) Quantitative relationship between hepatocytic neoplasms and islands of cellular alterations during hepatocarcinogenesis in the male F344 rats. *Am. J. Pathol., 119,* 171-174

Kitawaga, T. (1971) Histochemical analysis of hyperplastic lesions and hepatomas of the liver of rats fed 2-fluorenylacetamide. *Gann, 62,* 207-216

Kitagawa, T. (1976) Sequential phenotypic changes in hyperplastic areas during hepatocarcinogenesis in the rat. *Cancer Res., 36,* 2534-2539

Kitagawa, T., Imai, F. & Sato, K. (1980) Re-evaluation of γ-glutamyl transpeptidase activity in periportal hepatocytes of rats with age. *Gann, 71,* 362-366

Kuhlmann, W.D., Krishan, R., Kunz, W., Guenthner, T.M. & Oesch, F. (1981) Focal elevation of liver microsomal epoxide hydrolase in early preneoplastic stages and its behaviour in the further course of hepatocarcinogenesis. *Biochem. biophys. Res. Commun., 98,* 417-423

Kunz, W., Schaude, G., Schwartz, M. & Tennekes, H. (1982) *Quantitative aspects of drug-mediated tumour promotion in liver and its toxicological implications.* In: Hecker, E., Fusenig, N., Marks, F. & Kunz, W., eds, *Carcinogenesis — A Comprehensive Survey,* New York, Raven Press, pp. 111-125

Kunze, E. (1979) Development of urinary bladder cancer in the rat. *Curr. Top. Pathol., 67,* 145-232

Kunze, E. & Schauer, A. (1971) Enzyme-histochemical and autoradiographic investivations on dibutylnitrosamine-induced papillomas of the urinary bladder in the rat [in German]. *Z. Krebsforsch., 75,* 146-160

Laib, R.J., Klein, K.P. & Bolt, H.M. (1985a) The rat liver foci bioassay. I. Age-dependence of induction by vinyl chloride of ATPase-deficient foci. *Carcinogenesis, 6,* 65-68

Laib, R.J., Pellio, T., Wünschel, U.M., Zimmermann, N. & Bolt, H.M. (1985b) The rat liver foci bioassay. II. Investigations on the dose-dependent induction of ATPase-deficient foci by vinyl chloride at very low doses. *Carcinogenesis, 6,* 69-72

Levitt, M.H., Harris, C.C., Squire, R., Springer, S., Wenk, M., Mollelo, C., Thomas, D., Kingsbury, E. & Newkirk, C. (1977) Experimental pancreatic carcinogenesis. I. Morphogenesis of pancreatic adenosarcoma in the Syrian golden hamster induced by *N*-nitrosobis(2-hydroxypropyl)amine. *Am. J. Pathol., 88,* 5-22

Lipsky, M.M., Hinton, D.E., Goldblatt, P.J., Klaunig, J.E. & Trump, B.F. (1979) Iron negative foci and nodules in safrole-exposed mouse liver made siderotic by iron-dextran injection. *Pathol. Res. Pract.*, *164*, 178-185

Longnecker, D.S., Wiebkin, P., Schaeffer, B.K. & Roebuck, B.D. (1984) Experimental carcinogenesis in the pancreas. *Int. Rev. exp. Pathol.*, *26*, 177-229

Longnecker, D.S., Roebuck, B.D., Kuhlmann, E.T. & Curphey, T.J. (1985) Induction of pancreatic carcinomas in rats with *N*-nitroso(2-hydroxypropyl)(2-oxopropyl)amine: histopathology. *J. natl Cancer Inst.*, *74*, 209-217

Magee, P.N. & Barnes, J.M. (1962) Induction of kidney tumours in the rat with dimethylnitrosamine (*N*-nitrosodimethylamine). *J. Pathol. Bacteriol.*, *84*, 19-31

Moore, M.A. & Kitagawa, T. (1986) Hepatocarcinogenesis in the rat; the effect of the promoters and carcinogens *in vivo* and *in vitro*. *Int. Rev. Cytol.*, *101*, 125-173

Moore, M.A., Mayer, D. & Bannasch, P. (1982) The dose-dependence and sequential appearance of putative preneoplastic populations induced in the rat liver by stop experiments with *N*-nitrosomorpholine. *Carcinogenesis*, *3*, 1429-1436

Moore, M.A., Hacker, H.J., Kunz, H.W. & Bannasch, P. (1983a) Enhancement of NNM-induced carcinogenesis in the rat liver by phenobarbital: a combined morphological and enzyme histochemical approach. *Carcinogenesis*, *4*, 473-479

Moore, M.A., Hacker, H.J. & Bannasch, P. (1983b) Phenotypic instability in focal and nodular lesions induced in a short term system in the rat liver. *Carcinogenesis*, *4*, 595-603

Moore, M.A., Takahashi, M., Ito, N. & Bannasch, P. (1983c) Early lesions during pancreatic carcinogenesis induced in Syrian hamster by DHPN or DOPN. I. Histologic, histochemical and radioautographic findings. *Carcinogenesis*, *4*, 431-437

Moore, M.A., Takahashi, M., Ito, N. & Bannasch, P. (1983d) Early lesions during pancreatic carcinogenesis induced in the Syrian hamster by DHPN or DOPN. II. Ultrastructural findings. *Carcinogenesis*, *4*, 439-448

Nettesheim, P. (1976) Precursor lesions of bronchogenic carcinoma. *Cancer Res.*, *36*, 2654-2658

Newberne, P.M. (1976) Experimental hepatocellular carcinogenesis. *Cancer Res.*, *36*, 2573-2578

O'Connor, C.A. & Woodward, S.C. (1981) Industry's role in cancer research: anticipating regulatory problems. *Regul. Toxicol. Pharmacol.*, *1*, 316-334

Ogawa, K., Medline, A. & Farber, E. (1979) Sequential analysis of hepatic carcinogenesis. A comparative study of the ultrastructure of preneoplastic, malignant, prenatal, postnatal, and regenerating liver. *Lab. Invest.*, *41*, 22-35

Parodi, S., Taningher, M. & Santi, L. (1983) Induction of preneoplastic nodules: quantitative predictivity of carcinogenicity. *Anticancer Res.*, *3*, 393-400

Parsa, I., Longnecker, D.S., Scarpelli, D.G., Pour, P., Reddy, J.K. & Lefkowitz, M. (1985) Ductal metaplasia of human exocrine pancreas and its association with carcinoma. *Cancer Res.*, *45*, 1285-1290

Peraino, C., Staffeldt, E.F. & Ludeman, V.A. (1981) Early appearance of histochemically altered hepatocyte foci and liver tumors in female rats treated with carcinogens one day after birth. *Carcinogenesis*, *2*, 463-465

Peraino, C., Richards, W.L. & Stevens, F.J. (1983) *Multistage hepatocarcinogenesis*. In: Slaga, T.J., ed., *Mechanisms of Tumor Promotion*, Vol. 1, Boca Raton, FL, CRC Press, pp. 1-53

Peraino, C., Staffeldt, E.F., Carnes, B.A., Ludeman, V.A., Blomquist, J.A. & Vesselinovitch, S.D. (1984) Characterization of histochemically detectable altered hepatocyte foci and their relationship to hepatic tumorigenesis in rats treated once with diethylnitrosamine or benzo-[*a*]pyrene within one day after birth. *Cancer Res.*, 44, 3340-3347

Pereira, M.A. (1982) Rat liver foci bioassay. *J. Am. Coll. Toxicol.*, 1, 101-117

Pereira, M.A. & Stoner, G.D. (1985) Comparison of rat liver foci assay and strain a mouse lung tumor assay to detect carcinogens: a review. *Fundam. appl. Toxicol.*, 5, 688-699

Pereira, M.A., Savage, R.E., Jr, Herren, S.L. & Guion, C.W. (1982) Comparison of enhancement of γ-GTase-positive foci and induction of ornithine decarboxylase in rat liver by barbiturates. *Carcinogenesis*, 3, 147-150

Pitot, H.C. & Sirica, A.E. (1980) The stages of initiation and promotion in hepatocarcinogenesis. *Biochim. biophys. Acta*, 605, 191-215

Pitot, H.C., Barsness, L., Goldsworthy, F. & Kitagawa, T. (1978) Biochemical characterisation of stages of hepatocarcinogenesis after a single dose of diethylnitrosamine. *Nature*, 271, 456-458

Pour, P., Althoff, J. & Takahashi, M. (1977) Early lesions of pancreatic ductal carcinoma in the hamster model. *Am. J. Pathol.*, 88, 291-308

Pour, P., Runge, R., Birt, D., Ginfell, R., Lawson, T., Nagel, D., Wallcave, L. & Salmasi, S. (1981) Current knowledge of pancreatic carcinogenesis in the hamster and its relevance to the human disease. *Cancer*, 47, 1573-1587

Préat, V., de Gerlache, J., Lans, M., Tapor, H. & Roberfroid, M. (1985) Comparison of the biological effects of two nitrosamines in a triphasic protocol of rat hepatocarcinogenesis. *Teratog. Mutagenesis Carcinog.*, 6, 165-172

Pugh, T.D. & Goldfarb, S. (1978) Quantitative histochemical and autoradiographic studies of hepatocarcinogenesis in rats fed 2-acetylaminofluorene. *Cancer Res.*, 38, 4450-4457

Rabes, H.M. (1983) Development and growth of early preneoplastic lesions induced in the liver by chemical carcinogens. *J. Cancer Res. clin. Oncol.*, 106, 85-92

Rabes, H.M., Scholze, P. & Jantsch, B. (1972) Growth kinetics of diethylnitrosamine-induced enzyme-deficient 'preneoplastic' liver cell populations *in vivo* and *in vitro*. *Cancer Res.*, 32, 2577-2586

Radomski, J.L., Krischer, C. & Krischer, K.N. (1978) Histologic and histochemical preneoplastic changes in the bladder mucosae of dogs given 2-naphthylamine. *J. natl Cancer Inst.*, 60, 327-333

Rao, M.S. & Reddy, J.K. (1980) Histogenesis of pseudo-ductular changes induced in the pancreas of guinea pigs treated with *N*-methyl-*N*-nitrosourea. *Carcinogenesis*, 1, 1027-1037

Rao, M.S., Lalwani, N.D. & Reddy, J.K. (1984) Sequential histologic study of rat liver during peroxisome proliferator [4-chloro-6-(2,3-xylidino)-2-pyrimidinylthio]acetic acid (Wy-14,643)-induced carcinogenesis. *J. natl Cancer Inst.*, 73, 983-990

Reznik-Schüller, H. (1977a) Ultrastructural alterations of APUD cells during nitrosamine-induced lung carcinogenesis. *J. Pathol.*, 121, 79-82

Reznik-Schüller, H. (1977b) Sequential morphologic alterations in the bronchial epithelium of Syrian golden hamster during *N*-nitrosomorpholine-induced pulmonary tumorigenesis. *Am. J. Pathol.*, 89, 59-66

Rossi, L., Ravera, M., Repetti, G. & Santi, L. (1977) Long-term administration of DDT or phenobarbital-Na in Wistar rats. *Int. J. Cancer, 19*, 179-185

Sato, K., Kitahara, A., Satoh, K., Ishikawa, T., Tatematsu, M. & Ito, N. (1984) The placental form of glutathione S-transferase as a new marker protein for preneoplasia in rat chemical carcinogenesis. *Gann, 75*, 199-202

Schauer, A. & Kunze, E. (1968) Enzyme-histochemical and autoradiographic investigations at different stages of rat liver carcinogenesis induced by diethylnitrosamine [in German]. *Z. Krebsforsch., 70*, 252-266

Schauer, A. & Kunze, E. (1976) *Liver tumours of the rat*. In: Turusov, V.S., ed., *Pathology of Tumours in Laboratory Animals*, Vol. I, *Tumours of the Rat*, Part 2 (*IARC Scientific Publications No. 6*), Lyon, International Agency for Research on Cancer, pp. 41-72

Scherer, E. (1984) Neoplastic progression in experimental hepatocarcinogenesis. *Biochim. biophys. Acta, 738*, 219-236

Scherer, E. & Emmelot, P. (1975) Foci of altered liver cells induced by a single dose of diethylnitrosamine and partial hepatectomy: their contribution to hepatocarcinogenesis in the rat. *Eur. J. Cancer, 11*, 145-154

Scherer, E., Hoffmann, M., Emmelot, P. & Friedrick-Freska, H. (1972) Quantitative study on foci of altered liver cells induced in the rat by a single dose of diethylnitrosamine and partial hepatectomy. *J. natl Cancer Inst., 49*, 93-106

Schulte-Hermann, R., Ohde, G., Schuppler, J. & Timmermann-Trosiener, I. (1981) Enhanced proliferation of putative preneoplastic cells in rat liver following treatment with the tumor promoters phenobarbital, hexachlorocyclohexane, steroid compounds and nafenopin. *Cancer Res., 41*, 2556-2562

Schulte-Hermann, R., Roome, N., Timmermann-Trosiener, I. & Schuppler, J. (1984) Immunocytochemical demonstration of a phenobarbital-inducible cytochrome P-450 in putative preneoplastic foci of rat liver. *Carcinogenesis, 5*, 143-153

Solt, D. & Farber, E. (1976) New principle for the analysis of chemical carcinogenesis. *Nature, 263*, 702-703

Squire, R.A. & Levitt, M.H. (1975) Report of a workshop on classification of specific hepatocellular lesions in rats. *Cancer Res., 35*, 3214-3223

Stewart, H.L. (1975) *Comparative aspects of certain cancers*. In: Becker, F.F., ed., *Cancer. A Comprehensive Treatise*, Vol. 4, New York, Plenum, pp. 303-374

Stewart, H.L., Williams, G., Keysser, C.H., Lombard, L.S. & Montali, R.J. (1980) Histologic typing of liver tumors of the rat. *J. natl Cancer Inst., 65*, 179-206

Stiller, D. & Rauscher, H. (1971) Irreversible preneoplastic defect in alkaline phosphatase in cancerization transitional epithelium. *Exp. Pathol., 5*, 255-258

Takahashi, M., Arai, H., Kokubo, T., Furukawa, F., Kurata, Y. & Ito, N. (1980) An ultrastructural study of precancerous and cancerous lesions of the pancreas in Syrian golden hamsters induced by *N*-nitrosobis(oxopropyl)amine. *Gann, 71*, 825-831

Takahashi, M., Nagase, S., Takahashi, M., Kokubo, T. & Hayashi, Y. (1981) Changes of amylase during experimental pancreatic carcinogenesis in hamsters. *Gann, 72*, 615-619

Taper, H.S., Fort, L. & Brucher, J.-M. (1971) Histochemical activity of alkaline and acid nucleases in the rat liver parenchyma during *N*-nitrosomorpholine carcinogenesis. *Cancer Res., 31*, 913-916

Tatematsu, M., Takano, T., Hasegawa, R., Imaida, K., Nakanowatari, J. & Ito, N. (1980) A sequential quantitative study of the reversibility or irreversibility of liver hyperplastic nodules in rats exposed to hepatocarcinogens. *Gann, 71*, 843-855

Tatematsu, M., Nagamine, Y. & Farber, E. (1983) Redifferentiation as a basis for remodeling of carcinogen-induced hepatocyte nodules to normal appearing liver. *Cancer Res., 43*, 5049-5058

Teebor, G.W. & Becker, F.F. (1971) Regression and persistence of hyperplastic nodules induced by *N*-2-fluorenylacetamide and their relationship to hepatocarcinogenesis. *Cancer Res., 31*, 1-3

Trump, B.F., McDowell, E.M., Glavin, F., Barrett, L.A., Becci, P.J., Schürch, W., Kaiser, H.E. & Harns, C.C. (1978) The respiratory epithelium. III. Histogenesis of epidermoid metaplasia and carcinoma *in situ* in the human. *J. natl Cancer Inst., 61*, 563-575

Tsuda, H., Lee, G. & Farber, E. (1980) Induction of resistant hepatocytes as a new principle for a possible short term in-vivo test for carcinogens. *Cancer Res., 40*, 1157-1164

Vanderlaan, M., Fong, S. & King, E.B. (1982) Histochemistry of NADH diaphorase and γ-glutamyltranspeptidase in rat bladder tumors. *Carcinogenesis, 3*, 397-402

Vesselinovitch, S.D. (1980) *Infant mouse as a sensitive bioassay system for carcinogenicity of N-nitroso compounds*. In: Walker, E.A., Castegnaro, M., Griciute, L. & Börzsönyi, M., eds, N-*Nitroso Compounds: Analysis, Formation and Occurrence (IARC Scientific Publications No.31)*, Lyon, International Agency for Research on Cancer, pp. 645-655

Vesselinovitch, S.D. & Mihailovich, N. (1983) Kinetics of diethylnitrosamine hepatocarcinogenesis in the infant mouse. *Cancer Res., 43*, 4253-4259

Vesselinovitch, S.D., Rao, K.V.N. & Mihailovich, N. (1979) Neoplastic response of mouse tissues during perinatal age periods and its significance in chemical carcinogenesis. *Natl Cancer Inst. Monogr., 51*, 239-250

Vesselinovitch, S.D., Hacker, H.J. & Bannasch, P. (1985) Histochemical characterization of focal hepatic lesions induced by single diethylnitrosamine treatment in infant mice. *Cancer Res., 45*, 2774-2780

Ward, J.M. (1981) Morphology of foci of altered hepatocytes and naturally-occurring hepatocellular tumours in F344 rats. *Virchows Arch. Pathol. Anat., 390*, 339-345

Ward, J.M. (1984) *Morphology of potential preneoplastic hepatocyte lesions and liver tumors in mice and a comparison with other species*. In: Popp, J.A., ed., *Mouse Liver Neoplasia*, Washington DC, Hemisphere, pp. 1-26

Williams, G.M. (1980) The pathogenesis of rat liver cancer caused by chemical carcinogens. *Biochim. biophys. Acta, 605*, 167-189

Williams, G.M. (1982) Phenotypic properties of preneoplastic rat liver lesions and application to detection of carcinogens and tumour promoters. *Toxicol. Pathol., 10*, 3-10

Williams, G.M. & Watanabe, K. (1978) Quantitative kinetics of development of *N*-2-fluorenylacetamide-induced, altered hyperplastic hepatocellular foci resistant to iron accumulation and of their reversion of persistence following removal of carcinogen. *J. natl Cancer Inst., 61*, 113-121

Williams, G.M., Klaiber, M., Parker, S.E. & Farber, E. (1976) Nature of early appearing, carcinogen-induced liver lesions resistant to iron accumulation. *J. natl Cancer Inst., 57*, 157-165

REPORT 3

ASSAYS FOR INITIATING AND PROMOTING ACTIVITIES

Prepared by:

P. Bannasch, R.A. Griesemer (Rapporteurs),
F. Anders, R. Becker, J.R. Cabral, G. Della Porta (Chairman),
V.J. Feron, D. Henschler, N. Ito, R. Kroes, P.N. Magee,
B. McKnight, R. Montesano, N.P. Napalkov, S. Nesnow, M. Roberfroid,
T. Slaga, V.S. Turusov, J. Wilbourn and G.M. Williams

1. Introduction

Investigators interested in the natural history of various forms of experimentally induced cancers have developed models to examine the various stages of the carcinogenesis process and, in parallel, to attempt to identify the capacity of various carcinogens to affect those stages (Berenblum, 1974; Börzsönyi *et al.*, 1984). A number of studies have defined two distinct stages of the carcinogenesis process — initiation and promotion. Described briefly in this report are some experimental models in which these two stages have been studied and attempts have been made to identify agents with initiating and/or promoting activity. As a general rule, the initation stage requires only one or two treatments with a carcinogen, and the promotion stage is brought about by repeated treatments, administered after an initiator. Carcinogens are considered to possess both initiating and promoting activity.

The terms 'two-stage', 'initiator', 'initiation', 'initiating activity', 'promoter', 'promotion' and 'promoting activity' used in the description of these models are merely operational and should not be strictly interpreted mechanistically.

Although the assays described here are not conventional bioassays for chemical carcinogens, because they are limited to effects on single organs or involve administration in conjunction with other chemical substances, they measure the potential contributions of test substances to the stages of the carcinogenic process. It is questionable whether pure initiators exist, i.e., such substances can exert their carcinogenic activity only if their application is followed by a promoter treatment. Likewise, pure promoters probably do not exist either, as practically all of them are weakly carcinogenic if applied at a high dose for a long time.

For a long time, the phenomenon of initiation-promotion was observed in the mouse skin only, but during the last 10-15 years, it has been demonstrated in many other organs of rodents. Some of the initiation/promotion models most often used (for skin, liver, bladder, kidney, thyroid) and one that utilizes small fish are described briefly in this report. Other models are described elsewhere (Pitot, 1983; Börzsönyi et al., 1984; Turusov & Koblyakov, 1986). Recent studies have provided evidence that transforming gene (oncogenes) are implicated in the various stages of carcinogenesis (Land et al., 1983; Balmain, 1985).

2. Assays for initiating and promoting activity in rodent skin

Of the various experimental models for multistage carcinogenesis, the mouse skin model is the clearest example in which two distinct stages have been delineated. The terms 'initiation' and 'promotion' were originally proposed by Rous and Kidd (1941) and Friedwald and Rous (1944) on the basis of studies on the induction in rabbits of skin tumours with coal tar and trauma. At the same time, the stages of development of skin tumours in mice treated with benzo[a]pyrene (the initiator) and croton oil (the promoter) were defined more clearly (Berenblum, 1941; Mottram, 1944; Berenblum & Shubik, 1947, 1949). The requirements for initiating or promoting processes in tumour development were outlined by Berenblum (1974): (i) that the two actions not overlap in time; (ii) that neither initiator nor promoter alone be to any significant extent carcinogenic; (iii) that the increased cancer incidence be observed only when the promoter is administered after the initiator and not *vice versa*; (iv) that variation in the interval between initiation and promotion not affect the final tumour incidence; and (v) that the tumour incidence be related to the dose of initiator. In addition it was shown that lengthening the time interval between the administration of croton oil or 12-O-tetradecanoylphorbol 13-acetate (TPA) reduces the incidence of papillomas.

These requirements, which were first established with croton oil, were later confirmed with TPA, the active principle of croton oil (Hecker et al., 1964). Repetitive applications of high doses of croton oil or TPA without initiation by 7,12-dimethylbenz[a]anthracene (DMBA) may also produce skin tumours in mice (Hecker, 1967; Chouroulinkov & Lazar, 1974).

Mice are more sensitive than rats and hamsters to skin carcinogenesis either by the complete carcinogenesis protocol or by the initiation-promotion protocol. The complete carcinogenesis protocol in mice gives rise to a high number of papillomas, followed — depending on the dose of carcinogen — by a high incidence of squamous-cell carcinomas. The initiation-promotion protocol in mice gives rise to a very high multiplicity of papillomas, many of which regress, and, in general, a low incidence of squamous cell carcinomas. Another interesting observation is that the papillomas induced by initiation-promotion protocol are mainly of monoclonal origin, whereas many of the skin tumours induced by repeated doses of DMBA are polyclonal (Reddy & Fialkow, 1983).

Both the complete carcinogenesis and initiation-promotion protocols generally give rise in rats to basal-cell carcinomas; however, high doses of DMBA alone may provoke very high incidences of papillomas, keratoacanthomas and squamous-cell carcinomas. The complete carcinogenesis protocol in hamsters produces some squamous-cell carcinomas and melanomas, whereas the initiation-promotion protocol produces primarily melanomas.

There are both good qualitative and quantitative correlations between the complete carcinogenic and tumour-initiating activities of several chemical carcinogens in mouse skin (Slaga et al., 1982; Slaga & Fischer, 1983). It is thus important to test compounds not only for tumour-initiating and tumour-promoting activities but also for complete carcinogenic activities, since, for example, a substance weak in promoting or initiating activity may not be detected when tested as a complete carcinogen. Scribner and Scribner (1980) attempted to differentiate the initiating and promoting activity of carcinogens in mouse skin and found that dibenz[a,c]anthracene possesses strong initiating activity and much weaker promoting activity, while 7-bromomethylbenz-[a]anthracene has the inverse relation of these two properties. Mouse skin tumour promotion itself has been described to consist of several substages (Slaga, 1983). As shown by Hecker (1967) for phorbol esters, promoting activity is manifested at a much lower dose than carcinogenicity.

Over 100 polycyclic aromatic hydrocarbons (PAH), PAH derivatives and PAH metabolites are known to be mouse skin-tumour initiators or complete carcinogens, and one PAH, DMBA, is most frequently used as a standard initiator when chemical substances are tested for promoting activity. The mouse skin tumorigenesis bioassay has also been used to identify many other chemicals of a wide variety of structural classes with tumour-initiating and complete carcinogenic activity, including aldehydes, carbamates, epoxides, haloalkylethers, haloaromatics, haloalkylketones, hydroxylamines, lactones, nitrosamides, sulfonates, sultones and ureas. They include such well-known chemical carcinogens as aflatoxin B_1, bis(chloromethyl)ether, chloromethyl methyl ether, urethane, N-acetoxy-2-acetylaminofluorene, β-propiolactone, N-methyl-N'-nitro-N-nitrosoguanidine, 1,3-propanesultone, N-methyl-N-nitrosourea, triethylenemelamine, and 4-nitroquinoline-N-oxide. It is important to stress that the skin is not the main target organ for many of these chemicals.

Many diverse chemical agents exhibit skin-tumour promoting activity (Fujiki & Sugimura, 1983; Slaga, 1983), including the diterpenes (phorbol esters), indole alkaloids (teleocidin and lyngbyatoxin), polyacetate (aplysiatoxin), chrysarobin and anthralin. Van Duuren et al. (1978) reported a fairly extensive structure-activity study with anthraline and derivatives, and Boutwell and Bosch (1959) reported a similar study with weaker promoters — phenolic compounds. Benzo[e]pyrene, retinoic acid, 1-fluoro-2,4 dinitrobenzene and benzoyl peroxide, as well as other peroxides have also been found to have relatively good promoting activity (Slaga et al., 1981). A non-TPA-type terpene promoter has recently been described (Hakii et al., 1986).

There are reasonably large data bases on the following stocks and strains of mice with regard to two-stage skin carcinogenesis: SENCAR, CD-1, ICR/Ha Swiss, NMRI, BALB/c, C3H, C57BL/6 and DBA/2 (Boutwell, 1964; Slaga, 1983; Slaga & Fischer, 1983). In general, all the carcinogens and initiators that have been tested in these stocks and strains have given positive results. The major differences are in the sensitivity and latency to tumour development as well as in the ratio of benign to malignant tumours (Van Duuren, 1969; Van Duuren et al., 1973; Burns et al., 1978; Phillips et al., 1978; DiGiovanni et al., 1980; Stenback, 1980; Verma & Boutwell, 1980; Hennings et al., 1981; Slaga & Fischer, 1983; Reiners et al., 1984).

Although most skin tumour promoters tested have also given positive results in this test, some promoters do not seem to be effective in some stocks and strains of mice. The best example is the difference in the sensitivity to two-stage carcinogenesis of SENCAR and C57BL/6 mice (Reiners et al., 1984): SENCAR mice are very sensitive to two-stage skin carcinogenesis by benzo[a]pyrene and TPA, whereas C57BL/6 mice are very resistant; however, C57BL/6 mice do respond to complete carcinogenesis by benzo[a]pyrene, and, in this regard, they appear to be slightly more sensitive than SENCAR mice. This suggests that the promotional stages of complete and two-stage carcinogenesis are dissimilar and may have a genetic basis. In addition, differences in sensitivity to initiation and promotion between strains of mice may be due to alterations in the promotional stage of two-stage carcinogenesis. It has been found that benzoyl peroxide is an effective promoter in C57BL/6 and SENCAR mice (Reiners et al., 1984), but, for unknown reasons, TPA is not an effective promoter in C57BL/6 mice.

It is recommended that animals of each sex be used to detect possible sex differences in the activity of skin tumour initiators and promoters. Generally, 30-50 mice of each sex are used to obtain statistically significant results. Because the hair of mice is in a nongrowing phase between seven and nine weeks of age, mice of this age are commonly used. The dorsal skin is shaved at least two days before application of a tumour initiator; the promoter is generally applied one to two weeks after tumour initiation to allow time for the skin to recover from the initiating treatment. Although some investigators repeatedly shave the mice throughout the treatment period, consistent results can be obtained with one initial shaving. Furthermore, multiple shaving may cause wounding, which may alter the tumour response. In assays for initiating activity, a test substance is applied to the skin one or several times, followed by repeated applications of a known tumour promoter (TPA). In assays for promoting activity, a single application of 10-20 μg DMBA is used as an initiator, followed by repeated applications of the test substance.

Tumour promoters are generally applied repetitively (one to three times per week) throughout the experimental period. However, recent experiments suggest that prolonged treatment may not further increase the carcinoma yield over that which can be achieved by limited treatment. Twice-weekly application of TPA for five to ten weeks is sufficient to give a maximal yield of carcinomas (Verma & Boutwell, 1980; Diwan *et al.*, 1985). This protocol was based on experiments in which a known, potent initiator (DMBA) and a known, potent promoter (TPA) are used, however, and it is questionable whether the same results can be obtained if either of these compounds is replaced by a substance possessing weak activity. For this reason, it is advisable to use several (or at least two) dose groups: one in which the promoter is applied for one year and another in which it is applied for a shorter period.

Skin-tumour formation is recorded weekly, and papillomas greater than 2 mm in diameter are included in the cumulative total if they persist for one week or longer. The tumour diagnoses are verified histologically after each mouse dies or is killed. Estimates of the probability of surviving to various ages are estimated using actuarial methods (see Report 1, section 4.5) and presented for each group in graphical form. The total numbers of carcinoma- and papilloma-bearing animals in each group are reported, and, if tumour-free survival is similar among the treatment groups, the numbers of carcinoma- and/or papilloma-bearing animals are compared statistically, using methods described in Report 1 (section 4.5; see also Gart *et al.*, 1986).

If tumour-free survival differs substantially across treatment groups, or if interest centres on relating treatment to the age of tumour occurrence, times to first papilloma are compared using the statistical techniques of survival analysis as described by Gart *et al.* (1986; section 5.6). The regression model of Cox can be used to relate quantitative treatment descriptors such as concentration, frequency and duration of application to onset time (see section 6.3 of Gart *et al.*, 1986, for details). Models such as the multistage model that are derived from theories about how treatment descriptors should affect the time to tumour onset can be fitted, as well as models that take into account the effect of previous tumours in the comparison of subsequent tumour onset when there are multiple tumours per animal (see sections 6.3 and 7.3 of Gart *et al.*, 1986). Tumours at distant sites can also be evaluated if desired.

3. Assays for initiating and promoting activities in rodent liver

The pathogenesis of hepatic cancer has been studied mainly in rats and mice (Farber, 1956; Stewart & Snell, 1957; Bannasch, 1968; Newberne & Butler, 1969; Kitagawa & Sugano, 1973). These studies showed that the natural history of the development of these tumours occurs in different stages and that various lesions precede the appearance of tumours (see Report 2; Farber, 1973; Pitot & Sirica, 1980). It was shown in parallel (Peraino *et al.*, 1973) that chronic administration of phenobarbital to rats drastically increases the carcinogenicity of 2-acetylaminofluorene

(AAF), indicating that, as in mouse skin, initiating and promoting events are taking place. On the basis of these observations, various models were developed to identify the capacity of environmental agents to affect specifically these stages in the induction of liver tumours. The endpoints measured in these studies are in most instances the appearance of preneoplastic lesions.

As noted in Report 2, liver tumorigenesis in rodents appears to be preceded by the development of phenotypically altered, focal cellular populations, which are regarded as preneoplastic lesions (Williams, 1982; Bannasch *et al.*, 1984). The results of the assays are based on quantification of these preneoplastic lesions, as estimated by histochemical methods, including γ-glutamyl transpeptidase-positive and adenosine triphosphatase-deficient focal hepatocyte populations (Scherer & Emmelot, 1976; Kitagawa & Sugano, 1978; Ogawa *et al.*, 1979) and immunohistochemically detectable glutathione *S*-transferase placental-type positive foci (Ogiso *et al.*, 1985; Tatematsu *et al.*, 1985; Thamavit *et al.*, 1985). Stereological techniques for accurate quantification have been recommended (Scherer, 1981; Nychka *et al.*, 1984).

Some of the main assays are described briefly here. It is evident that their execution as well as interpretation of their results require considerable expertise and the exercising of scientific value judgement by the investigators. These in-vivo assays were originally designed to analyse the natural history of liver carcinogenesis, and only recently have attempts been made to use them as screening tests for chemical carcinogens.

Figure 1 shows schematically some of the assays used for detecting chemical carcinogens with initiating and promoting activities.

One assay for the detection of initiating activity is based on the protocol (Fig. 1A) described by Solt and Farber (1976). In this test, preneoplastic basophilic foci can be detected within four weeks when a single dose of *N*-nitrosodiethylamine (NDEA) is followed by administration in the diet of AAF (0.02%) for two weeks coupled with a partial hepatectomy. It has been proposed that the AAF treatment exerts a selective growth advantage on the NDEA-initiated cells over that of normal hepatocytes, and the partial hepatectomy functions as a generalized growth stimulus (Farber, 1980).

In a variation of this assay system (Fig. 1B), a single dose of test compound is given to rats 12-18 hours after partial hepatectomy, followed by feeding of AAF for two to four weeks, combined with administration of a single highly toxic dose of carbon tetrachloride administered at week 3. Animals are killed at week 5 and γ-glutamyl transpeptidase-positive liver-cell foci are measured (Tsuda *et al.*, 1980).

In these systems, known hepatocarcinogens as well as carcinogens previously shown to have a carcinogenic effect in tissues other than the liver, such as PAH, safrole and *N*-[4-(5-nitro-2-furyl)-2-thiazolyl]formamide (FANFT), all gave positive results, indicating that a wide spectrum of carcinogens is detectable in this assay. Both hepatocarcinogens and some extrahepatic carcinogens were capable of inducing liver

Fig. 1. Experimental design of some in-vivo assay systems for detecting initiating and promoting activities of carcinogens in rat liver

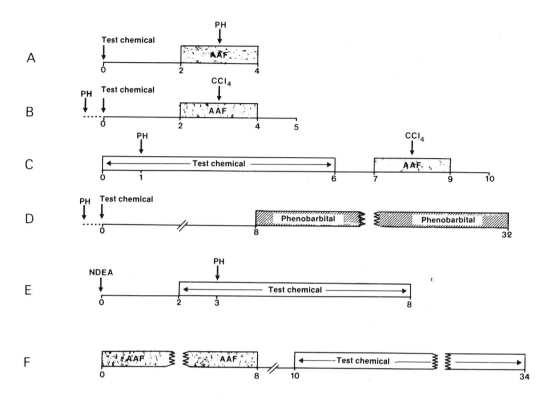

A, Solt and Farber (1976); B, Tsuda *et al.* (1980); C, Hasegawa *et al.* (1982a); D, Pitot *et al.* (1978); E, Ito *et al.* (1980); F, Numoto *et al.* (1984)

AAF, 2-acetylaminufluorene, 0.02% in the diet; phenobarbital, 0.05% in drinking-water; PH, partial hepatectomy; NDEA, *N*-nitrosodiethylamine, 10 mg/kg body weight; CCl_4, carbon tetrachloride, 2 ml/kg body weight. The numbers indicate experimental weeks.

cancer when the animals were kept under observation for a long period of time (8-13 months) (Solt *et al.*, 1983). For several pyrolysis products that also gave positive results in these assays (Hasegawa *et al.*, 1982a), subsequent long-term animal bioassays revealed unequivocal evidence for in-vivo tumorigenesis (Hosaka *et al.*, 1981; Ohgaki *et al.*, 1984; Takayama *et al.*, 1984).

A somewhat similar assay system (Fig. 1C) for detecting initiating potential involves feeding a test compound not as a single dose, but for six weeks, combined with partial hepatectomy, followed by AAF plus carbon tetrachloride as in the previous assay. The results obtained were almost identical to those seen after single-dose initiation and indicated that carcinogens that usually produce tumours in extrahepatic organs might be screened by this method (Hasegawa *et al.*, 1982b; Tatematsu *et al.*, 1983).

In all the assay systems so far described, the stimulation of the growth of initiated cells is provided by treatment with the known carcinogen, AAF, which probably has initiating and promoting activity, making it difficult to differentiate promoting from additive effects. However, an increase in the incidence of focal or nodular lesions was also observed by treatment with a choline-deficient diet or with orotic acid after exposure to various initiating agents (Shinozuka & Lombardi, 1980; Columbano *et al.*, 1982).

Pitot (1977) and Pitot *et al.* (1978) incorporated partial hepatectomy into a two-stage model (Fig. 1D). Rats were given a single dose of NDEA by gastric intubation 24 hours after partial hepatectomy. Eight weeks later a group of these animals was placed on a diet containing 0.05% phenobarbital for 24 weeks. Other investigators used benzo[*a*]pyrene instead of NDEA as the initiator after partial hepatectomy (Kitagawa *et al.*, 1980).

A similar protocol was used by Ito *et al.* (1980) (Fig. 1E), with some modifications by Lawns *et al.* (1983), to detect the promoting activity of carcinogens. In this assay, the test chemical was administered after an initial treatment with a single dose of NDEA (Fig. 1E) or multiple treatments with AAF (Fig. 1F) (Numoto *et al.*, 1984). Watanabe and Williams (1978) also reported that phenobarbital promoted the development of altered foci that were resistant to iron accumulation and increased the incidence of liver tumours. In place of AAF, a number of other chemical carcinogens have been used as initiators, including NDEA (Weisburger *et al.*, 1975; Nishizumi, 1976), 3'-methyl-4-dimethylaminoazobenzene (Kitagawa & Sugano, 1978; Peraino *et al.*, 1978) and 2-methyl-4-dimethylaminoazobenzene (Kitagawa *et al.*, 1979). DDT was also found significantly to enhance liver tumour development (Peraino *et al.*, 1975) when given in place of phenobarbital.

In principle, this general model bears some similarity to that developed for the skin (section 2), as shown by Williams and Furuya (1984) when dietary phenobarbital resulted in an enhancement of preneoplastic foci only when given after AAF. Partial hepatectomy is performed in order to increase the rate of cell proliferation in the liver of adult rats. The livers of neonatal and juvenile rats have a relatively high level of cell proliferation, and these may be used instead of partial hepatectomy for optimal sensitivity of the assay. For example, juvenile rats (32 days of age) were sensitive to the induction of γ-glutamyl transpeptidase-positive foci by NDEA combined with phenobarbital, without partial hepatectomy. While in the skin, a long (up to one year) delay in the application of promoter may not significantly reduce the incidence of papillomas, delaying the promotional regimen in the liver from nine to 29 days after the administration of NDEA resulted in a 50% decrease in the incidence of γ-glutamyl transpeptidase-positive foci (Herren-Freund *et al.*, 1986).

4. Two-stage models for the induction of neoplastic changes in the urinary bladder

Application of the two-stage model concept to urinary bladder carcinogenesis has allowed the development of assays for initiating and promoting activities of chemicals in this organ. Three assays have been developed in rats using different initiators; all three measure the production of bladder cancer.

The first assay was introduced by Hicks et al. (1975) and Hicks and Chowaniec (1977). Instillation of a subcarcinogenic dose of N-methyl-N-nitrosourea into rat bladder followed by oral administration of saccharin or cyclamate resulted in a high incidence of bladder cancer, while long-term administration of saccharin or cyclamate alone induced only a very low (or no increased) incidence of bladder cancer.

A second model was developed by Cohen et al. (1979) and Fukushima et al. (1981). FANFT was administered in the diet (0.2%) for four or six weeks as the initiator, followed by feeding with saccharin and tryptophan as possible promoting chemicals. Surviving rats were examined at the end of two years, and saccharin was reported to have promoting activity for the urinary bladder.

In a third model, nitrosobutyl(4-hydroxybutyl)amine (NBHBA) was given in drinking-water (0.05%) for four weeks as the initiator; then, test chemicals were given for the following 32 weeks, and the induction of bladder cancer within the experimental period scored. Many chemicals, including saccharin, sodium L-ascorbate and some antioxidants (butylated hydroxyanisole, butylated hydroxytoluene and sodium erythorbate), were found to be promoters of bladder carcinogenesis in this assay (Nakanishi et al., 1980; Fukushima et al., 1983a,b; Imaida et al., 1983; Ito et al., 1983; Fukushima et al., 1984).

These experimental assays have been used mainly for investigations of the promoting potential of test chemicals. Quantitative analysis of putative preneoplastic bladder lesions (see Report 2) has been included in the third model, and is potentially applicable, as is cytological examination of urine (see Report 2), to all urinary bladder assays. Enzyme changes reminiscent of those observed in altered foci in liver carcinogenesis, including focal loss of alkaline phosphatase (Kunze & Schauer, 1971; Ito, 1973; Hicks et al., 1978; Radomski et al., 1978), NADH diaphorase (Kunze & Schauer, 1971; Vanderlaan et al., 1982), γ-glutamyl transferase (Vanderlaan et al., 1982), and increased activity of nonspecific esterase and β-glucuronidase, have been reported in hyperplastic bladder lesions (Ito, 1973). However, these marker enzymes are not validated predictors of bladder carcinogenesis at the present time.

For quantitative analysis, bladder lesions are counted in histological sections stained with haematoxylin and eosin; the total length of basement membrane analysed is measured, and the numbers of preneoplastic and neoplastic lesions are expressed per 10 cm of basement membrane. Thus, for example, in experiments on the promoting effect of sodium citrate, the number of papillary or nodular hyperplasias/10 cm of basement membrane was 7.78 in a group receiving carcinogen (NBHBA) and promoter *versus* 0.46 in a group receiving carcinogen alone (Fukushima *et al.*, 1986).

Although increased urothelial proliferation has been induced by partial cystectomy (Tatematsu *et al.*, 1981; Fukushima *et al.*, 1982) and freezing ulceration (Shirai *et al.*, 1978), in neither case did the mucosal regeneration enhance urinary bladder carcinogenesis, either before or during carcinogen administration (Fukushima *et al.*, 1982). Unilateral ureteral ligation has been found, however, to enhance both urinary bladder epithelial proliferation and the incidence of ureteral and bladder cancers in rats (Ito *et al.*, 1971; Miyata *et al.*, 1984). Rats were given 0.05% NBHBA for two weeks as the initiator, and then test chemicals were administered for 22 weeks. Unilateral ligation of one ureter was performed at the end of week 3. From a wide range of compounds, some of them powerful carcinogens, NBHBA, N-nitrosoethylhydroxybutylamine, sodium-*o*-phenylphenate, saccharin, butylated hydroxyanisole, sodium L-ascorbate, sodium erythrobate and ethoxyquin, all significantly promoted preoplastic lesions induced in the urinary bladder (Miyata *et al.*, 1984, 1985).

5. Assays for initiating and promoting activities in other organs of rodents

5.1 *Lung*

In this model, male or female A/J mice with a predisposition to lung tumour formation are used, and urethane, 3-methylcholanthrene, benzo[*a*]pyrene, and N-nitrosodimethylamine are used as initiators. For example, a single intraperitoneal administration of 1000 mg/kg bw urethane, followed for three months by a diet containing 0.75% butylated hydroxytoluene induced two-fold increases in the number of lung adenomas per mouse, as compared to urethane alone. No increase in tumour multiplicity was noted when lower doses of urethane (<50 mg/kg bw) were administered before placing the animals on the butylated hydroxytoluene diet. Administration of this compound could be delayed for several months after urethane treatment without abolishing the promoting effect. Male A/J mice administered a single dose of 3-methylcholanthrene, N-nitrosodimethylamine or benzo[*a*]pyrene were placed for eight weeks on a 0.75% butylated hydroxytoluene diet and were killed four months after carcinogen administration. Butylated hydroxytoluene increased tumour multiplicity from four fold to ten fold over the values found in mice treated with carcinogens alone (Witschi & Morse, 1983). Under the same conditions, butylated hydroxytoluene was not effective in strains of mice that have low spontaneous incidences of lung tumours and are resistant to the carcinogenicity of urethane (Witschi, 1983).

Some enhancement of lung carcinogenesis was observed in AKR mice injected with N-nitrosodimethylamine when newborn animals were treated subsequently with phorbol (Armuth & Berenblum, 1972), and in strain A mice given urethane and then treated with saccharin (Theiss *et al.*, 1980).

Promotion of lung tumours in mice by administration of glycerol after initiation by 4-nitroquinoline-N-oxide has also been reported (Inayama, 1986). An increase in the incidence of lung carcinomas was observed in rats treated with N-nitrosobis(2-hydroxypropyl)amine followed by bleomycin (Shirai et al., 1984a).

5.2 Kidney

Various reports describe the promoting or enhancing activity of chemicals in kidneys of rats pretreated with N-nitroso-N-ethylhydroxyethylamine. This compound was given either in drinking-water for one or two weeks at concentrations of 500 or 1000 mg/l or in diet and followed by promoter administration for 30-35 weeks, at which time the rats were killed. Dark-cell and clear-cell renal tumours were distinguished microscopically and counted separately (Hiasa et al., 1984a). Nephrotoxic agents, such as basic lead acetate, β-cyclodextrin, DL-serine and trisodium nitriloacetate, were reported to exert a pronounced promoting activity with regard to renal carcinogenesis (Hiasa et al., 1982a, 1983a, 1984b). When the food additive potassium bromate was used as a promoter, a dose-related increase of the number of kidney dysplastic foci was found (Kurokawa et al., 1985).

A slightly modified model was used for testing the promoting activity of folic acid, N-(3,5-dichlorophenyl)succinimide, 2,3-dibromo-1-propanol phosphate and basic lead acetate. At week 3 of promoter treatment, the right kidney was removed to enhance cell proliferation, and the animals were killed 35 weeks after the start of promoter administration. All of the nephrotoxic agents except 2,3-dibromo-1-propanol phosphate were promoters of renal carcinogenesis (Shirai et al., 1984b).

5.3 Thyroid

In experimental models for two-stage carcinogenesis in rat thyroid, AAF, N-methyl-N-nitrosourea, N-nitrosobis(2-hydroxypropyl)amine, ^{131}I and irradiation are used as initiators, and goitrogenic chemicals and other treatments are used as promoters. Hall (1948) first reported two-stage carcinogenesis in the thyroid of rats given AAF followed by methylthiouracil treatment, resulting in the development of thyroid adenomas.

In other studies, N-methyl-N-nitrosourea was given either as a single intravenous injection (Ohshima & Ward, 1984; Milmore et al., 1982) or as multiple intraperitoneal injections (Tsuda et al., 1983). N-Nitrosobis(2-hydroxypropyl)amine was given either once intraperitoneally (Hiasa et al., 1984c) or subcutaneously for several weeks (Hiasa et al., 1982b). ^{131}I (3μCi intraperitoneally; Nadler et al., 1970) and local irradiation (300 rads X-rays; Nadler et al., 1969) have also been used as initiators.

Goitrogenic substances and treatments used as promoters of thyroid carcinogenesis include propyl- and methylthiouracil (Schaffer & Müller, 1980), 3-amino-1,2,4-triazole (Hiasa et al., 1982b), iodine deficiency (Nadler et al., 1969; Ohshima & Ward, 1984) and

4,4'-diaminodiphenylmethane (Hiasa et al., 1984c). Barbital and phenobarbital are also reported to be promoters of tumour development in the rat thyroid (Tsuda et al., 1983; Hiasa et al., 1982c, 1983b). Promoter treatment is usually started within two weeks after administration of an initiator has been stopped and continues for 18-33 weeks, after which the animals are killed.

At autopsy, thyroid glands are weighed and examined macroscopically. Serial sections are sometimes made of each thyroid, and tumours are counted in the largest serial section (Hiasa et al., 1984c) or in all sections. Established classifications of thyroid tumours in rats (Napalkov, 1976) can be used when evaluating the histological findings.

In the models described above, combined treatment with an initiator followed by a promoter resulted in a very high (up to 100%) incidence of thyroid neoplasms (carcinomas and adenomas), while a relatively low incidence of these tumours was usually found when an initiator was given alone; as a rule, no tumour was observed in groups receiving promoter alone. It should be noted that all (except perhaps 4,4'-diaminodiphenylmethane) of the goitrogenic chemicals and treatments that exert promoting activity on the thyroid have long been used at much higher doses to produce adenomas and carcinomas in this organ (Napalkov, 1976).

6. An assay for carcinogenic and promoting activity in small fish

6.1 Introduction

Several genera of small teleost fish serve increasingly as animal models in cancer research (Dawe et al., 1981; Hoover, 1984). Their small size, low maintenance costs and demonstrated response to carcinogens make them attractive alternatives to rodents (Matsushima & Sugimura, 1976; Anders et al., 1981; Dawe & Couch, 1984). The genus of fish considered in this report is *Xiphophorus*, within which the species *X. maculatus* (platyfish) and *X. helleri* (sword-tail, both from central America) are especially useful (Anders, 1967; Kallman, 1973; Rosen, 1979; Anders et al., 1984).

All pure bred *Xiphophorus* fish from wild populations have been nonsusceptible to induction of neoplasia by carcinogens. In contrast, up to 10% of the fish of certain interpopulational and interspecific hybrid populations that have been bred in the laboratory, are susceptible. Following treatment with chemical or physical carcinogens, they develop a large variety of neoplasms of neurogenic, epithelial and mesenchymal origin (Anders & Anders, 1978; Anders, 1981; Anders et al., 1983).

6.2 Oncogenes in Xiphophorus fish

Genetic, cytogenetic and molecular studies have assigned the capacity to develop neoplasia to a deregulated oncogene (*Tu*), which, in the case of the platyfish from Rio Jamapa, is located in a terminal Giemsa-staining band of the X-chromosome but may also be translocated to the Y-chromosome. *Tu* contains several copies of *c-src*, *c-erb* and probably *c-yes* that are homologues of retroviral oncogenes. These oncogenes are highly amplified in the developing neoplasms, and c-*src* levels may be elevated up to 50 times over that of nontumourous control tissues (Schartl et al., 1985).

Tu is normally under the control of population-specific systems of linked and non-linked regulatory genes, which, if eliminated by hybridization (replacement of chromosomes containing regulatory genes by chromosomes lacking these genes) or impaired by carcinogens, permit the *Tu* oncogene to become active. There are two sets of regulatory genes controlling *Tu* — the tissue-specific (mesenchymal, epithelial, neurogenic, pigment cell system-specific) and the compartment-specific (dorsal fin, tail fin, eye, meninges, side of the body) regulatory genes. Furthermore, a post-transformational acting gene that pushes the tumour cells to differentiate terminally has been specified (Anders et al., 1985a,b).

6.3 Development of tester strains for bioassays

A tester strain was developed for the detection of mutagenic carcinogens that takes advantage of the easy detectability of melanomas. The basis of the construction of this strain is a female platyfish in which the *Tu*-linked pigment cell-specific regulatory gene is impaired. In addition, the *Tu*-linked regulatory genes specific for the dorsal fin and the posterior part the body are impaired by mutation. Because of the remaining non-linked regulatory genes, no tumour develops; the animals may, however, exhibit some small pigmented spots.

Crosses of this platyfish with a sword-tail lacking these regulatory genes result in F_1 hybrids that develop melanoma. The melanomas consist mainly of well-differentiated, transformed pigment cells which are histologically similar to those of the maternal spots. Back-crosses of the F_1 hybrids with the sword-tail as the recurrent parent result in three types of segregant in the offspring: one-fourth develop benign melanoma, like most of the F_1; another fourth develop malignant melanoma consisting mainly of incompletely differentiated transformed pigment cells; and half develop neither spots nor melanomas. Further back-crosses of the fish carrying malignant melanoma with the sword-tail result always in 50% of melanoma-free and 50% of malignant melanoma-bearing segregants. Finally, the malignant melanoma-bearing hybrids contain a genome in which one chromosome comes from the platyfish and the other from the sword-tail. The chromosome originating from the platyfish contains the critical *Tu* oncogene complex with impaired regulatory genes. Whenever melanomas occur in these animals, they develop in both the dorsal and the posterior part of the body.

To establish the strain containing a single functioning regulatory gene that protects the fish from melanoma, the platyfish chromosome used normally in the crossings was replaced by a platyfish chromosome in which the pigment cell-specific regulatory gene is a normal one. Due to the active pigment cell-specific regulatory gene, no melanoma develops spontaneously in the hybrids. Crosses between two hybrids of the latter genotype were the basis for the establishment of a strain homozygous for this chromosome. Because each of the two copies of the *Tu* oncogene-complex is repressed by its own linked pigment cell-specific regulatory gene, which acts in the *cis* position only, the incidence of animals developing melanoma following treatment with carcinogens potentially doubles. These fish are highly suitable as test animals for mutagenic carcinogens.

Construction of the tester strain for detection of tumour promoters is also based on the use of hybrids that develop malignant melanoma spontaneously. In this case, a chromosome containing a mutant gene that codes for a block of pigment cell differentiation at the stage of stem melanoblasts (not yet competent for neoplastic transformation) was introduced into the system. In animals heterozygous for this gene ($+/g$), the de-repressed *Tu* oncogene-complex still mediates melanoma formation. In animals that are homozygous for this mutation (g/g), however, no melanoma develops. The block of differentiation at the stage of the stem melanoblasts exerted by the homozygous g mutation protects the fish from its own de-repressed oncogenes and, therefore, from melanoma formation (Anders *et al.*, 1985a,b).

6.4 Assays with the tester strains

A large variety of mutagenic and nonmutagenic agents has been tested in the two tester strains. Table 1 shows that tumour initiation (specified by somatic mutation) and tumour promotion (specified by cell differentiation; from the not yet competent to the competent stage for neoplastic transformation) can be distinguished in these strains. In the case of promotion, the cells give rise to a melanoma of multicellular origin within about three to five weeks. In the case of initiation, melanoma development starts with one melanoma of unicellular origin within a period of two to 12 months. The smallest initial stages of melanomas detected consisted of four cells. X-rays, N-methyl-N-nitrosourea and N-ethyl-N-nitrosourea, which are mutagens and carcinogens, may also trigger melanoma by promotion of cell differentiation.

7. References

Anders, A. & Anders, F. (1978) Etiology of cancer as studied in the platyfish-swordtail system. *Biochim. biophys. Acta*, *516*, 61-95

Anders, A., Anders, F., Lüke, W., Henze, M., Schartl, M. & Schmidt, C.-R. (1985a) Oncogenes during evolution, development and tumorigenesis [in German]. *Zellenergetik Zelldifferenzierung Ergebnis. Sonderforschungsber.*, *103*, 172-186

Anders, A., Dess, G., Nishimura, S. & Kersten, H. (1985b) *A molecular approach to the study of malignancy and benignancy in melanoma of* Xiphophorus. In: Bagnara, J., Klaus, S.N., Paul, E. & Schartl, M., eds, *Pigment Cell 1985, Biological, Molecular and Clinical Aspects of Pigmentation*, Tokyo, University of Tokyo Press, pp. 315-324

Anders, F. (1967) Tumour formation in platyfish-swordtail hybrids as a problem of gene regulation. *Experientia*, *23*, 1-10

Table 1. Tumour initiation (somatic mutation) and promotion (of cell differentiation) distinguished by tester strains of *Xiphophorus*[a]

Agent	Dose	Period	Initiation[b]		Promotion	
			No. of survivors	With tumours (%)	No. of survivors	With tumours (%)
N-Methyl-N-nitrosourea	10^5 μg/l	5 × 14 days 45 min	457	88(19)	36	18(50)
N-Ethyl-N-nitrosourea	10^5 μg/l	5 × 7 days 1 h	235	21(9)	21	2(10)
N-Nitrosodiethylamine	45×10^3 μg/l	2 × 1 week for 8 weeks	23	2(9)	24	0
TPA (in dimethyl sulfoxide)	1 μg/l/day	10 weeks	121	0	23	11(48)
4-O-Methyl TPA	μg/l/day	10 weeks	20	0	15	0
Dimethyl sulfoxide	2×10^6 μg/l/8 days	10 weeks	98	0	12	1(8)
Diazepam	500 μg/l/day	48 weeks	27	0	24	0
Phenobarbital	7.5×10^3 μg/l/day	10 weeks	25	0	11	7(64)
Cyclamate	10^6 μg/l	8 weeks	28	0	22	12(55)
Saccharin	2×10^6 μg/l	8 weeks	NT	NT	10	6(60)
5-Azacytidine	5×10^6 μg/l	10 weeks	21	0	23	3(13)
L-Ethionine	1.25×10^3 μg/l/day	8 weeks	47	0	12	7(58)
Actinomycin D	500 μg/l/day	8 weeks	18	0	16	0
Testosterone	20 μg/l/day	8 weeks	15	0	10	0
Methyltestosterone	2–20 μg/l/day	24 weeks	NT	NT	22	22(100)
5-Dihydrotestosterone	2–20 μg/l/day	24 weeks	113	0	30	30(100)
Methylandrostenolone	2–20 μg/l/day	24 weeks	NT	NT	11	11(100)
Oestrogen	2–20 μg/l/day	24 weeks	NT	NT	21	21(100)
Diethylstilboestrol	2–20 μg/l/day	48 weeks	NT	NT	5	0
17-Hydroprogesterone	2–20 μg/l/day	48 hours	NT	NT	16	0
Progesterone	2–20 μg/l/day	48 hours	NT	NT	9	0
X-rays	1000 rads	3 × 45 min/6-week interval	805	163(20)	8	0
					45	45(100)

[a]Representative examples of experiments which, when repeated, gave comparable results
[b]NT, not tested

Anders, F. (1981) The role of hereditary and environmental factors and the causation of malignant growth based on studies in *Xiphophorus* [in German]. *Klin. Wochenschr.*, *59*, 943-956

Anders, F., Schwab, M. & Scholl, E. (1981) *Strategy for breeding test animals of high susceptibility to carcinogens*. In: Stich, H.F. & San, R.H.C., eds, *Short-term Tests for Chemical Carcinogens*, New York, Springer, pp. 399-407

Anders, F., Schmidt, C.-R., Herbert, A., Anders, A. (1983) *Tests for carcinogens for the initiating and/or promoting activity in the assay system* Xiphophorus. In: *Testing of Chemicals for Carcinogenicity, Mutagenicity and Teratogenicity*, Munich, Ges. f. Strahlen- und Umweltforschung, pp. 253-274

Anders, F., Schartl, M., Barnekow, A. & Anders, A. (1984) *Xiphophorus* as an in-vivo model for studies on normal and defective control of oncogenes. *Adv. Cancer Res.*, *42*, 191-275

Armuth, V. & Berenblum, I. (1972) Systemic promoting action of phorbol in liver and lung carcinogenesis in AKR mice. *Cancer Res.*, *32*, 2259-2262

Balmain, A. (1985) Transforming ras oncogenes and multistage carcinogenesis. *Brit. J. Cancer*, *51*, 1-7

Bannasch, P. (1968) The cytoplasm of hepatocytes during carcinogenesis. Light and electron microscopic investigations of the nitrosomorpholine-intoxicated rat liver. *Recent Results Cancer Res.*, *19*, 1-100

Bannasch, P., Hacker, H.J., Klimek F. & Mayer, D. (1984) Hepatocellular glycogenosis and related pattern of enzymatic changes during hepatocarcinogenesis. *Adv. Enzyme Regul., 22, 97-121*

Berenblum, I. (1941) The cocarcinogenic action of croton resin. *Cancer Res.*, *1*, 44-48

Berenblum, I. (1974) *Carcinogenesis as a Biological Problem*. Amsterdam, Elsevier

Berenblum, I. & Shubik, P. (1947) A new quantitative approach to the study of chemical carcinogenesis in the mouse's skin. *Br. J. Cancer*, *1*, 383-391

Berenblum, I. & Shubik, P. (1949) The persistence of latent tumour cells induced in the mouse skin by a single application of 9,10-dimethyl-1,2-benzanthracene. *Br. J. Cancer*, *3*, 384-386

Börzsönyi, M., Day, N.E., Lapis, K. & Yamasaki, H., eds (1984) *Models, Mechanisms and Etiology of Tumour Production (IARC Scientific Publications No. 56)*, Lyon, International Agency for Research on Cancer

Boutwell, R.K. (1964) Some biological aspects of skin carcinogenesis. *Prog. exp. Tumor Res.*, *4*, 207-250

Boutwell, R.K. & Bosch, D.K. (1959) Tumor promoting action of phenol and related compounds for mouse skin. *Cancer Res.*, *19*, 413-419

Burns, F.J., Vanderlaan, M., Synder, E. & Albert, R.E. (1978) *Induction and progression kinetics of mouse skin papillomas*. In: Slaga, T.L., Sivak, A & Boutwell, R.K. eds, *Carcinogenesis*, Vol. 2, *Mechanisms of Tumor Promotion and Cocarcinogenesis*, New York, Raven Press, pp. 91-96

Chouroulinkov, I. & Lazar, P. (1974) Carcinogenic and cocarcinogenic action of 12-0-tetradecanoylphorbol-13-acetate (TPA) on mouse skin [in French]. *C.R. Acad. Sci. Paris Serie D*, *278*, 3027-3031

Cohen, S.M., Arai, M., Jacons, J.B. & Friedell, G.H. (1979) Promoting effect of saccharin and DL-tryptophan in urinary bladder carcinogenesis. *Cancer Res., 39,* 1207-1217

Columbano, R., Ledda, G.M., Rao, P.M., Rajalakshmi, S. & Sarma, D.S. (1982) Dietary orotic acid, a new selective growth stimulus for carcinogen altered hepatocytes in rat. *Cancer Lett.*, *16*, 191-196

Dawe, C.J. & Couch, J.A. (1984) Mouse versus minnow: the future of fish in carcinogenicity testing. *Natl Cancer Inst. Monogr.*, *65*, 223-235

Dawe, C.J., Harshbarger, J.C., Kondo, S., Sugimura, T. & Takayama, S., eds (1981) *Phyletic Approaches to Cancer. Proceedings of the 11th International Symposium of The Princess Takamatsu Cancer Research Fund, Tokyo, 1980*, Tokyo, Japan Scientific Societies Press

DiGiovanni, J., Slaga, T.J. & Boutwell, R.K. (1980) Comparison of the tumor initiating activity of 7,12-dimethylbenz[*a*]anthracene and benzo[*a*]pyrene in female SENCAR and CD-1 mice. *Carcinogenesis*, *1*, 381-389

Diwan, B.A., Ward, J.M., Henneman, J. & Wenk, M.L. (1985) Effects of short-term exposure to the tumor promoter, 12-0-tetradecanoylphorbol-13-acetate on skin carcinogenesis in SENCAR mice. *Cancer Lett.*, *26*, 177-184

Farber, E. (1956) Similarities in the sequence of early histological changes induced in the liver of the rat by ethionine, 2-acetylaminofluorene, and 3-methyl-4-dimethyl-aminoazobenzene. *Cancer Res.*, *16*, 142-148

Farber, E. (1973) Hyperplastic liver nodules. *Methods Cancer Res.*, *7*, 345-375

Farber, E. (1980) The sequential analysis of liver cancer induction. *Biochem. biophys. Acta*, *605*, 149-166

Friedwald, W.F. & Rous, P. (1944) The initiating and promoting elements in tumor production. An analysis of the effects of tar, benzopyrene and methylcholanthrene on rabbit skin. *J. exp. Med.*, *80*, 101-126

Fujiki, H. & Sugimura, T. (1983) New potent tumour promoters: teleocidin, lyngbiatoxin A and aplysiatoxin. *Cancer Surveys*, *2*, 539-556

Fukushima, S., Friedell, G., Jacobs, J.B. & Cohen, S.M. (1981) Effect of L-tryptophan and sodium saccharin on urinary tract carcinogenesis initiated by *N*-[4-(5-nitro-2-furyl)-2-thiazolyl]-formamide. *Cancer Res.*, *41*, 3100-3103

Fukushima, S., Hirose, M., Okuda, M., Nakanowatari, J., Hatano, A. & Ito, N. (1982) Effect of partial cystectomy on the induction of pre-neoplastic lesions in rat bladder initiated with *N*-butyl-*N*-4-(4-hydroxybutyl)nitrosamine followed by bladder carcinogens and promoters. *Urol. Res.*, *10*, 115-118

Fukushima, S., Imaida, K., Sakata, T., Okamura, T., Shibata, M. & Ito, N. (1983a) Promoting effects of sodium L-ascorbate on 2-stage urinary bladder carcinogenesis in rats. *Cancer Res.*, *43*, 4454-4457

Fukushima, S., Kurata, Y., Shibata, M., Ikawa, E. & Ito, N. (1983b) Promoting effect of sodium o-phenylphenate and o-phenylphenol on two-stage urinary bladder carcinogenesis in rats. *Gann*, *74*, 625-632

Fukushima, S., Kurata, Y., Shibata, M., Ikawa, E. & Ito, N. (1984) Promotion by ascorbic acid, sodium erythorbate and ethoxyquin of neoplastic lesions in rats initiated with *N*-butyl-*N*-(4-hydroxybutyl)nitrosamine. *Cancer Lett.*, *23*, 29-37

Fukushima, S., Thamavit, W., Kurata, Y., Ito, N. (1986) Sodium citrate: a promoter of bladder carcinogenesis. *Jpn. J. Cancer Res. (Gann)*, *77*, 1-4

Fukushima, S., Kurata, Y., Shibata, M., Ikawa, E. & Ito, N. (1984) Promotion by ascorbic acid, sodium erythorbate and ethoxyquin of neoplastic lesions in rats initiated with N-butyl-N-(4-hydroxybutyl)nitrosamine. *Cancer Lett.*, 23, 29-37

Fukushima, S., Thamavit, W., Kurata, Y., Ito, N. (1986) Sodium citrate: a promoter of bladder carcinogenesis. *Jpn. J. Cancer Res. (Gann)*, 77, 1-4

Gart, J., Krewski, D., Lee, P.N., Tarone, R.E., Wahrendorf, J. eds (1986) *Statistical Methods in Cancer Research,* Vol. III, *The Design and Analysis of Long-term Animal Carcinogenicity Experiments (IARC Scientific Publications No. 79)*, Lyon, International Agency for Research on Cancer (in press)

Hakii, H., Fujiki, H., Suganuma, M., Nakayasu, M., Tahira, T., Sugimura, T., Scheuer, P.J. & Christensen, S.B. (1986) Thapsigargin, a histamine secretagogue, is a non-12-O-tetradecanoyl-phorbol-13-acetate (TPA) type tumor promoter in two-stage mouse skin carcinogenesis. *J. Cancer Res. clin. Oncol.*, 111, 177-181

Hall, W.H. (1948) The role of initiating and promoting factors in the pathogenesis of tumours of the thyroid. *Br. J. Cancer*, 2, 273-280

Hasegawa, R., Tsuda, H., Ogiso, T., Ohshima, M. & Ito, N. (1982a) Initiating activities of pyrolysis products of L-lysine and soybean globulin assessed in terms of the induction of γ-glutamyl transpeptidase-positive foci in rat liver. *Gann*, 73, 158-159

Hasegawa, R., Tatematsu, M., Tsuda, H., Shirai, T., Hagiwara, A. & Ito, N. (1982b) Induction of hyperplastic liver nodules in hepatectomized rats treated with 3'-methyl-4-dimethyl- aminoazobenzene, benzo[a]pyrene or phenobarbital before or after exposure to N-2-fluorenylacetamide. *Gann*, 73, 264-269

Hecker, E. (1967) Phorbol esters from croton oil. Chemical nature and biological activities. *Naturwissenschaften*, 54, 282-284

Hecker, E., Bresch, H. & von Szczepanski, C. (1964) Cocarcinogen A 1 the first pure, highly active constituent of croton oil. *Angew. Chem.*, 76, 225-226

Hennings, H., Devor, D., Wenk, M.L., Slaga, T.J., Former, B., Colburn, N.H., Bowden. G.T., Elgjo, K. & Yuspa, S.H. (1981) Comparison of two-stage epidermal carcinogenesis initiated by 7,12-dimethylbenz[a]anthracene or N-methyl-N'-nitro-N-nitrosoguanidine in newborn and adult SENCAR and Balb/c mice. *Cancer Res.*, 211, 773-779

Herren-Freund, S.L., Pereira, M.A., Long, R.E., Klourez, U.M. (1986) The effect of single *versus* split doses of diethylnitrosamine on the induction of gamma-glutamyltranspeptidase-foci in the livers of adult and juvenile rats. *Carcinogenesis*, 7, 1107-1110

Hiasa, Y., Ohshima, M., Kitahori, Y., Konishi, N., Fujita, T. & Yuasa, T. (1982a) beta-Cyclodextrin: promoting effect on the development of renal tumular cell tumours in rats treated with N-ethyl-N-hydroxyethyl-nitrosamine. *J. natl Cancer Inst.*, 69, 963-967

Hiasa, Y., Ohshima, M., Kitahori, Y., Yuasa, T., Fujita, T. & Iwata, C. (1982b) Promoting effects of 3-amino-1,2,4-triazole on the development of thyroid tumors in rats treated with N-bis(2-hydroxypropyl)nitrosamine. *Carcinogenesis*, 3, 381-384

Hiasa, Y., Kitahori, Y., Ohshima, M., Fujita, T., Yuasa, T., Konishi, N. & Miyashiro, A. (1982c) Promoting effects of phenobarbital and barbital on development of thyroid tumors in rats treated with N-bis(2-hydroxypropyl)nitrosamine. *Carcinogenesis*, 3, 1187-1190

Hiasa, Y., Ohshima, M., Kitahori, Y., Fujita, T., Yuasa, T. & Miyashiro, A. (1983a) Basic lead acetate: promoting effect on the development of renal tubular cell tumors in rats treated with N-ethyl-N-hydroxyethylnitrosamine. *J. natl Caner Inst.*, 70, 761-765

Hiasa, Y., Kitahori, Y., Konishi, N., Enoki, N. & Fujita, T. (1983b) Effect of varying the duration of exposure to phenobarbital on its enhancement of N-bis(2-hydroxypropyl)nitrosamine induced thyroid tumorigenesis in male Wistar rats. *Carcinogenesis*, 4, 935-937

Hiasa, Y., Enoki, N., Kitahori, Y., Konishi, N. & Shimoyama, T. (1984a) DL-Serine: promoting activity on renal tumorigenesis by *N*-ethyl-*N*-hydroxyethyl-nitrosamine in rats. *J. natl Cancer Inst., 73,* 297-299.

Hiasa, Y., Kitahori, Y., Konishi, N., Enoki, N., Shimoyama, T. & Miyashiro, A. (1984b) Trisodium nitroloacetate monohydrate: promoting effects on the development of renal tubular cell tumours in rats treated with *N*-ethyl-*N*-hydroxyethylnitrosamine. *J. natl Cancer Inst., 72,* 483-489

Hiasa, Y., Kitahori, Y., Enoki, N., Konishi, N. & Shimoyama, T. (1984c) 4,4'-Diaminodiphenylmethane: promoting effect on the development of thyroid tumors in rats treated with *N*-bis(2-hydroxypropyl)nitrosamine. *J. natl Cancer Inst., 72,* 471-476

Hicks, R.M. & Chowaniec, J. (1977) The importance of synergy between weak carcinogens in the induction of bladder cancer in experimental animals and humans. *Cancer Res., 37,* 2943-2949

Hicks, R.M., Wakefield, J.St J. & Chowaniec, J. (1975) Evaluation of a new model to detect bladder carcinogens or co-carcinogens; results obtained with saccharin, cyclamate and cyclophosphamide. *Chem.-biol. Interactions, 11,* 225-233

Hicks, R.M., Chowaniec, J. & Wakefield, J.St J. (1978) *Mechanisms of tumor promotion and cocarcinogenesis.* In: Slaga, T.J. & Boutwell, R.K., eds, *Carcinogenesis,* Vol. 2, *Mechanisms of Tumor Promotion and Cocarcinogenesis,* New York, Raven Press, pp. 475-489

Hosaka, S., Matsushima, T., Hirono, I. & Sugimura, T. (1981) Carcinogenic activity of 3-amino-1-methyl-5H-phrido[4,3-b]-indole (Trp-P-2), a pyrolysis product of tryptophan. *Cancer Lett., 13,* 23-28

Hoover, K.L., ed. (1984) *Use of Small Fish Species in Carcinogenicity Testing (Natl Cancer Inst. Monograph No. 65),* Bethesda, MD, National Cancer Institute

Imaida, K., Fukushima, S., Shirai, T., Ohtani, M., Nakanishi, K. & Ito, N. (1983) Promoting activities of butylated hydroxyanisole and butylated hydroxytoluene on 2-stage urinary bladder carcinogenesis and inhibition of γ-glutamyl transpeptidase-positive foci development in the liver of rats. *Carcinogenesis, 4,* 895-899

Inayama, Y. (1986) Promoting action of glycerol in pulmonary carcinogenesis model using a single administration of 4-nitroquinoline 1-oxide in mice. *Jpn. J. Cancer Res. (Gann), 77,* 345-350

Ito, N. (1973) Experimental studies on tumors of the urinary system of rats induced by chemical carcinogens. *Acta pathol. Jpn, 23,* 87-109

Ito, N., Makiura, S., Yokota, Y., Kamamoto, Y., Hiasa, Y. & Sugiura, S. (1971) Effect of unilateral ureter ligation on development of tumors in the urinary system of rats treated with *N*-butyl-*N*-(4-hydroxybutyl)nitrosamine. *Gann, 62,* 359-365

Ito, N., Tatematsu, M., Nakanishi, K., Hasegawa, R., Takano, T., Imaida, K. & Ogiso, T. (1980) The effects of various chemicals on the development of hyperplastic liver nodules in hepatectomized rats treated with *N*-nitrosodiethylamine or *N*-2-fluorenylacetamide. *Gann, 71,* 832-842

Ito, N., Fukushima, S., Shirai, T. & Nakanishi, K. (1983) Effects of promoters of *N*-butyl-*N*-(4-hydroxybutyl)nitrosamine-induced urinary bladder carcinogenesis in the rat. *Environ. Health Perspect., 50,* 61-69

Kallman, K. (1973) *The platyfish,* Xiphophorus maculatus. In: King, R.C., ed., *Handbook of Genetics,* Vol. 4, New York, Plenum, pp. 81-132

Kitagawa, T. & Sugano, H. (1973) *Analytical and experimental epidemiology of cancer.* In: Nakahara, W., Hirayama, T., Nishioka, K. & Sugano, H., eds, *Time Table for Hepatocarcinogenesis in Rat,* Tokyo, University of Tokyo Press, pp. 91-108

Kitagawa, T. & Sugano, H. (1978) Enhancing effects of phenobarbital on the development of enzyme-altered islands and hepatocellular carcinomas initiated by 3'-methyl-4-dimethylaminoazobenzene or diethylnitrosamine. *Gann, 69,* 679-687

Kitagawa, T., Pitot, H.C., Miller, E.C. & Miller, J.A. (1979) Promotion by dietary phenobarbital of hepatocarcinogenesis by 2-methyl-*N*,*N*-dimethyl-4-aminoazobenzene in the rat. *Cancer Res.*, *39*, 112-115

Kitagawa, T., Kirakawa, T., Ishikawa, T., Nemoto, N. & Takayama, S. (1980) Induction of hepatocellular carcinoma in rat liver by initial treatment with benzo[*a*]pyrene after partial hepatectomy and promotion by phenobarbital. *Toxicol. Lett.*, *6*, 167-171

Kunze, E. & Schauer, A. (1971) Enzyme-histochemical and autoradiographic investigations on dibutylnitrosamine-induced papillomas of the urinary bladder in the rat [in German]. *Z. Krebsforsch.*, *75*, 146-160

Kurokawa, Y., Aoki, S., Imazawa, T., Hayashi, Y., Matsushima, Y. & Takamura, N. (1985) Dose-related enhancing effect of potassium bromate on renal tumorigenesis in rats initiated with *N*-ethyl-*N*-hydroxyethylnitrosamine. *Jpn. J. Cancer Res. (Gann)*, *76*, 583-589

Land, H., Parada, L.F. & Weinberg, R.A. (1983) Cellular oncogenes and multistep carcinogenesis. *Science*, *222*, 771-778

Lawns, M., de Gerlache, J., Taper, H.S., Préat, V. & Roberfroid, M.B. (1983) Phenobarbital as a promoter in the initiation/selection process of experimental rat hepatocarcinogenesis. *Carcinogenesis*, *4*, 141-144

Matsushima, T. & Sugimura, T. (1976) *Experimental carcinogenesis in small aquarium fishes.* In: Dawe, C.J., Scarpelli, D.G. & Wellings, S.R., eds, *Tumors in Aquatic Animals*, Vol. 20, Basel, S. Karger, pp. 367-379

Milmore, J.E., Chandrasekaran, V. & Weisburger, J.H. (1982) Effects of hyothyroidism on development of nitrosomethylurea-induced tumors of the mammary gland, thyroid gland, and other tissues. *Proc. Soc. exp. Biol. Med.*, *169*, 487-493

Miyata, Y., Fukushima, S., Imaida, K. & Ito, N. (1984) Promoting effects of unilateral ureteric ligation on two-stage urinary bladder carcinogenesis in rats. *Cancer Lett.*, *23*, 265-272

Miyata, Y., Fukushima, S., Hirose, M., Masui, T. & Ito, N. (1985) Modifying potentials of various environmental chemicals on *N*-butyl-*N*-(4-hydroxybutyl)nitrosamine-initiated urinary bladder carcinogenesis in rats with ureteric ligation. *Jpn. J. Cancer Res. (Gann)*, *76*, 828-834

Mottram, J.C. (1944) A developing factor in experimental blastogenesis. *J. Pathol.*, *54*, 181-187

Nadler, N.J., Mandavia, M.G. & Leblond, C.P. (1969) Influence of preirradiation on thyroid tumorigenesis by low iodine in the rat. *UICC Monogr. Ser.*, *12*, 125-130

Nadler, N.J., Mandavia, M. & Goldberg, M. (1970) The effect of hypophysectomy on the experimental production of rat thyroid neoplasm. *Cancer Res.*, *30*, 1909-1911

Nakanishi, K., Hagiwara, A., Shibata, M., Imaida, K., Tatematsu, M. & Ito, N. (1980) Dose-response of saccharin in induction of urinary bladder hyperplasia in Fischer 344 rats pretreated with *N*-butyl-*N*-(4-hydroxybutyl)nitrosamine. *J. natl Cancer Inst.*, *65*, 1005-1010

Napalkov, N.P. (1976) *Tumours of thyroid gland.* In: Turusov, V.S. ed., *Pathology of Tumours in Laboratory Animals*, Vol. I, *Tumours of the Rat, Part 2 (IARC Scientific Publications No. 6)*, Lyon, International Agency for Research on Cancer, pp. 239-272

Newberne, P.M. & Butler, W.H. (1969) Acute and chronic effects of aflatoxin on the liver of domestic and laboratory animals; a review. *Cancer Res.*, *29*, 236-250

Nishizumi, M. (1976) Enhancement of diethylnitrosamine hepatocarcinogenesis in rats by exposure to polychlorinated biphenyls or phenobarbital. *Cancer Lett.*, *2*, 11-16

Numoto, S., Furukawa, K., Furuya, K. & Williams, G.M. (1984) Effects of the hepatocarcinogenic peroxisome-proliferating hypolipidemic agents clofibrate and nafenopin on the rat liver cell membrane enzymes gamma-glutamyltranspeptidase and alkaline phosphatase and on the early stages of liver carcinogenesis. *Carcinogenesis, 5*, 1603-1611

Nychka, D., Pugh, T.D., King, J.H., Koen, H., Wahba, G., Chover, J. & Goldfarb, S. (1984) Optimal use of sampled tissue sections for estimating the number of hepatocellular foci. *Cancer Res., 44*, 178-183

Ogawa, K., Medline, A. & Farber, E. (1979) Sequential analysis of hepatic carcinogenesis. A comparative study of the ultrastructure of preneoplastic, malignant, prenatal, postnatal and regenerating liver. *Lab. Invest., 41*, 22-35

Ogiso, T., Tatematsu, M., Tamano, S., Tsuda, H. & Ito, N. (1985) Comparative effects of carcinogens on the induction of placental glutathione S-transferase-positive liver nodules in a short term assay and of hepatocellular carcinomas in a long-term assay. *Toxicol. Pathol. 13*, 257-265

Ohgaki, H., Matsukura, N., Morino, K., Kawachi, T., Sugimura, T. & Takayama, S. (1984) Carcinogenicity in mice of mutagenic compounds from glutamic acid and soybean globulin pyrolysates. *Carcinogenesis, 5*, 815-819

Ohshima, M. & Ward, J.M. (1984) Promotion of N-methyl-N-nitrosourea-induced thyroid tumors by iodine deficiency in F344/NCr rats. *J. natl Cancer Inst., 73*, 289-294

Peraino, C., Fry, R.J.M., Staffeldt, E. & Kisieleski, W.E. (1973) Effects of varying the exposure to phenobarbital on its enhancement of 2-acetylaminofluorene-induced hepatic tumorigenesis. *Cancer Res., 33*, 2701-2705

Peraino, C., Fry, R.J.M., Staffeldt, E. & Christopher, J.P. (1975) Comparative enhancing effects of phenobarbital, amobarbital, diphenylhydantoin, and dichlorodiphenyltrichloroethane on 2-acetylaminofluorene-induced hepatic tumorigenesis in the rat. *Cancer Res., 35*, 2884-2890

Peraino, C., Fry, R.J.M. & Grube, D.D. (1978) *Drug-induced enhancement of hepatic tumorigenesis*. In: Slaga, T.J., Sivak, A. & Boutwell, R.D., eds, *Carcinogenesis*, Vol. 2, *Mechanisms of Tumor Promotion and Cocarcinogenesis*, New York, Raven Press, pp. 421-432

Phillips, D.H., Grover, P.L, & Sims, P. (1978) The covalent binding of polycyclic hydrocarbons to DNA in the skin of mice of different strains. *Int. J. Cancer, 22*, 487-494

Pitot, H.C. (1977) The natural history of neoplasia. *Am. J. Pathol., 89*, 402-411

Pitot, H.C. (1983) Contributions to our understanding of the natural history of neoplastic development in lower animals to the cause and control of human cancer. *Cancer Surv., 2*, 519-537

Pitot, H.C. & Sirica, A.E. (1980) The stages of initiation and promotion in hepatocarcinogenesis. *Biochim. biophys. Acta, 605*, 191-215

Pitot, H.C., Barsness, L., Goldsworthy, T. & Kitagawa, T. (1978) Biochemical characterization of stages of hepatocarcinogenesis after a single dose of diethylnitrosamine. *Nature, 271*, 456-458

Radomski, J.L., Krischer, C. & Krischer, K.N. (1978) Histologic and histochemical preneoplastic changes in the bladder mucosae of dogs given 2-naphthylamine. *J. natl Cancer Inst., 60*, 327-333

Reddy, A.L. & Fialkow, P.J. (1983) Papillomas induced by initiation-promotion differ from those induced by carcinogen alone. *Nature, 304*, 69-71

Reiners, J.J., Jr, Nesnow, S & Slaga, T.J. (1984) Murine susceptibility to two-stage skin carcinogenesis is influenced by the agent used for promotion. *Carcinogenesis, 5*, 301-307

Rosen, D.E. (1979) Fishes from the uplands and intermontane basins of Guatemala: revisionary studies and comparative geography. *Bull. Am. Mus. nat. Hist., 162*, 267-376

Rous, P. & Kidd, J.G. (1941) Conditional neoplasms and subthreshold neoplastic states. *J. exp. Med.*, *73*, 299-390

Schaffer, R. & Miller, H.A. (1980) On the development of metastasizing tumors of the rat thyroid gland after combined administration of nitrosomethylurea and methylthiouracil. *Cancer Res. clin. Oncol.*, *96*, 281-285

Schartl, M., Schmidt, C.-R., Anders, A. & Barnekow, A. (1985) Elevated expression of the cellular *src* gene in tumors of different etiology in *Xiphorphorus*. *Int. J. Cancer*, *36*, 199-107

Scherer, E. (1981) Use of a programmable pocket calculator for the quantitation of precancerous foci. *Carcinogenesis*, *2*, 805-807

Scherer, E. & Emmelot, P. (1976) Kinetics of induction and growth of enzyme deficient islands involved in hepatocarcinogenesis. *Cancer Res.*, *36*, 2544-2554

Scribner, N.K. & Scribner, J.D. (1980) Separation of initiating and promoting effects of the skin carcinogen 7-bromomethylbenz[*a*]anthracene. *Carcinogenesis*, *1*, 97-100

Shinozuka, H. & Lombardi, B. (1980) Synergistic effect of a choline-devoid diet and phenobarbital in promoting the emergence of foci of γ-glutamyltranspeptidase-positive hepatocytes in the liver of carcinogen-treated rats. *Cancer Res.*, *40*, 3846-3849

Shirai, T., Cohen, S.M., Fukushima, S., Hananouchi, M. & Ito, N. (1978) Reversible papillary hyperplasia of the rat urinary bladder. *Am. J. Pathol.*, *91*, 33-48

Shirai, T., Masuda, A., Hirose, M., Ikawa, E. & Ito, N. (1984a) Enhancement of *N*-bis(2-hydroxypropyl)nitrosamine-initiated lung tumor development in rats by bleomycin and *N*-methyl-*N*-nitrosourethane. *Cancer Lett.*, *25*, 25-31

Shirai, T., Ohshima, M., Masuda, A., Tamano, S. & Ito, N. (1984b) Promotion of 2-(ethylnitrosamino)-ethanol-induced renal carcinogenesis in rats by nephrotoxic compounds: positive responses with folic acid, basic lead acetate and *N*-(3,5-dichlorophenyl)succinimide but not with 2,3-dibromo-1-propanol phosphate. *J. natl Cancer Inst.*, *72*, 477-482

Slaga, T.J. (1983) Overview of tumor promotion in animals. *Environ. Health Perspect.*, *50*, 3-14

Slaga, T.J. & Fischer, S.M. (1983) Strain differences and solvent effects in mouse skin carcinogenesis experiments using carcinogens, tumor initiators and promoters. *Prog. exp. Tumor Res.*, *26*, 85-109

Slaga, T.J., Klein-Szanto, A.J.P., Triplett, L.L., Yotti, L.P. & Trosko, J.E. (1981) Skin tumor promoting activity of benzoyl peroxide, a widely used free radical generating compound. *Science*, *213*, 1023-1025

Slaga, J.J., Fischer, S.M., Triplett, L.L. & Nesnow, S. (1982) Comparison of complete carcinogenesis and tumor initiation in mouse skin: tumor initiation-promotion, a reliable short-term assay. *J. Am. Coll. Toxicol.*, *1*, 83-99

Solt, D.B. & Farber, E. (1976) A new principle for the analysis of chemical carcinogenesis. *Nature*, *263*, 702-703

Solt, D.B., Cayama, E., Tsuda, H., Enomoto, K., Lee, G. & Farber, E. (1983) Promotion of liver cancer development by brief exposure to dietary 2-acetylaminofluorene plus partial hepatectomy or carbon tetrachloride. *Cancer Res.*, *43*, 188-191

Stenback, F. (1980) Skin carcinogenesis as a model system; observations on species, strain and tissue sensitivity to 7,12-dimethylbenz[*a*]anthracene with and without promotion from croton oil. *Acta pharmacol. toxicol.*, *46*, 89-97

Stewart, H.L. & Snell, K.C. (1957) The histopathology of experimental tumors of the liver of the rat. *Acta unio int. contra Cancrum*, *13*, 770-803

Takayama, S., Masuda, M., Mogami, M., Ohgaki, H., Sato, S. & Sugimura, T. (1984) Induction of cancers in the intestine, liver and various other organs of rats by feeding mutagens from glutamic acid pyrolysate. *Gann, 75*, 207-213

Tatematsu, N., Imaida, K., Fukushima, S., Arai, M., Mizutani, M. & Ito, N. (1981) Cytopathological effect of partial cystectomy on rats. *Acta pathol. Jpn, 31*, 535-543

Tatematsu, M., Hasegawa, R., Imaida, K., Tsuda, H. & Ito, N. (1983) Survey of various chemicals for initiating and promoting activities in a short-term in-vivo system based on generation of hyperplastic liver nodules in rats. *Carcinogenesis, 4*, 381-386

Tatematsu, M., Mera, Y., Ito, N., Satoh, K. & Sato, K. (1985) Relative merits of immunohistochemical demonstrations of placental, A, B, and C forms of glutathione S-transferase and histochemical demonstration of γ-glutamyl transferase as markers of altered foci during liver carcinogenesis in rats. *Carcinogenesis, 6*, 1621-1626

Thamavit, W., Tatematsu, N., Ogiso, T., Mera, Y., Tsuda, H. & Ito, N. (1985) Dose-dependent effects of butylated hydroxyanisole, butylated hydroxytoluene and ethoxyquin in induction of foci of rat liver cells containing the placental form of glutathione S-transferase. *Cancer Lett., 27*, 295-303

Theiss, J.C., Arnold, L.J. & Shimkin, M.B. (1980) Effect of commercial saccharin preparations on urethan-induced lung tumorigenesis in strain A mice. *Cancer Res., 40*, 4322-4324

Tsuda, H., Lee, G. & Farber, E. (1980) Induction of resistant hepatocytes as a new principle for a possible short-term in-vivo test for carcinogens. *Cancer Res., 40*, 1157-1164

Tsuda, H., Fukushima, S., Imaida, K., Kurata, Y. & Ito, N. (1983) Organ-specific promoting effect of phenobarbital and saccharin in induction of thyroid, liver, and urinary bladder tumors in rats after initiation with *N*-nitrosomethylurea. *Cancer Res., 43*, 3292-3296

Turusov, V.S. & Koblyakov, V.A. (1986) Stages of carcinogenesis and mechanisms of action of chemical carcinogens [in Russian]. *Itogi Nauk. Tek. Onkol., 15*, 6-76

Vanderlaan, M., Fong, S. & King, E.B. (1982) Histochemistry of NADH diaphorase and γ-glutamyl-transpeptidase in rat bladder tumors. *Carcinogenesis, 3*, 397-402

Van Duuren, B.L. (1969) Tumor promoting agents in two-stage carcinogenesis. *Prog. exp. Tumor Res., 11*, 31-68

Van Duuren, B.L., Sivak, A., Segal, A., Seidman, I. & Katz, C. (1973) Dose-response studies with a pure tumor-promoting agent, phorbol myristate acetate. *Cancer Res., 33*, 2166-2172

Van Duuren, B.L., Witz, G. & Goldschmidt, B.M. (1978) *Structure-activity relationships of tumor promoters and cocarcinogens and interaction of phorbol myristate acetate and related esters with plasma membranes*. In: Slaga, T.J., Sivak, A. & Boutwell, R.K., eds, *Carcinogenesis*, Vol. 2, *Mechanisms of Tumor Promotion and Cocarcinogenesis*, New York, Raven Press, pp. 491-507

Verma, A.K. & Boutwell, R.K. (1980) Effects of dose and duration of treatment with the tumor-promoting agent, 12-O-tetradecanoyl-phorbol-13-acetate on mouse skin. *Carcinogenesis, 1*, 271-276

Watanabe, K. & Williams, G.M. (1978) Enhancement of rat hepatocellular-altered foci by the liver tumor promoter phenobarbital: evidence that foci are precursors of neoplasms and that the promoter acts on carcinogen-induced lesions. *J. natl Cancer Inst., 61*, 1311-1314

Weisburger, J.H., Madison, R.M., Ward, R.J.M., Vignera, C. & Weisburger, E.K. (1975) Modification of diethylnitrosamine liver carcinogenesis with phenobarbital but not with immunosuppression. *J. natl Cancer Inst., 54*, 1185-1188

Williams, G.M. (1982) Phenotypic properties of preneoplastic rat liver lesions and application to detection of carcinogens and tumour promoters. *Toxicol. Pathol.*, *10*, 3-10

Williams, G.M. & Furuya, K. (1984) Distinction between liver neoplasm promoting and syncarcinogenic effects demonstrated by exposure to phenobarbital or diethylnitrosamine either before or after *N*-2-fluorenylacetamide. *Carcinogenesis*, *5*, 171-174

Witschi, H.P. (1983) Promotion of lung tumors in mice. *Environ. Health Perspect.*, *50*, 267-273

Witschi, M.R. & Morse, C.C. (1983) Enhancement of lung tumour formation in mice by dietary butylated hydroxytoluene: dose-time relationships and cell kinetics. *J. natl Cancer Inst.*, *71*, 859-866

SHORT-TERM ASSAYS TO PREDICT CARCINOGENICITY

Chairman: M.L. Mendelsohn
Vice-Chairman: M. Hofnung
Main Rapporteur: S. Venitt

REPORT 4

DNA DAMAGE AND REPAIR

Prepared by:
S. Venitt (Rapporteur), H. Bartsch, G. Becking,
R.P.P. Fuchs, M. Hofnung, C. Malaveille, T. Matsushima,
A.E. Pegg, M.R. Rajewsky, M. Roberfroid,
H.S. Rosenkranz (Chairman) and G.M. Williams

1. Introduction

Comprehensive reviews on the formation and detection of DNA adducts and DNA repair, and the relevance of these topics to carcinogenesis are available (Searle, 1984).

Carcinogenic chemicals of the type which undergo covalent binding to biological macromolecules do so either directly (as in the case of some alkylating agents) or after metabolic activation to a chemically reactive form, the so-called ultimate carcinogen. The reaction with DNA usually takes place at the level of the bases, although reactions with the sugar-phosphate backbone have also been described (formation of chain breaks or of phosphotriesters). A chemically modified base is usually referred to as a 'damaged base', a 'lesion', an 'adduct' or a 'premutagenic lesion'. Most long-term and short-term carcinogen screening tests measure endpoints (e.g., DNA repair synthesis, mutagenesis, various chromosomal anomalies, cytological alterations, in-vitro cell transformation) that are far removed from the events that induce the later biological effects. Measurement of covalent binding of chemicals to DNA in intact mammalian organisms or cultured cells can be used as the basis for short-term screening assays, provided that the detection of the DNA damage is sensitive and easy to perform. However, the quantity of adducts formed *in vivo* is very small, and they can be detected by purely chemical means only in certain cases, e.g., when they are highly fluorescent. Several techniques have been developed, namely (i) the use of radioactively labelled chemicals, (ii) the use of specific anti-adduct antibodies, and (iii) post-labelling techniques. The most commonly used technique involves radioactively labelled chemicals, the main limitation being its restriction to compounds that are available in such a form.

The vast majority of DNA lesions undergo error-free repair by excision repair or recombinational repair, and only a small fraction appear to lead to the production of mutations. In fact, mutagenesis can be viewed as the result of the processing of lesions by specific pathways (mutagenesis pathways), and not merely as the result of errors made during repair processes or DNA replication. These pathways are still largely unknown, even in bacteria. In the case of a forward mutation assay in bacteria, the frequency at which a given lesion is converted into a non-silent mutation is usually very low (10^{-2} to 10^{-3}) (Bichara & Fuchs, 1985), indicating that most lesions are removed from DNA in an error-free manner by various repair pathways without giving rise to phenotypic expression. As a consequence, frequent repair events can be monitored by means of biochemical assays (based on assays of incision, excision, resynthesis or ligation), whereas rare mutational events rely on strong selective endpoints (phenotypic detection). Thus far, tests that rely for detection of repair on assays for unscheduled DNA synthesis (UDS) have been adopted for screening purposes.

2. Detection of DNA damage

2.1 *Use of radioactively labelled chemicals to measure total binding to DNA*

Since the first unequivocal demonstration of the covalent binding of carcinogenic polycyclic aromatic hydrocarbons to DNA of skin of mice treated *in vitro* (Brookes & Lawley, 1964), it has generally been agreed that DNA is the critical target for covalent binding of carcinogens and their metabolites. Although there is generally more binding to proteins than to nucleic acids, only binding to DNA has been positively correlated with carcinogenic potency. Due to the low levels of binding of chemicals to DNA *in vivo*, this analysis is restricted primarily to compounds that are available in a radiolabelled form. Many studies have shown that binding of a carcinogen or its metabolite to DNA is related to the carcinogenic response only if the experiments are carried out *in vivo*, due to biases introduced by the lack of appropriate metabolic activation in most in-vitro systems. One way of expressing binding to DNA, proposed by Lutz (1979), is the 'covalent binding index' (CBI), defined as micromoles of chemical bound per mole of nucleotide/millimole of chemical administered per kg body weight of animal. However, this is an index of overall binding and takes no account of the relative proportions of the different adducts that contribute to total binding.

2.2 *Use of radioactively labelled chemicals in the analysis of adducts*

Reviews of studies of carcinogen-DNA adduct formation are available (e.g., Searle, 1984). The quantitative and qualitative analysis of DNA adducts requires degradation of DNA to the nucleotide, nucleoside or base level, followed by chromatography. Hydrolysis of DNA can be achieved enzymatically (usually using a combination of endonucleases, exonucleases and phosphatases, yielding nucleosides) or chemically (acid hydrolysis of DNA, giving free bases). Enzymatic hydrolysis procedures are usually unable fully to degrade DNA, and a core fraction containing modified oligonucleotides resistant to further hydrolysis is often obtained; the analysis of adducts is thus restricted to those degraded to the level of mononucleosides. Chemical hydrolysis is generally complete, but acid-labile adducts may be destroyed.

Chromatographic analysis of adducts usually involves two steps: crude separation of the bulk of unmodified nucleosides from the pool of carcinogen-nucleoside adducts, followed by reverse-phase high-performance liquid chromatography to resolve the various adducts. Individual adducts are identified by cochromatography with authentic (chemically synthesized) adducts, using ultraviolet spectrophotometry in combination with detection of radioactivity arising from adducts formed *in vivo*. An example of this approach is given by Warren (1984).

2.3 *Immunochemical detection of adducts* (see review by Kriek *et al*., 1984)

During the last few years, highly sensitive, specific immunoanalytical methods have been developed for the detection and quantification of DNA adducts. After some early studies with polyclonal antisera (Poirier *et al.*, 1977; Leng *et al.*, 1978; Müller & Rajewsky, 1978) monoclonal antibodies with very high antibody affinity constants (ranging from 5×10^8 to 3×10^{10} l/mol) and specificity were produced. Antibodies have now been produced to such diverse carcinogens as aromatic amines, polycyclic aromatic hydrocarbons, aflatoxins, methylating and ethylating agents, as well as to specific types of ultraviolet damage (thymidine dimers) and oxidative damage (thymine glycol) (Strickland & Boyle, 1981; Adamkiewicz *et al.*, 1982; Leadon & Hanawalt 1983; Adamkiewicz *et al.*, 1985).

Studies have also been published in which a polyclonal antibody to benzo[*a*]pyrene diolepoxide-I-modified DNA was used to quantify adduct levels in lung tissue and white blood cells of individuals with various exposures to benzo[*a*]pyrene (Perera *et al.*, 1982; Shamsuddin *et al.*, 1985). Cisplatin-DNA [*N*-7-d (GpG) diammine platinum intrastrand] adducts have been quantified in patients undergoing chemotherapy (Fichtinger-Schepman *et al.*, 1985; Poirier *et al.*, 1985). These studies demonstrate the feasibility of this approach and indicate that immunological methods are sufficiently sensitive for studies of human dosimetry. The list of monoclonal antibodies directed against specific DNA adducts is growing steadily, as is the number of different immunoanalytical procedures — competitive radioimmunoassay, enzyme immunoassays, immuno-slot-blot procedures, direct and indirect immunostaining and immuno-electron microscopy (De Murcia *et al.*, 1979; Müller & Rajewsky, 1980; van der Laken *et al.*, 1982; Nehls *et al.*, 1984). For example, O^6-ethyldeoxyguanosine formed in the DNA of cells exposed to ethylating *N*-nitroso compounds can be detected by radioimmunoassay using anti-O^6-ethyldeoxyguanosine monoclonal antibody at a molar ratio of 3×10^{-8} O^6-ethyl- to deoxyguanosine in a 2-mg sample of DNA (Adamkiewicz *et al.*, 1985). Using the immuno-slot-blot procedure, 0.3 fmol of the same alkylation product can be detected in a 3-μg sample of DNA (Nehls *et al.*, 1984). By direct immunofluorescence (using computer-based image analysis of electronically intensified fluorescence signals), less than 10^3 O^6-ethyldeoxyguanosine molecules per diploid genome can still be visualized in the nuclear DNA of individual cells (Nehls et al., 1984). Similar sensitivity is observed with monoclonal antibodies against O^6-methyldeoxyguanine (Wild et al., 1983).

Using monoclonal antibodies, it will be possible to quantify specific DNA adducts (resulting, e.g., from exposure to carcinogens or chemotherapeutic agents) in small human tissue samples. The use of anti-O^6-methyldeoxyguanosine monoclonal antibody permits detection of this DNA adduct in human oesophageal and stomach mucosa at a level of 25 fmol/mg DNA (Umbenhauer et al., 1985). Furthermore, it will be possible to measure adducts in different types of cells at different stages of differentiation and development. A further important problem that may be approached using immunostaining techniques is the determination, within defined cell populations, of subgroups of cells that are deficient in the repair of specific DNA lesions.

2.4 ^{32}P post-labelling detection of adducts (see review by Randerath et al., 1984)

The method of ^{32}P post-labelling developed by Randerath and coworkers (1981) involves the introduction of a ^{32}P label into DNA constituents after exposure of the DNA to nonradioactive, putative chemical carcinogens. DNA isolated from tissues exposed to chemicals are hydrolysed to deoxynucleoside 3′-monophosphates by incubation with micrococcal endonuclease and spleen phosphodiesterase. The digest is then treated with ^{32}P-ATP and T4 polynucleotide kinase to convert the monophosphates to 5′-^{32}P-labelled deoxynucleoside 3′,5′-diphosphates. This mixture of normal and chemically modified ^{32}P-labelled deoxynucleoside 3′,5′-diphosphates is separated by thin-layer chromatography. Another technique involves removal of the 3′-phosphate and subsequent analysis of the 5′-^{32}P-labelled deoxynucleoside monophosphates by reverse-phase chromatography (either thin-layer or high-performance liquid chromatography) (Reddy et al., 1984).

This technique has been used to detect the adducts formed in vivo in various tissues after exposure of mice or rats to a total of 28 compounds comprising seven arylamines and derivatives, three azo compounds, two nitroaromatics, 12 polycyclic aromatic hydrocarbons and four methylating agents. Of these 28 compounds, the 25 known carcinogens gave a positive result, while three compounds (anthracene, pyrene and perylene) that have not been shown to be carcinogenic in animals, gave negative results. Within the group of polycyclic aromatic hydrocarbons, a good correlation was observed between the previously described carcinogenic potency of individual compounds for mouse skin (Dipple, 1976; Phillips et al., 1979) and their binding level to mouse skin DNA (Reddy et al., 1984). This method is applicable directly to the detection and quantification of adducts in DNA from human sources and may thus contribute to the monitoring of human populations for exposure to genotoxic chemicals. It has already been used to detect adducts in buccal mucosa of betel-quid chewers (see Randerath et al., 1984) and in placental tissue of smokers and nonsmokers (Everson et al., 1986).

2.5 Physico-chemical methods for detection of adducts

Direct chemical techniques are available for the determination of a limited number of adducts. For example, photon-counting fluorescence spectroscopy has allowed the measurement of 7-methylbenzo[a]anthracene- and benzo[a]pyrene-DNA adducts in mouse skin without prior degradation of the DNA (Daudel et al., 1974, 1975a,b). Unfortunately, the poor resolution of conventional spectra does not allow the identification or resolution of adducts in samples from material obtained in situations with complex exposures. Recently, fluorescence-line-narrowed spectrometry employing a laser as the excitation source has been developed to distinguish samples with very similar conventional fluorescence spectra (Heisig et al., 1984). This technique allows analysis of DNA from humans exposed to mixtures of polycyclic aromatic hydrocarbons, but the sensitivity must be increased. Synchronous fluorescence spectrometry has recently been applied to human samples (Vahakangas et al., 1985). In this method, excitation and emission wavelengths are scanned simultaneously with a fixed wavelength difference, resulting in much simpler spectra. The sensitivity has been reported as one adduct per 10^7 bases. A possible problem is interference from fluorescent contaminants in the sample.

Adduct levels can be measured not only in surrogate or target cells and tissue, but also in urine. Such adducts may result from excision repair or spontaneous depurination of unstable adducts. Bennett et al. (1981) demonstrated that rats treated with aflatoxin B_1 excreted in their urine 48 hours later about 35% of the amount of adduct seen initially in the liver. Several studies of humans have also been reported: six of 45 urine samples from a district in Kenya with high aflatoxin exposure were shown to contain levels of aflatoxin adducts detectable by high-performance liquid chromatography and fluorescence spectroscopy (Autrup et al., 1983).

2.6 Enzyme-sensitive sites

Another approach to detecting adducts is measurement of the presence of sites sensitive to enzymes that incise DNA in regions of damage (Paterson, 1978). Damaged DNA is incubated with a purified endonuclease, and the number of sensitive sites is measured by the reduction in the size of DNA, usually revealed by alkaline sucrose gradient centrifugation. DNA damage caused by several types of carcinogen has been detected in this manner (Duker & Teebor, 1976; Heflich et al., 1977; Paterson, 1978). The major limitations of this technique at present are the requirement for the appropriate pure nuclease and the scarcity of information on the susceptibility of different types of chemical damage to incision by nucleases *in vitro*.

2.7 Alkaline sucrose gradients and alkaline elution

Another technique for detecting DNA damage is measurement of reduction in size of single-strand DNA. DNA breaks can occur as a result of the action of free radicals (at apurinic/apyrimidinic sites) and of alkyl phosphotriesters under alkaline conditions or as a result of incision by repair nucleases. The size of single-strand DNA can be determined by alkaline sucrose gradient centrifugation (McGrath & Williams, 1966; Lett *et al.*, 1967) or by alkaline elution from membrane filters (Kohn, 1979).

The alkaline elution technique, as developed by Kohn (1979), measures the elution rate of DNA through the pores of a membrane filter. The elution rate is a function of the molecular weight of the eluted DNA and is thus faster for lower molecular weight DNA. In the first part of the procedure, cells are deposited on a membrane filter and lysed by means of an anionic detergent at pH 10. Most of the cellular macromolecules wash through with the lysing solution. The native nuclear DNA is retained quantitatively on the filter. DNA with many double-strand breaks from dead cells will wash through at this point; DNA with many single-strand breaks from dead cells will wash through at the start of the subsequent elution.

In the second part of the procedure, eluting solution at pH 12 is pumped slowly through the filter, and fractions are collected to determine the time-course of DNA release. DNA is measured either by radioactivity or by fluorimetric assay. Alkaline elution is a rapid and highly sensitive technique for measuring single-strand breaks and has been applied to chemical screening with cultured cells (Swenberg *et al.*, 1976) and with animals (Parodi *et al.*, 1978; Petzold & Swenberg, 1978). Its application *in vivo* has been used to detect carcinogens (cycasin and 1,2-dimethylhydrazine) that had thus far escaped detection in assays conducted *in vitro*.

The advantages of this approach are that it yields equivalent results to alkaline sucrose gradients but is faster, simpler, cheaper and more sensitive. The in-vivo assay with fluorimetric measurement of DNA can be applied to a large spectrum of tissues from a variety of animal species treated *in vivo*, thus reflecting a diversity of pharmacokinetic and metabolic factors. In the conduct of this assay, proper controls are essential in order to distinguish true breaks from mechanical fragmentation of DNA or nucleolytic degradation. The latter must be stringently excluded by histological evaluation of tissues. One drawback of the approach *in vivo* (which requires one animal for each estimation) is that it requires knowledge of the toxicity (LD_{50}) of the test chemical in order to determine the test dose. Another present limitation is that it has not yet been established for which tissues, apart from liver, this technique can be recommended for screening. Since no generally agreed protocol for a test is currently available, the assay has not yet been extensively validated. Therefore, caution is urged in the interpretation of alkaline elution data in terms of assessing potential carcinogenicity, since this technique also reveals low levels of DNA breaks, due presumably to noncarcinogenic agents (Kohn, 1979).

3. Detection of DNA repair

As mentioned in the introduction, most DNA lesions are detected by means of repair mechanisms such as excision repair or recombinational repair. As far as excision repair is concerned, the different steps in this process (incision, excision, resynthesis, ligation) can be monitored at the biochemical level.

3.1 *Biophysical techniques*

Of the different repair assays that have been proposed (Table 1), none was considered to be useful for screening purposes in a report of the US Environmental Protection Agency Gene-Tox programme (Larsen *et al.*, 1982). The conclusion was that these tests, although valid for studying fundamental repair phenomena in eukaryotic cells, are time-consuming, expensive and require highly specialized skills. Nevertheless, with the development of techniques with increased sensitivity for the detection of DNA adducts, it is also possible to monitor repair by measuring the loss of adducts.

Table 1. Methods for studying excision DNA repair in cultured cells[a]

Steps involved	Technique
Incision in region of DNA damage	Alkaline sucrose gradients Alkaline elution
Excision of damaged region	Loss of damaged bases Loss of enzyme-sensitive sites
Resynthesis of excised region	^{3}H-Thymidine incorporation, autoradiography, liquid scintillation counting Isopycnic gradients Bromouracil photolysis Benzoylated naphthoylated DEAE cellulose chromatography
Rejoining of strand	Alkaline sucrose gradients Alkaline elution

[a]From Williams (1979)

Current knowledge about enzymatic removal of adducts from the DNA of a variety of organisms treated with a range of chemicals and the relevance of this topic to carcinogenesis has been reviewed by Lawley (1984) and by Osborne (1984).

3.2 *DNA repair synthesis*

Originally, DNA repair synthesis in mammalian cells was detected, by autoradiography after incorporation of ^{3}H-thymidine, as UDS, i.e., DNA synthesis outside the S-phase of the cell cycle (Rasmussen & Painter, 1966; Painter & Cleaver, 1969). This procedure is still the most frequently used technique for measuring repair synthesis. Autoradiographic measurement of repair is based on the ready distinction between light nuclear labelling due to ^{3}H-thymidine incorporation during repair and the heavy nuclear labelling indicative of replicative DNA synthesis. In this way, thymidine incorporation into S-phase cells can be excluded from the results.

UDS can also be measured by liquid scintillation counting of radioactive thymidine incorporation into the DNA of cells in which replicative DNA synthesis has been suppressed by hydroxyurea (Evans & Norman, 1968), arginine deficiency (Stich & San, 1970) or by a combination of both (Trosko & Yager, 1974). The hydroxyurea block has been used extensively (Cleaver, 1969; Lieberman et al., 1971; Smith & Hanawalt, 1976; Martin et al., 1978), but since replicative DNA synthesis is not arrested completely by hydroxyurea (Roberts et al., 1968; Smith & Hanawalt, 1976), there is always uncertainty about the significance of increases above the persisting background. In addition, the sensitivity of this approach may be limited for two reasons: (i) hydroxyurea may affect repair synthesis (Williams, 1977); and (ii) DNA damage inhibits residual replicative synthesis, and repair will therefore not become evident until the repair synthesis results in enough thymidine incorporation to equal the inhibited replicative synthesis and thereby exceed the control level. Another complication of this method is that suppression of replicative DNA synthesis by hydroxyurea can be interfered with by other chemicals (Brandt et al., 1972), resulting in restoration of replicative synthesis that can be confused with repair. Hydroxyurea has also been used to inhibit DNA synthesis for autoradiographic measurement of UDS (Lieberman et al., 1971), but this application suffers from the same uncertainty described above.

Several definitive techniques for assessing repair synthesis, which are not complicated by replicative synthesis, are available for use in rapidly dividing cultures. Incorporation of the thymidine analogue, 5-bromodeoxyuridine (BUdR), into regions of DNA repair synthesis does not significantly alter the density of parental DNA; this DNA can therefore be separated on density gradients from newly-replicated DNA, which is of high density due to extensive incorporation of BUdR. Additional incorporation of radioactively labelled thymidine or BUdR into the isolated parental DNA provides a measure of repair synthesis (Pettijohn & Hanawalt, 1964; Roberts et al., 1968).

Other techniques for estimating the extent of repair involve photolysis of BUdR incorporated during repair (Regan et al., 1971) and measurement of thymidine incorporation into DNA growing points, which are retained on benzoylated naphthoylated DEAE-cellulose columns because of their single-stranded regions (Scudiero et al., 1975). Despite their attractive features, these definitive techniques have not found wide use in screening because of their demanding technical requirements. Nevertheless, they could be used on a selective basis to confirm or extend results obtained with simpler methods. Furthermore, techniques that permit measurement of incorporation of precursors other than thymidine may be required to detect compounds that, under some conditions, produce specific types of repair (Hennings & Michael, 1976) or which might block thymidine utilization during repair synthesis.

3.3 *Measurement of ligation*

The final step in excision repair is the rejoining of the DNA strand by a ligase. Ligation cannot be measured directly, but the elongation-rejoining process can be followed by observing the restoration of fragmented DNA to greater length using gradient centrifugation techniques or alkaline elution. Provided that DNA prelabelled with a radioactive isotope is identified as fragmented and then restored to its normal sedimentation or elution profile, these techniques measure DNA repair reliably. The only limitation to their application is the requirement for prelabelled DNA, which, for cell culture, necessitates the use of replicating cell lines.

3.4 *Tests that employ unscheduled DNA synthesis (UDS)*

Following the excision of a damaged segment of DNA, the single-strand gap is filled, using the opposite strand as template. According to a report of the US Environmental Protection Agency Gene-Tox programme (Mitchell *et al.*, 1983), only three approaches are recommended: (i) autoradiographic measurement using human diploid fibroblasts; (ii) autoradiographic measurement using primary rat hepatocytes; and (iii) measurement using liquid scintillation counting of DNA extracted from human diploid fibroblasts.

(a) Diploid human fibroblasts

Both early-passage cultures of diploid human cells and human diploid cell lines have been used. WI-38 cells are the only line that has been used to a significant extent. These cultures have little, if any, enzymatic capacity for metabolic activation of chemicals and require the addition of exogenous metabolic activation systems (see Report 15).

(b) Rat hepatocytes

Cultures of freshly isolated hepatocytes offer two major advantages: (i) they are essentially non-dividing, and (ii) they are capable of metabolically activating a range of carcinogens (Williams, 1976, 1977, 1978, 1979; Casciano *et al.*, 1978; Michalopoulos *et al.*, 1978; Yager & Miller, 1978). Originally, only rat hepatocytes were validated (Williams *et al.*, 1982); but now hepatocytes from a variety of species (McQueen & Williams, 1983), including humans (Michalopoulos *et al.*, 1978), are used. The choice is subject to the same considerations as apply to metabolic activation in all in-vitro systems.

UDS tests measure a specific primary response to DNA damage, they appear to respond to most classes of chemicals (IARC, 1980; Mitchell *et al.*, 1983; Williams, 1985), and they are relatively rapid and economical. Induction of UDS has also been examined in an in-vivo/in-vitro system, in which UDS is determined in tissue culture in the presence of ^{3}H-deoxythymidine after administration of the test chemicals *in vivo*; a variety of carcinogens with carcinogenic organotropisms has been tested, with promising results (Furihata & Matsushima, 1986).

4. References

Adamkiewicz, J., Drosdziok, W., Eberhardt, W., Langenberg, U. & Rajewsky, M.F. (1982) *High-affinity monoclonal antibodies specific for DNA components structurally modified by alkylating agents.* In: Bridges, B.A., Butterworth, B.E. & Weinstein, I.B., eds, *Indicators of Genotoxic Exposure (Banbury Report 13)*, Cold Spring Harbor, NY, Cold Spring Harbor Laboratory, pp. 265-276

Adamkiewicz, J., Eberle, G., Huh, N., Nehls, P. & Rajewsky, M.F. (1985) Quantitation and visualization of alkyl deoxynucleosides in the DNA of mammalian cells by monoclonal antibodies. *Environ. Health Perspect.*, *62*, 49-55

Autrup, H., Bradley, K.A., Shamsuddin, A.K.M., Wakhisi, J. & Wasunna, A. (1983) Detection of putative adduct with fluorescence characteristic identical to 2,3-dihydro-2-(7^1-guanyl)-3-hydroxyaflatoxin B$_1$ in human urine collected in Murang'a District, Kenya. *Carcinogenesis*, *4*, 1193-1195

Bennett, R.A., Essigmann, J.M. & Wogan, G.N. (1981) Excretion of an aflatoxin-guanine adduct in the urine of aflatoxin B$_1$-treated rats. *Cancer Res.*, *41*, 650-654

Bichara, M. & Fuchs, R.P.P. (1985) DNA binding and mutation spectra of the carcinogen N-2-aminofluorene in *Escherichia coli*. A correlation between the conformation of the premutagenic lesion and the mutation specificity. *J. mol. Biol.*, *183*, 341-351

Brandt, W.N., Flamm, W.G. & Bernheim, N.J. (1972) The value of hydroxyurea in assessing repair synthesis of DNA in HeLa cells. *Chem.-biol. Interactions*, *5*, 327-339

Brookes, P. & Lawley, P.D. (1964) Evidence for the binding of polynuclear aromatic hydrocarbons to the nucleic acids of mouse skin: relation between carcinogenic power of hydrocarbons and their binding to DNA. *Nature*, *202*, 781-784

Casciano, D.A., Dan, J.A., Oldham, J.W. & Carie, M.D. (1978) 2-Acetylaminofluorene-induced unscheduled DNA synthesis in hepatocytes isolated from 3-methylcholanthrene treated rats. *Cancer Lett.*, *5*, 173-178

Cleaver, J.E. (1969) Repair replication of mammalian cell DNA: effects of compounds that inhibit DNA synthesis or dark repair. *Radiat. Res.*, *37*, 334-348

Daudel, P., Croisy-Delcey, M., Alonso-Verduras, C., Duquesne, M., Jaquignon, P., Markovits, P. & Vigny, P. (1974) Study with fluorescence of nucleic acids extracted from cultured cells treated with methyl-7-benzo[*a*]anthracene [in French]. *C.R. Acad. Sci. Ser. D*, *278*, 2249-2252

Daudel, P., Duquesne, M., Vigny, P., Grover, P.L. & Sims, P. (1975a) Fluorescence spectral evidence that benzo[*a*]pyrene-DNA products in mouse skin arise from diol-epoxides. *FEBS Lett.*, *57*, 250-253

Daudel, P., Kawamura, H., Croisy-Delcey, M. & Duquesne, M. (1975b) Fluorescence study of DNA treated with methyl-7-benzo[a]anthracene in the presence of a microsomal oxidation system [in French]. *C.R. Acad. Sci. Ser. D, 280*, 521-524

De Murcia, G., Lang, M.C.E., Freund, A.M., Fuchs, R.P.P., Daune, M.P., Sage, E. & Leng, M. (1979) Electron-microscopic visualization of N-acetoxy-N-2-acetylaminofluorene binding sites in Col E-1 DNA by means of specific antibodies. *Proc. natl Acad. Sci. USA, 76*, 6076-6080

Dipple, A. (1976) *Polynuclear aromatic carcinogens*. In: Searle, C.E., ed., *Chemical Carcinogens (ACS Monograph 182)*, Washington DC, American Chemical Society, pp. 245-314

Duker, N.J. & Teebor, G.W. (1976) Detection of different types of damage in alkylated DNA by means of human corrective endonuclease (correndonuclease). *Proc. natl Acad. Sci. USA, 73*, 2629-2633

Evans, R.G. & Norman, A. (1968) Radiation stimulated incorporation of thymidine into the DNA of human lymphocytes. *Nature, 217*, 455-456

Everson, R.B., Randerath, E., Santella, R.M., Cefalo, R.C., Avitts, T.A. & Randerath, K. (1986) Detection of smoking-related covalent DNA adducts in human placenta. *Science, 231*, 54-57

Fichtinger-Schepman, A.M.J., Baan, R.A., Luiten-Schuite, A., Van Dijk, M. & Lohman, P.H.M. (1985) Immunochemical quantitation of adducts induced in DNA by cis-diamminedichloroplatinum(II) and analysis of adduct-related DNA-unwinding. *Chem.-biol. Interactions, 55*, 275-288

Furihata, C. & Matsushima, T. (1986) Use of *in vivo/in vitro* unscheduled DNA synthesis for identification of organ-specific carcinogens. *CRC crit. Rev. Toxicol.* (in press)

Heflich, R.H., Dorney, D.J., Maher, V.M. & McCormick, J.J. (1977) Reactive derivatives of benzo[a]pyrene and 7,12-dimethylbenz[a])anthracene cause S_1 nuclease sensitive sites in DNA and 'UV-like' repair. *Biochem. biophys. Res. Commun., 77*, 634-641

Heisig, V., Jeffrey, A.M., McGlade, M.J. & Small, G.J. (1984) Fluorescence-line-narrowed spectra of polycyclic aromatic carcinogen-DNA adducts. *Science, 223*, 289-291

Hennings, H. & Michael, D. (1976) Guanine-specific DNA repair after treatment of mouse skin cells with N-methyl-N'-nitro-N-nitrosoguanidine. *Cancer Res., 36*, 2321-2325

IARC (1980) *IARC Monographs on the Evaluation of the Carcinogenic Risk of Chemicals to Humans*, Suppl. 2, *Long-term and Short-term Assays for Carcinogens: A Critical Appraisal*, Lyon, pp. 201-226

Kohn, K.W. (1979) DNA as a target in cancer chemotherapy: measurement of macromolecular DNA damage produced in mammalian cells by anticancer agents and carcinogens. *Meth. Cancer Res., 16*, 291-345

Kriek, E., Welling, M. & van der Laken, C.J. (1984) *Quantitation of carcinogen-DNA adducts by a standardized high-sensitive enzyme immunoassay*. In: Berlin, A., Draper, M., Hemminki, K. & Vainio, H. eds, *Monitoring Human Exposure to Carcinogenic and Mutagenic Agents (IARC Scientific Publications No. 59)*, Lyon, International Agency for Research on Cancer, pp. 297-305

van der Laken, C.J., Hagennaars, A.M., Hermsen, G., Kriek, E., Kuipers, A.J., Nagel, J., Scherer, E. & Welling, M. (1982) Measurement of O^6-ethyldeoxyguanosine and N-(deoxyguanosine-8-yl)-N-acetyl-2-aminofluorene in DNA by high-sensitive enzyme immunoassays. *Carcinogenesis, 3*, 569-572

Larsen, K.H., Brash, D., Cleaver, J.E., Hart, R.W., Maher, V.M., Painter, R.B. & Sega, G.A. (1982) DNA repair assays as tests for environmental mutagens — a report of the US EPA Gene-Tox Program. *Mutat. Res., 78*, 287-318

Lawley, P.D. (1984) *Carcinogenesis by alkylating agents*. In: Searle, C.E., ed., *Chemical Carcinogens (ACS Monographs 182)*, Washington DC, American Chemical Society, pp. 325-484

Leadon, S.A. & Hanawalt, P.C. (1983) Monoclonal antibody to DNA containing thymine glycol. *Mutat. Res., 112*, 191-200

Leng, M., Sage, E., Fuchs, R.P.P. & Daune, M.P. (1978) Antibodies to DNA modified by the carcinogen *N*-acetoxy-*N*-acetyl-2-aminofluorene. *FEBS Lett. 92*, 207-210

Lett, J.T., Caldwell, I.R., Dean, C.J. & Alexander, P. (1967) Rejoining of X-ray induced breaks in the DNA of leukaemia cells. *Nature, 214*, 790-792

Lieberman, M.W., Baney, R.N., Lee, R.E., Sell, S. & Farber E. (1971) Studies on DNA repair in human lymphocytes treated with proximate carcinogens and alkylating agents. *Cancer Res., 31*, 1297-1306

Lutz, W.K. (1979) In-vivo covalent binding of organic chemicals to DNA as a quantitative indicator in the process of chemical carcinogenesis. *Mutat. Res., 65*. 289-356

Martin, C.N., McDermid, A.C. & Garner, R.C. (1978) Testing of known carcinogens and noncarcinogens for their ability to induce unscheduled DNA synthesis in HeLa cells. *Cancer Res., 38*, 2621-2627

McGrath, R.A. & Williams, R.W. (1966) Reconstruction *in vivo* of irradiated *Escherichia coli* deoxyribonucleic acid; the rejoining of broken pieces. *Nature, 212*, 534-535

McQueen, C.A. & Williams, G.M. (1983) The use of cells from rat, mouse, hamster and rabbit in the hepatocyte primary culture/DNA-repair test. *Ann. N.Y. Acad. Sci., 407*, 119-130

Michalopoulos, G., Sattler, G.L., O'Connor, L. & Pitot, H.C. (1978) Unscheduled DNA synthesis induced by procarcinogens in suspensions and primary cultures of hepatocytes on collagen membranes. *Cancer Res., 38*, 1866-1871

Mitchell, A.D., Casciano, D.A., Meltz, M.L., Robinson, D.E., Sen, R.H.C., Williams, G.M. & von Halle, E.S. (1983) Unscheduled DNA synthesis tests — A report of the US Environmental Protection Agency Gene-Tox Program. *Mutat. Res., 123*, 363-410

Müller, R., & Rajewsky, M.F. (1978) Sensitive radioimmunoassay for detection of O^6-ethyldeoxyguanosine in DNA exposed to the carcinogen ethylnitrosourea *in vivo* or *in vitro*. *Z. Naturforsch., 33c*, 897-901

Müller, R. & Rajewsky, M.F. (1980) Immunological quantification by high-affinity antibodies of O^6-ethyldeoxyguanosine in DNA exposed to *N*-ethyl-*N*-nitrosourea. *Cancer Res., 40*, 887-896

Nehls, P., Adamkiewicz, J. & Rajewsky, M.F. (1984) Immuno-slot-blot: a highly sensitive immunoassay for the quantitation of carcinogen-modified nucleosides in DNA. *J. Cancer Res. clin. Oncol., 108*, 23-29

Osborne, M.R. (1984) *DNA interactions of reactive intermediates derived from carcinogens*. In: Searle C.E., ed., *Chemical Carcinogens (ACS Monographs 182)*, Washington DC, American Chemical Society, pp. 485-524

Painter, R.B. & Cleaver, J.E. (1969) Repair replication, unscheduled DNA synthesis and the repair of mammalian DNA. *Radiat. Res., 37*, 4151-4166

Parodi, S., Taningher, M., Santi, L., Cavanna, M., Sciaba, L., Maura, A. & Brambilla, G. (1978) A practical procedure for testing DNA damage *in vivo*, proposed for a pre-screening of chemical carcinogens. *Mutat. Res., 54*, 39-46

Paterson, M.C. (1978) Use of purified lesion-recognizing enzymes to monitor DNA repair *in vivo*. *Adv. Radiat. Biol., 7*, 1-53

Perera, F.P., Poirier, M.C., Yuspa, S.H., Nakayama, J., Jaretzki, A., Curnen, M.M., Knowles, D.M. & Weinstein, I.B. (1982) A pilot project in molecular cancer epidemiology: determination of benzo[a]pyrene-DNA adducts in animal and human tissues by immunoassays. *Carcinogenesis, 3*, 1405-1410

Pettijohn, D. & Hanawalt, P. (1964) Evidence for repair-replication of ultraviolet damaged DNA in bacteria. *J. mol. Biol., 9*, 395-410

Petzold, G.L. & Swenberg, J.A. (1978) Detection of DNA damage induced *in vivo* following exposure of rats to carcinogens. *Cancer Res., 38*, 1589-1598

Phillips, D.H., Grover, P.L. & Sims, P. (1979) A quantitative determination of the covalent binding of a series of polycyclic hydrocarbons to DNA in mouse skin. *Int. J. Cancer, 23*, 201-208

Poirier, M.C., Yuspa, S.H. Weinstein, I.B. & Blobstein, S. (1977) Detection of carcinogen-DNA adducts by radioimmunoassay. *Nature, 270*, 185-188

Poirier, M.C., Reed, E., Zwelling, L.A., Ozols, R.F., Litterst C.L. & Yuspa, S.H. (1985) Polyclonal antibodies to quantitate cis-diamminedichloro platinum(II)-DNA adducts in cancer patients and animal models. *Environ. Health Perspect., 62*, 89-94

Randerath, K., Reddy, M.V. & Gupta, R.C. (1981) ^{32}P-Labeling test for DNA damage. *Proc. natl Acad. Sci. USA, 78*, 6126-6129

Randerath, K., Randerath, E., Agrawal, H.P. & Reddy, M.V. (1984) *Biochemical (postlabelling) methods for carcinogen-DNA adduct analysis*. In: Berlin, A., Draper, M., Hemminki, K. & Vainio, H., eds, *Monitoring Human Exposure to Carcinogenic and Mutagenic Agents (IARC Scientific Publications No. 59)*, Lyon, International Agency for Research on Cancer, pp. 217-231

Rasmussen, R.E. & Painter, R.B. (1966) Radiation-stimulated DNA synthesis in cultured mammalian cells. *J. Cell Biol., 29*, 11-19

Reddy, M.V., Gupta, R.C., Randerath, E. & Randerath, K. (1984) ^{32}P-Postlabeling test for covalent DNA binding of chemicals *in vivo*: application to a variety of aromatic carcinogens and methylating agents. *Carcinogenesis, 5*, 231-243

Regan, J.D., Setlow, R.B. & Ley, R.D. (1971) Normal and defective repair of damaged DNA in human cells: a sensitive assay utilizing the photolysis of bromodeoxyuridine. *Proc. natl Acad. Sci. USA, 68*, 708-712

Roberts, J.J., Crathorn, A.R. & Brent, T.P. (1968) Repair of alkylated DNA in mammalian cells. *Nature, 218*, 970-972

Scudiero, D., Henderson, E., Norin, A. & Strauss, B. (1975) The measurement of chemically-induced DNA repair synthesis in human cells by BND-cellulose chromatography. *Mutat. Res., 29*, 473-488

Searle, C.E., ed. (1984) *Chemical Carcinogens (ACS Monographs 182)*, Washington DC, American Chemical Society

Shamsuddin, A.K.M., Sinopoli, N.T., Hemminki, K., Boesch, R.R. & Harris, C.C. (1985) Detection of benzo[a]pyrene: DNA adducts in human white blood cells. *Cancer Res., 45*, 66-68

Smith, C.A. & Hanawalt, P.C. (1976) Repair replication in human cells: simplified determination utilizing hydroxyurea. *Biochim. biophys. Acta, 432*, 336-347

Stich, H.F. & San, R.H.C. (1970) DNA repair and chromatid anomalies in mammalian cells exposed to 4-nitroquinoline 1-oxide. *Mutat. Res., 10*, 389-404

Strickland, P.T. & Boyle, J.M. (1981) Characterization of two monoclonal antibodies specific for dimerised and non-dimerised adjacent thymidines in single stranded DNA. *Photochem. Photobiol., 34*, 595-601

Swenberg, J.A., Petzold, G.L. & Harback, P.R. (1976) In-vitro DNA damage/alkaline elution assay for predicting carcinogenic potential. *Biochem. biophys. Res. Commun.*, 7, 732-738

Trosko, J.D. & Yager, J.D. (1974) A sensitive method to measure physical and chemical carcinogen-induced 'unscheduled DNA synthesis' in rapidly dividing eukaryotic cells. *Exp. Cell Res.*, 88, 47-55

Umbenhauer, D., Wild, C.P., Montesano, R., Saffhill, R., Boyle, J.M., Huh, N., Kirstein, U., Thomale, J., Rajewsky, M.F. & Lu, S.H. (1985) O^6-Methyldeoxyguanosine in oesophageal DNA among individuals at high risk of oesophageal cancer. *Int. J. Cancer*, 36, 661-665

Vahakangas, K., Trivers, G., Rowe, M. & Harris, C.C. (1985) Benzo[*a*]pyrene diol epoxide-DNA adducts detected by synchronous fluorescence spectrophotometry. *Environ. Health Perspect.*, 62, 101-104

Warren, W. (1984) *The analysis of alkylated DNA by high pressure liquid chromatography*. In: Venitt, S. & Parry J.M., eds, *Mutagenicity Testing — A Practical Approach*, Washington DC, IRL Press, pp. 24-44

Wild, C.P., Smart, G., Saffhill, R. & Boyle, J.M. (1983) Radioimmunoassay of O^6-methyldeoxyguanosine in DNA of cells alkylated *in vitro* and *in vivo*. *Carcinogenesis*, 4, 1605-1609

Williams, G.M. (1976) Carcinogen-induced DNA repair in primary rat liver cell cultures; a possible screen for chemical carcinogens. *Cancer Lett.*, 1, 231-236

Williams, G.M. (1977) The detection of chemical carcinogens by unscheduled DNA synthesis in rat liver primary cell cultures. *Cancer Res.*, 37, 1845-1851

Williams, G.M. (1978) Further improvement in the hepatocyte DNA repair test for carcinogens; detection of carcinogen biphenyl derivatives. *Cancer Lett.*, 4, 69-75

Williams, G.M. (1979) The status of in-vitro test systems utilizing DNA damage and repair for the screening of chemical carcinogens. *J. Assoc. off. anal. Chem.*, 62, 857-863

Williams, G.M. (1985) Identification of genotoxic and epigenetic carcinogens in liver culture systems. *Regul. Toxicol. Pharmacol.*, 5, 132-144

Williams, G.M., Laspia, M.F. & Dunkel, V.C. (1982) Reliability of the hepatocyte primary culture/DNA repair test in testing of coded carcinogens and non-carcinogens. *Mutat. Res.*, 97, 359-370

Yager, J.D. & Miller, J.A. (1978) DNA repair in primary cultures of rat hepatocytes. *Cancer Res.*, 38, 4385-4395

REPORT 5

SHORT-TERM ASSAYS USING BACTERIA

Prepared by:
S. Venitt (Rapporteur), H. Bartsch, G. Becking,
R.P.P. Fuchs, M. Hofnung, C. Malaveille, T. Matsushima,
M.R. Rajewsky, M. Roberfroid, and H.S. Rosenkranz (Chairman)

1. Introduction

The biochemistry and genetics of bacteria have been the subject of intensive study for many years. Despite their apparent simplicity compared with eukaryotic organisms, bacteria possess elaborate mechanisms for responding to DNA damage. Bacteria are widely used, therefore, both in fundamental studies of the mechanisms involved in the biological response to DNA damage (e.g., mutagenicity, DNA repair) and in short-term screening tests for potential carcinogens. Such studies have contributed to the widely accepted notion that DNA damage is involved in the process of carcinogenesis. However, certain cancers are induced by agents that do not appear to cause DNA damage, and such agents will not be detected by the types of short-term tests discussed in this report.

DNA damage and its repair can be studied by chemical, physical and immunochemical means, as discussed in Report 4. This report deals exclusively with monitoring bacterial responses to DNA damage. These responses include well-known cellular endpoints such as the induction of mutations, increased killing of DNA repair-deficient cells and the production of bacteriophage by lysogenic bacteria. More recent developments involve the use of molecular responses such as induction of specific genes, e.g., SOS genes.

All bacterial short-term tests consist of two components: (i) the target cell and (ii) the metabolizing system. The latter is used to convert certain compounds to DNA-reactive species — primarily electrophiles. The metabolizing system is supplied exogenously, although the endogenous capabilities of the target cell cannot be ignored and must be taken into consideration when interpreting results. The present report is concerned only with the target cell. Exogenously supplied activation systems are discussed in Report 15. Detailed guidelines on the conduct of short-term tests using bacteria have been published (Organisation for Economic Co-operation and Development, 1983; Venitt *et al.*, 1983; Tweats *et al.* 1984).

2. Mechanisms of bacterial responses to DNA damaging agents

Our knowledge of the response of bacteria to DNA-damaging agents has progressed considerably since the last report (IARC, 1980). In particular, the chemical nature of DNA lesions, enzymatic processing of these lesions, regulation of the synthesis of these enzymes and the mechanisms of mutagenesis are much better understood (Walker, 1984), allowing a deeper understanding of the scientific basis of bacterial tests, better interpretation of the responses of established tests and aids in the development of new ones.

A schematic view of the action of a typical DNA damaging agent (which causes a variety of DNA lesions) is shown in Figure 1. Depending both on its nature and its position — in particular with respect to the DNA replication fork — each lesion can undergo at least four fates, which are not mutually exclusive nor exclusive of other possibilities: (1) The lesion may be processed without mutagenic effects. This may occur by reversion of the lesion (due, for example to the enzyme photolyase) or by excision repair of the lesion (due, for example, to *uvr* endonuclease and other enzymes). (2) The lesion may provoke direct miscoding. (3) The lesion may induce the SOS response, which includes SOS mutagenesis. (4) The lesion may induce other responses, such as the adaptative response to alkylating agents, the response to oxidative agents and the heat-shock response.

Mutagenesis occurs mainly by passive mutagenesis (2) and active or SOS mutagenesis (3). Phage induction is predominantly a consequence of SOS induction (3), although it may also result from mutagenesis in the Cl gene [(2) and (3)]. Differential survival is due mainly to (1), (3) and (4). Thus, these physiological endpoints have several components and do not simply reflect the initial lesions.

Bacteria such as *Escherichia coli* and *Salmonella typhimurium* have complex genetic networks that respond to stress, which may include attack on the genome. Four of these networks have been studied intensively: (i) the adaptative system (*ada*), which responds to alkylating agents (Demple *et al.*, 1985); (ii) the SOS system, which responds to most DNA lesions (Walker, 1984); (iii) the system for oxidative stress (Christman *et al.*, 1985); and (iv) the heat-shock response (Grossman *et al.*, 1984). These networks share certain genes and are thus not totally independent of one another. Induction of genes belonging to these networks can be monitored as early signals of genetic damage. In comparison, phage induction, assays for survival and mutagenesis are indirect, late, global manifestations of cellular responses to DNA damage.

Assays based on the monitoring of SOS responses have been developed since the publication of the previous report (IARC, 1980); the regulation of the SOS system is therefore described briefly. There has been considerable progress in unravelling the molecular mechanisms of SOS responses (Walker, 1984). Thanks to in-vivo gene fusion and cloning techniques, knowledge of the genes involved and of their regulation is now

Fig. 1. Pathways allowing replication to progress through a DNA lesion

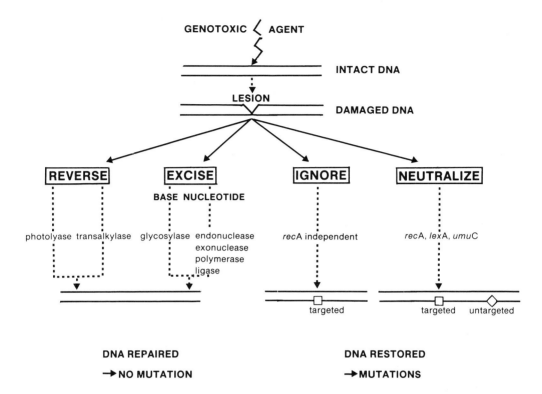

Reverse: The lesion is eliminated without base or nucleotide excision.
Excision: The lesion is eliminated through base or nucleotide excision.
Ignore: The lesion allows DNA replication to proceed.
Neutralize: Replication occurs after SOS induction.

Only the last two pathways are mutagenic. Tester strains detect mainly SOS mutagenesis, i.e., the fourth pathway; however, some compounds, such as 2-aminopurine, are mostly mutagenic through direct miscoding, i.e., the third pathway.

very detailed. Two genes play a key role: *lexA* encodes a repressor for all the genes of the system (over 15 are known); *recA* encodes a protein able to cleave (or to promote cleavage of) the *lexA* repressor upon activation by an SOS inducing signal. The *recA* protein also promotes the cleavage of the λCl repressor, the regulatory protein which maintains phage λ in its dormant state. The exact nature of the SOS inducing signal remains unknown; however, it is produced when DNA lesions perturb or stop DNA replication. Until recently, one of the easiest ways of detecting SOS induction was to assay for phage λ induction. There are now more direct and sensitive ways (see section 8.2).

3. Advantages and limitations of bacterial tests

Bacteria grow rapidly in simple, defined media; this is why bacterial short-term tests are among the simplest, quickest and most inexpensive to conduct. Their simplicity is such that commercially available self-contained kits have been devised for some tests.

Bacterial tests are quite flexible and can be used, for example, for:

(*a*) primary screening of chemicals for potential carcinogenicity;

(*b*) isolating biologically active compounds from complex mixtures (see Report 17);

(*c*) detecting proximate or ultimate metabolites from procarcinogens;

(*d*) monitoring human exposure to mutagens and carcinogens by testing body fluids or excreta (Report 14);

(*e*) studying the mechanism of the mutagenic effects of chemicals;

(*f*) host-mediated assays (Report 6); and

(*g*) analysis of the specific types of DNA damage caused by carcinogens and mutagens.

Although bacterial tests are designed to predict effects in eukaryotes, they have some inherent limitations due to the differences between bacteria and eukaryotic cells. For example, the DNA of eukaryotic cells, but not prokaryote cells, is packaged into chromosomes which are bounded by a nuclear membrane. Moreover, some responses to DNA damage may be different.

4. Criteria for selecting tests

Many short-term tests using bacteria have been developed. A comparative evaluation is given in Table 1, which lists those that have been used in several laboratories.

The following criteria should be considered in selecting a particular test:

(*a*) *reproducibility* within the same laboratory and between laboratories;

(*b*) *sensitivity*, on the basis of three parameters:

   (i) the minimum dose that will give a positive response

   (ii) the maximum dose that can be tested; and

   (iii) the possible range of variation of the minimal and maximal responses.

The maximal obtainable response may be limited by toxic effects on cells. Differences in sensitivity may be due to several factors, such as permeability and the repair capability of test organisms, experimental procedures (e.g., plate incorporation assay *versus* liquid-preincubation assay), metabolic activation systems, co-mutagens and cofactors in the activation system;

Table 1. A selective list of quantitative bacterial short-term assays the use of which has led to publication of data by more than one laboratory

Test name[a]	Year[b]	Indicator[c]	Organism[d]	Mechanism[e]	Validation[f]	Strains[h]	Simplicity[g]	Sensitivity[i]	Reference[j]
Ames	1975	his^+	S. typh.	rev	>>1000	>>100	++	+++	Maron & Ames (1983)
WP2	1976	$trp+$	E. coli	rev	>>100	>>10	++	+++	Venitt et al. (1984)
K12/343/113	1974	k	E. coli		>30	>3	+	+++	Mohn et al. (1984)
araR	1978	araR	S. typh.	dir	>40	>4	++	+++	Ruiz-Rubio et al. (1985)
Inductest	1976	phage	E. coli	CI cleav	>100	>5	++	+	Mamber et al. (1984)
SOS chromo-test	1982	β-gal, pho	E. coli	LexA cleav	>100	>5	+++	+++	Quillardet et al. (1985)
Rec assay		Growth inhibition	B. subtilis	DNA repair	>>100	>10	+++	++	Leifer et al. (1981)
PolA assay		Growth inhibition	E. coli	DNA repair	>>100	>10	+++	++	Leifer et al. (1981)

[a]Common name of the test
[b]Year of first publication
[c]Indicator or endpoint: his^+, his^+ colonies; trp^+, trp^+ colonies; araR, arabinose-resistant colonies; gal^+, gal^r colonies; phage, phage plaques; β-gal, β-galactosidase assay; pho, alkaline phosphatase assay
[d]Test organism: S. typh., Salmonella typhimurium; E. coli, Escherichia coli; B. subtilis, Bacillus subtilis
[e]Molecular mechanisms involved: rev, reversion assay; dir, direct mutagenesis assay; CI cleav, cleavage of the CI repressor for phage λ; LexA cleav, cleavage of the LexA repressor
[f]Rough lower estimate of the number of compounds examined; and rough lower estimate of the number of laboratories having published papers on the test: >, 'more than'; >>, 'much more than'
[g]Simplicity: estimation of the simplicity of the standard procedure, including the need for equipment: +, fairly simple; ++, simple; +++, very simple
[h]Strains: number of strains recommended for the standard test
[i]Sensitivity: ability to detect low amounts of genotoxic agents: +, fairly sensitive; ++, very sensitive; +++, extremely sensitive
[j]Reference: reference to a recent paper on the test
[k]Several indicators are available in this test, see text.

(*c*) *rapidity*: how long it takes to do a single experiment, and how many experiments (including controls and replicate testing) are needed to test a chemical completely;

(*d*) *simplicity*: the operations and equipment required for conducting routine standard procedures, the number of bacterial strains needed being an important factor;

(*e*) *complexity*: bacterial tests are relatively easy to conduct, but they required a certain level of technical expertise and in many cases cannot be reduced to simple routine procedures;

(*f*) *quantification*: the possibility of producing reproducible dose-response effects;

(*g*) *protocols*: well-defined standard protocols and appropriate internal controls;

(*h*) *cost and manpower*: cost considerations should not be made at the expense of quality; for example, omission of a single control (solvent or positive control) or lack of replicate testing can completely invalidate results;

(*i*) *existing data base*: the number of chemicals and chemical classes that have been tested, the number of laboratories using the test, and the number of years it has been used; and

(*j*) *knowledge of mechanism*: to assist in conduct and interpretation of a given test.

In addition, the choice of a test should be considered in relation to the purpose of testing.

5. Relationships between endpoints and carcinogenesis

The molecular mechanisms of the responses of bacteria to genotoxic agents are in some cases well understood, but knowledge is much more limited in the case of mammalian cells. Despite recent progress, our understanding of the molecular mechanisms of carcinogenesis is even more restricted. It is thus premature to use theoretical models for extrapolating from bacterial responses to carcinogenesis. In practice, the predictive value of bacterial tests, as for all short-term tests, is established empirically — by determining how closely results from bacterial tests accord with those gained from carcinogenicity tests using animals.

A mutation may be defined as a stable heritable change in a DNA nucleotide sequence, which may be detected as a phenotypic change. Heritable changes in nucleotide sequences can be due to base substitution (transition, transversions), frameshifts (deletions or additions of one or a few base pairs), large deletions, insertions and translocations. Few, if any, mutagens induce only one type of mutational change: rather, most mutagenic agents tend to exhibit a characteristic mutational spectrum which depends upon (i) the nature of the primary DNA alteration (e.g., modification of base, phosphate or sugar, strand break, incorporation of modified base), and (ii) the subsequent secondary effects of DNA repair and replication. The same mutagen may therefore produce different mutational spectra in organisms of different genetic background.

Mutations affecting the structure and function of cellular oncogenes (i.e., mutations that activate oncogenes) have been described, and this knowledge may well provide a mechanistic link between mutagenesis and carcinogenesis. There is, however, no evidence that any one specific type of mutation is involved in the generation of all cancers. On the contrary, oncogene activation may occur through a variety of genetic events. Therefore, there is *a priori* no reason to think that any one method of monitoring DNA damage is better than another.

Since mutagenic specificity has sometimes been thought to reflect carcinogenic potential, some aspects of this question are discussed below.

In a reversion assay, the target for mutagenesis — i.e., all the possible changes leading to reversion — is larger than the reversion event itself. The target is usually not, however, affected by all DNA-damaging agents (Fig. 1; Levin & Ames, 1985). In a direct mutagenesis assay, the aim is to detect inactivation of a gene, and it is generally believed that, since the target of forward mutagenesis is much larger, it should be affected by any DNA-damaging agent. Use of a complete battery of strains in a reversion assay should present a target large enough to detect a spectrum of mutations as wide as that detected by a forward mutagenesis assay.

Several systems for studying the exact nature of spontaneous or induced mutations and which involve extensive DNA sequencing have been developed (see, for example Koffel-Schwartz *et al.*, 1984; Wood & Hutchinson, 1984). Such systems are not meant for routine testing. They show that many mutagenic agents preferentially induce a specific type of mutagenic change (for example, ethylmethanesulphonate induces GC $\rightarrow$ AT transition; *N*-acetoxy-2-acetylaminofluorene induces -1 and -2 frameshifts); these changes are not the only ones to occur, however.

The proportions of different mutations may depend to a large degree on several modifying factors. For example, there are different pathways for the reversion of mutations, and the activity of these pathways is dependent on various factors, including the DNA sequence of the initial mutation and the genetic background of the strain. In addition, the reversion promoted by a mutagen will depend on the type of lesions produced and the genetic and physiological states of the cellular pathways for processing these lesions (Miller, 1983).

Thus, the notion of mutagenic specificity is relative. The most important concept is that these endpoints reveal DNA damage, and it is this damage that is very probably involved in carcinogenesis.

6. Validation of short-term tests

Short-term tests have been validated by comparing test results obtained with a variety of chemical classes of carcinogens and noncarcinogens. The sensitivity (the percentage of true positives), the specificity (the percentage of true negatives) and the accuracy (the percentage of correct matches) of short-term tests have been calculated in the following way (Cooper *et al.*, 1979):

$$\text{Sensitivity} = \frac{\text{No. of carcinogens found positive in test} \times 100}{\text{No. of carcinogens tested}}$$

$$\text{Specificity} = \frac{\text{No. of noncarcinogens found negative in test} \times 100}{\text{No. of noncarcinogens tested}}$$

$$\text{Accuracy} = \frac{\text{No. of correct test results} \times 100}{\text{No. of chemicals tested}}$$

The results of validation are generally reported both from the laboratory that originally established the test method as well as from other laboratories where it is used. However, the results are dependent on the prevalence of carcinogens among the tested chemicals, and they may therefore be biased by the selection of test chemicals. More accurate validations of short-term tests have been done on the basis of published data —for example, in the Gene-Tox Program of the US Environmental Protection Agency (ICPEMC, 1984; Palajda & Rosenkranz, 1985) — in which the results of tests performed according to standard protocols are reviewed. Discrepancies were found to be due mainly to the limited numbers of chemicals and chemical classes tested and differences in the purity of the chemicals used in different laboratories. A third method for validating short-term tests is national and international collaborative studies of the same batches of chemicals using the same test (e.g., de Serres & Ashby, 1981; Dunkel *et al.*, 1984, 1985).

The predictivity of an assay can be defined either as the probability that the tested chemical is a carcinogen — given that the test result is positive — or as the probability that the tested chemical is a noncarcinogen — given that the test result is negative. It has been suggested (Rosenkranz *et al.*, 1984; Chankong *et al.*, 1985) that these probabilities can be calculated from the known sensitivities and specificities of the assay. Obviously, the predictive value of assays should be considered before selecting an assay for screening potential carcinogens. It is important to note that sensitivities, specificities and predictivities can be computed only if the available data base includes a significant proportion of carcinogens and noncarcinogens.

The evaluation of bacterial assays in the Gene-Tox data base indicated that, for the chemicals tested, no single one of the widely used assays was ideal. Some of the assays were better predictors of noncarcinogens than of carcinogens and *vice versa*, and, in this respect, the bacterial assays were no different from mammalian assays for mutations and DNA damage (Pet-Edwards *et al.*, 1985). No chemical class could be identified that was exclusively either mutagenic or DNA damaging in bacteria or in mammalian cells.

7. Assays for mutagenesis

7.1 *The* Salmonella/*microsome test*

The *Salmonella*/mammalian microsome test — the 'Ames test' (Ames *et al.*, 1973a,b, 1975; Maron & Ames, 1983) is the most popular of the bacterial assays and has been validated on a large scale (Kier *et al.*, 1986).

The test measures reverse mutation from histidine auxotrophy to prototrophy in several specially constructed mutants of *Salmonella typhimurium*. A homogenate of rat liver (or other mammalian tissue) is added to the bacterial suspension as an approximation of mammalian metabolism (Ames *et al.*, 1973a; IARC, 1980). By 1983, data on more than 5000 chemicals tested using this method had been published (Maron & Ames, 1983). By now, the number of publications on chemicals tested in this way has grown again (e.g., Kier *et al.*, 1986).

Since the last report (IARC, 1980) on long-term and short-term screening assays, several interlaboratory trials have assessed the reproducibility of standard protocols for the *Salmonella*/microsome test (e.g., Venitt, 1982; Haworth *et al.*, 1983; Dunkel *et al.*, 1984; Venitt & Forster, 1985). The test has evolved, not so much in the protocol (see, for example, Maron & Ames, 1983; Venitt *et al.*, 1984), as in the genetic nature of the strains recommended.

In a reversion assay such as the *Salmonella*/microsome test, the response of each tester strain to mutagenic agents depends on the nature of the initial *his* mutation and on the capacity for processing DNA lesions (e.g., presence or absence of excision repair). A given tester strain usually responds to certain classes of compounds and not to others. It is thus critical to establish a battery which can detect the widest possible range of compounds with the minimum number of strains.

In the original minimal battery (Ames *et al.*, 1975), the use of five strains was recommended (Table 1). Each strain carried one of three different *his* mutations (*his*G46, *his*C3076 and *his*D3052), all of which are deficient in excision repair (*uvr*B) and have gained enhanced permeability to certain chemicals as the result of a deficiency in the cell envelope lipopolysaccharide (*rfa*). Two of the strains carry the plasmid pkM101, which was subsequently shown to harbour SOS genes (*muc* AB) involved in mutagenesis (Shanabruch & Walker, 1980).

Three important advances have been made: (1) the number of chemical classes that gives responses has been enlarged (e.g., mitomycin C, aldehydes and oxidative mutagens could not be detected previously under 'standard' assay conditions); (2) the change in nucleotide sequence responsible for the initial *his* mutation is now known (Tables 2 and 3); and (3) the nature of the reversion events is better understood (see below and Fig. 1).

Table 2. Batteries of tester strains recommended for routine screening in the Ames test

The first five strains were initially recommended (Ames *et al.*, 1975). The last four were recommended recently (Maron & Ames, 1983). Ancillary strains are available for special cases. All have a deficiency in cell envelope lipopolysaccharide (*rfa*)

Strain	*His*- mutation name	Repair	pKM101	Nature of mutation	Examples of classes of compounds detected
TA1535	*his*G46	ΔuvrB	-	AT $\rightarrow$ GC	
TA1537	*his*C3076	ΔuvrB	-	+1 near C...C	
TA1538	*his*D3052	ΔuvrB	-	-1 near CG...CG	
TA98	*his*D3052	ΔuvrB	+	-1 near CG	frameshift mutagens
TA100	*his*G46	ΔuvrB	+	AT $\rightarrow$ GC	
TA97	*his*D6610	ΔuvrB	+	+4 near CCC	frameshift mutagens, PR toxin
TA102	PAQ1 *his*G428/Δhis	+	+	GC $\rightarrow$ AT	Oxidants, X-rays, mytomycin C, ultraviolet radiation, bleomycin, hydrogen peroxide, quinones, streptomycin

The new minimal battery of strains recommended by Maron and Ames (1983) includes four strains, each of which carries a different *his* mutation (*his*D3052, *his*G46, h*is*D6610 and *his*G428) (Table 2). Two of the strains (TA100 and TA98) were included in the first battery. Two are new (TA97 and TA102) (Levin *et al.*, 1982; Maron & Ames, 1983). All strains are *rfa* and carry plasmid pKM101. TA102 (Levin *et al.*, 1984a) carries the *his*G428 mutation on a multicopy plasmid and is *uvr*+. This strain is sensitive to mitomycin C, X-rays and oxidative mutagens such as hydrogen peroxide, bleomycin and quinones (Chesis *et al.*, 1984; Levin *et al.*, 1984a). The availability of these new strains requires an assessment both of their performance with a wide variety of chemical classes and their interlaboratory reproducibility. The selection of an optimal battery of strains will require analysis of all data generated with these and the earlier strains.

Table 3. Sequence change in the strains recommended for routine screening in the Ames test.

The DNA sequence near the site of the initial mutation is indicated. The *his*G428 mutation is a chain terminating mutation of the ochre type. See text for details.

Mutation name	Nature of change	Precise sequence change
*his*G428	substitution GC → AT	WT -CAG AGC AAG CAA GAG CTG- mutant -CAG AGC AAG *T*AA (ochre)
*his*G46	substitution AT → GC	leu WT CTC mutant C*C*C (missense) pro
*his*D6610	frameshift +4	WT GTC ACC CCT GAA GAG ATG GCG mutant GTC AC*A CCC* TGA (opal)
*his*D3052	frameshift -1	-1 near CG CG CG CG
*his*C3076	frameshift +1	+1 near CCC

The initial *his* mutations represented in the new minimal battery of Maron and Ames (1983) represent four fundamental nucleotide changes: *his*D3052 is a -1 frameshift mutation, *his*G46 is an AT → GC transition, *his*D6610 is a +4 frameshift mutation and *his*G428 is a GC → AT transition (Tables 2 and 3). The reversion of these mutations to the original sequence covers a large spectrum of events: two basic frameshift events (+1 and -1) and the two possible transition events (GC → AT and AT → GC). It must be recalled that reversion can also be due to secondary mutation, i.e., intragenic and extragenic suppressors. The reversion test thus allows the detection of a much larger spectrum of genetic events than the simple restoration of original DNA sequences (Levin *et al.*, 1984b; Levin & Ames, 1985).

7.2 *Other tests employing* Salmonella typhimurium

(a) *Forward mutation assay using arabinose resistance*

In this system, forward mutations to L-arabinose resistance are detected (Ruiz-Vazquez *et al.*, 1978). Sensitivity is due to an *ara*D mutation that not only blocks the utilization of L-arabinose as a carbon source, but also leads to the accumulation of a toxic intermediate which inhibits bacterial growth when the arabinose operon is expressed. In this assay, resistance to L-arabinose may be due to mutations in at least three different genes: *ara*A, *ara*B and *ara*C (Pueyo & Lopez-Barea, 1979).

The usefulness of this assay has been extended by introduction of the *uvr*B mutation, an increase in cell permeability due to an *rfa* mutation and introduction of plasmid pKM101 (Pueyo, 1978, 1979; Whong *et al.*, 1981; Ruiz-Rubio & Pueyo, 1982; Xu *et al.*, 1984). This improved strain permits the detection of oxidative mutagens.

(b) Forward mutation assay using 8-azaguanine resistance

Skopek *et al.* (1978) have developed a forward mutation system employing strain TM35 derived from a spontaneous *his*$^+$ revertant of *S. typhimurium* TA1535. In this test, resistance to the purine analogue 8-azaguanine is used as a genetic marker.

7.3 Assays based on Escherichia coli WP2 derivatives

This test is based on the reversion of an *ochre* mutation at the *trp*E locus, yielding tryptophan auxotrophy. Several strains are available, including WP2, WP2 *uvr*A, WP2(pKM101) and WP2*uvr*A (pKM101). Of these, WP2*uvr*A (pKM101) has proved to be the most useful for detecting the mutagenic activity of a very wide variety of chemical classes (Matsushima *et al.*, 1981; Venitt & Crofton-Sleigh, 1981; Dyrby & Ingvardsen, 1983; Venitt *et al.* 1984; Araki *et al.*, 1985).

7.4 E. coli K12/343/113

The *E. coli* K12/343/113 test (Mohn & Ellenberger, 1977) is a versatile assay in that several different endpoints (forward mutation, reversion, phage induction, differential DNA repair) can be monitored. Recent developments have improved and extended the usefulness of the assay, and several tester strains with different capacities for processing DNA lesions (*uvr*B, *rec*A, *pol*A, *dam*3, pKm101), different permeability to large molecules and differing in certain metabolizing activities (nitroreductase deficiency, glutathione deficiency) are now available. Of the genetic indicators used in this test, the authors suggest that the 'forward-mutation detection systems to nalidixic-acid-resistance and to valine resistance and the frameshift specific nad$_{113}$ back mutation system seem to be sufficiently developed and calibrated to be used in routine experiments and this is also applicable to the differential DNA repair tests' (Mohn *et al.*, 1984).

7.5 Other tests

A forward mutation assay termed the 'replicative killing' (RK) test, using prototrophic bacteria has been described (Hayes *et al.*, 1983, 1984). It employs an *E. coli* strain with a defective thermoinducible lysogenic prophage. The ability of a chemical to increase the fraction of colony-forming cells at 42°C is taken as an indication of mutagenicity. The assay requires further evaluation (Hayes & Gordon, 1984).

8. Assays for SOS induction

One way of detecting DNA lesions is to monitor the induction of the SOS system, either directly or indirectly.

8.1 *Assays based on λCI repressor cleavage*

(a) *Phage production*

With mutagenesis, this is one of the oldest procedures for assaying for genotoxicity (Lwoff, 1953). The main advantage of this type of test is that it requires only one tester strain and gives a response overnight. However, the tests are indirect, since many steps are required before phage particles are produced. In addition, they are limited in sensitivity, because the first step in phage λ induction is cleavage of the λ repressor (the product of gene CI). This phage-encoded protein maintains the phage in its dormant state, and proteolytic cleavage of the repressor allows expression of phage functions. It is now known that the λ repressor is relatively resistant to cleavage compared with the *lex*A repressor.

A recent study of 46 chemicals (including 30 known animal carcinogens) confirmed this lack of sensitivity (Mamber *et al.*, 1984). To increase sensitivity, a procedure has been developed involving the use of a mutant of phage λ with a thermosensitive repressor (C1857) (Ho & Ho, 1979). However, this assay presents practical problems, since it is essential to maintain a constant temperature of exactly 35°C to obtain reproducible results.

(b) *Biochemical induction assay*

A biochemical (colorimetric) assay of phage λ induction has also been developed (Elesperu & White, 1983). This assay (BIA) is based on a tester strain containing a derivative of phage λ with a genetic fusion whereby gene *lac*Z, encoding the easily assayed *E. coli* enzyme β-galactosidase, is under the direct control of the C1 repressor. Cleavage of the C1 repressor results in β-galactosidase. The BIA is a simple, fast (<5 hours) and direct assay of λ repressor cleavage. A control for toxicity can be included by performing the assay in parallel with a strain bearing a C1857 thermosensitive repressor at 42°C. Since the β-galactosidase assay reflects λ repressor cleavage, it has an inherent limit in sensitivity (see section 8.1(*a*)).

8.2 *Direct assays based on* lex*A cleavage*

The expression of an SOS gene can be monitored directly by means of its fusion with *lac*Z, the structural gene for *E. coli* β-galactosidase. Two such tests have been developed but these tests must not be confused with 'repair tests', in which differential survival between a repair-proficient and repair-deficient is increased (see section 9). The relationship between SOS induction and differential survival is far from simple.

(a) *SOS chromotest*

The SOS chromotest (Quillardet et al., 1982) makes use of a *sfi*A-*lac*Z operon fusion in a *uvr*A, *rfa* strain of *E. coli* K12. Constitutive synthesis of alkaline phosphatase, an enzyme independent of the SOS system, serves as a control for toxicity. The assay takes only a few hours and is quantitative. The slope of the linear region of the dose-response curve is a single parameter (SOS inducing potency or SOSIP), which reflects the inducing ability of the test compound (Quillardet et al., 1982). Two extensive validation studies of the SOS chromotest have been published (Ohta et al., 1984; Quillardet & Hofnung, 1985; Quillardet et al., 1985).

(b) Umu *test*

The *umu* test involves an *umu-lac*Z gene fusion carried by a multicopy plasmid introduced into one of the tester strains developed by Ames and his co-workers, TA1535, which is *uvr*A and *rfa* (Oda et al., 1985). A validation study has been published. The *Salmonella*/microsome test, the SOS chromotest and the *umu* test appear to have similar sensitivity when this is expressed as the lowest amount of chemical detected.

9. Differential survival tests

The detection of DNA damaging agents by repair-deficient bacterial assays is based on the differential inhibition of growth of repair-proficient and repair-deficient bacterial strains. These tests employ paired, isogenic strains — one with normal DNA repair capability and one that is deficient in one or more repair pathways. In excision repair, although the incision-excision step can be mediated by different endonucleases (according to the nature of the lesion), the resynthesis step is common to all pathways and is mediated by DNA polymerase I, encoded by the gene *pol*A. *Pol*A mutants are therefore particularly suitable for this test system. Another gene that plays a central role in repair and mutagenesis is *rec*A, and *rec*A mutants are therefore also suitable for these test systems. Indeed, the two most widely used pairs of strains are the *E. coli* *pol*A$^+$/*pol*A$^-$ pair and the *Bacillus subtilis* *rec*$^+$/*rec*$^-$ pair.

There are several procedures for determining preferential inhibition of DNA repair-deficient strains:

(a) *The disc diffusion test*: the chemical is placed on a filter disc in a petri dish containing a growing lawn of bacteria. The diameter of the zone in which growth is inhibited is measured as a function of the quantity of the chemical spotted on the disc.

(b) *The suspension test*: a bacterial culture is exposed to a chemical, and the number of surviving bacteria is determined as colony-forming units, either as a function of time of exposure or as a function of the concentration of the chemical. As for other bacterial tests, appropriate systems for metabolic activation must be added.

Some of the unresolved problems that have been experienced with tests based on differential inhibition of bacterial growth are discussed by Tweats et al. (1984). These include (1) problems of testing insoluble or poorly diffusible materials with the disc-diffusion method, (2) lack of sensitivity when testing materials that are metabolized inefficiently *in vitro*, and (3) the occurrence of false-positive results in tests conducted without exogenous metabolism of indirectly-acting agents that normally require metabolic activation.

A promising new approach has been undertaken by Mohn and co-workers (Mohn *et al.*, 1984; Kerklaan *et al.*, 1985), using four strains of *E. coli* K12/343/113, one of which is wild-type with respect to DNA repair, and three others (*pol*A,*rec*A, *uvr*B/*rec*A). Each of the four strains carries a different amino acid mutation. This allows the use of different selective media for determining the survival of each DNA repair strain after treatment of a mixture of all four strains with the test chemical (see also Chapter 6, section 2).

10. Miscellaneous tests

10.1 *Bioluminescence test*

This test uses a dark mutant of luminous bacteria *(Photobacterium leiognathi)* (Ulitzur *et al.*, 1980; Ulitzur & Weiser, 1981; Ulitzur *et al.*, 1981) and is based on the capacity of a genotoxic agent to restore the ability of the photobacterium to emit light. The procedure is technically simple, measurement being performed on a scintillation counter. The test appears to be very sensitive and quantitative and detects a wide range of genotoxic agents. One important problem, however, is lack of knowledge about the mechanism involved in the restoration of photoluminescence.

10.2 *Tests for transpositions*

Transposable genetic elements are believed to play an important role in genetic rearrangements. Bacterial tests for transposition have been proposed, but no correlation has been found between effects on transposition and carcinogenicity (Data *et al.*, 1983; Shinder *et al.*, 1984).

10.3 *Multitest*

An assay using a single *E. coli* strain has been proposed to determine mutagenic, phage-inducing and recombinogenic activities. Quantitation and validation of this test are at an early stage (Toman *et al.*, 1985).

11. References

Ames, B.N., Durston, W.E., Yamasaki, E. & Lee, F.D. (1973a) Carcinogens are mutagens: a simple test system combining liver homogenates for activation and bacteria for detection. *Proc. natl Acad. Sci. USA, 70*, 2281-2285

Ames, B.N., Lee, F.D. & Durston, W.E. (1973b) An improved bacterial test system for detection and classification of mutagens and carcinogens. *Proc. natl Acad. Sci. USA, 70*, 782-786

Ames, B.N., McCann, J. & Yamasaki, E. (1975) Methods for detecting carcinogens and mutagens with the *Salmonella*/mammalian-microsome mutagenicity test. *Mutat. Res., 31*, 347-364

Araki, A., Muramatsu, M. & Matsushima, T. (1984) Comparison of mutagenicity of *N*-nitrosamines of *Salmonella typhimurium* TA100 and *Escherichia coli* WP2 *uvr*A/pKM101 using rat and hamster liver S9. *Gann, 75*, 8-16

Chankong, V., Haimes, Y.Y., Rosenkranz, H.S. & Pet-Edwards, J., (1985) The carcinogenicity prediction and battery selection method (CPBS). *Mutat. Res., 153*, 135-166

Chesis, P.L., Leven, D.E., Smith, M.T., Ernster, L. & Ames, B.N. (1984) Mutagenicity of quinones: pathways of metabolism activation and detoxification. *Proc. natl Acad. Sci. USA, 81*, 1696-1700

Christman, M.F., Morgan, R.W., Jacobson, F.S. & Ames, B.M. (1985) Positive control of a regulon for defenses against oxidative stress and some heat-shock protein in *Salmonella typhimurium*. *Cell, 41*, 753-762

Cooper, J.A., Saracci, R. & Cole, P. (1979) Describing the validity of carcinogen screening tests. *Br. J. Cancer, 39*, 87-89

Data, A.R., Randolph, B.W. & Rosner, J.L. (1983) Detection of chemicals that stimulate Tn9 transposition in *Escherichia coli* K 12. *Mol. gen. Genet., 189*, 245-250

Demple, B., Sedgwick, B., Robins, P., Totty, N., Waterfield, M.D. & Lindahl, T. (1985) Active site and complete sequence of the suicidal methyltransferase that counters alkylation mutagenesis. *Proc. natl Acad. Sci. USA, 82*, 2688-2692

Dunkel, V.C., Zeiger, E., Brusick, D., McCoy, E., McGregor, D., Mortelmans, K., Rosenkranz, H.S. & Simmon, V.F. (1984) Reproducibility of microbial mutagenicity assays. I. Tests with *Salmonella typhimurium* and *Escherichia coli* using a standardised protocol. *Environ. Mutag., 6* (Suppl. 2), 1-154

Dunkel, V.C., Zeiger, E., Brusick, D., McCoy, E., McGregor, D., Mortelmans, K., Rosenkranz, H.S. & Simmon, V.F. (1985) Reproducibility of microbial mutagenicity assays. II. Testing of carcinogens and non-carcinogens in *Salmonella typhimurium* and *Escherichia coli*. *Environ. Mutag., 7* (Suppl. 5), 1-248

Dyrby, T. & Ingvardsen, P. (1983) Sensitivity of different *E. coli* and *Salmonella* strains in mutagenicity testing calculated on the basis of selected literature. *Mutat. Res., 123*, 47-60

Elespuru, R.K. & White, R.J. (1983) Biochemical prophage induction assay: a rapid test for antitumor agents that interact with DNA. *Cancer Res., 43*, 2819-2830

Grossman, A.D., Erickson, J.W. & Gross, C.A. (1984) The htpR gene product of *E. coli* is a sigma factor for heat shock promoters. *Cell, 38*, 383-390

Haworth, S., Lawlor, T., Mortelmans, K., Speck, W. & Zeiger, E. (1983) Salmonella mutagenicity test results for 250 chemicals. *Environ. Mutagenesis, 1*, 3-142

Hayes, S. & Gordon, A. (1984) Validating RK test: correlation with Salmonella mutatest and SOS chromotest assay results for reference compounds and influence of pH and dose response on measured toxic, mutagenic effects. *Mutat. Res., 130*, 107-111

Hayes, S., Hayes, C., Taitt, E. & Talbert, M. (1983) *A simple forward selection for independently determining the toxicity and mutagenic effect on environmental chemicals: measuring replicative killing of* Escherichia coli *by an integrated fragment of bacteriophage λ DNA*. In: Kolber, A.R., Wong, T.K., Grant, L.D., de Woskin, R.S. & Hugues, T.J., eds, In Vitro *Toxicity Testing of Environmental Agents*, Part A, New York, Plenum Press, pp. 61-71

Hayes, S., Gordon, A., Sadowski, I. & Hayes, C. (1984) RK bacterial test for independently measuring chemical toxicity and mutagenicity: short-term forward selection assay. *Mutat. Res., 130*, 97-106

Ho, Y.L. & Ho, S.K. (1979) The induction of a mutant prophage λ in *Escherichia coli*: a rapid screening test for carcinogens. *Cancer Res., 99*, 257-264

IARC (1980) *IARC Monographs on the Evaluation of the Carcinogenic Risk of Chemicals to Humans*, Suppl. 2, *Long-term and Short-term Screening Assays for Carcinogens: A Critical Appraisal*, Lyon, pp. 85-106

ICPEMC (International Commission for Protection against Environmental Mutagens and Carcinogens) (1984) Report of the ICPEMC Task Group 5 on the differentiation between genotoxic and non-genotoxic carcinogens. *Mutat. Res. 133*, 1-49

Kerklaan, P.R.M., Bouter, S., van Eburg, P.A. & Mohn, G. (1985) Evaluation of the DNA-repair host-mediated assay. 1. Induction of repairable DNA damage in *E. coli* cells recovered from liver, spleen, lungs, kidneys, and the blood stream of mice treated with methylating carcinogens. *Mutat. Res.*, 148- 1-12

Kier, L.E., Brusick, D.J. Auletta, A.E., Von Halle, E.S., Brown, M.M., Simmon, V.F., Dunkel, V., McCann, J., Mortelmans, K., Prival, M., Rao, T.K., Ray, V. (1986) The *Salmonella typhimurium*/mammalian microsomal assay. A report of the US Environmental Protection Agency Gene-Tox Program. *Mutat. Res., 168*, 69-240

Koffel-Schwartz, N., Verdier, J.M., Bichara, M., Freund, A.M., Daune, M.P. & Fuchs, R.P.P. (1984) Carcinogen-induced mutation spectrum in wild-type, uvrA and umuC strains of *Escherichia coli. J. mol. Biol. 177*, 33-51

Leifer, Z., Kada, T., Mandel, M., Zeiger, E., Stafford, R. & Rosenkranz, H.S. (1981) An evaluation of tests using DNA repair-deficient bacteria for predicting genotoxicity and carcinogenicity. *Mutat. Res., 87*, 211-297

Levin, D.E. & Ames, B.N. (1985) Classifying mutagens as to their specificity in causing the six possible transitions and transversions: a simple analysis using the *Salmonella* mutagenicity assay. *Environ. Mutagenesis, 8*, 9-28

Levin, D.E., Yamasaki, E. & Ames, B.N. (1982) A new *Salmonella* tester strain for the detection of frameshift mutagens: A run of cytosines as a mutational hot-spot. *Mutat. Res., 94*, 315-330

Levin, D.E., Hollstein, N., Christman, M.F. & Ames, B.N. (1984a). Detection of oxidative mutagens in a new *Salmonella* tester strain (TA102). *Meth. Enzymol. 105*, 249-263

Levin, D.E., Marnett, L.J. & Ames, B.N. (1984b) Spontaneous and mutagen-induced deletions: mechanistic studies in *Salmonella* tester strain TA102. *Proc. natl Acad. Sci. USA, 81*, 4457-4461

Lwoff, A. (1953) Lysogeny. *Bacteriol. Rev., 17*, 269-337

Mamber, S.W., Bryson, V. & Katz, S.E. (1984) Evaluation of the *Escherichia coli* K12 inductest for detection of potential chemical carcinogens. *Mutat. Res., 130*, 141-151

Maron, D.M. & Ames, B.N. (1983) Revised methods for the *Salmonella* mutagenicity test. *Mutat. Res., 113*, 173-215

Matsushima, T., Takamoto, Y., Shirai, A., Sawamura, M. & Sugimura, T. (1981) *Reverse mutation test on 42 coded compounds with the* E. coli *WP2 system*. In: de Serres, F.J. & Ashby, J., eds, *Evaluation of Short-Term Tests for Carcinogens: Report of the International Collaborative Program*, New York, Elsevier, pp. 387-395

Miller, J.M. (1983) Mutational specificity in bacteria. *Ann. Rev. Genet., 17*, 215-238

Mohn, G. & Ellenberger, J. (1977) *The use of Escherichia coli, K12/343/313 (λ) as a multi-purpose indicator strain in various mutagenicity testing procedures.* In : Kilbey, B.J., Legator, M., Nichols, W. & Ramel, C., eds, *Handbook of Mutagenicity Test Procedures*, Amsterdam, Elsevier, pp. 95-118

Mohn, G., Kerklaan, P.R.M., van Zeeland, A.A., Ellenberger, J., Baan, R.A., Lohman, P.H.M. & Pons, F.W. (1984) Methodologies for the determination of various genetic effects in permeable strains of *E. coli* K-12 differing in DNA repair capacity. Quantification of DNA adduct formation, experiments with organ homogenates and hepatocytes, and animal-mediated assays. *Mutat. Res., 125*, 133-184

Oda, Y., Nakamura, S., Oki, I., Kato, T. & Shinagawa, H. (1985) Evaluation of the new system (*umu*-test) for the detection of environmental mutagens and carcinogens. *Mutat. Res., 147*, 219-229

Ohta, T., Nakamura, N., Moriya, M., Shirasu, Y. & Kada, T. (1984) The SOS function-inducing activity of chemicals mutagens in *Escherichia coli*. *Mutat. Res., 131*, 101-109

Organization for Economic Cooperation and Development (1983) *OECD Data Interpretation Guides for Initial Hazard Assessment*, Paris

Palajda, M., & Rosenkranz, H.S. (1985) Assembly and preliminary analysis of a genotoxicity data base for predicting carcinogens. *Mutat. Res., 153*, 79-134

Pet-Edwards, J., Chankong, V., Rosenkranz, H.S. & Haimes. Y.Y. (1985) Application of the carcinogenicity prediction and battery selection method (CPBS) to the Gene-Tox data base. *Mutat. Res., 153*, 187-20

Pueyo, C. (1978) Forward mutations to arabinose resistance in *Salmonella typhimurium* strains: a sensitive assay for mutagenicity testing. *Mutat. Res., 54*, 311-321

Pueyo, C. (1979) Natuban induces forward mutations to L-arabinose-resistance in *Salmonella typhimurium*. *Mutat. Res., 67*, 189-192

Pueyo, C. & Lopez-Barea, J. (1979) The L-arabinose resistance test with *Salmonella typhimurium* strain SV3 selects forward mutations at several *ara* genes. *Mutat. Res., 64*, 249-258

Quillardet, P. & Hofnung, M. (1985) The SOS chromotest, a colorimetric bacterial assay for genotoxins: procedures. *Mutat. Res., 147*, 65-78

Quillardet, P., Huisman, O., D'Ari, R. & Hofnung, M. (1982) SOS chromotest, a direct assay of induction of an SOS function in *Escherichia coli* K12 to measure genotoxicity. *Proc. natl Acad. Sci. USA, 79*, 5971-5975

Quillardet, P., De Bellecombe, C. & Hofnung, M. (1985) The SOS chromotest, a colorimetric bacterial assay for genotoxins: validation study with 83 compounds. *Mutat. Res., 147*, 79-95

Rosenkranz, H.S., Klopman, G., Chankong, V., Pet-Edwards, J. & Haimes, Y.Y. (1984) Prediction of environmental carcinogens: a strategy for the mid 1980's. *Environ. Mutagenesis, 6*, 231-258

Ruiz-Rubio, M. & Pueyo, C. (1982) Double mutants with both *his* reversion and *ara* forward mutation systems of *Salmonella*. *Mutat. Res., 105*, 383-386

Ruiz-Rubio, M., Alejandre-Duran, E. & Pueyo, C. (1985) Oxidative mutagens specific for AT base pairs induce forward mutations to L-arabinose resistance in *Salmonella typhimurium*. *Mutat. Res., 147*, 153-163

Ruiz-Vazquez, R., Pueyo, C. & Cerda-Almedo, E. (1978) A mutagen assay detecting forward mutations in an arabinose-sensitive strain of *Salmonella typhimurium*. *Mutat. Res., 54*, 121-129

de Serres, F.J. & Ashby, J., eds (1981) *Evaluation of Short-term Tests for Carcinogens: Report of the International Collaborative Programme*, New York, Elsevier

Shanabruch, W.G. & Walker, G.C. (1980) Localisation of the plasmid (pKM101) gene(s) involved in $recA^+ lexA^+$-dependent mutagenesis. *Mol. gen. Genet.*, *179*, 289-297

Shinder, G., Tourjman, S. & Dubow, M.S. (1984) *Bacteriophage* mu *DNA transposition to identify new classes of genotoxic agents*. In: Liu, D. & Dutka, B.J., eds, *Drug and Chemical Toxicology*, Vol 1,. *Toxicology Screening Procedures Using Bacterial System*, New York, Marcel Dekker, pp. 295-308

Skopek, T.R., Liber, H.L., Kroleswski, J.J. & Thilly, W.G. (1978) Quantitative forward mutation assay. *Proc. natl Acad. Sci. USA*, *75*, 410-414

Toman, Z., Dambly-Chaudiere, C., Tenenbaum, L. & Radman, M. (1985) A system for detection of genetic and epigenetic alterations in *Escherichia coli* induced by DNA-damaging agents. *J. mol. Biol.*, *186*, 97-105

Tweats, D., Bootman, J., Combes, R., Green, M. & Watkins, P. (1984) *Assays for DNA repair in bacteria*. In: Dean, B.J., ed., *Report of the UKEMS Sub-Committee on Guidelines for Mutagenicity Testing*, Part 2, Swansea, UK Environmental Mutagen Society pp. 5-25

Ulitzur, S. & Weiser, I. (1981) Acridine dyes and other DNA-intercalating agents induce the luminescence system of luminous bacteria and their dark variants. *Proc. natl Acad. Sci. USA*, *78*, 3338-3342

Ulitzur, S., Weiser, I. & Yannai, S. (1980) A new sensitive and simple bioluminescence test for mutagenic compounds. *Mutat. Res.*, *74*, 113-124

Ulitzur, S., Weiser, I. & Yannai, S. (1981) *Bioluminescence test of mutagenic agents*. In: DeLuca, M.A. & McElroy, W.D., eds, *Bioluminescence and Chemiluminescence*, New York, Academic Press

Venitt, S. (1982) UKEMS Collaborative Genotoxicity Trial. Bacterial mutation tests of 4-chloromethylbiphenyl, 4-hydroxymethylbiphenyl and benzylchloride. Analysis of data from 17 laboratories. *Mutat. Res.*, *100*, 91-109

Venitt, S. & Crofton-Sleigh, C. (1981) *Mutagenicity of 42 coded compounds in a bacterial assay using* Escherichia coli *and* Salmonella typhimurium. In: de Serres, F.J. & Ashby, J., eds, *Evaluation of Short-Term Tests for Carcinogens: Report of the International Collaborative Program*, New York, Elsevier, pp. 351-360

Venitt, S. & Forster R. (1985) *Bacterial mutagenicity assays: coordinators' report*. In: Parry, J.M. & Arlett, C.F., eds, *Comparative Genetic Toxicology: The Second UKEMS Study*, London, MacMillan Press, pp. 103-144

Venitt, S., Forster, R. & Longstaff, E. (1983) *Bacterial mutation assays*. In: Dean, B.J., ed., *Report of the UKEMS Sub-committee on Guidelines for Mutagenicity Testing*, Part 1, Swansea, UK Environmental Mutagen Society, pp. 5-40

Venitt, S., Crofton-Sleigh, C. & Forster, R. (1984) *Bacterial mutation assays using reverse mutation*. In: Venitt, S. & Parry, J.M., eds, *Mutagenicity Testing, a Practical Approach*, Chapter 3, Oxford, IRL Press, pp. 45-98

Walker, G.C. (1984) Mutagenesis and inducible responses to deoxyribonucleic acid damage in *Escherichia coli*. *Microbiol. Rev.*, *48*, 60-93

Whong, W.Z., Stewart, J. & Ong, T.M. (1981) Use of the improved arabinose-resistant assay system of *Salmonella typhimurium* for mutagenesis testing. *Environ. Mutagenesis*, *3*, 93-99

Wood, R.D. & Hutchinson, F. (1984) Non-targeted mutagenesis of unirradiated lambda phage in *Escherichia coli* host cells irradiated with ultraviolet light. *J. mol. Biol.*, *173*, 293-305

Xu, J., Whong, W.Z. & Ong, T.M. (1984) Validation of the *Salmonella* (SV50)/arabinose-resistant forward mutation assay system with 26 compounds. *Mutat. Res.*, *130*, 79-86

REPORT 6

THE HOST-MEDIATED ASSAY

Prepared by:
S. Venitt (Rapporteur), H. Bartsch, G. Becking,
R.P.P. Fuchs, M. Hofnung, C. Malaveille, T. Matsushima,
M.R. Rajewsky, M. Roberfroid and H.S. Rosenkranz (Chairman)

1. Introduction

The host-mediated assay has been evaluated by the US Environmental Protection Agency Gene-Tox Program (Legator *et al.*, 1982). McGregor (1980) made a short review of the place of the host-mediated assay in genetic toxicology.

2. Principles, scientific basis and description of the host-mediated assay

Short-term tests conducted *in vitro* which employ an exogenous metabolic system cannot take account of the patterns of absorption, distribution, metabolism and excretion of foreign compounds that are characteristic of mammals. The host-mediated assay was initially developed to overcome this deficiency (Gabridge & Legator, 1969). Later, it exploited the sensitivity of indicator organisms used in in-vitro short-term tests. In the conduct of the assay, a host animal is dosed with the test material and inoculated by a different route with suitable marker organisms. After various intervals, marker organisms are recovered and assayed for a given genetic endpoint. Control animals concurrently receive the vehicle only. Marker organisms include bacteriophage, bacteria, fungi and mammalian cells, and endpoints include DNA strand-breaks, point mutations, differential survival of repair-proficient and repair-deficient bacteria, recombinational events, deletions, recessive-lethal mutations and chromosomal aberrations. Although any animal species could be used, in practice, the majority of studies have employed rodents. The sites for injection of marker organisms include the peritoneal cavity, the vascular system, the testis and the intestinal tract. Organisms can be recovered from the peritoneal cavity, liver, spleen, lungs, blood and intestinal tract (Legator *et al.*, 1982). In addition, procedures have been reported for conducting host-mediated assays *via* the stomach of mice (Barale *et al.*, 1983) and by the transplacental route in Syrian hamsters (Quarles, 1981).

Recently, Mohn and his co-workers (Mohn et al., 1984; Kerklaan et al., 1985) have reported an improved protocol employing differential survival in bacteria as a rapid, sensitive endpoint for the host-mediated assay in mice. This system employs a pair of E. coli K-12 343 strains which differ greatly in DNA-repair capacity, each one of which also carries another marker which allows identification on selective medium, giving red (repair+) and white (repair-) colonies. Cultures of each strain are injected into the lateral tail-vein of the same mouse. Groups of inoculated mice are then treated intraperitoneally with the test material. After 2 to 4 hours, the mice are killed, and various organs (e.g., liver, spleen, lungs, kidney, stomach, intestine, colon and blood) are removed and homogenized. The homogenates are centrifuged and the bacteria resuspended in an appropriate medium. After overnight incubation, bacteria are plated out on selective medium, and, after further incubation, the surviving fraction of the repair-negative strain is calculated relative to the survival of the repair-positive strain.

Several reference mutagens (e.g., cisplatin, 4-nitroquinoline oxide, N-nitrosodimethylamine, mitomycin C, 1,2-dimethylhydrazine, N-methyl-N-nitrosourea, methyl methane sulfonate, hexamethylphosphoramide, 2-acetyl aminofluorene) give positive results in organs including the liver, lungs, spleen, kidneys and blood of treated mice.

3. Advantages and limitations of the assay

3.1 *Disadvantages*

Major disadvantages include: (*a*) lack of sensitivity, often requiring doses of the test chemical close to the LD_{50}; (*b*) low recovery of marker organisms from organs or sites other than the peritoneal cavity; and (*c*) large variations in results within groups of identically treated animals. However, the recent but as yet unvalidated system described by Mohn et al. (1984) appears to have overcome most of these difficulties.

3.2 *Advantages*

The host-mediated assay combines the sensitivity and utility of recently developed marker organisms with the full metabolic competence of the intact mammal, allowing studies of distribution and organ-specific activation of test compounds.

4. Validation and performance of the host-mediated assay

In a recent evaluation (Legator et al., 1982), results of host-mediated assays of 208 chemicals were evaluated. Of these, 133 were mutagenic in the assay with one or more indicators, and 76 chemicals, several of which are not considered to be carcinogenic, were not detected by any of the indicators. Of the 208 chemicals, 125 had been tested in carcinogenicity assays in rodents; 90 (72%) of the carcinogens were detected as mutagens in the host-mediated assay. Several of the carcinogens that were not detected may have given negative results because of improper selection of the indicator organism.

The group concluded that the host-mediated assay is an important test in mutagenicity/carcinogenicity research and that, by proper selection of protocols and indicators, valuable information can be gained that otherwise would be overlooked in strictly in-vitro assays.

5. References

Barale, R., Zucconi, D., Romaon, M. & Loprieno, N. (1983) The intragastric host-mediated assay for the assessment of the formation of direct mutagens *in vivo*. *Mutat. Res.*, *113*, 21-32

Gabridge, M.G. & Legator, M.S. (1969) A host-mediated microbial assay for the detection of mutagenic compounds. *Proc. Soc. exp. Biol. Med.*, *130*, 831-834

Kerklaan, P.R.M., Bouter, S., van Elburg, P.A. & Mohn, G. (1985) Evaluation of the DNA-repair host-mediated assay. 1. Induction of repairable DNA damage in *E. coli* cells recovered from liver, spleen, lungs kidneys, and the blood stream of mice treated with methylating carinogens. *Mutat. Res.*, *148*, 1-12

Legator, M.S., Bueding, E., Batzinger, R., Connor, T.H., Eisenstadt, E., Farrow, M.G., Ficsor, G., Hsie, A., Seed, J. & Stafford, R.S. (1982) An evaluation of the host-mediated assay and body-fluid analysis. A report of the US Environmental Protection Agency Gene-Tox Program. *Mutat. Res.*, *125*, 153-184

McGregor, D.B. (1980) The place of the host-mediated assay. *Arch. Toxicol.*, *46*, 111-121

Mohn, G., Kerklaan, P.R.M., van Zeeland, A.A., Ellenberger, J., Baan, R.A., Lohman, P.H.M. & Pons, F.-W. (1984) Methodologies for the determination of various genetic effects in permeable strains of *E. coli* K-12 differing in DNA repair capacity. Quantification of DNA adduct formation, experiments with organ homogenates and hepatocytes, and animal-mediated assays. *Mutat. Res.*, *125*, 153-184

Quarles, J.M. (1981) Transplacental host-mediated assay in the hamster as a rapid screening test for chemical carcinogens. *Biol. Res. Pregn.*, *2*, 188-194

REPORT 7
ASSAYS FOR GENETIC CHANGES IN MAMMALIAN CELLS

Prepared by:

B. Lambert (Rapporteur), E.H.Y. Chu,
L. De Carli, U.H. Ehling, H.J. Evans, M. Hayashi,
W.G. Thilly (Chairman) and H. Vainio

1. General considerations

1.1 *Genetic changes associated with neoplasia*

A number of different types of genetic change have been shown to occur in cells recovered from human tumours, as exemplified in Table 1. Whether these changes arise early or late in tumour development is not known, neither is their contribution to the various developmental stages during the multistep process of carcinogenesis. Nevertheless, the consistency with which these changes occur in certain types of tumours, and their specificity at the molecular or chromosomal level, provide a rational basis for the use of assays for genetic change in somatic cells in testing for potential carcinogens.

Table 1. Genetic changes in human tumour cells

Type of genetic	Example	Reference
Gene mutation	Human bladder tumour, C → T transversion in codon 12 of c-Ha-*ras*1	Reddy *et al.* (1982)
Chromosomal translocation	Burkitt's lymphoma, t(8;14), involving transposition of c-*myc*	Zech *et al.* (1976)
Chromosomal deletion	Retinoblastoma, inherited deletion of chromosome 13,q14	Yunis & Ramsay (1978); Cavenee *et al.* (1983)
Aneuploidy	Chronic lymphocytic leukaemia, trisomy 12	Gahrton *et al.* (1980)
Gene amplification	Neuroblastoma, 5-1000 × amplification of N-*myc*	Schwab *et al.* (1983)

However, assuming a primary role for genetic changes in tumour development, a number of assumptions are made in the use of these endpoints for testing for carcinogens. In addition, several prerequisites have to be fulfilled in order to interpret the results.

1.2 *Assumption of genetic equivalence*

In measuring genetic changes in mammalian cell populations, it is assumed that a mechanistic equivalence exists among various experimental systems. This equivalence may be manifested at different levels: (1) The types and mechanisms of mutational events, whether chromosomal or genic, should be the same in humans and other eukaryotes. Thus, various chromosomal mutations (e.g., numerical and structural) and nucleotide changes (e.g., base substitutions, deletions, additions) occur in all chromosome regions of all species. (2) In all organisms so far examined, mutations appear to be under genetic control — both direct and indirect. Accordingly, the occurrence of a mutant phenotype is affected by the genotype of cells, and involves genes which (i) control DNA-repair pathways, (ii) have an epistatic effect, (iii) modify DNA precursor pools, and (iv) others. As a consequence, differences in mutational response to internal and external stimuli have been observed for different species and even different strains of the same species. Even though the types and mechanisms of mutational events and their regulation appear to be equivalent in all eukaryotic organisms, the variable mutagenic response complicates testing for potential carcinogens. However, these variations are to be expected and must be taken into consideration both in the design of experiments and in the interpretation of results.

1.3 *Assumption of biochemical equivalence*

In using a test for chemically induced genetic changes as a surrogate for the assay of preneoplastic genetic changes in humans, one assumes that the biochemical conditions of assay are equivalent. Thus, independent of the mammalian cell type used, there is an underlying assumption that the test chemical will undergo biochemical processing as it would in the human body.

Since many chemicals of diverse structure are to be tested, this assumption is far-reaching and important in the area of short-term testing. In order for the outcome of a genetic assay to be considered an important indicator of carcinogenic risk, the biochemical processes of metabolism and the processing of DNA or chromatin reaction product should be similar to those encountered by the chemical in one or more human tissues.

The actual patterns of metabolism and of DNA repair for *any* chemical in *all* human tissues in which cancer occurs are not known. The patterns of metabolism (nature of metabolites and rate of production and degradation) in some human tissues are understood for only a few carcinogens, and there is as yet no useful information on the nature of DNA adducts and their rate of removal and relation to genetic change for any known carcinogen in humans. The existing data do not allow the unambiguous conclusion that the biochemical processes of metabolism and DNA repair in any given mammalian-cell assay or rodent species do or do not mimic the behaviour of a human tissue at risk for cancer.

Cultured cells exist under physiological conditions that are markedly different from those of cells *in vivo*, especially with regard to oxygen tension, hormonal and nutrient status and cellular interrelationships. Cells in culture are generally deficient in enzymes for xenobiotic metabolism. Differences among different rodents species in the ability to metabolize carcinogens are also well documented (see Report 15).

In summary, therefore, it may be fair to say that the only reason to expect that the genetic assays described in this section have any predictive value for human carcinogenesis is that there is a wide variety of biochemical possibilities within the tissues of humans. Thus, there is a finite chance that the biochemical conditions within cells and in the assays mimic those in one or more human tissues. It might equally be predicted that the biochemical differences between test systems and human tissues would mean that these systems have minimal value for identifying human carcinogens.

1.4 *Accuracy and precision in short-term assays*

The accuracy of assays is affected by various forms of bias. For gene mutation assays, the following sources of bias should be considered:

(*a*) *Phenotypic lag*: The period between chemical treatment and the expression of genetic changes varies among the phenotypes selected and among different cells used in the assay.

(*b*) *Selection of phenotypes*: The actual relationship between the selective conditions and the nature of mutants selected should be established.

(*c*) *Growth rates*: All mutants and nonmutants should grow in the same way under nonselective conditions, particularly in the presence of toxic agents.

(*d*) *Assumption of independence*: It is assumed that events leading to cell death and to mutation are independent.

(*e*) *Cell density-dependent artefacts*: The probability of colony formation is a function of cell density, for reasons that are not well understood.

The only known means for overcoming bias in mutation assays is to perform a reconstruction experiment with mutant cells identical to those which arise in the assay itself.

For chromosomal assays, the following sources of bias should be considered:

(a) *Phenotypic lag*: The period between chemical treatment and the expression of chromosomal changes varies for different endpoints, chemicals tested and cell types used in the assay.

(b) *Selection of phenotypes*: The optimal concentration of bromodeoxyuridine to be used in assays for sister chromatid exchange (SCE), for example, should be established.

(c) *Cell density-dependent artefacts*: In assays conducted *in vitro*, the cell density at the time of treatment with test chemicals might be important.

(d) *Definition of structural aberrations*: Inconsistent distinctions between gaps and breaks, for instance, result in inconsistency of results not only between laboratories but also within a laboratory.

(e) *Extremely high concentration of test chemical*: A number of nonspecific toxic events should be considered probable, especially when conducting assays *in vitro*.

(f) *Cytological preparation artefacts*: The quality of chromosome spreads is important both for aberration analysis and for chromosome counts.

(g) *Subjective observation*: For objective observation, preparations should be coded and scored without knowledge of the experimental treatment.

The precision of assays is also affected by random error. The protocol for the assay should be considered statistically. In order to achieve the desired precision of an assay, the expected confidence limits should be calculated. Experimental group size and the number of cells to be scored can be determined on the basis of calculated confidence limits. Knowledge of the background frequency of chromosomal anomalies in an individual test system is essential in order to calculate the effect of the chemical tested and to estimate the degree of precision of the system. Data from both positive and negative controls should be presented for each assay.

1.5 *Prediction of carcinogenicity*

Kawachi *et al.* (1980) compared the outcomes of a variety of short-term tests, including chromosomal assays, with results of carcinogenicity tests with the same chemical. For chromosomal aberrations *in vitro*, they found a 68% agreement with carcinogenicity for 58 chemicals. For aberrations *in vivo*, the agreement was 60% for 53 chemicals.

A recent progress report by Zeiger and Tennant (1986) contains a similar analysis of 95 chemicals that have been tested in the US National Cancer Institute/National Toxicology Program cancer bioassay programme using mice and rats as well as in in-vitro tests for aberrations and sister chromatid exchanges (SCE) in Chinese hamster ovary (CHO) cells. The concordance between aberrations and carcinogenesis was 49%, that between SCE and carcinogenesis, 56%, and that between aberrations and SCE, 64%. It must be emphasized that this is a subset of 210 chemicals, and that the choice of chemicals in the subset was influenced by known lack of concordance between data from the *Salmonella*/microsome assay and results from assays for carcinogenicity. Until the complete set has been analysed, caution should be exercised in interpreting the results.

It has been suggested that more short-term studies for genetic change should be carried out in animals *in vivo* (Ashby & Purchase, 1985). The data base for such genetic effects in experimental animals is limited, and additional information is required in order to relate the results to potential carcinogenicity.

It would be tempting to place greater importance on effects observed in humans *in vivo*, and chromosomal damage is the endpoint that can be checked most easily. Of the 20 chemicals for which there is evidence of induction of chromosomal damage in humans, nine are recognized human carcinogens (Table 2). It should be noted, however, that, in addition to the nine chemicals evaluated as definite human carcinogens, four of the remaining 11 have been evaluated as being probably carcinogenic to humans (Tomatis, 1986). The remaining chemicals are mainly those used in cancer chemotherapy, such as 1-(2-chloroethyl)-3-cyclohexyl-1-nitrosourea (CCNU), 6-mercaptopurine, methotrexate and bleomycins, and it is not yet established whether or not any of these compounds is a human carcinogen.

The most important aspect of in-vivo assays in mammals is that they have been conducted *in vivo*, i.e., that the data generated can be used to assess the significance to mammals of mutagenic and clastogenic effects previously observed only *in vitro*. These assays can therefore be used in a predictive sense in mammals in the following circumstances: (1) An agent shown to be either mutagenic and clastogenic *in vitro* is also shown to induce mutation or cytogenetic changes *in vivo*. From this, it can be inferred that the agent remains active *in vivo* and thus may also be capable of inducing cancer or heritable mutations in mammals. It follows, equally, that if such an agent fails to induce mutations or cytogenetic effects *in vivo*, then the chances of it being either a carcinogen or a heritable mutagen in mammals are reduced. (2) In proven cases of human exposure to a known mutagen and/or clastogen, it may be possible to monitor exposure by conducting cytogenetic or mutation analysis of the peripheral lymphocytes of exposed individuals. In such cases, a positive effect will be indicative of exposure but cannot, at present, be considered predictive of the future occurrence of cancer in the affected individual.

1.6 *Needs and future developments*

In mammalian cell genetics, three areas of research and development are strongly recommended:

(*a*) Understanding of the relationships between genetic changes observed *in vivo* to those seen *in vitro* requires carefully planned studies for measuring gene mutations, anomalies of chromosome structure and aneuploidy.

Table 2. Compounds or exposures evaluated within the *IARC Monographs* programme that induce chromosomal damage in humans[a]

**Adryamycin
*Arsenic and certain arsenic compounds
*Benzene
*Betel-quid with tobacco
***Bleomycins
*1,4-Butanediol dimethylsulfonate (myleran)
**CCNU [1-(2-chloroethyl)-3-cyclohexyl-1-nitrosourea]
**Chloroprene
*Cyclophosphamide
**Ethylene oxide
*Hexavalent chromium
*Melphalan
***6-Mercaptopurine
***Methotrexate
**Nitrogen mustard
***Norethynodrel
**Procarbazine
***Styrene
*Tobacco smoke
*Vinyl chloride

*Carcinogenic to humans
**Probably carcinogenic to humans
***The carcinogenicity of these compounds could not be adequately evaluated

[a]From Tomatis (1986)

(b) Measurements of genetic changes in mammalian cells are tedious and labour-intensive. Molecular probes and automated counting systems should be developed.

(c) Means for measuring the ability of a chemical to induce those genetic changes associated with the development of human cancer should be developed and applied. Rapidly expanding understanding of these precancerous events should be incorporated into assay design without delay.

2. Measurements of gene mutations in human cells

2.1 *Relationship of gene mutations to human cancer*

A number of different human tumours have been shown to have specific base-pair substitutions in the c-*ras* gene, which confers an ability on DNA from these tumours to transform mouse cell lines neoplastically. Thus, it is assumed that exposure of humans to chemicals that can cause such mutations would increase the probability of their developing cancer.

2.2 Relationship of gene mutations measured in human cells to mutations known to occur in human cancer

The human cell mutation assays that have been developed to date involve drug-resistant loci to detect a wide variety of genetic changes, including base-pair substitutions, frameshifts and large insertions and deletions. No assay has been developed to measure specific gene mutations that act on human proto-oncogenes. It is therefore an untested hypothesis that the existing gene mutation assays would detect the kinds of mutations that result in human neoplasia. The complete range of mutations involved in human cancer is unknown, so it cannot be stated with certainty how well assays for mutations at some human genes represent those that apparently precede some human cancers. Of course, mutation in humans is *per se* of public health interest, independent of cancer.

2.3 Relationship of biochemical processes in human cell mutation assays to those occurring in humans

Most researchers who use human cells do so because they wish to mimic as closely as possible the responses of the human body to mutagens. There are many important differences among species with regard to the metabolism of common environmental chemicals, and there are wide variations in the dynamics of xenobiotic metabolism among tissues of the same species. In particular, most human cells grown in culture show little if any of the drug-metabolizing complement of enzymes, especially cytochrome P450. Only one human cell system used in mutation studies (Crespi & Thilly, 1984) has been reported in which the cells retained this metabolic capacity.

It is clear that there are equally important biochemical deficiencies in routes of DNA repair among cells derived from different species, and the use of human cells is meant to serve as a better approximation of expected human responses than would rodent or bacterial cells. The quantitative and qualitative differences in DNA repair among different human tissues and between human cells *in vivo* and *in vitro* have not yet been well enough described to permit any meaningful summary. However, such differences are strongly suspected.

Sometimes overlooked is the relationship between mutation and concentration of human cells; when plotted, there are many examples of a marked departure from linearity. Experiments in which the lowest concentrations tested far exceed the concentrations experienced in real human exposures cannot therefore be applied reliably in estimates of absolute or relative human risk (Liber *et al.*, 1985).

2.4 Accuracy and precision in the performance of human cell mutation assays

A mutation assay consists of treating N_o cells containing m_o preexisting mutants and at some later time measuring the number of cells present, N_t, and the number of mutants, m_t/N_t. The change in the mutant fraction, $m_t/N_t - m_o/N_o$ is the result of the assay.

Constraints on interpretation of this estimate of induced mutant fraction arise from both random and systematic error.

(a) Random error

(i) *Treatment of the population*

When N_o cells containing m_o preexisting mutants are treated, the numbers of total and mutant cells that survive treatment represent the denominator and numerator of the mutant fraction that is to be determined, m/N.

Under circumstances in which mutants and nonmutants are equally susceptible to the toxic action of treatment, such that the surviving fraction is S, then N_oS total cells survive, m_oS preexisting mutants survive, and m_iS newly induced mutants survive, where m_i = the number of cells with potentially detectable mutations induced by treatment.

The resulting mutant fraction, $\dfrac{m}{N} = \dfrac{m_oS + m_iS}{N_oS}$, has an expected variance of m/N^2 (Leong et al., 1985).

In order to achieve a desired precision, the confidence limits expected for a particular experiment should be calculated and the contribution of each independent manipulation to the variance be calculated. The step of treatment resulting in a mutant fraction m/N with a variance of m/N^2 has 95% confidence limits of approximately

$$\frac{m}{N} + 2\sqrt{\frac{m}{N^2}} .$$

In many applications, a confidence interval of $\pm 20\%$ of the mean yields a sufficient degree of precision. To meet this requirement,

$$\frac{2\sqrt{\frac{m}{N^2}}}{\frac{m}{N}} \leqslant 0.2$$

$$\frac{\sqrt{m}}{m} \leqslant 0.1$$

or $m \geqslant 100$.

In practical terms, the product of the original cell number, N_o, the survival, S, and the observed mutant fraction, m/N, should be greater than 100 in order to assure a level of precision of $\pm 20\%$ for the mutant fraction immediately after treatment.

(ii) *Growth and dilution of populations after treatment*

Because induced genetic mutations are not immediately expressed as detectable phenotypic changes and because the ability of treated cells to form colonies at low cell densities is suppressed for some generations after treatment, it is necessary to permit several (one to as many as 14 have been reported) generations of growth before determining the mutant fraction. In many cases, the concomitant increase in cell number creates unwieldy numbers, so that a series of samples is taken, diluted in culture media and permitted to continue growing. The number of cells and mutants carried in these manipulations must be accounted for, as after the original treatment.

In the case of N cells and m mutants permitted to grow to tN cells and tm mutants before sampling and dilution by $1/t$, while the expected mutant fraction m/N is unchanged, the variance of this expectation would increase by $\left(1-\frac{1}{t}\right)\left(\frac{m}{N^2}\right)$

(Leong et al., 1985). If this process is repeated k times, each step would contribute equally to the variance of m/N that is,

$$\text{variance due to growth and sampling} = \frac{k\left(1-\frac{1}{t}\right)m}{N^2}.$$

(iii) *Determining the number of mutant and nonmutant cells in a sample*

Methods that depend on visual recognition of mutant cell colonies involve sampling the culture for N_f total cells, which have a plating efficiency, a, so that N_f colonies are observed under nonselective conditions.

Since $m \ll N$, the number of cells plated under selective conditions is much larger than N_f, defined here by factor g. Thus, the number of colonies seen under selective conditions is agm_f, where m_f is the number of mutant cells in a sample of N_f cells.

Thus, the observed mutant fraction in total and mutant colonies arranged as on Petri dishes is $\frac{m_f}{N_f} \cdot \frac{agm_f}{aN_f}\left(\frac{1}{g}\right)$

The variance contributed by this process is $\frac{agm_f}{(aN_f)^2}\left(\frac{1}{g^2}\right)$.

In the mutation assays with human lymphoblasts and lymphocytes reviewed here, colonies are arrayed in microtitre dishes, which complicates estimation of dispersion in this final step of mutation assay. Furth et al. (1981a) have offered a solution that has proven adequate in practice.

(iv) *Background and spontaneous mutants*

Preexisting or background mutants represent the 'noise' in most single-treatment mutation assays. A treated culture contains surviving preexisting mutants Sm_0 and surviving induced mutants Sm_1. Untreated cultures contain m_0 preexisting mutants, and $m_i = 0$.

The final mutant fraction determined for an untreated culture is m_0/N_0 with a total variance of

$$\frac{m_0}{N_0} + \frac{k\left(1-\frac{1}{t}m_0\right)}{N_0^2} + \frac{agm_0}{(aN_0)^2 g^2}.$$

Calculation of the induced mutant fraction,

$$\frac{m_i}{N_0} = \frac{m_0 + m_i}{N_0} - \frac{m_0}{N_0},$$

must be accompanied by calculation of the confidence limits of this derived variable:

$$V\frac{m_i}{N_0} = V\left(\frac{m_0 + m_i}{N_0}\right) \times V\frac{m_0}{N_0},$$

which is considerably greater than the variance of the mutant fraction found in the treated culture when m_0 is of the same magnitude as m_i.

It is desirable to test concentrations as low as possible in order to mimic the conditions of human exposure. The resulting small increases in M_i, the number of induced mutants, will obviously be easier to detect when contrasted with a low background. Thus, experimenters employ strategies such as negative selection and recloning of populations to minimize m_0, the background number of mutants.

During treatment and subsequent growth and dilution steps, new mutants can arise spontaneously. As attempts to study lower and lower fractions of induced mutants proceed, it becomes necessary to account for the spontaneous mutation rate by more precise and greater numbers of experiments.

(v) *Number of replicates*

Human and rodent cell mutation assays are tedious and require a fair degree of manual dexterity, time, patience and money. Thus, many reports have appeared in which multiple samples from the same treated or untreated cultures have been substituted for treatment and observation of independent cultures. Remarkably small confidence intervals accompany these substitutions, which appear to represent a misunderstanding of sources of random error in mutation assay protocols. When independent replicate experiments have been performed, the statistical variation in mutant fractions has been observed to correspond closely to the expected variation due to random error, as discussed here (Leong et al., 1985).

(b) Systematic error

Many of the errors that lead to inaccuracy in measurements of mutant fractions have been recognized and corrected, but others have not. Thus, appropriate reconstruction experiments must be carried out to test the entire assay procedure for systematic bias.

(c) Known sources of bias

(i) Phenotypic lag

When the coding region of a gene is mutated, aberrant and inactive proteins may appear. However, the proteins from the gene that were synthesized before mutation occurs can continue to define the cell phenotype until they are degraded or diluted by cell division. This is the case for mutation to phage resistance in *E. coli* and for 6-thioguanine resistance in human cell mutation assays. In this latter assay, loss of all, or nearly all, activity of previously existing hypoxanthine phosphoribosyltransferase (HPRT) must occur before cells will grow in the present of 1 μg/ml 6-thioguanine.

In human lymphoblast lines, the period between mutagen treatment and the expression of 6-thioguanine resistance has been reported to range from five to 14 days. It becomes necessary, therefore, to measure the length of phenotypic lag directly for each cell strain or line with each set of selective conditions.

(ii) Selection of phenotypes

The actual relationship between the rigour of the selective conditions (e.g., concentration of selective agent) and the nature of the mutants selected should be established, and those unlikely to be affected by minor errors in protocol should be adopted. For instance, while as little as 80 ng/ml 6-thioguanine yield the same number of mutant cells as 1 μg/ml in one human B-cell mutation assay, 60 ng/ml permit substantial growth of nonmutants, making the observations ambiguous.

(iii) Selective pressure

Mutants should be able to grow under normal nonselective conditions with the same doubling times as nonmutants. Similarly, treatment with toxic agents should affect mutants and nonmutants in the same way. For instance, one would not use an assay for mutation in the *hprt* gene to measure mutation induced by a substitute for that enzyme, such as 8-azahypoxanthine.

(iv) Assumption of independence

It is assumed that cell killing and mutation are statistically independent events. However, when an average cell has received multiple lethal hits at low survival levels, all observed mutants will not have received a lethal hit.

(v) *Density-dependent artefacts*

The probability of colony formation is a function of cell density, for reasons that are not all well understood. Formation of mutant colonies can be suppressed by the production of toxic metabolites in neighbouring cells, or they can be sustained by feeder effects from non-growing neighbours.

The final step in most mutation assays involves placing a low density of cells (about 100 cells/plate) under nonselective conditions to determine the number of colony-forming units. A high cell density (about 10^4-10^6/plate) is used under selective conditions to determine the number of mutant colony-forming units. Obviously, if the colony-forming efficiencies are not identical in each case, use of the ratio of colony formation under selective and nonselective conditions can lead to serious bias.

This bias can be further exacerbated in practice by failing to permit adequate recovery of colony-forming efficiency after treatment, since low-density colony-forming ability appears to be recovered more slowly than high-density colony-forming ability after toxic treatment. Gross overestimation of induced mutant fractions has resulted from neglect of this phenomenon.

(*d*) *Requirement for reconstruction experiments*

The only known means for overcoming bias in mutation assays is to perform a reconstruction experiment with mutant cells identical to those that arise in the assay itself.

In a typical experiment, a set of mutants is selected randomly, without regard to colony size or plate position. These mutants are grown separately to significant numbers, and are combined and added to a population of wild-type cells, so that the fraction of mutant cells is higher than would be induced in a typical experiment, e.g., 10^{-3}. This population is then permitted to grow for a number of generations (>20) with periodic determination of mutant fraction to test for any selective advantage or disadvantage for the mutants. It is important to select a large number of independent mutants in this kind of experiment, since the doubling time of clonal isolates can vary significantly from the average for the population as a whole.

Finally, the population of mutants and nonmutants is treated with a series of toxic mutagens, the action of which would not change the artificially high mutant fraction of the original culture. The treated population is handled by the identical mutation assay protocol. If the mutant fraction is unchanged by the protocol, the assay is apparently unbiased; if it is increased or decreased, the investigators should attempt to discover the magnitude of the bias and its variation among experiments.

Stratagems such as using mutant cells of independent strains or lines or single mutants from the same strain or line do not constitute an adequate reconstruction experiment.

2.5 Summary of observations in human cell cultures

(a) Human fibroblasts

The pioneering work of Puck *et al.* (1956) and Szybalski and Smith (1959) was followed by that of DeMars (1965, 1973), Albertini and DeMars (1970) and DeMars and Held (1972), who were the first to devise quantitative assays for gene mutations in human fibroblasts. Table 3 lists the work of that laboratory and of four laboratories from which significant contributions using human fibroblasts have been made.

(b) Human lymphoblast lines

The first human lymphoblast *hprt* mutation assays that took account of the long phenotypic lag were performed by Thilly *et al.* (1976). Several cell lines were later developed to make the assay easier, including the TK6, which permits use of the thymidine kinase (TK) locus which has only a two-day phenotypic lag (Skopek *et al.*, 1978), and the AHH1, which permits direct study of mutagens requiring metabolism by cytochrome P450 (Crespi & Thilly, 1984). Recently, the TK6 line has been used to study low doses of ionizing radiation (Grosovsky & Little, 1985). Table 4 summarizes these observations.

(c) Primary human T cells

The techniques of lymphoblast assays and T-cell expansion were combined by Albertini (1982). Morley and coworkers (Morley *et al.*, 1983; Sanderson *et al.*, 1984) and Vijayalaxmi and Evans (1984a) performed a series of quantitative studies in which the primary T-cell cultures were exposed to mutagens. Work on primary human T cells is summarized in Table 5.

2.6 hprt *mutant fractions among blood cells* in vivo

The same constraints apply to accurate and precise measurement of mutant fractions for populations of cells *in vivo* and *in vitro*. Thus, the cell population sampled must be of sufficient size to contain 100 mutant clone-forming units if a precision (95% confidence) of $\pm 20\%$ is desired.

Since only one forward-mutation detection system exists at present for a hemizygous marker, and since large hemizygous markers are the only kinds that yield mutant fractions of $>10^{-6}$, studies of mutant fractions *in vivo* have used the *hprt* locus and selection with ~ 1 μg/ml 6-thioguanine. In its simplest form, the assay involves either autoradiographic enumeration of DNA-synthesizing peripheral blood lymphocytes stimulated with phytohaemagglutinin, or enumeration in microwells of macroscopic T-cell colonies which form in the presence of allogenic stimulator cells and an undefined mixture of 'T-cell growth factors'. To our knowledge, three laboratories have reported the results of assays of 6-thioguanine-resistant blood cells, all using similar techniques.

Table 3. Studies of human fibroblast mutation[a]

Mutagen	Cell type[b]	LD$_{50}$	MD 10^{-5}	MD 10^{-4}	Selective marker[c]	Reference
Heat	Normal fibroblasts (NF)	--	35.6°C (during cultivation)	--	6TG^R	Simons (1982)
Azaserine						
X-rays	NF	80 rads	50 rads	--	6TG^R	de Ruijter & Simons (1980)
X-rays	NF	90 rads	35 rads	50 rads	6TG^R	Cox & Masson (1979)
X-rays	NF	90 rads	35 rads	--	6TG^R	Cox & Masson (1976)
X-rays	NF	1.3 grays	0.4 grays	--	6TG^R	Aust et al. (1984)
X-rays (250keV)	NF	100 rads	40 rads	--	6TG^R	Thacker & Cox (1976)
X-rays (carbon K ultrasoft)	NF	25 rads	30 rads	--	6TG^R	Goodhead et al. (1979)
^{60}C + X-rays	NF	100 rads	40 rads	--	6TG^R	Thacker & Cox (1976)
Helium ions					6TG^R	Cox & Masson (1976)
20 keV/μm	NF	65 rads	25 rads		6TG^R	
28 keV/μm	NF	55 rads	20 rads			
50 keV/μm	NF	30 rads	15 rads			
70 keV/μm	NF	25 rads	10 rads			
90 keV/μm	NF	20 rads	5 rads			
Ultraviolet radiation	NF	36 ergs/mm^2	--	--	--	Maher et al. (1975)
Ultraviolet radiation	NF	3 J/mm^2	1.2 J/m^2	1.6 J/m^2	8AG^R	Maher et al. (1979)

Table 3. (contd)

Mutagen	Cell type[b]	LD_{50}	MD 10^{-5}	MD 10^{-4}	Selective marker[c]	Reference
Ultraviolet radiation	NF	4.5 J/m²	2.5 J/m²	4.35 J/m²	6TG^R	Konze-Thomas et al. (1982)
Ultraviolet radiation	XP12BE (< 1% normal fibroblast repair line)	1 erg/mm²	--	--	--	Maher et al. (1975)
Ultraviolet radiation	XP12BE	0.1 J/m²	0.1 J/m²	0.15 J/m²	8AG^R	Maher et al. (1979)
Ultraviolet radiation	XP12BE	0.4 J/m²	0.2 J/m²	0.5 J/m²	6TG^R	Konze-Thomas et al. (1982)
Ultraviolet radiation	XP2BE (15-25% normal fibroblast repair)	5 erg/mm²	--	--	--	Maher et al. (1975)
Ultraviolet radiation	XP2BE	0.3 J/m²	0.15 J/m²	0.25 J/m²	8AG^R	Maher et al. (1979)
Ultraviolet radiation + 8-methoxypsoralen	NF	224 J/m² 10 µg/ml 8-methoxy-psoralen	--	--	--	Burger & Simons (1979)
	NF	2200 J/m² 0.25 µg/ml 8-methoxy-psoralen	1200 J/m² 0.25 µg/ml	--	--	
Sunlamp fluence	NF	10 kJ/m²	--	13 kJ/m²	6TG^R	Howell et al. (1984)
	HCMM/DNS	6 kJ/m²	--	7 kJ/m²	6TG^R	
Sunlamp fluence	NF	10 J/m² × 10⁻³	--	12.5 J/m² × 10⁻³	6TG^R	Patton et al. (1984)
	XP12BE	1 J/m² × 10⁻³	--	--	--	

Table 3. (contd)

Mutagen	Cell type[b]	LD$_{50}$	MD 10^{-5}	MD 10^{-4}	Selective marker[c]	Reference
Germicidal lamp fluence	NF	4.2 J/m^2	--	5.8 J/m^2	6TG^R	Patton et al. (1984)
	XP12BE	0.4 J/m^2	--	0.4 J/m^2	6TG^R	
N-Ethyl-N-nitrosourea	NF	1.35 mM × 3 h	0.25 mM × 3 h	0.55 mM × 3 h	6TG^R	Simon et al. (1981)
	NF (SV40)	0.75 mM × 3 h	--	--	--	
N-Ethyl-N-nitrosourea	NF (SL68)	1.25 mM × 1 h	--	0.1 mM × 1 h	DT^R	Drinkwater et al. (1982)
N-Ethyl-N-nitrosourea	XP12BE	0.25 mM × 3 h	0.01 mM × 3 h	0.3 mM × 3 h	6TG^R	Simon et al. (1981)
	XP12BE (SV40)	0.3 mM × 3 h	--	--	--	
N-Ethyl-N-nitrosourea	NF	1.2 mM × 1 h	--	0.5 mM × 1 h 0.2 mM × 1 h	6TG^R DT^R	Aust et al. (1984)
N-Methyl-N-nitrosourea	NF	0.33 mM × 3 h	--	--	--	Simon et al. (1981)
	XP12BE	0.0025 mM × 1 h	--	--	--	
4-Nitroquinoline 1-oxide	NF	190 nM × 2 h	--	--	--	Howell et al. (1984)
	HCMM/DNS-1	90 nM × 2 h	--	130 nM × 2 h	6TG^R	
	HCMM/DNS-2	50 nM × 2 h	--	--	--	
Benzo[a]pyrene	XP12BE	0.23 μM × 24 h 0.12 μM × 48 h	0.16 μM × 48 h	0.31 μM × 48 h	6TG^R	Aust et al. (1980)

Table 3. (contd)

Mutagen	Cell type[b]	LD$_{50}$	MD 10^{-5}	MD 10^{-4}	Selective marker[c]	Reference
Anti-benzo[a]pyrene diol epoxide	NF	0.11 μM × 2 h	--	--	--	Yang et al. (1982)
	XP12BE	0.015 μM × 2 h	--	0.02 μM × 2 h	6TG^R	
Anti-benzo[a]pyrene diol epoxide	NF	0.16 μM × 1 h	--	0.1 μM × 1 h	6TG^R	Aust et al. (1984)
				0.07 μM × 1 h	DT^R	
N-Methyl-N'-nitro-N-nitrosoguanidine	NF	0.0025 mM × 1 h	--	--	--	Simon et al. (1981)
	XP12BE	0.0025 mM × 1 h	--	--	--	
N-Methyl-N'-nitro-N-nitrosoguanidine	NF	3.5 μM × 10^6 × 4 h	--	1.5 μM × 10^6	AG^R	Jacobs & DeMars (1978)
N-Acetoxy-2-acetyl-aminofluorene	NF	1.25 μM × 3 h	--	--	--	Heflich et al. (1980)
	XP12BE	0.25 μM × 3 h	--	--	--	
N-Acetoxy-2-acetyl-aminofluorene	NF	0.8 μM × 3 h	0.3 μM × 1 h	--	DT^R + 6TG^R	Aust et al. (1984)
ICR-191	NF	0.25 μM × 1 h	0.05 μM × 1 h	--	DT^R	Aust et al. (1984)
	NF	0.25 μM × 1 h	--	0.14 μM × 1 h	6TG^R	
Aflatoxin B$_1$ dichloride	NF	8.2 nM × 1 h	2.8 nM × 1 h	--	6TG^R	Mahoney et al. (1984)
	XP12BE	2.7 nM × 1 h	--	2.5 nM × 1 h	6TG^R	

[a]This table is organized around the mutagen studied. Because the treatment necessary to induce a given mutant fraction seems a fair basis for comparing experiments, we have calculated, by interpolation, the dose necessary to induce a mutant fraction of 10^{-5} *hprt* mutants and that necessary to induce a 10^{-4} mutant fraction. Some high background levels make the lower calculation impossible.

[b]NF, normal fibroblasts; XP12BE, xeroderma pigmentosum cells; HCMM, hereditary cutaneous malignant melanoma; DNS, dysplastic naevus syndrome

[c]6TG, 6-thioguanine resistance; 8-azaguanine resistance; DT^R, diphtheria-toxin resistance.

Table 4. Studies of human lymphoblasts[a]

Mutagen	Cell line	LD$_{50}$	MD 10^{-5}	MD 10^{-4}	Selective marker[b]	Reference
^{125}I-Deoxyuridine	TK6	25 decays/cell; 22 h	10 decays/cell; 22 h	--	6TG^R	Liber et al. (1983)
^{3}H-Thymidine	TK6	330 decays/cell; 22 h	275 decays/cell; 22 h	--	6TG^R	Liber et al. (1983)
X-rays	TK6	60 rads	120 rads	--	6TG^R	Liber et al. (1983)
X-rays	TK6	--	150 rads	--	6TG^R	Grosovsky & Little (1985)
			5 rads/day; 30 days		6TG^R	
			10 rads/day; 13 days		6TG^R	
			130 rads/day (accum. dose)		6TG^R	
			130 rads (single acute exposure)		F$_3$TdR^R	
			145 rads (1-10 rads/day)		F$_3$TdR^R	
Heat (45°C)	HL(MIT-2)	45°C; 65 min	--	45°C; 9 min	6TG^R	Gilman & Thilly (1977)
N-Methyl-N-nitrosourea	HL(MIT-2)	1.4 μg/ml; 24 h	--	32 μg/ml; 24 h	6TG^R	Thilly et al. (1976)
N-Methyl-N-nitrosourea	TK6	5 μM; 24 h	4 μM; 24 h 6.2 μM; 24 h	-- --	F$_3$TdR^R 6TG^R	Penman et al. (1983)

Table 4. (contd)

Mutagen	Cell line	LD$_{50}$	MD 10^{-5}	MD 10^{-4}	Selective marker[b]	Reference
N-Methyl-N-nitrosourea	TK6	--	1 μM; 3 days 0.5 μM; 7 days 0.25 μM; 10 days 0.125 μM; 16.5 days 1 μM; 5.5 days 0.5 μM; 11.5 days 0.25 μM; 19 days 0.125 μM; 19 days	-- --	F$_3$TdR^R	Penman et al. (1983)
					6TG^R	
N-Methyl-N-nitrosourea	TK6	4 μM; 24 h	1.5 μM; 24 h		F$_3$TdR^R + 6TG^R	Liber & Thilly (1982)
N-Methyl-N-nitrosourea	MIT-2 WI-L2	8 μM; 24 h --	5 μM; 24 h 8 μM; 24 h	30 μM; 24 h --	6TG^R 6TG^R	Slapikoff et al. (1980)
N-Methyl-N-nitrosourea	MIT-2	41 μM; 24 h	37 μM; 24 h	--	6TG^R	Thilly et al. (1978)
N-Methyl-N-nitrosourethane	MIT-2	0.11 μM; 24 h	--	0.4 μM; 24 h	6TG^R	Penman et al. (1979)
Benzo[a]pyrene	HH-4	--	1.2 μg/ml; 3 h	--	6TG^R	Thilly et al. (1980)
Benzo[a]pyrene	AHH-1	--	7.6 μM; 24 h	--	6TG^R	Crespi & Thilly (1984)
Benzo[a]pyrene	TK6	24 μM; 3 h	--	15 μM; 3 h 21 μM; 3 h	F$_3$TdR^R 6TG^R	Liber & Thilly (1982)
Benzo[a]pyrene	HH-4 WI-L2	-- --	1.5 μg/ml; 3 h --	7 μg/ml; 3 h --	6TG^R 6TG^R	Skopek et al. (1979)
Cyclopenta[c,d]pyrene	HH-4	--	3.8 μg/ml; 3 h		6TG^R	Thilly et al. (1980)

Table 4. (contd)

Mutagen	Cell line	LD_{50}	MD 10^{-5}	MD 10^{-4}	Selective marker[b]	Reference
Cyclopenta[c,d]pyrene	AHH-1	--	300 nM; 24 h	--	6TG^R	Crespi & Thilly (1984)
Cyclopenta[c,d]pyrene	HH-4	--	4 µg/ml; 3 h	--	6TG^R	Skopek et al. (1979)
Fluoranthene	HH-4	--	0.5 µg/ml; 3 h	--	6TG^R	Thilly et al. (1980)
Fluoranthene	AHH-1	--	88 µM; 48 h	--	6TG^R	Crespi & Thilly (1984)
Kerosene soot	HH-4	--	9.2 µg/ml; 3 h	--	6TG^R	Thilly et al. (1980)
Kerosene soot	HH-4	--	9.5 µg/ml; 3 h	23 µg/ml; 3 h	6TG^R	Skopek et al. (1979)
Butylmethane sulfonate	MIT-2	21 mM; 24 h	0.1 mM; 24 h	--	6TG^R	Thilly et al. (1980)
Butylmethane sulfonate	MIT-2	0.9 mM; 24 h	0.5 mM; 24 h	--	6TG^R	Hoppe et al. (1978)
Butylmethane sulfonate	MIT-2	1.15 mM; 24 h	--	0.2 mM; 24 h	6TG^R	Slapikoff et al. (1978)
Butylmethane sulfonate	MGL8B-2	0.65 mM; 24 h	0.75 mM; 24 h	--	6TG^R	Thilly et al. (1980)
Butylmethane sulfonate	MGL8B-2	6.5 mM; 24 h	--	8 mM; 24 h	6TG^R	Slapikoff et al. (1978)
Butylmethane sulfonate	GM130	1.55 mM; 24 h	--	--	--	Thilly et al. (1980)
Butylmethane sulfonate	GM1	Thilly et al. (1980)				
Butylmethane sulfonate	WI-L2	4 mM; 24 h	--	6 mM; 24 h	6TG^R	Slapikoff et al. (1978)
Butylmethane sulfonate	TK6	2.5 mM; 1 h	1 mM; 1 h	--	F_3TdR^R + 6TG^R	Furth et al. (1981b)

Table 4. (contd)

Mutagen	Cell line	LD_{50}	MD 10^{-5}	MD 10^{-4}	Selective marker[b]	Reference
Butylmethane sulfonate	TK6	--	1 mM; 1 h	--	$F_3TdR^R + 6TG^R$	Liber & Thilly (1982)
ICR-191	MIT-2	0.6 µM; 24 h	--	1.75 µM; 24 h	$6TG^R$	Thilly et al. (1980)
ICR-191	MIT-2	2.4 µM; 24 h	--	0.3 µM; 24 h	$6TG^R$	DeLuca et al. (1977a)
ICR-191	MIT-2	1.1 µM; 24 h	0.1 µM; 24 h	1.8 µM; 24 h	$6TG^R$	Slapikoff et al. (1980)
ICR-191	GM130	--	0.1 µM; 24 h	1.2 µM; 24 h	$6TG^R$	
	WI-L2	0.4 µM; 24 h	0.25 µM; 24 h	--	$6TG^R$	
ICR-191	TK6	--	0.15 µM; 24 h	--	F_3TdR^R	Liber & Thilly (1982)
ICR-191	AHH-1	0.5 µg/ml; 24 h	--	0.42 µg/ml; 24 h	$6TG^R$	Crespi & Thilly (1984)
ICR-191	GM130	2.25 µM; 24 h	--	0.9 µM; 24 h	$6TG^R$	Thilly et al. (1980)
5-Chlorodeoxyuridine	MIT-2	37 µM; 24 h	6.25 µM; 24 h	--	$6TG^R$	Penman et al. (1976)
5-Bromodeoxyuridine	MIT-2	12 µM; 24 h	35 µM; 24 h	--	$6TG^R$	Penman et al. (1976)
5-Iododeoxyuridine	MIT-2	10 µM; 24 h	8 µM; 24 h	--	$6TG^R$	Penman et al. (1976)
N-Methyl-N'-nitro-N-nitrosoguanidine	MIT-2	5 ng/ml; 24 h	--	13.5 ng/ml; 24 h	$6TG^R$	Penman & Thilly (1976)
N-Methyl-N'-nitro-N-nitrosoguanidine	MIT-2	0.13 µM; 24 h	0.075 µM; 24 h	--	$6TG^R$	Thilly et al. (1978)

Table 4. (contd)

Mutagen	Cell line	LD_{50}	MD 10^{-5}	MD 10^{-4}	Selective marker[b]	Reference
N-Methyl-N'-nitro-N-nitrosoguanidine	MIT-2	100 nM; 24 h	30 nM; 24 h	275 nM; 24 h	6TG^R	Slapikoff et al. (1980)
9-Aminoacridine	MIT-2	5 μM; 24 h	--	--	--	DeLuca et al. (1977b)
Ethylmethane sulfonate	AHH-1	45 μM; 24 h	48 μM; 24 h	--	6TG^R	Crespi & Thilly (1984)
Ethylmethane sulfonate	TK6	32 μM; 24 h	15 μM; 24 h	140 μM; 24 h	F$_3$TdR^R+ 6TG^R	Penman et al. (1983)
		--	35 μM; 24 h			
		--	6 nM; 2 days	16 nM; 17.5 days	F$_3$TdR^R	
			8 nM; 2 days	--		
			4 nM; 4 days	--		
			2 nM; 6.5 days	--		
			1 nM; 9 days	--		
			16 nM; 1 day	--	6TG^R	
			8 nM; 2.5 days	--		
			4 nM; 4.5 days	--		
			2 nM; 7 days	--		
			1 nM; 11 days	--		
Ethylmethane sulfonate	MIT-2	0.35 mM; 24 h	--	--	--	Hoppe et al. (1978)
Aflatoxin B$_1$	AHH-1	13 μM; 48 h	3.8 μM; 48 h	--	6TG^R	Crespi & Thilly (1984)
2-Acetylaminofluorene	AHH-1	50 μM; 48 h	172 μM; 48 h	--	6TG^R	Crespi & Thilly (1984)
4-Nitroquinoline 1-oxide	TK6	420 nM; 24 h	35 nM; 24 h	--	F$_3$TdR^R	Penman et al. (1984)
		--	50 nM; 24 h	--	6TG^R	
	TK6		26 nM; 2 days	--	6TG^R	
			13 nM; 6.5 days	--	6TG^R	
			6.6 nM; 19 days	--	6TG^R	
			--	53 nM; 17 days	6TG^R	

Table 4. (contd)

Mutagen	Cell line	LD$_{50}$	MD 10^{-5}	MD 10^{-5}	Selective marker[b]	Reference
4-Nitroquinoline 1-oxide	TK6	--	0.05 μM; 24 h	--	6TG^R	Liber & Thilly (1982)
		0.08 μM; 24 h	--	F$_3$TdR^R		
Ultraviolet radiation	TK6	11 J/m^2	2.5 J/m^2	--	6TG^R	DeLuca et al. (1984)
Ultraviolet radiation	TK6	--	1.6 J/m^2	--	6TG^R	Liber & Thilly (1982)
			2.3 J/m^2		F$_3$TdR^R	
Ultraviolet radiation	TK6	7 J/m^2	2.2 J/m^2	--	6TG^R	DeLuca et al. (1983)
			2.5 J/m^2		F$_3$TdR^R	
Ultraviolet radiation	XPA-3	2.3 J/m^2	1 J/m^2	--	6TG^R	DeLuca et al. (1984)
Propylmethane sulfonate	MIT-2	0.5 mM; 24 h	0.2 mM; 24 h	--	6TG^R	Hoppe et al. (1978)
Methylmethane sulfonate	MIT-2	0.1 mM; 24 h	--	--	6TG^R	Hoppe et al. (1978)
N-Nitrosotaurocholic acid	TK6	--	0.55 mM; 3 h	--	F$_3$TdR^R	Puju et al. (1982)
N-Nitrosoglycocholic acid	TK6	--	0.04 μM; 3 h	--	F$_3$TdR^R	Puju et al. (1982)
β-Propiolactone	MIT-2	180 μM; 24 h	--	190 μM; 24 h	6TG^R	Penman et al. (1979)
Formaldehyde	TK6	90 μM; 2 h	125 μM; 2 h	--	F$_3$TdR^R	Goldmacher & Thilly (1973)

[a]This table is organized around the mutagen studied. Because the treatment necessary to induce a given mutant fraction seems a fair basis for comparing experiments, we have calculated, by interpolation, the dose necessary to induce a mutant fraction of 10^{-5} hprt mutants and that necessary to induce a 10^{-4} mutant fraction. Some high background levels make the lower calculation impossible.

[b]6TG, 6-thioguanine resistance; F$_3$TdR, trifluorothymidine resistance

Table 5. Studies of T-cell (TC) mutations[a]

Mutagen	Cell type	LD_{50}	MD 10^{-5}	MD 10^{-4}	Selective marker[b]	Reference
X-rays	TC/solid tumours	--	40 rads	--	$6TG^R$	Dempsey et al. (1985)
	TC/normal	--	60 rads	--	$6TG^R$	
	TC/lymphomas	--	60 rads	--	$6TG^R$	
X-rays	TC/normal	80 rads	--	--	--	Kutlaca et al. (1984)
	TC/carcinoma	57 rads	--	--	--	
	TC/Hodgkin's lymphoma	57 rads	--	--	--	
	TC/Hodgkin's non-lymphoma	75 rads	--	--	--	
X-rays	TC/normal	75 rads	--	--	--	Seshadri et al. (1983)
X-rays	TC/normal	80 rads	--	--	--	Kutlaca et al. (1982a)
X-rays	TC/normal	--	80 rads	--	$6TG^R$	Sanderson et al. (1984)
	TC/non-proliferating	240 rads	80 rads	--	$6TG^R$	
	TC/proliferating	120 rads	280 rads	--		
X-rays	TC/normal	--	40 rads	--	$6TG^R$	Vijayalaxmi & Evans (1984a)
Ultraviolet radiation	TC/solid tumours	--	105 erg/mm^2	--	$6TG^R$	Dempsey et al. (1985)
	TC/normal	--	75 erg/mm^2	--	$6TG^R$	
	TC/lymphomas	--	20 erg/mm^2	--	$6TG^R$	
Ultraviolet radiation	TC/normal	40 erg/mm^2	--	--	--	Kutlaca et al. (1984)
	TC/cancer	40 erg/mm^2	--	--	--	
	TC/lymphoma	40 erg/mm^2	--	--	--	

Table 5. (contd)

Mutagen	Cell type	LD$_{50}$	MD 10^{-5}	MD 10^{-4}	Selective marker[b]	Reference
Ultraviolet radiation	TC/60–90 years	28 erg/mm^2	--	--	--	Kutlaca et al. (1982b)
	TC/40–60 years	31 erg/mm^2	--	--	--	
	TC/17–40 years	29 erg/mm^2	--	--	--	
Ultraviolet radiation	TC/normal	28 erg/mm^2	--	--	--	Kutlaca et al. (1982a)
Ultraviolet radiation	TC/normal	--	50 erg/mm^2	--	6TG^R	Sanderson et al. (1984)
Bleomycin	TC/20 years	2 µg/ml; 3 days	--	--	--	Seshadri et al. (1979)
	TC/40 years	13 µg/ml; 3 days	--	--	--	
	TC/60 years	0.7 µg/ml; 3 days	--	--	--	
Mitomycin C	TC/20 years	0.2 µg/ml; 3 days	--	--	--	Seshadri et al. (1979)
	TC/40 years	0.17 µg/ml; 3 days	--	--	--	
	TC/60 years	0.17 µg/ml; 3 days	--	--	--	
	TC/80 years	0.015 µg/ml; 3 days	--	--	--	

[a]This table is organized around the mutagen studied. Because the treatment necessary to induce a given mutant fraction seems a fair basis for comparing experiments, we have calculated, by interpolation, the dose necessary to induce a mutant fraction of 10^{-5} hprt mutants and that necessary to induce a 10^{-4} mutant fraction. Some high background levels make the lower calculation impossible.

[b]6TG^R, 6-thioguanine resistance

Albertini and his colleagues (Strauss & Albertini, 1978) introduced the idea of recognizing and counting drug-resistant cells using ^{3}H-thymidine uptake after the wild-type population had been killed off by a selective agent. This method promised to be fast (no waiting for colony formation) and potentially applicable to any DNA or (if using ^{3}H-uracil) to RNA-synthesizing cell populations. A series of papers reported characterization of the apparent mutant fraction using this approach until it became clear that the vast majority of cells taking up ^{3}H-thymidine under the assay conditions were not in fact *hprt* mutants but were instead escaping the selective conditions by other means. This problem was addressed by Albertini (1982), who reported in this and following publications (e.g., Albertini *et al.*, 1982) the use of T-cell cloning techniques in indirect mutation assays. These involved essentially the same cell handling used in assays on human B-cell lines but added the newly-reported strategy for cloning and expanding T cells (Paul *et al.*, 1981; Albertini *et al.*, 1982).

In the studies of Albertini (1982, 1985) and Albertini *et al.* (1982), a range of 10^{-6} to 10^{-5} was noted for the *hprt* clone-forming mutants. Further studies have addressed the molecular nature of the mutant *hprt* genes and methods based on T-cell receptor genes for testing the independence of the isolated mutants. A variety of *hprt* mutants was shown to exist in blood samples, as determined by restriction-fragment digests and Southern blots of DNA isolated from individual T-cell mutants (Albertini *et al.*, 1985). Improvements in conducting the more rapid autoradiographic assay, which is shown to yield numerical results equivalent to the T-cell clonal assay, are reported by Albertini (1985).

Morley and coworkers (Kutlaca *et al.*, 1982a; Morley *et al.*, 1982) have studied differential sensitivity of lymphocytes to irradiation and looked for increased amounts of mutation in somatic cells as a function of age (Seshadri *et al.*, 1979; Kutlaca *et al.*, 1982b) using the autoradiographic technique with 6-thioguanine. Further studies (Morley *et al.*, 1983) showed that with a variation of the T-cell cloning protocols, mutation assays could be performed on T-cell samples *in vivo* using manipulative techniques previously applied in human B-cell assays. As noted in Table 5, the technique was used for studying induced mutations in isolated primary T cells.

A somewhat lower estimate of mutant accumulation with age was observed using the clonal assay, as compared to the autoradiography-based study (Trainor *et al.*, 1984). Studies showing decreased cloning efficiency with age (Chrysostomou *et al.*, 1984) have put into question use of the ratio of cloning efficiencies in the presence and absence of selective conditions when the two populations are present at greatly different densities; i.e., a systematic bias might occur owing to cell-density effects, which would raise the apparent *hprt* mutant fraction as a function of age. Resolution of this matter awaits appropriate reconstruction experiments, possibly using blood samples from Lesch-Nyhan heterozygotes depleted of *hprt* mutants by brief growth in aminopterin-containing (HAT) medium.

A significant increase in the *hprt* mutant fraction, as determined by colony formation in cells from patients undergoing radiotherapy and chemotherapy, has been reported (Dempsey *et al.*, 1985). It has also been found that a large fraction of the isolated *hprt* mutants carry significant DNA insertions or deletions, as detected by restriction enzyme digestion and Southern blotting (Turner *et al.*, 1985). In this important matter, data from the laboratories of Albertini and Morley are in close agreement.

Evans and Vijayalaxmi (1981) used the autoradiographic technique described by Albertini but employed 8-azaguanine instead of 6-thioguanine as the selective agent. However, the high frequency of variant cells that was reported (with background frequencies of $\sim 10^{-4}$ in normal donors) suggested that these variants were not mutant cells and cast doubt on use of this approach. The technical problems were discussed and partially overcome by use of clonal assays in subsequent work (Vijayalaxmi & Evans, 1984a,b).

3. Measurements of gene mutations in rodent cells

3.1 *Introduction*

The foundations of human genetics, radiation genetics and cancer biology have derived largely from experimental use of rodent species. The development of the concepts and techniques of the molecular and cell biology of cancer can be attributed principally to work with cultures of animal cells. Therefore, although the main goal of short-term tests for carcinogenicity of environmental chemicals is to reduce and prevent human cancers, contributions towards achieving this goal may not come solely from direct studies of the human organism and cells. On the contrary, there is every reason to expect that results from studies with somatic rodent cells both *in vivo* and *in vitro* will contribute and complement those from human studies.

This section of Report 7 is an updating of the information presented in Report 3 of the first IARC document on long-term and short-term tests (IARC, 1980). During the intervening years, several reviews on mutagenesis in mammalian cells in culture have appeared (Bradley *et al.*, 1981; Howard-Flanders, 1981; Hsie *et al.*, 1981; Cox, 1982; Chu, 1983; Clive *et al.*, 1983; Morrow, 1983; Chu *et al.*, 1984). The present report is therefore not intended to be an exhaustive review. Instead, it emphasizes the cellular and molecular basis of mutagenicity test systems using mammalian cells, and the relevance, limits and potential of the endpoints measured in relation to the detection of mutagens/carcinogens in the environment. In addition, assays for somatic mutations in mice *in vivo* are summarized, to allow a comparison with assays for somatic mutations *in vitro*.

3.2 *The nature of somatic variation* in vitro

Detection of genetic changes in cultured mammalian cells is considered to be more relevant to the identification of environmental chemicals that are potentially mutagenic or carcinogenic to humans than are similar assays on microorganisms. At the same time, assays employing mammalian cells offer the advantages of ease of handling, low cost and rapidity, compared with studies using whole animals. Early work on cultured cells produced cell lines suitable for genetic analysis, isolated and characterized a large array of mutant clones, and demonstrated the feasibility of mutagenicity testing. However, a recurring question is whether the heritable phenotypic variation observed in cultured mammalian cell populations is of genetic or epigenetic origin.

Most mutants, such as drug-resistant cells and auxotrophs, maintain stable phenotypes for long periods of time in the absence of selective agents. Although unstable mutants and high reversion frequencies have been reported, such phenotypic instability might be due to the employment of inappropriate conditions for isolating 'mutants' (IARC, 1980) or to the characteristics of the particular mutant allele.

Early studies by Harris (1971) and Mezger-Freed (1972) suggested that there was no correlation between mutation frequency and cell ploidy. Subsequent studies have indicated that such an association may exist (Chasin, 1973; McBurney & Whitmore, 1973; Hsie *et al.*, 1977; Morrow *et al.*, 1978).

The contention that somatic variation may arise at least in part from modulation of gene expression (rather than gene mutation) has received recent experimental support from studies on DNA methylation. For instance, treatment of human-mouse somatic cell hybrids with 5-azacytidine results in reactivation of genetically inactive human genes on the X chromosome (Mohandas *et al.*, 1981). Harris (1982) showed that brief exposure of 5-bromodeoxyuridine-resistant, thymidine kinase (TK)-deficient Chinese hamster cells to 5-azacytidine resulted in a massive conversion to a HAT^+ state (ability to grow in a medium containing hypoxanthine, aminopterin and thymidine), suggesting that the induction of revertants (and the original TK^- variant itself) might have resulted from changes in DNA methylation patterns. Similarly, 5-azacytidine treatment has been used to reactivate the expression of *hprt* by the inactive gene in Chinese hamster cells (Grant & Worton, 1982). Nonetheless, the correlation between DNA methylation and genetic inactivation observed in cultured cells and cell hybrids may not represent accurately and entirely the situation in the mammalian organism. Wolf and Migeon (1982) used cloned DNA fragments from human X chromosomes as probes and examined the restriction enzyme digestion pattern of DNA from placentas and cultured skin fibroblasts. They found that DNA methylation of the chromosome changes with replication, is not correlated with the number of X chromosomes or transcriptional activity, and is less stable and more prevalent than in the human X chromosome present in interspecific hybrids. They showed further that 5-azacytidine did not cause allelic reactivation of two heterozygous X-linked genes in normal female fibroblasts in culture. This is in contrast to the findings, mentioned earlier, in established cell lines and cell hybrids. Thus, DNA methylation may explain only in part the observed, heritable changes in the phenotypes of cultured somatic cells. Furthermore, this type of modification is not genetic because no alteration in genetic information is involved. However, because of the probably multifactorial origin of human cancers, assays that enable the study of epigenetic modulation of gene expression should be encouraged. It is important to elucidate the underlying molecular mechanisms as well as to determine the frequency of such phenocopies. It will be of interest to demonstrate whether the stable shift of phenotypes could lead to neoplastic transformation *in vitro* and to tumorigenesis *in vivo*.

A large body of experimental evidence (see reviews by Howard-Flanders, 1981; Chu *et al.*, 1984) strongly suggests a genetic basis for mammalian cell variation. A number of criteria has been proposed and applied to assess the mutational origin of phenotypic variation observed in cultured mammalian cells (Chu, 1974; Chu *et al.*, 1975; Siminovitch, 1976). These include: (*a*) random occurrence, (*b*) retention of stable phenotype in the absence of selection, (*c*) induction by mutagens in a dose-dependent manner, (*d*) mutagenic specificity, (*e*) conditional lethality, (*f*) interallelic complementation, (*g*) changes in the activity and physicochemical or immunological properties of specific gene products, (*h*) changes in the amino acid sequence of the gene product, and (*i*) changes in the nucleotide sequence of the gene. Some of the supporting evidence is circumstantial and indirect, and not all of the criteria have been met in every observed somatic variation. However, most of the published results of mutation assays using cultured rodent cells have considered these criteria. The recent demonstration of nucleotide changes in experimentally induced mutations at several loci in cultured rodent cells should dispel the doubt that variations in these cells could indeed be the result of gene mutations (Graf & Chasin, 1982; Meuth & Arrand, 1982; Fuscoe *et al.*, 1983; Nalbontoglu *et al.*, 1983; Goncalves *et al.*, 1984; King & Brookes, 1984).

3.3 *Mutation induction using selectable markers*

Numerous cell lines, genetic markers and selective systems are now available, and the list is still expanding. In order to perform reliable and efficient mutation assays for testing environmental agents that may be mutagens and carcinogens, consideration should be given to the types of cells used, markers, selective procedures and genetic predisposition to mutagenesis.

(*a*) *Choice of cell material*

Suitable cells for mutagenesis studies should exhibit the following characteristics: (i) a high plating efficiency (more than 50%), (ii) short generation time (less than 24 hours), (iii) stable karyotype at diploid or near diploid level, and (iv) suitable cell and colony morphology. Established cell lines used, such as V79 and CHO Chinese hamster cells and L5178Y mouse lymphoma cells, generally meet most of the above characteristics. The BHK cell line, consisting of fibroblasts from baby Syrian hamster kidneys, has the advantage of retaining the capacity for enzymatic activation of carcinogenic hydrocarbons (Stoker & Macpherson, 1964; Wigley *et al.*, 1979). Adult rat liver epithelial cells are used for mutagenicity assays for the same reason (Tong & Williams, 1980; Tong *et al.*, 1984). Primary low-passage fibroblasts are obtained by dissociation of cells of embryo, newborn or adult tissues. However, due to their limited lifespan and low plating efficiency, primary low-passage fibroblasts are not commonly used for mutagenicity testing.

(b) Genetic markers

The criteria to be considered in choosing genetic markers for detecting mutagenic chemicals are: (i) simple selection procedure, (ii) induction of either autosomal dominant or X-linked mutation or autosomal recessive mutation in heterozygotes (or hemizygotes), (iii) forward rather than reverse mutation, and (iv) low frequency of spontaneous mutation.

Isolation of drug-resistant mutants by forward mutations has been used widely and is the best method available for studying mutagenesis in mammalian cells, because of the ease with which single-step selection can be applied. The mutations used most commonly are resistance to 8-azaguanine or 6-thioguanine, which arise from a loss of HPRT activity; resistance to 5-bromodeoxyuridine and other thymidine analogues, associated with a loss of TK activity; resistance to 2,6-diaminopurine, due to a loss of adenine phosphoribosyltransferase (APRT) activity; and resistance to ouabain, due to alteration of the membrane-associated Na^+/K^+ ATPase activity.

Studies of cell hybridization between drug-resistant mutants and parental sensitive cells indicate that resistance to ouabain is a codominant trait. Mutation of the X-linked *hprt* locus is recessive. Resistance to 5-bromodeoxyuridine and to 2,6-diaminopurine is also dependent on recessive mutations, but suitable sublines heterozygous or hemizygous for the TK or *aprt* locus in V79, CHO or L5178Y cell lines have been developed.

Detection of reverse mutations is useful for (i) accurate estimation of the stability of the assay system, (ii) analysis of the genetic basis of the original mutations, (iii) testing mutagenic specificity, and (iv) detection of potential mutagens. Reverse mutation from $hprt^-$ to $hprt^+$ or from TK^- to TK^+ is possible using a medium containing HAT; reverse mutation from $aprt^-$ to $aprt^+$ can be selected in a medium containing adenine and alanosine or azaserine. However, forward mutations are preferred over reverse mutations in mutagenicity testing, because the former represent a broader spectrum of genetic lesions.

On the basis of studies with other organisms, the rates of spontaneous and induced mutations may differ among different loci. Evidence has been presented (Meuth *et al.*,

1979a) that in certain mutator strains of CHO cells, the spontaneous rates were five- to 50-fold higher than in parental cells for two markers (resistance to 6-thioguanine and to ouabain), but the rates for two other markers (reversion of proline-auxotrophy to prototrophy and forward mutation to emitine resistance) were unaffected. Certain agents such as X- and γ-rays fail to induce ouabain resistant mutants in mammalian cells, although they are capable of inducing 6-thioguanine resistant mutants in the same cell type (Arlett *et al.*, 1975; Chang *et al.*, 1978; Thacker *et al.*, 1978). It appears, therefore, that mutation rates may be locus-specific for mammalian cells as well. For more accurate assessment of the mutagenic potential of environmental chemicals, the use of several cell types and several genetic markers is advisable. The employment and further development of other selectable markers not reviewed here (e.g., resistance to α-amanitine: Lobban & Siminovitch, 1975) should be encouraged.

(c) Selection procedure

Selection of mutants in mammalian cells in culture involves a number of technical considerations, including the isolation protocol, rate of replication at the time of treatment, phenotypic expression time, cell density, concentration of selective agent and culture medium employed. The importance of these and other factors that may influence the mutant recovery has been discussed in previous reviews (IARC, 1980; Chu *et al.*, 1984). Details of the technical procedures are beyond the scope of this report. The practitioner should refer to the original papers, because factors that affect mutant yield have been shown to vary for both biological and technical reasons.

The mutagenic/carcinogenic effect of many chemicals is dependent on the formation of reactive electrophilic intermediates, mainly by the activity of microsomal mixed-function oxygenases (see Report 15). Most cell lines currently used for mutagenesis studies, such as V79, CHO and L5178Y cells, are not capable of activating procarcinogens or promutagens. Inclusion of a metabolic activation system in a test system is therefore essential. Three mutagenesis assays using mammalian cells and metabolic activation systems have been developed: (i) a microsome-mediated assay, (ii) a cell-mediated assay, and (iii) a host-mediated assay. In addition, adult rat liver epithelial cells are used to detect genotoxic compounds by mutation at the *hprt* locus (Tong & Williams, 1980; Tong *et al.*, 1984).

(d) Genetic predisposition to mutagenesis

The progression of a primary DNA lesion to a substantial alteration in gene expression is a complex process, which acts as a sieve that allows only a proportion of the DNA lesions to proceed toward the final product as a mutant clone. Many of the enzymes of DNA replication and repair contribute to the fate of primary lesions, and the specificity of these enzymes is responsible for the diversity of mutational responses.

Cultured skin fibroblasts from patients with the hereditary disorder xeroderma pigmentosum exhibited increased sensitivity to ultraviolet radiation (UV) and increased induced mutation rates compared with normal fibroblasts (Maher et al., 1976, 1977). The recent development of in-vitro techniques has allowed the isolation of mutagen-sensitive and DNA repair-defective mutants of mouse and Chinese hamster cells (for citations, see Chu et al., 1984). Results of comparative studies indicate that not only are mammalian DNA repair mechanisms different in many respects from those of bacteria and other organisms, but also that some human cells studied are quite different from mouse and Chinese hamster cells in their DNA excision repair (Regan & Setlow, 1973). Therefore, there is an urgent need to isolate human mutants and to use them, in combination with the rodent mutants, for comparing species; this will aid predictions of the potential mutagenic effects of environmental chemicals on the basis of studies on the whole organism carried out in animals, but not in humans.

Mechanisms other than DNA repair can influence mutagenesis in mammalian cells. For instance, a V79 cell mutant selected for its resistance to aphidicolin, a specific inhibitor of DNA polymerase-α, is characterized by slow growth, sensitivity to UV, and hypersensitivity to UV mutagenesis (Liu et al., 1982, 1983). Imbalance in intracellular DNA precursor pools as a result of either an excessive supply of exogenous nucleosides (Bradley & Sharkey, 1978; Peterson et al., 1978; Peterson & Peterson, 1979) or a genetic alteration (Meuth, 1981; Meuth et al., 1979a,b; Weinburg et al., 1981; Li & Chu, 1982; Li et al., 1983), have led to a significant elevation in the number of mutations. These and other hypermutable cell strains will be very useful in increasing sensitivity for detecting certain classes of mutagens.

3.4 *Nature of the data and interpretation of results*

(a) *Qualitative considerations*

Since the demonstration of experimental mutagenesis in cultured Chinese hamster cells (Chu & Malling, 1968; Kao & Puck, 1968), there has been a vast expansion in the number of cell types and genetic markers used, as well as in the amount of data generated. As implied in the previous section, biological differences in species and tissue of origin and genetic loci chosen, among other factors, could be responsible for the large differences found in mutation rates in cultured cells. Therefore, for mutagenicity testing, consideration must be taken of the biological consistency of the cell material and of cell culture conditions that could modify the behaviour of the same cell type. For a particular cell system chosen, biological variability can further be minimized by (i) cell cloning, (ii) monitoring and control of heterogeneity in test cell populations, (iii) standardization of culture conditions, and (iv) other measures. The reader is referred to the publication of the UK Environmental Mutagen Society (Dean, 1983, 1984), for example, for detailed experimental protocols.

(b) Quantitative considerations

In a mutagenicity testing system using cultured cells, cell population size, total expected number of spontaneous and induced mutants per survivor, number of replications, number of dose levels, mean lethal dose (LD_{50}) of test agents, and other factors must be defined, pretested and standardized before large-scale experiments are initiated. Appropriate statistical analysis should be made to determine the level of experimental error and significance of differences. Reconstruction experiments are sometimes necessary to demonstrate the appropriate conditions for maximal recovery of mutants. The levels of spontaneous forward and reverse mutations (when applicable) at a given genetic locus must be determined before use of the test system. These factors are discussed in detail in section 2.4.

The linearity of dose-effect curves has not yet been determined for very low doses of chemicals. In both CHO and L5178Y cell systems, dose-effect relationships within a certain range of concentrations have been obtained for many known mutagens (Hsie *et al.*, 1975; Clive *et al.*, 1979).

In the L5178Y mouse cell lymphoma assay, two phenotypes — small and large colonies (Hozier *et al.*, 1985; Moore *et al.*, 1985a,b; Sawyer *et al.*, 1985) — are scored as representing induced genetic change. However, to the best knowledge of the group, no paper has appeared in which a quantitative assay involving *small* colonies of trifluorothymidine-resistant cells has been tested for experimental bias; i.e., no reconstruction experiment has been reported. As a result, it is not possible to conclude that the apparent rise in frequency of these small colonies among surviving cells represents induction of genetic change.

3.5 Evaluation and future development of rodent cell culture test systems

(a) Utility of assays for detection of chemicals with potential carcinogenic activity

As reviewed previously (IARC, 1980), over 200 chemicals have been tested in a variety of assay systems for mammalian cell mutagenesis. Although the list has expanded in the intervening years, it is difficult to make a useful comparison of test results, because of heterogeneity in the biological materials used and the experimental protocols employed in different laboratories. It is imperative to reduce such experimental variability by the various measures outlined above. It would be important to demonstrate similarities and differences in mutagenic response among a variety of mammalian cell types, between cells of different species origin (e.g., mouse *versus* human) and between mammalian cells and other organisms. When these requirements are met, it will be helpful to compare the ability of cultured cells to differentiate between carcinogens and noncarcinogens.

(b) Analysis of mammalian cell mutations at the molecular level

Thus far, we have discussed the assessment of somatic mutations of cultured mammalian cells at the phenotypic level. Recent technical advances have permitted analysis of somatic cell mutants at the protein or nucleic acid levels. Several current studies in progress are: (i) amino acid sequence analysis of *hprt*, (ii) multilocus somatic mutation assay by one-dimensional gel electrophoresis, and (iii) multilocus somatic mutation assay by two-dimensional gel electrophoresis, and (iv) DNA alterations in mammalian cell mutants by analysis of altered restriction fragment pattern or nucleotide sequence (for review, see Chu *et al.*, 1984). Although these molecular approaches to mammalian cell mutagenesis are of enormous scientific interest and significance, their applications for large-scale screening of environmental mutagens/carcinogens must await further development.

(c) New assumptions in the evaluation of cellular toxicity

The preceding sections have focused on the induction by environmental agents of mutations at the general level (point mutations) in mammalian cells and the use of mutagenicity tests to prescreen for environmental carcinogens. In a broader sense, mutations include alterations at both the genic and chromosomal levels. The forward mutation assay systems mentioned earlier, which indicate loss of a particular cellular function, detect a broad spectrum of genetic lesions, encompassing base-pair substitutions, deletions, insertions, chromosome deletions, structural rearrangements and chromosome loss.

As was pointed out in the earlier report (IARC, 1980), mammalian cells in culture have the additional advantage that several endpoints can be measured simultaneously in the same cell system, to detect specific locus mutation. These endpoints include neoplastic transformation, chromosomal changes and DNA damage and repair. The usefulness of cultured mammalian cell systems as assays for detecting 'carcinogens' can be extended on the basis of several concepts outlined recently by Trosko (1984).

Carcinogenesis appears to consist of initiation and promotion phases, and it is clear from experimental studies that the mechanisms of initiation are quite different from those of promotion. The mechanisms of initiation can involve mutagenesis, whereas promotion appears to involve nonmutagenic or epigenetic processes leading to clonal expansion of initiated cells. In-vitro assays designed to detect initiators are usually termed assays for 'genotoxic' activity, i.e., for detecting genic and chromosomal mutations. However, not all mutagenic chemicals interact with or damage DNA. For instance, agents that cause chromosome number imbalance (aneuploidy) probably interact with the membrane or mitotic spindle fibre but do not damage DNA (Barrett *et al.*, 1981; Kaufman, 1983; Kafer, 1984; Tsutsui *et al.*, 1984). Aneuploidy has been shown to be associated with neoplasia in humans and animals (Klein, 1981, 1983; Yunis, 1983).

Another class of genetic alteration that does not involve nucleotide sequence is gene amplification. Gene amplification occurs during normal development and can be induced experimentally (Schimke, 1982, 1984a,b; Heilbronn et al., 1985); it has been shown to be involved in the etiology of several human and animal tumours. Several agents, such as UV and tumour promoters, have been shown to induce gene amplification in mammalian cell cultures.

Concern about the role of epigenetic modulation of gene expression by a variety of agents that may ultimately lead to cancer formation has been discussed earlier. The effect of certain promotors can be assayed in cultured cells, as indicated by the inhibition of cell-to-cell communication (Yotti et al., 1979; Trosko et al., 1983; see Report 10).

In short, in addition to genotoxic effects that may be measured using the existing mammalian cell systems, especially with high doses, it is imperative to develop methods that would reveal other adverse effects in the broad sense of cellular toxicity. Improvement of existing tests and development of new tests for investigating the effects of chronic and low-level exposure to potentially hazardous chemicals in the environment would be important. (See Report 10 for further discussion on the detection of tumour-promoting activity.)

3.6 *Somatic mutations* in vivo

(a) *Mouse lymphocytes*

Recently, a clonogenic assay has been developed to measure thioguanine-resistant (HPRT$^+$) spleen lymphocytes in mice (Jones et al., 1985a,b). Lymphocytes are cultured in 96-well microtitre plates; and proliferation is initiated by the mitogen concanavalin-A and is supported thereafter by conditioned medium containing interleukin-2. The spontaneous frequency of HPRT$^+$ cells ranged from $1-3 \times 10^{-6}$. *In vivo*, intraperitoneal injection of mice with N-ethyl-N-nitrosourea 15 days before in-vitro culture resulted in a linear dose-related increase in mutant cells. This work represents a major new experimental approach for comparing mammalian somatic mutation *in vivo* and *in vitro*.

(b) *Mouse spot test*

The mouse spot test for coat colour variation was developed by Russell and Major (1957) for assaying radiation; it was first employed for testing chemical mutagenicity by Fahrig in 1975. The method permits detection of a spectrum of genetic alterations, ranging from gross chromosomal damage to gene mutations that result in the inactivation or loss of the gene product at a set of heterozygous loci controlling hair pigmentation.

The test consists of treatment *in utero* of mouse embryos which are heterozygous for several recessive coat-colour alleles. Treatment occurs during the time of migration of the melanocyte precursor cells from the neural crest to the dermis. Alteration or loss of a wild-type allele at one of the heterozygous coat-colour loci in one of the melanocyte precursor cells leads to expression of the recessive allele. The resulting mutant clone can be observed when the offspring are two weeks of age. Each offspring examined represents a population of clones derived from treated pigment precursor cells. The mutant clones are recognized by external examination as coloured patches, consisting of hairs of brownish, yellowish or greyish colour. The procedure is therefore a relatively inexpensive and rapid test for screening somatic mutations in mice *in vivo*. Various crosses have been used to produce the heterozygous embryos (Neuhäuser-Klaus, 1981; Braun *et al.*, 1982; Hart, 1985).

(i) *Relevance of the test*

In comparison with the methods discussed for screening other types of genetic changes in animals *in vivo*, this method is relatively inexpensive and fast. Progress has been made in differentiating between induced mutations and mitotic recombination (Fahrig & Neuhäuser-Klaus, 1985). Since the target cells of the mouse spot test are in the embryo, the agent or its metabolites must cross the placenta. In addition, any reduction in the average litter size indicates embryotoxicity or toxic effects on the mother. Since the embryos are exposed at mid-gestation, simple external examination of the newborns can yield ancillary information on the teratogenicity of an agent.

(ii) *Disadvantages*

The disadvantage of the test is that the altered clone cannot be confirmed by breeding tests, so that classification of any spot as doubtful cannot be verified. By microscopic examination, however, Fahrig and Neuhäuser-Klaus (1985) have classified hairs from spots into the loci at which the mutations occurred.

(iii) *Results*

The Gene-Tox Program reviewed results from seven laboratories, testing a total of 30 substances, including three solvents. Of the 27 compounds, 16 gave positive, six negative and five inconclusive results (Russell *et al.*, 1981). Of the five compounds classified as inconclusive, two — caffeine and hycanthone — have since been classified mutagenic (Styles & Penman, 1985).

A new tabulation of Styles and Penman (1985) gives the results for 60 agents tested in the mouse spot test. The data were compared with those of a bacterial mutation assay (*Salmonella*/microsome test) and of lifetime rodent bioassays. The authors concluded:

'The performance of the spot test as an in-vivo complementary assay to the in-vitro bacterial mutagenesis test reveals that of 60 agents, 38 were positive in both systems, 6 were positive only in the spot test, 10 were positive only in the bacterial test and 6 were negative in both assays.

'The spot test was also considered as a predictor of carcinogenesis; 45 chemicals were carcinogenic of which 35 were detected as positive by the spot test and 3 out of 6 non-carcinogens were correctly identified as negative.

'If the results are regarded in sequence, i.e., that a positive result in a bacterial mutagenicity test reveals potential that may or may not be realized *in vivo*, then 48 chemicals were mutagenic in the bacterial mutation assay of which 38 were active in the spot test and 31 were confirmed as carcinogens in bioassays. Twelve chemicals were non-mutagenic to bacteria of which six gave positive responses in the spot test and five were confirmed as carcinogens.

'These results provide strong evidence that the mouse-coat spot-test is an effective complementary test to the bacterial mutagenesis assay for the detection of genotoxic chemicals and as a confirmatory test for the identification of carcinogens. The main deficiency at present is the paucity of data from the testing of non-carcinogens.'

The predictive value of the mouse spot test for specific locus mutations is unknown because there is an insufficient data base for comparison. Therefore, it would be interesting to test more completely in the specific locus experiment agents such as benzo[*a*]pyrene, caffeine, hycanthone and *N*-methyl-*N*'-nitro-*N*-nitrosoguanidine, which give positive results in the spot test.

4. Assays of structural chromosomal changes, sister chromatid exchange and micronucleus formation

4.1 *Introduction*

Changes in the structure or number of chromosomes occur spontaneously at low frequency in somatic and germ cells of both animals and plants. The consequences of such aberrations will depend on the type of change, and on the cell type and tissue in which they occur.

In somatic cells, chromosomal alterations usually lead to reproductive cell death; however, certain chromosomal changes that permit cell survival and proliferation may, in rare situations, induce or take part in neoplastic development. Consequently, chromosomal changes have been used as endpoints to monitor exposure of human populations to potential carcinogens (Evans & Lloyd, 1978), and to develop test systems for the detection of chemical agents with potential carcinogenic activity (IARC, 1980).

Most cytogenetic test systems are based on the detection of chromosomal alterations by analysis of metaphase cells under the light microscope. In some eukaryotes (e.g., yeast), in which the small size of the chromosomes makes microscopic analysis impossible, chromosomal alterations are studied by formal genetic analysis of suitable gene and centromere markers (Zimmermann et al., 1984). Recently, human restriction-fragment-length polymorphism markers have become available which allow the detection of certain chromosomal alterations by molecular methods (White, 1982). In this report, the following chromosomal endpoints in mammalian somatic cells are discussed: structural chromosomal aberrations, micronucleus formation and sister chromatid exchange (SCE). Aneuploidy is considered only in relation to micronucleus formation, since this type of effect is discussed in section 5. Gene conversion, mitotic recombination and segregation, and nondisjunction are chromosomal events that may be of importance in carcinogenesis. These endpoints are readily analysed in yeast systems (Report 11).

4.2 *Principles and scientific basis of the test systems*

Changes in the normal morphology or staining pattern of chromosomes are thought to reflect changes at the molecular level. Indeed, by combination of high-resolution chromosome banding and recombinant DNA methods, the close relationship between rearrangements at the chromosomal and molecular levels in several mammalian tumours has been demonstrated (reviewed by Klein, 1981; Klein & Klein, 1985).

Chromosomal anomalies can arise through several different targets and pathways. Many chemical carcinogens and their active metabolites are electrophilic compounds which may react with and damage DNA (Miller, 1978). Such damage may result in changes in chromosomal structure that can be classified into a number of distinct categories, e.g., deletions and rearrangements of chromosomes, or SCE.

Another potential target for the attack of chemical carcinogens is the spindle apparatus. Disturbance of spindle function may lead to nondisjunction and irregular segregation of the chromosomes, resulting in aneuploidy (reviewed by Onfelt & Ramel, 1979). Acentric chromosome fragments, as well as whole chromosomes which lag at anaphase, may not be incorporated into any of the daughter nuclei and therefore appear as micronuclei in the postmitotic interphase cell (Jenssen, 1982; Heddle et al., 1983).

The biochemical and molecular events leading to chromosomal changes are not known. Chemical agents that interact with DNA, and especially those which give rise to DNA strand breaks and DNA adducts, are usually effective in causing chromosomal damage or SCE, but these effects can also be caused by, for example, inhibitors of enzymes involved in DNA repair processes (Kihlmann & Natarajan, 1984), or disturbance in the pools of bases and nucleosides (Anderson et al., 1981; Kunz, 1982). Moreover, structural chromosomal aberrations and SCE seem to arise by at least partly different mechanisms (Wolff, 1982a).

Results of extensive studies of some human autosomal recessive conditions (e.g., Fanconi's anaemia, Bloom's syndrome and ataxia telangiectasia), all of which are associated with an increased incidence of neoplasia and 'spontaneous' chromosomal aberrations or SCE (reviewed by Arlett & Lehmann, 1978), suggest that a number of so far unknown genes involved in DNA replication and/or DNA repair are of great importance for the maintenance and stability of the constitution of normal chromosomes.

Cells at different stages of their cycle may differ greatly in their sensitivity to induction of chromosomal alterations. Moreover, the frequency with which these events are observed depends on the time of sampling of the cell population after treatment, which further indicates the complexity of the processes involved.

(a) Structural chromosomal aberrations

Chemical compounds that induce structural chromosomal aberrations are designated as 'clastogens'. In general, these compounds give rise to primary DNA lesions, some of which are further processed during the replication and/or repair of DNA into secondary lesions, which develop into the various types of chromosomal aberrations visible in the metaphase cell. In principle, the classification of chromosomal aberrations is based on the structural unit involved, i.e., the whole chromosome or the single chromatid, and the type of morphological alteration, i.e., breaks (producing deletions and fragments) or rearrangements (producing inversions, translocations and exchanges within or between chromosomes or chromatids). Present knowledge of the development of the various types of aberration, as well as useful examples of the most common classification schemes, have been reviewed (IARC, 1980; Archer *et al.*, 1981; Scott *et al.*, 1983; Harnden & Klinger, 1985). In the present context, only general aspects of the use of chromosomal aberrations as an endpoint in test systems are considered.

Studies with ionizing radiation have shown that the unit of breakage and reunion in the chromatin is the single chromatid (reviewed by Evans, 1977). Hence, aberrations formed in unreplicated G_1 chromosomes will be duplicated during the S-phase and appear as chromosome-type aberrations during the subsequent mitosis. Aberrations formed after DNA replication, i.e., during or after the S-phase, involve only one of the chromatids and will be recorded as chromatid aberrations.

In contrast to ionizing radiation, the vast majority of known chemical clastogens are S-dependent, which means that they induce aberrations only in cells that pass the S-phase after exposure. Moreover, the predominant type of aberration seen in the first post-treatment mitosis of these cells is of the chromatid type. Thus, it is not the primary DNA lesion *per se*, but rather the processing of the lesion during the S-phase that leads to aberrations. Erroneous processing, such as misreplication or misrepair, could lead to deletions as a result of persistent breaks in the chromatid,Çor to rearrangements and exchanges resulting from faulty rejoining or broken chromatids.

The yield of chromosomal aberrations is often found to be higher when the cells are exposed to clastogens in the early S-phase than during the G_1-phase. This is probably due to the limited time available for error-free repair of the primary DNA lesions in cells entering the S-phase shortly after exposure. Consequently, agents that interfere with the rate or fidelity of DNA replication and repair can influence the final yield of chromosomal aberrations (Preston, 1982; Kihlmann & Natarajan, 1984). Such interactions may occur during testing of mixtures rather than pure compounds. Another consequence is that one could optimize the sensitivity of a test system using chromosomal aberrations as the endpoint by exposing synchronous cells in the early S-phase. However, there are important exceptions to this general rule, in that a few chemical clastogens (e.g., bleomycin, cytosine arabinoside and some compounds that generate free radicals) and ionizing radiation are S-independent and induce aberrations during all phases of the cell cycle.

The majority of cells with aberrations in metaphase may not survive cell division. Unbalanced chromosome-type aberrations are frequently lethal, because both chromatids are affected and both daughter cells suffer from structural chromosome damage. A chromatid deletion will result in a structural aberration in one of two daughter cells, the other being normal. Reciprocal chromatid exchanges may result in homozygosity in a certain proportion of the daughter cells (if homologous chromosomes were involved) or hemizygosity and duplication (if non-homologous chromosomes were involved) of the exchanged segment. Very little is known about the ability of different chemicals to induce these types of derived, stable aberrations, which may be of great importance in cancer induction.

(b) Sister chromatid exchange

SCE represents a symmetrical exchange of chromatid segments within a single chromosome. Such an exchange does not produce an altered chromosome morphology, and the SCE is not detectable unless the sister chromatids are differentially stained or labelled. The points of exchange in the two sister chromatids appear to occur cytologically at the same locus, but the precision of exchange at the molecular level is not known. So far, there is no evidence that SCE *per se* is a cytotoxic or lethal event. Nevertheless, it has been used extensively as an endpoint in cytogenetic research and mutagenicity testing, and these studies have recently been reviewed (Sandberg, 1982; Wolff, 1982b; Tice & Hollaender, 1984).

The detection of SCE involves pretreatment of cells with agents that are incorporated into DNA and which themselves induce SCE, e.g., bromodeoxyuridine (BUdR) and ^{3}H-thymidine (Latt, 1981). Thus, SCE induction by chemical compounds is generally studied against an artificially increased background level, and the presence of BUdR in DNA or precursor pools may interact with the SCE-inducing effect of other

DNA lesions. However, studies of ring chromosomes in which pretreatment with BUdR or ^{3}H-thymidine is unnecessary have indicated that SCE do occur with a low, spontaneous frequency in normal cells (reviewed by Geard, 1984). This conclusion is supported by recent studies with monoclonal antibodies directed against BUdR-containing DNA. This method, which allows the detection of SCE at very low levels of BUdR substitution, should be useful in evaluating interactions between BUdR-induced and mutagen-induced SCEs (Pinkel et al., 1985).

A large number of chemical compounds have been found to increase SCE frequency in plant, animal and human cells in vivo and in vitro (reviewed by Perry, 1980; Latt et al., 1981; Takehisa, 1982). The great majority of these compounds also give rise to DNA and chromosomal damage, as well as gene mutation. Alkylating agents and other electrophilic compounds are generally very potent SCE inducers, whereas base analogues, intercalating agents and compounds that generate free radicals usually cause little if any increase in the frequency of SCEs. Many attempts have been made to determine whether certain types of DNA lesions are more frequently involved in SCE induction than others, e.g., thymidine dimers, O^6-alkylation of guanine, interstrand DNA cross-links, DNA-protein cross-links (reviewed by Wolff, 1982a). In general, none of these studies has given conclusive results, and it is still not known whether SCE results from breakage and reciprocal reunion of DNA strands at the site of the lesion, or if it is a result of events taking place at sites distant from the location of the damage. In the former case, it would be assumed that the type of lesion as well as the repair mechanisms elicited are of critical importance in SCE formation. In the second case, the critical events could be related more closely to mechanisms of replication, the size of the replicon and the action of topo-isomerases.

Since SCE is an S-phase-dependent event, SCE-inducing agents increase SCE frequency only if the cells are allowed to pass through the S-phase after exposure. Treatment of cells in the early S-phase generally induces more SCEs than treatment in G_1 or late S (Lambert et al., 1984). These observations suggest that SCE-inducing damage can be removed before the S-phase, and that the formation of SCE is closely linked to the replication process. This interpretation is further supported by results from studies of repair- and replication-deficient cells from patients with xeroderma pigmentosum, Fanconi's anaemia and Bloom's syndrome (Evans, 1982). The removal of SCE-inducing damage has also been demonstrated in animal and human cells in vivo, by a reduction in SCE frequency with time between exposure and the sampling of the cells for analysis (reviewed by Lambert et al., 1982; Tucker et al., 1986). Hence, the time during the cell cycle at which treatment occurs, and the time at which the cells are harvested for analysis, are important determinants of the final yield of SCE.

Chemically induced SCEs (as well as the BUdR-induced background SCEs) are distributed evenly among the chromosomes — roughly in proportion to the length of the chromosome arms. However, preferential location of SCEs in centromere and telomere regions, as well as at or near the junctions between light- and dark-staining chromosome bands, have been reported (Latt, 1974). The biological significance of this apparent nonrandom location of SCEs is not known but has tentatively been related to sequence rearrangements involved in DNA amplification (Latt *et al.*, 1986).

Studies of the relationship between chromosomal aberrations and SCE suggest that the two endpoints have different mechanisms:

(i) chromatid breaks are uncommon at sites of SCE and *vice versa*;

(ii) some agents induce many chromosomal aberrations but few SCEs, e.g., ionizing radiation and bleomycin;

(iii) many agents induce SCE at low, subtoxic concentrations, which produce few if any chromatid or chromosomal aberrations; and

(iv) the correlation between chromosomal aberrations and SCE in the various repair-deficient and chromosome breakage syndromes is inconsistent.

As most clastogenic agents also induce SCE, it is likely that many common types of primary DNA lesion lead to either or both of the two endpoints, depending on the mechanisms involved.

A linear relationship has been established between the induced frequencies of SCE and gene mutation in V79 and CHO cells, and in human lymphocytes, for a limited number of chemical mutagens and gene loci (reviewed by Carrano & Thompson, 1982). The slope of the regression line differs for different chemical compounds, indicating that some are more efficient inducers of SCE relative to mutation than others. Moreover, several compounds and physical agents induce an increase in the frequency of SCE in the absence of mutation and *vice versa*. Nevertheless, these results suggest a similarity between gene mutation and SCE formation with regard to the types of lesions involved, or the processing of the lesions. Further general observations supporting a relationship between gene mutation and SCE are that:

(i) both endpoints are S-phase-dependent;

(ii) they are detectable at low toxicity and high cell survival;

(iii) many SCE-inducing agents are mutagens; and

(iv) the frequencies of spontaneous SCE and gene mutation are relatively similar over the entire genome.

Even though the major proportion of induced SCEs probably do not represent recombinational or mutational events, it is likely that some fraction of the exchanges occur at non-homologous loci (Stringer *et al.* 1985). Such 'unequal' SCEs could represent mutational events, small chromosomal deletions or be involved in DNA amplification. However, the possible relationship between SCE and other endpoints of tentative importance in carcinogenesis (e.g., mitotic recombination and cell transformation) will probably not be elucidated until further knowledge has been obtained about the mechanism(s) of SCE formation.

(c) *Micronuclei*

Micronuclei arise from chromosome fragments or whole chromosomes that are excluded from the daughter nuclei following cell division. They are usually located in the cytoplasm, where they form single, sharply demarcated, rounded bodies with similar staining properties but a smaller size (1/5-1/20) than the main nucleus. The identification of micronuclei in the interphase of suitably spread and stained preparations is relatively simple, although artefactual structures with similar appearance have been reported occasionally (Jenssen, 1982). To distinguish micronuclei from these artefacts, fluorescent staining methods have been used (Hayashi *et al.*, 1983; MacGregor *et al.*, 1983).

Several in-vitro and in-vivo systems have been developed to study micronuclei, including peripheral lymphocytes and bone-marrow cells from humans, and meiotic germ cells and peripheral-blood cells from rodents (reviewed by Heddle *et al.*, 1983; Choy *et al.*, 1985). The most frequently used in-vivo assay is the mouse bone-marrow test, which takes advantage of enucleated polychromatic erythrocytes to facilitate detection of micronuclei in the absence of a main nucleus.

The induction of micronuclei is related either to chromosomal aberrations, resulting in the formation of acentric fragments, or to dysfunction of the spindle apparatus, giving rise to lagging chromosomes at anaphase. The clastogenic effect is not easily distinguised from spindle dysfunction, unless the DNA content and size of the micronuclei, as well as the timing of the response, are measured.

Like chromosomal aberrations, micronuclei induction is greatly influenced by the time of treatment and sampling in relation to the cell cycle. As micronuclei are detected in the interphase cell, it is essential that the cells in the analysed population be allowed to divide between treatment and analysis. The majority of clastogens gives rise to lesions during the G_1- and S-phases, leading to the formation of micronuclei in the subsequent cell cycle. It is therefore common to allow for one cell cycle between the treatment and the harvesting of cells. At shorter sample intervals, micronuclei may escape detection: longer sampling times may lead to loss or dilution of micronucleated cells. Micronuclei induced by dysfunction of the spindle or by S-phase-independent clastogens can be expected to show up at shorter sampling intervals than micronuclei induced by S-phase-dependent chemicals. Multiple sampling times in relation to treatment are therefore often mandatory.

Centric fragments, chromosomal rearrangements without concomitant production of acentric fragments, and nondisjuction, in which the chromosomes remain associated with the spindle, are not usually detected in the micronucleus assay. Hence, the counting of micronuclei gives little or no information in addition to that which can be obtained from the analysis of metaphase cells for chromosomal aberrations. These limitations are to some extent compensated for by the simplicity of the micronucleus assay and its wide applicability to in-vivo testing, e.g., in human exfoliated cells (Stich & Rosin, 1984).

4.3 Description of the test systems

Cells from many different organisms, including humans, have been used to study the clastogenic effects of potential chemical carcinogens *in vitro* as well as *in vivo*. The most common in-vitro assays use CHO cells or human peripheral lymphocytes (Natarajan & Obe, 1982; Obe & Beek, 1982; Evans, 1984). The rodent bone-marrow system is the most commonly used in-vivo system (Kliesch *et al.*, 1981; Heddle *et al.*, 1984). Many chemicals have been tested in these assays, and the results have recently been compiled and evaluated; guidelines for the procedures, and outlines of the critical steps involved, have been presented in several recent publications (Latt *et al.*, 1981; Preston *et al.*, 1981; Heddle *et al.*, 1983; Scott *et al.*, 1983; Topham *et al.*, 1983; Perry *et al.*, 1984; Dean & Danford, 1984). Only points of principal importance are discussed here.

(a) In-vitro systems

Culture conditions can greatly influence the sensitivity of the response in in-vitro systems. The choice of appropriate doses and treatment times is facilitated by the results of an initial study to determine the cytotoxic effect of the chemical under various exposure conditions. The mitotic index and rate of cell proliferation are useful indicators of cytotoxicity.

Two types of treatment are commonly used. One is addition of the compound for a limited time (1-2 hours) to a partially synchronized cell population, followed by washing and addition of fresh medium. This procedure is preferable in studies of cells exposed at different stages of the cell cycle. When continuous treatment is used, the compound is usually added to a nonsynchronously replicating cell population and is allowed to remain for the duration of the culture time. Cells in all stages of the cell cycle are thus exposed; the dose depends on the stability of the test compound. Knowledge about the half-life and possible degradation products of the compound in the medium is helpful in choosing the proper treatment protocol.

Three doses of the test compound covering a ten-fold concentration range, the highest dose having a detectable cytotoxic effect, are usually required. The fixation time has to be chosen with regard to the delay in cell proliferation that is often caused by the test compound. The use of several fixation times has the advantage that the analysed cell population can be enriched in cells exposed at different stages in the cell cycle.

CHO cells are maintained in exponential growth by regular subculturing, which prevents the cells from reaching confluence. Under these conditions, the cell cycle time is about 12-16 hours, and the chromosomal constitution remains relatively stable, with a modal chromosome number close to the normal diploid number. Partial synchronization of CHO cells can be obtained by shaking off and subculturing the mitotic cells. This is a useful procedure for exposing cells at defined stages in the cell cycle. Mitotic shake-off is also used frequently to enrich metaphase cells at harvest.

Human lymphocytes are usually obtained from healthy volunteers. Due to possible variation between individuals, it is advisable to set up at least two parallel cultures from different individuals for each compound tested. The predominant fraction of the circulating lymphocytes are in a resting stage, G_0, and it is necessary to induce mitotic activity in the cell population with mitogens such as phytohaemagglutinin (PHA). T-lymphocytes, which constitute the major lymphocyte subpopulation, respond to PHA stimulation and reach the S-phase after about 20-30 hours. Mitotic activity usually begins at about 40 hours and reaches a maximum at 70-90 hours after PHA stimulation. The cell-cycle time is highly variable, but averages 18-24 hours, except for the first cell cycle, which is longer. Moreover, initiation of the first cell cycle may vary between different cells in a culture. Hence, only partial synchronization is obtained after PHA stimulation, and only for the first cell cycle.

Whole-blood, buffy-coat and purified mononuclear cells can be used for short-term lymphocyte cultures. However, the growth kinetics and medium requirements differ considerably between these systems; results from one type of culture are not always comparable with those of another. Serum is a less well-defined component of the growth medium, and its content of proteins and radical scavengers may protect against the clastogenic effect of certain compounds. Cells are therefore sometimes treated in the absence of serum. Other confounding factors in the medium include nucleotide precursors. Alterations in precursor pools may influence not only the rate of replication, but also the frequency of chromosomal aberrations and SCE.

A serious disadvantage of cell cultures is their limited capacity for metabolic conversion of potential chemical carcinogens into active metabolites or ultimate carcinogens. Moreover, detoxification processes may be decreased *in vitro*, creating treatment conditions that are artificial and very different from those *in vivo*. The introduction of rodent liver microsomal preparations, together with the appropriate coenzymes, has to some extent circumvented this problem. However, the use of these microsomal preparations to simulate metabolism *in vivo* is itself problematic and adds a further dimension of complexity to the test systems (see Report 15 and Bartsch *et al.*, 1982).

Microsomal preparations are variable in composition and activity, depending on, for example, species, strain, organ, inducing agent, concentration and exposure time used. Hence, there is urgent need for better definition of the optimal conditions for metabolic activation *in vitro*, and for the exploration of alternative methods for this purpose.

At present, the microsomal activation system is a necessary component of all mammalian cytogenetic in-vitro assays. When used as suggested in the various test protocols (e.g., Dean, 1983, 1984), the system works according to expectation for a large number of indirect mutagens and carcinogens. Nevertheless, there is no recommended, simple, uniform procedure for the use of metabolic activation systems under the various conditions of testing for cytogenetic effects in mammalian cell cultures. Recent work has indicated that a number of alternative 'extrahepatic' metabolic pathways are of importance in the activation and detoxification of mutagens and carcinogens (Rydström et al., 1983). These problems are discussed in Report 15.

(b) In-vivo systems

Many tissues from various rodent species have been used to study the induction of chromosomal aberrations, SCE and micronuclei *in vivo*. A major requirement for the studies is that the chemical or its metabolite(s) reach the tissue, which should also have a high rate of cell proliferation. The bone marrow of rats and mice is most frequently used because of its high proportion of dividing cells and the relatively simple procedures for sampling and preparation of cells.

The bone marrow contains several subpopulations of cells, which divide asynchronously and have different cell-cycle times. It is important that the interval between administration of the test compound and sampling of cells permit expression of the endpoints under study. Thus, for the analysis of chromosomal aberrations, the cells are harvested at first post-treatment division, i.e., usually less than 24 hours after treatment. In studies of SCE, two rounds of replication must occur before sampling, and a sufficient concentration of BUdR must be present during at least one of the replications to allow differential staining of the chromatids (see also section 4.2 (b)).

The induction of cytogenetic endpoints is often associated with a delay in cell proliferation and prolongation of the cell cycle. Preliminary studies must therefore be carried out to find the suitable dose range and to assess the bone-marrow toxicity of high doses. This is an important part of the study, as the limited sensitivity of in-vivo assays makes it necessary to test compounds at or close to the maximum tolerated dose, while at the same time extensive distortion of the marrow-cell population and inhibition of the proliferation rate must be avoided. In addition, preliminary toxicity studies could provide important information about the possible differential sensitivity of the various marrow-cell subpopulations, which may lead to spurious conclusions.

The design of the study must take into consideration the possibilities of inadequate or slow metabolism of the test chemical, or high reactivity or rapid inactivation of its metabolite(s), which may prevent the active compound from reaching the bone marrow during the exposure time. The route of administration may vary and is usually chosen with regard to the use of the compound, human exposure routes, previous toxicity data, and the solubility, stability and metabolic pathways of the compound. Hence, prior knowledge of the metabolism and pharmacokinetics of the test compound is a great advantage.

In conclusion, very few general recommendations can be made for in-vivo tests with regard to dose range, timing of treatment and sampling or route of administration because of the versatility and complexity of these systems. Nevertheless, guidelines for the most common test situations, including single, multiple and chronic dose regimens and suitable sampling times for chromosomal aberrations, SCE and micronuclei have recently been published (Heddle *et al.*, 1983; Topham *et al.*, 1983; Heddle *et al.*, 1984). If strictly followed, the suggested protocols should provide reliable results for many compounds and test conditions.

(c) Advantages and disadvantages of cytogenetic test systems

The cytogenetic test systems can be adjusted to a variety of species, tissues and cells *in vitro* and *in vivo*. It is possible to use the same cell type or tissue *in vitro* as well as *in vivo*, and several of the endpoints can be scored in the same system simultaneously (Hayashi *et al.*, 1984; Almássy & Holmberg, 1985) or under very similar conditions (Kliesch *et al.*, 1981; Jones *et al.*, 1985a). The endpoints are specific and sensitive for the purpose of detecting agents with the capacity to cause DNA damage and various types of chromosomal abnormalities, and are likely to be of great relevance to predicting carcinogenic activity as well. Moreover, all of the endpoints are applicable to the monitoring of human populations, which is a valuable complement to the evaluation of epidemiological studies.

The main disadvantages of cytogenetic tests are the amount of work and time needed to conduct them and the considerable subjectivity involved in the analyses. Recent development of sensitive methods for in-situ hybridization, in combination with detection of specific DNA probes by fluorescence staining, may considerably facilitate cytogenetic analysis and improve the specificity of the technique (Pinkel *et al.*, 1986). Moreover, the development of rapid, automated systems is promising (Hutter & Stohr, 1982; Fantes *et al.*, 1983; Philip & Lundsteen, 1985; Gray & Langlois, 1986). Hence the problem of microscopic scoring and analysis of a large number of cells and chromosomes may soon be diminished. The subjectivity involved in the analysis is obvious, but is usually well controlled in experienced laboratories and does not seem to be a major problem. Of greater importance is the difference between laboratories in the interpretation of the various types of chromosomal aberration, e.g., gaps *versus* breaks. The consistent use and publication of historical, positive and negative control data is helpful in dealing with this problem.

In-vitro tests have the obvious advantage of being sensitive, and applicable to a number of different experimental requirements. Their main disadvantage is the need for supplementation with metabolic activation, which almost doubles the amount of work and still provides only limited guidance to the possible effects *in vivo*. The chromosomal aberration tests yield important information about the types of aberrations induced, which can help in deducing the possible mechanisms of action of a compound. The potential sensitivity of the test is very high, but exploitation of this potential is limited because of the work needed to analyse a large number of metaphases. SCE assays are extremely sensitive to a large number of compounds of the types that produce DNA adducts and cross-links, and respond to some nonclastogenic agents as well. The test is less laborious but cannot replace the chromosomal aberration test, as several potent clastogens cause little or no increase in SCE. However, the major disadvantage of SCE assays is that the mechanism leading to this endpoint is unknown, and its biological significance uncertain. Positive responses in SCE assays should therefore be followed up by studies of the induction of chromosomal aberrations.

Cytogenetic assays conducted *in vivo* have the advantage of generating endpoints in the intact mammal, which are generally considered to be more relevant than in-vitro results to predict carcinogenicity, and possibly also in human risk evaluation. However, in-vivo tests are more laborious and less sensitive than the corresponding in-vitro assays. As cells in culture often tolerate higher doses of a test compound than the intact animal, in-vitro tests are usually more effective in detecting weak mutagens, which might be missed in in-vivo tests. A negative response *in vivo* therefore calls for a complementary test *in vitro*.

The rodent bone-marrow micronucleus test is particularly attractive because of its relative simplicity. The endpoint is easily recognized, its biological significance is comparatively clear, and it is sensitive to impairment of the spindle apparatus, an effect that is not easily detected in other assays. The number of cells than can be scored is almost unlimited, and the spontaneous background level is usually very low. Less training in cytogenetics is needed to score for micronuclei than for chromosomal aberrations and SCE. However, the micronucleus test, like the in-vivo chromosomal aberration assay, seems to have low sensitivity for detecting many agents that require metabolic activation or which are inactivated rapidly. The mechanistic information obtained in the micronucleus test is obviously much less than that from chromosomal aberration analysis, but this may not be a serious problem in screening studies.

4.4 *Nature of the data and interpretation of the results*

(a) *Qualitative considerations*

Of major importance for the evaluation of data obtained in cytogenetic short-term tests is the investigators' practical experience of the conduct of such tests and the scientific interpretation of the results. Hence, compilation and evaluation of data collected from negative- and positive-control experiments is worthwhile. The analysis of cytogenetic endpoints is associated with a degree of uncertainty because of, for example, variability between slide readers, the quality of preparations, and use of different staining methods. Highly standardized procedures for scoring and analysis, as well as for the preparation of slides should reduce these problems. To avoid personal bias, the slides must always be coded before analysis. In general, the use of chromosome banding techniques does not increase the sensitivity of the chromosomal aberration assay, but it may improve the degree of resolution and provide valuable mechanistic information.

The analysis of SCE and micronucleated cells is usually not problematic, but the classification of chromatid gaps and breaks is a source of great variation between readers and laboratories (Bianchi et al., 1982; Brøgger et al., 1984). The significance of gaps is not clear (Brøgger, 1982), but they are generally considered to be associated with breaks. Nevertheless, the two classes of aberration should be recorded and tabulated separately.

(b) Quantitative considerations

Observations from a large number of cells or animals must usually be pooled. It is therefore important to present the data in such a way that statistical calculations can be performed by an independent evaluator. The relatively low incidence of cytogenetic endpoints under many test conditions poses difficulties for statistical analysis. Nonparametric statistical methods and methods suggested in recommended protocols (Dean, 1983, 1984) should then be used. Additional problems associated with statistical evaluation are considered by Tice and Hollaender (1984) and in Report 16.

Results of cytogenetic tests are best regarded as qualitative rather than quantitative. The criteria for classifying an agent as giving positive or negative results in a particular assay should be established prior to initiation of the study, giving consideration to the following points:

(i) possible inconsistencies in actual and historical control data and unexpected results in the positive control;

(ii) the sensitivity of the assay, i.e., the minimal response required to obtain the desired statistical significance at a particular dose or in a particular group of treated animals, and the number of cells that must be analysed in order to achieve that level of significance;

(iii) whether a response should be considered positive if there is no clear dose-effect relationship or only when the data combined from all doses show a significant increase; and

(iv) the justification for a repeat study. For obvious reasons, it is particularly important to avoid repeating in-vivo studies; however, this may be necessary due to technical failures or inconclusive results, which emphasizes the importance of the points mentioned above.

If the data do not satisfy the criteria for a positive response, the result should be considered negative. Even so, the study may show unexpected responses which call for repeat or follow-up experiments with appropriate modifications.

The final decision as to whether the response should be considered positive or negative must be related to the actual testing conditions. More general conclusions about the ability of an agent to induce cytogenetic damage require demonstrations of the reproducibility of the results and (preferably) additional tests in other cell systems and with other endpoints.

4.5 *Relevance of chromosomal endpoints to cancer*

Distinctions between different types of genetic alterations may not always be clear. A chromosomal translocation or a small (intragenic) chromosomal deletion could, for example, affect the cell in the same way as a gene mutation. Gene amplification may be associated with chromosomal changes in the form of homogeneously staining regions or double minute chromosomes. SCE is not known to involve rearrangement of DNA or chromosomes but could very well be associated with such events, as well as with mutation. Micronuclei may represent chromosomal fragments due to clastogenic damage and additional chromosomes resulting from spindle dysfunction. One common denominator of all of the chromosomal endpoints is that they can be induced by a wide variety of DNA-damaging agents, many of which have been shown to be carcinogens.

(*a*) *Chromosomal alterations in human neoplasias*

Recent advances in methods for culturing tumour cells and the use of high-resolution banding of chromosomes in combination with molecular genetic techniques have improved both the identification of the chromosomal alterations associated with human malignancies and knowledge of the activation of cellular oncogenes (reviewed by Yunis, 1983). Such information has just begun to emerge, and generalizations may therefore be premature. Nevertheless, the results demonstrate convincingly that malignant cells of most human neoplasias contain chromosomal abnormalities which are usually complex and involve structural as well as numerical changes. In addition, a number of consistent and specific chromosomal aberrations have been detected in various types of human malignant cells (reviewed by Mitelman, 1983).

A limited number of chromosomal break-points seem to be common to the aberrations found in human tumours of different origin, and these breakpoints appear to cluster in a relatively few chromosomal regions (Mitelman, 1983, 1984), some of which are close to the locations of human cellular oncogenes (Rowley, 1983). In some leukaemias and lymphomas, specific chromosomal translocations have been shown to result in the transposition and subsequent activation of the associated oncogene (reviewed by Klein & Klein, 1985). The activation of an oncogene and enhancement of its expression in experimental systems as well as in human tumours frequently involves chromosomal rearrangements, such as deletions, insertions, transpositions and translocations as well as amplification, resulting in homogeneously staining regions and double minute chromosomes (reviewed by Varmus, 1984).

Specific, constitutional chromosomal aberrations have been detected in a few inherited conditions associated with an increased cancer incidence, such as retinoblastoma (Cavenee *et al.*, 1983), aniridia-Wilms' tumour (references given by Solomon, 1984) and hereditary renal carcinoma (Cohen *et al.*, 1979). Expression of the malignant phenotype in retinoblastoma and Wilms' tumour is associated with additional chromosomal changes involving the homologue of the constitutionally deleted chromosome. In at least half of retinoblastoma cases and two-thirds of Wilms' tumour cases, the changes appear to be caused by chromosome loss or mitotic recombination, resulting in hemi- or homozygosity of the constitutionally deleted chromosome (Cavenee et al., 1985). Similar mechanisms seem to be involved in other tumours (Hansen el al., 1985; Koufos et al., 1985).

In the human autosomal recessive conditions Bloom's syndrome, ataxia telangiectasia and Fanconi's anaemia, increased incidence of malignancy is associated with an increased spontaneous frequency of nonspecific chromosomal aberrations and SCE in somatic cells (reviewed by Arlett & Lehmann, 1978). Cells from these individuals, as well as from patients with xeroderma pigmentosum, another inherited cancer-prone condition associated with defective DNA repair mechanism(s), show abnormal responses with regard to the induction of chromosomal aberrations and SCE when exposed to clastogens or DNA-damaging agents *in vitro* (reviewed by Lehmann, 1982). These observations suggest that an enhanced frequency of nonspecific chromosomal aberrations in somatic cells may increase the probability of developing neoplasia.

In spite of the obvious impact of these observations on current understanding of cancer development, it is still an open question whether chromosomal changes are primary steps related to the induction of transformation, or later steps related to further development of the transformed phenotype and progression of neoplastic growth. Moreover, certain malignancies may develop along a pathway that does not involve genetic alterations at all. Nevertheless, it is evident that most human malignancies and cancer-prone conditions involve or are associated with some more or less specific chromosomal abnormalities, suggesting an important role of these endpoints in carcinogenesis.

(b) Chromosomal endpoints in short-term tests

Even though many of the chromosomal aberrations detected in short-term tests are probably lethal to cells, it is likely that chemical clastogens also induce a low frequency of non-lethal deletions or rearrangements, which are converted into stable aberrations in subsequent cell cycles. Further research to clarify the relationship between gross structural aberrations and the small, balanced types of aberrations likely to be of relevance in tumour development is needed.

There is no direct link between SCE and carcinogenesis. However, SCE causes no or little reduction in cell viability and is strongly correlated to mutation. A fraction of SCEs may be 'unequal', giving rise to deletion or duplication mutations in the progenitor cells (Stringer et al., 1985). Moreover, it has been suggested that SCE is associated with homologous recombination, which may lead to expression of recessive genes by making them homozygous or by altering gene expression *via* changes in the pattern of DNA methylation (Radman et al., 1982). These suggestions must be considered purely speculative as long as the molecular and enzymatic mechanisms involved in SCE formation are unknown. Accordingly, the relevance of SCE as an endpoint in short-term tests has to be based on its high sensitivity for detecting DNA-damaging, carcinogenic, clastogenic and mutagenic chemicals, and a number of weak carcinogens as well. Hence, even though SCE could be a biologically unimportant by-product of carcinogenic damage, it must be regarded as a very appropriate endpoint to be utilized in test systems to detect chemical carcinogens.

4.6 Conclusions

Four strands of evidence confirm the relationship between chromosomal alterations and cancer development. The first emerges from recent studies of oncogene activation and chromosomal rearrangements in human malignancies. The second comes from studies of some human childhood cancers associated with a constitutional chromosome deletion. The third relates to observations of an increased frequency of chromosomal alterations and enhanced sensitivity to clastogens and SCE-inducing agents in some human cancer-prone conditions. The fourth is the strong positive correlation between the carcinogenicity and the clastogenicity or SCE-inducing ability of a large number of chemicals in short-term tests. Taken together, these observations suggest that chromosomal alterations are extremely valuable and highly relevant endpoints for the detection of potential carcinogens.

5. Assays of aneuploidy in mammalian cells

5.1 *Association of numerical chromosomal changes with reproductive pathology and tumours*

Constitutional variations in the human chromosome number are clearly related to infertility and to congenital or hereditary diseases associated with malformations and mental defects. Polyploidy and aneuploidy are thought to be among the major causes of embryonic loss. Data on chromosome number at the very early stages of development can be obtained by analysing products of in-vitro fertilization (Angell *et al.*, 1983). Studies of spontaneous abortions indicate a frequency of about 50% for all chromosomal abnormalities, and most of these (70%) are aneuploidies. The best estimates on the incidence of numerical chromosomal abnormalities in human populations come from surveys of newborn babies. The overall frequency of aneuploidy is one in 650 live births (Chandley, 1982). Analysis of the human sperm chromosome complement by in-vitro fertilization of hamster eggs (Rudak *et al*, 1978; Martin, 1983) can be used to establish the frequency of aneuploidy among the male gametes. This frequency has been reported to be 1.8% and 3.1% for hyperhaploidy and hypohaploidy, respectively (Martin, 1985). Similar results have been obtained by other groups using similar techniques (Brandiff *et al.*, 1984, 1985). These data indicate a substantial contribution of numerical chromosome errors to the pathology of reproduction and development in humans.

Constitutional aneuploidies are also sometimes associated with an increased risk of developing specific types of cancer. Examples include leukaemia in patients with Down's syndrome, breast cancer in patients with Klinefelter's syndrome and gonadoblastoma in individuals with an abnormal sex chromosome constitution (reviewed by Sandberg, 1980).

There is increasing evidence that aneuploidies and the processes from which they originate may also play an important role in the emergence and progression of neoplasia. Firstly, the occurrence of nonrandom trisomies in some tumours, such as trisomy-15 in murine T-cell lymphomas, is a strong indication that a gene dosage mechanism operates in the development of malignancy (Yunis, 1983). Secondly, changes in chromosome numbers, tending mostly towards hyperploidy, may be important events in the evolution of tumour-cell populations (Wake et al., 1982; Larizza & Schirrmacher, 1984). Mitotic nondisjunction may lead to the expression of a recessive allele that controls neoplastic transformation, as in retinoblastoma and Wilms' tumour (Cavenee et al., 1983; Solomon, 1984).

5.2 *Rationale of aneuploidy assay systems and their equivalence to tumours developing in the organism*

Errors in the distribution of chromosome numbers in mammals can occur during either meiosis or mitosis. These two processes exhibit common targets to aneuploidy-inducing agents, namely, the centromere and the kinetochore on the chromosomes, the microtubules in the spindle, and the associated centrioles and the centrosome of the spindle apparatus. These cell constituents can be considered specific targets, since they are integral parts of the apparatus for chromosome duplication, association and movement during cell division. Nonspecific targets both in meiosis and mitosis are membranes, as sites of oxidative damage, and intermediate filaments of the vimentin and cytokeratin type (Aubin et al., 1980; Harris, 1975). Moreover, spindle disturbances may be caused by various agents that interact with different cellular processes, such as the calcium-sequestering system, oxidative phosphorylation and sulfhydryl balance. Point mutations (i.e., changes in DNA sequence restricted to one or a few neucleotides, not necessarily located on the centromere but distributed everywhere in the genome) must also be considered a possible cause of polyploidy and aneuploidy. Several mutants for α- and β-tubulin genes have been isolated in *Drosophila melanogaster* (Raff, 1984), which show phenotypes with specific alterations in microtubule-mediated processes. In mammalian cells, temperature-sensitive mutants characterized by various mitotic abnormalities have been reported (Wang et al., 1983).

Given the specific and nonspecific mechanisms that generate numerical chromosome variations, a wide variety of chemical agents are possible aneuploidy inducers. They may include substances that interact directly with DNA or with constituents of the mitotic apparatus or which act through intermediary molecules. Because of the possibility of indirect effects, knowledge of the toxicity of a compound to be screened for induction of aneuploidy is an essential step in the assay.

In testing for the potential carcinogenicity of chemicals, mitotically dividing populations appear to be the only relevant material for analysing aneuploidy. Tests can be carried out either *in vivo* or *in vitro*. In in-vivo tests, the substance is administered to an animal, and the mitoses are examined in direct preparations or after culture; in in-vitro tests, the whole procedure is accomplished in cell culture. Observations on mitotic effects can be made at two different levels: (i) by examining the morphology of chromosomes (c-mitoses), the appearance of the spindle (e.g., birefringence) or metaphase and anaphase configurations (multipolar mitoses); and (ii) by counting the chromosomes in metaphase or scoring anaphases with lagging chromosomes. The first type of observation is adequate only for qualitative evaluations, whereas the second provides quantitative results. Anaphase analysis can give only supplementary information. The counting of chromosomes in metaphase appears to be the most reliable method for measuring mitotic aneuploidy, as it enables the presence or absence of a whole chromosome to be detected. DNA probes consisting of chromosome-specific sequences are a promising tool for the development of new and more specific tests (e.g., Pinkel *et al.*, 1986).

The basic constituents of the mitotic apparatus, including both chromosomal and extrachromosomal structures, and the mechanisms of action, are thought to be the same in all cells of all tissues, whether of normal or of malignant origin. Differences between species are expected to be irrelevant. Therefore, measurement of aneuploidy rates in isolated cells derived either from the exposed organism or cultured *in vitro* should be equivalent to measuring effects that may be involved in tumour development and progression. Biochemical equivalence among different cell types and different species of animals is subject to the same considerations as have already been discussed for other genetic effects in mammalian cell systems (see section 1).

5.3 *Description of the test systems and nature of the data*

(a) *In-vivo cell systems*

Only mitotic cell systems are considered, because they are particularly relevant to carcinogenicity assays. Theoretically, any cell type from tissues which divide *in vivo* or are capable of proliferation *in vitro* after removal from an organism can be used to reveal changes in chromosome numbers following exposure to a test chemical. Transplantable tumours can be exploited for this purpose, as was done in the case of structural chromosomal aberrations (Wobus *et al.*, 1978). However, in view of the chromosomal instability of these tumours and their prevailing hyperploid constitution, they must be considered poor indicators of numerical chromosomal aberrations. The most commonly used cell types in studies on cytogenetic damage induced *in vivo* are peripheral-blood lymphocytes and bone-marrow cells. Lymphocytes are the most suitable material for investigations on human populations exposed to mutagens or carcinogens. They are not commonly used in in-vivo assays with experimental animals, although they should be considered a useful source of material, in view of the need to relate data from in-vitro and in-vivo assays. Test procedures with bone-marrow cells are basically those already described in section 4.

(i) *Bone marrow metaphases and/or anaphases*

The protocols used to detect structural chromosomal aberrations in bone marrow can be adapted to the requirements of an aneuploidy test by introducing a few modifications in exposure to colcemid and hypotonic treatment. The duration of colcemid treatment, either *in vivo* or *in vitro*, during the short culture period can be doubled, because, within certain limits, chromosome contraction does not affect chromosome counts and the number of mitoses that can be scored is also higher. Hypotonic treatment should be done in such a way as to prevent excessive chromosome scattering.

(ii) *Bone-marrow micronuclei in polychromatic erythrocytes*

This test is used primarily for detecting structural chromosomal aberrations, as described in the previous chapter. Its use for aneuploidy testing relies on the fact that an appreciable fraction of micronuclei consists of whole chromosomes. However, the criteria used to discriminate between two subsets of micronuclei, based on morphometric analysis or on determination of DNA content, do not appear to be fully satisfactory.

(b) *In-vitro cell systems*

A great variety of cultured mammalian cell types have been used in cytogenetic tests to evaluate mutagenic/carcinogenic potential of chemicals (e.g., lymphocytes, fibroblasts, epitheliocytes). Euploidy is not an absolute requirement for the detection of structural chromosomal aberrations, but the constancy of the chromosome set under basal conditions is by definition essential for analysing numerical chromosomal abnormalities induced by chemicals. For this reason, it is necessary to use homodiploid cell populations, as in B-lymphocyte or T-lymphocyte cell cultures or in fibroblast cell strains with a definite life span. Human cells are preferable because they vitiate problems of extrapolation from animals and also because they show considerable chromosome stability *in vitro*.

The most highly standardized systems for the study of chromosomal damage induced in human cells *in vitro* are blood lymphocyte cultures and fibroblast cultures. The lack of metabolizing capacities in such cultures requires the addition of a $9000 \times g$ supernatant fraction (S9) of liver from rats (see Report 15).

There is no validated procedure for detecting specialized aneuploidies, namely, the effects of nondisjunction for specific chromosomes. A method has been proposed, based on detection of the Y chromosome on Q-banded metaphases from human diploid cells in primary culture — either PHA-stimulated lymphocytes or fibroblasts (Tenchini et al., 1983). This method takes advantage of the intense brilliance of the q12 band in the long arm of the Y chromosome, which allows it to be easily recognized. The presence of additional copies of this chromosome can thus be readily recognized. The mitoses need not be karyotyped, and the analysis is therefore exactly comparable with that of the Barr body in resting nuclei. Treatment with test compounds, cell cultures and chromosomal preparations are done according to standard methods (see section 4.3). The only modification concerns hypotonic treatment, which can be milder than usual to avoid excessive spread of chromosomes in the metaphase plates. A method similar to the 'double Y method' uses the late replicating X in females as a marker; the presence of multiple Xs can be revealed by autoradiography. However, the relatively complex technical procedures required by this method make it less practicable. In general, test systems based on the detection of non-disjunctional events involving individual chromosomes offer the advantage of a higher target specificity. A serious limitation lies in the size of cell samples to be scored, which would require the use of automated systems for scoring mitoses (Shafer et al., 1985).

5.4 *Qualitative and quantitative considerations*

Any assessment of the accuracy and precision of aneuploidy tests is hindered by the scarcity of data and by the lack of adequate numerical analysis in existing reports. Variations in chromosome counts may arise from two sources: random errors and systematic errors. Systematic error is related primarily to factors inherent in in-vitro culture, especially in cytotoxicity assays, and to the quality of the cytological preparations.

Because of difficulties in discriminating between generalized toxic effects and the induction of spindle disturbances and errors in chromosome distribution, it is important to obtain cytotoxicity data before undertaking an actual assay. The cytotoxicity of a potential carcinogen can be determined in preliminary experiments on permanent cell lines, either in the presence or in the absence of S9 fraction, by using suitable parameters such as plating efficiency, growth rate and mitotic index.

At least three doses of the chemical must be included in the test, ranging from subtoxic to nontoxic levels. A dose-response relationship should be sought in order to establish the positive effect of a potential carcinogen on chromosome distribution at mitosis.

Significant endpoints are spindle appearance, state of contraction of the chromosomes, chromosome counts of metaphases and lagging chromosomes at anaphases. Whereas the first two indicators are of a qualitative nature, the latter two permit quantitative evaluations. The counts should be done on at least 100-200 mitoses per dose, depending on the spontaneous background frequency, when numerical variation is being considered for any element of the chromosome complement (generalized aneuploidy), and on 1000-2000 mitoses when a specific pair of chromosomes is being dealt with. The data can be expressed as a percentage of abnormal mitoses when only two classes are recognized, corresponding to cells with an exactly diploid number of chromosomes and cells with a hyperploid constitution. This implies that cells with missing chromosomes are not considered because of the high risk of losses due to technical artefacts. The significance of the differences observed can be evaluated statistically on $\log(x + 10)$ transformed values with Student's 't' test. When the procedures for preparing chromosomes are properly controlled, hypodiploid cells can be included in the results, thus generating three classes — hypodiploid, diploid and hyperploid. The most convenient statistical treatment in this case is a chi-square determination on a 2×3 table.

5.5 *Assessment of the test systems*

Table 6 gives a list of chemicals for which positive results have been obtained in assays for aneuploidy or spindle disturbances. All the data obtained are reported, whether on a qualitative or a quantitative basis and with or without a clear dose-response relationship. Twelve of the compounds listed in the table can safely be recognized as aneuploidy inducers, in that they cause a significant increase in hyperploidy. They are used in Table 7 for a comparison between data on aneuploidy and on carcinogenicity.

This comparison was restricted to compounds for which aneuploidy induction was determined quantitatively by chromosome counts. All information about carcinogenicity was derived from the *IARC Monographs* (IARC, 1974a,b, 1976a,b, 1979, 1982), except in two cases, where additional information was available from the US National Toxicology Program (1985a,b). Due to the limited number of chemicals screened, the selection criteria and the high proportion of missing data, no correlation value can be calculated from these comparisons. However, it can be seen that most of the compounds reported as carcinogenic gave a positive response in tests for mitotic errors and/or induction of numerical chromosome changes. At present, therefore, data on aneuploidy must be considered as complementary information, which may be helpful in the evaluation of the carcinogenic potential of a chemical agent.

The relative advantages and disadvantages of the in-vivo and in-vitro systems used for aneuploidy tests and their relative sensitivity and efficiency, are closely comparable with those for assays of structural chromosomal aberrations. These have been discussed in detail in a previous report (IARC, 1980) and in section 4 of this report. Moreover, any critical evaluation of screening methods for chemicals capable of inducing numerical chromosome variation is hampered by lack of validation of the testing procedures. However, attention is drawn to some features that may be pertinent to aneuploidy tests *per se*. These concern comparisons (i) between diploid and quasidiploid or heteroploid systems, which are characteristic of primary cultures or cell strains and permanent cell lines, respectively; and (ii) between scoring for generalized and specialized aneuploidy.

Table 6. Chemicals that have given positive results in tests for aneuploidy or spindle disturbance[a]

Chemical	Cells[b]	Effects[c]	Reference
Asbestos (chrysotile)	SHE	AP, PP	Oshimura et al. (1984)
Asbestos (crocidolite)	SHE	PP, anaphase lag	Hesterberg & Barrett (1985); Oshimura et al. (1984)
Benomyl	HPBL	AP	Gupta & Legator (1976); Tenchini et al. (1983)
Benzo[a]pyrene	SHE	AP	Benedict et al. (1972)
Benz[a]anthracene	SHE	AP	Benedict et al. (1972)
Carbaryl (1-naphthyl-N-methyl carbamate)	V79	AP, PP	Onfelt & Klasterska (1983)
Colcemid	HPBL	AP	Tenchini et al. (1983)
	BHK21	AP, PP	Barass (1982)
	DON	AP, PP	Hsu et al. (1983a)
	CHW	AP	Cox et al. (1976)
	SHE	AP, PP	Tsutsui et al. (1984)
Diazepam	DON	cm, mm, AP	Hsu et al. (1983b)
	JOK-1	cm	Andersson et al. (1981)
Diethylstilboestrol	SHE	AP	Tsutsui et al. (1983)
Diethylstilboestrol dipropionate		cm, abnormal mitoses	Danford & Parry (1982); Parry et al. (1982)
9,10-Dimethyl-1,2-benz[a]anthracene	SHE	AP	Benedict et al. (1972)
Distamicin A	HPBL	AP	Tenchini et al. (1983)
Griseofulvin	HPBL	AP, PP	Larizza et al. (1974)
	PtK$_1$	cm	Mullins & Snyder (1979)
N-Methyl-N'-nitro-N-nitrosoguanidine	CHO	AP, PP	Bempong (1979)
Mysoline	HPBL	AP, PP	Bishun et al. (1975)
Oestradiol	HSM, HSL, HSF	AP	Lycette et al. (1970)
Phenytoin	HPBL	AP, PP	Bishun et al. (1975)
Rhodamine B	MF	AP, PP	Lewis et al. (1981)
TPA (12-O-tetradecanoyl phorbol 13-acetate)	MPE	AP	Dzarlieva & Fusenig (1982)
Vinblastine	DON	AP	Hsu et al. (1983a)
	CH	AP, PP, mm	Palyi (1976)
Zarotin	HPBL	AP, PP	Bishun et al. (1975)
Ethidium bromide	DON, TCH-2352	mm, er, abnormal centrioles	McGill et al. (1974)
Gentian violet	CHO	mm	Au et al. (1978)
Halothane	V79	cm, mm	Sturrock & Nunn (1976)
Isopropyl(N-3-chlorophenyl) carbamate	3T3	abnormal mitoses	Oliver et al. (1978)
Joduron	HyCH	cm	Schmid & Bauchinger (1976)
Mercuric chloride	HPBL	cm	Verschaeve et al. (1984)
Methylmercury	HPBL	cm	Verschaeve et al. (1984)
Metronidazole	V79	PP, er	Korbelick & Horvat (1980)
Nitrous oxide	HeLa	mm, cm	Brinkley & Rao (1973)

Table 6. (contd)

Chemical	Cells[b]	Effects[c]	Reference
Nocodazole	HeLa, CHO, WI38, L(NCTC-929)	mitotic arrest	Zieve et al. (1980a)
Olivetol	HPBL	anaphasic lag, unequal segregation, mm	Morishima et al. (1976)
Potassium dichromate	HEp-2	cm, PP	Majone (1977)
Sodium ortho-vanadate	PtK$_1$	inhibition of chromosome movement	Cande & Wolniak (1978)

[a] Modified from Galloway & Ivett (1986)

[b] Continuous cell lines: CH, CHO, CHW, V79, DON, TCH-2352; HyCH: Chinese hamster; HeLa, HEp-2: human carcinomas; JOK-1: human leukaemia; MPE, L(NCTC-929), 3T3: mouse; PtK$_1$: marsupial kidney. SHE, BHK21: Syrian hamster; HSM, HSL, HSF: human synovial fluid; MF: *Muntiacus muntjac* fibroblasts

[c] Cells in short-term cultures or cell strains: HSBP, W138: human fibroblasts; HPBL: human peripheral blood lymphocytes

[d] AP, aneuploidy; PP, polyploidy; cm, c-mitoses; er, endoreduplications; mm, multipolar mitoses

Table 7. Mammalian cell systems

Chemical	Response to aneuploidy test	Carcinogenicity[a]	Reference
Asbestos (chrysotile)	+	+ (animal, human)	IARC (1982); US National Toxicology Program (1985a)
Benomyl	+	0	
Carbaryl	+	i	IARC (1976a)
Colcemid	+	0	
Colchicine	+	0	
Cyproheptadine hydrochloride	-	0	
Diazepam	+	0[b]	
Diethylstilboestrol	+	+ (animal, human)	IARC (1982)
9,10-Dimethyl-1,2-benz[a]anthracene	+	i (animal)[c]	US National Toxicology Program (1985b)
Ethylene thiourea	-	+ (animal)	IARC (1974a)
Griseofulvin	+	+ (animal)	IARC (1976b)
N-Methyl-N'-nitro-N-nitrosoguanidine	+	+ (animal)	IARC (1974b)
Mitomycin C	-	+ (animal)	IARC (1976b)
Testosterone	-	+ (animal)	IARC (1979)
TPA (12-O-tetradecanoyl-phorbol 13-acetate)	+	0	
Vinblastine	+	i (animal, human)	IARC (1982)

[a] +, positive in aneuploidy tests, carcinogenic; 0, not reported; i, inadequate evidence

[b] Oxazepam, a major metabolite of diazepam, was carcinogenic in mice (IARC, 1977)

[c] Tested in combination with 2,3,7,8-tetrachlordibenzo-p-dioxin

Established human or other mammalian cell lines are easier to handle than primary cultures and diploid cell strains, and are therefore particularly helpful in ascertaining the clonal nature of induced aneuploidy and for toxicity studies based on plating efficiencies. However, instability of the chromosomal complement in most of the permanent cell lines is a serious hindrance to their use in aneuploidy tests.

Generalized aneuploidy, which is the most commonly used endpoint, has the advantage of having a variety of exploitable targets, and therefore the size of the mitotic samples to be analysed can be limited. Even so, generalized aneuploidy appears to be less indicative than specialized aneuploidy, because single disjunctional errors are considered to be one of the possible steps in cancer development.

5.6 *Conclusions*

Aneuploidy is an important indicator of genetic damage, reflecting alterations of structure and function necessary to maintain chromosomal dosage. Because of the considerable impact of numerical chromosomal aberrations on human health, much effort should be expended on developing specific test methods. Studies on aneuploidy-inducing chemicals are still fragmentary, and there is as yet no uniform methodological approach to the problem. The data reported are often inconclusive. Since no definitive protocols are available for aneuploidy assays in mammalian cells, further tests need to be developed and validated. The group made the following recommendations:

(*a*) The effect of any test compound on the mitotic index should be carefully analysed in order to find the proper dose range, before initiating a aneuploidy assay.

(*b*) In-vitro treatments with the addition of an appropriate metabolizing mixture are complementary to in-vivo studies and readily provide data for dose-response relationships.

(*c*) Efforts should be concentrated on human diploid cell systems such as fibroblast strains and B- and T-lymphocyte cultures, because they exhibit high karyotypic stability and allow results to be more directly extrapolated to evaluation of human hazard.

(*d*) In addition to chromosome counts, qualitative investigations should be done on spindle configurations (using the novel techniques based on fluorescent antibodies) and on chromosome morphology.

(*e*) New methods should be sought, based on molecular probes and automated scanning systems, to replace present laborious procedures which employ microscopic examination of individual mitotic preparations.

6. References

Albertini R.J. (1982) *Studies with T-lymphocytes: an approach to human mutagenicity testing.* In: Bridges, B.A., Butterworth, B.F. & Weinstein, I.B., eds, *Indicators of Genotoxic Exposure (Banbury Report 13)*, Cold Spring Harbor, NY, CSH Press, pp. 343-412

Albertini, R.J. (1985) Somatic gene mutations *in vivo* as indicated by the 6-thioguanine resistant T lymphocytes in human blood. *Mutat. Res. 150*, 411-422

Albertini, R.J. & DeMars, R. (1970) Diploid azaguanine-resistant mutants of cultured human fibroblasts. *Science, 169*, 482-485

Albertini, R.J., Castle, K.L. & Borcherding, W.R. (1982) T cell cloning to detect the mutant 6-thioguanine resistant lymphocytes present in human peripheral blood. *Proc. natl Acad. Sci. USA, 79*, 6617-6621

Albertini, R.J., O'Neill, J.P., Nickles, J.A., Heintz, N.H. & Kelleher, P.C. (1985) Alterations of the *hprt* gene in human *in vivo*-derived 6-thioguanine resistant T lymphocytes. *Nature, 316*, 369-371

Almássy, Z. & Holmberg, M. (1985) Simultaneous detection of SCE and Q-bands on human chromosomes by a double-staining technique. *Cancer Genet. Cytogenet., 16*, 153-156

Anderson, D., Richardson, C.R. & Davies, P.J. (1981) The genotoxic potential of bases and nucleosides. *Mutat. Res., 91*, 265-272

Andersson, L., Lehto, V.-P., Stenman, S., Bradley, R.A. & Virtanen, I. (1981) Diazepam induces mitotic arrest at prometaphase by inhibiting centriolar separation. *Nature, 291*, 247-248

Angell, R.R., Aitken, R.J., van Look, P.F.A., Lumsden, M.A. & Templeton, A.A. (1983) Chromosome abnormalities in human embryos after in-vitro fertilization. *Nature, 303*, 336-338

Archer, P.G., Bender, M., Bloom, A.D., Brewen, J.G., Carrano, A.V. & Preston, R.J. (1981) *Guidelines for cytogenetic studies in mutagen-exposed human populations.* In: Bloom, A.D., ed., *Guidelines for Studies of Human Populations Exposed to Mutagenic and Reproductive Hazards*, New York, March of Dimes Birth Defects Foundation, pp. 1-35

Arlett, C.F. & Lehmann, A.R. (1978) Human disorders showing increased sensitivity to the induction of genetic damage. *Ann. Rev. Genet., 12*, 95-115

Arlett, C.F., Turnbull, D., Harcourt, S.A., Lehmann, A.R. & Colella, C.M. (1975) A comparison of the 8-azaguanine and ouabain-resistance systems for the selection of induced mutant Chinese hamster cells. *Mutat. Res., 33*, 261-278

Ashby, J. & Purchase, I.F.H. (1985) Significance of the genotoxic activities observed *in vitro* for 35 of 70 NTP non-carcinogens. *Environ. Mutagenesis, 7*, 747-758

Au, W., Pathak, S., Collie, C.J. & Hsu, T.C. (1978) Cytogenetic toxicity of gentian violet and crystal violet on mammalian cells *in vitro*. *Mutat. Res., 58*, 269-276

Aubin, J.E., Osborn, M., Franke, W.W. & Weber, K. (1980) Intermediate filaments of the vimentin-type and the cytokeratin-type are distributed differently during mitosis. *Exp. Cell Res., 129*, 149-165

Aust, A.E., Falahee, K.J., Maher, V.M. & McCormick, J.J. (1980) Human cell-mediated benzo[*a*]-pyrene cytotoxicity and mutagenicity in human diploid fibroblasts. *Cancer Res., 40*, 4070-4075

Aust, A.E., Drinkwater, N.R., Debien, K., Maher, V.M. & McCormick, J.J. (1984) Comparison of the frequency of diphtheria toxin and thioguanine resistance induced by a series of carcinogens to analyze their mutational specificities in diploid human fibroblasts. *Mutat. Res., 125*, 95-104

Barass, N.C. (1982) The incidence of spontaneous and radiation-induced chromosome damage in a trisomic variant of a diploid mammalian cell line. *Prog. Mutat. Res.*, *4*, 85-98

Barrett, J.C., Wong, A. & McLachlan, J.A. (1981) Diethylstilbestrol induces neoplastic transformation without measurable gene mutation at two loci. *Science*, *212*, 1402-1404

Bartsch, H., Kuroki, T., Roberfroid, M. & Malaveille, C. (1982) *Metabolic activation systems in vitro for carcinogen/mutagen screening tests.* In: de Serres, F.J. & Hollaender, A., eds, *Chemical Mutagens, Principles and Methods for their Detection*, Vol. 7, New York, Plenum, pp. 95-161

Bempong, M.A. (1979) Mutagenicity and carcinogenicity of N-methyl-N'-nitro-N-nitrosoguanidine. I. Induction of chromosome aberrations and mitotic anomalies in Chinese hamster ovary cells. *J. environ. Pathol. Toxicol.*, *2*, 633-656

Benedict, W.F., Gielen, J.E. & Nebert, D.W. (1972) Polycyclic hydrocarbon-produced toxicity, transformation, and chromosomal aberrations as a function of aryl hydrocarbon hydroxylase activity in cell cultures. *Int. J. Cancer*, *9*, 435-451

Bianchi, M., Bianchi, N.O., Brewen, J.G., Buckton, K.E., Fabry, L., Fischer, P., Gooch, P.C., Kucerova, M., Leonard, A., Mukherjee, R.N., Mukherjee, U., Nakai, S., Natarajan, A.T., Obe, G., Palitti, F., Pohl-Ruling, J., Schwarzacher, H.G., Scott, D., Sharma, T., Takahashi, E., Tanzarella, C. & Van Buul, P.P. (1982) Evaluation of radiation-induced chromosomal aberrations in human peripheral blood lymphocytes *in vitro*. Results of an IAEC-coordinated programme. *Mutat. Res.*, *96*, 233-242

Bishun, N.P., Smith, N.S. & Williams, D.C. (1975) Chromosomes and anticonvulsant drugs. *Mutat. Res.*, *28*, 141-143

Bradley, M.O. & Sharkey, N.A. (1978) Mutagenicity of thymidine to cultured Chinese hamster cells. *Nature*, *274*, 607-608

Bradley, M.O., Bhuyan, B., Francis, M.C., Langenbach, R., Peterson, A. & Huberman, E. (1981) Mutagenesis by chemical agents in V79 Chinese hamster cells: a review and analysis of the literature. *Mutat. Res.*, *87*, 81-142

Brandiff, B., Gordon, L., Ashworth, L., Watchmaker, G., Carrano, A. & Wyrobek, A. (1984) Chromosomal abnormalities in human sperm: comparisons among four healthy men. *Human. Genet.*, *66*, 193-201

Brandiff, B., Gordon, L., Ashworth, L., Watchmaker, G., Moore, D., II, Wyrobek, A.J. & Carrano, A.V. (1985) Chromosomes of human sperm: variability among normal individuals. *Human Genet.*, *70*, 18-24

Braun, R., Russell, L.B. & Schöneich, J. (1982) Workshop on the practical application of the mammalian spot test in routine mutagenicity testing of drugs and other chemicals. *Mutat. Res.*, *97*, 155-161

Brinkley, B.R. & Rao, P.N. (1973) Nitrous oxide: effects on the mitotic apparatus and chromosome movement in HeLa cells. *J. Cell Biol.*, *58*, 96-106

Brøgger, A. (1982) The chromatid gap — a useful parameter in genotoxicology? *Cytogenet. Cell Genet.*, *33*, 14-19

Brøgger, A., Norum, R., Hansteen, I.L., Clausen, K.O., Skardal, K., Mitelman, F., Kolnig, A.M., Strömbeck, B., Nordenson, I., Andersson, G., Jakobsson, K., Mäkipaakkanen, J., Norppa, H., Järventaus, H. & Sorsa, M. (1984) Comparison between five Nordic laboratories on scoring of human lymphocyte chromosome aberrations. *Hereditas*, *100*, 209-218

Burger, P.M. & Simons, J.W.I.M. (1979) Mutagenicity of 8-methoxypsoralen and long-wave ultraviolet irradiation in diploid human skin fibroblasts. *Mutat. Res.*, *63*, 371-380

Cande, W.Z. & Wolniak, S.M. (1978) Chromosome movement in lysed mitotic cells is inhibited by vanadate. *J. Cell Biol.*, *79*, 573-580

Carrano, A.V. & Thompson, L.H. (1982) Sister chromatid exchange and single gene mutation. *Cytogenet. Cell Genet.*, *33*, 57-61

Cavenee, W.K., Dryja, T.P., Phillips, R.A., Benedict, W.F., Godbout, R., Gallie, B.L., Murphree, A.L., Strong, L.C. & White, R.L. (1983) Expression of recessive alleles by chromosomal mechanisms in retinoblastoma. *Nature*, *305*, 779-784

Cavenee, W.K., Hansen, M.F., Nordenskjöld, M., Kock, E., Maumenee, I., Squire, J.A., Phillips, R.A. & Gallie, B.L. (1985) Genetic origin of mutations predisposing to retinoblastoma. *Science*, *228*, 501-503

Chandley, A.C. (1982) *The Origin of Aneuploidy. Human Genetics*, Part B, *Medical Aspects*, New York, Alan R. Liss, pp. 337-347

Chang, C.-C., Trosko, J.E. & Akera, T. (1978) Characterization of ultraviolet light-induced ouabain-resistant mutations in Chinese hamster cells. *Mutat. Res.*, *51*, 85-98

Chasin, L.A. (1973) The effect of ploidy on chemical mutagenesis in cultured Chinese hamster cells. *J. cell. Physiol.*, *82*, 299-308

Choy, W.N., MacGregor, J.T., Shelby, M.D. & Maronpot, R.R. (1985) Induction of micronuclei by benzene in $B6C3F_1$ mice: Retrospective analysis of peripheral blood smears from the NTP carcinogenesis bioassay, *Mutat. Res.*, *143*, 55-59

Chrysostomou, A., Seshadri, R. & Morley, A.A. (1984) Decreased cloning of lymphocytes from elderly individuals. *Scand. J. Immunol.*, *19*, 293-296

Chu, E.H.Y. (1974) Induction and analysis of gene mutations in cultured mammalian somatic cells. *Genetics*, *78*, 115-132

Chu, E.H.Y. (1983) Mutation system in cultured mammalian cells. *Ann. N.Y. Acad. Sci.*, *407*, 221-230

Chu, E.H.Y. & Malling, H.V. (1968) Mammalian cell genetics. II. Chemical induction of specific locus mutations in Chinese hamster cells *in vitro*. *Proc. natl Acad. Sci. USA*, *61*, 1306-1312

Chu, E.H.Y., Sun, N.C. & Chang, C.C. (1975) *Genetic markers associated with hamster chromosomes.* In: Richmonds, C.R., Peterson, D.F., Mullaney, P.P. & Anderson, E.C., eds, *Mammalian Cells: problems and Probes.*, Springfield, VA, National Technical Information Service, US Department of Commerce, pp. 228-238

Chu, E.H.Y., Li, I-C. & Fu, J. (1984) *Mutagenesis studies with cultured mammalian cells: Problems and prospects.* In: Chu, E.H.Y. & Generoso, W.M., eds, *Mutation, Cancer and Malformation*, New York, Plenum, pp. 315-336

Clive, D.K., Johnson, D., Spector, J.F.S., Batson, A.G. & Brown, M.M.M. (1979) Validation and characterization of the L5178Y TK+/- mouse lymphoma assay system. *Mutat. Res.*, *59*, 61-108

Clive, D., McCuen, R., Spector, J.F.S., Piper, C. & Mavournin, K.H. (1983) Specific gene mutation in LS178y cells in culture. A report of the US Environmental Protection Agency Gene-Tox Program. *Mutat. Res.*, *115*, 225-251

Cohen, A.J., Li, F.P., Berg, S., Marchetto, D.J., Tsai,S., Jacobs, S.C. & Brown, R.S. (1979) Hereditary renal-cell carcinoma associated with a chromosomal translocation. *New Engl. J. Med.*, *301*, 592-595

Cox, D.M., Birnie, S. & Tucker, D.N. (1976) The in-vitro isolation and characterization of monosomic sublines derived from a colcemid-treated Chinese hamster cell population. *Cytogenet. Cell Genet.*, *17*, 18-25

Cox, R. (1982) *Mechanisms of mutagenesis in cultured mammalian cells*. In: Sugimura, T., Kondo, S. & Takebe, H., eds, *Environmental Mutagens and Carcinogens*, New York, Alan R. Liss, pp. 157-166

Cox, R. & Masson, W.K. (1976) X-ray induced mutation to 6-thioguanine resistance in cultured human diploid fibroblasts. *Mutat. Res.*, *37*, 125-136

Cox, R. & Masson, W.K. (1979) Mutation and inactivation of cultured mammalian cells exposed to beams of accelerated heavy ions. III. Human diploid fibroblasts. *Int. J. Radiat. Biol.*, *36*, 149-160

Crespi, C.L. & Thilly, W.G. (1984) Assay for gene mutation in a human lymphoblast line, AHH-1, competent for xenobiotic metabolism. *Mutat. Res.*, *128*, 221-230

Danford, N. & Parry, J.M. (1982) Abnormal cell division in cultured human fibroblasts after exposure to diethylstilboestrol. *Mutat. Res.*, *103*, 379-383

Dean, B.J., ed. (1983) *Report of the UKEMS Sub-committee on Guidelines for Mutagenicity Testing*, Part 1, Swansea, UK Environmental Mutagen Society

Dean, B.J., ed. (1984) *Report of the UKEMS Sub-committee on Guidelines for Mutagenicity Testing*, Part 2, London, UK Environmental Mutagen Society

Dean, B.J. & Danford, N. (1984) *Assays for the detection of chemically-induced chromosome damage in cultured mammalian cells*, In: Venitt S. & Parry, J.M., eds, *Mutagenicity Testing: A Practical Approach*, Oxford, IRL Press, pp. 187-232

DeLuca, J.G., Kaden, D.A., Krolewski, J., Slopek, T.R. & Thilly, W.G. (1977a) Comparative mutagenicity of ICR-191 to *S. typhimurium* and diploid human lymphoblasts. *Mutat. Res.*, *46*, 11-18

DeLuca, J.G., Krolewski, J., Skopek, T.R., Kaden, D.A. & Thilly, W.G. (1977b) 9-Aminoacridine— a frameshift mutagen for *Salmonella typhimurium* TA 1537 inactive at the *hgprt* locus in human lymphoblasts. *Mutat. Res.*, *42*, 327-330

DeLuca, J.G., Weinstein, L. & Thilly, W.G. (1983) Ultraviolet light-induced mutation of diploid human lymphoblasts. *Mutat. Res.*, *107*, 347-370

DeLuca, J.G., Kaden, D.A., Komives, E.A. & Thilly, W.G. (1984) Mutation of xeroderma pigmentosum lymphoblasts by far-ultraviolet light. *Mutat. Res.*, *128*, 47-57

DeMars, R. (1965) *Investigations in human genetics with cultivated human cells: a summary of present knowledge*. In: Sonneborn, T.M., ed, *The Control of Human Heredity and Evolution*, New York, MacMillan Co. pp. 48-79

DeMars, R. (1973) Mutation studies with human fibroblasts. *Environ. Health Perspec.*, *6*, 127-136

DeMars, R. & Held, K. (1972) The spontaneous azaguanine-resistant mutants of diploid human fibroblasts. *Humangenetik*, *16*, 87-110

Dempsey, J.L., Seshadri, R.S. & Morley, A.A. (1985) Increased mutation frequency following treatment with cancer chemotherapy. *Cancer Res.*, *45*, 2873-2877

Drinkwater, N.R., Corner, R.C., McCormick, J.J. & Maher, V.M. (1982) An *in-situ* assay for induced diphtheria-toxin-resistant mutants of diploid human fibroblasts. *Mutat. Res.*, *106*, 277-289

Dzarlieva, R.T. & Fusenig, N.E. (1982) Tumor promoter 12-*O*-tetradecanoylphorbol-13-acetate enhances sister chromatid exchanges and numerical and structural chromosome aberrations in primary mouse epidermal cell cultures. *Cancer Lett.*, *16*, 7-17

Evans, H.J. (1977) *Molecular mechanisms in the induction of chromosome aberrations.* In: Scott, D., Bridges, B.A. & Sobels, F.H., eds, *Progress in Genetic Toxicology*, Amsterdam, Elsevier, pp. 57-74

Evans, H.J. (1982) *Sister chromatid exchanges and disease states in man.* In: Wolff, S., ed., *Sister Chromatid Exchange*, New York, John Wiley & Sons, pp. 183-228

Evans, H.J. (1984) *Human peripheral blood lymphocytes for the analysis of chromosome aberrations in mutagen tests.* In: Kilbey, B.J., Legator, M., Nichols, W. & Ramel, C., eds, *Handbook of Mutagenicity Test Procedures*, 2nd ed., Amsterdam, Elsevier, pp. 405-427

Evans, H.J. & Lloyd, D.C., eds (1978) *Mutagen Induced Chromosome Damage in Man*, Edinburgh, Edinburgh University Press

Evans, H.J. & Vijayalaxmi (1981) Induction of 8-azaguanine resistance and sister chromatid exchange in human lymphocytes exposed to mitomycin C and X-rays *in vitro. Nature, 292,* 601-605

Fahrig, R. (1975) A mammalian spot test: induction of genetic alterations in pigment cells of mouse embryos with X-rays and chemical mutagens. *Mol. gen. Genet., 138,* 309-314

Fahrig, R. & Neuhaüser-Klaus, A. (1985) Similar pigmentation characteristics in the specific-locus and the mammalian spot test. A way to distinguish between induced mutation and recombination. *J. Hered., 76,* 421-426

Fantes, J.A., Green, D.K., Elder, J.K., Malloy, P. & Evans, H.J. (1983) Detecting radiation damage to human chromosomes by flow cytometry. *Mutat. Res., 119,* 161-168

Furth, E.E., Thilly, W.G., Penman, B.W., Liber, H.L. & Rand, W.M. (1981a) Quantitative assay for mutation in diploid human lymphoblasts using microtiter plates. *Anal. Biochem., 110,* 1-8

Furth, E.E., Thilly, W.G., Penman, B.W., Liber, H.L. & Rand, W.M. (1981b) Quantitative assay for mutation in diploid human lymphoblast lines. *Mutat. Res., 54,* 193-196

Fuscoe, J.C., Fenwick, R.G., Jr, Ledbetter, D.H. & Caskey, C.T. (1983) Detection and amplification of the HGPRT locus in Chinese hamster cells. *Mol. cell. Biol., 3,* 1086-1096

Gahrton, G., Robert, K.-H., Friberg, K., Zech, L. & Bird, A.G. (1980) Extra chromosome 12 in chronic lymphocytic leukaemia. *Lancet, i,* 146-147

Galloway, S.M. & Ivett, J.L. (1986) Chemically-induced aneuploidy in mammalian cells in culture. *Mutat. Res., 167,* 89-105

Geard, C.R. (1984) *Ring chromosomes and sister chromatid exchanges.* In: Tice, R.R. & Hollaender, A., eds, *Sister Chromatid Exchanges*, New York, Plenum, pp. 91-101

Gilman, M.Z. & Thilly, W.G. (1977) Cytotoxicity and mutagenicity of hyperthermia for diploid human lymphoblasts. *J. thermal Biol., 2,* 95-99

Goldmacher, V.S. & Thilly, W.G. (1983) Formaldehyde is mutagenic for cultured human cells. *Mutat. Res., 116,* 417-422

Goncalves, O., Drobetsky, E. & Meuth, M. (1984) Structural alterations of the *aprt* locus induced by deoxyribonucleoside triphosphate pool imbalances in Chinese hamster ovary cells. *Mol. Cell Biol., 4,* 1792-1799

Goodhead, D.T., Thacker, J. & Cox, R. (1979) Effectiveness of 0.3 keV carbon ultrasoft X-rays for the inactivation and mutation of cultured mammalian cells. *Int. J. Radiat. Biol., 36,* 101-104

Graf, L.J. & Chasin, L.A. (1982) Direct demonstration of genetic alterations at the dihydrofolate reductase locus by gamma irradiation. *Mol. Cell Biol., 2,* 93-96

Grant, S.G. & Worton, R.G. (1982) 5-Azacytidine-induced reactivation of HPRT on the inactive X chromosome in diploid Chinese hamster cells. *Am. J. human Genet., 34*, 171A

Gray, J.W. & Langlois, R.G. (1986) Chromosome classification and purification using flow cytometry and sorting. *Ann. Rev. Biophys. biophys. Chem., 15*, 195-235

Grosovsky, A.J. & Little, J.B. (1985) Evidence for linear response for the induction of mutations in human cells by X-ray exposures below 10 rads. *Proc. natl Acad. Sci. USA, 82*, 2092-2095

Gupta, A.K. & Legator, M.S. (1976) *Chromosome aberrations in cultured human leukocytes after treatment with fungicide 'Benlate'*. In: *Proceedings of the Symposium on Mutagenicity Carcinogenicity Teratogenicity Chemistry, M.S. University, Baroda, India*, pp. 95-103

Hansen, M.F., Koufos, A., Gallie, B.L., Phillips. R.A., Fogstad, O., Brögger, A., Gedde-Dahl, T. & Cavenee, W. (1985) Osteosarcoma and retinoblastoma: a shared chromosomal mechanism revealing recessive predisposition. *Proc. natl Acad. Sci. USA, 82*, 6216-6220

Harnden, D.C. & Klinger, H.P., eds (1985) *An International System for Human Cytogenetic Nomenclature*, Basel, Karger

Harris, M. (1971) Mutation rates in cells at different ploidy levels. *J. Cell Physiol., 78*, 177-184

Harris, M. (1982) Induction of thymidine kinase in enzyme-deficient Chinese hamster cells. *Cell, 29*, 483-492

Harris, P. (1975) The role of membranes in the organization of the mitotic apparatus. *Exp. Cell Res., 94*, 409-425

Hart, J. (1985) The mouse spot test: results with a new cross. *Arch. Toxicol., 58*, 1-4

Hayashi, M., Sofuni, T. & Ishidate, M. (1983) An application of acridine orange fluorescent staining to the micronucleus test. *Mutat. Res., 120*, 241-247

Hayashi, M., Sofuni, T. & Ishidate, M. (1984) Kinetics of micronucleus formation in relation to chromosomal aberrations in mouse bone marrow. *Mutat. Res., 127*, 129-137

Heddle, J.A., Hite, M., Kirkhart, B., Mavournin, K., MacGregor, J.T., Newell, G.W. & Salamone, M.F. (1983) The induction of micronuclei as a measure of genotoxicity. A report of the US Environmental Protection Agency Gene-Tox Program. *Mutat. Res., 123*, 61-118

Heddle, J.A., Stuart, E. & Salamone, M.F. (1984) *The bone marrow micronucleus test*. In: Kilbey, B.J., Legator, M., Nichols, W. & Ramel, C., eds, *Handbook of Mutagenicity Test Procedures*, Amsterdam, Elsevier, pp. 441-457

Heflich, R.H., Hazard, R.M., Lommel, L., Scribner, J.D., Maher, V.M. & McCormick, J.J. (1980) A comparison of the DNA binding, cytotoxicity and repair synthesis induced in human fibroblasts by reactive derivatives of aromatic amide carcinogens. *Chem.-biol. Interactions, 29*, 43-56

Heilbronn, R., Schlehofer, J.R., Yalkinoglu, A.O. & zur Hausen, H. (1985) Selective DNA-amplification induced by carcinogens (initiators): evidence for a role of proteases and DNA polymerase alpha. *Int. J. Cancer, 36*, 85-91

Hesterberg, T.W. & Barrett, J.C. (1985) Induction by asbestos fibers of anaphase abnormalities: mechanism for aneuploidy induction and possibly carcinogenesis. *Carcinogenesis, 6*, 473-475

Hoppe, H., IV, Skopek, T.R., Liber, H.L. & Thilly, W.G. (1978) Alkyl methane sulfonate mutation of diploid human lymphoblasts and *Salmonella typhimurium*. *Cancer Res., 38*, 1595-1600

Howard-Flanders, P. (1981) Mutagenesis in mammalian cells. *Mutat. Res., 86*, 307-327

Howell, J.N., Greene, M.H., Corner, R.C., Maher, V.M. & McCormick, J.J. (1984) Fibroblasts from patients with hereditary cutaneous malignant melanoma are abnormally sensitive to the mutagenic effect of simulated sunlight and 4-nitroquinoline 1-oxide. *Proc. natl Acad. Sci. USA, 81*, 1179-1183

Hozier, J., Sawyer, J., Clive, D. & Moore, M.M. (1985) Chromosome 11 aberrations in small colony L5178Y TK-/- mutants early in their clonal history. *Mutat. Res., 147*, 237-242

Hsie, A.W., Brimer, P.A., Mitchell, T.J. & Gosslee, D.G. (1975) The dose-response relationship for ethyl methanesulfonate-induced mutations at the hypoxanthine-guanine phosphoribosyl transferase locus in Chinese hamster ovary cells. *Somatic Cell Genet., 1*, 247-261

Hsie, A.W., Brimer, P.A., Machnoff, R. & Hsie, M.H. (1977) Further evidence for the genetic origin of mutations in mammalian somatic cells: the effects of ploidy level and selection stringency on dose-dependent chemical mutagenesis to purine analogue resistance in Chinese hamster ovary cells. *Mutat. Res., 45*, 271-282

Hsie, A.W., Casciano, D.A., Cough, D.B., Krahn, D.F., O'Neill, J.P. & Whitfield, B.L. (1981) The use of Chinese hamster ovary cells to quantify specific locus mutations and to determine mutagenicity of chemicals. *Mutat. Res., 86*, 193-214

Hsu, T.C., Shirley, L.R. & Takanari, H. (1983a) Cytogenetic assays for mitotic poissons: the diploid Chinese hamster cell system. *Anticancer Res., 3*, 155-160

Hsu, T.C., Liang, J.C. & Shirley, L.R. (1983b) Aneuploidy induction by mitotic arrestants. Effects of diazepam on diploid Chinese hamster cells. *Mutat. Res., 122*, 201-209

Hutter, K.-J. & Stohr, M. (1982) Rapid detection of mutagen induced micronucleated erythrocytes by flow cytometry. *Histochemistry, 75*, 353-362

IARC (1974a) *IARC Monographs on the Evaluation of Carcinogenic Risk of Chemicals to Man*, Vol. 7, *Some Anti-thyroid and Related Substances, Nitrofurans and Industrial Chemicals*, Lyon, pp. 54, 245

IARC (1974b) *IARC Monographs on the Evaluation of Carcinogenic Risk of Chemicals to Man*, Vol. 4, *Some Aromatic Amines, Hydrazine and Related Substances, N-Nitroso Compounds and Miscellaneous Alkylating Agents*, Lyon, p. 183

IARC (1976a) *IARC Monographs on the Evaluation of Carcinogenic Risk of Chemicals to Man*, Vol. 12, *Some Carbamates, Thiocarbamates and Carbazides*, Lyon, p. 37

IARC (1976b) *IARC Monographs on the Evaluation of Carcinogenic Risk of Chemicals to Man*, Vol. 10, *Some Naturally Occurring Substances*, Lyon, pp. 153, 171

IARC (1977) *IARC Monographs on the Evaluation of Carcinogenic Risk of Chemicals to Man*, Vol. 13, *Some Miscellaneous Pharmaceutical Substances*, Lyon, pp. 57, 58

IARC (1979) *IARC Monographs on the Evaluation of the Carcinogenic Risk of Chemicals to Humans*, Vol. 21, *Sex Hormones*, Lyon, p. 519

IARC (1980) *IARC Monographs on the Evaluation of the Carcinogenic Risk of Chemicals to Humans*, Suppl. 2, *Long-term and Short-term Screening Assays for Carcinogens: A Critical Appraisal*, Lyon

IARC (1982) *IARC Monographs on the Evaluation of the Carcinogenic Risk of Chemicals to Humans*, Suppl. 4, *Chemicals, Industrial Processes and Industries Associated with Cancer in Humans (IARC Monographs, Volumes 1-29)*, Lyon, pp. 52, 79, 88, 184, 202, 249

Jacobs, L. & DeMars, R. (1978) Quantification of chemical mutagenesis in diploid human fibroblasts: induction of azaguanine-resistant mutants by N-methyl-N'-nitro-N-nitrosoguanidine. *Mutat. Res., 53*, 29-53

Jenssen, D. (1982) The induction of micronuclei. In: Sandberg, A.A., ed., *Sister Chromatid Exchange*, Chapter 4, New York, Alan R. Liss, pp. 47-63

Jones, I.M., Burkhart-Shultz, K. & Carrano, A.V. (1985a) A study of the frequency of sister chromatid exchange and of thioguanine resistant cells in mouse spleen lymphocytes after in-vivo exposure to ethylnitrosourea. *Mutat. Res.*, *143*, 245-249

Jones, I.M., Burkhart-Schultz, K. & Carrano, A.V. (1985b) A method to quantify spontaneous and in-vivo induced thioguanine-resistant mouse lymphocytes. *Mutat. Res.*, *147*, 97-105

Kafer, E. (1984) Disruptive effects of ethyl alcohol on mitotic chromosome segregation in diploid and haploid strains of *Aspergillus nidulans*. *Mutat. Res.*, *135*, 53-75

Kao, F.-T. & Puck, T.T. (1968) Genetics of somatic mammalian cells. VII. Induction and isolation of nutritional mutants in Chinese hamster cells. *Proc. natl Acad. Sci. USA*, *60*, 1275-1281

Kaufman, M.H. (1983) Ethanol-induced chromosomal abnormalities at conception. *Nature*, *302*, 258-260

Kawachi, T., Yahagi, T., Kada, T., Tazima, Y., Ishidate, M., Sasaki, M. & Sugiyama, T. (1980) *Cooperative programme on short-term assays for carcinogenicity in Japan*. In: Montesano, R., Bartsch, H. & Tomatis, L., eds, *Molecular and Cellular Aspects of Carcinogen Screening Tests (IARC Scientific Publications No. 27)*, Lyon, International Agency for Research on Cancer, pp. 323-330

Kihlmann, B.A. & Natarajan, A.T. (1984) *Potentiation of chromosomal alterations by inhibitors of DNA repair*. In: Collins, A., Downes, C.S. & Johnson, R.T., eds, *DNA Repair and its Inhibition (Nucleic Acids Symposium Series No. 13)*, Oxford, IRL Press, pp. 319-339

King, H.W.S. & Brookes, P. (1984) On the nature of the mutations induced by the diol epoxide of benzo[a]pyrene in mammalian cells. *Carcinogenesis*, *5*, 965-970

Klein, G. (1981) The role of gene dosage and genetic transpositions in carcinogenesis. *Nature*, *294*, 313-318

Klein, G. (1983) Specific chromosomal translocations and the genesis of B-cell-derived tumors in mice and man. *Cell*, *32*, 258-260

Klein, G. & Klein, E. (1985) Evolution of tumours and the impact of molecular oncology. *Nature*, *315*, 190-195

Kliesch, U., Danford, N. & Adler, I.D. (1981) Micronucleus test and bone marrow chromosome analysis. A comparison of two methods *in vivo* for evaluating chemically induced chromosome alterations. *Mutat. Res.*, *80*, 321-332

Konze-Thomas, B., Hazard, R.M., Maher, V.M. & McCormick, J.J. (1982) Extent of excision repair before DNA synthesis determines the mutagenic but not the lethal effect of UV radiation. *Mutat. Res.*, *94*, 421-434

Korbelik, M. & Horvat, D. (1980) The mutagenicity of nitroaromatic drugs. Effect of metronidazole after incubation in hypoxia *in vitro*. *Mutat. Res.*, *78*, 201-207

Koufos, A., Hansen, M.F., Copeland, N.G., Jenkins, N.A., Lampkin, B.C. & Cavenee, W.K. (1985) Loss of heterozygosity in three embryonal tumours suggest a common pathogenetic mechanism. *Nature*, *316*, 330-334

Kunz, B.A. (1982) Genetic effects of deoxyribonucleotide pool imbalances. *Environ. Mutagenesis*, *4*, 695-725

Kutlaca, R., Alder, S.J., Seshadri, R. & Morley, A.A. (1982a) Radiation sensitivity of human lymphocytes. *Mutat. Res.*, *94*, 125-131

Kutlaca, R., Seshadri, R. & Morley, A.A. (1982b) Effect of age on sensitivity of human lymphocytes to radiation. A brief note. *Mech. Aging Devel.*, *19*, 97-101

Kutlaca, R., Seshadri, R. & Morley, A.A. (1984) Sensitivity of lymphocytes from patients with cancer to X- and ultraviolet radiation. *Cancer, 54*, 2952-2955

Lambert, B., Lindblad, A., Holmberg, K. & Francesconi, D. (1982) *The use of sister chromatid exchange to monitor human populations for exposure to toxicologically harmful agents.* In: Wolff, S., ed., *Sister Chromatid Exchange*, New York, John Wiley & Sons, pp. 149-182

Lambert, B., Sten, M. & Hellgren, D. (1984) Removal and persistence of SCE-inducing damage in human lymphocytes *in vitro. Basic Life Sci., 29*, 647-662

Larizza, L. & Schirrmacher, V. (1984) Somatic cell fusion as a source of genetic rearrangement leading to metastatic variants. *Cancer Metastasis Rev., 3*, 193-222

Larizza, L., Simoni, G., Tredici, F. & De Carli, L. (1974) Griseofulvin: A potential agent of chromosomal segregation in cultured cells. *Mutat. Res., 25*, 123-130

Latt, S.A. (1974) Localization of sister chromatid exchanges in human chromosomes. *Science, 185*, 74-76

Latt, S.A. (1981) Sister chromatid exchange formation. *Ann. Rev. Genet., 15*, 11-55

Latt, S.A., Allen, J., Bloom, S.E., Carrano, A., Falke, E., Kram, D., Schneider, E., Schreck, R., Tice, R., Whitfield, B. & Wolff, S. (1981) Sister chromatid exchanges: A report of the Gene-Tox Program. *Mutat. Res., 87*, 17-62

Latt, S.A., Shiloh, Y., Sakai, K., Brodeur, G., Donlon, T., Korf, B., Shipley, J., Bruns, G., Heartlein, M., Kanda, N., Kohl, N., Alt, F. & Seeger, R. (1986) *Novel DNA rearrangement phenomena associated with DNA amplifications in human neuroblastomas and neuroblastoma cell lines.* In: Ramel, C., Lambert, B. & Magnusson, J., eds, *Genetic Toxicology of Environmental Chemicals. Part A: Basic Principles and Mechanisms of Action*, New York, Alan R. Liss, (in press)

Lehmann, A.R. (1982) *Xeroderma pigmentosum, Cockayne syndrome and ataxia-telangiectasia: Disorders relating DNA repair to carcinogenesis.* In: Bodmer, W.F., ed., *Inheritance of Susceptibility to Cancer in Man*, Oxford, Oxford University Press, pp. 93-118

Leong, P.-M., Thilly, W.G. & Morgenthaler, S. (1985) Variance estimation in single-cell mutation assays: Comparison to experimental observations in human lymphoblasts at 4 gene loci. *Mutat. Res., 150*, 403-410

Lewis, I.L., Patterson, R.M. & McBay, H.C. (1981) The effects of Rhodamine B on the chromosomes of *Muntiacus muntjac. Mutat. Res., 88*, 211-216

Li, I.-C. & Chu, E.H.Y. (1982) Direct selection of mammalian cell mutants deficient in thymidylate synthetase (Abstract no. 22002) *J. Cell Biol., 95*, 447a

Li, I.-C., Fu, J., Hung, Y.-T. & Chu, E.H.Y. (1983) Estimation of mutation rates in cultured mammalian cells. *Mutat. Res., 111*, 253-262

Liber, H.L. & Thilly, W.G. (1982) Mutation assay at the thymidine kinase locus in diploid human lymphoblasts. *Mutat. Res., 94*, 467-485

Liber, H.L., LeMotte, P.K. & Little, J.B. (1983) Toxicity and mutagenicity of X-rays and [^{125}I]dUrd or [^{3}H]TdR incorporated in the DNA of human lymphoblast cells. *Mutat. Res., 111*, 387-404

Liber, H.L., Danheiser, S.L. & Thilly, W.G. (1985) *Mutation in single cell systems induced by low-level mutagen exposure.* In: Woodhead, A.D., Shellabarger, C.J. & Bond, V.P., eds, *Assessment of Risk from Low-Level Exposure to Radiation and Chemicals: A Critical Overview*, New York, Plenum, pp. 169-204

Liu, P.K., Chang, C.-C. & Trosko, J.E. (1982) Association of mutator activity with UV sensitivity in an aphidicolin-resistant mutant of Chinese hamster V79 cells. *Mutat. Res., 106*, 317-332

Liu, P.K., Chang, C.-C., Trosko, J.E., Dube, D.K., Martin. G.M. & Loeb, L.A. (1983) Mammalian mutator mutant with an aphidicolin-resistant DNA polymerase alpha. *Proc. natl Acad. Sci. USA, 80*, 797-801

Lobban, P.E. & Siminovitch, L. (1975) Alpha-amanitin resistance: a dominant mutation in CHO cells. *Cell, 4*, 167-172

Lycette, R.R., Whyte, S. & Chapman, C.J. (1970) Aneuploid effect of oestradiol on cultured human synovial cells. *N.Z. med. J., 72*, 114-117

MacGregor, J.T., Wehr, C.M. & Langlois, R.G. (1983) A simple fluorescent staining procedure for micronuclei and RNA in erythrocytes using Hoechst 33258 and pyronin Y. *Mutat. Res., 120*, 269-275

Maher, V.M., Birch, N., Otto, J.R. & McCormick, J.J. (1975) Cytotoxicity of carcinogenic aromatic amines in normal and xeroderma pigmentosum fibroblasts with different DNA repair capabilities. *J. natl Cancer Inst., 54*, 1287-1294

Maher, V.M., Quellette, L.M., Curren, R.D. & McCormick, J.J. (1976) Frequency of ultraviolet light-induced mutations is higher in xeroderma pigmentosum variant cells than in normal human cells. *Nature, 261*, 593-595

Maher, V.M., McCormick, J.J., Grover, P.L. & Sims, P. (1977) Effect of DNA repair on the cytotoxicity and mutagenicity of polycyclic hydrocarbon derivatives in normal and xeroderma pigmentosum fibroblasts. *Mutat. Res., 44*, 313-326

Maher, V.M., Domey, D.J., Mendrala, A.L., Konze-Thomas, B. & McCormick, J.J. (1979) DNA excision-repair processes in human cells can eliminate the cytotoxic and mutagenic consequences of ultraviolet irradiation. *Mutat. Res., 62*, 311-323

Mahoney, E.M., Ball, J.C., Swenson, D.H., Richmond, D., Maher, V.M. & McCormick, J.J. (1984) Cytotoxicity and mutagenicity of aflatoxin dichloride in normal and repair deficient diploid human fibroblasts. *Chem.-biol. Interactions, 50*, 59-76

Majone, F. (1977) Effects of potassium dichromate on mitosis of cultured mammalian cells. *Caryologia, 30*, 469-481

Martin, R.H. (1983) A detailed method for obtaining preparations of human sperm chromosomes. *Cytogenet. Cell Genet., 35*, 252-256

Martin, R.H. (1985) *Chromosomal abnormalities in human sperm.* In: Dellarco, V.L., Voytek, P.E. & Hollaender, A., eds, *Aneuploidy, Etiology and Mechanisms*, Part 2, *Etiological Aspects of Human Aneuploidy*, New York, Plenum, pp. 91-102

McBurney, M.W. & Whitmore, G.F. (1973) Selection for temperature-sensitive mutants of diploid and tetraploid mammalian cells. *J. Cell Physiol., 83*, 69-74

McGill, M., Pathal, S. & Hsu, T.C. (1974) Effects of ethidium bromide on mitosis and chromosomes: a possible material basis for chromosome stickiness. *Chromosoma, 47*, 157-167

Meuth, M. (1981) Role of deoxynucleoside triphosphate pools in the cytotoxic and mutagenic effects of DNA alkylating agents. *Somatic Cell Genet., 7*, 89-102

Meuth, M. & Arrand, J.E. (1982) Alterations of gene structure in ethyl methane sulfonate-induced mutants of mammalian cells. *Mol. Cell Biol., 2*, 1459-1462

Meuth, M., L'Heureux-Huard, N. & Trudel, M. (1979a) Characterization of a mutator gene in Chinese hamster ovary cells. *Proc. natl Acad. Sci. USA, 76*, 6505-6509

Meuth, M., Trudel, M. & Siminovitch, L. (1979b) Selection of Chinese hamster cells auxotrophic for thymidine by 1-beta-D-arabinofuranosyl cytosine. *Somatic Cell Genet., 5*, 303-318

Mezger-Freed, L. (1972) Effect of polyploidy and mutagens on bromodeoxyuridine resistance in haploid and diploid frog cells. *Nature New Biol.*, *235*, 245-246

Miller, E.C. (1978) Some current perspectives on chemical carcinogenesis in humans and experimental animals. *Cancer Res.*, *38*, 1479-1496

Mitelman, F. (1983) Catalogue of chromosome aberrations in cancer. *Cytogenet. Cell Genet.*, *36*, 1-2

Mitelman, F. (1984) Restricted number of chromosomal regions implicated in aetiology of human cancer and leukaemia. *Nature*, *310*, 325-327

Mohandas, T., Sparkes, R.S. & Shapiro, L.J. (1981) Reactivation of an inactive human X chromosome: evidence for X inactivation by DNA methylation. *Science*, *211*, 393-396

Moore, M.M., Clive, D., Howard, B.E., Batson, A.G. & Turner, N.T. (1985a) In-situ analysis of trifluorothymidine-resistant (TFT^r) mutants of L5178Y$^{+/-}$ mouse lymphoma cells. *Mutat. Res.*, *151*, 147-159

Moore, M.M., Clive, D., Hozier, J.C., Howard, B.E., Batson, A.G., Turner, N.T. & Sawyer, J. (1985b) Analysis of trifluorothymidine-resistant (TFT^r) mutants of L5178Y/TK$^{+/-}$ mouse lymphoma cells. *Mutat. Res.*, *151*, 161-174

Morishima, A., Henrich, R.T., Jou, S. & Nahas, G.G. (1976) *Errors of chromosome segregation induced by Olivetol, a compound with the structure of C-ring common to cannabinoids: formation of bridges and multipolar divisions.* In: Nahas, G.G., ed., *Marihuana: Chemistry, Biochemistry and Cellular Effects*, New York, Springer, pp. 265-271

Morley, A.A., Cox, S., Wigmore, D., Seshadri, R. & Dempsey, J.L. (1982) Enumeration of thioguanine-resistant lymphocytes using autoradiography. *Mutat. Res.*, *95*, 363-375

Morley, A.A., Trainor, K.J., Seshadri, R. & Ryall, R.G. (1983) Measurement of in-vivo mutations in human lymphocytes. *Nature*, *302*, 155-156

Morrow, J. (1983) *Eukaryotic Cell Genetics*, New York, Academic Press

Morrow, J., Stocco, D. & Barron, E. (1978) Spontaneous mutation rate to thioguanine resistance is decreased in polyploid hamster cells. *J. Cell Physiol.*, *96*, 81-85

Mullins, J.M. & Snyder, J.A. (1979) Effects of griseofulvin on mitosis in PtK$_1$ cells. *Chromosoma*, *72*, 105-113

Nalbantoglu, J., Goncalves, O. & Meuth, M. (1983) Structure of mutant alleles at the *aprt* locus of Chinese hamster ovary cells. *J. mol. Biol.*, *167*, 575-594

Natarajan, A.T. & Obe, G. (1982) *Mutagenicity testing with cultured mammalian cells: cytogenetic assays.* In: Heddle, J.A. ed., *Mutagenicity, New Horizons in Genetic Toxicology*, New York, Academic Press, pp. 172-213

Neuhäuser-Klaus, A. (1981) An approach towards the standardization of the mammalian spot test. *Arch. Toxicol.*, *48*, 229-243

Obe, G. & Beek, B. (1982) *The human leukocyte test system.* In: de Serres, F.J. & Hollaender, A., eds, *Chemical Mutagens, Principles and Methods for their Detection*, Vol. 7, New York, Plenum, pp. 337-400

Oliver, J.M., Krawiec, J.A. & Berlin, R.D. (1978) A carbamate herbicide causes microtubule and microfilament disruption and nuclear fragmentation in fibroblasts. *Exp. Cell Res.*, *116*, 229-237

Onfelt, A. & Klasterska, I. (1983) Spindle disturbances in mammalian cells. II. Induction of viable aneuploid/polyploid cells and multiple chromatid exchanges after treatment of V79 Chinese hamster cells with carbaryl. Modifying effects of glutathione and S9. *Mutat. Res.*, *119*, 319-330

Onfelt, A. & Ramel, C. (1979) Some aspects of the organization of microfilaments and microtubules in relation to non-disjunction, *Environ. Health Perspec., 31*, 45-52

Oshimura, M., Hesterberg, T.W., Tsutsui, T. & Barrett, J.C. (1984) Correlation of asbestos-induced cytogenetic effects with cell transformation of Syrian hamster embryo cells in culture. *Cancer Res., 44*, 5017-5022

Palyi, I. (1976) Mechanisms of spontaneous and induced heteroploidization and polyploidization. *Acta morphol. Acad. Sci. Hung., 24*, 307-315

Parry, E.M., Danford, N. & Parry, J.M. (1982) Differential staining of chromosomes and spindle and its use as an assay for determining the effect of diethylstilboestrol on cultured mammalian cells. *Mutat. Res., 105*, 243-252

Patton, J.D., Rowan, L.A., Mendrala, A.L., Howell, J.N., Maher, V.M. & McCormick, J.J. (1984) Xeroderma pigmentosum fibroblasts including cells from XP variants are abnormally sensitive to the mutagenic and cytotoxic action of broad spectrum simulated sunlight. *Photochem. Photobiol., 39*, 37-42

Paul, W.E., Sredni, B. & Schwartz, R.H. (1981) Long term growth and cloning of non-transformed lymphocytes. *Nature, 254*, 697

Penman, B.W. & Thilly, W.G. (1976) Concentration-dependent mutation of diploid human lymphoblasts by methylnitronitrosoguanidine: the importance of phenotypic lag. *Somatic Cell Genet., 2*, 325-330

Penman, B.W., Wong, M. & Thilly, W.G. (1976) Mutagenicity of 5-halo-deoxyuridines to diploid human lymphoblasts. *Life Sci., 19*, 563-568

Penman, B.W., Hoppe, H., IV & Thilly, W.G. (1979) Concentration-dependent mutation by alkylating agents in human lymphoblasts and *Salmonella typhimurium*: *N*-methyl-*N*-nitrosourethane and β-propiolactone. *J. natl Cancer Inst., 63*, 903-907

Penman, B.W., Crespi, C.L., Komives, E.A., Liber, H.L. & Thilly, W.G. (1983) Mutation of human lymphoblasts exposed to low concentrations of chemical mutagens for long periods of time. *Mutat. Res., 108*, 417-436

Perry, P.E. (1980) *Chemical mutagens and sister chromatid exchange*. In: de Serres, F.J. & Hollaender, A., eds, *Chemical Mutagens, Principles and Methods for their Detection*, Vol. 6, New York, Plenum, pp. 1-39

Perry, P., Henderson, L. & Kirkland, D. (1984) *Sister chromatid exchange in cultured cells*. In: Dean, B.J., ed., *Report of the UK Environmental Mutagen Society Sub-committee on Guidelines for Mutagenicity Testing*, Part 2, London, UK Environmental Mutagen Society, pp. 89-121

Peterson, A.R. & Peterson, H. (1979) Facilitation by pyrimidine deoxyribonucleosides and hypoxanthine of mutagenic and cytotoxic effects of monofunctional alkylating agents in Chinese hamster cells. *Mutat. Res., 61*, 319-331

Peterson, A.R., Landolph, J.R., Peterson, H. & Heidelberger, C. (1978) Mutagenesis of Chinese hamster cells is facilitated by thymidine and deoxycytidine. *Nature, 276*, 508-510

Philip, J. & Lundsteen, C. (1985) Semiautomated chromosome analysis. *Clin. Genet., 27*, 140-146

Pinkel, D., Thompson, L.H., Gray, J.W. & Vanderlaan, M. (1985) Measurement of sister chromatid exchanges at very low bromodeoxyuridine substitution levels using monoclonal antibody in Chinese hamster ovary cells. *Cancer Res., 45*, 5795-5798

Pinkel, D., Straume, T. & Gray, J.W. (1986) Cytogenetic analysis using quantitative, high sensitivity fluorescence hybridization. *Proc. natl Acad. Sci. USA, 83,* 2934-2938

Preston, R.J. (1982) The use of inhibitors of DNA repair in the study of the mechanisms of induction of chromosome aberrations. *Cytogenet. Cell Genet., 33,* 20-26

Preston, R.J., Au, W., Bender, M.A., Brewen, J.G., Carrano, A.V., Heddle, J.A., McFee, A.F., Wolff, S. & Wassom, J.S. (1981) Mammalian in vivo and in vitro cytogenetic assays. A report of the US EPA's Gene-Tox Program. *Mutat. Res., 87,* 143-188

Puck, T.T., Marcus, P.I. & Cieciura, S.J. (1956) Clonal growth of mammalian cells *in vitro. J. exp. Med., 103,* 273-284

Puju, S., Shuker, D.E.G., Bishop, W.W., Falchuk, K.R., Tannenbaum, S.R. & Thilly, W.G. (1982) Mutagenicity of *N*-nitroso bile acid conjugates in *Salmonella typhimurium* and diploid human lymphoblasts. *Cancer Res., 42,* 2601-2604

Radman, M., Jeggo, P. & Wagner, R. (1982) Chromosomal rearrangement and carcinogenesis. *Mutat. Res., 98,* 249-264

Raff, E.C. (1984) Genetics of microtubule systems. *J. Cell Biol., 99,* 1-10

Reddy, E.P., Reynolds, R.K., Santos, E. & Barbacid, M. (1982) A point mutation is responsible for the acquisition of transforming properties by the T24 human bladder carcinoma oncogene. *Nature, 300,* 149-150

Regan, J.D. & Setlow, R.B. (1973) *Repair of chemical damage to human DNA.* In: Hollaender, A., ed., *Chemical Mutagens: Principles and Methods for Their Detection.,* Vol. 3, New York, Plenum, pp. 151-170

Rowley, J.D. (1983) Human oncogene locations and chromosome aberrations. *Nature, 301,* 290-291

Rudak, E., Jacobs, P.A. & Yanagimachi, R. (1978) Direct analysis of the chromosome constitution of human spermatozoa. *Nature, 274,* 911-913

de Ruijter, Y.C.E.M. & Simons, J.W.I.M. (1980) Determination of the expression time and the dose-response relationship for mutations at the HGPRT locus induced by X-irradiation in human diploid skin fibroblasts. *Mutat. Res., 69,* 325-332

Russell, L.B. & Major, M.H. (1957) Radiation-induced presumed somatic mutations in the house mouse. *Genetics, 42,* 161-175

Russell, L.B., Selby, P.B., von Halle, E., Sheridan, W. & Valcovic, L. (1981) Use of the mouse spot test in chemical mutagenesis: interpretation of past data and recommendations for future work. *Mutat. Res., 86,* 355-379

Rydström, J., Montelius, J. & Bengtsson, M., eds (1983) *Extrahepatic Drug Metabolism and Chemical Carcinogenesis,* Amsterdam, Elsevier

Sandberg, A.A. (1980) *The Chromosomes in Human Cancer and Leukemia,* New York, Elsevier, pp. 462-463

Sandberg, A.A., ed. (1982) *Sister Chromatid Exchange,* New York, Alan R. Liss

Sanderson, B.J.S., Dempsey, J.L. & Morley, A.A. (1984) Mutations in human lymphocytes: Effect of X- and UV-irradiation. *Mutat. Res., 140,* 223-227

Sawyer, J., Moore, M.M., Clive, D. & Hozier, J. (1985) Cytogenetic characterization of the L5178Y $TK^{+/-}$ 3.7.2C mouse lymphoma cell line. *Mutat. Res., 147,* 243-253

Schimke, R., ed. (1982) *Gene Amplification,* Cold Spring Harbor, NY, CSH Press

Schimke, R.T (1984a) Gene amplification, drug resistance and cancer. *Cancer Res., 44,* 1735-1742

Schimke, R.T. (1984b) Gene amplification in cultured animal cells. *Cell, 37*, 705-713

Schmid, S. & Bauchinger, M. (1976) The cytogenetic effect of an X-ray contrast medium on Chinese hamster cell cultures. *Mutat. Res., 34*, 295-298

Schwab, M., Alitalo, K., Klempnauer, K.H., Varmus, H.E., Bishop, J.M., Gilbert, F., Brodeur, G., Goldstein, M. & Trent, J. (1983) Amplified DNA with limited homology to the *myc* cellular oncogene is shared by human neuroblastoma cell lines and a neuroblastoma tumour. *Nature, 305*, 245-248

Scott, D., Danford, N, Dean, B, Kirkland, D. & Richardson, C. (1983) *In-vitro chromosome aberration assays.* In: Dean, B.J., ed., *Report of the UK Environmental Mutagen Society Sub-committee on Guidelines for Mutagenicity Testing*, Part 1, Swansea, UK Environmental Mutagen Society, pp. 41-64

Seshadri, R.S., Morley, A.A., Trainor, K.J. & Sorrell, J. (1979) Sensitivity of human lymphocytes to bleomycin increases with age. *Experientia, 35*, 233-234

Seshadri, R., Sutherland, G.R., Baker, E., Kutlaca, R., Wigmore, D. & Morley, A.A. (1983) SCE, X-radiation sensitivity and mutation rate in multiple sclerosis. *Mutat. Res., 110*, 141-146

Shafer, D.A., Mandelberg, K.I. & Falek, A. (1985) *Computer automation of metaphase finding, sister chromatid exchange and chromosome damage analysis.* In: de Serres, F.J., ed., *Chemical Mutagens*, Vol. 10, New York, Plenum

Siminovitch, L. (1976) On the nature of heritable variation in cultured somatic cells. *Cell, 7*, 1-11

Simon, L., Hazard, R.M., Maher, V.M. & McCormick, J.J. (1981) Enhanced cell killing and mutagenesis by ethylnitrosourea in xeroderma pigmentosum cells. *Carcinogenesis, 2*, 567-570

Simons, J.W.I.M. (1982) Effect of temperature on mutation in cultured human skin fibroblasts. *Mutat. Res., 92*, 417-426

Skopek, T.R., Liber, H.L., Penman, B.W. & Thilly, W.G. (1978) Isolation of a human lymphoblastoid line heterozygous at the thymidine kinase locus: possibility for a rapid human cell mutation assay. *Biochem. biophys. Res. Commun., 84*, 411-16

Skopek, T.R., Liber, H.L., Kaden, D.A., Hites, R.A. & Thilly, W.G. (1979) Mutation of human cells by kerosene soot. *J. natl Cancer Inst., 63*, 309-312

Slapikoff, S.A., Andon, B.M. & Thilly, W.G. (1978) Comparison of toxicity and mutagenicity of butyl methanesulfonate among human lymphoblast lines. *Mutat. Res., 54*, 193-196

Slapikoff, S.A., Andon, B.M. & Thilly, W.G. (1980) Comparison of toxicity and mutagenicity of methylnitrosourea, methylnitronitrosoguanidine and ICR-191 among human lymphoblast lines. *Mutat. Res., 70*, 365-371

Solomon, E. (1984) Recessive mutation in aetiology of Wilms' tumour. *Nature, 309*, 111-112

Stich, H.F. & Rosin, M.P. (1984) Micronuclei in exfoliated human cells as a tool for studies in cancer risk and cancer intervention. *Cancer Lett., 22*, 241-253

Stoker, M. & Macpherson, I.A. (1964) Syrian hamster fibroblast cell line BHK21 and its derivatives. *Nature, 203*, 1355-1357

Strauss, G.H. & Albertini, R.J. (1978) *6-Thioguanine resistant lymphocytes in human peripheral blood.* In: Scott, D., Bridges, B.A. & Sobels, F.H., eds, *Progress in Genetic Toxicology: Developments in Toxicology and Environmental Sciences*, Vol. 2, Amsterdam, Elsevier, p. 327

Stringer, J.R., Kuhn, R.M., Newman, J.L. & Meade, J.C. (1985) Unequal homologous recombination between tandemly arranged sequences stably integrated into cultured rat cells. *Mol. Cell Biol., 5*, 2613-2622

Sturrock, J.E. & Nunn, J.F. (1976) Synergism between haloethane and nitrous oxide in the production of nuclear abnormalities in the dividing fibroblasts. *Anesthesiology, 44*, 461-471

Styles, J.A. & Penman, M.G. (1985) The mouse spot test. Evaluation of its performance in identifying chemical mutagens and carcinogens. *Mutat. Res., 154*, 183-204

Szybalski, W. & Smith, J. (1959) Genetics of human cell lines. I. 8-Azaguanine resistance, a selective 'single-step' marker. *Proc. Soc. exp. Biol. Med., 101*, 662-666

Takehisa, S. (1982) *Induction of sister chromatid exchanges by chemical agents*. In: Wolff, S., ed., *Sister Chromatid Exchange*, New York, John Wiley & Sons, pp. 87-147

Tenchini, M.L., Mottura, A., Velicogna, M., Pessina, M., Rainaldi, G. & De Carli, L. (1983) Double Y as an indicator in a test of mitotic non-disjunction in cultured human lymphocytes. *Mutat. Res., 121*, 139-146

Thacker, J. & Cox, R. (1975) Mutation induction and inactivation in mammalian cells exposed to ionizing radiation. *Nature, 258*, 429-431

Thacker, J., Stephens, M.A. & Stretch, A. (1978) Mutation to ouabain resistance in Chinese hamster cells: induction by ethyl methanesulfonate and lack of induction by ionizing radiation. *Mutat. Res., 51*, 255-270

Thilly, W.G., DeLuca, J.G., Hoppe, H., IV & Penman, B.W. (1976) Mutation of human lymphoblasts by methylnitrosourea. *Chem.-biol. Interactions, 15*, 33-50

Thilly, W.G., DeLuca, J.G., Hoppe, H., IV & Penman, B.W. (1978) Phenotypic lag in mutation to 6-thioguanine resistance in diploid human lymphoblasts. *Mutat. Res., 50*, 137-144

Thilly, W.G., DeLuca, J.H., Furth, E.E., Hoppe, H., IV, Kaden, D.A., Krolewski, J.J., Liber, H.L., Skopek, T.R., Slapikoff, S.A., Tizard, R.J. & Penman, B.W. (1980) *Gene-locus mutation assays in diploid human lymphoblast lines*. In: de Serres, F.J. & Hollaender, A., eds, *Chemical Mutagens*, Vol. 6, New York, Plenum, pp. 331-364

Tice, R.R. & Hollaender, A., eds (1984) *Sister Chromatid Exchanges, 25 Years of Experimental Research*, New York, Plenum

Tomatis, L. (1986) *Relation between mutagenesis, carcinogenesis and teratogenesis — experience from the IARC Monographs Programme*. In: Ramel, C., Lambert, B. & Magnusson, J., eds, *Genetic Toxicology of Environmental Chemicals*, New York, Alan R. Liss, pp. 3-12

Tong, C. & Williams, G.M. (1980) Definition of conditions for the detection of genotoxic chemicals in the adult rat-liver epithelial cell/hypoxanthine-guanine phosphoribosyl transferase (ARL/-HGPRT) mutagenesis assay. *Mutat. Res., 74*, 19

Tong, C., Teland, S. & Williams, G.M. (1984) Differences in responses of four adult rat liver epithelial cell lines to a spectrum of chemical mutagens. *Mutat. Res., 130*, 53-61

Topham, J., Albanese, R. Bootman, J., Scott, D. & Tweats, D. (1983) *In-vivo cytogenetic assays*. In: Dean, B.J., ed., *Report of the UK Environmental Mutagen Society Sub-committe on Guidelines for Mutagenicity Testing*, Part 1, Swansea, UK Environmental Mutagen Society, pp. 119-141

Trainor, K.J., Wigmore, D.J., Chrysostomou, A., Dempsey, J.L., Seshadri, R. & Morley, A.A. (1984) Mutation frequency in human lymphocytes with age. *Mech. Ageing Devel., 27*, 83-86

Trosko, J.E. (1984) A new paradigm is needed in toxicology evaluation, *Environ. Mutagenesis, 6*, 767-769

Trosko, J.E.. Chang, C.C. & Medcalf, A. (1983) Mechanisms of tumour promotion: potential role of intercellular communication. *Cancer Invest., 1*, 511-526

Tsutsui, T., Maizumi, H., McLachlan, J.A. & Barrett, J.C. (1983) Aneuploidy induction and cell transformation by diethylstilbestrol: a possible chromosomal mechanism in carcinogenesis. *Cancer Res.*, *43*, 3814-3821

Tsutsui, T., Maizumi, H. & Barrett, J.C. (1984) Colcemid-induced neoplastic transformation and aneuploidy in Syrian hamster embryo cells. *Carcinogenesis*, *5*, 89-93

Tucker, J.D., Strout, C.L., Christensen, M.L. & Carrano, A.V. (1986) Sister chromatid exchange induction and persistence in peripheral blood and spleen lymphocytes of mice treated with ethylnitrosourea. *Environ. Mutagenesis* (in press)

Turner, D.R., Morley, A.A., Haliandros, M., Kutlaca, R. & Sanderson, B.J. (1985) In-vivo somatic mutations in human lymphocytes frequently result from major gene alterations. *Nature*, *315*, 343-345

US National Toxicology Program (1985a) *Toxicology and Carcinogenesis Studies of Chrysotile Asbestos (CAS No. 12001-29-5) in F344/N Rats (Feed Studies)* (*Tech. Rep. Ser. No. 295*), Washington DC, US Department of Health and Human Services

US National Toxicology Program (1985b) *Carcinogenesis Bioassay of 2,3,7,8-Tetrachlorodibenzo-p-dioxin (CAS No. 1746-01-6) in Swiss-Webster Mice (Dermal Study)* (*Tech. Rep. Ser. No. 201*), Washington DC, US Department of Health and Human Services

Varmus, H.E. (1984) The molecular genetics of cellular oncogenes. *Ann. Rev. Genet.*, *18*, 553-612

Verschaeve, L., Kirsch-Volders, M. & Susanne, C. (1984) Mercury-induced segregational errors of chromosomes in human lymphocytes and in Indian muntjac cells. *Toxicol. Lett.*, *21*, 247-253

Vijayalaxmi & Evans, H.J. (1984a) Induction of 6-thioguanine-resistant mutants and SCEs by 3 chemical mutagens (EMS, ENU and MMC) in cultured human blood lymphocytes. *Mutat. Res.*, *129*, 283-289

Vijayalaxmi & Evans, H.J. (1984b) Measurement of spontaneous and X-irradiation-induced 6-thioguanine resistant human blood lymphocytes using a T-cell cloning technique. *Mutat. Res.*, *125*, 87-94

Wake, N., Isaacs, J. & Sandberg, A.A. (1982) Chromosome changes associated with progression of the Dunning R-3327 rat prostatic adenocarcinoma system. *Cancer Res.*, *42*, 4131-4142

Wang, R.J., Wissinger, W., King, E.J. & Wang, G. (1983) Studies on cell division in mammalian cells. VII. A temperature-sensitive cell line abnormal in centriole separation and chromosome movement. *J. Cell Biol.*, *96*, 301-306

Weinburg, G., Ullman, B. & Martin, D.W., Jr (1981) Mutator phenotypes in mammalian cell mutants with distinct biochemical defects and abnormal deoxyribonucleoside triphosphate pools. *Proc. natl Acad. Sci. USA*, *78*, 2447-2451

White, R. (1982) *DNA polymorphism: new approaches to the genetics of cancer*. In: Bodmer, W.F., ed., *Inheritance of Susceptibility to Cancer in Man*, Oxford, Oxford University Press, pp. 175-186

Wigley, C.B., Newbold, R.F., Ames, J. & Brookes, P. (1979) Cell-mediated mutagenesis in cultured Chinese hamster cells by polycyclic hydrocarbons: mutagenicity and DNA reaction related to carcinogenicity in a series of compounds. *Int. J. Cancer*, *23*, 691-696

Wobus, A.M., Schöneich, J. & Thieme, R. (1978) The effect of the mode of administration of nitrogen mustard and cytosine arabinoside on the production of chromosomal aberrations in mouse bone marrow and ascites tumour cells. *Mutat. Res.*, *58*, 67-77

Wolf, S.F. & Migeon, B.R. (1982) Studies of X chromosome DNA methylation in normal human cells. *Nature*, *295*, 667-671

Wolff, S. (1982a) *Chromosome aberrations, sister chromatid exchanges, and the lesions that produce them*. In: Wolff, S., ed., *Sister Chromatid Exchange*, New York, John Wiley & Son, pp. 41-57

Wolff, S., ed. (1982b) *Sister Chromatid Exchange*, New York, John Wiley & Son

Yang, L.L., Maher, V.M. & McCormick, J.J. (1982) Relationship between excision repair and the cytotoxic and mutagenic effect of the 'anti' 7,8-diol-9,10-epoxide of benzo[a]pyrene in human cells. *Mutat. Res.*, 94, 435-447

Yotti, L.P., Chang, C.C. & Trosko, J.E. (1979) Elimination of metabolic cooperation in Chinese hamster cells by a tumor promoter. *Science*, 206, 1089-1091

Yunis, J. (1983) The chromosomal basis of human neoplasia. *Science*, 221, 227-236

Yunis, J.J. & Ramsay, N. (1978) Retinoblastoma and subband deletion of chromosome 13. *Am. J. Dis. Childh.*, 132, 161-163

Zech, L., Haglund, U., Nilsson, K. & Klein, G. (1976) Characteristic chromosome abnormalities in biopsies and lymphoid-cell lines from patients with Burkitt and non-Burkitt lymphomas. *Int. J. Cancer*, 17, 47-56

Zeiger, E. & Tennant, R.W. (1986) *Mutagenesis, clastogenesis, carcinogenesis; expectations, correlations, and relations*. In: Ramel, C., Lambert, B. & Magnusson, J., eds, *Genetic Toxicology of Environmental Chemicals*, New York, Alan R. Liss, pp. 75-84

Zieve, G.W., Turnbull, D., Mullins, J.M. & McIntosh, J.R. (1980) Production of large numbers of mitotic mammalian cells by use of the reversible microtubule inhibitor nocodazole. *Exp. Cell Res.*, 126, 397-405

Zimmermann, F.K., von Borstel, R.C., von Halle, E.S., Parry, J.M., Siebert, D., Zetterberg, G., Barale, R. & Loprieno, N. (1984) Testing of chemicals for genetic activity with *Saccharomyces cerevisiae*: a report of the US Environmental Protection Agency Gene-Tox Program. *Mutat. Res.*, 133, 199-244

REPORT 8

ASSAYS FOR GERM-CELL MUTATIONS IN MAMMALS

Prepared by:

U.H. Ehling (Rapporteur), E.H.Y. Chu,
L. De Carli, H.J. Evans, M. Hayashi, B. Lambert,
D. Neubert, W.G. Thilly (Chairman) and H. Vainio

1. Introduction

A not insubstantial proportion of human ill health and disease results from the inheritance of gene or chromosomal mutations. In many instances, a mutational change at a single locus may exist within a population and be transmitted to a proportion of individuals; in others, the mutation may have newly arisen. The rates at which such spontaneous mutations occur have been estimated for only a very limited number of human genes in which the mutation results in a well defined anomaly and is readily detectable. Chromosomal mutations that involve changes in chromosome number, or gross changes in structure, are readily detectable by cytogenetic screening, and their prevalence in various adult and newborn populations is well established. Moreover, they occur frequently in abortuses (up to 50% of abortions in the first trimester are chromosomally abnormal), and they are a major contributory factor to human reproductive failure.

Although information on spontaneous mutation at the level of chromosomes and, to a much lesser extent, genes is available for some human populations, there is virtually no information on *induced* heritable mutations. A limited number of studies on the progeny of populations exposed to known mutagens has been undertaken, but no substantial data on mutation at the gene level have been obtained; there is a similar paucity of information for induced chromosomal mutations.

The development of techniques to visualize chromosomes in human germ cells *in vitro* has opened up the possibility of determining chromosomal mutation rates in sperm and egg cells of individuals exposed to mutagens. A number of groups (Martin, 1985; Brandriff *et al.*, 1986) have reported on the types and frequencies of chromosomal mutations observed in human sperm complements visualized in hamster oocytes or in human zygotes produced *in vitro* (Rudak et al., 1978; Angell et al., 1983). These studies have demonstrated a high spontaneous chromosomal mutation frequency in germ cells from normal individuals. Moreover, in a recent report (Martin et al., 1985), chromosomal anomalies in spermatozoa of males exposed to X-rays for therapeutic purposes have been analysed and their frequencies shown to increase with the level of X-ray exposure.

Despite these recent developments in our knowledge of mutation frequencies in humans, it is inevitable that information on the mutagenic potential of environmental chemical agents can be obtained only through studies on relevant animal systems. Germ-cell mutations in mammals are screened in order to evaluate the ability of a substance to induce genetic changes that are transmitted through the germ line, for the following reasons: to confirm the results of other short-term tests; to determine the factors that affect the mutation process *in vivo*, e.g., dose, dose rate, dose fractionation, differential spermatogenic response and sex differences; and for risk estimation.

Mutational damage can be classified into three different categories: (1) Mendelian mutants; (2) chromosomal aberrations; and (3) irregularly inherited disorders. To date, no test system has been developed for screening irregularly inherited disorders. This review is therefore restricted to Mendelian mutations and chromosomal aberrations.

2. Mendelian mutations

This category includes recessive specific-locus mutations, dominant-cataract mutations and mutations detected by biochemical methods. The largest data base is available for specific-locus mutations.

2.1 *Specific-locus mutations*

The most efficient method for studying induced germ-cell mutations in mammals is the mouse specific-locus test, reviewed for the Gene-Tox Program by Russell et al. (1981). Consequently, the principles of mammalian germ-cell radiation mutagenesis have been developed from the experimental results obtained using the specific-locus test (Ehling & Favor, 1984). The mutagenic activity of 25 compounds has been tested in this assay (Table 1).

The occurrence of mutation is a rare event. When employing laboratory mammals as experimental organisms, in order to have enough mutational events for conclusive experimental results, the experimental protocol must allow the rapid, simple examination of large numbers of animals in which a newly occurring mutation might possibly be expressed.

Table 1. Agents evaluated for the induction of specific-locus mutations[a]

Compound	Classification
Benzo[a]pyrene	inconclusive
Butylated hydroxytoluene	inconclusive
Caffeine	inconclusive
5-Chlorouracil	inconclusive
Coal-liquid A	inconclusive
Cyclophosphamide	mutagenic
Diethylsulfate	inconclusive[b]
7,12-Dimethylbenz[a]anthracene	inconclusive
Ethyl methanesulfonate	mutagenic
N-Ethyl-N-nitrosourea	mutagenic
5-Fluorouracil	inconclusive
Hycanthone	inconclusive
ICR-170	inconclusive
Isopropyl methanesulfonate	inconclusive
Methyl methanesulfonate	mutagenic
N-Methyl-N'-nitro-N-nitrosoguanidine	inconclusive
Mitomycin C	mutagenic
Monocrotaline	inconclusive
Myleran	inconclusive
N-Nitrosodiethylamine	inconclusive
Procarbazine	mutagenic
Propyl methanesulfonate	mutagenic
Sodium bisulfite	inconclusive
Triethylenemelamine	mutagenic
Wheat + 200 krad	inconclusive

[a] From Russell et al. (1981)

[b] Ehling and Neuhäuser-Klaus (1985) demonstrated that the compound is mutagenic.

(a) Description of the test

A specific-locus test consists of treatment of parental mice homozygous for a wild-type set of marker loci. The treated homozygous wild-type mice are mated with a tester stock that is homozygous-recessive at the marker loci. Resultant F_1 offspring are expected to be heterozygous at the marker loci and to express the wild-type phenotype. In the event of a mutation from the wild-type allele at any of the heterozygous loci, the F_1 offspring would express the recessive phenotype.

The standard specific-locus test, developed by W.L. Russell (1951), employs a tester stock homozygous recessive at seven loci (*a, non-agouti*; *b, brown*; c^{ch}, *chinchilla*; *d, dilute*; *p, pink-eyed dilution*; *s, piebald*; *se, short ear*) controlling coat pigmentation, intensity or pattern as well as the size of the external ear. The recessive phenotypes at each of the seven loci are easily recognizable and, with the exception of intermediate alleles, characteristic for the locus. Since the recessive phenotypes are easily recognizable, examination procedures are simple and fast. This allows the screening of large numbers of F_1 animals with a minimal demand on resources. That the recessive phenotypes are often characteristic for the locus at which a mutation occurs reduces the effort necessary for confirming the presence of a newly occurring mutation by a test of allelism at the suspected locus.

(b) Relevance of the test

A mammalian germ-cell mutational assay is the most relevant experimental system for detecting possible mutagenic activity of test agents to human germ cells. The decision to conduct a germ-cell mutation test in mammals is costly in resources and time. It is therefore logical that the most efficient assay be the method of choice, to minimize the investment required and to maximize the probability of an unambiguous result. The specific-locus test is, at present, the most efficient germ-cell mutation assay in mice because of the simple experimental procedures and the large data base on historical control animals.

A mutation from the wild-type dominant to a recessive allele may be the result of any mutational event in which there is loss or inactivation of the wild-type gene product, such as a 'null' or nonsense mutation, or an alteration at the active site of the gene product. These mutations may be the result of a variety of DNA lesions, e.g., a single base-pair substitution, insertion or deletion, or a deletion of a small or larger DNA base sequence. Thus, the specific-locus test screens for a broad spectrum of mutations. Since the mutations are recovered and transmitted, it is possible to characterize the mutational events at the genetic (Russell, 1971) and molecular levels (Rinchik *et al.*, 1985).

A large body of experimental results already exists which enables comparison of results for a compound not previously assayed in a mammalian germ-cell test system with data on chemicals that have already been studied and with the wealth of information on ionizing radiation. The development of systems that screen for other endpoints (see discussion of dominant and biochemical mutational tests in sections 2.2 and 2.3) also requires comparative results; therefore, combined experiments, in which mutations to specific loci as well as other genetic endpoints are screened in the same experimental animals (see discussion of the multiple endpoint system in section 4), would appear useful.

(c) Disadvantages

Mouse specific-locus experiments indicate mutations to *recessive* alleles at a selected set of seven loci. Thus, the qualities that are responsible for the high efficiency of the specific-locus test may limit its relevance to the genetic endpoints that pose the most genetic hazard to humans. This dilemma could be solved by conducting combined experiments in which both specific-locus mutations and *dominant* mutations are screened simultaneously in the same experimental animals.

(d) Results of specific-locus experiments

The results obtained with this method up to 1980 are summarized in Table 1. In addition, Russell *et al.* (1984) reported that ethylene oxide did not induce specific-locus mutations in spermatogonia. In this experiment, male mice were exposed in an inhalation chamber to ethylene oxide in air at a concentration of (generally) 255 ppm. After accumulating total exposures of 101 000 or 150 000 ppm per hour in 16-23 weeks, a total of 71 387 offspring was observed. The spermatological stem-cell mutation rate at each exposure level, as well as the combined result, did not differ significantly from the historical control frequency. At the lower and higher exposure levels, the results rule out (at the 5% significance level) an induced frequency that is, respectively, 0.97 and 6.33 times the spontaneous rate; the combined results rule out a multiple of 1.64. Further results have been published for female mice. N-Ethyl-N-nitrosourea (ENU) (Ehling, 1984a) and triethylenemelamine (Cattanach, 1982) induce specific-locus mutations in the first seven weeks after treatment, whereas procarbazine induces mutations only in later matings (Ehling, 1984a).

Table 1 shows that eight chemicals out of 25 are clearly mutagenic, whereas for 17 inconclusive results are obtained. The high frequency of inconclusive results is attributable to two main reasons: (i) small sample size and (ii) lack of coverage of the whole spermatogenic cycle.

Because of the differential spermatogenic response, all gametogenic stages must be tested (Ehling & Neuhäuser-Klaus, 1984) before concluding that a compound is not mutagenic to mice. For example, diethylsulfate was classified in Table 1 as 'inconclusive'. However, Ehling and Neuhäuser-Klaus (1985) clearly demonstrated that diethylsulfate is mutagenic in spermatozoa and spermatids. Similarly, only 507 offspring derived from treated post-spermatogonial germ-cell stages were tested with isopropyl methanesulfonate, and only 911 offspring derived from post-spermatogonial germ-cell stages after treatment with myleran. In addition, these offspring were conceived immediately after treatment or 21 days post-pretreatment. Therefore, the most sensitive germ-cell stages for the induction of mutations by these compounds were not sampled (Ehling, 1978). For a definitive evaluation, the differential spermatogenic response and the sample size have to be taken into account. Many different factors affect the sensitivity to mutation induction, especially by chemical mutagens (Ehling & Favor, 1984).

It is interesting to note that the nature of mutations induced in post-spermatogonial germ-cell stages is different from the mutations induced in spermatogonia. The high frequency of homozygous lethal mutations induced in post-spermatogonial germ-cell stages suggests that small deficiencies are the main cause. This observation also explains the correlation between the induction of dominant-lethal and specific-locus mutations by cyclophosphamide, ethyl methanesulfonate and methyl methanesulfonate (Ehling & Neuhäuser-Klaus, 1984). The relatively low frequency of homozygous lethal mutations suggests that mutations induced by ENU, mitomycin C and procarbazine in spermatogonia may be due mainly to base-pair changes or small deficiencies and do not involve large deficiencies.

2.2 Dominant mutations

Although mutations to recessive visible alleles can be screened more efficiently by the specific-locus test in mice (Ehling et al., 1985), the study of mutations to dominant visible alleles is important. Dominant mutations may represent a different class of DNA lesion at the molecular level than recessive mutations (Favor, 1983, 1984, 1986). Furthermore, since, by definition, newly occurring dominant mutations are expressed immediately when heterozygous in an outbred population, they represent an important endpoint in the assessment of human mutagenic hazards (Ehling, 1984a).

(a) Description of the test

In principle, the search for newly occurring dominant mutations is simple: one recovers variant phenotypes in the F_1 generation following mutagenic treatment. However, in practice, a systematic search for newly occurring dominant mutations requires a carefully designed experimental protocol. The phenotypes by which dominant visible mutations are identified, must be carefully defined in order to reduce observational bias and to fix the number of loci at which dominant mutations are screened. The examination procedures must be simple and fast so that experiments of the required size can be carried out. Finally, phenotypic variants identified as suspected mutants must be subjected to genetic confirmation (Ehling, 1984b). The dominant-cataract mutation test fulfills these requirements and, to date, represents the most practicable experimental design for estimating the induced mutation rate to dominant alleles in mammalian germ cells (Ehling et al., 1982, 1985; Favor, 1986). Other experimental systems for detecting dominant mutations include those for dominant visible (Schlager & Dickie, 1971; Searle & Beechy, 1986), histocompatibility (Bailey & Kohn, 1965) and skeletal (Ehling, 1966; Selby & Selby, 1977) mutations. A screen for dominant visible mutations may be subject to variability due to observational bias, whereas, although in the histocompatibility and skeletal systems observational bias is reduced, the examination procedures are not appropriate to screen the large number of progeny required in a search for newly occurring mutations.

(b) Relevance of the test

The aim of such tests is to estimate the rate of mutation rate to dominant alleles and to characterize the recovered mutations. If the suggestion is correct that mutations to dominant alleles represent a different spectrum of DNA lesions than mutations to recessive alleles (Favor, 1983, 1984), then it is vital that information on mammalian mutagenesis include results of mutations to dominant alleles. The dominant-cataract test, as designed, is very efficient, since it screens, in the same experimental animals, for both recessive specific-locus and dominant cataract mutations. By imposing the criterion that phenotypic variants suspected of being caused by dominant mutation be subjected to a genetic confirmation test, the results obtained define unambiguous germinal mutations in mammals. Moreover, genetic characteristics of the mutations, such as fertility and penetrance effects, may be estimated. A combination of the dominant-cataract test and the specific-locus test represents the most complete attempt to analyse potential mutagenic exposure in mammalian germ cells and to estimate the genetic risk of such exposure to the human population.

(c) *Disadvantages*

As with any mammalian germinal mutagenesis test, experiments are costly and relatively labour intensive. The dominant-cataract test represents the best attempt at developing a system that screens for dominant mutations and borrows from the elegance of the specific-locus test method. It is not proposed as a test to determine the potential mutagenicity of an agent, but is a test for quantifying dominant mutations that may pose a human genetic hazard. Furthermore, it can be used to determine if mutations to dominant alleles represent a different spectrum of DNA lesions from those of recessive alleles.

(d) *Results of dominant-mutation tests*

To date, results from dominant-cataract mutation tests are available for radiation, for ENU and for procarbazine. Table 2 lists the results after spermatogonial treatment. Mutants were recovered in all experiments and have been maintained for further characterization. Further experiments have been conducted in combination with specific-locus biochemical markers, which enable direct comparison among the genetic endpoints.

Since the mutation rate to dominant-cataract alleles is lower than that to recessive specific-locus alleles, it has been suggested that the spectrum of DNA lesions resulting in a dominant mutation may be more restricted than that resulting in a recessive mutation (Kratochvilova, 1981; Ehling *et al.*, 1982; Favor, 1983, 1984; Ehling *et al.*, 1985). This hypothesis is based on the premise that a recessive allele could result from a wide variety of mutational events leading to loss or inactivation of the gene product. A dominant mutation, by comparison, is more likely to result in an altered gene product which interferes with the normal gene product in the way that a heterozygote results in a mutant phenotype (Kacser & Burns, 1981). ENU is more efficient than ionizing radiation in inducing dominant-cataract mutations, but the opposite is true for recessive specific-locus mutations, and this may reflect a difference in the spectrum of

Table 2. Mutation frequencies of different genetic endpoints following spermatogonia treatment of (101/El $\times$ C3H/El)F_1 hybrid male mice[a]

Treatment	Specific-locus mutations[b]	Dominant-cataract mutations	Protein-charge mutations	Enzyme-activity mutations
Historical control	19(13)/227 805	1/22 594	0/5 812	0/3 610
N-Ethyl-N-nitrosourea				
160 mg/kg	35(32)/ 8 658	14/ 6 435	NT[c]	NT
	52(50)/ 13 018	NT	1/1 892	14/3 093
250 mg/kg	64(58)/ 9 766	17/ 9 352	3/4 254	4(3)[d]/ 505
Procarbazine				
600 mg/kg	4/ 13 071	1/12 056	1/7 506[e]	1/2 974
X-irradiation				
3 + 3 Gy	18/ 10 054	3(2)[d]/ 8 889	NT	NT
	11/8 085	0/ 6 662	0/1 003	1/3 388

[a]From Ehling et al. (1985)

[b]Number of independent mutational events in parentheses

[c]Not tested

[d]Presumed cluster (two mutations were recovered in the same litter and express the same phenotype; however, a test for allelism has not yet been conducted.)

[e]Includes results of the present multiple endpoint experiment (0/5 725) as well as results of an independent experiment (1/1 781)

DNA lesions induced by different agents. Radiation induces a higher frequency of small deletions, which would result in a higher frequency of recessive mutations due to a loss of the gene products; in contrast, ENU induces a higher frequency of intragenic changes, and, as expected, a relatively higher frequency of dominant mutations is observed. This hypothesis is also supported by the homozygous viability of the recovered mutations. For both recessive specific-locus and dominant-cataract mutations, a much higher frequency of homozygous lethal mutations was observed after treatment with radiation than with ENU (Kratochvilova, 1981; Ehling & Favor, 1984; Favor, 1984).

Finally, results from the dominant-cataract mutation test provide a basis for estimating human genetic risk due to mutagenic exposure. Data on mutation rates from the dominant-cataract system can be extrapolated to estimate the expected total number of dominant mutations in the first generation following exposure employing the direct risk estimation procedure (Ehling, 1984a, 1985). Furthermore, genetic characterization of recovered dominant mutations, including fertility and penetrance effects (Favor, 1984), provides useful additional information for estimating the total risk due to induced dominant mutations beyond the first generation following exposure. Fertility effects in recovered mutants can be used to calculate the expected total number of carriers of a newly induced mutation before its eventual elimination from the population. In combination with penetrance effects, these can be used to estimate the total number of affected individuals before an induced dominant mutation is lost from the population (Favor, 1986). For example, it was shown that the average reduction in fertility of the dominant-cataract mutations recovered after treatment with 250 mg/kg ENU was 0.12, and that 8.3 mutation carriers per newly induced mutation would be expected before eventual loss of the mutation from the population. Of the mutations recovered, 25% were shown to have a reduced penetrance, the average penetrance value being 0.39. The total number of affected individuals per newly-occurring dominant mutation before its elimination from the population can thus be estimated as:

Affected individuals = $8.3 \times (0.75 + 0.25 \times 0.39) = 7.0$

2.3 Biochemical mutations

Like dominant visible mutations, biochemical mutations are also relevant to inherited human diseases, and there is homology between animal and human enzyme systems. Genetic diseases resulting from inherited biochemical mutations are well documented (McKusick, 1983). Recovered biochemical mutations reflect closely the direct gene product and therefore provide a possibility for characterizing DNA alterations resulting in a particular phenotypic change.

(a) Description of the tests

Two basic systems have been used to recover biochemical mutations in mice: (i) electrophoretic-mobility variants (Soares, 1979; Johnson & Lewis, 1981; Pretsch & Charles, 1984; Peters et al., 1986) and (ii) enzyme-activity variants (Feuers et al., 1982; Pretsch & Charles, 1984). The electrophoretic-mobility system can be used to screen mutational events, such as base-pair substitutions and small deletions or insertions, which cause a change in the net charge of the enzyme molecule. Additionally, large deletions or mutations at regulatory loci which result in a 'null' mutation may be detected. As with the test for dominant visible mutations, the experimental protocol employed to screen for electrophoretic-mobility variants is critical. For example, 'null' mutations are detectable when the parental strains employed differ for the allelic form of the enzyme under study. The resulting F_1 offspring should codominantly express both alleles, and, for a single-locus-monomer enzyme, a double-band phenotype should be observed. The absence of an allelic band indicates the occurrence of a 'null' mutation. Should the parental strains not differ in the allelic form of the enzyme studied, then one electrophoretic-mobility band would be visualized in the F_1 offspring. The occurrence of a 'null' mutation would result in a less intensely stained band, which may or may not be evident. Consequently, loci at which the parental strains differ are more effective in detecting mutations (Lewis & Johnson, 1986).

Johnson and Lewis (1981) have shown that after spermatogonial treatment of mice with ENU, approximately equal numbers of mutations occur that cause an alteration in the electrophoretic mobility of an enzyme or a 'null' expression of the enzyme. Of the four 'null' mutations recovered, three were recovered at loci for which electrophoretic differences existed between the genotypes of the strains of parental mice, while one provisional 'null' mutation was recovered at a locus in which the parental strains were of the same allelic form. Peters et al. (1986) demonstrated a three-fold higher incidence of 'null' mutations than of electrophoretic-mobility variants after spermatogonial treatment with ENU. Radiation exposure of Drosophila melanogaster, which could be expected to induce a higher frequency of deletions, yielded a five-fold higher frequency of 'null' mutations than of mutations that resulted in an alteration of the electrophoretic mobility of an enzyme molecule (Mukai & Cockerham, 1977).

In order to search systematically for enzyme-activity mutations, the criterion of an outlier, presumed to be due to a mutation, must be precisely defined. Feuers et al. (1980) proposed a lower limit of 50% wild-type activity for a one-locus (structural) enzyme system and 25% wild-type activity for a two-locus enzyme system. These values are based on the assumption that the mutations screened are 'nulls'. For a mutation that causes increased activity, an increase of two standard deviations above the normal activity is employed. In a similar experimental protocol, outliers were defined as individuals not included in the interval, defined by the mean ± three standard deviations, of a population of F_1 individuals examined for a particular enzyme on a particular day (Pretsch & Charles, 1984; Ehling et al., 1985). Once outliers have been identified, Feuers et al. (1982) analyse for changes in the catalytic properties of the enzyme (K_m and V_{max}), and, if this second analysis also indicated a variation from the normal value, presumed mutants were subjected to a genetic confirmation test (progeny of the suspected mutant were examined). Pretsch and Charles (1984) re-analysed the samples from identified outliers, and, when the outlier value was confirmed, presumed mutants were subjected to a genetic confirmation test. The experimental protocol of Pretsch and Charles has the distinct advantage that enzyme activities in erythrocytes are studied. This biopsy procedure allows the rapid, simple examination of live animals —critical for a mammalian germ-cell mutation test.

Results for spermatogonial treatment of mice with radiation, ENU and procarbazine (Table 3) show a total of 11 confirmed mutations with decreased activity and eight with increased activity (Ehling et al., 1985). Peters et al. (1986), employing the test protocol of Pretsch and Charles (1984), have recovered five enzyme-activity mutants after spermatogonial treatment with ENU, two with loss of activity and three with increased activity. After post-spermatogonial treatment with ethylmethane sulfonate, eight mutations, all with decreased enzyme activity, were recovered (Bishop & Feuers, 1982).

Table 3. Observed and presumed mutation rate per locus ($\times 10^{-5}$) for different genetic endpoints following spermatogonial treatment of $(101/E1 \times C3H/E1)F_1$ hybrid mice[a]

Treatment	Specific-locus mutations[b]	Dominant-cataract mutations	Protein-charge mutations	Enzyme-activity mutations
Historical control	1.2	0.1		
N-Ethyl-N-nitrosourea				
160 mg/kg	57.8	7.3	NT[b]	NT
	57.1	NT	2.3	37.7
250 mg/kg	93.6	6.1	3.1	66.0
Procarbazine				
600 mg/kg	4.4	0.3	0.6	2.8
X-irradiation				
3 + 3 Gy	22.8	0.6		2.5

[a]From Ehling et al. (1985)
[b]Not tested

(b) Relevance of the test

Systematic screening for biochemical mutations in experimental mammals extends the field of mammalian mutagenesis to new dimensions: (i) An additional genetic endpoint is studied which allows a comparison of the effects of a mutagen in different germinal mutation test systems. (ii) Hereditary diseases in humans due to gene homology and inborn metabolic disorders can be compared with recovered biochemical mutations in experimental mammals; such mutations can also be compared with the large body of data on mutation rates to electrophoretic mobility variants in humans (Harris et al., 1974; Neel et al., 1980a,b). (iii) The molecular characterization of recovered biochemical mutations may contribute to an understanding of the mechanism of mutational events in mammalian germ cells. Recovered electrophoretic-mobility variants have a higher probability of characterization, since the changes probably occur in the structural gene. The characterization of haemoglobin mutations (Popp et al., 1983; Peters et al., 1986) is an example of the potential of this system.

(c) Disadvantages

This system is the most labour-intensive mammalian germ-cell mutagenicity test. It is not intended to be used as a screen for potential mutagenic activity but should be employed as an additional genetic endpoint for assessing proven mutagens, with the possibility of characterizing the recovered mutagenic events.

The ultimate expression of an enzyme is complicated (Paigen, 1971), and caution should be exercised in the interpretation of results. For example, post-translational modification of an enzyme may be controlled by loci other than the structural locus, so that an electrophoretic-mobility variant might not necessarily represent a mutation at the structural gene. The level of enzyme activity ultimately realized is even more complex, the end result depending upon the inherent catalytic properties of the structural gene as well as on enzyme synthesis, degradation and localization. A mutation at any of the possible loci controlling these processes would result in the expression of an enzyme-activity variant. The relatively high mutation rate per locus for enzyme activity mutation, as calculated by Ehling et al. (1985), may in fact be due to an underestimation of the number of loci screened. Obviously, genetic variants should be developed to dissect enzyme systems genetically. Since experiments to screen for induced enzyme-activity mutants have been initiated only recently, the systems are not well understood. However, using the recovered mutants, these systems will be elucidated for the benefit of both biochemical geneticists and those who wish to study the process of mutation in mammalian germ cells.

(d) Results of biochemical mutation tests

The frequencies of recovered electrophoretic-mobility and enzyme-activity mutations recovered after treatment with ENU, procarbazine or radiation are presented in Table 2, from a combined experiment in which specific-locus and dominant-cataract mutations were also screened, and these are therefore directly comparable. Seven loci were scored for specific loci and 23 for electrophoretic-mobility loci; for dominant-cataract and enzyme-activity mutations, it was assumed that 30 and 12 loci, respectively, were screened.

The smaller number of animals screened in the two biochemical mutation tests than in the specific-locus and dominant-cataract tests reflects the greater effort required. The electrophoretic-mobility mutation test requires a much larger expenditure of resources — in particular, animal housing — than the enzyme activity mutation test, because the electrophoretic-mobility test assays for liver enzymes and therefore requires a liver biopsy, which can be done only when animals are approximately eight weeks of age. Since the enzyme-activity assay is for erythrocyte enzymes, a blood biopsy can be taken at weaning; thus, animals are weaned, and assayed and classified within a few days of weaning.

Fewer biochemical mutations were recovered after procarbazine and radiation treatment than after treatment with ENU, as in specific-locus and dominant-cataract tests.

Estimates of mutation rates per locus are presented in Table 3 and represent the only systematic comparison of mutation rates to different genetic endpoints in mouse germ cells. The mutation rate to specific-locus alleles is consistently highest, although that for enzyme-activity mutations is of the same order of magnitude; mutation rates to dominant-cataract and electrophoretic-mobility alleles are an order of magnitude lower. Caution must be exercised since (i) the electrophoretic-mobility results are for loci at which allelic differences do not exist between the parental strains employed, and (ii) the number of loci screened in the enzyme-activity mutation test may be underestimated and result in an overestimation in the calculated per-locus mutation rate. Improvements in the tests may be expected, making comparative results for mutation rates to different genetic endpoints even more meaningful.

3. Chromosomal mutations

Two well-established methods exist for studying chromosomal mutations in mammals: the dominant-lethal mutation assay and the heritable translocation test. Both tests were evaluated in the Gene-Tox Program (Generoso et al., 1980a; Green et al., 1985). Sex-chromosome loss and nondisjunction can also be used to study chromosomal mutation, but, since no new data have been published (IARC, 1980), this review covers only new data on the dominant-lethal mutation and heritable translocation tests.

3.1 Dominant-lethal mutation assay

The term 'dominant lethal' is used to describe embryonic death resulting from chromosomal breakage in parental germ cells. Any induced change that affects the viability of the germ cells themselves or which renders the gametes incapable of participating in fertilization is excluded. To study the cytogenetic basis of chemically-induced dominant lethals, Brewen et al. (1975) collected fertilized ova from females mated to young adult male mice after treatment with methyl methanesulfonate. The ova were collected from day 1 to day 23 after treatment. The types of aberrations observed were predominantly double fragments (presumably isochromatid deletions), chromatid interchanges and some chromatid deletions, as well as a shattering effect of the male chromosomal complement with 100 mg/kg methyl methanesulfonate during the peak sensitivity of dominant-lethal induction. These data strongly suggest that chromosomal aberrations observed at the first cleavage division of zygotes are the basis of methyl methanesulfonate-induced dominant lethality. In general, it can be concluded that most dominant lethals probably result from multiple chromosomal breaks in the germ cells.

The effectiveness of the blood-testis barrier (Setchell, 1970) has sometimes been suggested to invalidate the dominant lethal assay. On the contrary, this assay is one of the few test systems that provides information about compounds that can cross the blood-testis barrier. Such information is of great importance for the evaluation of chemical mutagens.

When a chemical substance has penetrated the blood-testis barrier, it can be subjected to enzymic activation processes in the different tissues of the gonad or may be detoxified. Then, the chemical or its metabolic products may interact with DNA, and the resulting damage may be involved in different DNA repair processes. Finally, a sperm develops which may or may not carry a mutation. Generoso et al. (1979) demonstrated that the yield of chemically-induced dominant lethal mutations in male mice depends on the genotype of the female with which the males were mated. This has been interpreted as a genetic difference in DNA-repair capability.

(a) Description of the test

The method consists of sequential mating between treated or untreated male mice and untreated females. Mating usually occurs at night, and conception can be recognized the following morning by the presence of a vaginal plug, which is a convenient means of timing a pregnancy. Pregnant females are killed on day 14-16 of pregnancy, and the corpora lutea, representing the number of ova shed, are counted and the uterine contents scored for early and late deaths and living fetuses. The induction of dominant lethals is determined by increases in pre- and post-implantation loss of zygotes in the experimental group over that in the control group. This simple procedure is an essential advantage of the test method. A test protocol has been developed on the basis of a collaborative study involving nine laboratories (Ehling et al., 1978).

(b) Relevance of the test

Chemical mutagens can be characterized by the differential sensitivity of various germ-cell stages to the induction of dominant lethals. Since the frequency of dominant lethals can change drastically in a 24-h mating interval, it is mandatory to use a sequential mating schedule of only a few days. A review of the differential induction of dominant lethals was published by Ehling (1977).

The greatest advantage of the test is its sensitivity. With a relatively small sample size, mutagenic activity can be detected in post-spermatogonial germ-cell stages. For example, the sample size required to detect mutagenic activity in NMRI-Kisslegg and $(101 \times C3H)F_1$ mice (Vollmar, 1977) are shown in Table 4. The data were taken from a total of 7000 untreated control animals, and a type 1 error of $\alpha = 0.05$ and an equally large type 2 error of $\beta = 0.05$ were assumed.

With a sample size of 15-19 fertilized $(101 \times C3H)F_1$ hybrid female mice per mating interval, one can easily detect the effect of 40 mg/kg bw methyl methanesulfonate. With a sample size of 45 animals one can detect a significant increase in dominant-lethal frequency with 10 mg/kg bw. Increasing the sample size to 160 females per mating interval did not further increase the sensitivity of the assay. It was not possible to detect a significant increase in mutation frequency after treatment with 5 mg/kg bw (Ehling, 1977).

(c) Disadvantages

Since calculations are based on comparisons of treated and control groups with respect to the ratio of live embryos to corpora lutea, one cannot discriminate between

Table 4. Sample sizes required to demonstrate a statistically significant increase in the dominant lethal assay in male mice[a]

Mutagenic effect $\Delta\ (\%)$[b]	Genotype	
	NMRI-Kisslegg	$(101 \times C3H)F_1$
10	70	45
15	27	19
20	22	15

[a] From Vollmar (1977)

[b] Lowering of the probability by $\Delta\ \%$ that a live implant will arise from an ovulated oocyte

preimplantation loss and unfertilized ova. Moreover, no formula, by itself, can achieve such a differentiation. For exact determination of dominant-lethal frequency, it is necessary to determine the rate of fertilization of ova (Kratochvilova, 1978). It must be mentioned that a decreased frequency of fertilization is also an indication of a possible hazard.

Another disadvantage of the method is that, due to germinal selection, it would be extremely difficult to detect dominant lethal mutations in spermatogonia.

(d) Results of dominant-lethal mutation tests

Green *et al.* (1985) reviewed 450 papers for the Gene-Tox Program. Of these, 305 papers were rejected because of insufficient quality. The remaining 145 papers covered 139 chemicals, 65 of which were classified as positive after evaluating all publications. The category of 'positive' included responses of a borderline nature. Although 99 chemicals were classified as negative, there is considerable overlap of chemicals in the two categories, due to conflicting conclusions of independent investigations for the same chemical, which accounts for the incongruity between the total number of chemicals tested and the numbers considered positive and negative.

The difficulty in evaluating the published data on dominant lethal tests is due partly to insufficient statistical treatment of the data. Knowledge of the sample size used to detect an increase and a proper statistical evaluation of the data are mandatory for correct evaluation of the dominant lethal assay.

3.2 Heritable translocation test

Reciprocal translocations are exchanges of segments between nonhomologous chromosomes which do not alter the polarity of the exchanged material (with respect to centromere-telomere orientation) and do not result in an alteration in the number of chromosomes. A cell carrying a reciprocal translocation contains a balanced genetic complement and can undergo cell division without loss of genetic material. The heritable translocation method is thus a test for detecting transmitted chromosomal damage that depends on breakage (IARC, 1980).

Screening for translocation heterozygotes has been carried out in two ways: (i) by testing progeny for fertility effects (semi- or complete sterility), followed by cytological analysis of those that show either of these characteristics; and (ii) by analysing all male progeny cytologically.

A detailed protocol of the test and suggestions for improvements were published by Generoso *et al.* (1980a).

(a) Relevance of the test

In contrast to the dominant lethal test, the translocation test recovers heritable defects. Because the heritable translocation test screens for transmissible exchanges, it is more costly and time-consuming than procedures that score for nonheritable aberrations. It is therefore essential that the conditions of the test be optimized. Generoso *et al.* (1978) computed the minimum numbers of progeny needed for different spontaneous rates for testing an agent in order to detect various translocation rates.

(b) Disadvantages

The main obstacle in the development of the heritable translocation test for use in wide-scale testing is the lack of knowledge of whether historical controls can be used in tests for significance. Questions of whether stocks of mice differ in the frequency of spontaneous occurrence and whether laboratory conditions affect the spontaneous occurrence need to be settled. At present, the available information on the spontaneous frequency of heritable translocations is inadequate for the purpose of evaluating the usefulness of historical controls (Generoso *et al.*, 1980a).

(c) Results of heritable translocation tests

In the Gene-Tox Report (Generoso *et al.*, 1980a), a total of 47 publications were evaluated; 29 were judged to contain adequate information to classify whether or not a given chemical induced heritable translocations. Data were available for 32 compounds, but were not adequate for 15 compounds. Of the remaining 17 compounds, a clear-cut determination of positive or negative effects was made for 14 compounds, while data for three compounds were still inconclusive (Table 5). In addition, it has been shown that ethylene oxide significantly increased the frequency of translocations (Generoso *et al.*, 1980b).

For the nine compounds that were tested for specific-locus mutations (Table 1) and for heritable translocations (Table 5), there is agreement in the classification in eight cases; isopropyl methanesulfonate was classified as inconclusive in the specific-locus

Table 5. Agents evaluated for the induction of heritable translocations[a]

Compound	Classification
Caffeine	inconclusive
Captan	negative
Cyclophosphamide	mutagenic
Ethyl methanesulfonate	mutagenic
Isopropyl methanesulfonate	mutagenic
6-Mercaptopurine	negative
Methyl methanesulfonate	mutagenic
Mitomycin C	mutagenic
Nitrogen mustard	inconclusive
Procarbazine	mutagenic
Sodium bisulfite	negative
Sodium calcium salt	negative
Sodium nitrite-treated pork	inconclusive
Trenimon	mutagenic
Triethylenemelamine	mutagenic
Tris(1-aziridinyl)phosphine oxide	mutagenic
Tris(1-aziridinyl)phosphine sulfide	mutagenic

[a]From Generoso et al. (1980a).

test and mutagenic in the heritable-translocation test. This classification is probably due to incomplete testing, there being no specific-locus data for spermatids.

4. Conclusions

The specific-locus test is the most efficient of the germ-cell mutation tests in mice, because a set of seven marker loci were chosen at which newly occurring mutations are easily screened. The dominant-cataract mutation test represents results for dominantly expressed mutations, which are important in terms of hazard to humans. Some biochemical mutations share this relevance, could be characterized at the molecular level, and may contribute to an understanding of the mechanism of mutation induction in germ cells of mammals. Since no single mutation test in mice offers all these advantages, the best strategy is to carry out combined experiments in which mutations to different genetic endpoints are scored simultaneously in the same experimental animals. The dominant-cataract mutation test was so developed to score for both specific-locus and dominant-cataract mutations, allowing a systematic comparison of the mutation frequency for recessive and dominant alleles following mutagenic treatment (Kratochvilova, 1981; Ehling et al., 1982; Favor, 1983). In addition, biochemical mutations have been screened in the same experimental groups (Pretsch & Charles, 1984). This coordinated effort to determine the mutation rate to specific-locus, dominant-cataract and biochemical alleles has been termed the 'multiple endpoint approach' (Ehling et al., 1985) and has allowed comparison of induced mutation rates to a variety of genetic endpoints systematically in a single group of experimental animals. Additional advantages of a multiple endpoint approach are that (i) the number of loci that can be scored is increased to approximately 70 — seven specific, 30 dominant-cataract, 23 electrophoretic and 12 enzyme-activity loci; (ii) recessive, dominant and codominant biochemical mutations are scored; and (iii) further characterization of mutagenic events to four different genetic endpoints is possible.

Apart from the scientific reasons discussed above, there are practical reasons which justify a multiple endpoint approach for studies of germ-cell mutations in mice. Namely, one maximizes the return of an admittedly infrequent mutational event for a given investment of resources, animal-house space and initial handling of mice up to and through weaning. The next goal is to develop a battery of germ-cell mutation tests which screens for variants suspected of being due to a mutation as simply and rapidly as possible, so that presumed mutations may be identified with minimal extra investment of space and time. At present, this battery consists of the specific-locus, dominant-cataract and erythrocyte-enzyme activity mutation tests.

The well-established chromosomal mutation tests, the dominant-lethal assay and the heritable translocation test should be used mainly for confirming the results of nonmammalian short-term tests. The development of techniques to measure mutation rates in humans and other mammals should be encouraged. Studies to analyse gene mutations and chromosomal structure and numbers in human sperm are needed.

5. References

Angell, R.R., Aitken, R.J., van Hook, P.F.A., Lumsden, M.A. & Templeton, A.A. (1983) Chromosome abnormalities in human embryos after in-vitro fertilisation. *Nature, 303,* 336-338

Bailey, D.W. & Kohn, H.I. (1965) Inherited histocompatibility changes in progeny of irradiated and unirradiated inbred mice. *Genet. Res., 6,* 330-340

Bishop, J.B. & Feuers, R.J. (1982) Development of a new biochemical mutation test in mice based upon measurement of enzyme activities. II. Test results with ethyl methanesulphonate (EMS). *Mutat. Res., 95,* 273-285

Brandiff, B., Gordon, L., Ashworth, L.K., Watchmaker, G. & Carrano, A.V. (1986) *Detection of chromosome abnormalities in human sperm.* In: Ramel, C., Lambert, B. & Magnusson, J., eds, *Genetic Toxicology of Environmental Chemicals,* New York, Alan R. Liss, pp. 469-476

Brewen, J.G., Payne, H.S., Jones, K.P. & Preston, R.J. (1975) Studies on chemically induced dominant lethality. I. The cytogenetic basis of MMS-induced dominant lethality in post-meiotic male germ cells. *Mutat. Res., 33,* 239-250

Cattanach, B.M. (1982) Induction of specific-locus mutations in female mice by triethylenemelamine (TEM). *Mutat. Res., 104,* 173-176

Ehling, U.H. (1966) Dominant mutations affecting the skeleton in offspring of X-irradiated male mice. *Genetics, 54*, 1381-1389

Ehling, U.H. (1977) Dominant lethal mutations in male mice. *Arch. Toxicol., 38*, 1-11

Ehling, U.H. (1978) *Specific-locus mutations in mice*. In: Hollaender, A. & de Serres, F.J., eds, *Chemical Mutagens*, Vol. 5, New York, Plenum, pp. 233-256

Ehling, U.H. (1984a) *Methods to estimate the genetic risk*. In: Obe, G., ed., *Mutations in Man*, Berlin (West), Springer, pp. 292-318

Ehling, U.H. (1984b) Variants and mutants. *Mutat. Res., 127*, 189-190

Ehling, U.H. (1985) *Induction and manifestation of hereditary cataracts*. In: Woodhead, A.D., Shellabarger, C.J., Pond, V. & Hollaender, A., eds, *Assessment of Risk from Low-Level Exposure to Radiation and Chemicals*, New York, Plenum, pp. 345-367

Ehling, U.H. & Favor, J. (1984) *Recessive and dominant mutations in mice*. In: Chu, E.H.Y. & Generoso, W.M., eds, *Mutation, Cancer and Malformation*, New York, Plenum, pp. 389-428

Ehling, U.H. & Neuhäuser-Klaus, A. (1984) *Dose-effect relationships of germ-cell mutations in mice*. In: Tazima, Y., Kondo, S. & Kuroda, Y., eds, *Problems of Threshold in Chemical Mutagenesis*, Tokyo, Kokusai-bunken Printing Co., pp. 15-25

Ehling, U.H. & Neuhaüser-Klaus, A. (1985) *Diethylsulfate (DES) induced specific locus mutations in male mice*. In: *Genetic Effects of Environmental Chemicals*, Commission of the European Communities, pp. 14-16

Ehling, U.H., Machemer, L., Buselmaier, W., Dycka, J., Frohberg, H., Kratochvilova, J., Lang, R., Lorke, D., Müller, D., Peh, J., Röhrborn, G., Roll, R., Schulze-Schencking, M. & Wiemann, H. (1978) Standard protocol for the dominant lethal test on male mice. *Arch. Toxicol., 39*, 173-185

Ehling, U.H., Favor, J., Kratochvilova, J. & Neuhäuser-Klaus, A. (1982) Dominant cataract mutations and specific-locus mutations in mice induced by radiation or ethylnitrosourea. *Mutat. Res., 92*, 181-192

Ehling, U.H., Charles, D.J., Favor, J., Graw, J., Kratochvilova, J., Neuhäuser-Klaus, A. & Pretsch, W. (1985) Induction of gene mutations in mice: the multiple endpoint approach. *Mutat. Res., 150*, 393-401

Favor, J. (1983) A comparison of the dominant cataract and recessive specific-locus mutation rates induced by treatment of male mice with ethylnitrosourea. *Mutat. Res., 110*, 367-382

Favor, J. (1984) Characterization of dominant cataract mutations in mice: penetrance, fertility and homozygous viability of mutations recovered after 250 mg/kg ethylnitrosourea paternal treatment. *Genet. Res., 44*, 183-197

Favor, J. (1986) *A comparison of the mutation rates to dominant and recessive alleles in germ cells of the mouse*. In: Ramel, C., Lambert, B. & Magnusson, J., eds, *Genetic Toxicology of Environmental Chemicals*, Part B, *Genetic Effects and Applied Mutagenesis*, New York, Alan R. Liss, pp. 519-528

Feuers, R.J., Delonchamp, R.R., Casciano, D.A., Burkhart, J.G. & Mohrenweiser, H.W. (1980) Assay for mouse tissue enzymes: levels of activity and statistical variation for 29 enzymes of liver or brain. *Analyt. Biochem., 101*, 123-130

Feuers, R.J., Bishop, J.B., Delongchamp, R.R. & Casciano, D.A. (1982) Development of a new biochemical mutation test in mice based upon measurement of enzyme activities. I. Theoretical concepts and basic procedure. *Mutat. Res., 95*, 263-271

Generoso, W.M., Cain, K.T., Huff, S.W. & Gosslee, D.G. (1978) *Heritable-translocation test in mice.* In: Hollaender, A. & de Serres, F.J., eds, *Chemical Mutagens*, Vol. 5, New York, Plenum, pp. 55-77

Generoso, W.M., Cain, K.T., Krishna, M. & Huff, S.W. (1979) Genetic lesions induced by chemicals in spermatozoa and spermatids of mice are repaired in the egg. *Proc. natl Acad. Sci. USA, 76,* 435-437

Generoso, W.M., Bishop, J.B., Gosslee, D.G., Newell, G.W., Sheu, C.-J. & von Halle, E. (1980a) Heritable translocation test in mice. *Mutat. Res., 76,* 191-215

Generoso, W.M., Cain, K.T., Krishna, M., Sheu, C.W. & Gryder, R.M. (1980b) Heritable translocation and dominant-lethal mutation induction with ethylene oxide in mice. *Mutat. Res., 73,* 133-142

Green, S., Auletta, A., Fabricant, J., Kapp, R., Manandhar, M., Sheu, C.-J., Springer, J. & Whitfield, B. (1985) Current status of bioassays in genetic toxicology — the dominant lethal assay. *Mutat. Res., 154,* 49-67

Harris, H., Hopkinson, D.A. & Robson, E.B. (1974) The incidence of rare alleles determining electrophoretic variants: data on 43 enzyme loci in man. *Ann. Hum. Genet., 37,* 237-253

IARC (1980) *IARC Monographs on the Evaluation of the Carcinogenic Risk of Chemicals to Humans*, Suppl. 2, *Long-term and Short-term Screening Assays for Carcinogenesis: A Critical Appraisal*, Lyon

Johnson, F.M., & Lewis, S.E. (1981) Electrophoretically detected germinal mutations induced in the mouse by ethylnitrosourea. *Proc. natl Acad. Sci. USA, 78,* 3138-3141

Kacser, H. & Burns, J.A. (1981) The molecular basis of dominance. *Genetics, 97,* 639-666

Kratochvilova, J. (1978) Evaluation of pre-implantation loss in dominant-lethal assay in the mouse. *Mutat. Res., 54,* 47-54

Kratochvilova, J. (1981) Dominant cataract mutations detected in offspring of gamma-irradiated male mice. *J. Hered., 72,* 302-307

Lewis, S.E. & Johnson, F.M. (1986) *The nature of spontaneous and induced electrophoretically detected mutations in the mouse.* In: Ramel, C., Lambert, B. & Magnusson, J., eds, *Genetic Toxicology of Environmental Chemicals*, Part B, *Genetic Effects and Applied Mutagenesis*, New York, Alan R. Liss, pp. 359-366

Martin, R.H. (1985) *Chromosomal abnormalities in human sperm.* In: Dellareo, V.L., Voytek, P.E. & Hollaender, A., eds, *Aneuploidy, Etiology and Mechanisms*, Part 2, *Etiological Aspects of Human Aneuploidy*, New York, Plenum, pp. 91-102

McKusick, V.A. (1983) *Mendelian Inheritance in Man*, 6th Ed., Baltimore, The Johns Hopkins University Press

Mukai, T. & Cockerham, C.C. (1977) Spontaneous mutation rates at enzyme loci in *Drosophila melanogaster*. *Proc. natl Acad. Sci. USA, 74,* 2514-2517

Neel, J.V., Mohrenweiser, H.W. & Meisler, M.H. (1980a) Rate of spontaneous mutation at human loci encoding protein structure, *Proc. natl Acad. Sci. USA, 77,* 6037-6041

Neel, J.V., Satoh, C., Hamilton, H.B., Otake, B., Goriki, K., Kageoka, T., Fujita, M., Neriishi, S. & Asakawa, J. (1980b) Search for mutations affecting protein structure in children of atomic bomb survivors: preliminary report. *Proc. natl Acad. Sci. USA, 77,* 4221-4225

Paigen, K. (1971) *The genetics of enzyme realization.* In: Recheigl, M., Jr, ed., *Enzyme Synthesis and Degradation in Mammalian Systems*, Basel, Karger, pp. 1-46

Peters, J., Ball, S.T. & Andrews, S.J. (1986) *The detection of gene mutations by electrophoresis and quantitative assay and their analysis*. In: Ramel, C., Lambert, B. & Magnusson, J., eds, *Genetic Toxicology of Environmental Chemicals*, Part B, *Genetic Effects and Applied Mutagenesis*, New York, Alan R. Liss, pp. 367-374

Popp, R.A., Bailiff, E.G., Skow, L.C., Johnson, F.M. & Lewis, S.E. (1983) Analysis of a mouse α-globin gene mutation induced by ethylnitrosourea. *Genetics*, *105*, 157-167

Pretsch, W. & Charles, D.J. (1984) *Detection of dominant enzyme mutants in mice: model studies for mutations in man*. In: Berlin, A., Draper, M., Hemminki, K. & Vainio, H., eds, *Monitoring Human Exposure to Carcinogenic and Mutagenic Agents* (*IARC Scientific Publications No. 59*), Lyon, International Agency for Research on Cancer, pp. 361-369

Rinchik, E.M., Russell, L.B., Copeland, N.G. & Jenkins, N.A. (1985) The dilute-short ear (d-se) complex of the mouse: lessons from a fancy mutation. *Trends Genet.*, *1*, 170-176

Rudak, E., Jacobs, P.A. & Yanagimachi, R. (1978) Direct analysis of the chromosome constitution of human spermatozoa. *Nature*, *274*, 911-913

Russell, L.B. (1971) Definition of functional units in a small chromosomal segment of the mouse and its use in interpreting the nature of radiation-induced mutations. *Mutat. Res.*, *11*, 107-123

Russell, L.B., Selby, P.B., von Halle, E., Sheridan, W. & Valcovic, L. (1981) The mouse specific-locus test with agents other than radiations. Interpretation of data and recommendations for future work. *Mutat. Res.*, *86*, 329-354

Russell, L.B., Cumming, R.B. & Hunsicker, P.R. (1984) Specific-locus mutation rates in the mouse following inhalation of ethylene oxide, and application of the results to estimation of human genetic risk. *Mutat. Res.*, *129*, 381-388

Russell, W.L. (1951) X-ray-induced mutations in mice. *Cold Spring Harbor Symp. Quant. Biol.*, *16*, 327-336

Schlager, G. & Dickie, M.M. (1971) Natural mutation rates in the house mouse. Estimates for five specific loci and dominant mutations. *Mutat. Res.*, *11*, 89-96

Searle, A.G. & Beechy, C.V. (1986) *The role of dominant visibles in mutagenicity testing*. In: Ramel, C., Lambert, B. & Magnusson, J., eds, *Genetic Toxicology of Environmental Chemicals*, Part B, *Genetic Effects and Applied Mutagenesis*, New York, Alan R. Liss, pp. 511-518

Selby, P.B. & Selby, P.R. (1977) Gamma-ray-induced dominant mutations that cause skeletal abnormalities in mice. I. Plan, summary of results and discussion. *Mutat. Res.*, *43*, 357-375

Soares, E.R. (1979) TEM-induced gene mutations at enzyme loci in the mouse. *Environ. Mutagen.*, *1*, 19-25

Vollmar, J. (1977) Statistical problems in mutagenicity tests. *Arch. Toxicol.*, *38*, 13-25

REPORT 9

MAMMALIAN CELL TRANSFORMATION IN CULTURE

Prepared by:

J.C. Barrett (Rapporteur), T. Kakunaga,
T. Kuroki (Chairman), D. Neubert, J.E. Trosko,
J.M. Vasiliev, G.M. Williams and H. Yamasaki

1. Introduction

Cell transformation can be defined as the induction in cultured cells of certain phenotypic alterations that are related to neoplasia. They include morphological transformation, focus formation on cell monolayers, indefinite lifespan ('immortalization'), growth in agar and altered growth potential on plastic surfaces (Heidelberger *et al.*, 1983; Barrett *et al.*, 1984). These heritable phenotypic changes have been shown, in certain cell types, to be associated either directly, or as part of a multistep process, with neoplastic conversion of cells, as measured by their ability to form tumours in animals. Carcinogens can induce stable cell transformation that is heritable and which should be distinguished from similar *reversible* phenotypic changes, which may be induced by alterations to the cell environment, e.g., enhancement of colony formation in agar by certain growth factors (Peehl & Stanbridge, 1981).

A variety of short-term, mammalian cell-culture assays for examining the biological effects of putative carcinogens are available. Some cell transformation assays have high predictive ability for the detection of known carcinogens (IARC, 1980; Heidelberger *et al.*, 1983; Kuroki & Sasaki, 1985). Use of cell transformation assays to study chemical carcinogenesis allows analysis of the cellular events in neoplasia, independent of certain postulated host effects such as immunological surveillance and cell-tissue interactions.

1.1 *Advantages and disadvantages*

In contrast to several other short-term tests, cell transformation assays are not predicated on a theoretical correlation between the endpoint and carcinogenesis; rather, the endpoints of transformation assays can be related to the neoplastic conversion of a cell — a central, but not the only, aspect of carcinogenesis (see Report 3). Cell transformation assays are based on the assumption that neoplastic transformation is induced in cells in culture by the same mechanisms as neoplastic alteration of cells *in vivo*. Cell transformation assays have detected chemicals, many of which are known carcinogens, that are not detected readily in gene mutation assays (Barrett *et al.*, 1984), suggesting that these assays detect relevant, carcinogen-induced events other than gene mutations. Certain human carcinogens that are not detected in other short-term tests have been shown to induce cell transformation: for example, diethylstilboestrol (Pienta, 1980; Barrett *et al.*, 1981), arsenic (DiPaolo & Casto, 1979; Lee *et al.*, 1985), benzene (Amacher & Zelljadt, 1983) and asbestos (DiPaolo *et al.*, 1982; Hesterberg & Barrett, 1984).

The cellular mechanisms of carcinogenesis can also be addressed experimentally using cell transformation assays (Land *et al.*, 1983; Barrett *et al.*, 1984), and these assays can be used to detect agents that may have tumour promoting activity (see Report 10).

The disadvantages of cell transformation assays are that specialized technical expertise and facilities are needed. Some variability in results occurs due mainly to differences between batches of fetal bovine serum, a required but undefined component in most mammalian cell culture systems (IARC/NCI/EPA Working Group, 1985). Another disadvantage is the subjective scoring criteria adopted in some assays; this is discussed further in sections 2-4 of this report. The time required for these assays (two to six weeks, depending on the system) is protracted compared with that needed for assays that employ prokaryotes. In general, several months may be required to complete and repeat assays on a single chemical. In addition, if confirmation is required that cells transformed by a chemical are tumorigenic *in vivo*, several additional months (up to one year) are required.

The state of development of cell transformation assays is not as well advanced as that of some other short-term tests in terms of established protocols, validation and determination of interlaboratory reproducibility. Many systems employ early-passage cells or epithelial cells, which retain the ability to activate a variety of chemicals metabolically; however, an optimal *exogenous* metabolic activation system for cell transformation assays (when this is necessary) has still not been established.

1.2 Interpretation of cell transformation assays: implications of the multistep nature of carcinogenesis and involvement of oncogenes

There is now considerable evidence that neoplastic development of cells in culture is a multistep process (Barrett, 1985). Normal fibroblast or epithelial cells in culture have a limited lifespan and, after a varying number of cell generations, undergo a process of senescence in which the cells cease to proliferate and then die. One key step in the neoplastic progression of cells in culture is escape from senescence and the acquisition of an indefinite lifespan, a process termed 'immortalization' (Barrett & Ts'o, 1978a;

Newbold *et al.*, 1982). Rodent cells in culture, especially mouse and rat cell cultures, are immortalized spontaneously (Ponten, 1971), and can also be induced by carcinogens and certain oncogenes (Newbold *et al.*, 1982). Cells that have undergone immortalization are usually aneuploid and are considered to be partially transformed or in an intermediate stage in the multistep process of neoplastic development (Barrett, 1980). This hypothesis is supported by observations that immortal cells (e.g., NIH3T3, BALBc/3T3 and C3H 10T½) are readily transformed to grow in agar and produce tumours when transfected with certain oncogenes (e.g., *ras*), which are ineffective in normal, diploid cells (Land *et al.*, 1983; Hsiao *et al.*, 1984; Oshimura *et al.*, 1985; Thomassen *et al.*, 1985).

These differences between normal cell cultures, which undergo senescence, and established cell lines, which are immortalized, have implications in understanding the biological significance of different cell transformation systems and in interpreting differing responses of such systems to chemical carcinogens. Furthermore, it is possible that the process of neoplastic development requires more than two steps, which implies that immortal cell lines may differ in the ways in which they convert to the neoplastic state.

2. Syrian hamster embryo cell transformation assay

The first quantitative cell transformation system for detecting potential chemical carcinogens used Syrian hamster embryo cells and was developed by Berwald and Sachs (1965). Following exposure to chemical carcinogens, these early-passage, diploid cells form morphologically altered colonies within one to eight days after treatment. The appearance of such colonies is normally the basis for the assay. This system has been extensively studied in a number of laboratories with over 200 carcinogens and noncarcinogens.

2.1 *Description of the test system*

Usually, this assay is performed as follows: primary cell cultures are prepared from inbred or random-bred pregnant Syrian hamsters (*Mesocricetus auratus*) at 12-14 days of gestation. The fetuses are removed, trypsinized and plated at high density in medium containing fetal bovine serum (Berwald & Sachs, 1965; Huberman & Sachs, 1966; DiPaolo *et al.*, 1971). The cells are grown to near confluence and are used directly (i.e., second passage) or can be cryopreserved in liquid nitrogen and used in the third passage (Pienta, 1980; DiPaolo *et al.*, 1981). The advantage of the latter procedure is that primary cultures can be screened for their response to reference compounds, and a highly susceptible batch of cells can be chosen for multiple assays to minimize the variations in response that may occur with cells derived from different litters of embryos.

For the clonal transformation assay, freshly prepared cells or replated cells from cryopreserved cultures are plated on a feeder layer of lethally irradiated cells. The feeder cells, which can be either homologous Syrian hamster embryo cells or heterologous rat embryo cells (Huberman & Sachs, 1966; DiPaolo et al., 1971), serve to support the growth and increase the colony forming ability of the target cells. The feeder cells can be omitted if increased numbers of target cells are plated (Barrett & Tso, 1978b) or if the medium is supplemented with growth factors (Evans & DiPaolo, 1982). After an overnight incubation to allow the cells to attach, the cultures are exposed to different doses of test chemical and to appropriate positive and negative control treatments. Single cells at clonal density are exposed to the carcinogen and allowed to form colonies for seven to nine days. The cells are then fixed, stained (usually with Giemsa) and examined for morphological changes. The normal cells grow in a parallel, orientated manner, with little criss-crossing of the cells on each other. In contrast, transformed cells are characterized by increased basophilia, criss-crossed growth and an increased nuclear:cytoplasmic ratio (Berwald & Sachs, 1965; DiPaolo et al., 1971; Pienta, 1980; Tu et al., 1986). The cells in the colonies are either sparse and separated or densely packed. Transformation can be scored by similar criteria in each type of colony (Huberman & Sachs, 1966).

Exposure to the chemical can either last for the entire period of colony formation (i.e., seven to nine days) or can be discontinued after defined periods (e.g., one to three days). The latter procedure may result in lower transformation frequencies but has the advantages of demonstrating the stability of the induced phenotypic change and also of allowing subsequent treatments with possible cocarcinogenic or promoting substances.

A modification of this assay (Casto et al., 1977) is to plate the cells at higher, nonclonal densities and to grow them to confluence, allowing the development of transformed foci of cells, which are scored 20-30 days after treatment. This assay, although requiring more time to execute, may reduce some of the ambiguity in scoring transformed cells. The reproducibility of this modification of the assay is, however, not well established in other laboratories, and the clonal assay described above is more widely used (Heidelberger et al., 1983).

The numbers of morphologically transformed and total colonies are determined, and the transformation frequency (i.e., % transformed colonies) is expressed as the (number of transformed colonies — total number of colonies) × 100. The relative survival (% of control) of each treatment group, e.g., (total colonies in treated group —total colonies in control (usually solvent treated) group) × 100, is also determined.

In the execution of the assay, it is essential to include known positive and negative chemicals as controls. The background frequency of spontaneously arising morphological transformants in control cultures (untreated or solvent treated) is generally low. No spontaneous transformation has been reported in many experiments (DiPaolo et al., 1971; Pienta, 1980), but frequencies of 0.01-0.05% have been reported (Barrett &

Ts'o, 1978a; Jones et al., 1984; Barrett & Lamb, 1985; Tu et al., 1986). Following exposure to a wide variety of carcinogens, induced morphological transformation frequencies of 1-10% of the surviving colonies have been observed (Berwald & Sachs, 1965; Huberman & Sachs 1966; DiPaolo et al., 1971).

2.2 *Nature of data and interpretation of results*

Many carcinogens and noncarcinogens have been tested for their ability to produce morphological transformation in this system (Dunkel et al., 1977; Umezawa et al., 1978; Pienta, 1980; Heidelberger et al., 1983). These include representative alkylating agents, polycyclic aromatic hydrocarbons, nitrosamines and nitrosamides, aromatic amines, aminoazo dyes, hormones, metal salts and a number of miscellaneous compounds. Positive correlations between the results of this assay and carcinogenicity have been reported, but further validation is needed (IARC, 1980; Heidelberger et al., 1983).

This system has been extended to include methods for the metabolic activation of indirect carcinogens. Thus, several chemicals, namely, auramine, N-nitrosodiethylamine, 3-methoxy-4-aminoazobenzene, natulan, 2-nitrofluorene, para-rosaniline and urethane, which gave negative results in the standard test, gave positive results when hamster liver homogenates and appropriate cofactors were added (Pienta, 1980). Similarly, if irradiated hamster hepatocytes are incorporated into the bioassay to provide metabolic activation (Poiley et al., 1979; Pienta, 1980), 2-nitrofluorene, 4-aminoazobenzene and N-nitrosodiethylamine give positive results. Pregnant females can also be tested with the test chemical and morphological transformation scored in embryo cell cultures derived from treated mothers. This in-vivo/in-vitro assay is useful for studying metabolism and transplacental carcinogenesis (DiPaolo et al., 1973; Inui et al., 1978).

The Syrian hamster embryo cell transformation assay has been used to detect a number of known human carcinogens not readily detected in other short-term assays and chemicals that do not have detectable activity in gene mutation assays, including diethylstilboestrol (Barrett et al., 1981), sodium bisulfite (DiPaolo et al., 1981), benzene (Amacher & Zelljadt, 1983), asbestos (DiPaolo et al., 1982; Hesterberg & Barrett, 1984) and arsenic (DiPaolo & Casto, 1979; Lee et al., 1985).

2.3 *Assessment of the test system*

The Syrian hamster embryo cell transformation assay has been studied for over 20 years. The technical difficulties in growing the cells and establishing optimal culture conditions to monitor morphological transformation quantitatively has limited its use. Nevertheless, over 20 laboratories have published results using this system, and more chemicals have been tested in this assay than in any other cell transformation system. However, validation, standardization, and interlaboratory comparisons of the assay are limited. Future improvements of the culture conditions, development of serum-free media, and selective assays for transformed cells are needed.

The clonal assay does not select for transformed cells; hence, the morphological appearance of each surviving colony, normal and transformed, must be scored. When the frequency of spontaneous transformation is low, induction of transformation by a chemical can be demonstrated with only a small number of positive colonies in the treated group (Pienta, 1980). This allows a qualitative assessment of a chemical; a quantitative analysis of dose-response relationship requires scoring of considerably more colonies (Barrett et al., 1984).

Morphological transformation of Syrian hamster embryo cells, as detected by the colony assay, represents the first step in a multistep, progressive process of neoplastic development of the cells (Barrett & Ts'o, 1978b). Initial carcinogen-induced changes can be studied, as well as the effects of tumour promoters (see Report 10). Further changes, including immortality and anchorage-independent growth, are required for neoplastic transformation of Syrian hamster cells (Barrett & Ts'o, 1978b; Newbold et al., 1982). Thus, multiple steps in the carcinogenic process can be detected in these cells and the mechanisms of later stages of tumour progression studied (Barrett, 1985). Different classes of viral oncogenes (e.g., *myc* and *ras*) are active in the neoplastic transformation of these cells (Oshimura et al., 1985; Thomassen et al., 1985). Future studies to identify and characterize carcinogen-induced oncogenes activated in different stages of the neoplastic transformation of these cells are needed which may aid in understanding carcinogen-induced changes in carcinogenesis.

The major advantages of this system are that (i) normal, diploid, early-passage cells are used, (ii) a number of carcinogens not readily detected in other systems give positive results in this system, and (iii) the cells intrinsically have a wider range of metabolic capabilities than most fibroblast cell lines.

The major limitations of this assay are (i) the subjectivity of the endpoint scored, (ii) the sensitivity of the assay to culture variables, especially serum, (iii) the variability of primary cultures of embryos, (iv) the technical difficulty of the assay, and (v) the effort required to obtain dose-response data.

3. BALB/c 3T3 cell transformation assay

The BALB/c 3T3 cell line was originally developed by Aaronson and Todaro (1968). This cell line has acquired some properties associated with neoplastic cells, e.g., aneuploidy and immortality, but retains other properties associated with non-neoplastic cells, such as flat morphology, a high sensitivity to density-inhibition of growth and movement, anchorage-dependence of growth, and failure to form tumours in animals. Following treatment with chemical or physical carcinogens, a minor population of cells lose these non-neoplastic characteristics and form foci, which consist of disorientated, piled up cells, growing over the untransformed cell monolayer. The focus-forming cells are usually tumorigenic upon injection into syngeneic or nude mice. Thus, this system measures only part of the multistep process of neoplastic transformation.

The BALB/c 3T3 cell line was initially used to study neoplastic transformation by oncogenic viruses. Neoplastic transformation of BALB/c 3T3 clones by chemical carcinogens was first reported in the early 1970s (Kakunaga & Kamahora, 1970; DiPaolo et al., 1972; Kakunaga, 1972, 1973). Since these original clones of BALB/c 3T3 no longer exist, or were very sensitive to unknown factors in serum, new subclones were isolated from early passages of the BALB/c 3T3-A31 clone (Kakunaga & Crow, 1980), and these have been used to study neoplastic transformation by chemicals and irradiation in many laboratories. A variety of chemicals have been tested in this system (Kuroki & Sasaki, 1985).

3.1 *Description of the test system*

The methods and protocols for the transformation assay using BALB/c 3T3 subclones have been standardized (IARC/NCI/EPA Working Group, 1985; Kakunaga & Yamasaki, 1985). Briefly, 10^4 target cells per 60 mm dish are plated onto at least 20 dishes for each concentration of chemical to be tested, and 24 hours later the test chemical is added. After three days' treatment, the medium is removed. Two dishes of each set are trypsinized and the surviving fraction determined from the cloning efficiency of the treated and control populations. The remaining dishes are washed, replenished with fresh medium without the test chemical and incubated for an additional four weeks. The medium is changed twice a week during this incubation period. At the end of the incubation period, the dishes are fixed and stained. Transformed foci which show deep basophilia, dense multilayering of cells, random cell orientation at any part of the focus edge and invasion into the surrounding contact-inhibited monolayer are scored as positive transformed foci. A detailed prescription for scoring transformed foci of BALB/c 3T3 subclones is available (IARC/NCI/EPA Working Group, 1985). The transformation frequency for each treatment is expressed as the total number of transformed foci per total replicate dishes and per surviving cell and as the number of dishes with foci.

The BALB/c 3T3 populations now being used for transformation assays are from a clone designated BALB/c 3T3-A31-1 and its subclones, A31-1-1 and A31-1-13 (Kakunaga & Crow, 1980).

3.2 *Nature of the data and interpretation of results*

Over 72 chemicals have been tested in this system (see Heidelberger et al., 1983; Kuroki & Sasaki, 1985). There is a reasonably good correlation between transformation and in-vivo carcinogenicity. Polycyclic aromatic hydrocarbons are metabolized in BALB/c 3T3 subclones and induce transformation. Metal carcinogens and aromatic amines also give positive results. For 28 known carcinogens tested, the sensitivity (calculated as the number of carcinogens giving positive results as a percentage of the total carcinogens tested) is 67.8%. Unfortunately, there are insufficient data on noncarcinogens: of three tested, two were negative.

3.3. Assessment of the test system

Since BALB/c 3T3 cells are an established cell line with high cloning efficiency, a large cell population with a relatively homogeneous phenotype can be derived from a single parent cell. In contrast to primary cell cultures, a large stock of cells can be prepared and used in different laboratories for transformation assays with a reasonable degree of reproducibility.

The high plating efficiency of BALB/c 3T3 subclones (>30%) allows calculation of induced transformation frequency per surviving cell; however, this determination is difficult (IARC/NCI/EPA Working Group, 1985).

Expression of the transformed phenotype is influenced by culture media and serum. Since serum contains variable amounts of growth factors and related factors, different batches of serum give different experimental results. This affects the frequencies of the induced and spontaneous transformation and significantly influences the effects of cell density on transformation frequency. The transformation frequency per treated cell or surviving cell may increase with decreasing numbers of cells plated per dish (Little, 1979), a phenomenon in part ascribed to serum factors. By selecting a particular batch of serum, it is possible to obtain a constant transformation frequency within the range of $5 \times 10^2 - 2 \times 10^4$ cells plated per 60-mm dish (Kakunaga, 1985). Thus, a batch of serum should be tested before use.

In order to assess the transforming potential of a chemical, it is essential to obtain a dose-related response for transformation. In contrast to other systems (see sections 2 and 4), many laboratories using BALB/c 3T3 cells as target cells have reported increases in transformation frequencies which are dose-related.

The cells from the transformed focus (which is the assay endpoint for the BALB/c 3T3 transformation system) usually produce tumours when transplanted subcutaneously into syngeneic or nude mice, indicating the relevance of the endpoint to neoplastic conversion of these cells. Cells from nontransformed areas of the dishes remain nontumorigenic (Kakunaga, 1985).

Wide variations of transformed phenotype have been a major problem common to all cell transformation systems. In general, scoring transformed foci is easier and more reproducible than scoring morphologically altered colonies (see section 2).

BALB/c 3T3 subclones have limited capacities to metabolize chemicals. They have relatively high arylhydrocarbon hydroxylase activities, but several classes of carcinogens, such as nitrosamines, aromatic amines, cyclic esters and azo dyes, are not metabolized in BALB/c 3T3 subclones. Addition of exogenous metabolic systems or cocultivation with hepatocytes can resolve this problem to some extent, but the technique is not completely established (Schectman, 1985; Sivak & Tu, 1985).

BALB/c 3T3 subclones used for transformation assays provide relatively homogeneous target-cell populations. However, spontaneous transformation occurs at frequencies of approximately 10^{-7} to 10^{-6} per cell division, depending on the batch of serum. Since the transformed cells have a growth advantage, it is necessary to use stocks of target cells with reduced numbers of spontaneously transformed cells.

4. C3H 10T1/2 cell assay

C3H 10T1/2 is a cell line derived from C3H mouse embryos by a protocol of subculturing the cells at 10-day intervals by plating 0.5×10^5 cells onto a 60-mm dish (Reznikoff et al., 1973). Like BALB/c 3T3 cells, these cells shows a high sensitivity to density-inhibition of cell growth. Treatment with carcinogens results in the loss by a certain fraction of the cells of sensitivity to contact inhibition and formation of a dense focus of cells with a criss-crossed arrangement (transformed foci). Although C3H 10T1/2 cells are not tumorigenic in appropriate hosts, they are aneuploid and produce tumours when inoculated *in vivo* by attachment to a solid surface (Boone & Jacobs, 1976). These findings suggest that the cells are partially transformed.

4.1. *Description of the test system*

The transformation assay is conducted by the original procedures described by Reznikoff et al. (1973), with slight modifications. An IARC/NCI/EPA Working Group has reported a recommended protocol for the C3H 10T1/2 cell assay (IARC/NCI/EPA Working Group, 1985; Kakunaga & Yamasaki, 1985). It is essential to use cells with a low background of spontaneous transformation, which is usually achieved by the use of prescreened serum and a cell stock with less than 15 passages. The assay system should be controlled by the inclusion of proper negative (or solvent) and positive controls (e.g., 48-hour treatment with 2.5 μg 3-methylcholanthrene). Serum should be selected to make the assay highly sensitive to known carcinogens. The transformation assay is performed by exposing C3H 10T1/2 cells, seeded at 2×10^3 cells/60-mm dish, to test compounds at several dose levels for 48 h. Treated cells are maintained for six to eight weeks and fixed and stained for determination of the presence of transformed foci. A cytotoxicity assay is performed in parallel by plating 200 cells/60-mm dish, incubating for two weeks and determining the number of surviving colonies in control and treated cultures.

Transformation is scored quantitatively by counting morphologically altered foci. Reznikoff et al. (1973) categorized foci into types I, II and III according to the degree of morphological aberration. Transformed foci that display the aberrant phenotypic characteristics designated types II and III should be counted independently. The differences between type-II and type-III foci are related to the degree of morphological aberrations associated with each. Type-III foci have the following properties: dense, multilayered, basophilic, random orientation at focus edge, invasion into the monolayer, predominantly spindle-shaped transformed cells. Type-II foci are distinguishable from type-III foci primarily by their more ordered and defined edges. In addition, they are dense, multilayered and less basophilic than type-III foci (IARC/NCI/EPA Working Group, 1985).

Note should be taken of unusual transformed foci (e.g., corded, banded, poorly attached), especially if they represent a substantial fraction of the foci, and caution should be exercised in including these in the total number of transformed foci. They should also be reported independently of the standard foci.

Focus size depends on the culture conditions and duration of the assay. Generally, foci of less than 1 to 2 mm should not be scored; however, small foci with striking type-III transformed morphology could be counted as transformed at the discretion of the investigator. For the purpose of identifying carcinogens, it is recommended to score type-II and -III foci as transformed because the cells derived from them produce tumours when injected into immunosuppressed syngeneic mice (IARC/NCI/EPA Working Group, 1985).

The transformation frequency (i.e., the number of transformed foci per treated cell after correcting for cell killing) is dependent upon the initial number of cells at risk. This complicates the quantification of the frequency of transformation of these cells and suggests that this conversion is not a one-step process (Reznikoff *et al.*, 1973; Haber *et al.*, 1977; Fernandez *et al.*, 1980; Kennedy *et al.*, 1980). Several explanations for these results with C3H 10T1/2 cells have been suggested (Fernandez *et al.*, 1980; Kennedy *et al.*, 1980; Backer *et al.*, 1982; Mordan *et al.*, 1983; Kennedy *et al.*, 1984). Because of this problem, dose-response data and comparative results can be obtained only if the same numbers of surviving cells are plated per dish.

In order to detect a wide variety of chemicals, a liver fraction or metabolically competent epithelial cells must be included in the transformation assay to provide an exogenous source of metabolic activation. (See Report 15.)

4.2 *Nature of the data and data base*

Sixty-seven chemicals have been tested with C3H 10T1/2 cells, including 26 known carcinogens and five noncarcinogens (Kuroki & Sasaki, 1985), showing a sensitivity of 80.8% (21/26) and a specificity of 60.0% (3/5). In general, results with polycyclic aromatic hydrocarbons show a good correlation with carcinogenicity *in vivo*. Several purine and pyrimidine derivatives were assayed with C3H 10T1/2 cells, but their in-vivo carcinogenicity has not been tested. One of the disadvantages of the C3H 10T1/2 cell system is that most alkylating agents cannot be detected efficiently by the standard protocol (Peterson *et al.*, 1981). Modifications of the assay conditions, such as use of synchronous cultures, treatment of cells five days after seeding or subsequent exposure to a phorbol ester tumour promotor, are required for detecting transforming activity of N-methyl-N'-nitro-N-nitrosoguanidine (Landolph, 1985).

Two-stage transformation can be studied in C3H 10T1/2 cells (see Report 10).

4.3 *Advantages and disadvantages of the test system*

Because C3H 10T1/2 cells are an established cell line, a cloned, uniform cell population can be obtained easily. Spontaneous transformation is rarer than in BALB/c 3T3 cells. Morphological transformation can be scored more easily than in Syrian hamster cells (see section 2). C3H 10T1/2 cells retain significant amounts of inducible cytochrome P448, and therefore polycyclic hydrocarbons induce transformation in the absence of exogenous metabolic activation.

A potential disadvantage of the C3H 10T1/2 cell assay is that the cells are aneuploid and immortal and are therefore abnormal. Long periods of cultivation (up to six to eight weeks) and frequent changes of medium and the requirement of a large number of dishes are also disadvantageous. Difficulty in obtaining transformation by alkylating agents is a limitation for testing a wide variety of chemicals. Compared with the BALB/c 3T3 system, it is more difficult to obtain dose-response relationships. Further studies are needed for validation, especially using noncarcinogens, for which few data are available.

5. Other cell transformation assays

In addition to the assays described above, a number of other cell transformation systems have been developed. The Working Group did not feel, however, that these tests were either as well-established or reproducible as the assays discussed above. However, some of them offer considerable promise or have been used occasionally; therefore, a summary of the current status of these assays is presented below.

5.1 *Assays with virus-infected fibroblasts*

Fischer rat-embryo cells infected with the nontransforming retrovirus Rauscher leukaemia virus (RLV) are susceptible to chemically induced neoplastic transformation (Freeman *et al.*, 1970; Price *et al.*, 1971, 1972). Fischer rat-embryo cell lines with fewer than 60 population doublings are resistant to spontaneous and chemically induced transformation, whereas the same cells chronically infected with RLV have increased sensitivity to transformation by a variety of chemical carcinogens (Price & Mishra, 1980). Traul *et al.* (1981) derived a cell line infected with RLV ($2FR_450$), which is now used in a short-term assay for chemically induced morphological transformation (Dunkel *et al.*, 1981; Suk, 1985). This assay system is currently being validated (Suk, 1985). The major limitations of the system are that it is impossible to quantify transformation on a per-cell basis and the lack of definition of the mechanism of viral enhancement of carcinogen-induced transformation (Tennant *et al.*, 1985). Results in this system cannot be attributed entirely to the action of the chemical alone; however, it may provide insights into viral-chemical interactions.

The induction of transformed foci of Syrian hamster embryo cells by simian adenovirus SA7 is enhanced by concurrent exposure to a wide variety of potential chemical carcinogens (Casto et al., 1973, 1974; Hatch et al., 1985). This assay is currently being validated as a screen for morphological transformation and has the advantage of being highly quantitative and very sensitive. The disadvantage of the assay is that it measures the enhancement by chemicals of transformation by the virus, and its relevance to non-viral-mediated transformation is uncertain.

5.2 *Neoplastic transformation of human cells*

One of the basic assumptions in short-term tests with nonhuman cellular systems is that the basic mechanisms of chemical action are the same as in human cells. This assumption can be tested, and, indeed, in some cases, e.g., gene mutations (see Report 7), many similarities are observed. However, significant differences unrelated to metabolic activation of carcinogens have been observed in the ability to induce neoplastic transformation of rodent *versus* human cells in culture (see below). These differences may be only technical but more likely reflect undefined intrinsic differences. Further studies on the molecular mechanisms of human and rodent cells are needed to solve this problem.

(a) Principles, scientific basis and description of the test

The systems for studying neoplastic transformation of human cells in culture can be divided into two groups, based on whether or not complete neoplastic transformation is used as the assay endpoint. One group of systems measures the process of conversion of normal, diploid cells to fully tumorigenic cells, while the other group measures specific phenotypic changes possibly related to the neoplastic change but which alone are insufficient for neoplastic transformation (Kakunaga, 1981; McCormick et al., 1985).

In the first type of assay, which measures complete neoplastic conversion, normal, diploid human fibroblasts or epithelial cells are exposed to a single dose of a carcinogen and then continuously propagated or exposed to multiple doses of carcinogens, with some interval of continuous propagation in culture. Cultures are propagated for varying periods of time (five weeks to three months) until morphologically altered cells or transformed foci appear. Morphologically altered cells are isolated and tested for other transformed phenotypes including anchorage-independent growth and tumorigenicity in nude mice. Tumorigenicity is sometimes, but rarely, achieved with low efficiency in these assays (Kakunaga, 1981).

The second group of assays, which rely on phenotypic markers, can be divided into two subgroups, based on which phenotype in the multistep process of neoplastic transformation is chosen as the assay endpoint. One subgroup uses diploid cells as target cells and anchorage-independent growth as the assay endpoint. The cells are treated with carcinogens, allowed to divide to express the carcinogen-induced changes and are then plated in soft agar. The numbers of colonies formed in soft agar are scored, and the transformation frequency is expressed as the number of agar colonies per cells plated (McCormick et al., 1985). The second subgroup uses flat-revertants of a human osteosarcoma cell line (Rhim et al., 1975) or fibroblasts immortalized by SV40 or Ela oncogene (Rhim et al., 1986) as target cells. Multiple transformed phenotypes, including morphological alterations, anchorage-independent growth and tumorigenicity are used as assay endpoints. These systems are similar to rodent systems using spontaneously immortalized cells (e.g., BALB/c 3T3 and C3H 10T1/2 cells) as target cells. However, the induction of transformation in these human cell systems has not been quantified.

(b) *Nature of the data and interpretation of results*

The frequency of complete neoplastic transformation of human cells by chemical carcinogens or radiation is very low (Kakunaga, 1978; Namba, 1985), due probably to the extremely low probability of human cells acquiring immortality, a property that is considered to be an essential step in the progress to neoplasia (Newbold et al., 1982; Barrett, 1985).

Induction of anchorage-independent growth of normal human diploid cells by chemical carcinogens has been reported from several laboratories (Freedman & Shin, 1977; Milo & DiPaolo, 1978; Silinskas et al. 1981; McCormick et al., 1985). Optimal conditions for obtaining dose-response curves for transformation have been described by Silinskas et al. (1981). Spontaneous transformation occurs with variable frequency, and the stability of this transformed phenotype is unclear, since isolated colonies do not grow with a high frequency in agar after isolation. The number of chemicals tested with this assay is small, and there are large variations in transformation frequencies between experiments and between laboratories. Furthermore, the biological significance of the assay endpoint is unclear. The cells that compose the colonies in soft agar show no other transformed phenotype, are senescent and have not been shown to progress to complete neoplastic conversion.

The induction of tumorigenic conversion of immortalized human cell lines has been reported by one laboratory (Rhim et al., 1975, 1986) using various cells as targets. The number of chemicals tested is very small and quantification of transformation frequency has never been attempted.

(c) *Assessment of the test systems*

None of the human cell transformation assays can readily be used to test chemicals. The anchorage-independent growth assay is possibly quantifiable, but many variables are still undefined in the execution of this assay, and its relevance to the neoplastic progression of these cells is not yet understood (McCormick et al., 1985). No evidence has been obtained indicating that the human diploid cells that grow in soft agar have a higher propensity to become tumorigenic. The extremely low probability of human cells acquiring immortality suggests that immortalization is the rate-limiting step in the neoplastic transformation of human cells.

Tests of human cells for tumorigenicity, quantification of tumorigenic conversion of human cell lines, and development of an assay system for the immortalization step are desirable.

5.3 *Epithelial cell transformation assays*

Epithelial cell systems have been of interest because most human cancers are of epithelial origin and because epithelial cells are the organ-specific functional cell types in many organs.

Human and rodent epithelial cells from many tissues have been studied, including liver (Montesano *et al.*, 1973; Williams *et al.*, 1973; Schaeffer & Heintz, 1978; Shimada *et al.*, 1983), bladder (Hashimoto & Kitagawa, 1974; Summerhayes *et al.*, 1981), epidermis (Elias *et al.*, 1974; Colburn *et al.*, 1978; Fusenig *et al.*, 1978; Slaga *et al.*, 1978; Ananthaswamy & Kripke, 1981; Kulesz-Martin *et al.*, 1985), kidney (Borland & Hard, 1974), mammary gland (Richards & Nandi, 1978), salivary gland (Wigley, 1979) and trachea (Pai *et al.*, 1983; Thomassen *et al.*, 1983).

Advances have been made in the development of quantitative assays for carcinogen-induced transformation of mouse keratinocyte lines (Kulesz-Martin *et al.*, 1985) and of primary rat tracheal epithelial cells (Thomassen *et al.*, 1983); the latter system can be used for in-vivo-in-vitro experiments in which cells are exposed in an intact trachea *in vivo* and transformation is assessed quantitatively *in vitro* (Nettesheim & Barrett, 1984).

A significant feature of epithelial cell systems is that they may possess carcinogen activation that is tissue-specific. However, quantification of the results obtained with these systems is still not satisfactory; determination of the transformation frequencies on a per-cell basis is limited to a few systems. In addition, standardized protocols need to be designed and validated.

An established cell line derived from the kidney of a weanling Syrian hamster, designated BHK (for baby hamster kidney), is available (IARC, 1980). Cells of this line are immortal, quasidiploid and generally grow in agar with a low frequency. The basis for this cell transformation assay is the increase in frequency of cells with anchorage-independent growth following treatment with the test chemical. BHK cells have been used as a short-term assay for potential carcinogens (Styles, 1977; IARC, 1980). The major difficulty with this assay is that the spontaneous background level is variable and can become quite high, presumably due to selection for the transformed cells. This problem has resulted in difficulties in reproducibility. Only when a subclone with a low background level is isolated and maintained will this assay be useful. At present, this assay is not routinely used for screening chemicals.

6. References

Aaronson, S.A. & Todaro, G.J. (1968) Development of 3T3-like lines from BALB/c mouse embryo cultures: transformation susceptibility to SV 40. *J. Cell. Physiol.*, 72, 141-148

Amacher, D.E. & Zelljadt, I. (1983) The morphological transformation of Syrian hamster embryo cells by chemicals reportedly non-mutagenic to *Salmonella typhimurium*. *Carcinogenesis*, 4, 291-295

Ananthaswamy, H.N. & Kripke, M.L. (1981) In-vitro transformation of primary culture of neonatal BALB/c mouse epidermal cells with ultraviolet-B radiation. *Cancer Res.*, 41, 2882-2890

Backer, J.M., Boerzig, M. & Weinstein, I.B. (1982) When do carcinogen treated 10T½ cells acquire the commitment for forming transformed foci? *Nature*, 299, 458-460

Barrett, J.C. (1980) A preneoplastic stage in the spontaneous neoplastic transformation of Syrian hamster embryo cells in culture. *Cancer Res.*, 40, 91-94

Barrett, J.C. (1985) *Cell culture models of multistep carcinogenesis*. In: Likhachev, A., Anisimov, V. & Montesano, R., eds, *Age-Related Factors in Carcinogenesis (IARC Scientific Publications No. 58)*, Lyon, International Agency for Research on Cancer, pp. 181-202

Barrett, J.C. & Lamb, P.W. (1985) *Tests with the Syrian hamster embryo cell transformation assay*. In: Ashby, J., de Serres, F.J., Draper, M., Ishidate, M., Margolin, B.H., Matter, B.E. & Shelby, M.D., eds, *Progress in Mutation Research*, Vol. 5, Amsterdam, Elsevier, pp. 623-628

Barrett, J.C. & Ts'o, P.O.P. (1978a) The relationship between somatic mutation and neoplastic transformation. *Proc. natl Acad. Sci. USA*, 75, 3297-3301

Barrett, J.C. & Ts'o, P.O.P. (1978b) Evidence for the progressive nature of neoplastic transformation *in vitro*. *Proc. natl Acad. Sci. USA*, 75, 3761-3765

Barrett, J.C., Wong, A. & McLachlan, J.A. (1981) Diethylstilbestrol induces neoplastic transformation of cells in culture without measurable somatic mutation at two loci. *Science*, 212, 1402-1404

Barrett, J.C., Hesterberg, T.W. & Thomassen, D.G. (1984) Use of cell transformation systems for carcinogenicity testing and mechanistic studies of carcinogenesis. *Pharmacol. Rev.*, 36, 53S-70S

Berwald, Y. & Sachs, L. (1965) In-vitro transformation of normal cells to tumor cells by carcinogenic hydrocarbons. *J. natl Cancer Inst.*, 35, 641-661

Boone, C.W. & Jacobs, J.B. (1976) Sarcomas routinely produced from putatively non-tumorigenic BALB/3T3 and C3H10T½ cells by subcutaneous inoculation attached to plastic plates. *J. supramol. Struct.*, 5, 131-137

Borland, R. & Hard, G.C. (1974) Early appearance of 'transformed' cells from the kidneys of rats treated with a 'single' carcinogenic dose of dimethylnitrosamine (DMN) detected by culture *in vitro*. *Eur. J. Cancer*, 10, 177-184

Casto, B.C., Pieczynski, W.J. & DiPaolo J.A. (1973) Enhancement of adenovirus transformation by pretreatment of hamster cells with carcinogenic polycyclic hydrocarbons. *Cancer Res.*, 3, 819-824

Casto, B.C., Pieczynski, W.J. & DiPaolo, J.A. (1974) Enhancement of adenovirus transformation by treatment of hamster embryo cells with diverse chemical carcinogens. *Cancer Res.*, 34, 72-78

Casto, B.C., Janosko, N. & DiPaolo, J.A. (1977) Development of a focus assay model for transformation of hamster cells *in vitro* by chemical carcinogens. *Cancer Res.*, 37, 3508-3515

Colburn, N.H., Bruegge, W.F.V., Bates, J. & Yuspa, S.H. (1978) Epidermal cell transformation *in vitro*. *Carcinogenesis*, 2, 257-271

DiPaolo, J.A. & Casto, B.C. (1979) Quantitative studies of in-vitro morphological transformation of Syrian hamster cells by inorganic metal salts. *Cancer Res. 39*, 1008-1013

DiPaolo, J.A., Donovan, P.J. & Nelson, R.L. (1971) In-vitro transformation of hamster cells by polycyclic hydrocarbons: factors influencing the number of cells transformed. *Nature-New Biol., 230*, 240-242

DiPaolo, J.A., Takano, K. & Popescu, N.C. (1972) Quantitation of chemically induced neoplastic transformation of BALB/3T3 cloned cell lines. *Cancer Res., 32*, 2686-2695

DiPaolo, J.A., Nelson, R.L., Donovan, P.J. & Evans. C.H. (1973) Host-mediated in-vivo-in-vitro assay for chemical carcinogenesis. *Arch. Pathol., 95*, 380-385

DiPaolo, J.A., DeMarinis, A.J. & Doniger, J. (1981) Transformation of Syrian hamster embryo cells by sodium bisulfite. *Cancer Lett., 12*, 203-308

DiPaolo, J.A., DeMarinis, A.J. & Doniger, J. (1982) Asbestos and benzo[*a*]pyrene synergism in the transformation of Syrian hamster embryo cells. *J. environ. Pathol. Toxicol., 5*, 535-543

Dunkel, V.C., Wolff, J.S. & Pienta, R.J. (1977) In-vitro transformation as a presumptive test for detecting chemical carcinogens. *Cancer Bull., 29*, 167-174

Dunkel, V.C., Pienta, R.J., Sivak, A. & Traul, K.A. (1981) Comparative neoplastic transformation responses of Balb/3T3 cells, Syrian hamster embryo cells, and Rauscher murine leukemia virus-Fischer 344 rat embryo cells to chemical carcinogens. *J. natl Cancer Inst., 67*, 1303-1315

Elias, P.M., Yuspa, S.H., Gullino, M., Morgan, D.L., Bates, R.R. & Lutzner, M.A. (1974) In-vitro neoplastic transformation of mouse skin cells: morphology and ultrastructure of cells and tumors. *J. invest. Dermatol., 62*, 569-581

Evans, C.H. & DiPaolo, J.A. (1982) Equivalency of endothelial cell growth supplement to irradiated feeder cells in carcinogen-induced morphological transformation of Syrian hamster embryo cells. *J. natl Cancer Inst., 68*, 127-131

Fernandez, A., Mondal, S. & Heidelberger, C. (1980) Probabilistic view of the transformation of cultured C3H 10T½ mouse embryo fibroblasts by 3-methylcholanthrene. *Proc. natl Acad. Sci. USA, 77*, 7272-7276

Freedman, V. & Shin, S.I. (1977) Isolation of human diploid cell variants with enhanced colony-forming efficiency in semisolid medium after a single-step chemical mutagenesis: brief communication. *J. natl Cancer Inst., 58*, 1873-1875

Freeman, A.E., Price, P.J., Igel, H.J., Young, J.C., Maryak, J.M. & Huebner, R.J. (1970) Morphological transformation of rat embryo cells induced by diethylnitrosamine and murine leukemia viruses. *J. natl Cancer Inst., 44*, 65-78

Fusenig, N.E., Amer, S.M., Bonkamp, P. & Worst, P.K.M. (1978) Characteristics of chemically transformed mouse epidermal cells *in vitro* and *in vivo*. *Bull. Cancer, 65*, 271-280

Haber, D.A., Fox, D.A., Dyan, W.S. & Thilly, W.G. (1977) Cell density dependence of focus formation in the C3H 10T½ transformation assay. *Cancer Res., 37*, 1644-1648

Hashimoto, Y. & Kitagawa, H.S. (1974) In-vitro neoplastic transformation of epithelial cells of rat urinary bladder by nitrosamines. *Nature, 252*, 497-499

Hatch, G.G., Anderson, T.M., Lubet, R.A., Nims, R.W., Schechtman, L.M., Spalding, J.W. & Tennant, R.W. (1985) *Chemical enhancement of SA7 virus transformation of hamster embryo cells: interlaboratory testing of diverse chemicals*. In: Barrett, J.C. & Tennant, R.W., eds, *Carcinogenesis — A Comprehensive Survey*, Vol. 9, *Mammalian Cell Transformation, Mechanisms of Carcinogenesis and Assays for Carcinogens*, New York, Raven Press

Heidelberger, C., Freeman, A.E., Pienta, R.J., Sivak, A., Bertram, J.S., Casto, B.C., Dunkel, V.C., Francis, M.W., Kakunaga, T., Little, J.B. & Schechtman, L.M. (1983) Cell transformation by chemical agents: a review and analysis the literature. *Mutat. Res., 114*, 283-385

Hesterberg, T.W. & Barrett, J.C. (1984) Dependence of asbestos- and mineral dust-induced transformation of mammalian cells in culture on fiber dimension. *Cancer Res., 44*, 2170-2180

Hsiao, W.-L.W., Gattoni-Cell, S. & Weinstein, I.B. (1984) Oncogene transformation of C3H 10T½ cells is enhanced by tumor promoters. *Science, 226*, 552-555

Huberman, E. & Sachs, L. (1966) Cell susceptibility to transformation and cytotoxicity by the carcinogenic hydrocarbon benzo[*a*]pyrene. *Proc. natl Acad. Sci. USA, 56*, 1123-1129

IARC (1980) *IARC Monographs on the Evaluation of the Carcinogenic Risk of Chemicals to Humans*, Suppl. 2, *Long-term and Short-term Screening Assays for Carcinogens: A Critical Appraisal*, Lyon, pp. 185-199

IARC/NCI/EPA Working Group (1985) Cellular and molecular mechanisms of cell transformation and standardization of transformation assays of established cell lines for the prediction of carcinogenic chemicals: overview and recommended protocols. *Cancer Res., 45*, 2395-2399

Inui, N., Nishi, Y. & Taketomi, M. (1978) Chromosome breakage and neoplastic transformation of Syrian golden hamster embryonic cells in tissue culture by transplacental application of 2-(2-furyl)-3-(5-nitro-2-furyl)acrylamide (AF-2). *Mutat. Res., 58*, 331-338

Jones, C.A., Callaham, M.F. & Huberman, E. (1984) Enhancement of chemical carcinogen-induced cell transformation in hamster embryo cells by $1\alpha,25$-dihydroxycholecalciferol, the biologically active metabolite of vitamin D_3. *Carcinogenesis, 5*, 1155-1159

Kakunaga, T. (1972) *Discussion on 'The link between oncogenicity of 4NQO and 4NPO derivatives, induction of DNA lesions and enhancement of viral transformation' by Stitch, H.F*. In: Nakahara, W., Takayama, S., Sugimura, T. & Odashima, S., eds, *Topics in Chemical Carcinogenesis*, Tokyo, University of Tokyo Press, pp. 31-43

Kakunaga, T. (1973) A quantitative system for assay of malignant transformation by chemical carcinogens using a clone derived from BALB/3T3. *Int. J. Cancer, 12*, 463-473

Kakunaga, T. (1978) Neoplastic transformation of human diploid fibroblast cells by chemical carcinogens. *Proc. natl Acad. Sci. USA, 75*, 1334-1338

Kakunaga, T. (1981) *Approaches towards the development of human transformation assay system*. In: Mishra, N.K., Dunkel, V. & Mehlman, M.A., eds, *Advances in Environmental Toxicology; Mammalian Cell Transformation by Chemical Carcinogens*, Princeton Junction, NJ, Senate Press, pp. 355-382

Kakunaga, T. (1985) *Critical review of the use of established cell lines for in-vitro cell transformation*. In: Kakunaga, T. & Yamasaki, H., eds, *Transformation Assay of Established Cell Lines: Mechanisms and Application (IARC Scientific Publications No. 67)*, Lyon, International Agency for Research on Cancer, pp. 55-73

Kakunaga, T. & Crow, J.D. (1980) Cell variants showing differential susceptibility to ultraviolet light-induced transformation. *Science, 209*, 505-507

Kakunaga, T. & Kamahora, J. (1970) *Process of neoplastic transformation of cultured mammalian cells by chemical carcinogens* (abstract). In: *Proceedings of the 29th Symposium of the Japanese Cancer Association*, Tokyo, Japanese Cancer Association, pp. 42-48

Kakunaga, T. & Yamasaki, H., eds (1985) *Transformation Assay of Established Cell Lines: Mechanisms and Application (IARC Scientific Publications No. 67)*, Lyon, International Agency for Research on Cancer

Kennedy, A.R., Fox, M., Murphy, G. & Little, J.B. (1980) On the relationship between X-ray exposure and malignant transformation in C3H 10T½ cells. *Proc. natl Acad. Sci. USA, 77*, 7262-7266

Kennedy, A.R., Cairns, J. & Little, J.B. (1984) The timing of the steps in transformation of C3H 10T½ cells by X-irradiation. *Nature, 307*, 85-86

Kulesz-Martin, M., Yoshida, M.A., Prestine, L., Yuspa, S.H. & Bertram, J.S. (1985) Mouse cell lines for improved quantitation of carcinogen-induced altered differentiation. *Carcinogenesis, 6*, 1245-1254

Kuroki, T. & Sasaki, K. (1985) *Relationship between in-vitro cell transformation and in-vivo carcinogenesis based on available data on the effects of chemicals.* In: Kakunaga, T. & Yamasaki, H., eds, *Transformation Assay of Established Cell Lines: Mechanisms and Application (IARC Scientific Publications No. 67)* Lyon, International Agency for Research on Cancer, pp. 93-118

Land, H., Parda, L.F. & Weinberg, R.A. (1983) Cellular oncogenes and multistep carcinogenesis. *Science, 222*, 771-778

Landolph, J.R. (1985) *Chemical transformation in C3H 10T½ Cl8 mouse embryo fibroblasts: historical background, assessment of the transformation assay, and evolution and optimization of the transformation assay protocol.* In: Kakunaga, T. & Yamasaki, H., eds, *Transformation Assay of Established Cell Lines: Mechanisms and Application (IARC Scientific Publications No. 67)* Lyon, International Agency for Research on Cancer, pp. 185-198

Lee, T.C., Oshimura, M. & Barrett, J.C. (1985) Comparison of arsenic-induced cell transformation, cytotoxicity, mutation and chromosome effects in Syrian hamster embryo cells in culture. *Carcinogenesis, 6*, 1421-1426

Little, J.B. (1979) Quantitative studies of radiation transformation with the A31-1-1 mouse BALB/3T3 cell line. *Cancer Res., 39*, 1474-1480

McCormick, J.J., Kateley-Kohler, S. & Maher, V.M. (1985) *Factors involved in quantitating induction of anchorage independence in diploid human fibroblasts by carcinogens.* In: Barrett, J.C. & Tennant, R.W., eds, *Carcinogenesis — A Comprehensive Survey*, Vol. 9, *Mammalian Cell Transformation: Mechanisms of Carcinogenesis and Assays for Carcinogens*, New York, Raven Press, pp. 233-247

Milo, G. & DiPaolo, J.A. (1978) Neoplastic transformation of human diploid cells *in vitro* after chemical carcinogen treatment. *Nature, 275*, 130-132

Montesano, R., Saint Vincent, L. & Tomatis, L. (1973) Malignant transformation *in vitro* of rat liver cells by dimethylnitrosamine and N-methyl-N'-nitro-N-nitrosoguanidine. *Br. J. Cancer, 28*, 215-220

Mordan, L.J., Martner, J.E. & Bertram, J.S. (1983) Quantitative neoplastic transformation of C3H/10T½ fibroblasts: dependence upon the size of the initiated cell colony at confluence. *Cancer Res., 43*, 4062-4067

Namba, M. (1985) *Neoplastic transformation of human diploid fibroblasts (KMST-6) by treatment with Co-60 gamma rays.* In: Barrett, J.C. & Tennant, R.W., eds, *Carcinogenesis — A Comprehensive Survey*, Vol. 9, *Mammalian Cell Transformation: Mechanisms of Carcinogenesis and Assays for Carcinogens*, New York, Raven Press, pp. 217-231

Nettesheim, P. & Barrett, J.C. (1984) Tracheal epithelial cell transformation: a model system for studies on neoplastic progression. *CRC crit. Rev. Toxicol., 12*, 215-239

Newbold, R.F., Overell, R.W. & Connell, J.R. (1982) Induction of immortality is an early event in malignant transformation of mammalian cells by carcinogens. *Nature, 299*, 633-635

Oshimura, M., Gilmer, T.M. & Barrett, J.C. (1985) Nonrandom loss of chromosomes 15 in Syrian hamster tumors induced by v-Ha-*ras* plus *v-myc* oncogens. *Nature, 316*, 636-639

Pai, S.B., Steele, V.E. & Nettesheim, P. (1983) Neoplastic transformation of primary tracheal epithelial cell cultures. *Carcinogenesis, 4*, 369-374

Peehl, D.M. & Stanbridge, E.J. (1981) Anchorage-independent growth of normal human fibroblasts. *Proc. natl Acad. Sci. USA, 78*, 3053-3057

Peterson, A.R., Landolph, J.R., Peterson, H., Spears, C.P. & Heidelberger, C. (1981) Oncogenic transformation and mutation of C3H/10T½ clone 8 mouse embryo fibroblasts by alkylating agents. *Cancer Res., 41*, 3095-3099

Pienta, R.J. (1980) *Transformation of Syrian hamster embryo cells by diverse chemicals and correlation with their reported carcinogenic and mutagenic activities*. In: de Serres, F.J., ed., *Chemical Mutagens*, Vol. 6, New York, Plenum, pp. 175-202

Poiley, J.A., Raineri, R. & Pienta, R. (1979) The use of hamster hepatocytes to metabolize carcinogens in an in-vitro bioassay. *J. natl Cancer Inst., 63*, 519-524

Pontén, J. (1971) Spontaneous and virus induced transformation in cell culture. *Virol. Monog., 8*, 1-253

Price, P.J. & Mishra, N.K. (1980) *The use of Fischer rat embryo cells as a screen for chemical carcinogens and the role of the nontransforming type 'C' RNA tumor viruses in the assay*. In: Mishra, N., Dunkel, V. & Mehlman, M., eds, *Advances in Modern Environmental Toxicology*, Vol. 1, *Mammalian Cell Transformation by Chemical Carcinogens*, Princeton Junction, NJ, Senate Press, pp. 213-239

Price, P.J., Freeman, A.J., Lane, W.T. & Huebner, R.J. (1971) Morphological transformation of rat embryo cells by the combined action of 3-methylcholanthrene and Rauscher leukemia virus. *Nature-New Biol., 230*, 144-146

Price, P.J., Suk, W.A. & Freeman, A.E. (1972) Type C RNA tumor viruses as determinants of chemical carcinogenesis: effects of sequence treatment. *Science, 177*, 1003-1004

Reznikoff, C.A., Bertram, J.S., Brankow, D.W. & Heidelberger. C. (1973) Quantitative and qualitative studies of chemical transformation of cloned C3H mouse embryo cells sensitive to postconfluence inhibition of cell division. *Cancer Res., 33*, 3239-3249

Rhim, J.S., Kim, C.M., Arnstein, P., Huchner, R.J., Weisburger, E.K. & Nelson Rees, W.A. (1975) Transformation of human osteosarcoma cells by a human chemical carcingen. *J. natl Cancer Inst., 55*, 1291-1295

Rhim, J.S., Fujita, K., Arnstein, P. & Aaronson, S.A. (1986) Neoplastic conversion of human keratinocytes by adenovirus-12-SV40 and chemical carcinogens. *Science, 232*, 385-388

Richards, J. & Nandi, S. (1978) Neoplastic transformation of rat mammary cells exposed to 7,12-dimethylbenz[a]anthracene or *N*-nitrosomethylurea in cell culture. *Proc. natl Acad. Sci. USA, 75*, 3836-3840

Schaeffer, W.I. & Heintz, N.H. (1978) A diploid rat liver cell culture. *In Vitro, 14*, 418-427

Schechtman, L.M. (1985) *Metabolic activation of procarcinogens by subcellular enzyme fractions in C3H 10T½ and BALB/c 3T3 cell transformation systems*. In: Kakunaga, T. & Yamasaki, H., eds, *Transformation Assay of Established Cell Lines: Mechanism and Application (IARC Scientific Publications No. 67)*, Lyon, International Agency for Research on Cancer, pp. 137-162

Shimada, T., Furukawa, K., Kreiser, D.M., Camein, A. & Williams, G.M. (1983) Induction of transformation by six classes of chemical carcinogens in adult rat liver epithelial cells. *Cancer Res., 43*, 5087-5092

Silinskas, K.C., Kateley, S.A., Tower, J.E., Maher, V.M. & McCormick, J.J. (1981) Induction of anchorage-independent growth in human fibroblasts by propane sultone. *Cancer Res.*, *41*, 1620-1627

Sivak, A. & Tu, A.S. (1985) *Use of rodent hepatocytes for metabolic activation in transformation assays.* In: Kakunaga, T. & Yamasaki, H., eds, *Transformation Assay of Established Cell Lines: Mechanisms and Application (IARC Scientific Publications No. 67)*, Lyon, International Agency for Research on Cancer, pp. 121-135

Slaga, T.J., Viaje, A., Bracken, W.M., Buty, S.G., Miller, D.R., Fischer, S.M., Richter, C.K. & Dumont, J.N. (1978) In-vitro transformation of epidermal cells from newborn mice. *Cancer Res.*, *38*, 2246-2252

Styles, J.A. (1977) A method of detecting carcinogenic organic chemicals using mammalian cells in culture. *Br. J. Cancer*, *36*, 558-563

Suk, W.A. (1985) *Interaction between nontransforming retroviruses and chemicals in cells* in vitro: *association with progression to neoplastic phenotypes.* In: Barrett, J.C. & Tennant, R.W., eds, *Carcinogenesis — A Comprehensive Survey*, Vol. 9, *Mammalian Cell Transformation: Mechanisms of Carcinogenesis and Assays for Carcinogens*, New York, Raven Press, pp. 423-436

Summerhayes, I.C., Cheng, Y.E., Sun, T.T. & Chen, L.B. (1981) Expression of keratin and vimentin intermediate filaments in rabbit bladder epithelial cells at different stages of benzo[*a*]pyrene-induced neoplastic progression. *J. Cell Biol.*, *90*, 63-69

Tennant, R.W., Stasiewicz, S. & Spalding, J.W. (1985) *Comparative evaluation of three mammalian cell transformation assay systems.* In: Barrett, J.C. & Tennant, R.W., eds, *Carcinogenesis - A Comprehensive Survey*, Vol. 9, *Mammalian Cell Transformation: Mechanisms of Carcinogenesis and Assays for Carcinogens*, New York, Raven Press, pp. 399-410

Thomassen, D.G., Gray, T.E., Mass, M.J. & Barrett, J.C. (1983) A high frequency of carcinogen-induced early, preneoplastic changes in rat tracheal epithelial cells in culture. *Cancer Res.*, *43*, 5956-5963

Thomassen, D.G., Gilmer, T.G., Annab, L.A. & Barrett, J.C. (1985) Evidence for multiple steps in neoplastic transformation of normal and preneoplastic Syrian hamster embryo cells following transfection with Harvey murine sarcoma virus oncogene (v-Ha-*ras*). *Cancer Res.*, *45*, 726-732

Traul, K.A., Hind, R.J., Wolff, J.S. & Korol, W. (1981) Chemical carcinogenesis *in vitro*. An improved method for chemical transformation in Rauscher leukemia virus-infected rat embryo cells. *J. appl. Toxicol.*, *1*, 32-37

Tu, A., Hallowell, W., Pallotta, S., Snak, A., Lubet, R.A., Curren, R.D., Avery, M.D., Jones, C., Sedita, B.A., Huberman, E., Tennant, R., Spalding, J. & Kouri, P.A. (1986) An interlaboratory comparison of transformation in Syrian hamster embryo cells using model and coded chemicals. *Environ. Mutagen.*, *6*, 77-98

Umezawa, K., Hirakawa, T., Tanaka, M., Katoh, Y. & Takayama, S. (1978) Statistical evaluation of Pienta's in-vitro carcinogenesis assay. *Toxicol. Lett.*, *2*, 23-27

Wigley, C.B. (1979) *Transformation in vitro of adult mouse salivary gland epithelium: a system for studies on mechanisms on initiation and promotion.* In: Frank, L.M. & Wigley, C.B., eds, *Neoplastic Transformation in Differentiated Epithelial Cell Systems* In Vitro, London, Academic Press, pp. 3-36

Williams, G.M., Elliot, J.M. & Weisburger, J.H. (1973) Carcinoma after malignant conversion *in vitro* of epithelial-like cells from rat liver following exposure to chemical carcinogens. *Cancer Res.*, *33*, 606-612

REPORT 10

IN-VITRO ASSAYS THAT MAY BE PREDICTIVE OF TUMOUR-PROMOTING AGENTS

Prepared by:

*J.C. Barrett (Rapporteur), T. Kakunaga,
T. Kuroki (Chairman), D. Neubert, J.E. Trosko,
J.M. Vasiliev, G.M. Williams and H. Yamasaki*

1. Introduction

The process of tumour promotion can be defined operationally only from experiments conducted *in vivo* (see Report 3); therefore, tests conducted *in vitro* can never predict completely the tumour-promoting activity for all chemicals. In theory, if the molecular, biochemical and cellular mechanisms involved in tumour promotion *in vivo* were known, then in-vitro tests could be developed to assess the ability of a chemical to induce one or more of the critical changes involved in tumour promotion. Such a test would indicate that a chemical with these critical properties might have tumour-promoting activity under certain conditions. Since promotion of neoplasia may involve different mechanisms in different tissues, no one in-vitro or even in-vivo assay can suffice to define the tumour-promoting activity or potential of all chemicals.

The demonstrations of Sivak and Van Duuren (1967, 1970) showed the possibility of using cell culture systems to identify promoters. Since that time, a number of other in-vitro assays have been studied as models for identifying tumour promoters. These have been based on possible theoretical mechanisms of tumour promotion, including (but not limited to) enhancement of cell transformation, modulation of the neoplastic phenotype, inhibition of cell differentiation, stimulation of cell proliferation (by nontoxic mechanisms, e.g., hormones, and by cell killing and compensatory hyperplasia), effects on the cell membrane, modulation of immune responses and induction of genetic effects. Proceedings of meetings on mechanisms of tumour promotion are available (Slaga *et al.*, 1978; Hecker *et al.*, 1982; Börzsönyi *et al.*, 1984).

Some tumour-promoting agents have been shown to induce genetic effects in certain systems. 12-0-Tetradecanoylphorbol 13-acetate (TPA) has been reported to induce sister chromatid exchanges in cultured mammalian cells (Kinsella & Radman, 1978; Gentil *et al.*, 1980; Nagasawa & Little, 1981), aneuploidy in yeast (Parry *et al.*, 1981), gene amplification (Varshavsky, 1981; Hayashi *et al.*, 1983), DNA strand breaks in leucocytes (Birnboim, 1982) and clastogenic effects in lymphocytes (Emerit & Cerruti, 1982). Parry *et al.* (1981) observed induction of aneuploidy in yeast with several agents other than phorbol esters, and Kinsella (1982) found induction of numerical chromosomal abnormalities in V79 cells with nonphorbol promoters. Most of these genetic effects have been attributed to the generation of free radicals by phorbol esters (Nagasawa & Little, 1981; Troll & Wiesner, 1985), which may explain why sister chromatid exchanges have not been observed in some studies (Loveday & Latt, 1979; Popescu *et al.*, 1980; Connell & Duncan, 1981; Kinsella, 1982). Tests for genetic effects have not been used to identify new promoters since many tumour promoters exhibit little or no genetic activity. Nevertheless, genetic assays for possible carcinogenic agents are well established and any new chemical should of course be assayed for mutagenic activity in order to evaluate its potential carcinogenicity. Such assays are not specific for tumour promoters.

Many phenotypic properties of cultured cells have been shown to be affected by phorbol ester promoters, e.g., modulation of cell differentiation, cell adhesion, aggregation and effects on the cytoskeleton (see Diamond *et al.*, 1978; Weinstein *et al.*, 1979; Diamond, 1984; Kennedy, 1984; Yamasaki, 1984a). Several biochemical effects can also be observed to occur rapidly *in vivo*, e.g., induction of ornithine decarboxylase activity in mouse skin (Boutwell, 1983) and altered agglutination of rat bladder cells (Kakizoe *et al.*, 1981). In general, these assays detect only phorbol esters and, with the exception of teleocidin (Fujiki *et al.*, 1981), have not been applied to the identification of new promoters.

Only two groups of assays have been studied which can be used to detect nongenetic effects of tumour promoters and which have shown the effects of a wide variety of chemicals that are active as tumour promoters: (i) enhancement of cell transformation; and (ii) inhibition of cell-to-cell communication. None of the assays in these two classes is well established in terms of standardized protocols, validation with coded compounds or determination of interlaboratory reproducibility. Nevertheless, encouraging results obtained with a spectrum of chemicals warrant further investigation of these assays.

2. Enhancement of cell transformation by two-stage protocol

2.1 *Introduction*

The enhancement of cell transformation by tumour-promoting agents forms the basis of an assay. Generally, a two-stage protocol is used in which cells are exposed first to an initiator and then, at various times after exposure, to a tumour-promoting agent. This approach was first described for rat embryo fibroblasts (Lasne *et al.*, 1974) and later extended to C3H 10T1/2 cell lines (Mondal *et al.*, 1976, Frazelle *et al.*, 1983),

Syrian hamster embryo cells (Poiley et al., 1979; Popescu et al., 1980; Rivedal & Sanner, 1981), BALB/c 3T3 cell lines (Sivak & Tu, 1980; Hirakawa et al., 1982), and to virally-infected rat embryo cells (Traul et al., 1981) and Syrian hamster embryo (Fisher et al., 1981) assays.

TPA has also been reported to induce irreversible growth in soft agar of a mouse epidermal cell line (JB-6) (Colburn et al., 1979) and of adenovirus-transformed rat embryo cell lines (Fisher et al., 1979).

2.2 Syrian hamster embryo-cell transformation assay

(a) Description of the assay

Hamster embryo cells are sequentially exposed to an initiating agent (for example, benzo[a]pyrene, N-hydroxy-2-acetylaminofluorene, 4-nitroquinoline 1-oxide, 3-methylcholanthrene or N-methyl-N'-nitro-N-nitrosoguanidine) and subsequently to the putative promoting agent. As in the conventional transformation test, morphological transformation (defined as altered colony morphology, including criss-crossing and piling up of cells) is used as the endpoint. (See Report 9.)

(b) Data obtained and interpretation of results

A number of investigators have studied the effect of TPA in this system (Popescu et al., 1980; Rivedal & Sanner, 1982). In general, treatment with 0.05-0.1 μg TPA enhanced morphological transformation following exposure of the cells to an initiator. Phorbol ester compounds with tumour-promoting activity in mouse skin enhanced the formation of morphologically transformed colonies, while nonpromoting analogues did not (Rivedal & Sanner, 1982). This effect was reversible. If TPA was removed, the morphology returned to normal within 24-48 h. In these studies, the morphologically transformed colonies resulting from enhancement by TPA were not tested for their ability to grow in soft agar or for tumorigenicity in hamsters. Thus, the transformed state induced by TPA in culture has not been related to promotion of the neoplastic conversion of these cells.

The effect of metal salts has been studied in this system (Rivedal & Sanner, 1981). Chromic chloride and zinc chloride alone did not induce transformation and did not enhance the transformation frequency when tested in combination with benzo[a]pyrene. Nickel sulfate and cadmium acetate initiated morphological transformation when TPA or benzo[a]pyrene was used as the second agent in the sequence. Both the nickel and cadmium salts also showed a promotion-like effect when the cells were first exposed to benzo[a]pyrene. It was concluded from these data that these metal salts are more potent as promoters than as initiators.

Several studies have provided information on the mechanism(s) by which transformation is enhanced. Popescu et al. (1980) showed that enhancement of transformation by TPA in hamster embryo cells was related more to the dose of carcinogen than to the dose of TPA. Also, TPA given before the carcinogen did not enhance transformation. It was also shown that TPA did not affect the frequency of sister chromatid exchange in normal or malignant hamster cells.

So far, only a few substances have been tested in this assay, and it is therefore difficult to draw any conclusions about its predictive value.

2.3 *BALB/c 3T3 and C3H 10T1/2 cell transformation assay systems*

(a) Description of the assay

Two-stage cell transformation has been demonstrated with focus assays using C3H 10T1/2 and BALB/c 3T3 cells (Mondal *et al.*, 1976; Sivak & Tu, 1980; Hirakawa *et al.*, 1982; Enomoto & Yamasaki, 1985). Treatment of cells with a subthreshold dose of a carcinogenic agent and subsequent exposure to a tumour promoter markedly increased the number of transformed foci.

Cells are treated with a carcinogenic agent on day 1 at a subthreshold dose which produces only a few transformed foci (e.g., one focus in 10 to 12 dishes). The subthreshold dose of carcinogen varies according to the stock of cells and other experimental conditions and should be carefully chosen. After treatment with the carcinogen, the cells are exposed to a tumour promoter for a given period — usually two weeks — by changing the medium containing the putative tumour promoter once or twice a week. Transformation is scored after an appropriate period of cultivation, e.g., four to six weeks for BALB/c 3T3 cells and six to eight weeks for C3H 10T1/2 cells. The assay system should be controlled for treatment with carcinogen alone and promoter alone, and with adequate positive and negative controls.

(b) Data obtained and interpretation of results

Two-stage transformation experiments with C3H 10T1/2 and BALB/c 3T3 cells have been done with a wide variety of initiating agents and promoting agents (Kennedy, 1984). Initiating agents studied include X-rays, ultraviolet light, neutrons, polycyclic aromatic hydrocarbons, *N*-methyl-*N'*-nitro-*N*-nitrosoguanidine and formaldehyde. The promoting agents include TPA (and other phorbol analogues), nickel sulfate, cadmium acetate, epidermal growth factor, cigarette smoke extract, and 2,3,7,8-tetrachlorodibenzo-*para*-dioxin (TCDD) (Kennedy, 1984; Abernethy *et al.*, 1985). Saccharin was reported to enhance the transformation of C3H 10T1/2 cells (Heidelberger & Mondal, 1982) but to give negative results for the transformation of BALB/c 3T3 cells (Sivak & Tu, 1980).

The action of promoting agents on cell transformation has been studied in greatest detail with TPA. Treatment with TPA alone has little effect, and the enhancement of carcinogen-induced changes is usually reversible at first (Diamond, 1984; Kennedy, 1984). The effects of TPA depend on the type of initiating agent used. Enhancement of transformation of C3H 10T1/2 cells by chemical carcinogens was observed only when low doses of initiators were used, and no further increase was seen at doses that induced significant yields of transformation by themselves. Increasing doses of physical initiating agents, however, such as ultraviolet light, X-rays and neutrons, followed by TPA treatment, resulted in dose-dependent yields of transformed foci (Kennedy, 1984).

Demonstration of the promoting effects of TPA on C3H 10T1/2 cells is highly variable and depends on the batch of serum used (Frazelle *et al.*, 1983), unlike TPA-mediated promotion of BALB/c 3T3 cells (Kakunaga, 1985).

2.4 Use of a mouse epidermal cell line (JB6) and adenovirus-infected rat embryo cells

The effects of TPA on other cell types are significantly different from those observed in the systems discussed above. TPA alone induces irreversible, anchorage-independent growth of a mouse epidermal cell line (JB-6) and of adenovirus-transformed rat embryo cell lines (Colburn *et al.*, 1979; Fisher *et al.*, 1979). Other promoting agents that are active in the JB-6 promotion assay include mezerein, certain detergents, cigarette smoke condensate, benzoyl peroxide, di(2-ethylhexyl)phthalate and epidermal growth factor (Colburn *et al.*, 1980; DiWan *et al.*, 1985; Gindhart *et al.*, 1985). A role for reactive oxygen has recently been suggested in promotion of transformation of JB6 by TPA (Nakamura *et al.*, 1985). One interesting feature of this system is that promotion-resistant and promotion-sensitive variants of JB6 cells exist, and two genes involved in this sensitivity have recently been identified (Colburn *et al.*, 1982; Lerman *et al.*, 1986).

3. Block of junctional intercellular communication

3.1 Principles and scientific basis of test system

Communication between cells is thought to be essential for the coordination of cell proliferation and differentiation (Loewenstein, 1979). One form of cell-to-cell communication is the transfer of low-molecular-weight molecules and ions, which is believed to occur through the gap-junction organelles of membranes (termed junctional communication) (Gilula *et al.*, 1972; Pitts & Simms, 1977; Loewenstein, 1981). Junctional intercellular communication occurs between cells *in vivo* and *in vitro* and is influenced by developmental and hormonal factors and by physiological changes in levels of carbon dioxide and of calcium ions by pH and by exogenous chemicals (Gilula *et al.*, 1972; Loewenstein, 1981).

Several tumour-promoting analogues of phorbol esters have been shown to inhibit junctional communication between various types of cells in culture, as measured by metabolic cooperation (Murray & Fitzgerald, 1979; Yotti *et al.*, 1979), electrical coupling (Enomoto *et al.*, 1981) and dye transfer (Friedman & Steinberg, 1982; Fitzgerald *et al.*, 1983; Enomoto *et al.*, 1984). Furthermore, TPA (a potent tumour promoter) decreased the appearance of gap junctions on the membranes of cells in culture and *in vivo* (Yansey *et al.*, 1982; Finbow *et al.*, 1983; Kalimi & Sirsat, 1984). In addition, other tumour-promoting stimuli, such as partial hepatectomy and skin wounding, are also known to decrease gap junctions (Loewenstein, 1979).

Numerous studies have shown that cells derived from tumours and transformed cells have a reduced level of communication or no capacity for intercellular communication (Loewenstein, 1979). Other tumour cells retain a normal capacity for communication (Weinstein et al., 1976). Results from cell culture studies have suggested a good correlation between blocked intercellular communication and the enhancement of cell transformation. For example, phorbol ester-mediated inhibition of intercellular communication was associated with enhancement of morphological transformation of BALB/c 3T3 cells (Enomoto & Yamasaki, 1985) and of Syrian hamster embryo cells (Rivedal et al., 1985); however, no such correlation was reported in the C3H 10T1/2 cell transformation system (Dorman et al., 1983). Furthermore, chemically transformed BABL/c 3T3 cells were able to communicate between themselves, but not with their surrounding nontransformed counterparts (Enomoto & Yamasaki, 1984). A clonal cell line of BALB/c 3T3 with high susceptibility to chemical induction of cell transformation showed decreased communication capacity when confluent, whereas a cell line with low susceptibility did not (Yamasaki et al., 1985). These results, together with the observation that some tumour-promoting agents block communication, suggest that blockage of intercellular communication is a possible determinant of tumour promotion (Williams, 1981; Trosko et al., 1982a; Yamasaki et al., 1984).

Since gap-junctional communication is an essential process in tissues of most organs, its inhibition *in vivo* may have consequences other than on carcinogenesis — for example, in teratogenesis, reproductive dysfunction and neurotoxicity (Loewenstein, 1979; Trosko et al., 1982b; Warner et al., 1984; Welsch & Stedman, 1984).

3.2 *Description of methods for measuring junctional communication*

(a) *Metabolic cooperation between $HPRT^+$ and $HPRT^-$ Chinese hamster V79 cells*

In cell culture, junctional communication can be measured by a variety of different methods (Loewenstein, 1979). Of these, metabolic cooperation (Subak-Sharpe et al., 1969; Cox et al., 1970; Hooper, 1982) is widely used and can be used to assay the transfer of various metabolites. For example, when hypoxanthineguanine phosphoribosyl transferase-deficient ($HPRT^-$) cells are cocultured with metabolically competent ($HPRT^+$) cells in the presence of 6-thioguanine, the latter phosphoribosylate the synthetic purine to a toxic metabolite which is then transferred to the $HPRT^-$ cells, resulting in cell death (Fig. 1).

Detailed protocols for the use of the Chinese hamster V79 assay have been described previously (Jone et al., 1985). A small number (100) of HPR^- V79 cells are plated and cocultured with a large number of $HPRT^+$ cells. After 3-4 hours, the test chemical or solvent is added together with 6-thioguanine and the cells are incubated for three days.

Fig. 1. Postulated mechanism by which metabolic cooperation occurs in V79 cells at the *hprt* locus

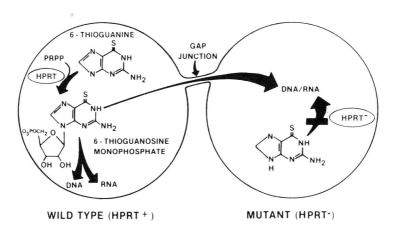

6-Thioguanine-resistant cells die in the presence of 6-thioguanine if they are in physical contact with 6-thioguanine-sensitive cells. From Trosko *et al.* (1982a). PRPP, 5-phosphoribosyl-1-pyrophosphate

Cultures are refed with fresh medium containing only 6-thioguanine after the third day, and at day 7 or 8 the cultures are fixed and stained and the surviving colonies are scored.

(*b*) *Nature of the data and interpretation of results of HPRT$^{+/-}$Chinese hamster V79 metabolic cooperation assays*

Generally, a good qualitative correlation between promoting activity and inhibition of intercellular exchange of molecules is evident (Table 1). Some chemicals that have not been demonstrated to be promoters also inhibited cell-cell communication — for example, diazepam; it may be that this agent has not been adequately tested for promoting activity in the appropriate tissue.

Discordant results from different laboratories for the same compounds may be due to different doses or assay conditions (Kinsella, 1982; Dorman & Boreiko, 1983; Jone *et al.*, 1985), indicating the importance of developing standard assay conditions for this system. Cell types that possess specific metabolic capabilities, or receptors for specific chemicals, may be needed to detect the activity of some chemicals.

Several new promoters have been identified by measurement of inhibition of intercellular exchange of molecules (Trosko *et al.*, 1981; Jensen *et al.*, 1982; Telang *et al.*, 1982; Williams & Numoto, 1984).

As part of the metabolic cooperation assay, the cytotoxic effect of test chemicals must be determined by a colony-forming assay using nontoxic or only slightly toxic doses. If the recovered 6-thioguanine-resistant colonies are smaller than the controls (solvent only), cytostatic activity is indicated.

Table 1. Some reported tumour-promoting agents, other than phorbol esters, which have been tested in the metabolic cooperation assay

Chemical	Reported target organ in vivo	Cell type used	Reported results	Reference
DDT	liver	RLE	+	Williams et al. (1981)
		V79	+	Umeda et al. (1982)
		HFC	+	Davidson et al. (1985a,b)
Chlordane	liver	V79	−	Trosko et al. (1982a)
		HPC/RLE	+	Telang et al. (1982)
Hepatchlor	liver	HPC/RLE	+	Telang et al. (1982)
		V79	+	Umeda et al. (1982)
Anthralin	skin	V79	+	Umeda et al. (1980)
			+	Trosko et al. (1982a)
Phenobarbital	liver	V79	−	Umeda et al. (1980)
		V79	+	Trosko et al. (1982a)
		HPC/RLE	+	Williams (1980)
Polybrominated biphenyls	liver	V79	+	Trosko et al. (1982a)
		HPC/RLE	+	Williams & Numata (1984)
Saccharin	bladder	V79	+	Trosko et al. (1980)
			−	Umeda et al. (1980)
			+	Welsch & Stedman (1984)
Phenol	skin	V79	−	Chen et al. (1984)
			−	Malcolm et al. (1985)
Deoxycholic acid	colon	V79	−	Noda et al. (1981)
Lithocholic acid	colon	V79	+	Noda et al. (1981)
Di(2-ethylhexyl)-phthalate	skin (2nd stage) liver	V79	+	Malcolm & Mills (1983)
Benzoyl peroxide	skin	V79	+	Slaga et al. (1981)
		HEK	+	Lawrence et al. (1984)
Certain fractions of cigarette smoke condensates	skin	V79	+	Hartman & Rosen (1983)

[a]RLE, rat liver epithelial cells; HFC, human fibroblast culture; HPC, rat hepatocyte primary culture; HEK, human epidermal keratinocytes

The assay must be performed under controlled, standardized conditions of temperature, pH, time of addition of the test chemicals and 6-thioguanine, using clones of 6-thioguanine-sensitive and 6-thioguanine-resistant cells which have high plating efficiencies, and 6-thioguanine-sensitive cells with a low background frequency of 6-thioguanine-resistant cells.

The data should be analysed statistically. The Student-Newman Keuls test has been used. A statistically significant recovery ($p < 0.01$) twice that of the background level, with a dose-related increase, is considered positive.

In principle, there are several other ways in which an increase in the number of 6-thioguanine-resistant colonies (i.e., a positive response in the assay) could occur *without* inhibition by the chemical of metabolic cooperation. These include induction of new 6-thioguanine-resistant mutants in the treated population and inhibition of the *hprt* enzyme. The former would be unlikely, but the mutagenic activity of the chemical should be considered. If a chemical directly inhibits the *hprt* enzyme in 6-thioguanine-sensitive cells, the observed response is the recovery of thousands of surviving cells in the dish rather than of discrete resistant colonies; this is an artefactual result that can easily be seen.

Experience to date has indicated that three classes of chemicals give positive results in the assay. Some chemicals, such as TPA, are active at the ng/ml level, which suggests a receptor-mediated response. Other chemicals (e.g., DDT), which are lipophilic, work at the μg/ml level, while others are active only at mg/ml levels. In general, the concentration at which different chemicals inhibit intercellular communication *in vitro* seems to correlate with the in-vivo potency of their tumour-promoting activity. For example, TPA is a potent promoter of skin tumours in certain strains of mice (Hecker, 1978), DDT is a promoter of rat liver tumours, but at higher doses (Peraino *et al.*, 1975), while saccharin is a relatively weak promoter of bladder tumours, only at very high doses (Hicks *et al.*, 1973).

(c) Metabolic cooperation assays other than V79 $HPRT^{+/-}$

Other systems are available to study the transfer of metabolites between cultured cells (Pitts & Simms, 1977). For example, phorbol esters and other tumour-promoting agents inhibit cooperation between rat hepatocytes and rat liver epithelial $HPRT^-$ cells (Williams, 1980) and between argininosuccinate lyase-deficient and argininosuccinate synthetase-deficient human or Chinese hamster cells (Davidson *et al.*, 1984).

An electrophysiological method can be used for studying junctional intercellular transfer of ions (Loewenstein, 1979). Although this method was successfully used to demonstrate that phorbol esters inhibit the transfer even of ions between cultured cells (Enomoto *et al.*, 1981), this assay is not easy to set up in laboratories that are not equipped for electrophysiology and cannot be used as a routine test.

Another assay for measuring junctional communication is the 'dye-transfer' assay, in which tracer molecules are microinjected into cells and their spread is monitored (Loewenstein, 1979). At present, a fluorescent probe, Lucifer Yellow CH, is being used widely (Stewart, 1978; Enomoto *et al.*, 1984). This assay has one great advantage as a possible test for tumour-promoting agents, namely, that the tracer dyes can be injected into any type of cell, making it possible to study the organ specificity of tumour-promoting agents (Yamasaki, 1984b).

3.3 *Assessment of assays of effects of chemicals on cell-to-cell communication*

Several assays designed to measure gap-junctional communication have been used to detect chemicals that promote tumours in several organs in several species (Table 1). This implies that the biochemical regulation of gap-junctional communication might be similar in most organs or species.

The possibility of missing potential tumour promoters that need specific receptors or metabolism is one limitation of these assays. Addition of a metabolizing system might not overcome the deficiencies of this assay with respect to drug metabolism, since the components of these cell-free systems interfere with both lipophilic test chemicals and with the cell membranes of the target cells. Primary hepatocytes, keratinocytes and other differentiated cells with metabolic capacities could be cocultured with the target Chinese hamster cell (or other cell type) used to measure gap-junctional intercellular communication.

Future development of these in-vitro assays may involve other techniques for measuring gap-junctional function: e.g., dye transfer, antibodies to gap-junctional proteins, and techniques employing human cells. These methods have the advantage that they can be applied to many cell types *in vitro* and *in vivo* at different cell densities.

4. References

Abernethy, D.J., Greenlee, W.F., Huband, J.C. & Boreiko, C.J. (1985) 2,3,7,8-Tetrachorodibenzo-*p*-dioxin (TCDD) promotes the transformation of C3H/10T½ cells. *Carcinogenesis, 6,* 651-653

Birnboim, H.C. (1982) DNA strand breakage in human leukocytes exposed to a tumor promoter phorbol myristate acetate. *Science, 215,* 1247-1249

Börzsönyi, M., Day, N.E., Lapis, K. & Yamasaki, H., eds (1984) *Models, Mechanisms and Etiology of Tumour Promotion (IARC Scientific Publications No. 56)*, Lyon, International Agency for Research on Cancer

Boutwell, R.K. (1983) *Biology and biochemistry of the two-step model of carcinogenesis.* In: Meyskens, F.L., ed., *Modulation and Mediation of Cancer by Vitamins*, Basel, Karger, pp. 2-9

Chen, T.H., Kavanagh, T.D., Chang, C.C. & Trosko, J.E. (1984) Inhibition of metabolic cooperation in Chinese hamster V79 cells by various organic solvents and simple compounds. *Cell Biol. Toxicol., 1,* 155-171

Colburn, N.H., Former, B.F., Nelson, K.A. & Yuspa, S.H. (1979) Tumour promoter induces anchorage independence irreversibly. *Nature, 281,* 589-591

Colburn, N.H., Koehler, B.A., & Nelson, K.J. (1980) A cell culture assay for tumor-promoter-dependent progression toward neoplastic phenotype: detection of tumor promoters and promotion inhibitors. *Teratog. Carcinog. Mutagenesis, 1,* 87-96

Colburn, N.H., Wendel, E. & Srinivas, L. (1982) Responses of preneoplastic epidermal cells to tumor promoters and growth factors: use of promoter-resistant variants for mechanism studies. *J. cell. Biochem., 18,* 261-270

Connell, J.R. & Duncan, S.J. (1981) The effect of non-phorbol promoters as compared with phorbol myristate acetate on sister chromatid exchange induction in cultured Chinese hamster cells. *Cancer Lett., 11,* 351-356

Cox, R.P., Krauss, M.R., Balis, M.E. & Dancis, J. (1970) Evidence for transfer of enzyme product as the basis of metabolic cooperation between tissue culture fibroblasts of Lesch-Nyhan disease and normal cells. *Proc. natl Acad. Sci. USA*, *67*, 1573-1579

Davidson, J.S., Baumgarten, I. & Harley, E.H. (1984) Metabolic cooperation between argininosuccinate synthetase-deficient and argininosuccinate lyase-deficient fibroblasts. *Exp. Cell Res.*, *150*, 367-368

Davidson, J.S., Baumgarten, I. & Harley, E.H. (1985) Use of a new citrulline incorporation assay to investigate inhibition of intercellular communication by 1-1-1-trichloro-2,2-bis(p-chlorophenyl)ethane in human fibroblasts. *Cancer Res.*, *45*, 515-519

Diamond, L. (1984) Tumor promoters and cell transformation. *Pharmacol. Ther.*, *26*, 89-145

Diamond, L., O'Brien, T. & Rovera, G. (1978) *Tumor promoters inhibit terminal cell differentiation in culture*. In: Slaga, T.J., Sivak, A. & Boutwell, R.K., eds, *Carcinogenesis*, Vol. 2, *Mechanisms of Tumor Promotion and Cocarcinogenesis*, New York, Raven Press, pp. 335-341

DiWan, B.A., Ward, J.M., Rice, J.M., Colburn, N.H. & Spangler, E.F. (1985) Tumor promoting effects of di(2-ethylhexyl)phthalate in JB6 mouse epidermal cells and mouse skin. *Carcinogenesis*, *6*, 343-347

Dorman, B.H. & Boreiko, C.J. (1983) Limiting factors of the V79 cell metabolic cooperation assay for tumour promoters. *Carcinogenesis*, *4*, 873-877

Dorman, B.H., Butterworth, B.E. & Boreiko, C.J. (1983) Role of intercellular communication in the promotion of C3H/10T½ cell transformation. *Carcinogenesis*, *4*, 1109-1115

Emerit, I. & Cerutti, P.A. (1982) Tumor promoter phorbol 12-myristate 13-acetate induces a clastogenic factor in human lymphocytes. *Proc. natl Acad. Sci. USA*, *79*, 7509-7513

Enomoto, T. & Yamasaki, H. (1984) Lack of intercellular communication between chemically transformed and surrounding non-transformed BALB/c 3T3 cells. *Cancer Res.*, *44*, 5200-5203

Enomoto, T. & Yamasaki, H. (1985) Phorbol ester-mediated inhibition of intercellular communication in BALB/c 3T3 cells: relationship to enhancement of cell transformation. *Cancer Res.*, *45*, 2681-2688

Enomoto, T., Sasaki, Y., Shiba, Y., Kanno, Y. & Yamasaki, H. (1981) Tumour promoters cause a rapid and reversible inhibition of the formation and maintenance of electrical cell coupling in culture. *Proc. natl Acad. Sci. USA*, *78*, 5628-5632

Enomoto, T., Martel, N., Kanno, Y. & Yamasaki, H. (1984) Inhibition of cell communication between BALB/c 3T3 cells by tumor promoters and protection by cAMP. *J. Cell Physiol.*, *121*, 323-333

Finbow, M.E., Shuttleworth, J., Hamilton, A.E. & Pitts, J.D. (1983) Analysis of vertebrate gap junction protein. *EMBO J.*, *2*, 1479-1486

Fisher, P.B., Dorsch-Häsler, K., Weinstein, I.B. & Ginsberg, H.S. (1979) Tumour promoters enhance anchorage-independent growth of adenovirus-transformed cells without altering the integration pattern of viral sequences. *Nature*, *281*, 591-594

Fisher, P.B., Mufson, R.A., Weinstein, I.B. & Little, J.B. (1981) Epidermal growth factor, like tumor promoters, enhances viral and radiation-induced cell transformation. *Carcinogenesis*, *2*, 183-187

Fitzgerald, D.J., Knowles, S.E., Ballard, F.J. & Murray, A.W. (1983) Rapid and reversible inhibition of junctional communication by tumor promoters in a mouse cell line. *Cancer Res.*, *43*, 3614-3618

Frazelle, J.H., Abernethy, D.J. & Boreiko, C.J. (1983) Factors influencing the promotion of transformation in chemically-initiated C3H/10T½ C18 mouse embryo fibroblasts. *Carcinogenesis*, *4*, 709-715

Friedman, E.A. & Steinberg, M. (1982) Disrupted communication between late-stage premalignant human colon epithelial cells by 12-O-tetradecanoylphorbol-13-acetate. *Cancer Res.*, *42*, 5096-5105

Fujiki, H., Mori, M., Nakayasu, M., Terada, M., Sugimura, T. & Moore, R.E. (1981) Indole alkaloids: dihydroteleocidin B, teleocidin and lyngbyatoxin A as members of a new class of tumor promoters. *Proc. natl Acad. Sci. USA*, *78*, 3872-3876

Gentil, A., Renault, G. & Margot, A. (1980) The effect of the tumor promoter 12-0-tetradecanoyl-phorbol-13-acetate (TPA) on UV- and MNNG-induced sister chromatid exchange in mammalian cells. *Int. J. Cancer*, *26*, 517-521

Gilula, N.B., Reeves, R.O. & Steinbach, A. (1972) Metabolic coupling, ionic coupling and cell contacts. *Nature*, *253*, 262-265

Gindhat, T.D., Srinivas, L. & Colburn, N.H. (1985) Benzoyl peroxide promotion of transformation of JB6 mouse epidermal cells: Inhibition by ganglioside GT but not by retinoic acid. *Carcinogenesis*, *6*, 309-311

Hartman, T.G. & Rosen, J.D. (1983) Inhibition of metabolic cooperation by cigarette smoke condensate and its fractions in V-79 Chinese hamster lung fibroblasts. *Proc. natl Acad. Sci. USA*, *80*, 5305-5309

Hayashi, K., Fujiki, H. & Sugimura, T. (1983) Effects of tumor promoters on the frequency of metallothionen I gene amplification in cells exposed to cadmium. *Cancer Res.*, *43*, 5433-5436

Hecker, E. (1978) *Structure-activity relationships in diterpene esters irritant and cocarcinogenic to mouse skin*. In: Slaga, T.J., Sivak, A. & Boutwell, R.K., eds, *Carcinogenesis*, Vol. 2, *Mechanisms of Tumor Promotion and Cocarcinogenesis*, New York, Raven Press, pp. 11-48

Hecker, E., Kunz, W., Fusenig, N.E., Marks, F. & Thielmann, H.W., eds (1982) *Carcinogenesis — A Comprehensive Survey*, Vol. 7, *Cocarcinogenesis and Biological Effects of Tumor Promoters*, New York, Raven Press

Heidelberger, C. & Mondal, S. (1982) *Effects of tumour promoters on the differentiation of C3H/10T½ mouse embryo fibroblasts*. In: Hecker, E., Fusenig, N.E., Kunz, W., Marks, F. & Thielmann, H.W., eds, *Carcinogenesis — A Comprehensive Survey*, Vol. 7, *Cocarcinogenesis and Biological Effects of Tumour Promoters*, New York, Raven Press, pp. 391-394

Hicks, R.M., Wakefield, J.St J. & Chowaniec, J. (1973) Co-carcinogenic action of saccharin in the chemical induction of bladder cancer. *Nature*, *243*, 347-349

Hirakawa, T., Kakunaga, T., Fujiki, H. & Sugimura, T. (1982) A new tumor-promoting agent, dihydroteleocidin B, markedly enhances chemically induced malignant cell transformation. *Science*, *216*, 527-529

Hooper, M.L. (1982) Metabolic cooperation between mammalian cells in culture. *Biochim. biophys. Acta*, *651*, 85-103

Jensen, R.K., Sleight, S.D., Goodman, J.I., Aust, S.D. & Trosko, J.E. (1982) Polybrominated biphenyls as promoters in experimental hepatocarcinogenesis. *Carcinogenesis*, *3*, 1183-1186

Jone, C., Trosko, J.E., Aylsworth, C.F., Parker, L. & Chang, C.C. (1985) Further characterization of the in-vitro assay for inhibitors of metabolic cooperation in the Chinese hamster V79 cell line. *Carcinogenesis*, *6*, 361-366

Kakizoe, K., Komatsu, H., Niijima, T., Kawachi, T. & Sugimura, T. (1981) Maintenance by saccharin of membrane alterations of rat bladder cells induced by subcarcinogenic treatment with bladder carcinogens. *Cancer Res.*, 4702-4705

Kakunaga, T. (1985) *Critical review of the use of established cell lines for in-vitro cell transformation.* In: Kakunaga, T. & Yamasaki, H., eds, *Transformation Assay of Established Cell Lines: Mechanisms and Application (IARC Scientific Publications No. 67)*, Lyon, International Agency for Research on Cancer, pp. 55-73

Kalimi, G.H. & Sirsat, S.M. (1984) Phorbol esters tumor promoter affects the mouse epidermal gap junctions. *Cancer Lett.*, 22, 343-350

Kennedy, A.R. (1984) *Promotion and other interactions between agents in the induction of transformation* in vitro *in fibroblasts*. In: Slaga, T.J., ed., *Mechanisms of Tumor Promotion*, Vol. 3, *Tumor Promotion and Carcinogenesis* in vitro, Boca Raton, FL, CRC Press, pp. 13-50

Kinsella, A.R. (1982) Elimination of metabolic cooperation and the induction of sister chromatid exchanges are not properties common to all promoting or co-carcinogenic agents. *Carcinogenesis*, 3, 499-503

Kinsella, A.R. & Radman, M. (1978) Tumor promoter induces sister chromatid exchanges: relevance to mechanisms of carcinogenesis. *Proc. natl Acad. Sci. USA*, 75, 6149-6153

Kurata, M., Hirose, K. & Umeda, M. (1982) Inhibition of metabolic cooperation in Chinese hamster cells by organochlorine pesticides. *Gann*, 73, 217-221

Lasne, C., Gentil, A. & Chouroulinkov, I. (1974) Two-stage malignant transformation of rat fibroblasts in tissue culture. *Nature*, 247, 490-491

Lawrence, N.J., Parkinson, E.K. & Emmerson, A. (1984) Benzoyl peroxide interferes with metabolic cooperation between cultured human epidermal keratinocytes. *Carcinogenesis*, 5, 419-421

Lerman, M.I., Hegamyer, G.A. & Colburn, N.H. (1986) Cloning and characterization of putative genes that specify sensitivity to neoplastic transformation by tumor promoters. *Int. J. Cancer*, 37, 293-302

Loveday, K.S. & Latt, S.A. (1979) The effect of a tumor promoter, 12-O-tetradecanoylphorbol-13-acetate (TPA) on sister chromatid exchange formation in cultured Chinese hamster cells. *Mutat. Res.*, 67, 343-348

Loewenstein, W.R. (1979) Junctional intercellular communication and the control of growth. *Biochim. biophys. Acta*, 560, 1-65

Loewenstein, W.R. (1981) Junctional intercellular communication: the cell-to-cell membrane channel. *Physiol. Res.*, 61, 829-913

Malcolm, A.R. & Mills, L.J. (1983) Inhibition of metabolic cooperation between Chinese hamster V79 cells by tumor promoters and other chemicals. *Ann. N.Y. Acad. Sci.*, 407, 448-450

Malcolm, A.R., Mills, L.J. & McKenna, E.J. (1985) Effects of phorbol myristate acetate, phorbol dibutyrate, ethanol, dimethylsulfoxide, phenol and seven metabolites of phenol on metabolic cooperation between Chinese hamster V79 lung fibroblasts. *Cell Biol. Toxicol.*, 1, 269-283

Mondal, S., Brankow, D.W. & Heidelberger, C. (1976) Two stage chemical oncogenesis in cultures of C3H/10T1/2 cells. *Cancer Res.*, 36, 2254-2260

Murray, A.W. & Fitzgerald, D.J. (1979) Tumour promoters inhibit metabolic cooperation in coculture of epidermal and 3T3 cells. *Biochem. biophys. Res. Commun.*, 91, 395-401

Nagasawa, H. & Little, J.B. (1981) Factors influencing the induction of sister chromatid exchanges in mammalian cells by 12-O-tetradeacanoyl phorbol-13-acetate. *Carcinogenesis*, 2, 601-607

Nakamura, Y., Colburn, N.H. & Ginhart, T.D. (1985) Role of reactive oxygen in tumor promotion: implications of superoxide anion in promotion of neoplastic transformation in JB-6 cells by TPA. *Carcinogenesis*, 6, 229-235

Noda, K., Umeda, M. & Ono, T. (1981) Effects of various chemicals including bile acids and chemical carcinogens on the inhibition of metabolic cooperation. *Gann, 72*, 772-776

Parry, J.M., Parry, E.M. & Barrett, J.C. (1981) Tumor promoters induce mitotic aneuploidy in yeast. *Nature, 294*, 263-265

Peraino, C., Fry, R.J., Staffeldt, E. & Christopher, J.P. (1975) Comparative enhancing effects of phenobarbital, amobarbital, diphenylhydantoin, and dichlorodiphenyltrichloroethane on 2-acetylaminofluorene-induced hepatic tumorigenesis in the rat. *Cancer Res., 35*, 2884-2890

Pitts, J.D. & Simms, J.W. (1977) Permeability of junctions between animal cells. *Exp. Cell Res., 104*, 153-163

Poiley, J.A., Raineri, R. & Pienta, R.J. (1979) Two-stage malignant transformation in hamster embryo cells. *Br. J. Cancer, 39*, 8-14

Popescu, N.C., Amsbaugh, S.C. & DiPaolo, J.A. (1980) Enhancement of N-methyl-N'-nitro-N-nitrosoguanidine transformation of Syrian hamster cells by a phorbol diester is independent of sister chromatid exchanges and chromosome aberrations. *Proc. natl Acad. Sci. USA, 77*, 7282-7286

Rivedal, E. & Sanner, T. (1981) Metal salts as promoters of in-vitro morphological transformation of hamster embryo cells initiated by benzo[a]pyrene. *Cancer Res., 41*, 2950-2953

Rivedal, E. & Sanner, T. (1982) Promotional effect of different phorbol esters on morphological transformation of hamster embryo cells. *Cancer Lett., 17*, 1-8

Rivedal, E., Sanner, T., Enomoto, T. & Yamasaki, H. (1985) Inhibition of intercellular communication and enhancement of morphological transformation of Syrian hamster embryo cells by TPA. Use of TPA-sensitive and TPA-resistant cell lines. *Carcinogenesis, 6*, 899-902

Sivak, A. & Tu, A.S. (1980) Cell culture tumor promotion experiments with saccharin, phorbol myristate acetate and several common food materials. *Cancer Lett., 10*, 27-32

Sivak, A. & Van Duuren, B.L. (1967) Phenotype expression of transformation: induction in cell culture by a phorbol ester. *Science, 157*, 1443-1449

Sivak, A. & Van Duuren, B.L. (1970) A cell culture system for the assessment of tumor-promoting activity. *J. natl Cancer Inst., 44*, 1091-1097

Slaga, T.J., Klein-Szanto, A.J., Triplett, L.L., Yotti, L.P. & Trosko, J.E. (1981) Skin tumor-promoting activity of benzoyl peroxide, a widely used free radical-generating compound. *Science, 213*, 1023-1025

Slaga, T.J., Sivak, A. & Boutwell, R.K., eds (1978) *Carcinogenesis — A Comprehensive Survey*, Vol. 2, *Mechanisms of Tumor Promotion and Cocarcinogenesis*, New York, Raven Press

Stewart, W.W. (1978) Functional connections between cells are revealed by dye-coupling with a highly fluorescent naphthalimide tracer. *Cell, 14*, 741-759

Subak-Sharpe, J.H., Burk, R.R. & Pitts, J.D. (1969) Metabolic cooperation between biochemically marked mammalian cells in tissue culture. *Cell Sci., 4*, 353-367

Telang, S., Tong, C. & Williams, G.M. (1982) Epigenetic membrane effects of a possible tumor promoting type on cultured liver cells by the nongenotoxic organochlorine pesticides chlordane and heptachlor. *Carcinogenesis, 3*, 1175-1178

Traul, K.A., Hink, R.J., Jr, Kachevsky, V. & Wolff, J.S., III (1981) Two stage carcinogenesis *in vitro*: transformation of 3-methylcholanthrene-initiated Rauscher murine leukemia virus-infected rat embryo cells by diverse tumor promoters. *J. natl Cancer Inst., 66*, 171-176

Troll, W. & Weisner, R. (1985) The role of oxygen radicals as a possible mechanism of tumor promotion. *Ann. Rev. Pharmacol. Toxicol.*, *25*, 509-528

Trosko, J.E., Dawson, B., Yotti, L.P. & Chang, C.C. (1980) Saccharin may act as a tumour promoter by inhibiting metabolic cooperation between cells. *Nature*, *285*, 109-110

Trosko, J.E., Dawson, B. & Chang, C.C. (1981) PBB inhibits metabolic cooperation in Chinese hamster cells *in vitro*: its potential as a tumor promoter. *Environ. Health Perspect.*, *37*, 179-182

Trosko, J.E., Yotti, L.P., Warren, S.T., Tsushimoto, G. & Chang, C. (1982a) *Inhibition of cell-cell communication by tumor promoters*. In: Hecker, E., Kunz, W., Fusenig, N.E., Marks, F. & Thielemann, H.W., eds, *Carcinogenesis — A Comprehensive Survey*, Vol. 7, *Cocarcinogenesis and Biological Effects of Tumour Promoters*, New York, Raven Press, pp. 565-585

Trosko, J.E., Chang, C.C. & Netzloff, M. (1982b) The role of inhibited cell-cell communication in teratogenesis. *Teratog. Carcinog. Mutagenesis*, *2*, 31-45

Umeda, M., Noda, K. & Ono, T. (1980) Inhibition of metabolic cooperation in Chinese hamster cells by various compounds including tumour promoters. *Gann*, *71*, 614-620

Varshavsky, A. (1981) Phorbol ester dramatically increases incidence of methotrexate-resistant mouse cell: possible mechanisms and relevance to tumour promotion. *Cell*, *25*, 561-572

Warner, A.E., Guthrie, S.C. & Gilula, N.B. (1984) Antibodies to gap-junctional protein selectively disrupt junctional communication in the early amphibian embryo. *Nature*, *311*, 127-131

Weinstein, I.B., Lee, L.S., Fisher, P.B., Mufson, R.A. & Yamasaki, H. (1979) Action of phorbol esters in cell culture: mimicry of transformation, altered differentiation and effects on cell membranes. *J. Supramol. Struct.*, *12*, 195-208

Weinstein, R.S., Merk, F.B. & Alroy, J. (1976) The structure and function of intercellular junctions in cancer. *Adv. Cancer Res.*, *23*, 23-89

Welsch, F. & Stedman, D.B. (1984) Inhibition of metabolic cooperation between Chinese hamster V79 cells by structurally diverse teratogens. *Teratog. Carcinog. Mutagenesis*, *4*, 285-301

Williams, G.M. (1980) Classification of genotoxic and epigenetic hepatocarcinogens using liver culture assays. *Ann. N.Y. Acad. Sci.*, *349*, 273-282

Williams, G.M. (1981) Liver carcinogenesis: the role for some chemicals of an epigenetic mechanism of liver tumor promotion involving modification of the cell membrane. *Food Cosmet. Toxicol.*, *19*, 577-583

Williams, G.M. & Numoto, S. (1984) Promotion of mouse liver neoplasms by the organochlorine pesticides chlordane and heptachlor in comparison to dichlorodiphenyltrichloroethane. *Carcinogenesis*, *5*, 1689-1696

Williams, G.M., Telang, S. & Tong, C. (1981) Inhibition of intercellular communication between liver cells by the liver tumour promoter 1,1,1-trichloro-2,2-bis(p-chlorophenyl)ethane. *Cancer Lett.*, *11*, 339-344

Yamasaki, H. (1984a) *Modulation of cell differentiation by tumor promoters*. In: Slaga, T.J., ed., *Mechanisms of Tumor Promotion*, Vol. 4, *Cellular Responses to Tumor Promoters*, Boca Raton, FL, CRC Press, pp. 1-26

Yamasaki, H. (1984b) In-vitro approaches to identify tumor-promoting agents: cell transformation and intercellular communication. *Food Addit. Contam.*, *1*, 179-187

Yamasaki, H., Enomoto, T. & Martel, N. (1984) *Intercellular communication, cell differentiation and tumour promotion*. In: Börzsönyi, M., Lapis, K., Day, N.E. & Yamasaki, H., eds, *Models, Mechanisms and Etiology of Tumour Promotion (IARC Scientific Publications No. 56)*, Lyon, International Agency for Research on Cancer, pp. 217-238

Yamasaki, H., Enomoto, T., Shiba, Y., Kanno, Y. & Kakunaga, T. (1985) Intercellular communication capacity as a possible determinant of transformation sensitivity of BALB/c 3T3 clonal cells. *Cancer Res.*, *45*, 637-641

Yancey, S.B., Edens, J.E., Trosko, J.E., Chang, C.C. & Revel, J.P. (1982) Decreased incidence of gap junctions between Chinese hamster V79 cells upon exposure to the tumor promoter 12-O-tetradecanoyl phorbol-13-acetate. *Exp. Cell Res.*, *139*, 329-340

Yotti, L.P., Chang, C.C. & Trosko, J.E. (1979) Elimination of metabolic cooperation in Chinese hamster cells by a tumour promoter. *Science*, *206*, 1089-1091

REPORT 11

ASSAYS FOR GENETIC CHANGES IN FUNGI

Prepared by:
E. Moustacchi, A. Carere, G. Morpurgo,
C. Ramel (Chairman) and F.E. Würgler (Rapporteur)

1. Gene mutation and allied effects

1.1 *Introduction*

International collaborative studies have shown that assays using yeasts (i) are promising predictors of carcinogenic activity (de Serres & Hoffmann, 1981); (ii) compare favourably in performance with other eukaryotic test systems in correctly identifying selected chemical carcinogens (Parry *et al.*, 1985); (iii) can detect a number of mutagens that give negative results in prokaryote screening tests (paraffins, olefins, actinomycin D); and (iv) are cheap to perform. For these reasons, many genetic toxicology laboratories have adopted *Saccharomyces cerevisiae* as a test organism.

The budding yeast *S. cerevisiae* and the fission yeast *Schizosaccharomyces pombe*, together with the moulds *Neurospora crassa* and *Aspergillus nidulans*, respond to environmental mutagens at a variety of genetic endpoints. Among these are reversions (base substitution and frameshifts), forward mutations, intragenic (conversion) and intergenic mitotic and meiotic recombination (crossing-over), chromosomal aberrations, chromosome loss and nondisjunction (aneuploidy). In addition, point mutations and deletions of mitochondrial DNA can be assayed in *S. cerevisiae*.

A number of reviews of genotoxicity assays using these lower eukaryotes are available (IARC, 1980; Lemontt, 1980; Haynes & Kunz, 1981; Lawrence, 1982; Käfer *et al.*, 1982; Scott *et al.*, 1982; Brockman *et al.*, 1984; Zimmermann *et al.*, 1984a; Moustacchi, 1985; Parry *et al.*, 1985). These reviews cover the scientific basis of the test systems, descriptions of assays for the different genetic endpoints, the role of metabolic activation and of DNA repair functions in mutagenesis and recombination. To date, 492 chemicals have been evaluated in *S. cerevisiae* (Zimmermann *et al.*, 1984b), together with 41 other industrial compounds (Dean *et al.*, 1985) and 20 compounds identified in pulp mill effluents (Nestmann and Lee, 1985). The *N. crassa* system has been used to evaluate 102 compounds (Brockman *et al.*, 1984) and the *A. nidulans* system, 124 (Scott *et al.*, 1982).

The advent of recombinant DNA techniques has provided a number of new approaches for the study of the mechanisms of mutagenesis and recombination, particularly in *S. cerevisiae*. Systems are being developed in which the nature of the induced genetic events can be defined at the level of the single nucleotide. Although these techniques are not yet used routinely, they undoubtedly aid in understanding the underlying mechanisms and may consequently enhance the predictive value of assays (see, for example, Ernst *et al.*, 1985).

1.2 *Principles and scientific basis of the test systems*

The tester strains used in assays of environmental agents give genetic responses that are detectable on selective media and can be used qualitatively in spot tests and quantitatively in plating tests to establish dose-effect relationships. The simplest tests make use of minimal media for detecting reversions from auxotrophy to prototrophy (e.g., reverse mutation at *his 1-7*, *cyc 1-115* or *ilv 1-92*, and gene conversion at *try-5* and *ade-2* with noncomplementing alleles in *S. cerevisiae*). In addition, media supplemented with a given antibiotic or inhibitor can be used to detect a change from sensitivity to resistance (e.g., forward mutation in haploids; mitotic crossing-over between gene and centromere in diploids; resistance to canavanine, cycloheximide, ethionine, 5-fluorophenyl-alanine). Such selective media allow rapid spot tests (three to five days) of one or several compounds at different concentrations with a small number of plates. Compounds that give positive results are then tested by the treat-and-plate method, taking into account the effect of the chemical on cell viability at a range of concentrations. Such quantitative comparisons are also essential for studying, for example, interactions between compounds and for defining the role of growth conditions in the mutagenic effect.

1.3. *Description of the test systems*

(*a*) Saccharomyces cerevisiae

A number of test strains have been constructed for routine screening of environmental mutagens/carcinogens (for reviews see IARC, 1980; Zimmermann *et al.*, 1984a). The most popular are the diploid strains of *S. cerevisiae*, which allow simultaneous screening of mitotic crossing-over, gene conversion and reverse mutation in the same strain (e.g., strain D7; Zimmermann *et al.*, 1975), and the haploid strains, which permit definition of the molecular nature of induced mutations (e.g., strain XV185-14C for detection of *ochre*, missense and frameshifts reversions; Mehta & von Borstel, 1981; von Borstel *et al.*, 1981). Table 1 lists the genotypes of strains most commonly used. Moreover, *S. cerevisiae*, a facultative aerobe, permits detection of mitochondrial point mutations and deletions when plated on suitable hydrocarbon substrates. Respiratory-deficient cells (mit^-, syn^- and rho^-) survive in the presence of glucose but cannot use glycerol and Krebs-cycle intermediates as energy sources, as do wild-type respiratory-competent cells. To the knowledge of the Working Group, no important new yeast tester strain for routine screening has been introduced since publication of the reviews cited above.

Table 1. Genotype of *S. cerevisiae* strains most commonly used for detecting chemically-induced chromosome malsegration

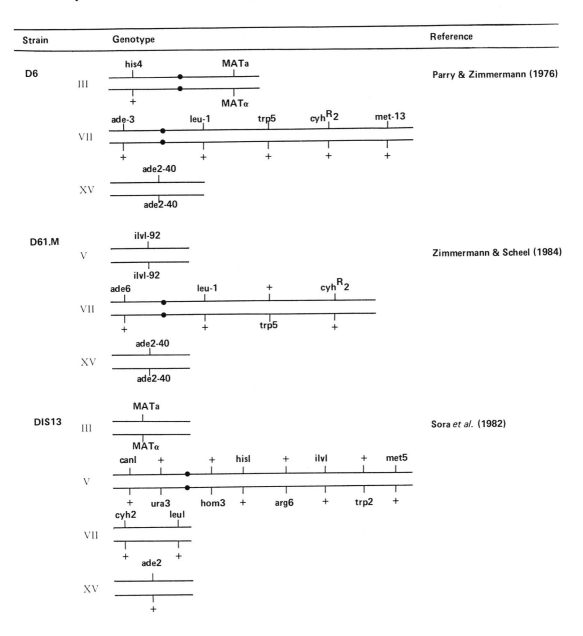

The genetic map of *S. cerevisiae*, on which 568 genes have been located, has been published (Mortimer & Schild, 1985).

(*b*) Schizosaccharomyces pombe

This species has many of the useful characteristics of *S. cerevisiae*, although it seems to be less sensitive, and, being *petite*-negative, cannot be used for mitochondrial mutagenesis studies. Gene mutation systems (forward and reverse mutations) have been developed and calibrated against directly and indirectly-acting mutagens (Loprieno *et al.*, 1983). Strain P1, which allows detection of forward mutation at five loci has been used most extensively.

(*c*) Neurospora crassa

N. crassa allows the detection of both forward and reverse gene mutations by assays for biochemical or morphological changes in treated cells. Up to now, the most commonly used assay is that measuring forward mutations in the *ade-3* region of the genome in either homokaryons or heterokaryons. The use of two-component heterokaryons is recommended, because it allows estimation of the frequency of recessive lethal damage over the entire genome. Indeed, in the heterokaryotic conidia, the arrangement of chromosomes is similar to that in a diploid cell, i.e., one haploid set of chromosomes of each genotype occupies the same cytoplasm rather than the same nuclcus, as in yeast.

In this system, recessive lethal mutations can result from either point mutations or chromosomal deletions. Detailed descriptions of experimental procedures, strains and genotypes are given by Brockman *et al.* (1984).

(*d*) Aspergillus nidulans

   (i) *Gene mutations*

Suitable haploid strains of this fungus (e.g., strains 35 and 18) have been employed for the detection of mutations induced by a fairly high number of chemical agents. The most widely used assays are two forward-mutation systems: (1) *8-azaguanine resistance*, which detects completely recessive point mutations in a single locus specifying a permease for the transport of the purine into the cell; and (2) *suppression of meth-G 1 (methionine requirement)*, which detects recessive point mutations in at least five genes that can suppress the requirement for methionine. The strain normally used is the haploid strain 35. Both systems have been shown to detect base-substitution as well as frame-shift types of mutagens.

   (ii) *Mitotic crossing-over*

Well marked heterozygous diploid strains (e.g., P1) have been employed to detect chemically induced somatic segregation (crossing-over and aneuploidy). In particular, mitotic crossing-over can be efficiently detected by using selective (*p*-fluorophenylalanine resistance) or nonselective systems. Induced somatic segregation is detected in pale green strains heterozygous for a conidia colour marker (yA2-/yA2$^+$) and for several nutritional requirements on both arms of the same or other chromosomes. Strains and the test methods are described in detail by Käfer *et al.* (1982) and Scott *et al.* (1982).

Yeast and mould test systems thus represent simple assays of genetic recombination and gene conversion. The systems available in *S. cerevisiae* respond significantly to as little as 25 rads of ionizing radiation.

1.4 *Nature of data, qualitative and quantitative interpretation*

(a) *General considerations*

Beside producing genetic changes, genotoxic compounds also kill cells. Consequently, comparisons between controls and treated samples are based on numbers of genetically changed cells per survivors as well as per total numbers of treated cells. Proper evaluation requires (Zimmerman *et al.*, 1984a):

(i) presentation of experimental data (actual numbers of colonies, numbers of plated cells) in independently repeated experiments or several parallel cultures, and not only transformed data;

(ii) establishment of dose-effect relationships (different concentrations for a given exposure time within limits of solubility and a given concentration for a range of treatment times within limits of the stability of the compound). Doubling of frequency over that in parallel controls is generally accepted as the lower limit of detection. A dose-dependent increase in the type of colonies that indicate the chosen genetic endpoint is good evidence for a positive response. If the increase is relative, i.e., the absolute numbers of genetic events decrease in response to dose, there is the possibility of selective survival as opposed to true induction of genetic events;

(iii) checking of compounds giving a negative response under a variety of conditions (e.g., stationary *versus* growing cells, low toxicity levels, small dose increments, metabolic activation with an appropriate positive control).

(b) *Quantitative comparison of the genetic activity of environmental agents*

For the evaluation of substances possibly hazardous to human health, it is essential to determine quantitatively how mutagenic various agents are in a given assay system. In addition, knowledge of the relative mutability of different strains is important in the design of short-term tests. Among the different methods for treating experimental data that have been proposed, comparison of mutation (or recombination) yields (mutants or recombinant per cell treated) seems to be one of the most informative (Eckardt & Haynes, 1980; Kunz *et al.*, 1985). For quantifying genetic effects, two parameters can be used: (i) mutant frequency (M), i.e., induced mutants per surviving cells; and (ii) mutant yield (Y), i.e., induced mutants per treated cells.

The dose dependence for $M(x)$ and $Y(x)$ can be described as survival on the basis of a Poisson distribution: $M(x)$ reflects the probability of the induction of genetic events in the DNA, whereas $Y(x)$ expresses simultaneously the genetic *and* lethal effects of an agent. In other words, $Y(x)$ describes the probability of occurrence of viable mutants in the population of treated cells. Plots of Y *versus* -log $S(x)$ (or lethal hits) can be used to determine the relative mutagenic efficiency (RME) of different agents as well as the relative mutability (RMT) of different strains to the same mutagen. Plots of Y *versus* dose also provide an unambiguous way of assessing the relative mutational sensitivities (RMS) or mutational resolutions (RMR) of different strains against a given mutagen.

This approach has been applied to measurements of mutagenesis induced by ultraviolet radiation and 4-nitroquinoline oxide in both repair-proficient and excision-deficient haploid strains of *S. cerevisiae* (Eckardt & Haynes, 1980). These two agents have the same mutagenic efficiency in yeast, and a linear relationship between the doses of ultraviolet light and 4-nitroquinoline oxide was constructed for mutation and killing. Moreover, it was shown that, although the wild-type strain is 15 times more mutable than the excision deficient strain, the latter is six times more sensitive as a detector of mutants at low doses; however, this increase in sensitivity of detection comes at a cost of mutational resolution. Application of this analysis to induction by ultraviolet radiation of mitotic recombination in wild-type and excision-repair-deficient diploid strains of yeast led to similar conclusions (Kunz *et al.*, 1985).

(c) Test performance

Data on 492 chemicals tested in *S. cerevisiae* have been evaluated (Zimmermann *et al.*, 1984b). A variety of genetic endpoints, including forward and reverse mutations, intra- and intergenic recombination and mitochondrial mutations were considered, for stationary and for exponentially growing mitotic cells and for meiotic cells. According to this evaluation, 249 compounds gave positive results for at least one genetic endpoint. In *S. pombe*, of 54 compounds tested for forward and/or reverse mutations, 47 gave positive results (Loprieno *et al.*, 1983). In *N. crassa*, of 102 compounds tested in one or more gene mutation tests, 94 gave positive and eight gave negative results (Brockman *et al.*, 1984). In *A. nidulans*, of 123 compounds tested, 51 gave positive results in haploid strains used for detection of gene mutation, while 42 gave positive results in diploids used for somatic segregation and recessive lethals (Scott *et al.*, 1982).

The proportions of 'positive' chemicals in the different systems may, however, be influenced by the following factors: (i) nonrandom choice of the chemicals (selection of compounds known to be mutagenic and/or carcinogenic); (ii) the criteria for judging a result as positive or negative; and (iii) the overall number of compounds tested and the limited number of chemical classes.

Several international validation studies aimed at selecting the most useful tests for screening have included yeast systems (de Serres & Ashby, 1981; Ashby et al., 1985). The first study comprised 30 assay systems in which 42 compounds were assayed (14 carcinogen/noncarcinogen pairs were included among the 42 chemicals tested). The second study was limited to ten compounds that were reported to be inactive or very difficult to detect in the 'Salmonella typhimurium/microsome assay, which were tested in eight types of assay. The results of these two international studies have recently been submitted to mathematical analysis using the multivariate method (Benigni & Giuliani, 1985; Benigni, 1986), and it was concluded that the S. cerevisiae XVI85-Ac mutation test performed remarkably well in predicting carcinogenicity. A further report (Ashby et al., 1985) concluded that although 'the Salmonella assay must remain the preferred primary assay... owing to the extensive data base available... appropriate yeast assays present a useful method of detecting gene-mutation and on occasion can be more sensitive than the Salmonella test.'

(d) Advantages and disadvantages of the test systems

The efficacy of microorganisms for assaying putative mutagens/carcinogens potentially hazardous to man is now accepted. Unicellular fungi that can be cultured in either haploid or diploid stable states and can be used for versatile genetic responses in rapid and inexpensive tests remain useful, since no one test system has yet been found to be 100% accurate. Fungi can be used to detect not only cell viability and reverse and forward mutations, which can easily be assayed in bacteria, but also genetic recombination, gene conversion and chromosomal malsegregation in both mitotic and meiotic cells. The nucleosomal organization of chromosomes, the presence of a nuclear membrane and of an organized chondriome, the potential for endogenous metabolic activation, and the presence of many genes involved in DNA repair, are properties in common with mammalian cells. One disadvantage of A. nidulans is that suitable exogenous metabolic activation has not yet been optimized. With the recent enormous developments in the molecular biology of yeast, the accumulation of knowledge on DNA repair genes (including their cloning and sequencing) and indications of the presence of inducible error-prone repair functions (Siede & Eckardt, 1984), it is clear that yeast is one of the few organisms, other than bacteria and viruses, in which mutagenic mechanisms can be elucidated rapidly at the molecular level.

Quantification of responses remains a crucial problem in genetic toxicology. Extrapolation of data from high to low levels, the existence of thresholds, doubling of doses over the spontaneous background, and the shape of dose-response curves are obvious matters of concern. The utility of fungal test systems for such quantitative evaluations is appreciable; however, it should be kept in mind that unicellular fungi are often less sensitive than bacteria in terms of the minimal concentrations required to elicit a response, due probably to powerful metabolic detoxification and to the cell membrane barrier. The last factor, which constitutes a serious disadvantage, is likely to be resolved by the use of agents that change cell membrane permeability without affecting cell growth and viability, as well as by the recent isolation of general permease mutants. Chemical treatment of spheroplasts followed by regeneration of whole cells in suitable medium, may be an alternative approach. The modified antibiotic, polymyxin B nonapeptide, which enhances the sensitivity of *S. cerevisiae* strains to several drugs, without itself being toxic, seems a promising agent (Boguslawski, 1985). The release of proteins into growth medium and rapid cell lysis are recent concerns in the industrial use of yeast. Newly developed strains remain to be explored for their sensitivity to chemical compounds.

1.5 *New developments*

In order rapidly to predict the nature of DNA lesions induced by newly identified chemicals of industrial or therapeutic interest, their cytotoxicity in wild-type and in mutants defective in specific steps of DNA repair can be checked. For instance, *S. cerevisiae* mutants exist that are specifically defective in the repair of DNA strand breaks (*rad 52* type), or of DNA interstrand cross-links (*pso2* type), or of bulky adducts on one DNA strand (*rad 3* type). If one such mutant is found to be clearly more sensitive to cell killing than the wild type by a compound of which the mode of action is unknown, it can be argued that the agent preferentially induces the type of lesion left unrepaired by the mutant, which can then be verified biochemically. This approach has been successfully used to predict the nature of DNA lesions induced *in vivo* by antitumour drugs such as neocarzinostatin (Moustacchi & Favaudon, 1982), an ellipticine derivative (BD40; Moustacchi *et al.*, 1983) and cisplatin (Hannan *et al.*, 1984). In all these cases, the cytotoxic effects observed in yeast concord with the characterization of DNA lesions and with the mutagenic/carcinogenic potential of the drugs. Moreover, the influence of a number of parameters on cellular cytotoxicity, such as the requirement for cell growth during incubation, the role of oxygen, interaction between compounds or with radiation, can be checked rapidly in the set of mutants and wild-type strains examined in parallel.

Genetic responses at specific stages in the cell cycle can be examined using conditional mutants blocked in genes controlling the different steps of cell division (*cdc* mutants). Indeed, temperature-sensitive heteroallelic strains with a *cdc* gene will not grow at the restrictive temperature, and only spontaneous and induced convertants in this gene can give rise to colonies. This system has been successfully used with 254 nm ultraviolet light and ionizing radiation to define the sensitive stages of the cell cycle for gene conversion (Fabre *et al.*, 1984), and the approach could be extended to chemicals.

Methods for supplying metabolic activation can be improved. Yeast strains possess the major activating system, i.e., the cytochrome P450-linked enzymes. Three factors are known to play a role in modulating cellular concentration levels: (i) the stage of growth, (ii) the genetic background of the strain and (iii) the amount of glucose in the medium (for a recent discussion, see von Borstel *et al.*, 1985). Owing to the presence of this endogenous activation system, certain compounds give positive results in yeast but are not mutagenic in other microbial systems. However, not all activating reactions can be performed by the yeast cytochrome P450 complex, and addition of rat liver microsomal mix is necessary in certain cases to activate promutagens in yeast assays. The role of parameters such as pH, temperature and ionic strength, which are known to have a strong influence on enzyme activities, has been investigated, as discussed below.

pH can, by itself, affect certain genetic endpoints in both growing and nongrowing yeast cells. For instance, with the commonly used D7 strain, a deviation to pH 5.8 from the normal pH 6.2 may enhance gene conversion frequency (*trp5* locus) by a factor of ten, with no effect on the frequency of reversion at the *ilv 1-92* locus (Nanni *et al.*, 1984). Thus, pH should be rigorously controlled in yeast assays, especially if gene conversion is examined. The pH of treatment solutions obviously influences the potencies of a variety of direct mutagens (for recent examples, see Rosin, 1984; Whong *et al.*, 1985) and is critical when exogenous metabolic activation is required.

The effect of temperature on enzyme stability and exogenous metabolic activity has also been examined (Mannironi *et al.*, 1985). The reliability and sensitivity of the tests conducted in yeast D7 strain can be improved by raising the incubation temperature to 42°C, which is optimal for the activity of the monooxygenase system and does not affect the spontaneous background of genetic events.

1.6 *Relevance of the genetic endpoints studied in lower eukaryotes to cancer*

Malignant transformation is considered to be a multistep process. The initiation phase is generally accepted to result from gene mutation in a broad sense (point mutation, chromosomal deletion or chromosomal rearrangements). There are good reasons to believe that not only the base sequence but also the chromatin structure play a role in both the frequency and the location of lesions in DNA (linker *versus* nucleosome, for instance). Moreover, once induced, the accessibility of the lesions to repair enzymes appears to differ according to the state of the chromatin (replicating *versus* nonreplicating regions, actively transcribed genes *versus* inactive genes).

Lower eukaryotes (in particular, yeast) share a number of features with mammalian cells, including activation-detoxification systems, a similar chromatin structure (nucleosomal organization; histones and nonhistones; nuclear proteins), a nuclear membrane, a stable diploid state, a cell cycle with distinct phases (G_1, S, G_2 and M), repair networks of high complexity, and cellular oncogenes. This suggests that extrapolation of data, at least for the mutagenic step, from yeast to human cells is likely to be valid in most cases.

Following this initial step, development of the malignant phenotype is thought by some to involve further chromosomal rearrangements resulting, among other events, in homozygosity of the mutated sites (Radman et al., 1982). Indeed, if mutations that occur in tumours are (like most mutations) recessive, they will not be expressed in diploid cells. Mitotic recombination, which can lead to homozygosity of recessive alleles, will result in the expression of such mutant genes and may consequently be involved in the carcinogenic process. Many chromosomal alterations associated with malignancies have been described (see Report 7). Monitoring of the recombinogenic potential of environmental chemicals, using yeasts and moulds in parallel with mutagenesis, is therefore relevant to cancer.

There is no consistent correlation between tumour-promoting activity and the ability of an agent to induce mitotic recombinations in yeast *per se* (Kunz et al., 1980; Fahrig, 1984); however, tumour promoters were found to be corecombinogenic in combination with other agents (Fahrig, 1984).

The covalent binding of carcinogens with nuclear DNA is thought to be the initial, critical event in the malignant transformation of cells by chemical carcinogens. However, mitochondrial DNA may be modified to a much greater extent than nuclear DNA (Backer & Weinstein, 1980; Magana-Schwencke et al., 1982; Niranjan et al., 1982). Because there is no excision repair in mitochondrial DNA (Clayton et al., 1974; Waters & Moustacchi, 1974), mitochondrial DNA-carcinogen adducts are known to persist for long periods of time after treatment. The possibility cannot be excluded that such alterations of the metabolic respiratory chain interfere directly (or more likely indirectly) with the stability of the nuclear genome. For instance, it is known that some of the enzymic steps of nuclear DNA repair are ATP-dependent; a reduction in this energy source due to mitochondrial alterations may in turn lead to a defect in the elimination of nuclear DNA damage. The mitochondrial 'petite' mutation, which is easily detectable in yeast, may serve as an indicator of such effects of carcinogenic compounds.

1.7 *Conclusions*

Although tests employing fungi are difficult to relate directly to human health hazards, short-term tests with these eukaryotic cells may serve:

    (i) to set priorities for the selection of chemicals to be included in long-term studies;

    (ii) to help in designing new molecular structures with reduced mutagenic activity by retaining the desired effect but abolishing the mutagenic effect; and

    (iii) to contribute to a better understanding of the basic molecular mechanisms underlying the mutagenic process.

2. Aneuploidy

2.1 *Introduction*

Recent progress in the understanding of the mechanisms of malignant transformation points to a role of specific chromosomal rearrangements and aneuploidy in tumour induction and progression (Onho, 1974; Yunis, 1983; Klein & Klein, 1984; Barrett *et al.*, 1985; Evans, 1985; Cavenee *et al.*, 1986). Fungal genetic systems provide sensitive tools for studying both the chemical induction and mechanisms of chromosomal malsegregation leading to aneuploidy in meiosis and mitosis.

The most commonly used fungi in studies of aneuploidy are the yeast *S. cerevisiae* (Zimmermann *et al.*, 1984a) and the moulds *A. nidulans* (Käfer *et al.*, 1982) and *N. crassa* (Brockman *et al.*, 1984). Genetic systems in fungi have the following advantages relevant to the study of aneuploidy:

(i) ease of culturing and handling;

(ii) organization of genetic material in typical eukaryotic chromosomes;

(iii) availability of selective biochemical and morphological markers;

(iv) packaging of the meiotic products into tetrads;

(v) suitability for mitotic segregation (e.g., parasexual cycle in *A. nidulans*); and

(vi) they may provide information on the mechanisms of aneuploidy induction.

Conversely, they have the following disadvantages:

(i) possible differences in the organization of meiotic/mitotic apparatus with respect to mammalian cells (e.g., intranuclear mitosis, absence of centrioles, mitotic pairing of chromosomes);

(ii) possible differences in critical targets (e.g., tubulin);

(iii) differences in fungal *versus* mammalian metabolism; and

(iv) the use of exogenous mammalian activation systems, which has been standardized only in a few systems.

The importance of the differences between fungal and mammalian cells is not yet clear, owing mainly to the lack of adequate comparative studies in mammalian and nonmammalian systems. Recently, several chemical agents have been found that specifically or primarily induce aneuploidy; these include agents known to interfere with targets other than DNA (e.g., tubulin, spindle microtubules assembly, centrioles/centrosomes, kinetochores, membranes). Such agents would be classified as inactive in most of the mutagenicity tests that are currently used, which detect primary or secondary effects on DNA. As a consequence, suitable systems aimed primarily at detecting aneuploidy are urgently needed to complement other studies.

2.2 Aneuploidy systems in yeasts

(a) Principles and scientific basis of the tests

The yeast most commonly used in tests for aneuploidy is *S. cerevisiae*, which is widely used for basic genetic research (Sherman & Lawrence, 1974) as well as for screening environmental chemical mutagens (Zimmermann, 1975; Zimmermann *et al.*, 1984a).

Chromosomal malsegregation leading to aneuploidy can be studied by means of genetic systems in appropriate diploid strains of yeast during both mitosis and meiosis. The most commonly used strains in the screening of environmental chemicals are D6 (Parry & Zimmermann, 1976) and D61.M (Zimmermann & Scheel, 1984) for mitotic chromosome loss (e.g., monosomy, from the 2n to 2n-1 condition), and DIS13 (Sora *et al.*, 1982) for chromosome gain (e.g., disomy, from the 2n to 2n+1 condition) and for diploidization in spores produced during meiotic cell division. The genotypes of these three strains are shown in Table 1.

The following strains have also been described for assay of chemically induced chromosomal malsegregation: Strain LBL1 (Esposito *et al.*, 1982) is a haploid that is disomic for chromosome VII; constructed to detect chromosome VII loss, it has been used to study the effects of ultraviolet light. Strains g632 and g551 are diploid strains described by Kunz and Haynes (1982) constructed to detect mitotic chromosome loss. Strains XD72 and XD79, described by Dixon (1983), detect mitotic chromosome loss. Strain D9J2, described by Parry *et al.* (1979a,b), detects meiotic disomy produced during the first meiotic division and has been used to study the effects of *p*-fluorophenylalanine, ultraviolet light and X-rays.

(i) Strain D6

This strain was developed by Parry and Zimmerman (1976) for the routine detection of mitotic chromosome loss; since then, it has been used extensively to test a large number of environmental chemicals (Parry, 1977; Parry *et al.*, 1979a,b; Parry & Sharp, 1981; Parry, 1982; Kelly & Parry, 1983; Ferguson & Parry, 1984; Parry *et al.*, 1985).

The test assay is based on a selective genetic system specifically aimed at detecting monosomy (e.g., chromosome VII loss) in mitotic cells, Strain D6 forms red colonies and requires adenine for growth because of the presence of the *ade2-40* allele in a homozygous condition on chromosome XV; it is also sensitive to cycloheximide because of the presence, in a heterozygous condition on chromosome VII, of the recessive cyh_2^R resistance allele. The loss of the chromosome VII homologue carrying the dominant wild-type allelles of this group of linked markers leads to the production of cells that, on high glucose medium containing 2 ppm cycloheximide, form white (owing to the expression of the *ade3* allele) and cycloheximide-resistant colonies (owing to the expression of the selective cyh_2^R allelle), which also express the nutritional alleles *leu-1, trp-5* and *met-13*. Presumptive monosomic cells must always be checked for the presence of the centromere-linked *leu-1* allele, since white, cycloheximide-resistant colonies could arise by other genetic events. The spontaneous frequency of the presumptive monosomic colonies has been estimated to be in the range of 4.18×10^{-6} — 4.18×10^{-7}. Technical details of the experimental protocol have been fully described (Parry & Parry, 1984).

(ii) *Strain D61.M*

This strain, which is very similar to D6, was described recently by Zimmermann and Scheel (1984). It detects chemically-induced mitotic chromosome loss and has been used to test about 20 chemicals.

The genetic principle of this system is practically the same as that for D6. Loss of the chromosome VII homologue carrying the wild-type alleles $ade6^+$, $leu\text{-}1^+$ and cyh_2^{R+} leads to the production of colonies that are white and adenine-dependent (owing to the expression of the recessive *ade6* allele) and are cycloheximide-resistant (due to the expression of the recessive cyh_2^R allele) (see Table 1). The *ade6* allele is much closer to the centromere than is the *ade3* allele; as a consequence, a double crossover event on both sides of the centromere is less frequent than with D6. As with D6, presumptive monosomics (i.e., white, cyh_2^R must always be checked for the centromere-linked *leu-1* marker.

Due to the presence of the homoallelic *ilv1-92* marker strain, D61.M can also be used for testing the induction of reverse gene mutations.

(iii) *Strain DIS13*

This strain was constructed by Sora *et al.* (1982) for the detection of meiotic aneuploidy (disomy: n + 1) and diploidization (2n) by a selective genetic system. Results obtained for about 12 chemical agents have been described (Sora & Bianchi, 1982; Sora *et al.*, 1982, 1983).

As shown in Table 1, this diploid strain is heterozygous for six markers of the right arm and two markers of the left arm of chromosome V, and heterozygous for the recessive allele *cyh2* (resistance to cycloheximide) on chromosome VII. By using a medium containing cycloheximide and the required nutrients — except those for the markers on the right arm of chromosome V — three main types of spores produced during meiosis can be selected: (1) diploid spores (2n) caused by failure of the II meiotic division (estimated spontaneous frequency, 0.54×10^{-4} per viable spore); (2) haploid spores that are disomic (n + 1) for chromosome V and are caused by nondisjunction in I or II meiotic division, with an overall spontaneous frequency of 0.95×10^{-4} per viable spore; and (3) haploid spores derived from the presence of five crossovers on chromosome V, so that all of the dominant alleles occur on the same chromatid. The last event is very unlikely, because it is expected to occur with a spontaneous frequency of 2.8×10^{-7} per viable spore, in accordance with the average crossover frequency reported by Mortimer and Schill (1980); this frequency has been confirmed in strain DIS 13 by Sora *et al.* (1982) after chemical treatments in seven independent experiments. The system has been used exclusively in the absence of an exogenous mammalian metabolic activation system.

(b) *Evaluation of the test systems*

(i) *Strains D6 and D61.M*

Interpretation of most data obtained with the mitotic aneuploidy system in these strains is difficult. The main problem is that the same phenotype of the presumptive monosomic colonies can arise by several genetic events, and it is difficult to discriminate among them. In particular, in strain D6, a comparison of the spontaneous frequency of the presumptive monosomics (estimated range, $4.18 \times 10^{-6} - 4.18 \times 10^{-7}$) with the expected frequency of double crossing-over on both sides of the centromere (about 1×10^{-7}), as computed from the observed spontaneous frequencies of single cross-over events (Parry & Zimmermann, 1976), suggests that a large fraction of the presumptive monosomics might well be due to the occurrence of a coincident double crossing-over. In strain D61.M, as the *ade-6* marker is much closer to the centromere than the *ade-3* marker of strain D6, the expected frequency of double crossing-over is considerably lower ($4 \times 10^{-9} - 4 \times 10^{-8}$). Moreover, presumptive monosomics (i.e., white, cyh_2^R) could be produced by the induction of 'petite' mutations in pre-existing cyh_2^R cells, due to the fact that respiratory deficient mutants are unable to develop the red pigment of *ade-2* mutants (Eckardt & Haynes, 1977). The spontaneous frequency of petite mutants in *S. cerevisiae* has been estimated to be about 1%; however this value might be increased enormously by chemical agents, including cycloheximide. Furthermore, increases in the number of colonies that are cycloheximide-resistant might occur by forward mutations at the $cyh_2^{R^+}$ allele (Brusick, 1972) or by mitotic crossing-over, while white adenine-requiring colonies might occur by forward mutations at the wild-type $ade-3^+$ locus. The last event seems to occur rather frequently (Eckardt & Haynes, 1977). In addition, strong selection for cells carrying both an *ade-2* and an *ade-X* mutation in adenine-requiring white colonies has been observed (Roman, 1956); since there is a requirement for growth during treatment, selection might play an important role. More generally, it is likely that chemical agents might affect, in different and unpredictable ways, various genetic events (aneuploidy, crossing over, gene conversion and petite mutations), making interpretation of the real nature of the mitotic segregants difficult if experiments are not carefully planned and evaluated.

Finally, it is rather surprising that the observed frequencies of the presumptive mitotic aneuploids, either spontaneous or induced, is so low; indeed, for D6, the spontaneous values estimated are in the range of gene mutation or gene conversion frequencies, while in all other organisms and even in *S. cerevisiae* in the case of loss of chromosome III in mitosis (Campbell *et al.*, 1975), these frequencies are in the order of

$1 \times 10^{-3} - 1 \times 10^{-4}$. Considering that in most organisms, and presumably also in yeasts, aneuploids have strong selective disadvantages, the most likely conclusion is that this system reveals only a small fraction of aneuploids. The frequencies seen in *S. cerevisiae*, either spontaneous or induced, are of the same order of magnitude as those obtained in *A. nidulans* using selective methods. In the latter organism, the true frequencies of aneuploidy are two to three orders of magnitude higher; it is very likely, therefore, that the results with *S. cerevisiae* are affected by strong selection factors before their detection. Mainly for these reasons, most of the results so far obtained with strain D6 have been considered inconclusive (Resnick *et al.*, 1986).

Although most of the problems discussed for D6 hold also for D61.M, the results obtained so far with the latter strain have generally been accepted with more confidence, because they have been obtained under controlled experimental conditions e.g., by checking the presence of the *leu-1* marker and by testing, in independent experiments, the induction of other genetic events like mutation, gene conversion and crossing-over.

(ii) *Strain DIS13*

The major advantage of the selective system based on the use of this strain is its ability to detect, very precisely, the induction of meiotic aneuploidy (disomy) as well as meiotic diploidization, giving some useful information on mechanisms. Strain DIS13 has been very useful in detecting disomic spores induced by agents such as benomyl, methyl methanesulfonate and cyclophosphamide as well as diploid spores induced by other chemicals (e.g., caffeine and bleomycin) (Sora & Bianchi, 1982; Sora *et al.*, 1982, 1983). One disadvantage is that, like other meiotic systems, it is a rather laborious assay, requiring extensive genetic analysis in order to detect spores that are disomic for chromosome V. Furthermore, as with all meiotic systems, a major problem is the possible interference of ascospore formation by the use of high doses of test chemicals. Moreover, this system, like other selective methods (e.g., those with strain D6 and D61.M), is likely to underestimate the true frequency of aneuploidies. Finally, this system has been used only in the absence of exogenous mammalian metabolic activation systems.

2.3 *Results obtained in* S. cerevisiae *aneuploidy systems*

A fairly large number of chemicals has been tested in *S. cerevisiae* aneuploidy systems, the majority in strain D6 and some (about 20) in strain D61.M, while only a few have been tested in strain DIS13. The data on many compounds have been evaluated in a report of the US Environmental Protection Agency Gene-Tox Program (Zimmermann *et al.*, 1984b), and all these data have recently been re-evaluated by a US Environmental Protection Agency Committee (Resnick *et al.*, 1986), which also examined some other compounds.

Table 2 gives the results obtained with 21 chemicals evaluated as positive (19) and negative (2), together with the Chemical Abstracts Services number, the strain used, degree of evidence for carcinogenicity and references. For various reasons (e.g., lack of adequate controls), the results for most chemicals so far tested at the mitotic level with strain D6 have been considered inconclusive and are therefore not reported.

Table 2. Agents evaluated in *S. cerevisiae* aneuploidy tests

Chemical	CAS No.	Strain	Response	Carcinogenicity[a]	References
Acetone	67-64-1	D61.M	+	0	Zimmermann et al. (1984a)
Acetonitrile	75-05-8	D61.M	+	0	Zimmermann et al. (1985a)
5-Azacytidine	320-67-2	D61.M	-	0	Zimmermann & Scheel (1984)
Benomyl	17804-35-2	DIS 13	+	1[b]	Sora et al. (1982)
Cyclophosphamide	50-18-0	DIS 13	+	+ (IARC, 1981)	Sora et al. (1983)
Diethyl ketone	96-22-0	D61.M	+	0	Zimmermann et al. (1985a)
Ethyl acetate	141-78-6	D61.M	+	0	Zimmermann et al. (1985a)
Ethylmethane sulfonate	62-50-0	D6	+	+ (IARC, 1974a)	Parry et al. (1979a)
p-Fluorophenyl-alanine	60-17-9	Several	+	0	Strömnaes (1968)
2-Methoxy ethyl acetate	110-49-6	D61.M	+	0	Zimmermann et al. (1982)
Methyl acetate	79-20-9	D61.M	+	0	Zimmermann et al. (1982)
Methylbenzimidazole carbamate	10605-21-7	D61.M	+	1[c]	Zimmermann et al. (1984a)
Methyl ethyl ketone	78-93-3	D61.M	+	0	Zimmermann et al. (1982)
Methylmethane sulfonate	66-27-3	DIS 13	+	+ (IARC, 1974a)	Sora et al. (1982)
Mezerein	34807-41-5	D6	-	+[d]	Parry et al. (1981)
Oncodazole	31430-18-9	D61.M	+	0	Zimmermann et al. (1984a)
Phenyl acetonitrile	140-29-4	D61.M	+	0	Zimmermann et al. (1985b)
Pivalinic acid nitrile		D61.M	+	0	Zimmermann et al. (1985b)
Pyridine	110-86-1	D61.M	+	0	Zimmermann et al. (1985b)
12-O-Tetradecanoyl phorbol 13-acetate	16561-29-8	D6	+	+[e]	Parry et al. (1981)
Valerolactone	108-29-2	D61.M	+	0	Zimmermann et al. (1985b)

[a]Classified mainly according to IARC (1972-1984) and Haseman et al. (1984); +, carcinogenic; 1, limited evidence; 0, not reported

[b]Dupont Report 1-81, 1981; Haskell Lab. R. 20-82, 1982

[c]Dupont Report 4-81, 1981; Haskell Lab. R. 70-82, 1982

[d]Slaga et al. (1980)

[e]Berenblum (1975)

Three agents (benomyl, cyclophosphamide and methyl methanesulfonate) gave positive results in the meiotic system in strain DIS13; one chemical (*p*-fluorophenylalanine) was assayed in several strains used to detect mitotic monosomy (Strömnaes, 1968); two (benomyl and *p*-fluorophenylalanine) are classical spindle poisons, while methyl methanesulfonate and cyclophosphamide are well-known genotoxic agents able to induce many other types of genetic events.

The remaining 17 chemicals listed in Table 2 were tested in the mitotic systems in strain D6 (3) and strain D61.M (14) and are those in which the *leu-1* marker was always checked; moreover, in most of the agents assayed with strain D61.M, the induction of other genetic events (e.g., gene mutation, gene conversion and crossing-over) has been tested in independent experiments. These chemicals are therefore likely to induce mitotic aneuploidy in yeast.

Most agents reported as giving positive results in strain D61.M (e.g., methylbenzimidazole carbamate, nocodazole, acetone and other polar solvents) seem to act mainly on tubulin (Zimmermann *et al.*, 1984b, 1985a,b).

2.4 *Correlation between genetic activity and carcinogenicity*

Table 2 indicates the following conclusions:

(i) For five of 21 compounds there is sufficient evidence or limited evidence of carcinogenicity, while for the remaining 16 agents no data on carcinogenicity were found.

(ii) Two agents for which there is sufficient (cyclophosphamide and methyl methanesulfonate) evidence of carcinogenicity were assayed and found to give positive results in the meiotic system in strain DIS13 (Sora *et al.*, 1982, 1983).

(iii) Three chemicals for which there is sufficient or limited evidence of carcinogenicity (ethyl methanesulfonate, 12-*O*-tetradecanoylphorbol 13-acetate and mezerein) were assayed in the mitotic system in strain D6 (Parry *et al.*, 1979a, 1981); the first two gave positive results, while mezerein was found negative.

(iv) It is obvious that this data base is too limited to allow any significant evaluation of the predictive value for carcinogenic activity of the *S. cerevisiae* aneuploidy systems.

2.5 *Test systems with* Neurospora crassa

(a) *Principles and scientific basis of the test*

N. crassa is a haploid (n=7) eukaryotic microorganism widely used in genetic research (Perkins *et al.*, 1982) as well as in environmental chemical mutagenesis (Brockman *et al.*, 1984). Its meiotic chromosomes look, and behave, like typical eukaryotic chromosomes (Barry, 1972). It was therefore an obvious candidate for studying chemically-induced aneuploidy at the meiotic level.

The first evidence of aneuploidy in *N. crassa* was provided by Pettinger (1954), who detected aneuploids as prototrophic pseudo-wild colonies (PWT) from ascospores of crosses between parents bearing several linked complementing nutritional markers. The PWT colonies were shown to be heterokaryons composed of the parental genotypes for only one chromosome pair. By genetic analysis, Pettinger inferred that the PWTs were derived from meiotic disomics (n+1), which were unstable and broke down to two nuclear components during vegetative growth.

Subsequently, Threlkeld and Stolz (1970) examined the mechanisms of meiotic aneuploidy by carrying out tetrad analysis of crosses heterozygous for different complementary *pan-2* mutants. In the presence of limiting amounts of pantothenate, the colour of ascospores of genotype *pan-2* is lighter than normal; thus, the majority of asci contained eight light spores. It was shown that among the possible sources of dark spores — mutation, recombination and aneuploidy — the latter was the most frequent. By studying the ascospore patterns for several additional markers in the aneuploid-containing asci, three main meiotic routes to aneuploidy were suggested: (1) nondisjunction at the first meiotic division; (2) nondisjunction at the second meiotic division (detected only in case of a crossing-over between the *pan-2* locus and the centromere); and (3) precocious (equational) division of one or both centromeres at the first meiotic division.

On this basis, Griffiths and Delange (1977) constructed strains specifically designed to study chemically-induced aneuploidy. Their system is based on the selection of n+1 prototrophic colonies, which arise from meiotic nondisjunction of linkage group 1, and uses PWT frequency as an index of aneuploidy frequency. The system, which has been reviewed in detail (Griffiths, 1982; Brockman *et al.*, 1984; Griffiths *et al.*, 1986) has been used to test a very large number of chemicals. The genotypes of the specially constructed strains 1-41-5 (used as the conidial male parent) and strain 1-348 (used as the protoperithecial female strain) is reported in Table 3. The spontaneous frequency of PWTs is low — about 6×10^{-5} ascospores — with a range rarely extending above 1.0×10^{-4}.

One experiment takes five weeks to perform. The aneuploidy frequency is usually expressed as the number of PWT colonies per total viable ascospores (10^5); in some cases, the number of disomics per number of treated ascospores has been used.

For each experiment, only the highest dose compatible with adequate fertility is usually analysed; therefore, only in a few experiments (e.g. those with *p*-fluorophenylalanine) have dose-response curves been obtained. In the *N. crassa* system, as in other fungal aneuploidy systems, there is only a relatively small effect of positive agents in the optimal dose range, which is usually very narrow. In *N. crassa*, a six- to ten-fold increase over the control is the best that has been achieved using *p*-fluorophenylalanine (Griffiths & Delanges, 1977), which remains probably the most consistently potent

Table 3. Genotypes of *N. crassa* and *S. brevicollis* strains commonly used for detecting aneuploidy

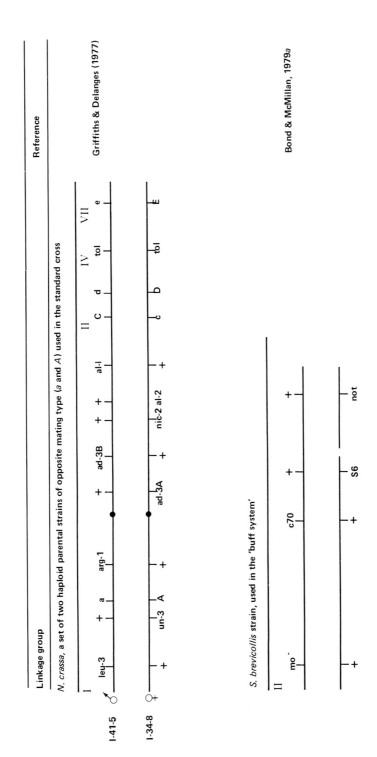

chemical agent so far tested in *N. crassa* and which is often used as a positive control. An exogenous mammalian metabolic system has not been used in conjunction with this assay system.

(b) Evaluation of the test system

There are several problems with the *N. crassa* selective meiotic system, due mainly to insufficient validation of chemically induced PWTs and its inability to distinguish between aneuploidy and other genetic events (e.g., recombination and mutation). Pettinger (1954) and Threlkeld and Stolz (1970) showed, by genetic analysis, that among the possible causes of the PWT phenotype (arg^+, ad^+, nic^+), i.e., mutation, recombination and aneuploidy, the last event (disomy of chromosome I) was the most frequent. However, in the case of chemically-induced PWT, this type of genetic analysis has been done only with *p*-fluorophenylalanine. With this agent, known to induce only aneuploidy, Griffiths and Delanges (1977) ruled out, by genetic analysis, the possibility that PWTs are triple crossovers. Since then, PWT frequency was automatically taken as an index of aneuploidy frequency and used to screen a large number of environmental chemicals, without taking note of their possible recombinogenic activity as a cause of PWT. In fact, with the exception of *p*-fluorophenylalanine, it is practically impossible to know the cause of the PWT induced by the chemicals so far tested.

Other serious limitations of this system are that (1) there is a high variability of spontaneous PWT frequencies and consequently high variance among replicates, even more pronounced in case of positive results; (2) high doses of test chemicals cannot be used because they are often incompatible with adequate fertility; (3) only small increases (usually from two- to three-fold) over the controls are obtained; (4) among the chemicals evaluated as positive, important classes (e.g., alkylating agents) are lacking; moreover, powerful mutagens (e.g., *N*-methyl-*N'*-nitro-*N*-nitrosoguanidine, methyl methanesulfonate) expected to induce aneuploidy, and classical nondisjunctional agents acting on the spindle (e.g., benomyl and chloral hydrate) give negative or inconclusive results; (5) the majority of agents evaluated as positive do not induce aneuploidy in other systems; some of them (e.g., sulfacetamide and trimethoprim) are folic acid antagonists known to be recombinogenic agents in *A. nidulans*; caffeine has been shown (Sora & Bianchi, 1982) to be a powerful inducer of meiotic diploidization due to failure of the II meiotic division in *S. cerevisiae*; trifluralin has been shown (Bond & McMillan, 1979b) to act mainly on the premeiotic extra-replication ascus class in *Sordaria*; and (6) as with all meiotic systems, exogenous mammalian metabolic activation systems have not been used.

(c) Results obtained

More than 100 chemicals have been tested in the *N. crassa* PWT selective system. The data on 70 were evaluated in a report of the US Environmental Protection Agency Gene-Tox Program (Brockman *et al.*, 1984), and all these data have recently been re-evaluated by a US Environmental Protection Agency committee (Griffiths et al., 1986), which also examined another 39 chemical agents. Table 4 shows the results, together with the Chemical Abstracts Services number and degree of evidence for carcinogenicity, when available. As stated above, except for p-fluorophenylalanine, there is serious doubt about the true nature of the PWTs induced by agents reported as positive.

2.6 *Test systems with* Sordaria brevicollis

(a) *Principles and scientific basis of the test*

S. brevicollis is a heterothallic ascomycete fungus which shares many of the useful characteristics of *N. crassa*, including an eight-spore ascus containing the products of a single meiosis.

In this fungus, a genetic system, known as the 'buff' system, was set up by Bond (1976) to study the induction and mechanisms of meiotic aneuploidy (see Table 3). The system uses two ascospore colour mutants, *c70* and *S6*, which are complementing alleles of the *b1* (buff ascospore colour) locus and form the basis for detecting aneuploidy (disomy) for linkage group II. In order to verify the disomic nature of the aneuploids the following flanking markers were used: *mo⁻* (morphological mutant), subsequently substituted by *met-1* (methionine requirement) and *not-1* (previously called *nic-1* and conferring nicotinamide or tryptophan requirement). In most cases, the two hetero-alleles *c70* and *S6* are phenotypically distinguishable, the first one conferring a slightly darker colour on the spore than the second. From crosses between the two *b1* alleles, the majority of the asci contain eight-colour mutant spores; occasionally, asci containing black spores are formed, which are disomics (n+1) for linkage group II, and are black because both parental complementary *b1* alleles are present in the same nucleus. By carrying out genetic analysis of cultures derived from the black spores, Bond was able to distinguish at least three principal causes of meiotic malsegregation: nondisjunction at the first meiotic division, nondisjunction at the second meiotic division, and premeiotic errors such as extrareplication of one or other of the marked homologues, resulting in formation of pseudo-wild types without accompanying spore abortion. According to Bond's calculations, the frequencies at the I and II meiotic divisions are 4.25×10^{-4} and 4.35×10^{-4}, respectively.

The buff system has been used to study the effects of two chemicals, *p*-fluoro-phenylalanine (Bond & McMillan, 1979a) and the herbicide trifluralin (Bond & McMillan, 1979b). *p*-Fluorophenylalanine gave a positive result (a ten-fold increase over the control being the highest effect observed), and a dose-response effect was also apparent. Most of the effect was due to increases in nondisjunction at both the I and II divisions; division II nondisjunction seemed to be increased to a greater extent than that at I division, but the data did not permit a definitive conclusion. Trifluralin showed a positive, but weaker and more variable effect, without a dose response; moreover, the main effect was on the extrareplication ascus class.

Table 4. Agents evaluated in the *Neurospora crassa* PWT system

Chemical	CAS No.	Response	Carcinogenicity[a]	References
Positive response				
Amethopterin	59-05-2	+	0	Griffiths (1979); Brockman *et al.* (1984)
m-Aminophenol	591-27-5	+	0	Griffiths (1979)
Atrazine	1912-24-9	+	0	Griffiths (1979)
Bromacil	314-40-9	+	0	Griffiths *et al.* (1986)
Caffeine	58-08-2	+	0	Griffiths (1979); Brockman *et al.* (1984)
2,4-Diaminoanisole sulfate	39156-41-7	+	+ (IARC, 1982a)	Griffiths (1979); Brockman *et al.* (1984)
2,4-Diaminotoluene	95-80-7	+	+ (IARC, 1978)	Griffiths (1979)
p-Fluorophenylalanine	60-17-3	+	0	Griffiths & Delanges (1977); Brockman *et al.* (1984)
Sulfacetamide	144-80-9	+	0	Griffiths (1979); Brockman *et al.* (1984)
Trifluralin	1582-09-8	+	1 (Griesemer & Cueto, 1980)[b]	Griffiths (1979); Brockman *et al.* (1984)
Trimethoprim	738-70-5	+	0	Griffiths (1979); Brockman *et al.* (1984)
Negative response				
2-Acetylaminofluorene	53-96-3	−	+ (Arcos & Argus, 1974)	Griffiths *et al.* (1986)
Acridine orange	65-61-2	−	1 (IARC, 1978)	Griffiths *et al.* (1986)
Alloxan	50-71-5	−	0	Griffiths (1979, 1982) Brockman *et al.* (1984)
Ametryne	834-12-8	−	0	Griffiths *et al.* (1986)
2-Aminofluorene	153-78-6	−	+ (National Cancer Institute, 1983)	Griffiths *et al.* (1986)
Aminopterin	54-62-6	−	0	Griffiths (1979, 1982) Brockman *et al.* (1984)
3-Amino-1,2,4-triazole	61-82-5	−	+ (IARC, 1982a)	Griffiths (1980); Brockman *et al.* (1984)
Amphotericin B	1397-89-3	−	0	Griffiths *et al.* (1986)
Azure A	531-53-3	−	0	Griffiths *et al.* (1986)
Azure B	531-55-5	−	0	Griffiths *et al.* (1986)
Azure C	531-57-7	−	0	Griffiths *et al.* (1986)
Benzo[*a*]pyrene	50-32-8	−	+ (IARC, 1984)	Griffiths *et al.* (1986)
Cyclic AMP	60-92-4	−	0	Griffiths *et al.* (1986)
Cytoxan	50-18-0	−	0	Griffiths *et al.* (1986)
2,4-Diaminotoluene	95-80-7	−	+ (IARC, 1978)	Griffiths *et al.* (1986) Brockman *et al.* (1984)

Table 4. (contd)

Chemical	CAS No.	Response	Carcinogenicity[a]	References
Diazinon	333-41-5	-	0	Griffiths et al. (1986); Brockman et al. (1984)
Dichloran	99-30-9	-	0	Griffiths et al. (1986); Brockman et al. (1984)
Diethylstilboestrol	56-33-1	-	+ (IARC, 1982a)	Griffiths (1981); Brockman et al. (1984); Griffiths et al. (1986)
Eptam	759-94-4	-	0	Griffiths (1979, 1982); Brockman et al. (1984)
Ethanol	64-17-5	-	+ (Kissen & Kaley, 1974)	Griffiths (1981); Brockman et al. (1984)
DL-Ethionine	67-21-0	-	+ (National Cancer Institute, 1983)	Griffiths (1979, 1982); Brockman et al. (1984)
Ethylene glycol	107-21-1	-	0	Griffiths (1979, 1982); Brockman et al. (1984)
4-Fluorotryptophan	25631-17-8	-	0	Griffiths et al. (1986)
5-Fluorouracil deoxyriboside	50-91-9	-	0	Griffiths (1979, 1982); Brockman et al. (1984)
Formaldehyde	50-00-0	-	+ (IARC, 1982b)	Griffiths et al. (1986)
Haloprogin	777-11-7	-	0	Griffiths (1980); Brockman et al. (1984)
Maleic hydrazide	123-33-1	-	- (IARC, 1974)	Griffiths (1979, 1982); Brockman et al. (1984)
Metronidazole	443-48-1	-	+ (IARC, 1977a)	Griffiths (1981); Brockman et al. (1984)
1-Naphthol	90-15-3	-	0	Griffiths (1979, 1982); Brockman et al. (1984)
Neutral red	553-24-2	-	0	Griffiths (1980); Brockman et al. (1984)
Nitrilotriacetic acid	139-13-9	-	+ (National Cancer Institute, 1977; Goyer et al. 1981)	Griffiths et al. (1986)
4-Nitroquinoline oxide	56-57-5	-	+ (Arcos & Argus, 1974)	Griffiths et al. (1986)
Nystatin	1400-61-9	-	0	Griffiths (1979, 1982); Brockman et al. (1984)
Orthene	30560-19-1	-	0	Griffiths (1979, 1982); Brockman et al. (1984)
Phenyl mercuric nitrate	55-68-5	-	0	Griffiths (1981); Brockman et al. (1984)
Quercetin	117-39-5	-	1 (IARC, 1983b)	Griffiths (1981); Brockman et al. (1984)
Resorcinol	108-46-3	-	1 (IARC, 1977b)	Griffiths (1979, 1982); Brockman et al. (1984)

Table 4. (contd)

Chemical	CAS No.	Response	Carcino-genicity[a]	Reference
Saccharin	81-07-2	-	l (IARC, 1982a)	Griffiths (1981); Brockman et al. (1984)
Sarcosine	107-97-1	-	0	Griffiths (1979, 1982); Brockman et al. (1984)
Simazine	122-34-9	-	0	Griffiths et al. (1986)
Stilphostrol	522-40-7	-	0	Griffiths (1979, 1982); Brockman et al. (1984)
Terbacil	5902-51-2	-	0	Griffiths et al. (1986)
o-Toluidine	95-53-4	-	+ (IARC, 1982a)	Griffiths et al. (1986)
Urethane	51-79-6	-	+ (IARC, 1974a)	Griffiths et al. (1986)
Vinblastine sulfate	143-67-9	-	l (IARC, 1982a)	Griffiths (1979, 1982); Brockman et al. (1984)
Inconclusive result				
Actidione	66-81-9		0	Griffiths (1979)
Chloral hydrate	302-17-0		0	Griffiths (1980)
Colchicine	64-86-8		0	Griffiths (1981)
Dimethylsulfoxide	67-68-5		0	Griffiths (1981)
Enterovioform	130-26-7		0	Griffiths (1979)
Ethylmethane sulfonate	62-50-0		+ (IARC, 1974a)	Griffiths (1980)
5-Fluorouracil	51-21-8		i (IARC, 1982a)	Griffiths (1979)
Hycanthone	3105-97-3		+ (IARC, 1977a)	Griffiths (1980)
8-Hydroxyquinoline	148-24-3		- (National Toxicology Program, 1985a)	Griffiths (1980)
N-Methyl-N'-nitro-N-nitrosoguanidine	70-25-7		+ (IARC, 1974b)	Griffiths (1981)
Mitomycin C	50-07-7		+ (National Cancer Institute, 1983)	Griffiths (1979)
p-Nitrobiphenyl	92-93-3		+ (IARC, 1974b)	Griffiths (1979)
4-Nitro-o-phenylene-diamine	99-56-9		- (National Cancer Institute, 1979a)	Griffiths (1979)
o-Phenylenediamine	95-54-5		+ (Weisburger et al. 1978)	Griffiths (1979)
p-Phenylenediamine	55-68-5		- (National Cancer Institute, 1979b)	Griffiths (1980)
Stannous fluoride	7783-47-3		+ (National Toxicology Program, 1982a)	Griffiths (1980)
Sulfanilamide	63-74-1	0		Griffiths (1979)

[a] Classified mainly according to IARC (1972-1984), NCI: Griesemer & Cueto (1980) and NTP: Haseman et al. (1984); +, carcinogenic; l, limited evidence; i, inadequate evidence; -, not carcinogenic; 0, not reported

[b] The commercial sample contained 0.1-0.2% impurities, including N-nitrosodipropylamine.

A second method, known as the 'grey' system, very similar to that described by Bond (1976), can be used in *S. brevicollis* (Fulton & Bond, 1983). In this case, the spore-colour mutations are *c31*, *rw25*, *yS121*, *yS18* and *B9*, heteroallelic of the *grey-6* (g6) locus on linkage group IV. Linked markers are also available at *met-2*, *ura-1* and *pdx-1*, conferring requirements for methionine, uracil and pyridoxine, respectively. These mutations are proximal to *grey-6* and, as for the buff system, are useful for identifying the complementing alleles when recovered from disomic spores. Comparing the two systems (the buff and the grey), Bond and McMillan found similar proportions of the errors that give rise to disomics for linkage II and IV.

(b) Evaluation of the test system

The *S. brevicollis* buff system is well suited for studying the induction of meiotic aneuploidy (disomy). The most important advantage of this system is its ability to differentiate the mechanisms of aneuploidy induction. As with other meiotic aneuploidy tests based on extensive genetic analysis, this test system is slow and complicated. Furthermore, there is the problem, common to all meiotic systems, of the lack of suitable methods for studying the effects of exogenous mammalian metabolic activation systems; finally, up to now it has been used with only two chemical agents. Nevertheless, studies with the *S. brevicollis* meiotic system should be encouraged.

2.7 *Test systems with* Aspergillus nidulans

(a) Principles and scientific basis of the tests

A. nidulans is a homothallic ascomycete widely exploited in genetic research (Clutterbuck, 1977) and in environmental chemical mutagenesis (Käfer *et al.*, 1982).

Wild-type strains are usually haploid; however, stable diploids are produced with low frequency as a spontaneous event by the fusion of haploid nuclei in heterokaryotic hyphae during the parasexual cycle (Pontecorvo, 1956). These diploids, which can easily be isolated from heterokaryons between differently marked haploid strains, are fairly stable and rarely produce mitotic segregants as the result of spontaneous mitotic crossing-over or chromosomal malsegregation. Mitotic aneuploids are usually found at a frequency of 1 - 2%.

Several well marked strains that have been widely employed in test systems specifically aimed at detecting chemicals able to induce mitotic segregation (crossing-over and aneuploidy) are shown in Table 5. Genotypes of other, less commonly used *A. nidulans* diploid strains are given by Käfer *et al.* (1982). The genetic systems are usually based on the expression of recessive markers, present in heterozygous condition on both arms of a well marked chromosome, in the parent diploid. Both selective and nonselective methods have been used for screening purposes.

(i) *Selective methods*

Recessive markers conferring resistance to *p*-fluorophenylalanine (*fpaA1* in strain P1) and pimaricin (*pimB10* in strain D7) have been used to select stable homozygous-

Table 5. Genotypes of *Aspergillus nidulans* strains commonly used for detecting chromosomal malsegration

Diploid	Genotype	Reference
P1	I: suAadE20 riboA / + + + fpaA anA + pabaA ya2 + adE20 biA / + + + + + + + proA + + + + III: sC12 / + phenA2 + IV: + methG + V: nicA2 / + + + pyroA4 lysB5 VII: nicB8 / +	Bignami et al. (1974) Morpurgo et al. (1979)
	I: suAadE20 pabaA ya2 adE20 + / + + + + biA II: AcrA + + / + wA2 cnxE16 III: sC12 / + phenA2 IV: + methG + V: nicA2⁻ / + + + pyroA4 + nicB8 VIII: riboB2 / + VI: sB3 / + lacA / + choA + chaA	Harsanyi et al. (1977)
	I: suAadE20 yA2 adE20 + / + + + + biA II: + / wA3 III: + / galA VI: pyroA4 / + facA303 / + sB3 VII: nicB8 / + VIII: riboB2 / +	Azevedo et al. (1977)
	I: suAadE20 + pabaA + ya2 adE20 + / + + + + + biA II: AcrA + / AcrA wa3 thiA4 cnxE16 III: galA IV: pyroA4 V: facA303 VI: sB3 VIII: riboB2 VII: nicB8 / + cnxB2 / +	Kappas (1978)
11	I: fpaB37 galD5 suAadE20 + riboA anA + pabaA ya2 adE20 biA / + + + + + sulA + + + + adE20 + II: AcrA wA3 III: ActA IV: + pyroA4 + facA303 V: + facA303 VIII: sDB5 fwA2 / + riboB2 chaA VI: lacA sB3 VII: CliA2 choA	Käfer et al. (1976)

Table 5 (contd)

Diploid	Genotype	Reference

19	I: fpaB37 galD5 suAadE20 riboA + yA2 adE20 +
	  + + + + anA + adE20 biA
	II: + + III + + IV pyroA4 V facA303 VI +
	  AcrA ActA + + + sB3
	VII: + + VIII + chaA +
	  AcrA•choA + fwA2 +
		Käfer et al. (1976)

20	I: fpaB37 galD5 suAadE20 + riboA anA pabaA yA2 adE20
	  + + + sulA + + + + +
	II: + III + IV + V facA303
	  AcrA ActA pyroA4 +
	V: + + VII sDB5 fwA2 +
	  nicA2 + + sB3 + chaA
	VI: + +
	  lacA sB3
		Käfer et al. (1976)

29	I: fpaB37 suAadE20 + riboA + pabaA yA2 adE20
	  + + sulA + anA + + +
	II: + + III + + IV biA
	  AcrA + wA cnxE16 galA ActA pyroA4
	  + + + methG +
	V: + + VI + VII + +
	  nicA2 facA303 sB3 oliC2 malA choA
	  + + + + + +
	VIII: fwA2 +
	  + chaA
		Käfer (1985)

30	I: fpaB37 suAadE20 riboA pabaA yA2 adE20
	  + + + + + +
	II: + + III + + IV + pyroA4
	  AcrA + wA cnxE16 galA ActA + methG
	V: + + VI + VII + malA +
	  nicA2 facA303 sB3 oliC2 + choA
	  + + sBA3 + +
	VIII: fwA2 +
	  + chaA
		Käfer (1985)

D7	I: pimB10 proA1 pabaA1 yA2 + IV pyroA4 + + + + biA1 +	de Bertoldi et al. (1980)

and hemizygous-resistant segregants from heterozygous-sensitive diploids (Morpurgo, 1962; Bignami et al., 1974; de Bertoldi et al., 1980). The analysis of biochemical markers and the identification of conidial colour markers makes it possible to distinguish whole-chromosome segregation (in haploids and nondisjunctional diploids) from mitotic crossing-over and other genetic events like breakage/deletion and point mutation, which are much rarer. These selective procedures have the advantage of being very simple and quick; however, while they are well suited for studying the induction of mitotic crossing-over, they cannot be considered fully reliable for detecting mitotic aneuploidy for at least two reasons: (1) possible interaction of the selective agents (which are themselves aneuploidy inducers) with the test chemicals; and (2) inhibition, due to the presence of high concentrations of selective agents, of the growth of abnormal segregants, including aneuploids, which are therefore recovered to a lower extent than with nonselective methods. Indeed, the spontaneous frequency of segregants for chromosome I in strain P1 varies from about $2-3 \times 10^{-3}$ to 1×10^{-5} when measured with nonselective and selective methods, respectively. Selective methods have therefore been used mainly to detect mitotic crossing-over.

(ii) *Nonselective methods*

There are two main types of nonselective test systems. The first are purely genetic test systems that detect stable euploid segregants derived from appropriate heterozygous diploids and identify them as products of crossing-over or as products of chromosomal malsegregation leading to aneuploidy or haploidy. This type of test is that most commonly used for screening up to now. The second type comprises tests that detect aneuploids in treated diploid and haploid strains as unstable, abnormally growing segregants which can be classified according to their characteristic phenotypes and patterns of segregation in newly generated segregants.

(iii) *Nonselective tests based on the genetic analysis of euploid mitotic segregants*

Diploid strains heterozygous for conidial colour and for several biochemical markers are usually used. Stable euploid segregants produced by chromosomal malsegregation (haploids and nondisjunctional diploids) and mitotic crossovers are visually identified as sectors, patches or spots in treated colonies and subsequently isolated and analysed. Segregation of the *yA2* marker, on the right arm of chromosome I, is frequently studied. Heterozygous, pale-green strains (e.g., P1) produce dark-green ($yA2^+/yA2^+$; $yA2^+/$ O) and yellow ($yA2^-/yA2^-$; $yA2^-/$ O) homozygous or hemizygous sectors, which are then classified as nondisjunctional diploids, haploids and crossovers according to their nutritional requirements. With the P1 strain, spontaneous frequencies of visually detectable yellow segregants are usually around 1% (crossovers) and 0.2-0.3% (chromosomal malsegregants).

Other colour markers frequently used together with *yA2* are *wA3* (white conidia) on the left arm of chromosome II (e.g., in strains developed by Azevedo *et al.*, 1977 and by Kappas, 1978) and *chaA* (chartreuse) and *fwA2* (fawn), usually in repulsion on the right arm of chromosome VIII for the detection of twin spots as the result of reciprocal crossing-over (e.g., in strains developed by Käfer *et al.*, 1976).

(iv) *Tests based on identification of aneuploids among abnormals from treated diploids*

Well-marked diploid strains, carrying alleles on both the homologues of each of the eight *A. nidulans* chromosome pairs, have been constructed by Käfer (diploids 29 and 30, Table 5). These strains are used with nonselective procedures to carry out detailed and conclusive analyses, aimed at identifying the primary genetic events induced rather than euploid stable secondary segregants originating from aneuploid types. The original change induced in the genotype of abnormally growing aneuploids is identified through their phenotypic characterization and replating, followed by the study of patterns of segregation in newly generated euploid sectors (Käfer, 1961, 1984).

(v) *Tests based on visual identification of aneuploids among abnormals from treated haploids*

Less frequently used tests involve haploid strains, based on the phenotypic characterization of disomics for each of the eight chromosome pairs (Käfer & Upshall, 1973). These assays can quickly provide conclusive evidence of the nature of the primary genetic event induced (aneuploidy *versus* chromosomal damage), because gross chromosomal aberrations and deletions, which could lead in diploid strains to chromosomal malsegregation as a secondary event, are expected to be lethal in haploid strains. Although relatively simple as an experimental procedure (Käfer, 1984), this type of test requires considerable experience, because the identification and classification of the observed presumptive disomics is based only on their characteristic morphology.

Technical details of the experimental protocols for all the *A. nidulans* aneuploidy systems have been described fully (Käfer *et al.*, 1982; Scott & Käfer, 1982; Gualandi & Morpurgo, 1984).

(b) Evaluation of the test systems

A great advantage of *A. nidulans* is that, by using the same strain, induction of the two main genetic events responsible for somatic segregation, namely, crossing-over and chromosomal malsegregation leading to aneuploidy, can be analysed simultaneously and with great precision.

For the reasons already mentioned, selective systems cannot be considered fully reliable for detecting mitotic aneuploidy, although they are well suited for studying the induction of mitotic crossing-over. Conversely, quick tests based on the detection of stable euploid segregants (nondisjunctional diploids and haploids) with nonselective procedures can be used efficiently for the initial detection of mitotic aneuploidy.

A unique feature in comparison with other tests in lower eukaryotes is that genetic analysis of *A. nidulans* can be carried out with great accuracy and detail, disclosing the exact nature of the genetic event induced and differentiation between primary aneuploidy and partial trisomy or monosomy resulting from structural chromosomal aberrations. This possibility is of major importance when exposure to a test chemical increases the frequency of all types of euploid segregants (crossovers, nondisjunctional diploids and haploids). In these cases, one cannot usually rule out the possibility that semilethal mutations or chromosomal damage (e.g., breakage/deletion), rather than aneuploidy, is the primary genetic damage induced and that the observed malsegregants are the result of secondary spontaneous events, as was shown to be the case for radiation-induced malsegregants (Käfer, 1985). The overall results of both types of nonselective tests indicate that, in the case of chemicals that can induce only whole-chromosome-type segregants (haploids and/or nondisjunctional diploids) but no crossovers, induction of aneuploidy is the primary genetic event.

A major weakness (at least at present) of *A. nidulans* aneuploidy tests is the lack of validated methods for efficient use of exogenous mammalian metabolic systems; so far, only one promutagen (cyclophosphamide) has been shown to be activated by a liver microsomal fraction. *A. nidulans*, like yeast and *N. crassa*, is endowed with intrinsic metabolic activities which have been shown to activate a certain number of pro-mutagens (*N*-nitrosamines, dimethylhydrazine, procarbazine, cyclophosphamide) (Bignami *et al.*, 1981). However, the experimental procedure has not yet been optimized and standardized.

Another shortcoming of assays using *A. nidulans* is the laborious nature of the experimental procedures. For example, when nonselective procedures are used, large numbers of colonies (1000 to 2000) must be inspected to obtain enough mitotic segregants (usually detected at less than 1% among untreated colonies) for significant comparisons of segregation frequencies in treated and untreated colonies, especially for weak effects (three- to four-fold increases). So far, only small numbers of colonies (about 200) have usually been analysed. Thus, most of the 'negative' results reported in Table 6 should be more properly evaluated as 'inconclusive'. The procedures described for identifying aneuploids by replating and further analysis of segregated sectors (Käfer, 1961, 1984) are cumbersome and require a high level of experience.

Finally, the absence of adequate standardization of experimental procedures seriously hinders interlaboratory comparisons.

(c) Results of tests for chromosomal malsegregation and aneuploidy in A. nidulans

Data on 76 chemicals tested in *A. nidulans* aneuploidy systems have been selected from the scientific literature (the main sources being two reviews by Käfer *et al.*, 1982, 1986) and are shown in Table 6. Of these, 45 compounds were evaluated as positive and 31 as negative. Only two positive agents (cyclophosphamide and iodochlorohydroxy-quinoline) were tested with selective procedures, while the remaining 43 agents were

Table 6. Agents evaluated in tests for mitotic malsegregation and aneuploidy in *A. nidulans*

Chemical	CAS No.	Carcinogenicity[a]	Assay code[b]	Other endpoints scored[c] Point mutation	Crossing-over	Reference
Positive response						
Acrylonitrile	107-13-1	+ (IARC, 1979a)	1a		-	Carere *et al.* (1985)
Actynomycin D	50-76-0	1 (IARC, 1982a)	1a		-	Kappas (1978)
Allyl chloride	107-05-1	0	1a	-		Crebelli *et al.* (1984)
Amphotericin B	1397-89-3	0	1a	-		Bellincampi *et al.* (1980)
Benomyl	17804-35-2	0	1a	-		Hastie (1970)
Benzene	71-43-2	+ (IARC, 1982b)	1a		-	Carere *et al.* (1984)
Botran	99-30-9	0	1a		+	Kappas (1978)
Carbon tetrachloride	56-23-5	+ (IARC, 1979b)	1a	±	+	Gualandi (1984)
Chloral hydrate	302-17-0	0	1a	±	-	Singh & Sinha (1976); Crebelli *et al.* (1985)
2-Chloroethanol	107-07-3	- (National Toxicology Program, 1985b)	1a	-	-	Crebelli *et al.* (1984)
Chloroneb	2675-77-6	0	1a		-	Azevedo *et al.* (1977)
Chlorpropham	101-21-3	i (IARC, 1976a)	1a	-	-	Gualandi & Bellincampi (1981)
Cyclophosphamide (+S9)	50-18-0	+ (IARC, 1981)	1b	+	+	de Bertoldi *et al.* (1980)
1,2-Dibromoethane	540-49-8	+ (IARC, 1977b)	1a	+	-	Crebelli *et al.* (1984)
2,2-Dichloroacetaldehyde	79-02-7	0	1a	+		Crebelli *et al.* (1984)
1,2-Dichloroethane	540-59-0	+ (IARC, 1979b)	1a		-	Crebelli *et al.* (1984)
2,2-Dichloroethanol	598-38-9	0	1a	±		Crebelli *et al.* (1984)
Dichlorvos	62-73-7	i (IARC, 1979b)	1a	+	+	Bignami *et al.* (1977)
Diepoxybutane	1464-53-5	+ (IARC, 1976b)	1a		+	Putrament (1967)

Table 6. (contd)

Chemical	CAS No.	Carcinogenicity[a]	Assay code[b]	Other endpoints scored[c]		Reference
				Point mutation	Crossing-over	
Econazole	27220-47-9	0	1a	NT	NT	Bellincampi et al. (1980)
Ethanol	64-17-5	+ (Kissin & Kaley, 1974)[d]	2		±	Harsanyi et al. (1977)
Fenarimol	60168-88-9	0	1a	NT	NT	Bellincampi et al. (1980)
Formaldehyde	50-00-0	+ (IARC, 1982b)	1a		+	Fratello et al. (1960)
Griseofulvin	126-07-8	+ (IARC, 1976c)	1a	NT	NT	Kappas & Georgopoulos (1974)
Indole-3-acetic acid	87-51-4	1 (Pesonen, 1950; Dunning & Curtis, 1958)	1a		+	Kappas (1983)
Indole-3-butyric acid	133-32-4	1 (Pesonen, 1950)	1a		+	Kappas (1983)
Iodochlorohydroxyquinoline	130-26-7	0	1b		−	Bignami et al. (1974)
Linoleic acid peroxide	12673-50-6	+ (Cutler & Schneider, 1974)	1a	+	+	Gualandi (1984)
Methylbenzimidazole carbamate	10605-21-7	0	1a	NT	NT	Kappas et al. (1974)
2,3-Methoxycarbonyl thioureidoaniline	27079-29-4	0	1a	NT	NT	Kappas et al. (1974)
Methyl methanesulfonate	66-27-3	+ (IARC, 1974a)	1a	+	+	Gualandi et al. (1979)
N-Methyl-N'-nitro-N-nitrosoguanidine	70-25-7	+ (IARC, 1974a)	2		+	Shanfield & Käfer (1971)
Methyl thiophanate	23564-05-8	0	1a	NT	NT	Kappas et al. (1974)
Miconazole	22916-47-8	0	1a	NT	NT	Bellincampi et al. (1980)
Nitrogen mustard	51-75-2	+ (IARC, 1975)	1a		−	Fratello et al. (1960)
4-Nitroquinoline-1-oxide	56-57-5	+ (Arcos & Argus, 1974)	1a	+	+	Morpurgo et al. (1979)
4-Fluorophenylalanine	60-17-3	0	1a	NT	NT	Morpurgo (1961)
Pimaricin	7681-93-8	0	1a	NT	NT	Bellincampi et al. (1980)

Table 6. (contd)

Chemical	CAS No.	Carcinogenicity[a]	Assay code[b]	Other endpoints scored[c] Point mutation	Crossing over	Reference
Safrole	94-59-7	+ (IARC, 1976c)	1a	–	–	Carere et al. (1985)
Sodium deoxycholate	320-95-4	0	2	NT	NT	Assinder & Upshall (1982)
Thiabenzadole	148-79-8	0	1a	NT	NT	Kappas et al. (1974)
Thiophanate	23564-06-9	0	1a	NT	NT	Kappas et al. (1974)
Thiram	137-26-8	i (IARC, 1976a)	2	NT	NT	Upshall & Johnson (1981)
2,2,2-Trichloroethanol	115-20-8	0	1a	+	–	Crebelli et al. (1985)
1,1,2-Trichloroethylene	79-01-6	+ (National Toxicology Program, 1982b)	1a	+	–	Crebelli et al. (1985)
Negative response						
Benzimidazole	51-17-2	0	1a			Kappas et al. (1974)
Bleomycin	11056-06-7	i (IARC, 1981)	1a			Demopoulos et al. (1982)
Captan	113-06-2	l (IARC, 1983a)	1a			Bignami et al. (1977)
Carboxin	290-72-2	0	1a			Georgopoulos et al. (1975)
Chlorfurazol	3615-21-2	0	1a			Kappas et al. (1974)
Chloroform	67-66-3	+ (IARC, 1979b)	1a			Crebelli et al. (1984); Gualandi (1984)
Cycloheximide	66-81-9	0	1a			Georgopoulos et al. (1975)
Daconil	1897-45-6	0	1a			Georgopoulos et al. (1975)
Dalapon	75-99-0	0	1a			Bignami et al. (1977)
Diethylhexylphthalate	117-81-7	+ (IARC, 1982b)	1a			Carere et al. (1985)
Diethylstilboestrol	56-53-1	+ (IARC, 1979c)	1a			Carere et al. (1985)
1,2-Dichloropropane	78-87-5	+ (National Toxicology Program, 1985c)	1a			Crebelli et al. (1984);
Dinobuton	973-21-7	0	1a			Bignami et al. (1977)
Dimethirimol	5221-53-4	0	1a			Georgopoulos et al. (1975)
Dodine	2439-10-3	0	1a			Bignami et al. (1977)

Table 6. (contd)

Chemical	CAS No.	Carcinogenicity[a]	Assay code[b]	Other endpoints scored[c]		Reference
				Point mutation	Crossing-over	
Ethidium bromide	1239-45-8	0	1a			Bonatelli & Azevedo (1977)
Hexamethylphosphoramide	680-31-5	+ (IARC, 1977b)	1a			Carere et al. (1985)
Ioxynil	1689-83-4	0	1a			Bignami et al. (1977)
Mecoprop	93-65-2	0	1a			Bignami et al. (1977)
Neburon	555-37-3	0	1a			Bignami et al. (1977)
Nystatin		0	1a			Georgopoulos et al. (1975)
Phenobarbital	57-30-7	+ (IARC, 1977a)	1a			Carere et al. (1985)
Picloram	1918-02-1	l (National Cancer Institute, 1978)	1a			Bignami et al. (1977)
Plondrel	5131-24-8	0	1a			Georgopoulos et al. (1975)
Polyoxind	19396-06-6	0	1a			Georgopoulos et al. (1975)
Sulfanilamide	63-74-1	0	2			Käfer et al. (1976)
Sulfallate	95-06-7	+ (IARC, 1983a)	1a			Morpurgo et al. (1977)
o-Toluidine	95-53-4	+ (IARC, 1982c)	1a			Carere et al. (1985)
Triarimol	26766-27-8	0	1a			Georgopoulos et al. (1975)
4,5,6-Trichloro-2-trifluoromethyl-benzimidazole	2338-27-4	0	1a			Kappas et al. (1974)
Zineb	12122-67-7	i (IARC, 1976a)	1a			Georgopoulos et al. (1975)

[a]Classified mainly according to IARC (1972-1984) and NCI (Haseman et al., 1984); +, carcinogenic; -, noncarcinogenic; l, limited evidence; i, inadequate evidence; 0, not reported

[b]1a, detection of euploid malsegregants (nondisjunction diploids and haploids) with nonselective procedure; 1b, detection of euploid malsegregants with selective procedure; 2, isolation and genetic analysis of aneuploids

[c]-, ineffective; ±, borderline; +, active; NT, not tested

[d]Epidemiological evidence only

assayed with the more informative and reliable nonselective systems. All data for the negative agents were obtained with nonselective systems.

In the heterogeneous group of positives, some agents are structurally or metabolically related. For instance, chloral hydrate and 2,2,2-trichloroethanol are the main metabolites of 1,1,2-trichloroethylene; 2,2-dichloroacetaldehyde and 2,2-dichloroethanol are the main metabolites of dichlorvos; benomyl, methylthiophanate and methylbenzimidazole carbamate are three closely related fungicides in that the first two are metabolically converted to the third. Another fungicide, thiabenzadole, is related to these three compounds, at least in mechanism of action, as shown by studies of cross-resistance of *A. nidulans* benomyl-resistant strains (Hastie & Georgopoulos, 1971).

Of the positive agents, 11 had no activity and five only borderline activity in inducing crossing-over and/or gene mutations in *A. nidulans*.

On the basis of the overall positive results and of current knowledge of the function of the mitotic apparatus, at least five mechanisms for the generation of mitotic aneuploidy in *A. nidulans* can be hypothesized:

(i) interaction of spindle poisons with microtubules and/or the disturbance of tubulin assembly/disassembly, as in the case of benomyl, methylbenzimidazole carbamate, griseofulvin, and perhaps *p*-fluorophenylalanine, chloral hydrate and ethanol;

(ii) alteration of membrane integrity or permeability by membrane active agents (e.g., the polyene antibiotics amphotericin B and pimaricin, as well as the phenethylimidazole compounds econazole and miconazole);

(iii) nonspecific interaction with cellular hydrophobic compartments, which could be the cause of the genotoxic activity of several organic solvents;

(iv) interference of free radicals with structures of the mitotic apparatus (e.g., —SH groups of microtubules, as suggested for carbon tetrachloride and linoleic acid lipoperoxide); and

(v) interaction of electrophilic reactants with DNA or chromosomal structures (e.g., centromere, kinetochore), important for correct chromosomal segregation.

The last class includes well-known alkylating agents (e.g., *N*-methyl-*N'*-nitro-*N*-nitrosoguanidine, methyl methanesulfonate, 1,2-dibromoethane) that can induce gene mutations and/or structural chromosomal aberrations; in these cases, the induced aneuploidy may be not a primary genetic event but only a secondary one, favoured by the initial induction of gross chromosomal aberrations. In *A. nidulans*, one can identify the primary genetic event by using nonselective methods (Käfer, 1961, 1984).

Some of the chemicals reported as 'negative' should more properly be evaluated as 'inconclusive', mainly because of the low number of scored colonies, as noted above.

Other important factors, such as inadequate metabolic activation and differences in the sensitivity of targets other than DNA, may also be involved.

(*d*) *Correlation between the genetic activity of chemicals and their carcinogenicity*

In total, 23 out of 33 agents (69.6%) for which there is sufficient or limited evidence of carcinogenicity were positive in *A. nidulans*; 69% (18 out of 26) of agents with sufficient evidence were found to be aneuploidizing. Conversely, 10 out of 33 (30%) agents for which there is sufficient or limited evidence of carcinogenicity were reported to be unable to induce aneuploidy in *A. nidulans*; the same proportion is also obtained when taking into account only the carcinogenic chemicals for which there is sufficient evidence but were negative in *A. nidulans* (8 out of 26).

Of the 45 chemicals evaluated as positive in *A. nidulans* aneuploidy systems, the most interesting are agents like actinomycin D, benzene, carbon tetrachloride, ethanol, griseofulvin and linoleic acid lipoperoxide, for which there is evidence of carcinogenicity (sufficient or limited), but which are difficult or impossible to detect in most mutagenicity tests measuring DNA damage and/or mutation.

Mainly on the basis of their activity in inducing other genetic events (e.g., crossing-over and gene mutation) in *A. nidulans*, the 23 aneuploidizing agents for which there is sufficient or limited evidence of carcinogenicity can be classified into at least two main groups: (1) those able to induce all genetic events studied (e.g., *N*-methyl-*N'*-nitro-*N*-nitrosoguanidine, methyl methanesulfonate, nitrogen mustard, 4-nitroquinoline-1-oxide, formaldehyde), for which DNA and chromosomal structures critical for proper chromosomal segregation (e.g., centromere, kinetochore) are the most likely targets; and (2) those able to induce only or mainly mitotic malsegregation; for some of them (e.g., ethanol, griseofulvin and maybe trichloroethylene), the spindle apparatus seems to be the only or the main target involved, while for others (e.g., benzene, carbon tetrachloride, linoleic acid lipoperoxide), other mechanisms (e.g., free-radical formation) may be involved.

Bearing in mind the limited value of the quantitative estimates presented here (due mainly to the small number and nonrandom choice of chemicals), a relatively high proportion of carcinogens seem to be able to induce mitotic chromosomal malsegregation in *A. nidulans*. Most interesting are those carcinogens usually described as 'false negatives' because they are difficult or impossible to detect in bacterial mutagenicity tests. This stresses the importance of using test systems specifically aimed at detecting aneuploidy when studying the correlation between carcinogenicity and mutagenicity of chemical agents.

2.8 *Conclusions*

None of the fungal genetic systems so far used to study aneuploidy fully meets the criteria usually adopted for validation (e.g., the extent to which a particular genetic endpoint is measured, the number and types of chemicals tested, standardization of protocols, interlaboratory comparisons). However, some of them, when properly used, can be considered well suited for studying induction as well as mechanisms of meiotic and mitotic aneuploidy.

The PWT meiotic system in *N. crassa* presents two major problems: (1) insufficient genetic validation in the case of chemical induction; and (2) inability to distinguish not only the mechanisms of aneuploidy but even between aneuploidy and other genetic events (e.g., recombination) without extensive genetic analysis. Of more than 100 chemicals tested, only one (*p*-fluorophenylalanine) has been shown to induce PWTs by meiotic aneuploidy; for the additional ten chemicals considered to give positive results in this system, the lack of careful genetic analysis prohibits interpretation of the real nature of their effects. Of interest is the *S. brevicollis* 'buff system', which allows unambiguous identification of meiotic aneuploids and also provides information on mechanisms. This test system has so far been used to test only a few chemical agents.

The system set up in strain DIS13 of the yeast *S. cerevisiae* seems well suited, from a genetic point of view, for studying chemically-induced meiotic aneuploidy (disomy). It is, however, rather laborious and tedious, requiring extensive genetic analysis; furthermore, it has so far been used only in the absence of exogenous mammalian activation systems. Finally, it has been used with too few chemicals, only three of which (methyl methanesulfonate, benomyl and cyclophosphamide) have been considered to give positive results. Data obtained with *S. cerevisiae* systems based on the use of strains D6 and D6.1M to study mitotic aneuploidy (monosomy) are difficult to interpret, and they require careful internal controls. The major problem is that the same phenotype of the presumptive monosomics can arise from several genetic events (e.g., crossing-over, gene conversion, gene mutations, petites), and it is difficult to discriminate between them after chemical induction. Mainly owing to lack of adequate internal controls, of the many chemicals so far tested in strains D6, only two (methyl methanesulfonate and 12-*O*-tetradecanoylphorbol 13-acetate) have been considered able to induce aneuploidy in this system. Although many of these problems also hold for strain D6-1M, the results obtained with 12 chemicals are considered to be due to monosomy, because they were obtained in better controlled experiments.

The nonselective systems in *A. nidulans* for studying mitotic aneuploidy seem genetically to be well suited for this purpose. On the one hand, the procedures that provide the most evidence, even on the nature of the primary event(s) induced, are cumbersome and require great experience. On the other hand, with the most commonly used, simpler, but less informative procedures, large numbers of plates have to be prepared and inspected. The main problem with both types of procedure is that adequate methods for the use of exogenous mammalian activation systems have not yet been optimized. Even with such limitations, 45 chemicals have been considered to give positive results in *A. nidulans* mitotic systems

3. References

Arcos, J.C. & Argus, M.F. (1974) *Chemical Induction of Cancer*, Vol. 2, New York, Academic Press

Ashby, J., De Serres, F.J., Draper, M., Ishidate, M., Jr, Margolin, B.H., Matter, B. & Shelby, M.D. (1985) Evaluation of short-term tests for carcinogens. *Progr. Mutat. Res.*, 5, 117-174

Assinder, S.J. & Upshall, A. (1982) Mitotic aneuoploidy induced by sodium deoxycholate in *Aspergillus nidulans. Mutat. Res.*, 93, 101-108

Azevedo, J.L., Santana, E.P. & Bonatelli, R., Jr (1977) Resistance and mitotic instability to chloroben and 1,4-oxathiin in *Aspergillus nidulans. Mutat. Res.*, 48, 163-172

Backer, J.M. & Weinstein, I.B. (1980) Mitochondrial DNA is a major cellular target for a dihydrodiol-epoxide derivative of benzo[a]pyrene. *Science*, 209, 297-299

Barrett, J.C., Oshimura, H., Tanaka, N. & Tsutsui, T. (1985) *Role of aneuploidy in early and late stages of neoplastic progression of Syrian hamster embryo cells in culture*. In: Dellarco, V.L., Woytek, P.E. & Hollaender, A., eds, *Aneuploidy: Etiology and Mechanisms*, New York, Plenum, pp. 523-538

Barry, E.G. (1972) Meiotic chromosome behaviour of an inverted insertional translocation in *Neurospora. Genetics*, 71, 53-62

Bellincampi, D., Gualandi, G., La Monica, E., Poley, C. & Morpurgo, G.P. (1980) Membrane-damaging agents cause mitotic non-disjunction in *Aspergillus nidulans. Mutat. Res.*, 79, 169-172

Benigni, R. (1986) The second international collaborative study on comparative mutagenesis revisited: analysis by multi-variate statistical methods. *Mutagenesis*, 1, 185-190

Benigni, R. & Giuliani, A. (1985) Cluster analysis of short-term tests: a new methodological approach. *Mutat. Res.*, 147, 139-151

Berenblum, I. (1975) *Sequential aspects of chemical carcinogenesis: skin*. In: Becker, F.F., ed., *Cancer: A Comprehensive Treatise*, Vol. 1, New York, Plenum, pp. 323-344

de Bertoldi, M., Griselli, M. & Barale, R. (1980) Different test assay in *Aspergillus nidulans* for the evaluation of mitotic gene conversion, crossing-over and non disjunction. *Mutat. Res.*, 74, 303-324

Bignami, M., Morpurgo, G., Pagliani, R., Carere, A., Conti, G. & Di Giuseppe, G. (1974) Non-disjunction and crossing-over induced by pharmaceutical drugs in *Aspergillus nidulans. Mutat. Res.*, 26, 159-170

Bignami, M., Aulicino, F., Vercich, A., Carere, A. & Morpurgo, G. (1977) Mutagenic and recombinogenic pesticides in *Aspergillus nidulans. Mutat. Res.*, 46, 395-402

Bignami, M., Conti, G., Crebelli, R. & Carere, A. (1981) Growth-mediated metabolic activation of promutagens in *Aspergillus nidulans. Mutat. Res.*, 80, 265-272

Boguslawski, G. (1985) Effects of polymyxin B sulfate and polymyxin B nonapeptide on growth and permeability of the yeast *Saccharomyces cerevisiae. Mol. gen. Genetics*, 199, 401-505

Bonatelli, R., Jr & Azevedo, J.L. (1977) Effect of ethidium bromide in diploid and duplication strains of *Aspergillus nidulans. Experientia*, 33, 311-312

Bond, D.J. (1976) A system for study of meiotic non-disjunction using *Sordaria brevicollis. Mutat. Res.*, 37, 213-220

Bond, D.J. & McMillan, L. (1979a) The effect of p-fluorophenylalanine on the frequency of aneuploid meiotic products in *Sordaria brevicollis. Mutat. Res.*, 60, 221-224

Bond, D.J. & McMillan, L. (1979b) Meiotic aneuploidy: its origins and induction following chemical treatment in *Sordaria brevicollis. Environ. Health Perspect.*, 31, 67-74 von Borstel, R.C., Shahin, M.M. & Mehta, R.D. (1981) *Protocol for a haploid yeast revision test for assaying mutagens.* In: Stich, F. & San, R.H.C., eds, *Short-term Tests for Chemical Carcinogens,* New York, Springer, pp. 171-174

von Borstel, R.C., O'Connell, D.F., Mehta, R.D. & Hennig, U.G.G. (1985) Modulation in cytochrome P-420 and P-450 content in *Saccharomyces cerevisiae* according to physiological conditions and genetic background. *Mutat. Res., 150,* 217-224

Brockman, H.E., de Serres, F.J., Ong, T.M., De Marini, D.M., Katz, A.J., Griffiths, A.J.F. & Stafford, R.S. (1984) Mutation tests in *Neurospora crassa, Mutat. Res., 113,* 87-134

Brusick, D.J. (1972) Induction of cycloheximide-resistants in *Saccharomyces cerevisiae* with N-methyl-N'-nitro-N-nitrosoguanidine and ICR-170. *J. Bacteriol., 109,* 1134-1138

Campbell, D.A., Fogel, S. & Lusnak, K. (1975) Mitotic chromosome loss in a disomic haploid of *Saccharomyces cerevisiae. Genetics, 79,* 383-396

Carere, A., Bellincampi, D., Conti, G., Conti, L., Crebelli, R., Gualandi, G. & Morpurgo, G. (1984) Genotoxic activity of selected chemical carcinogens in *A. nidulans. Mutat. Res., 147,* 287-288

Carere, A., Conti, G., Conti, L. & Crebelli, R. (1985) Assay in *Aspergillus nodulans* for the induction of forward-mutation in haploid strain 35 and for mitotic non-disjunction, haploidization and crossing-over in diploid strain P1. *Progr. Mutat. Res., 5,* 307-312

Cavenee, W.B., Koufos, A. & Hansen, M.F. (1986) Recessive mutant genes predisposing to human cancer. *Mutat. Res., 168,* 3-14

Clayton, D.A., Doda, J.N. & Friedberg, E.C. (1974) The absence of pyrimidine dimer repair mechanism in mammalian mitochondria. *Proc. natl Acad. Sci. USA, 71,* 2777

Clutterbuck, A.J. (1977) *The genetics of conidiation in* Aspergillus nidulans. In: Smith, J.E. & Paterman, J.A., eds, *Genetics and Physiology of Aspergillus,* London, Academic Press, pp. 305-317

Crebelli, R., Conti, G. & Carere, A. (1984) Induction of somatic segregation by halogenated aliphatic hydrocarbons in *Aspergillus nidulans. Mutat. Res., 138,* 33-38

Crebelli, R., Conti, G., Conti, L. & Carere, A. (1985) Mutagenicity of trichloroethylene, trichloroethanol and chloral hydrate in *Aspergillus nidulans. Mutat. Res., 155,* 105-111

Cutler, M.G. & Schneider, R. (1974) Tumours and hormonal changes produced in rats by subcutaneous injections of linoleic acid hydroperoxide. *Food Cosmet. Toxicol., 12,* 451-459

Dean, B.J., Brooks, T.M., Hodson-Walker, G. & Hutson, D.H. (1985) Genetic toxicology testing of 41 industrial chemicals. *Mutat. Res., 153,* 57-77

Demopoulos, N.A., Kappas, A. & Pelecanos, M. (1982) Recombinogenic and mutagenic effects of the antitumor antibiotic bleomycin in *Aspergillus nidulans. Mutat. Res., 102,* 51-57

Dixon, N. (1983) *A Yeast Screening System for the Detection of Mutation, Recombination and Aneuploidy (US Department of Energy Document LDL-16686),* Washington DC, US Department of Energy

Dunning, W.F. & Curtis, M.R. (1958) The role of indole in incidence of 2-acetylaminofluorene-induced bladder cancer in rats. *Proc. Soc. exp. Biol. Med., 99,* 91-95

Eckardt, F. & Haynes, R.H. (1977) Induction of pure and sectored mutant clones in excision proficient and deficient strain of yeast. *Mutat. Res., 43,* 327-338

Eckardt, F. & Haynes, R.H. (1980) Quantitative measures of mutagenicity and mutability based on mutation yield data. *Mutat. Res., 74*, 439-458

Ernst, J.F., Hampsey, D.M. & Sherman, F. (1985) DNA sequences of frameshift and other mutations induced by ICR-170 in yeast. *Genetics, 111*, 233-241

Esposito, M.S., Maleas, D.T., Bjornstad, K.A. & Bruschi, C.V. (1982) Simultaneous detection of change in chromosome number, gene conversion and intergenic recombination during meiosis of *Saccharomyces cerevisiae*: spontaneous and ultraviolet light induced events. *Curr. Genet., 6*, 5-11

Evans, H.J. (1985) *Neoplasia and cytogenetic abnormalities*. In: Dellarco, V.L., Woytek, P.E. & Hollander, A., eds, *Aneuploidy: Etiology and Mechanisms*, New York, Plenum, pp. 165-178

Fabre, F., Boulet, A. & Roman, H. (1984) Gene conversion at different points in the mitotic cycle of *Saccharomyces cerevisiae*. *Mol. gen. Genetics, 195*, 139-143

Fahrig, R. (1984) Genetic mode of action of cocarcinogens and tumor promoters in yeast and mice. *Mol. gen. Genetics, 194*, 7-14

Ferguson, L.R. & Parry, J.M. (1984) Mitotic aneuploidy as a possible mechanism for tumor promoting activity in bile acids. *Carcinogenesis, 5*, 447-451

Fratello, B., Morpurgo, G. & Sermonti, G. (1960) Induced somatic segregation in *Aspergillus nidulans*. *Genetics, 45*, 785-800

Fulton, A.M. & Bond, D.J. (1983) An investigation into the origins of meiotic aneuploidy using ascus analysis. *Genet. Res., 41*, 165-175

Georgopoulos, S.G., Kappas, A. & Hastie, A.C. (1975) Induced sectoring in diploid *Aspergillus nidulans* as a criterium of fungitoxicity by interference with hereditary processes. *Phytopathology, 66*, 217-220

Goyer, R.A., Falk, H.L., Hogan, M., Feldman, D.D. & Richter, W. (1981) Renal tumors in rats given trisodium nitrilotriacetic acid in drinking water for two years. *J. natl Cancer Inst., 66*, 869-880

Griesemer, R.A. & Cueto, C., Jr (1980) *Toward a classification scheme for degrees of experimental evidence for the carcinogenicity of chemicals for animals*. In: Montesano, R., Bartsch, H. & Tomatis, L., eds, *Molecular and Cellular Aspects of Carcinogen Screening Tests (IARC Scientific Publications No. 27)*, Lyon, International Agency for Research on Cancer, pp. 259-281

Griffiths, A.J.F. (1979) Neurospora prototroph selection system for studying aneuploid production. *Environ. Health Prospect., 31*, 75-80

Griffiths, A.J.F. (1980) *Progress Report (NIEHS 263-77-C0604 cc)*, Washington DC, National Institute of Environmental Health Science

Griffiths, A.J.F. (1981) *Neurospora and environmentally induced aneuploidy*. In: Stich, H.F. & San, R.H.C. eds, *Short-term Tests for Chemical Carcinogens*, New York, Springer, pp. 187-199

Griffiths, A.J.F. (1982) *Short-term tests for chemicals that promote aneuploidy*. In: Hollaender, A. & de Serres, F.J., eds, *Chemical Mutagens: Principles and Methods for their Detection*, Vol. 7, New York, Plenum, pp. 189-210

Griffiths, A.J.F. & Delange, A.M. (1977) p-Fluorophenylalanine increases meiotic non-disjunction in a *Neurospora* system. *Mutat. Res., 46*, 345-354

Griffiths, A.J.F., Brockman, H.E., DeMarini, D.M. & de Serres, F. (1986) The efficacy of *Neurospora* in detecting agents that cause aneuploidy. *Mutat. Res., 167*, 35-45

Gualandi, G. (1984) Genotoxicity of the free-radical producers CCl_4 and lipoperoxide in *Aspergillus nidulans*. *Mutat. Res., 136*, 109-114

Gualandi, G. & Bellincampi, D. (1981) Induced gene mutation and mitotic non-disjunction in *Aspergillus nidulans*. *Toxicol. Lett.*, 9, 389-394

Gualandi, G. & Morpurgo, G. (1984) *Methods for detecting the induction of mitotic chromosomal mis-distribution in* Aspergillus nidulans. In: Kilbey, B., Legator, M., Nichols, W. & Ramel, C., eds, *Handbook of Mutagenicity Test Procedures*, 2nd ed., Amsterdam, Elsevier, pp. 707-720

Gualandi, G., Bellincampi, D. & Puppo, S. (1979) MMS induction of different types of genetic damage in *Aspergillus nidulans*: a comparative analysis in mutagenesis. *Mutat. Res.*, 62, 255-266

Hannan, M.A., Zimmer, S.G. & Hazle, J. (1984) Mechanisms of cisplatin (cis-diamminodichloroplatinum II)-induced cytotoxicity and genotoxicity in yeast. *Mutat. Res.*, 127, 23-30

Harsanyi, Z., Granek, I.A. & Mackenzie, D.W.R. (1977) Genetic damage induced by ethyl alcohol in *Aspergillus nidulans*. *Mutat. Res.*, 48, 51-74

Haseman, J.K., Crawford, D.D., Huff, J.E., Boorman, G.A. & McConnell, E.E. (1984) Results from 86 two-year carcinogenicity studies conducted by the National Toxicology Program. *J. Toxicol. environ. Health*, 14, 621-639

Hastie, A.C. (1970) Benlate-induced instability of *Aspergillus* diploids. *Nature*, 226, 771

Hastie, A.C. & Georgopoulos, S.C. (1971) Mutational resistance to fungitoxic benzimidazole derivatives in *Aspergillus nidulans*. *J. gen. Microbiol.*, 67, 371-373

Haynes, R.H. & Kunz, B.A. (1981) *DNA repair and mutagenesis in yeast*. In: Strathern, J.N., Jones, E.W. & Broach, J.R., eds, *The Molecular Biology of the Yeast* Saccharomyces cerevisiae, Cold Spring Harbor, NY, CSH Press, pp. 317-414

IARC (1974a) *IARC Monographs on the Evaluation of Carcinogenic Risk of Chemicals to Man*, Vol. 7, *Some Anti-thyroid and Related Substances, Nitrofurans and Industrial Chemicals*, Lyon, pp. 253-260, 245-251, 111-140

IARC (1974b) *IARC Monographs on the Evaluation of Carcinogenic Risk of Chemicals to Man*, Vol. 4, *Some Aromatic Amines, Hydrazine and Related Substances, N-Nitroso Compounds and Miscellaneous Alkylating Agents*, Lyon, pp. 173-179, 183-195, 113-117

IARC (1975a) *IARC Monographs on the Evaluation of Carcinogenic Risk of Chemicals to Man*, Vol. 9, *Some Aziridines, N-, S- and O-Mustards and Selenium*, Lyon, pp. 193-207

IARC (1976a) *IARC Monographs on the Evaluation of Carcinogenic Risk of Chemicals to Man*, Vol. 12, *Some Carbamates, Thiocarbamates and Carbazides*, Lyon, pp. 55-68, 225-236, 245-257

IARC (1976b) *IARC Monographs on the Evaluation of Carcinogenic Risk of Chemicals to Man*, Vol. 11, *Cadmium, Nickel Some Epoxides, Miscellaneous Industrial Chemicals and General Considerations on Volatile Anaesthetics*, Lyon, pp. 115-123

IARC (1976c) *IARC Monographs on the Evaluation of Carcinogenic Risk of Chemicals to Man*, Vol. 10, *Some Naturally Occurring Substances*, Lyon, pp. 153-161, 231-244

IARC (1977a) *IARC Monographs on the Evaluation of Carcinogenic Risk of Chemicals to Man*, Vol. 13, *Some Miscellaneous Pharmaceutical Substances*, Lyon, pp. 91-100, 157-181, 113-122

IARC (1977b) *IARC Monographs on the Evaluation of the Carcinogenic Risk of Chemicals to Man*, Vol. 15, *Some Fumigants, the Herbicides 2,4-D and 2,4,5-T, Chlorinated Dibenzodioxins and Miscellaneous Industrial Chemicals*, Lyon, pp. 155-175, 211-222

IARC (1978) *IARC Monographs on the Evaluation of the Carcinogenic Risk of Chemicals to Man*, Vol. 16, *Some Aromatic Amines and Related Nitro Compounds — Hair Dyes, Colouring Agents and Miscellaneous Industrial Chemicals*, Lyon, pp. 145-152, 83-95

IARC (1979a) *IARC Monographs on the Evaluation of the Carcinogenic Risk of Chemicals to Humans*, Vol. 19, *Some Monomers, Plastics and Synthetic Elastomers, and Acrolein*, Lyon, pp. 73-113

IARC (1979b) *IARC Monographs on the Evaluation of the Carcinogenic Risk of Chemicals to Humans*, Vol. 20, *Some Halogenated Hydrocarbons*, Lyon, pp. 371-399, 429-448, 97-127, 401-427

IARC (1979c) *IARC Monographs on the Evaluation of the Carcinogenic Risk of Chemicals to Humans*, Vol. 21, *Sex Hormones (II)*, Lyon, pp. 173-231

IARC (1980) *IARC Monographs on the Evaluation of the Carcinogenic Risk of Chemicals to Humans*, Suppl. 2, *Long-term and Short-term Screening Assays for Carcinogens: A Critical Appraisal*, Lyon, pp. 135-155

IARC (1981) *IARC Monographs on the Evaluation of the Carcinogenic Risk of Chemicals to Humans*, Vol. 26, *Some Antineoplastic and Immunosuppressive Agents*, Lyon, pp. 165-202, 97-113

IARC (1982a) *IARC Monographs on the Evaluation of the Carcinogenic Risk of Chemicals to Humans*, Suppl. 4, *Chemicals, Industrial Processes and Industries Associated with Cancer in Humans (IARC Monographs, Volumes 1 to 29)*, Lyon

IARC (1982b) *IARC Monographs on the Evaluation of the Carcinogenic Risk of Chemicals to Humans*, Vol. 29, *Some Industrial Chemicals and Dyestuffs*, Lyon, pp. 345-389, 93-148, 269-294

IARC (1982c) *IARC Monographs on the Evaluation of the Carcinogenic Risk of Chemicals to Humans*, Vol. 27, *Some Aromatic Amines, Anthraquinones and Nitroso Compounds, and Inorganic Fluorides Used in Drinking-water and Dental Preparations*, Lyon, pp. 155-175

IARC (1983a) *IARC Monographs on the Evaluation of the Carcinogenic Risk of Chemicals to Humans*, Vol. 30, *Miscellaneous Pesticides*, Lyon, pp. 295-318, 283-291

IARC (1983b) *IARC Monographs on the Evaluation of the Carcinogenic Risk of Chemicals to Humans*, Vol. 31, *Some Food Additives, Feed Additives and Naturally Occurring Substances*, Lyon, pp. 213-229

IARC (1984) *IARC Monographs on the Evaluation of the Carcinogenic Risk of Chemicals to Humans*, Vol. 32, *Polynuclear Aromatic Compounds, Part 1, Chemical, Environmental and Experimental Data*, Lyon, pp. 211-224

Käfer, E. (1961) The processes of spontaneous recombination in vegetative nuclei of *Aspergillus nidulans*. *Genetics, 46*, 1581-1609

Käfer, E. (1984) Disruptive effects of ethyl alcohol on mitotic chromosome segregation in diploid and haploid strains of *Aspergillus nidulans*. *Mutat. Res., 135*, 53-75

Käfer, E. (1985) Aspergillus *tests which distinguish induced primary aneuploidy from secondary non-disjunction*. In: *Fourth International Conference on Environmental Mutagens, Stockholm, 24-28 June 1985, Abstract Book*, p. 38

Käfer, E. & Upshall, A. (1973) The phenotypes of the eight disomics and trisomics of *Aspergillus nidulans*. *J. Hered., 64*, 35-38

Käfer, E., Marshall, P. & Cohen, G. (1976) Well-marked strains of *Aspergillus* for tests of environmental mutagens: identification of induced mitotic recombination and mutation. *Mutat. Res., 38*, 141-146

Käfer, E., Scott, B.R., Dorn, G.L. & Stafford, R.S. (1982) *Aspergillus nidulans*: systems and results of tests for chemical induction of mitotic segregation and mutation. I. Diploid and duplication assay systems. A report of the US EPA Gene Tox Program. *Mutat. Res., 98*, 1-48

Käfer, E., Scott, B.R. & Kappas, A. (1986) Systems and results of tests for chemical induction of mitotic malsegregation and aneuploidy in *Aspergillus nidulans. Mutat. Res., 167*, 9-34

Kappas, A. (1978) On the mechanism of induced somatic recombination by certain fungicides in *Aspergillus nidulans. Mutat. Res., 51*, 189-197

Kappas, A. (1983) Genotoxic activity of plant growth-regulating hormones in *Aspergillus nidulans. Carcinogenesis, 4*, 1409-1411

Kappas, A. & Georgopoulos, S.G. (1974) Interference of griseofulvin with the segregation of chromosomes at mitosis in diploid *Aspergillus nidulans. J. Bacteriol., 119*, 334-335

Kappas, A., Georgopoulos, G. & Hastie, A.C. (1974) On the genetic activity of benzimidazole and thiophanate fungicides on diploid *Aspergillus nidulans. Mutat. Res., 26*, 17-27

Kelly, D. & Parry, J.M. (1983) Metabolic activation of cytochrome P-450/P-448 in the yeast *Saccharomyces cerevisiae. Mutat. Res., 108*, 147-159

Kissin, B. & Kaley, M.M. (1974) *Alcohol and cancer.* In: Kissin, B. & Begleiter, H., eds, *The Biology of Alcoholism*, Vol. 3, *Clinical Pathology*, New York, Plenum, pp. 481-511

Klein, G. & Klein, E. (1984) Oncogene activation and tumor progression. *Carcinogenesis, 5*, 429-435

Kunz, B.A., Hannan, M.A. & Haynes, R.H. (1980) Effect of tumor promoters on ultraviolet-light-induced mutation and mitotic recombination in *Saccharomyces cerevisiae. Cancer Res., 40*, 2323-2329

Kunz, B.A., Eckardt, F. & Haynes, R.H. (1985) Analysis of non-linearities in frequency curves for UV-induced mitotic recombination in wild type and excision-repair-deficient strains of yeast. *Mutat. Res., 151*, 235-242

Lawrence, C.W. (1982) Mutagenesis in *Saccharomyces cerevisiae. Adv. Genet., 21*, 173-274

Lemontt, J.F. (1980) *Genetic and physiological factors affecting repair and mutagenesis in yeast.* In: Generoso, W.M., Shelby, M.D. & de Serres, F.J., eds, *DNA Repair and Mutagenesis in Eucaryotes*, New York, Plenum, pp. 85-120

Loprieno, N., Barale, R., von Halle, E.S. & von Borstel, R.C. (1983) Testing of chemicals for mutagenic activity with *Schizosaccharomyces pombe. Mutat. Res., 115*, 215-223

Magana-Schwencke, J.A.P., Henriques, P., Chanet, R. & Moustacchi, E. (1982) The fate of 8-methoxypsoralen photoinduced crosslinks in nuclear and mitochondrial yeast DNA: comparison of wild-type and repair-deficient strains. *Proc. natl Acad. Sci. USA, 79*, 1722-1726

Mannironi, C., Cundari, E., Bauer, C., Del Carratore, R., Bronzetti, G., Corsi, C., Nieri, R. & Paolino, M. (1985) Study of the optimal temperature of the liver microsomal assay with mice S9 fractions. *Mutat. Res., 147*, 231-235

Mehta, R.D. & von Borstel, R.C. (1981) *Mutagenic activity of 42 encoded compounds in the haploid yeast reversion assay, strain XV185-14C.* In: de Serres, F.J. & Ashby, J., eds, *Evaluation of Short-term Tests for Carcinogens*, New York, Elsevier, pp. 414-423

Morpurgo, G. (1961) Somatic segregation induced by p-fluorophenylalanine. *Aspergillus Newsl., 2*, 10

Morpurgo, G. (1962) Quantitative measurements of induced somatic segregation in *Aspergillus nidulans. Sci. Rep. Ist. Sup. Sanità, 2*, 234

Morpurgo, G., Aulicino, F., Bignami, M., Conti, L. & Velcich, A. (1977) Relationship between structure and mutagenicity of dichlorvos and other pesticides. *Accad. Naz. Lincei, 63*, 693-701

Morpurgo, G., Bellincampi, D., Gualandi, G., Baldinelli, L. & Serlupi Crescenzi, O. (1979) Analysis of mitotic non-disjunction with *Aspergillus nidulans. Environ. Health Perspect., 31*, 81-95

Mortimer, R.K. & Schild, D. (1980) The genetic map of *Saccharomyces cerevisiae*. *Microbiol. Rev.*, 44, 519-571

Mortimer, R.K. & Schild, D. (1985) Genetic map of *Saccharomyces cerevisiae*. Edition 9. *Microbiol. Rev.*, 49, 181-212

Moustacchi, E. (1986) *DNA repair in yeast: control and biological consequences*. In: Lett, J., ed., *Advances in Radiation Research*, New York, Academic Press (in press)

Moustacchi, E. & Favaudon, V. (1982) Cytotoxic and mutagenic effects of neocarzinostatin in wild type and repair deficient yeasts. *Mutat. Res.*, 104, 87-94

Moustacchi, E., Favaudon, V. & Bisagni, E. (1983) Likelihood of the new antitumor drug 10-(sigma-diethylaminopropyl(BD-40)) a pyridopyrroloisoquinoline derivative, to induce DNA strand breaks *in vivo* and its non-mutagenicity in yeast. *Cancer Res.*, 43, 3700-3706

Nanni, N., Bauer, C., Cundari, E., Corsi, C., Del Carratore, R., Nieri, R., Paulini, M., Crewshaw, J. & Bronzetti, G. (1984) Studies of genetic effects in the D7 strain of *Saccharomyces cerevisiae* under different conditions of pH. *Mutat. Res.*, 139, 189-192

National Cancer Institute (1977) *Bioassays of Nitrilotriacetic Acid (NTA) and Nitrilotriacetic Acid, Trisodium Salt, Monohydrate ($Na_3NTA.H_2O$) for Possible Carcinogenicity (NCI Tech. Rep. Ser. No.6; DHEW Publ. No.(NIH)77-806)*, Bethesda, MD

National Cancer Institute (1978) *Bioassay of Picloram for Possible Carcinogenicity (NCI Tech. Rep. Ser. No.23; DHEW Publ. No.(NIH)78-823)*, Bethesda, MD

National Cancer Institute (1979a) *Bioassay of p-Phenylenediamine Dihydrochloride for Possible Carcinogenicity (NCI Tech. Rep. Ser. No.174; DHEW Publ. No.(NIH)79-1730)*, Bethesda, MD

National Cancer Institute (1979b) *Bioassay of 4-Nitro-o-phenylenediamine for Possible Carcinogenicity (NCI Tech. Rep. Ser. No.180; DHEW Publ. No. (NIH)1736)*, Bethesda, MD

National Cancer Institute (1983) *Survey of Compounds which Have Been Tested for Carcinogenic Activity (PHS Publication No. 149), Literature for the Years 1974-1975 on 575 Compounds*, Washington DC, US Government Printing Office

National Toxicology Program (1982a) *Carcinogenesis Bioassay of Stannous Chloride in F344/N Rats and B6C3F_1/N Mice (Feed Study) (NTP Tech. Rep. Ser. No. 231; NIH Publ. No.82-1787)*, Research Triangle Park, NC

National Toxicology Program (1982b) *Carcinogenesis Bioassay of Trichloroethylene in F344/N Rats and B6C3F_1/N Mice (Gavage Study) (NTP Tech. Rep. Ser. No.243; NIH Publ. No.82-1799)*, Research Triangle Park, NC

National Toxicology Program (1985a) *Toxicology and Carcinogenesis Studies of 8-Hydroxyquinoline in F344/N Rats and B6C3F_1 Mice (Feed Studies) (NTP Tech. Rep. Ser. No.276; NIH Publ. No.85-2532)*, Research Triangle Park, NC

National Toxicology Program (1985b) *Toxicology and Carcinogenesis Studies of 2-Chloroethanol (Ethylene Chlorohydrin) in F344/N Rats and Swiss CD-1 Mice (Dermal Studies) (NTP Tech. Rep. Ser. No.275; NIH Publ. No.86-2531)*, Research Triangle Park, NC

National Toxicology Program (1985c) *Toxicology and Carcinogenesis Studies of 1,2-Dichloropropane in F344/N Rats and B6C3F_1 Mice (Gavage Study) (NTP Tech. Rep. Ser. No.263; NIH Publ. No.85-2523)*, Research Triangle Park, NC

Nestmann, E.R. & Lee, E.G.H. (1985) Genetic activity in *Saccharomyces cerevisiae* of compounds found in effluents of pulp and paper mills. *Mutat. Res.*, 155, 53-60

Niranjan, B.G., Bhat, N.K. & Avadhani, N.G. (1982) Preferential attack of mitochondrial DNA by aflatoxin B_1 during hepatocarcinogenesis. *Science*, *215*, 73-75

Onho, S. (1974) *Aneuploidy as a possible means employed by malignant cells to express recessive phenotypes*. In: German, J., ed., *Chromosomes and Cancer*, New York, John Wiley & Sons, pp. 77-94

Parry, J.M. (1977) *The detection of chromosome non-disjunction in the yeast* Saccharomyces cerevisiae. In: Scott, D., Bridges, B.A. & Sobels, F.H., eds, *Progress in Genetic Toxicology*, Amsterdam, Elsevier, pp. 223-229

Parry, J.M. (1982) Assay of the induction of mitotic crossing-over and aneuploidy in yeast by BBC, 4-CMB and 4-HMB. *Mutat. Res.*, *100*, 139-143

Parry, J.M. & Parry, E.M. (1984) *The detection of induced chromosome aneuploidy using strains of the yeast* Saccharomyces cerevisiae. In: Kilbey, B.J., Legator, M., Nichols, W. & Ramel, C., eds, *Handbook of Mutagenicity Test Procedures*, Amsterdam, Elsevier, pp. 689-706

Parry, J.M. & Sharp, D.C. (1981) Induction of mitotic aneuploidy in the yeast strain D6 by 42 coded compounds. *Progr. Mutat. Res.*, *1*, 468-480

Parry, J.M. & Zimmermann, F.K. (1976) The detection of monosomic colonies produced by mitotic chromosome non-disjunction in the yeast *Saccharomyces cerevisiae*. *Mutat. Res.*, *36*, 49-66

Parry, J.M., Sharp, D. & Parry, E.M. (1979a) Detection of mitotic and meiotic aneuploidy in the yeast *Saccharomyces cerevisiae*. *Environ. Health Perspect.*, *31*, 97-111

Parry, J.M., Sharp, D., Tippins, R.S. & Parry, E.M. (1979b) Radiation induced mitotic and meiotic aneuploidy in the yeast *Saccharomyces cerevisiae*. *Mutat. Res.*, *61*, 37-55

Parry, J.M., Parry, E.M. & Barrett, J.C. (1981) Tumor promoters induce mitotic aneuploidy in yeast. *Nature*, *294*, 263-265

Parry, J.M., Arni, P., Brooks, T., Carere, A., Ferguson, L., Heinisch, J., Inge Vechtomov, S., Loprieno, N., Nestmann, E. & von Borstel, R. (1985) Summary report on the performance of the yeast and *Aspergillus* assay. *Progr. Mutat. Res. 5*, 25-46

Perkins, D.D., Radford, A., Newmayer, B. & Bjorkman, M. (1982) Chromosomal loci of *Neurospora crassa*. *Microbiol. Rev.*, *46*, 426-570

Pesonen, S. (1950) Tumors resulting from parthenogenesis induced by plant hormones in albino rats. *Acta endocrinol.*, *5*, 409-412

Pettinger, T.H. (1954) The general incidence of pseudowild types in *Neurospora crassa*. *Genetics*, *39*, 326-342

Pontecorvo, G. (1956) The parasexual cycle. *Ann. Rev. Microbiol.*, *10*, 393-400

Putrament, A. (1967) *Diepoxybutane induced mitotic recombination in* Aspergillus nidulans. In: *Proceedings of the Symposium on Mutational Process*, Prague, Czechoslovak Academy of Sciences, pp. 107-114

Radman, M., Jeggo, P. & Wagner, R. (1982) Chromosomal rearrangements and carcinogenesis. *Mutat. Res.*, *98*, 249-264

Resnick, M.A., Mayer, V.W. & Zimmermann, F.K. (1986) The detection of chemically induced aneuploidy in *Saccharomyces cerevisiae*: an assessment of mitotic and meiotic systems. *Mutat. Res.*, *167*, 47-60

Roman, H. (1956) A system selective for mutations affecting the synthesis of adenine in yeast. *C.R. Trav. Lab. Carlsberg Ser. Physiol.*, *26*, 299-314

Rosin, M.P. (1984) The influence of pH on the convertogenic activity of plant phenolics. *Mutat. Res.*, *135*, 109-113

Scott, B.R. & Käfer, E. (1982) Aspergillus nidulans — *an organism for detecting a range of genetic damage*. In: de Serres, F.J. & Hollaender, A., eds, *Chemical Mutagens, Principles and Methods for their Detection*, Vol. 7, New York, Plenum, pp. 447-479

Scott, B.R., Dorn, G.L., Käfer, E. & Stafford, R.S. (1982) *Aspergillus nidulans*: systems and results of tests for chemical induction of mitotic segregation and mutation. II. Haploid assay systems and overall response of all systems. *Mutat. Res.*, 98, 49-94

de Serres, F.J. & Ashby, J., eds (1981) Evaluation of Short-term Tests for Carcinogens, Reports of the International Program. *Prog. Mutat. Res.*, 1

de Serres, F.J. & Hoffmann, G.R. (1981) Summary report on the performance of yeast assays. *Prog. Mutat. Res.*, 1, 68-76

Shanfield, F. & Käfer, E. (1971) Chemical induction of mitotic recombination in *Aspergillus nidulans*. *Genetics*, 67, 209-219

Shermann, F. & Lawrence, C.W. (1974) Saccharomyces. In: King, R.C. ed., *Handbook of Genetics*, Vol. 1, New York, Plenum, pp. 359-393

Siede, W. & Eckardt, F. (1984) Inducibility of error-prone DNA repair in yeast? *Mutat. Res.*, 129, 3-11

Singh, M. & Sinha, U. (1976) Chloral hydrate induced haploidization in *Aspergillus nidulans*. *Experientia*, 15, 1144-1145

Slaga, T.J., Fischer, S.M., Nelson, K. & Gleason, G.L. (1980) Studies on the mechanism of skin tumor promotion: evidence for several stages in promotion. *Proc. natl Acad. Sci. USA*, 77, 3659-3663

Sora, S. & Bianchi, M. (1982) Propanol, atenol and trifluoroperazine reduce the spontaneous occurrence of meiotic diploid products in *Saccharomyces cerevisiae*. *Mol. cell. Biol.*, 2, 1299-1303

Sora, S., Lucchini, G. & Magni, G.E. (1982) Meiotic diploid progeny and meiotic non disjunction in *Saccharomyces cerevisiae*. *Genetics*, 101, 17-33

Sora, S., Crippa, M. & Lucchini, G. (1983) Disomic and meiotic products in *Saccharomyces cerevisiae*. Effect of vincristine, vinblastine, adriamycin, bleomycin, mitomycin C and cyclophosphamide. *Mutat. Res.*, 107, 249-264

Strömnaes, O. (1968) Genetic changes in *Saccharomyces cerevisiae* grown on media containing D,L-p-fluorophenylalanine. *Hereditas*, 59, 197-220

Threlkeld, S.F.H. & Stolz, J.M. (1970) A genetic analysis of non-disjunction and mitotic recombination in *Neurospora crassa*. *Genet. Res.*, 16, 29-35

Upshall, A. & Johnson, P.E. (1981) Thiram-induced abnormal chromosome segregation in *Aspergillus nidulans*. *Mutat. Res.*, 89, 297-301

Waters, R. & Moustacchi, E. (1974) The fate of ultraviolet-induced pyrimidine dimers in the mitochondrial DNA of *Saccharomyces cerevisiae* following various postirradiation cell treatments. *Biochim. biophys. Acta*, 366, 241-253

Weisburger, E.K., Rusfield, A.B., Hambuerger, F., Weisburger, J.H., Borger, E., van Dongen, C.G. & Chu, K.C. (1978) Testing of twenty-one environmental aromatic amines or derivatives for long term toxicity or carcinogenicity. *J. environ. Pathol. Toxicol.*, 2, 325-356

Whong, W.Z., Ong, T.M. & Brockman, H.E. (1985) Effect of pH on the mutagenic and killing potencies of ICR-170 in *ade 3* tests of *Neurospora crassa*. *Mutat. Res.*, 142, 19-22

Yunis, J.J. (1983) The chromosomal basis of human neoplasia. *Science*, 221, 227-236

Zimmermann, F.K. (1975) Procedures used in the induction of mitotic recombination and mutation in the yeast *Saccharomyces cerevisiae*. *Mutat. Res.*, 31, 71-86

Zimmermann, F.K. & Scheel, I. (1984) Genetic effects of 5-azacytidine in *Saccharomyces cerevisiae*. *Mutat. Res.*, *139*, 21-24

Zimmermann, F.K., Kern, R. & Rasenberger, H. (1975) A yeast strain for simultaneous detection of induced crossing over, mitotic gene conversion, and reverse mutation. *Mutat. Res.*, *28*, 381-388

Zimmermann, F.K., Mayer, V.W. & Parry, J.M. (1982) Genetic toxicology studies using *Saccharomyces cerevisiae*. *J. appl. Toxicol.*, *2*, 1-10

Zimmermann, F.K., von Borstel, R.C., von Halle, E.S., Parry, J.M., Siebert, D., Zetterberg, G., Barale, R. & Loprieno, N. (1984a) Testing of chemicals for genetic activity with *Saccharomyces cerevisiae*: a report of the US Environmental Protection Agency Gene-Tox Program. *Mutat. Res.*, *133*, 199-244

Zimmermann, F.K., Mayer, V.W. & Scheel, I. (1984b) Induction of aneuploidy by oncodazole (nocodazole), an antitubulin agent, and acetone. *Mutat. Res.*, *141*, 15-18

Zimmermann, F.K., Mayer, V.W., Scheel, I. & Resnick, M.A. (1985a) Acetone, methyl ethyl ketone, ethyl acetate, acetonitrile and other polar aprotic solvents are strong inducers of aneuploidy in *Saccharomyces cerevisiae*. *Mutat. Res.*, *149*, 339-351

Zimmermann, F.K., Groschel-Stewart, U., Scheel, I. & Resnick, M.A. (1985b) Genetic change may be caused by interference with protein-protein interactions. *Mutat. Res.*, *150*, 203-210

REPORT 12

ASSAYS FOR GENETIC ACTIVITY IN *DROSOPHILA MELANOGASTER*

Prepared by:

F.E. Würgler (Rapporteur), C. Ramel (Chairman),
E. Moustacchi & A. Carere

1. Introduction

Drosophila melanogaster is a eukaryotic organism in which induced mutations can be detected in germ cells as well as in somatic cells *in vivo*. The short generation time of ten days, inexpensive culture media, and a large number of well-defined genetic assays are the principal advantages of using *D. melanogaster* for studying genotoxic effects of chemicals. Multipurpose stocks allow the detection in gonadal and/or somatic tissues of intragenic mutations, deletions, duplications, translocations, inversions, chromosome loss, chromosome gain and mitotic recombination. For some assays, standardized test protocols are available: the sex-linked recessive lethal mutation assay (Lee *et al.*, 1983; Würgler *et al.*, 1984; Organization for Economic Cooperation and Development, 1984), and the test for heritable translocations (Valencia *et al.*, 1984; Würgler *et al.*, 1984). For other assays, different protocols and test strains are used, including tests for chromosome loss and nondisjunction (Valencia *et al.*, 1984; Würgler *et al.*, 1984; Zimmering *et al.*, 1986) and the somatic mutation and recombination tests ('SMART') (Würgler & Vogel, 1986)

Within the past five years, the standard assays, in particular the tests for sex-linked recessive lethal mutations and for reciprocal translocations, have been validated further and used in large-scale testing programmes such as the International Programme on Chemical Safety (IPCS) collaborative study on in-vitro assays (Ashby *et al.*, 1985) and the collaborative study on in-vivo assays (Ashby *et al.*, 1986) and the US National Toxicology Program (Zeiger, 1983). Two Gene-Tox reports contain critical evaluations of all data published up to about 1980 for the standard *Drosophila* tests (Lee *et al.*, 1983; Valencia *et al.*, 1984). In addition, a number of new assays have been developed, and for some, in particular those using somatic cells, extensive validation studies have been initiated (Fujikawa *et al.*, Graf *et al.*, Kondo *et al.*, Vogel *et al.*, unpublished data). The present status of the assays using somatic cells of *Drosophila* has recently been reviewed (Würgler & Vogel, 1986). As exemplified by the use of *Drosophila* in the recent IPCS study, a conceptual change in the strategy of mutagenicity testing with this organism is occurring in that larvae rather than adult flies are exposed to test agents.

In addition, background information on *Drosophila* has improved substantially in the fields of xenobiotic metabolism, DNA repair (Boyd *et al.*, 1983) and the analysis of the molecular nature of induced genetic variations. Research on the potential role of mobile elements in spontaneous and induced mutagenesis has yielded interesting results.

Of critical importance is the presence in larvae and adults of versatile enzyme activities which can metabolize xenobiotics. This allows the detection of procarcinogens which require activation to mutagenic metabolites (Baars, 1980; Vogel, 1982). Recently, the genetic control of a number of cytochrome P450-dependent activities has been determined (Hällström, 1986a). In the near future, this will allow improvements in the metabolic performance of certain tester strains.

Several lines of evidence on the metabolism of xenobiotics in *Drosophila* lead to the conclusion that the results obtained in mutagenicity assay systems using this organism are likely to be highly relevant to the prediction of adverse genetic effects in mammals. However, it should not be concluded that data obtained in insects can be extrapolated quantitatively to mammals, including humans.

2. **Principles and scientific basis of** *Drosophila* **assays**

2.1 *General principles*

Routine screening assays in germ cells and somatic cells are based on the segregation of visible genetic markers and scoring of particular phenotypes. These techniques give only limited information on the changes at the DNA and chromosome level. Cytogenetic studies of chromosomal aberrations (Slizinska, 1969) and sister chromatid exchanges are possible (Gatti *et al.*, 1980), although they are not used routinely.

2.2 *Genetic endpoints*

(*a*) *Sex-linked recessive lethals*

The kinds of genetic damage that may be expressed as recessive lethals include intragenic changes (point mutations) and small and large rearrangements (translocations, inversions, duplications and multilocus deficiencies (Auerbach, 1962)). Among sex-linked recessive lethals induced by ethyl methanesulfonate (EMS), methyl methanesulfonate (MMS) and ICR-170, the proportion of gross structural aberrations is probably below 10% (Fahmy & Fahmy, 1961; Carlson *et al.*, 1967; Lifschytz & Falk, 1969; Lim & Snyder, 1974; Liu & Lim, 1975; Lefevre, 1981a), but is considerably higher for X-rays (30-40%; Lefevre, 1981b), for the bifunctional alkylating agent diepoxybutane and for the trifunctional triethylene-melamine (Fahmy & Bird, 1952; Lim & Snyder, 1974; Olsen & Green, 1982).

Temperature-sensitive (*ts*) recessive lethals, which are assumed to represent mainly base-pair substitutions of the missense type (Lee, 1976; Suzuki *et al.*, 1976; Kramers, 1977), have been recovered after exposure to EMS, mitomycin C and gamma-rays (Suzuki *et al.*, 1967; Kaufman & Suzuki, 1974) and to the presumed frame-shift mutagens ICR-170 and hycanthone (Carlson *et al.*, 1967; Woodruff & Gander, 1974; Kramers, 1977). The highest proportion of *ts* mutations, about 10%, was found among EMS-induced recessive lethals. Additional evidence for the presence of base-pair substitutions is provided by the fact that out of 16 EMS-induced inactivating mutations at the *adh* (alcohol dehydrogenase) locus, 11 still produced identifiable protein (Schwartz & Sofer, 1976). Shukla and Auerbach (1980), however, claim that the majority of EMS-induced visibles are deletions.

Of critical importance is the induction of deletions, since they are, as nonrevertable mutations, of great relevance with respect to human genetic risk (Shukla & Auerbach, 1980). A compound studied intensively by complementation mapping in this respect is hycanthone methanesulfonate. Multilocus deletions occurred at a frequency of 54% among lethals in the proximal region of the X chromosome, and at a rate of 7% among lethals in other regions. This result is in agreement with the data on *Neurospora crassa* (Ong & de Serres, 1980). The overall fraction of deletions is, however, not large enough to account for the deficit in base-pair substitutions, which can be inferred from the large difference in *ts* mutation induction between hycanthone methanesulfonate and EMS. Therefore, it is reasonable to assume that a sizeable portion of hycanthone methanesulfonate-induced lethals is of the frame-shift type, although the possibility of small intralocus deficiencies arising by a different mechanism cannot be excluded.

Reardon *et al.* (1986) have analysed diepoxybutane-induced mutants determined to have lesions affecting expression of the *rosy* (*ry*) locus. Of the 21 mutants analysed genetically, five were putative deficiencies involving *ry* and adjacent lethal loci. However, molecular analysis confirmed that only two of these five putative deficiencies were in fact deletions detectable by the methods used. The remaining 16 were viable as homozygotes, suggesting that their lesions were confined to the *ry* locus. Seven of these 16 intragenic mutants were determined to be deletions of genetic material, as evidenced by altered restriction patterns relative to the wild-type patterns. Thus, nine of 21 (43%) diepoxybutane-induced mutants are due to deletions ranging in size from approximately 50 base pairs to more than eight kilobase pairs. Most of the deletions (7/9) are intragenic and less than 150 base pairs in size; it seems that most, if not all, affect coding rather than regulatory sequences.

Four formaldehyde-generated alcohol dehydrogenase-negative mutants have been cloned and sequenced (Benyajati *et al.*, 1983). All four mutants bear small deletions within the gene, ranging in size from six to 34 base pairs. Two of the deletions lie within a 65-base-pair intervening sequence and are accompanied by other aberrations. The other two are within the protein coding region of the gene. Some of the aberrations may be explained by a slipping mispairing mechanism. Among groups of mutants at the *scute* and *white* loci, this molecular approach allowed the recognition of intralocus insertions as well as deletions, which were mostly of spontaneous origin and induced by X-rays, respectively (Carramolino *et al.*, 1982; Zachar & Bingham, 1982).

(b) Chromosomal aberrations

Heritable translocations have been analysed cytologically only very occasionally (Vogel & Natarajan, 1979a).

In crosses that detect complete and partial chromosome losses, the information with respect to the actual chromosomal configuration is less precise than with heritable T2:3 translocations. As an alternative to direct cytological analyses (Traut & Scheid, 1970), it is possible to test defective Y chromosomes with a loss of one visible marker for the presence or absence of other marker genes. Zimmering and Kammermeyer (1982) showed that in males treated with *N*-nitrosodiethylamine and procarbazine and crossed to *mus302* females, 18/19 y^+ marker losses from the former and 50/50 from the latter are associated with a simultaneous loss of one (or more) adjacent loci, suggesting deletion and consequent chromosomal breakage.

(c) Mobile elements

More than 5% of the total genome of *Drosophila* consists of transposable elements (Spradling & Rubin, 1981, 1982).

When males from a strain of *Drosophila* that has multiple copies of the *P* family of transposable elements integrated into its genome (Bingham *et al.*, 1982) ('*P*' strain) are crossed to females of strains lacking those elements ('*M*' strains), the F_1 progeny manifest a number of aberrant traits, known collectively as hybrid dysgenesis (Kidwell *et al.*, 1977). Progeny of the reciprocal cross (*P* females × *M* males) are normal. The dysgenic syndrome is characterized by a high degree of genomic instability, presumably caused by enhanced tranposition of *P* elements. Temperature-sensitive sterility, male recombination, high frequency of chromosomal rearrangements, and increased frequencies of lethal and visible (often unstable) mutations are frequently observed in dysgenic hybrids. The response to selection, phenotypic variation, and realized heritability were all increased in dysgenic lines artificially selected for high and low abdominal bristle number (Mackay, 1984). A nondysgenic population was reported to respond more rapidly to malathion selection than a dysgenic one (Morton & Hall, 1985). The role of dysgenesis in determining the level of spontaneous mutations and its implications for evolutionary biology are at present still under discussion (O'Hare, 1985; Simmons & Karess, 1985).

One very peculiar finding was that *P*-element activity is restricted to the germ line, where most transposon-mediated mutations occur in premeiotic germ cells, that is, in early stages of germ cells (Green, 1967; Bingham, 1981; Rasmuson *et al.*, 1981).

Recently, Rubin (see O'Hare, 1985) presented evidence that the restriction to the germ line is due to different splicing of P transcripts in somatic and germ cells.

Knowledge of the nature of spontaneous mutations compared with chemically induced mutations in regard to risk estimates is important. Mutations induced by mutation/recombination (MR) mutator systems have been reported to induce high frequencies of unstable locus-specific visible mutations (primarily at the *singed* bristle and *rasberry* eye colour loci; Green, 1977). MR-induced lethals occur at many sites along the X chromosome (in contrast to the locus specificity of visible mutations); some, but not all of these sites at which recessive lethals arise in the MR-system are the same as those known to be targets for X-ray-induced lethals (Eeken et al., 1985). With in-situ hybridization, it was demonstrated that the majority of MR-induced lethals is associated with a particular mobile DNA sequence, the P element; i.e., they arose as a result of transposition.

In general, transposable elements are very insensitive to activation by mutagens. Datta et al. (1983) reported that, among more than 100 chemical compounds tested to stimulate the Tn9 transposition in *Escherichia coli*, only seven were active. The nature of the active compounds suggests the involvement of lipid or membrane in the transposition process. In *Drosophila*, Sobels and Eeken (see Simmons & Karess, 1985) reported that three chemical mutagens had no measurable effect on P-element activity, but X-rays seemed to enhance the frequency of P-mediated chromosomal breakage. In contrast, the unstable *white-zeste* system (see section 3.3(c)) is sensitive to a wide spectrum of chemical mutagens (for review, see Würgler & Vogel, 1986). Recently, Ryo et al. (1985) reported the discovery of a second unstable w^+ strain which can be reverted by X-rays and EMS in early larval stages. It will be important to learn why the transposing element in these systems is activated by chemicals, while most of the P elements seem to be refractory to chemical induction.

Overall, these studies indicate that there are, most probably, distinct differences between the mutational profiles and the nature of mutations arising spontaneously and those induced by physical and chemical mutagens. In the light of this finding, the 'doubling-dose' concept currently used for genetic risk evaluation needs revising (Eeken et al., 1985).

(d) Mitotic recombination

Chemically induced recombination in *Drosophila* can be studied in the mitotic divisions of premeiotic germ line cells as well as in somatic cells. Products of mitotic recombination in the germ line can be classified with respect to the chromosomal region in which the exchange took place (Hannah-Alava, 1968). So far, only a few chemicals, such as formaldehyde, mustard gas, dihydroxydimethyl peroxide (Sobels & van Stennis, 1957) and EMS (Miglani & Mohindra, 1985), have been tested for the induction of recombination in males. A cytological analysis of some cross-over products indicated that EMS induces not only recombination, but also inversions and deletions. In order to link the mitotic recombination seen in somatic cells (see section 3.3) with the well-known mutagenic effects in male germ cells, comparative studies with induced recombination in the somatic and germ line cells would be very useful.

In somatic cells (e.g., eye, wing, abdomen), recombinational events are scored on the basis of the resulting twin spots. For ionizing radiation, it has been demonstrated that induced mitotic recombination takes place in both eu- and heterochromatin, but preferentially in heterochromatin (for review, see Becker, 1976).

Single spots in somatic cells can originate from, for example, gene mutations, deletions or gene conversion. The relative frequencies of these changes are at present unknown. The mechanism by which an individual single spot observed in somatic cells originates and the molecular nature of the change are unknown. There is some indirect evidence that certain chemicals might induce genetic changes independent of recombination. In inversion-heterozygote individuals, single *mwh* spots are observed in the wing assay after exposure of larvae to senkirkine (Frei, unpublished data), hexamethyl phosphoramide (Szabad & Bennettova, 1985), isopropyl methanesulfonate (Katz, 1984), N-nitrosomorpholine (Surian et al., 1985) and 5-azacytidine (Katz, 1985). Since recombination between inversion heterozygous chromosomes is lethal (Merriam & Garcia-Bellido, 1972), the spots recovered from inversion heterozygous larvae are suggestive of the induction of recombination-independent mechanisms, but do not constitute final proof.

2.3. *Metabolic activation*

(a) *Xenobiotic metabolizing enzymes*

Initially, only indirect evidence for versatile xenobiotic metabolism in *Drosophila* was available. Evidence has accumulated, predominantly from mutagenesis studies with the sex-linked recessive lethal test, that *Drosophila* can perform essential activation steps which convert procarcinogens into reactive electrophilic species. Chemicals that can be activated include synthetic chemical carcinogens (alkaryltriazenes, nitrosamines, hydrazo- and azoxyalkanes, halo-olefins and halo-alkanes) and naturally-occurring carcinogens (aflatoxins, pyrrolizidine alkaloids, certain nitrosamines). For a review, see Vogel (1982).

During the last ten years, knowledge concerning cytochrome P450-dependent metabolism in *Drosophila* has increased from practically nothing to the analysis of about 15 reactions in several different strains. The influence of seven enzyme inducers has been at least partly characterized (Baars et al., 1977; Baars, 1980; Hällström & Grafström, 1981; Hällström et al., 1982, 1983; Waters et al., 1983; Hällström et al., 1984; Zijlstra et al., 1984; Hällström, 1986a).

The presence of cytochrome P450 in adult flies and larvae has been shown by means of reduced carbon monoxide-bound difference-spectra (Kulkarni et al., 1976; Baars et al., 1977; Zijlstra et al., 1979). Spectral analysis also revealed the presence of cytochrome b_5. Enzymes responsible for phase-II reactions are also present, e.g., epoxide hydrolase, glutathione S-transferase, glucuronyltransferase and phosphotransferase (Baars et al., 1980, 1983; Jansen et al., 1984, 1986). No sulfotransferase activity has yet been detected (Baars, 1980). Glucuronidation does not seem to occur in insects (Yang, 1976). Using enzyme inhibitors, Zijlstra and Vogel (1985) showed that monoamine oxidases might be involved in the bioactivation and deactivation of mutagens in *Drosophila*.

Substantial genetic variation is observed in most cytochrome P450-dependent reactions (Hällström et al., 1984; Zijlstra et al., 1984). These intrastrain variations and hybrids and segregants from crosses between strains with different cytochrome P450-dependent activities have allowed analysis of the genetic regulation of these enzyme activities. Table 1 presents a summary of the present state of these analyses (Hällström, 1986a). The wide range of high constitutive activities in insecticide-resistant strains determined by the 2-65 cM gene indicates that the resistance is achieved by a regulatory gene mutation, as has also recently been proposed for the house fly. A closely-linked, second chromosome gene determines high biphenyl 4-hydroxylation and aminopyrine demethylation in the *Oregon R(R)* strain. It is not possible to separate the two proposed regulatory genes on the basis of the genetic analysis. They are judged to be separate genes due to the difference in dominance, the increase in the MW 56 000 protein present only in *Oregon R(R)* and the close correlation of the activities within but not between groups of *Drosophila* strains.

In addition to the two regulatory genes, three presumed structural genes have been identified on the second and third chromosome. At present, therefore, at least five genes determining cytochrome P450 activity have been identified in *Drosophila*. The as yet unlocalized demethylation of ethylmorphine and benzphetamine are strongly correlated, but they show no correlation with the activities determined by the genes that have been localized. They seem to represent one or more additional gene(s).

The cytochrome P450 system in *Drosophila* is complex. This is underlined by the fact that the different reactions seem to be unrelated functionally. The 65 cM gene regulates both *O*- and *N*-dealkylations as well as aromatic and aliphatic hydroxylations. Different *N*-demethylations are regulated by different genes, and the reactions at different positions of the same substrate (e.g., biphenyl) are not genetically coupled.

A major problem in *Drosophila* testing is the low capacity for detecting most aromatic hydrocarbons and amines. The induction of specific metabolic reactions by polycyclic aromatic hydrocarbons and the activation of carcinogens like benzo-[*a*]pyrene and 2-aminofluorene are regulated in mammals by the *Ah* locus *via* a cytosolic protein receptor (reviewed by Eisen et al., 1983). Table 2 lists five activities that are inducible by polycyclic aromatic hydrocarbons in mammals and regulated by

Table 1. Genetic localization of some cytochrome P450-dependent activities in *Drosophila melanogaster*[a]

Approximate localization	Dominance	Suggested function	Associated activities and features	Reference
2-65 cM	Complete	Regulatory	*p*-Nitroanisole *O*-demethylation	Hällström & Blanck (1985); Hällström (1986b)
			Biphenyl 3-hydroxylation	Hällström & Blanck (1985); Hällström (1986b)
			Vinyl chloride metabolism	Hällström *et al.* (1982)
			N-Nitrosodimethylamine toxicity	Hällström *et al.* (1982)
			N-Nitrosodimethylamine mutagenicity	Vogel (1980)
			N-Nitrosodimethylamine demethylation	Waters *et al.* (1983)
			Protein MW 54 000	Hällström & Blanck (1985); Hällström (1986b)
			Insecticide resistance	Kikkawa (1961)
2-63 cM	Semidominant	Regulatory	Biphenyl 4-hydroxylation	Hällström & Blanck (1985); Hällström (1986b)
			Aminopyrine *N*-demethylation	Hällström (unpublished data)
			Protein MW 56 000	Hällström & Blanck (1985); Hällström (1986b)
At or to the left of 2-48 cM	Semidominant	Structural	Biphenyl 4-hydroxylation	Hällström & Blanck (1985); Hällström (1986b)
3-51 cM	Semidominant	Structural	Benzo[*a*]pyrene hydroxylation	Hällström & Blanck (1985); Hällström (1986b)
3-58 cM	Semidominant	Structural	7-Ethoxycoumarin *O*-demethylation	Hällström & Blanck (1985); Hällström (1986b)
Not localized			Ethylmorphine *N*-demethylation[b]	Hällström (unpublished data)
			Benzphetamine *N*-demethylation[b]	Hällström (unpublished data)

[a]From Hällström (1986a)

[b]Ethylmorphine *N*-demethylation and benzphetamine *N*-demethylation metabolism are not genetically localized, but the activities are highly correlated in nine strains studied, suggesting a common regulation, and are not correlated with any other activity studied.

the *Ah* locus but which are are low and uninducible in *Drosophila*, indicating perhaps that the responsible cytochrome P450 forms are present in low amounts and are constitutive in this organism. Surprisingly, the presence of the *Ah* receptor has been demonstrated in *Drosophila* (Bigelow *et al.*, 1985), but its biological role is probably not to regulate polycyclic aromatic hydrocarbon induction of cytochrome P450. Since a complex physiological response to these compounds is observed in mammals, there might be other functions for the receptor in *Drosophila*, and the coupling of the receptor to cytochrome P450 might have appeared later in evolution.

Table 2. Metabolic reactions with low, apparently constitutive activity and little strain variation in *Drosophila* (Hällström, 1986a); whereas in mammals, these activities are inducible by aromatic hydrocarbons and regulated by the *Ah* locus.

Biphenyl 2-hydroxylation

Benzo[*a*]pyrene dihydrodiol formation

7-Ethoxyresurfin *O*-deethylation

2-Acetylaminofluorene *N*-hydroxylation

N,N-Dimethylaniline *N*-oxygenation

The small genetic variation in the metabolism of polycyclic aromatic hydrocarbons observed so far makes it difficult to find test strains especially suited for the detection of aromatic hydrocarbons and amines. On the basis of available comparisons (nine strains, nine reactions), the substrain *Oregon R* selected for DDT resistance (*Oregon R(R)*) is the only strain with an overall high metabolic capacity (Hällström, 1986a).

(b) Location of xenobiotic metabolizing activities in adult flies

Insects do not possess a specific organ like the mammalian liver for detoxification, but several tissues appear to be involved. These are the fat bodies, the Malpighian tubules, various parts of the digestive tract, and certain parts of the gonadal tissue (Casida, 1969). Most triazenes, nitrosamines, azoxy- and hydrazo-alkanes, oxazaphosphorines, halo-alkanes and halo-olefines show peak mutagenic activity in metabolically active stages of spermatogenesis, i.e., in various spermatid stages, while spermatocytes are highly susceptible to killing (Vogel, 1975). Studies by Tates (1971) indicate that these stages have a highly developed endoplasmic reticulum.

A second piece of evidence supporting intragonadal activation comes from dose-effect studies with some nitrosamines and procarbazine (Blijleven & Vogel, 1977). With these compounds a steep increase in the induction of sex-linked recessive lethals is found with low doses, followed by a plateau. This is in marked contrast to the flat curves obtained for metabolically inert sperm. These observations can be explained by the formation of short-lived metabolites which act only in the cell in which they are formed. Spermatozoa that do not possess an endoplasmic reticulum must have been attacked by more stable material migrating from the surrounding tissue to those cells.

Recently, Förster and Würgler (1984) showed that aflatoxin B_1 is metabolized *in vitro* in explanted testes from different *Drosophila* strains. Benzo[*a*]pyrene hydroxylation activity has been found in different parts of the body: head, thorax, abdomen, testis. In males, inducibility of this activity by phenobarbital was similar in the head, thorax and abdomen, but significantly lower in testis (Zijlstra *et al.*, 1984).

(c) *Drug metabolizing activities in larvae*

By growing larvae on *N*-nitrosodiethylamine-containing medium, it was shown that (i) larvae possess enzymes for its metabolic activation; and (ii) mutation induction occurs in first, second and third instar larvae, indicating mixed-function oxidase activity in all three stages of larval development (Vogel, 1977). The observation that larval tissues contain a variety of drug metabolizing activities has been substantially extended during the last few years, based on the mutagenicity studies with somatic cells. So far no significant difference between larvae and adults in activation potential with respect to specific classes of chemicals has been demonstrated (for review, see Würgler & Vogel, 1986). But studies of the enzyme distribution pattern in larvae and adult flies have established that the spectral and enzyme features of the larval cytochromes differ considerably from those present in adult flies (Hällström *et al.*, 1983; Vogel *et al.*, 1983b; Zijlstra *et al.*, 1984). These observations indicate the occurrence of developmental modifications in the cytochrome P450 system in *Drosophila*, which might reflect the need for different isozymes and rapid increases in specific forms in the two developmental stages due to differences in feeding behaviour and hormonal status. In adult flies, sex pheromones are probably important substrates that are not present in larvae. It is also obvious that the process of metamorphosis demands dramatic alterations in the cytochrome P450 system (Porsch-Hällström, 1984).

(d) *Exclusion of microbial activation in the gut*

For nitroheterocyclic compounds, it is adequately established that the reduction of the nitro group is the first step in the formation of reactive metabolites (Swaminathan & Lower, 1978). Bacteria, in particular, are very efficient in carrying out reduction of nitroheterocyclics: in *E. coli*, three nitroreductase systems have been described (McCalla *et al.*, 1975). To establish the possible role of intestinal flora of larvae and flies in the activation process, the use of germ-free cultures has been studied. An easy method for rearing germ-free *Drosophila* lines was developed by Molnar and Kiss (1980). Blijleven (1979) tested the activation of *N*-nitrosodiethylamine and procarbazine. Kramers (1982) tested two nitroheterocyclic compounds in germ-free *Drosophila* and came to the conclusion that activation of the compounds is carried out by the flies rather than by the bacterial flora.

(e) *DNA repair*

Photoreactivation, excision repair and postreplication repair have been studied in detail in *Drosophila* (Boyd *et al.*, 1983).

The possibility of improving the sensitivity of the different assays should be explored. Systematic studies of repair-defective strains are feasible in *Drosophila* (e.g., Smith & Dusenbery, 1985; Dusenbery, 1986; Smith *et al.*, 1986). The use of very specific repair defects, e.g., *mu-2*, which potentiates the induction of terminal deficiencies (Mason *et al.*, 1984), should also be included. Attempts to improve the efficiency of stocks used in screens for chemically-induced mutations and chromosomal breakage in *Drosophila* may also include the use of hybrid dysgenic mutator lines (Yanopoulos *et al.*, 1980; Eeken & Sobels, 1981; Woodruff & Brodberg, 1983)

(f) Tumours

A long and unsuccessful search for *Drosophila* mutants with truly neoplastic growth created scepticism as to the ability of these cells to become neoplastic. It was only in 1967 that malignant and benign neoplasms were found in *Drosophila*. In the short time since the initial discovery of the first recessive-lethal mutant with true neoplasms, many additional mutants with many different alleles have been found. Neoplastic growth has often been observed in atelotypic tissue sublines derived from wild-type embryos and imaginal discs cultured *in vivo* in female adult abdomens. Developmental, histological and fine-structural studies of these neoplastic tissue sublines, and of neoplasms of genetic origin, unequivocally demonstrate their neoplastic nature (for reviews, see Gateff, 1978, 1982).

Drosophila cells can become neoplastic as long as they retain their ability to divide. Larval and most adult cells of *Drosophila*, like the neurons and neutrophilic leucocytes of vertebrates, do not divide and are thus incapable of neoplastic transformation. Neoplastic cells of *Drosophila* share many features with vertebrate neoplastic cells. The main characteristics of vertebrate malignant cells are: (i) fast autonomous growth; (ii) invasiveness; (iii) metastasis; (iv) transplantability; (v) host lethality; (vi) loss of function; and (vii) loss of structure. Malignant cells of *Drosophila* show striking similarities, except for the ability to metastasize.

The most prominent expression of malignancy is a highly increased autonomous growth rate. Compared to their wild-type counterparts, malignant *Drosophila* cells *in situ* as well as in culture *in vivo* similarly exhibit a manifold increase in their cell division rate.

Like vertebrate malignant cells, those from *Drosophila* are also invasive. Their invasive behaviour depends, however, to a great extent on certain structural tissue characteristics. In contrast to vertebrate tissues, which consist of many cell layers, most *Drosophila* tissues are made up of monolayered cells. The apical cell surfaces are folded into microvilli and the basal surfaces are covered by a rigid basement membrane. It is this intact basement membrane which prevents invasion of the monolayer epithelium by malignant cells. However, mutants that cause abnormalities of the basement membranes, in addition to neoplasms, show typical invasion of organs by neoplastic cells. For instance, in the mutant *l(1)bwn*, which develops neoplastic imaginal discs, the envelope of the brain-ventral-ganglion complex is defective, and invasion of neoplastic imaginal disc cells into the brain-ventral-ganglion complex is observed regularly. The nervous system, the gonads and the adult thoracic muscles are the only organs in which the cells are not arranged in monolayers. Typical invasion by neoplastic cells is therefore observed primarily in these organs.

3. Description of the assay system

3.1 *The sex-linked recessive lethal test*

(a) Test procedure

Usually, the assay is begun by exposing mature adult males to the test compound; in this way, all the different stages of spermatogenesis are exposed to the chemical. Treated males are mated to tester females, the F_1 progeny are inbred, and in the F_2 exposed X chromosomes are hemizygous in males. If a lethal is induced, 50% of the F_2 males will die (for details, see IARC, 1980; Lee *et al.*, 1983; Würgler *et al.*, 1984).

The test for sex-linked recessive lethals is at present the best validated *Drosophila* mutagenicity assay. It detects predominantly gene mutations and small deletions. Since 600-800 genes are tested simultaneously on every X chromosome (Abrahamson *et al.*, 1980), induced mutation frequencies as low as $3.3-2.5 \times 10^{-6}$ per locus per generation can be detected. Therefore, it is possible to test chemicals at concentration levels that could result in a doubling of the spontaneous mutation rate of an average human gene (5×10^{-6} to 5×10^{-7} per locus per generation; Cavalli-Sforza & Bodmer, 1971). A recent description of the technical aspects of the assay is given by Würgler *et al.* (1984).

A more versatile but more complicated scheme suitable for measuring recessive lethals induced in male as well as female germ lines has been developed by Lee (1976) and Vogel (1984).

(b) Spontaneous frequencies of recessive lethals

In large-scale mutagenicity screening programmes, large sets of control data on spontaneous mutation frequencies are collected. Interest in the problem of variability of spontaneous lethal frequencies was stimulated by a paper by Kilbey *et al.* (1981), who reported spontaneous frequencies ranging from 0-0.33% and proposed that in exposed series the significance should be calculated against a frequency of 0.4% instead of the concurrent control. Gocke *et al.* (1982) argued against this procedure, because the gain of avoiding misclassification of some nonmutagenic substances as mutagens might well be counterbalanced by the resulting misclassification of weak mutagens as non-mutagens. Their own experience with a constant protocol and control-runs every two months over a period of three years indicated no seasonal variation in the spontaneous recessive lethal frequency. In a total of over 280 000 chromosomes tested, the average spontaneous frequency was 0.25%. They suggest using either concurrent controls or historical controls from the laboratory conducting the test.

In contrast to the report of Gocke *et al.* (1982), in the most recent study a variation with time was found for the spontaneous sex-linked lethal mutation frequency (Mason *et al.*, 1985). Examination of approximately 1.7×10^6 X chromosomes in three laboratories suggested two sources of variation in the spontaneous mutation frequency. Firstly, the underlying mutation rate may change with time, suggesting genetic drift among the genes controlling the spontaneous mutation rate. In one set of experiments, the recessive lethal frequency increased from 0.2% to 0.4% in six years. Secondly, there may be seasonal variation in the mutation frequency even under apparently controlled conditions, suggesting that uncontrolled environmental factors may play an important role in determining the mutation frequency. Therefore, Mason *et al.* (1985) concluded that concurrent controls should be used in mutation experiments, except in cases in which the frequency in the untreated control is so low as to be unmeasurable in a single experiment or the control frequency can be shown not to vary with time.

(c) Test performance

The Gene-Tox report (Lee *et al.*, 1983) is based on 421 compounds for which sufficient data were available (Table 3). Of these, 198 compounds were found to be positive and 46 negative, and 177 chemicals could not be classified as positive or negative. There were 62 compounds that could be classified as positive or negative for both carcinogenesis and mutagenesis. Of these, there was agreement between the classification of carcinogenesis and mutagenesis for 56 (50 positive and six negative), i.e., 90% would have been correctly classified as carcinogens from the recessive-lethal test alone. Of 55 compounds classified as carcinogens, 50 were positive and five negative (50/55), giving 91% agreement among carcinogens. Of the seven compounds that were negative in the test for carcinogenesis, six were negative and one positive for mutagenesis (6/7), giving 86% agreement among noncarcinogens. This correlation of mutagenesis with carcinogenesis is limited, in that neither test has been applied to a random sample of compounds: there is no consistent pattern in the selection of compounds tested except that investigators have been successful in selecting compounds that give positive results. The criteria for a positive response are easier to meet than those for a negative response. Therefore, a nonmutagen is more likely to be classified as inadequately tested (177 compounds in the Gene-Tox report) than is a mutagen, yet the correlation with mutagenesis is high among both carcinogens and noncarcinogens. For the outcomes of recent large-scale studies, including the recessive-lethal assay, see section 6.

(d) Delayed mutations

A high proportion of the mutations induced in postmeiotic germ cells by nitrogen mustard (Auerbach, 1951) and other chemical mutagens (Browning & Altenburg, 1961; Carlson & Oster, 1962; Mathew, 1964; Alderson, 1965; Jenkins, 1967; Abeleva *et al.*, 1969; Lee *et al.*, 1970) are fixed only after one or more cell divisions have occurred. Following fertilization of the egg by a sperm (possibly carrying mutagen-induced premutational lesions), cleavage proceeds with synchronous nuclear divisions to form a blastoderm consisting of a total of about 4000 nuclei after 12 cleavage divisions. During the ninth cleavage division, three to seven nuclei migrate into the polar region of the embryo to become the primordial germ cells. Delay in fixation of a mutation during

Table 3. Numbers of compounds in the Drosophila Gene-Tox data base evaluated for mutagenicity and correlated with carcinogenicity (Lee et al., 1983; Valencia et al., 1984)

Test	Mutagenicity[a]	Carcinogenicity[a]				
		+	-	?	NI	Total
Recessive lethals	+	50	1	0	147	198
	-	5	6	0	35	46
	inc 1	5	1	3	45	54
	inc 2	11	1	1	67	80
	inc 3	3	0	0	40	43
	Total	74	9	4	334	421
Heritable translocations	+	19	0	2	6	27
	-	2	0	1	5	8
	(+)	4	0	0	4	11
	inc	4	1	1	5	8
	(-)	2	0	0	5	7
	Total	31	1	4	25	61
Clastogenesis	+	15	1	0	10	26
	-	4	0	2	7	13
	(+)	1	0	1	1	3
	(inc	4	0	4	14	22
	(-)	2	0	1	9	12
	Total	26	1	8	41	76

[a] +, positive; -, negative; inc, inconclusive; inc 1, at least 3000 chromosomes tested, but these tests did not meet the criteria for either a positive or a negative conclusion; inc 2, between 1000 and 3000 chromosomes tested; inc 3, less than 1000 chromosomes tested; (+), inconclusive, but looks positive; (-), inconclusive, but looks negative; ?, questionable; NI, not indicated by Griesemer and Cueto (1980)

cleavage could be due to modification of DNA, so that during successive cell divisions the DNA replication is error prone. The premutational lesions must escape the repair processes that are active shortly after entry of sperm into the egg. The nonrepaired lesions could produce both stable mutant and stable nonmutant nuclei during cleavage. High frequencies of somatic mosaics are expected, and these have been demonstrated by Lee et al. (1970) for EMS-induced *yellow* mutants in somatic tissues. Transmission of an induced mutation to progeny will occur only if nuclei containing a stable mutation or a mixture of nuclei with and without the stable mutation (produced gonadal mosaics) are incorporated into the germ line. Lee et al. (1976) point out that about 13% of mosaic embryos have gonads complete for the recessive lethal and, therefore, will give a positive test when scored for X-chromosomal recessive lethals in the F_2 generation. Furthermore, the proportion of recessive lethals detected from mosaic F_1 embryos depends upon the level of exposure, as shown with EMS by Epler (1966). Because chemical mutagens may differ in their ability to cause mosaics, a fraction of which will be scored as complete in the F_2 generation, testing of F_3 to detect fractional mutations is essential when recessive lethals are used as a comparative measure of dose. Lee (1976) devised a test protocol that enables a clear distinction of mosaic lethals from detrimentals and also safeguards against classification errors due to sex-chromosome nondisjunction.

A systematic study of monofunctional alkylating agents revealed a decreasing yield of delayed mutations in passing from MMS, EMS and diethylstiboestrol through N-nitrosodiethylamine and N-ethyl-N-nitrosourea (ENU) (Vogel & Natarajan, 1979a,b). It is striking that those agents relatively inefficient in inducing chromosomal breakage (the nitrosamines) gave high ratios of F_2-lethals to F_3-lethals. The remaining compounds tended to induce more delayed mutations. Thus, the ability of these chemicals to induce delayed mutations and their efficiency in causing chromosomal breakage seem to be positively related. Vogel and Natarajan (1979a,b) concluded that the ability to produce delayed mutations correlated with preference for alkylation of base nitrogens, especially the guanine N-7, the major product of DNA reacted with MMS, EMS, diethyl sulfate or dimethyl sulfate (Lawley & Brookes, 1963; Lawley & Shah, 1972; Sun & Singer, 1975; Swenson & Lawley, 1978). Alkylations at the N-7 and N-3 positions of purine bases are known to cause depurination of DNA (Lawley & Brookes, 1963; Lawley, 1974). Depurination has been associated with strand breakage (Brookes & Lawley, 1961, 1963), misincorporation of nucleotides (Shearman & Loeb, 1977), chromosomal breakage and occasional base-pair deletions, which may occur as delayed effects (Oeschger & Hartman, 1970; Freese, 1971; Auerbach, 1976).

It is interesting that hexamethylphosphoramide, a compound that does not detectably bind to the O-6 or N-7 positions of guanine, is not very efficient in producing delayed mutations. The frequencies of recessive lethals in the F_2 generation were four- to seven-fold higher than the percentages of mutations detected in the F_3 generation (Vogel et al., 1985).

(e) Germinal selection

The fact that most directly- and indirectly-acting mutagens produce lower frequencies of recessive lethals in premeiotic stages has generally been explained by 'germinal selection'. This term, introduced by Muller (1954), indicates elimination of cells that are deficient in a gene required for development of a spermatogonial cell into a spermatozoa. This phenomenon has been investigated by comparing X-linked with chromosome II recessive lethals, since, for the latter, germinal selection would not be expected to occur. In fact, for all mutagens tested, the ratio of chromosome II lethals to X-chromosome lethals (which is a measure of germinal selection) was higher in spermatogonia than in mature sperm (Purdom, 1957; Shukla & Auerbach, 1980; Vogel et al., 1982). Data on radiation have been reviewed by Sankaranarayanan and Sobels (1976). The difference was small for N-nitrosodiethylamine (Vogel, 1982; Vogel et al., 1982), which induces very few chromosomal aberrations (Vogel & Leigh, 1975; Vogel & Natarajan, 1979a,b). For EMS and diepoxybutane, however, the difference was large enough to suggest that, as a rough estimate, three-quarters of the spermatogonia carrying induced X-chromosomal lethals are eliminated before reaching the mature sperm stage (Shukla & Auerbach, 1980; Vogel et al., 1982). It has been suggested that germinal selection would apply mainly to chromosomal aberrations, and especially deletions.

Yoshikawa et al. (1984a,b) demonstrated that (i) there is a strong germinal selection against ENU-induced lethal-bearing X chromosomes in spermatogonia, irrespective of the dose of ENU; (ii) studies with a chromosome II lethal system using the Cy/Pm method show promise for constructing a realistic dose-response relationship in spermatogonial cells; and (iii) the shape of the dose-response curve for ENU-induced chromosome II lethals in the spermatogonia of Drosophila is very similar to that recorded for specific-locus mutations in spermatogonial stem cells of mice.

Only the study of autosomal lethals will give a realistic picture of intrinsic mutability throughout the cell cycle, since only autosomal lethals allow a comparison of the sensitivity to a given mutagen among all germ-cell stages.

3.2 Tests for chromosomal aberrations

(a) Introduction

Routine screening tests for the following types of chromosomal aberrations are available: autosomal reciprocal translocations, partial Y-chromosome loss, loss of rod and ring X chromosomes. The literature on chemical induction of chromosomal aberrations and clastogenic effects in Drosophila has been reviewed in a Gene-Tox report (Valencia et al., 1984). The tests for heritable translocations are discussed below. The aneuploidy tests (chromosome loss and nondisjunction) are evaluated in Report 13.

(b) Heritable translocations

The most commonly used test for heritable translocations detects induced mutations by phenotypic markers on separate chromosomes which appear to be linked (pseudolinkage). Eye colour and body colour may be used as suitable markers to study the induction of reciprocal translocations between autosomal chromosomes 2 and 3 (T2:3). Woodruff and Brodberg (1983) have shown that the use of females with attached XY chromosomes in screens for chemically induced reciprocal translocations allows for the recovery of breakage events involving the Y chromosome and autosomes.

In the Gene-Tox data base, the reviewing panel found acceptable data for 61 chemicals (Valencia et al., 1984) (Table 3). Of these, 27 could be classified as positive for the induction of heritable translocations, eight were classified as nonmutagens for the induction of translocations, and the data for the remaining 26 compounds were inconclusive or too small a number of chromosomes had been tested. The conclusive data for 33 compounds were compared with the results in the recessive lethal assay: the agreement between the two tests is good (78.8%; 26 positive in both assays; two negative in both assays). The discordant five compounds (D-lysergic acid, vinyl chloride, procarbazine, naltrexone and hycanthone methanesulfonate) are positive in the recessive lethal assay but negative for the induction of translocations. *N*-Nitrosodiethylamine, which is positive in the recessive lethal assay and classified inconclusive in the translocation assay, should be added to the group of mutagens with a low potential for the induction of chromosomal aberrations.

In a correlation of heritable translocations with mammalian carcinogenicity, 19 of the 31 carcinogens were deemed mutagenic, two were not mutagenic and for ten there were inconclusive data.

3.3 *Somatic mutation and recombination tests (SMART)*

In practice, the standard *Drosophila* assays for the detection of sex-linked recessive lethals and autosomal reciprocal translocations involve at least two generations of flies. These assays are therefore time-consuming and tedious. For this reason, it was recently suggested that one-generation test systems using somatic cells might improve the detection capacity and the versatility of *Drosophila* assays (Vogel et al., 1985; Würgler & Vogel, 1986). Three different assays were developed that have reached reasonable levels of validation. Two assays are somatic mutation and recombination tests (SMART): (1) the wing mosaic system based on the wing-hair markers *multiple wing hairs* (*mwh*) and *flare* (*flr*) (Graf et al., 1984; Würgler et al., 1985), and (2) the *white/white-coral* eye mosaic system using two alleles at the *white* locus (Vogel, 1985). The genetic principle of these assays consists of the phenotypic expression of recessive markers in cell clones derived from mutagen-exposed heterozygous cells. Among the genetic events leading to such phenotypic expression are gene mutations, deletions, mitotic recombination and possibly gene conversion. A third test, the unstable *white-zeste* eye mosaic system (Rasmuson et al., 1984; Fujikawa et al., 1985), detects induced genetic events related to an insertion sequence in the region of the *white* gene on the X chromosome.

(a) Multiple wing hairs/flare wing mosaic system

The markers used alter the phenotypic expression of the hairs on the wing blade. This combination of markers used in transheterozygous larvae allows the detection of mutagenic as well as recombinogenic activity of chemicals. Mitotic recombination between *flr* and the centromere results in two genetically marked daughter cells, one homozygous for *mwh* and the other homozygous for *flr*. During larval and pupal development, these two cells produce two adjacent cell clones. During metamorphosis, the cells differentiate into wing blade cells that express the *mwh* and those that express the *flr* phenotype; the mutant clones become visible as a twin spot. Mitotic recombination between *mwh* and *flr* leads to a *mwh* single spot. In addition, single spots (*mwh* or *flr*) are produced by nonrecombinogenic events, e.g., gene mutation, deletion. Clones induced early in the development of the larvae result in larger spots than those induced later. Therefore, the type and size of each spot is recorded. The following types of spots are evaluated separately: (i) small single spots consisting of only one or two cells; (ii) large single spots; and (iii) twin spots. The frequencies and sizes of the *mwh* clones may be used to calculate the mean clone induction frequency per cell cycle per 10^5 cells (Szabad *et al.*, 1983; Würgler & Vogel, 1986). Further details are given by Graf *et al.* (1984) and Würgler and Vogel (1986).

(b) White/white-coral eye mosaic system

The *white/white-coral* system was initially developed by Becker (1957) in an attempt to study the developmental processes in the eye imaginal discs. This system has been evaluated in detail by Vogel and coworkers (Vogel, 1985).

Advantage is taken of the ability to distinguish the expression of two different alleles at the *white* eye locus on a *sepia*-eye-colour background. Female larvae of constitution w/w^{co} at the *white* locus are exposed to mutagens. The distal position of the *white* gene at the tip of the X chromosome far away from the centromere (approximately 70 centimorgans) optimizes the possibility of recovering spots due to mutagen-induced somatic recombination. Such an exchange leads to one daughter cell homozygous for w^{co} and to another one homozygous for w. The adjacent clones derived from these two cells form a w/w^{co} twin spot in the differentiated eye consisting of an area of ommatidia with dark colour (*white-coral*) and an adjacent area with white colour, contrasting with the intermediate colour of the heterozygous cells in the rest of the eye. In addition to these twin spots, single mosaic light (or white) spots can also be scored. Dark spots are not scored because they are supposed to include mostly variegation effects (Vogel, unpublished data). Details of the test are given by Vogel (1985) and Würgler and Vogel (1986).

(c) Unstable white-zeste eye mosaic system

The test system is based on the recessive *zeste* mutation and the unstable *white* locus described by Rasmuson and Green (1974). Normally, the *zeste* mutation manifests its phenotype of lemon-yellow eye colour when the normal *white* allele is present in two copies, as in normal females and in males with a duplication of the entire w^+ locus, or at least the proximal part of it. When the *white* allele is present in a single copy, as in normal males and in females heterozygous for a deletion of the *white* locus, a red eye colour is produced (Jack & Judd, 1979). The *zeste-white* interaction reveals that the unstable *white* locus can occur in two different states: in males, it can either suppress *zeste*, and thus act as a normal single w^+ allele, or it can let the *zeste* phenotype appear, in which case it acts as a duplication of w^+. Shifts between the two states occur spontaneously with a certain frequency and can be induced by a number of mutagens (Rasmuson et al., 1984; Würgler & Vogel, 1986). This phenomenon seems to be associated with a 1-kilobase insertion at the proximal end of the *white* region (Rasmuson et al., 1981). Integrated at this position, the element seems to activate the function of the region and gives red eye pigmentation. Although some molecular studies on spontaneous and induced changes are being performed (Rasmuson et al., 1984; Gubb et al., 1985; Ising, unpublished data), the exact nature of the changes at the DNA level is not yet clear. Rasmuson (1985a,b) and Ryo et al. (1985) interpret their findings on the somatic phenotypic change from zeste to red as resulting from small deletions, i.e., excision of a DNA element responsible for expression of the *zeste* phenotype.

(d) Performance of somatic assays
(i) Detection of mutagens and promutagens

A compilation of all data available by mid-1985 revealed that 46 compounds have been tested with the unstable *white-zeste*, 57 with the *white/white-coral* and 92 with the wing assay (Würgler & Vogel, 1986). Promutagens like *N*-nitrosodimethylamine and *N*-nitrosodiethylamine, which act *via* short-lived genotoxic metabolites, are active in all three assays, indicating that cells of the eye and wing imaginal discs contain xenobiotic-activating enzymes. This is also substantiated by the activity of aflatoxin B_1 in the unstable *white-zeste* and the wing assay. Since validation studies with both direct and indirect mutagens are still in progress, the spectrum of activities cannot yet be discussed in detail; nevertheless, it is clear that a wide spectrum of promutagens is detected in the somatic assays (Würgler & Vogel, 1986).

(ii) Comparison of the three somatic assays

A comparison of the three assays is necessarily restricted because only a limited number of chemicals have been tested in at least two of these assays. Nevertheless, some general conclusions can be drawn:

(1) Powerful reference mutagens such as the directly-acting mono- and polyfunctional alkylating agents are detected without difficulty in somatic assays.

(2) Unstable, directly-acting mutagens are detected in the wing and the *white/white-coral* assay using feeding as a simple route of administration. A good example is β-propiolactone, which has a half-life of about 3.5 h in aqueous solution.

(3) The detection of intercalating agents, several of which have been studied in the wing test, seems to be more difficult. The limited experience gained so far indicates that intercalators of the acridine type induce only small single spots, if any. Acridine orange induced only large single spots at the concentrations tested but no significant frequency

of twin spots, as would have been expected on the basis of its ability to induce homozygosity (mitotic recombination and gene conversion) in the yeast *Saccharomyces cerevisiae* (Simmons, 1979). In the wing assay, ICR-170, a molecule with an intercalating moiety and an alkylating side chain, is capable of inducing all three types of spots in the wing assay (Graf et al., 1984; Katz, unpublished data). The antibiotics adriamycin and daunomycin, which are active in both the wing and the *white/white-coral* test, are also thought to act through an intercalation mechanism, similar to that described for acridines (Vig, 1977).

(4) It is of particular practical interest that mutagens which do not directly interact with DNA, such as methotrexate (an antifolate which leads to disturbances in the nucleotide pools), are detected in the somatic assays, which, by their nature, detect effects induced in proliferating cells. This is true for the *white/white-coral* (Vogel et al., unpublished data) as well as the wing assay (Würgler et al., 1983; Fujikawa, unpublished data).

(5) Carcinogenic peroxisome proliferators, such as di(2-ethylhexyl)phthalate, are thought to act as indirect genotoxic agents (Reddy et al., 1980, 1982). It has been proposed that they act through excessive production of hydrogen peroxide and the associated oxidative damage. With the *Drosophila* stocks used so far, di(2-ethylhexyl)phthalate and clofibrate gave negative or marginal effects (Vogel et al., 1985; Frei, unpublished data).

(6) A major problem in *Drosophila* mutagenicity testing has been the generally weak mutagenic effectiveness of carcinogenic aromatic amines and polycyclic hydrocarbons. In the last international study (Vogel et al., 1985), 2-acetylaminofluorene and benzo[a]pyrene appeared convincingly to have genotoxic effects, at least in somatic cells, although the effect was small in relation to their carcinogenic potency and mutagenic action in other test systems. This is not surprising bearing in mind the low capacity for biotransformation of aromatic compounds by *Drosophila*. It is known that polyaromatic hydrocarbons and aromatic amines and amides have a far weaker effect in *Drosophila* than in mammals. Notable exceptions are those polyaromatic hydrocarbons substituted with methyl groups, such as 7,12-dimethylbenz[a]anthracene (DMBA), 7,8,12-trimethylbenz[a]anthracene and 9,10-dimethylanthracene. This presumably is due to a difference in the cytochrome P450 species responsible for the activation of these compounds (Vogel et al., 1985).

Somatic assays such as the *white/white-coral eye* test and the *multiple wing hairs/flare wing* test are interesting because they include among the genetic effects, in addition to gene mutations, deletions, chromosomal aberrations as well as mitotic recombination and possibly gene conversion. The unstable *white-zeste* eye system allows the inclusion of a mutagen-sensitive transposing element in the analysis. Because of the well-recognized potential usefulness of *Drosophila* somatic assays for screening carcinogens and mutagens, their validation should be completed with high priority.

These validation studies should include the quantitative aspect of establishing dose-response curves as well as the study of defined noncarcinogens (Shelby & Stasiewicz, 1984; Ashby & Purchase, 1985).

Another aspect of the validation of the somatic assays is mechanistic studies. The origin only of twin spots is established to be from mitotic recombination. Single spots may, depending on the marker system used, also result from mitotic recombination, but there are good reasons, based on theoretical considerations and experience with other *Drosophila* assays, to suppose that gene mutations, deletions and chromosomal aberrations are also important in the formation of single spots. Specific additional assays in which recombination is excluded should demonstrate, for some model compounds, the relative frequency of gene mutations (including *ts* mutants indicative of base-pair substitutions) and induced deletions (e.g., by studying the *white-ivory* (w^i) system which reverts specifically by deletions; Green, 1962; Ryo *et al.*, 1985; Green *et al.*, 1986).

The simultaneous determination of induced mitotic recombination in somatic cells and male germ-line cells would allow determination of the relative sensitivity of the two tissues. In this way, the somatic and germ-cell assays could be calibrated, and a certain degree of extrapolation from somatic to germ cells might become possible.

(iii) *Comparison of somatic assays with germ-cell assays*

Eighty compounds have been tested in both the sex-linked lethal assay and at least one somatic assay (Würgler & Vogel, 1986). Of these, 55 (69%) are positive in both. Only two compounds (captan and 2,4,5-trichlorophenoxyacetic acid) have been found positive in germ cells but negative in one somatic assay (the unstable *white-zeste* system). These compounds will be retested in the wing and/or the eye assay. Eight compounds (cytosine arabinoside, benzyl chloride, 4-chloromethylbiphenyl, 2,4-dichlorophenoxyacetic acid, methotrexate, safrole, strychnine and vincristine) are positive in somatic cells but negative in the germ-cell assay. Another seven compounds (hydrazine, Trp-P-1, Trp-P-2, acrylonitrile, 2-(2-furyl)-3-(5-nitro-2-furyl)acrylamide, hydroxylamine and 5-fluorouracil) gave positive results in the somatic assays but inconclusive results with the recessive lethal assay in germ cells. Many antimetabolites give discordant results, perhaps because predominantly postmeiotic stages were tested in the recessive lethal assays, and strongly S-phase-dependent mutagenic activities could, therefore, not be detected. Strychnine is a representative of a class of chemicals which induce recombination but no other genotoxic effect (Würgler, 1986). This has been confirmed by studies in yeast (Fahrig, unpublished data).

A direct comparison between genotoxic effects in the eye imaginal discs and in larval gonads was reported by Vogel (1984), using the following model mutagens: MMS, ENU, bleomycin, DMBA and 9,10-dimethylanthracene. The frequency of sex-linked recessive lethals in germ cells was recorded, and effects in somatic cells were studied with the *white/white-coral* test. In the recessive lethal test, ENU was strongly mutagenic and was the only reference mutagen readily detectable in both sexes, whereas the identification of MMS and bleomycin as mutagens caused severe problems, since both compounds were either ineffective or showed only marginal activity at rather high doses. DMBA was more active in testes, 9,10-dimethylanthracene more so in the female germ line. Except for ENU, which induces predominantly point mutations, and for 9,10-dimethylanthracene for which a few clusters were found, no large clusters of mutants of common origin were observed.

This pattern of effects described for the recessive lethal assay was quite different from the markedly high response of somatic tissue to the five chemicals studied. The most striking difference between the two assays was the clear response of somatic tissues to bleomycin and MMS. A drastic difference in detection capacity between the larval germ-line assays and somatic mutation tests is therefore evident.

(iv) *Comparison of* Drosophila *somatic assays with yeast assays*

In a comparison of the somatic assays in *Drosophila* with the homozygosity assay in yeast for detecting mitotic recombination and gene conversion (Würgler & Vogel, 1986), 75% of the chemicals tested in both systems (34/48) give concordant results. Eight compounds gave discordant results: six were positive in *Drosophila* but negative in yeast, and two differed in the opposite direction. One positive and three negative compounds in yeast gave inconclusive or marginally positive results in *Drosophila*. These should be tested further in *Drosophila* to obtain conclusive decisions about their activity in the somatic assays.

(v) *Comparison of* Drosophila *somatic assays with the mouse spot test*

A total of 23 compounds was tested in both the somatic assays in *Drosophila* and the mouse spot test (Würgler & Vogel, 1986). Of these, 17 chemicals were positive in both assays. Four compounds (*N*-nitrosodiethylamine, *N*-methyl-*N*-nitrosourea, trenimon and diethyl sulfate) were positive in *Drosophila* somatic assays but negative in the mouse spot test. Two chemicals (*N*-nitrosodimethylamine and hycanthone methanesulfonate) which gave inconclusive results in the mouse spot test were found to be positive in *Drosophila*.

(vi) *Comparison of* Drosophila *somatic assays with in-vivo carcinogenicity data*

Among the 76 carcinogens tested in somatic assays, 67 compounds (88%) were found to be mutagenic in at least one somatic assay (Würgler & Vogel, 1986). Three gave marginal effects, and six (8%) were not detected by the somatic assays. A set of noncarcinogens, as defined by negative results in animals of each sex in a two-species long-term assay (Shelby & Stasiewicz, 1984), are still to be tested in the *Drosophila* somatic assays.

Of particular interest are the studies of Kondo and his coworkers (Yoo *et al.*, 1985), who tested nine carcinogenic amino acid pyrolysis products (which need metabolic activation) in the *Drosophila* wing assay. The reported values of carcinogenic potency in the mouse tests for seven of the nine compounds are well correlated with the mutagenic potency values obtained in the *Drosophila* wing assay, but not with those obtained in the *Salmonella typhimurium* assay.

4. Treatment conditions and dosimetry

4.1 *Feeding*

In *Drosophila* mutation assays, one of the commonly used methods of administering a test chemical is by allowing adult flies to imbibe sucrose solutions containing the test chemical. In most cases, the standard feeding technique as described by Vogel and Luërs (1974) is used. In the case of relatively volatile compounds (e.g., chloroprene and 1-chlorobutadiene; Vogel, 1979), the procedure has to be slightly changed.

In feeding procedures, the amount of the test solution ingested by the flies depends upon the palatability of the treatment solution, due to the nature and concentration of the test chemical and its toxic and repellent effects. Quantitative assays to measure the response of *Drosophila* males to repellents (Falk & Atidia, 1975) and attractants (Ford & Tomkins, 1985) have been developed. Aaron and Lee (1977) have demonstrated quantitatively an inhibitory effect of high EMS concentrations (25 mM) on the uptake of feeding solutions by adult males. An approximate dose of a chemical ingested by the flies can be determined on the basis of the uptake of ^{14}C-sucrose (MacDonald & Luker, 1980; Gollapudi *et al.*, 1985).

4.2 *Injection*

For screening, injection is a more tedious technique than feeding and is, therefore, the method of second choice. In the US National Toxicology Program, only those compounds that give negative results in *Drosophila* with feeding are tested by injection (see section 6). Injection techniques are available for larvae, but they are not used in mutagenicity studies (Würgler & Vogel, 1986).

4.3 *Inhalation*

Inhalation should be an efficient route of application, because the tracheae and tracheoles transport gaseous compounds directly to the target tissues, in particular to imaginal discs in larvae. The applicability of different inhalation techniques has been demonstrated for vinyl chloride (Verburgt & Vogel, 1977; Magnusson & Ramel, 1978), halothane (Kramers & Burm, 1979), urethane (Namura, 1979), 1,2-dichlorethane and methyl bromide (Kramers *et al.*, 1985a,b).

4.4 *Solvents*

When a compound is not soluble in water, an initial solution is made in some other solvent, such as ethanol, Tween 80, Tween 60 or dimethyl sulfoxide. For polycyclic aromatic hydrocarbons and aromatic amines, mixtures of solvents must be used (Vogel *et al.*, 1983a). It is recommended that the use of dimethyl sulfoxide be avoided when possible, since it can interfere with enzyme induction (Magnusson *et al.*, 1979) and with the activities necessary for promutagen conversion (Sosnowski *et al.*, 1976; Yahagi *et al.*, 1977; Younes *et al.*, 1979). Zijlstra and Vogel (1984) suggested that the choice of the solvent can be crucial: DMBA was ineffective when dissolved in oil/dimethylformamide, whereas a special fat emulsion of DMBA gave high mutation frequencies.

Some strains are sensitive to the induction of genetic damage by ethanol, e.g., excision repair-deficient larvae used for the *mei-9* wing spot test showed an increased frequency of spots when exposed to ethanol (Würgler *et al.*, 1985; Graf, unpublished data). The effect is due to the mutagenic metabolite acetaldehyde (Obe & Ristow, 1977; Graf, unpublished data). Fujikawa *et al.* (1983) added heat-stable test compounds dissolved in ethanol to hot medium, heated to 80°C and stirred for 5-10 min to allow the ethanol to evaporate.

5. Storage effects

If mutagen-exposed mature sperm are stored in females and are used for the insemination of eggs only after many days, a time-dependent increase in chromosomal aberrations ('storage effect') may be observed. Technically, the inseminated females are used either for many successive broods or are kept on sugar-agar medium, to prevent them from laying eggs, for ten or more days before they are allowed to oviposit eggs on normal yeast-containing medium. Storage effects have been observed for heritable 2-3 translocations as well as for ring-X chromosome losses (e.g., Vogel & Natarajan, 1979a,b). Pronounced storage effects are observed with monofunctional alkylating agents, and studies of stored sperm are essential for the detection of chromosomal rearrangements induced by these agents (Vogel & Natarajan, 1979a). It is thought that storage effects result from the delayed opening of latent breaks. Lee (1978), studying storage effects on dominant lethals induced by EMS, and Ryo *et al.* (1981) studying the same phenomenon with MMS-induced recessive lethals, have concluded that apurinic sites may be responsible, at least in part, for increases in this kind of genetic damage. Depurination of alkylated DNA is a slow process; 7-methylguanines, the major DNA adducts induced by MMS, are spontaneously lost by depurination with a half-life of three to six days at 37°C (Lawley, 1974), and the apurinic sites are converted into single-strand breaks with a half-life of 20-100 h at 37°C (Lindahl & Ljungquist, 1975).

The majority of chemical mutagens induce a number of different DNA lesions, and particular types of lesions may lead to specific genetic endpoints. In addition, different types of lesions may respond differentially to storage. That these phenomena can result in complex differences between the action of functionally related compounds was shown by recent studies of the following cross-linking agents: cisplatin (Woodruff et al., 1980; Brodberg et al., 1983), hexamethylphosphoramide, 2,5-bis(methoxyethoxy)-3,6-bisethyleneimino-1,4-benzoquinone (A 139) and tris(1-aziridinyl)phosphine sulfide (Thiotepa) (Vogel et al., 1985).In addition to single adducts, these compounds form different types of cross-links (interstrand and intrastrand DNA-DNA and DNA-protein cross-links). Differential effects of excision-repair deficiencies and storage on the frequency of induced complete loss of ring chromosomes and partial loss of rod chromosomes led to the following hypothesis (Vogel et al., 1985): one type of cross-link (which one remains to be determined) leads to sister chromatid exchanges and is responsible for those ring-X losses which result from such exchanges. Another type, or types, of cross-link leads to chromosomal aberrations (measured as partial chromosome loss) and to that fraction of the ring-X losses that results from a chromosomal aberration-type of event. Induction of the two mechanisms by individual agents depends differently on storage and excision repair.

6. Experience in mutagenicity screening

Drosophila assays were included in international collaborative studies on the evaluation of short-term tests for carcinogens (de Serres & Ashby, 1981; Ashby et al., 1985, 1986).They are also part of the US National Toxicology Program (Zeiger & Drake, 1980; Zeiger, 1983). At present, 168 chemicals have been tested for mutagenicity in *Drosophila* (Woodruff et al., 1984; Valencia et al., 1985; Woodruff et al., 1985; Yoon et al., 1985; Zimmering et al., 1985). Each compound was tested first for the induction of sex-linked recessive lethals by feeding the chemical in a solution of 5% aqueous sucrose; if it was not mutagenic by this route, it was tested by injection in a solution of 0.7% aqueous sodium chloride. If found inactive, the compound was not tested further. If it was mutagenic, the compound was tested for induction of reciprocal translocations using the exposure protocol and germ-cell stage that yielded positive results in the recessive lethal assay. Among the 168 chemicals tested, 38 were positive, 22 inconclusive and 108 negative in the recessive lethal assay. Among the 38 positive compounds, 14 induced heritable translocations; the remaining ten were negative.

In the most recent study (Woodruff et al., 1985), 11 of 48 chemicals that were found not to be mutagenic after feeding were mutagenic after injection. This ratio is very much dependent on the sample of chemicals chosen, because a similar relationship did not hold in previous studies — only one of 72 chemicals not mutagenic after feeding was mutagenic after injection. Overall, 24 of 158 chemicals were mutagenic after feeding, 17 of 151 were mutagenic when injected, and three were found to be mutagenic after both types of administration.

For 56 of the 168 chemicals tested, information on carcinogenicity was available. With this set of 56 chemicals, the agreement between mutagenicity in *Drosophila* and carcinogenicity in rodents is rather poor: 19/56 (34%) of the carcinogens are mutagens in *Drosophila*. However, there are two reasons for questioning the validity of this comparison. First, many of these 56 chemicals are insoluble or poorly soluble in water, and adult feeding and injection techniques necessitate the use of aqueous solutions, while corn oil gavage was used to treat rodents. A few chemicals were dissolved in oil for injections to *Drosophila*, thus avoiding aqueous solutions; however, the mortality rate was very high in controls because of leakage after injections. If chemicals that are insoluble at a concentration of 100 ppm in water are excluded from the comparison, only 17 chemicals are left in the sample, which is too few to make the comparison meaningful. The second reason for questioning the validity of this comparison is the specificity shown in the carcinogenesis bioassays. Among the nonmutagenic carcinogens, 72% were specific to one species (rat or mouse), and the majority are specific to a single organ (usually the liver). Clearly, data on more chemicals are necessary before comparisons of this type can be made with confidence.

Wild *et al.* (1983) studied 76 artificial flavouring substances for mutagenicity in the *Salmonella*/mammalian microsome assay, the sex-linked recessive-lethal test in *Drosophila*, and the micronucleus test in mouse bone marrow. It is noteworthy that 68 (89%) of the flavouring agents showed no effect in any of the tests. Four compounds were mutagenic in the *Salmonella* test, and two of these also induced a significant increase in sex-linked recessive-lethal mutations. One compound showed marginal activity in the *Salmonella* assay, but was negative in *Drosophila* and the mouse. One compound negative in *Salmonella* was positive in *Drosophila*; two further compounds appeared to be weakly mutagenic in *Drosophila* only. None of the flavouring substances induced micronuclei, i.e., cytogenetic damage in the bone marrow of mice. However, the authors conclude that this cannot be interpreted as a general absence of mutagenic effects in mice. The potential human health hazard associated with any of the artificial flavouring substances cannot be assessed without further special studies. This clearly underlines the fact that *Drosophila* assays are valid screening assays but cannot be used directly in human risk evaluation.

Drosophila mutagenicity assays are used to test pesticides (including insecticides) for their ability to induce lethals and chromosome losses, for several reasons. Large numbers and amounts of pesticides are released into the environment daily and, hence, may be a potential human mutagenic hazard, and *Drosophila* tests are relatively easy to perform on eukaryotes *in vivo*. However, an insect is obviously not an ideal test organism for mutagenicity testing of insecticides; mutagenic insecticides may cause rapid development of resistance in target insects. Four organophosphorus compounds gave negative results in the standard recessive-lethal assay (Benes & Sram, 1969; Vogel, 1974), but one of them, tested in an insecticide-resistant strain tolerating 35 times higher concentrations, showed a significant induction of lethals. Velazquez et al. (1984) reported that the insecticide endosulfan induces lethals as well as complete chromosomal losses but no partial Y chromosomal losses in Drosophila.

An intrinsic problem of insecticide testing in Drosophila is the limited concentration range that can be used. Woodruff et al. (1983) tried to solve this problem by assaying for complete ring-X chromosomal loss and partial Y chromosomal loss in crosses of exposed males to repair-deficient females, thus taking advantage of maternal effects. Of the 13 pesticides tested, one induced significant ring-X chromosomal loss; no pesticide induced a significant increase in partial chromosomal loss. That the positive compound failed to induce partial losses may be connected with the fact that premutational lesions and mutagenic mechanisms differ, at least partly, between these two endpoints. The results of Kramers (1985) on trenimon-induced losses indicate that partial losses are observed at higher doses than are ring-X losses. There is still controversy concerning the practical usefulness of repair-defective females for chromosome-loss assays, because the spontaneous mutation frequencies are increased and often vary from experiment to experiment. In addition, the enhanced sensitivity of repair-deficient mutants to cell killing and the problems associated with their reduced fertility should not be ignored (Vogel et al., 1983a).

Drosophila assays were included in the two most recent international collaborative studies (Vogel et al., 1985, 1986). In the first study, ten compounds were tested in the three somatic assays. Four carcinogens, hexamethylphosphoramide, ortho-toluidine, safrole and acrylonitrile, as well as the noncarcinogen caprolactam, were positive in the Drosophila assays, whereas benzoin, diethylstilboestrol and phenobarbital were considered to be inactive. No clear-cut decision could be made in the cases of benzene and di(2-ethylhexyl)phthalate. In the second study, in which the two carcinogens benzo[a]pyrene and 2-acetylaminofluorene and the two noncarcinogens pyrene and 4-acetylaminofluorene were tested, germ-line assays and somatic assays were compared. The most striking result was that none of the four chemicals was detected as a mutagen after treatment of male larvae. However, 2- and 4-acetylaminofluorene produced recessive lethals, which occurred in clusters in exposed female larvae. Benzo[a]pyrene and pyrene were nonmutagenic in female premeiotic cells. The three somatic assays gave positive results with two carcinogens (benzo[a]pyrene and 2-acetylaminofluorene), whereas pyrene was nonmutagenic. 4-Acetylaminofluorene produced weakly positive responses in the white/white-coral and in the wing test but was ineffective in the white-zeste system. The performance of the zeste-white mosaicism assay was particularly interesting, because both benzo[a]pyrene and 2-acetylaminofluorene, but not pyrene or 4-acetylaminofluorene, were considerably more cytotoxic in the absence of excision repair. This finding suggests the involvement of efficient repair of benzo[a]pyrene- and 2-acetylaminofluorene-related DNA lesions in repair-competent strains. The satisfactory performance of the somatic-cell assays suggests that they are promising new assays for predicting mammalian genotoxicity. However, more definite conclusions regarding the general applicability, the advantages and the limitations of this group of assays must await the results of extensive calibration studies yet to be completed in several laboratories.

In their studies with gaseous mutagens, Kramers et al. (1985a) showed that prolonged exposure can lead to the detection of lower concentrations of mutagens. This suggests that *Drosophila* could act as an in-situ monitor for the presence of mutagens in ambient air. Donner et al. (1983) exposed flies for several weeks at different locations in a rubber factory where concentrations of pollutants were known to occur and observed a slight but significant increase in the frequency of lethals at several locations. In preliminary experiments carried out by Kramers et al. (unpublished data), flies were placed in an automobile traffic tunnel for two to three weeks, resulting in a sex-linked recessive lethal frequency of $14/7079 = 0.20\%$, which is about the same as that of the laboratory control. Clearly, the approach has to be extended to a larger variety of polluted areas to be able to evaluate the possibilities and restrictions of in-situ application of mutagenicity tests with *Drosophila*, but the approach should be explored further.

7. Statistics

7.1 *Significance tests and sample size*

In general, it has been recommended that the significance of induced mutation frequencies in *Drosophila* assays be tested using either a one-tailed Fisher exact test (Fisher, 1935) or an approximation to the Fisher exact test — the conditional binomial test tabulated by Kastenbaum and Bowman (1970). Recently, Margolin et al. (1983) studied two statistical methods for their applicability to mutagenicity experiments that produce binomial responses from a control group and a single treated group. Attention was focused on experiments with (1) group-sample sizes greater than 500 and (2) a probability of less than 0.05 that a binary observation from any experimental unit is 'positive'. In addition, it was assumed that historical control data would not be included in the statistical analysis. The first test was the conditional binomial test (Kastenbaum & Bowman, 1970). The second was based on a standard normal approximation to the distribution of the difference between two sample proportions. A formula was presented for each analysis that relates the associated probability of detecting a mutagen to the mutant frequency and sample size of the two groups. On the basis of extensive results, it was concluded that the normal test is the preferred analysis for experiments in which the ratio of the two sample sizes is between 0.80 and 1.25. On the further assumption that an experiment is to be conducted with equal experimental group sample sizes, recommendations were made for the size of this common sample needed to achieve a specific effect.

The special problem of distinguishing between a negative and an inconclusive data set has been addressed by Selby and Olson (1981). Two-step procedures as proposed by them and by Frei (unpublished data) allow the classification of an experimental result as positive, weak, inconclusive or negative.

7.2 Historical controls

Several laboratories hold large sets of historical control data, especially for the sex-linked recessive-lethal assay and for the heritable-translocation assay (Gocke *et al.*, 1982; Woodruff *et al.*, 1984; Mason *et al.*, 1985). Mason *et al.* (1985) consider that there is reason to believe that long-term mutagen screening projects would benefit from the use of historical control data in combination with concurrent control data. Some problems related to the use of historical controls include variation over time and need to quantify the variability among experiments. Even when control frequencies are demonstrated to be reproducible, procedures for combining historical and concurrent control data must be determined. Statistical procedures for dealing with these problems are not yet available.

7.3 Clusters

The phenomenon of clustering (e.g., if, with the test for sex-linked recessive lethals, several lethal chromosomes are obtained from the same male) often creates uncertainties, particularly if weak mutagens are studied. The presence of clusters (suspected groups of identical mutations) can be judged on the basis of a comparison of the actual numbers of mutations obtained from an individual male with the expected number based on a Poisson distribution, taking into account the mean mutation frequency observed in the test series. A graphical procedure for detecting clusters has been published by Würgler and Graf (1985).

Two contrasting situations are encountered, depending on whether gonia or postmeiotic stages are studied. In postmeiotic cells, a mutagen cannot induce clusters; therefore, if clusters are observed, they represent premeiotic spontaneous events. By eliminating such pre-existing clusters from the data, the variance can be decreased, thereby increasing the sensitivity of the test system. Overestimation of mutation frequencies will be avoided, and, in borderline cases, the detection capacity of the assay will be increased. If mitotically dividing gonial cell populations are analysed (e.g., by treatment of larvae), mutated as well as nonmutant cells will multiply after induction of a mutation. Engels (1979) showed that when clustering is pronounced, the unweighted average mutation rate is a more efficient estimator than the usual average weighted by family size. When estimating mutation rates, therefore, all members of a cluster should be counted, but the family and cluster sizes need to be taken into account in determining the precision of the estimate.

8. Assessment of the test systems

(1) The general advantages of *Drosophila* as a eukaryotic organism for use in short-term tests for carcinogens are the short generation time (10-14 days), the small size of the animals, high fertility (leading to a large number of progeny), and inexpensive culture techniques. *Drosophila* assays are thus economical of time, space and money.

(2) The extended use of *Drosophila* in basic genetics, molecular genetics, radiation genetics and genetic toxicology has led to a unique collection of mutants and tester strains. In addition, the fact that there are only four chromosomes in the genome, one of which is small, greatly helps the construction of new assays and the improvement of old ones.

(3) *Drosophila* assays allow the detection of a large array of genetic endpoints: gene mutations, deletions, duplications, translocations, complete and partial chromosome loss, chromosome gain, recombination and effects on transposing elements.

(4) *Drosophila* is a useful organism for in-vivo studies, taking into account certain pharmacological and pharmacokinetic effects. Of particular importance is the versatile xenobiotic metabolism, which is in many respects comparable to extrahepatic metabolism in mammals. Analysis of the genetic control of the cytochrome P450 systems will allow the construction of tester strains with optimized metabolic capacity.

(5) Several assays have been validated using large numbers of compounds: recessive lethals, 421; heritable translocations, 61; clastogenesis, 76; nondisjunction, 44; SMART assays using the wing or the eye, over 130.

(6) The most important progress in recent years stems from the development and validation of the SMART assays. In contrast to the earlier germ-line assays, these tests are inexpensive and highly flexible. With reasonable experimental effort, it is possible to test different routes of exposure (feeding, inhalation and, eventually, injection), different exposure schedules (chronic or acute) and different doses.

(7) Some *Drosophila* assays are quite sensitive: in the recessive-lethal assay, 600-800 loci are tested per X chromosome, allowing studies of exposure levels that are expected to lead to a doubling of the spontaneous gene mutation frequency in humans (5×10^{-6} to 5×10^{-7} per gene per generation). In somatic assays, several thousand cells in every eye and more than 20 000 cells on every wing are assayed. In the wing assay, rough estimates of the induction frequency of genetic alterations per cell division can be obtained.

(8) The correct classification of carcinogens and noncarcinogens in the different assays is as follows: 90% (56/62) in the sex-linked recessive-lethal test; 90.5% (19/21) in the heritable-translocation test; 75% (15/20) in clastogenesis assays; and 88% (67/76) in SMART assays.

(9) Disadvantages of *Drosophila* assays include the following.

(a) Pharmacokinetics are different from those of mammals (e.g., adults and larvae do not have an organ in which the xenobiotic metabolism is concentrated, as in the mammalian liver).

(b) Certain metabolic differences have been found, e.g., in the metabolism of benzo[a]pyrene.

(c) Molecular dosimetry in target cells remains to be explored further and has to be developed to a point where it can be included in routine testing.

(d) The role played by mobile genetic elements in spontaneous and induced genetic events remains to be detemined in detail.

(e) Data obtained in *Drosophila* cannot be used directly in quantitative human risk evaluation.

(10) The somatic assays and the sex-linked recessive-lethal test detect gene mutations that are thought to play a critical role in the initiation step(s) of carcinogenesis. Mitotic recombination, translocations and aneuploidy are genetic events that lead potentially to the phenotypic expression of recessive mutations by creating homozygosity (or hemizygosity) or changes in the control of gene expression during the post-initiation stages of carcinogenesis.

9. Acknowledgement

We thank R.L. Dusenbery, J. Mason, P.D. Smith, J. Szabad, R. Woodruff and S. Zimmering for sending manuscripts and printers' proofs of unpublished articles and their kind permission to include this information in the present report.

10. References

Aaron, C.S. & Lee, W.R. (1977) Rejection of ethyl methanesulfonate feeding solution by *Drosophila melanogaster* adult males. *Drosophila Inform. Serv.*, 52, 64-65

Abeleva, E.A., Burychenko, G.M. & Potekhina, N.A. (1969) Nature of the distribution of the frequencies of mutants in the progeny of *Drosophila melanogaster* males subjected to the influence of methyl methanesulfonate and ethyl methanesulfonate as a criterion for the nature of mutation. *Genetika*, 5, 95-102

Abrahamson, S., Würgler, F.E., deJong, C. & Meyer, H.U. (1980) How many loci on the X-chromosome of *Drosophila melanogaster* can mutate to recessive lethals? *Environ. Mutagen.*, 2, 447-453

Alderson, T. (1965) Chemically induced delayed germinal mutation in *Drosophila*. *Nature*, 207, 164-166

Ashby, J. & Purchase, I.F.H. (1985) Significance of the genotoxic activities observed *in vitro* for 35 of 70 NTP noncarcinogens. *Environ. Mutagen.*, 7, 747-758

Ashby, J., de Serres, F.J., Draper, M., Ishidate M., Jr, Margolin, B.H., Matter, B.E. & Shelby, M.D. (1985) Evaluation of short-term tests for carcinogens. Report of the International Programme on Chemical Safety's collaborative study on in-vitro assays. *Progr. Mutat. Res.*, 5

Ashby, J., de Serres, F.J., Draper, M., Ishidate M., Jr, Margolin, B.H., Matter, B.E. & Shelby, M.D. (1986) Evaluation of short-term tests for carcinogens. Report of the International Programme on Chemical Safety's collaborative study on in-vivo assays. *Progr. Mutat. Res.*, 6 (in press)

Auerbach, C. (1951) Problems in chemical mutagenesis. *Quant. Biol.*, 16, 199-213

Auerbach, C. (1962) *Mutation: An Introduction to Research on Mutagenesis*, Edinburgh, Oliver & Boyd

Auerbach, C. (1976) *Mutation Research: Problems, Results and Perspectives*, London, Chapman & Hall

Baars, A.J. (1980) Biotransformation of xenobiotics in *Drosophila melanogaster* and its relevance for mutagenicity testing. *Drug Metab. Rev.*, 11, 191-221

Baars, A.J., Zijlstra, J.A., Vogel, E. & Breimer, D.D. (1977) The occurrence of cytochrome P-450 and aryl hydrocarbon hydroxylase activity in *Drosophila melanogaster* microsomes, and the importance of this metabolizing capacity for the screening of carcinogenic and mutagenic properties of foreign compounds. *Mutat. Res.*, 44, 257-268

Baars, A.J., Blijleven, W.G.H., Mohn, G.R., Natarajan, A.T. & Breimer, D.D. (1980) Preliminary studies on the ability of *Drosophila* microsomal preparations to activate mutagens and carcinogens. *Mutat. Res.*, 72, 257-264

Baars, A.J., Jansen, M. & Breimer, D.D. (1983) *Drug metabolizing enzymes in Drosophila melanogaster, in relation to genotoxicity testing*. In: Rydstroem, J., Montelius, J. & Bengtsson, M., eds, *Extrahepatic Drug Metabolism and Chemical Carcinogenesis*, Amsterdam, Elsevier, pp. 243-244

Becker, H.J. (1957) On X-ray-induced mosaic spots and mutations resulting in defective eyes in *Drosophila* and the developmental physiology of the eye [in German]. *Z. Vererbungsl.*, 88, 333-373

Becker, H.J. (1976) *Mitotic recombination*. In: Ashburner, M. & Novitski, E., eds, *Genetics and Biology of Drosophila*, Vol. 1C, New York, Academic Press, pp. 1020-1087

Benes, V. & Sram, R. (1969) Mutagenic activity of some pesticides in *Drosophila melanogaster*. *Ind. Med.*, 38, 50-53

Benyajati, C., Place, A.R. & Sofer, W. (1983) Formaldehyde mutagenesis in *Drosophila*. Molecular analysis of ADH-negative mutants. *Mutat. Res.*, 111, 1-7

Bigelow, S.W., Zijlstra, J.A., Vogel, E.W. & Nebert, D.W. (1985) Measurement of the cytosolic *Ah* receptor among four strains of *Drosophila*. *Arch. Toxicol.*, 56, 219-225

Bingham, P.M. (1981) A novel dominant allele at the *white* locus of *Drosophila melanogaster* is mutable. *Quant. Biol.*, 45, 519-525

Bingham, P.M., Kidwell, M.G. & Rubin, G.M. (1982) The molecular basis of *P-M* hybrid dysgenesis: the role of the *P* element, a *P*-strain-specific transposon family. *Cell*, 29, 995-1004

Blijleven, W.G.H. (1979) Metabolic activation of DEN and procarbazine in germ-free *Drosophila*. *Mutat. Res.*, 63, 413-415

Blijleven, W.G.H. & Vogel, E. (1977) The mutational spectrum of procarbazine in *Drosophila melanogaster*. *Mutat. Res.*, 45, 47-59

Boyd, J.B., Harris, P.V., Presley, J.M. & Narachi, M. (1983) *Drosophila melanogaster: a model eukaryote for the study of DNA repair*. In: Friedberg, E.C. & Bridges, B.A., eds, *Cellular Response to DNA Damage*, New York, Alan R. Liss, pp. 107-123

Brodberg, R.K., Lyman, R.F. & Woodruff, R.C. (1983) The induction of chromosome aberrations by cis-platinum(II)diamminodichloride in *Drosophila melanogaster*. *Environ. Mutagen.*, *5*, 285-297

Brookes, P. & Lawley, P.D. (1961) The reaction of mono- and di-functional alkylating agents with nucleic acids. *Biochem. J.*, *80*, 496-503

Brookes, P. & Lawley, P.D. (1963) Effects of alkylating agents on T2 and T4 bacteriophages. *Biochem. J.*, *89*, 138-144

Browning, L.S. & Altenburg, E. (1961) Gonadal mosaicism as a factor in determining the ratio of visible to lethal mutations in *Drosophila*. *Genetics*, *46*, 1317-1321

Carlson, E.A. & Oster, I.I. (1962) Comparative mutagenesis of the dumpy locus in *Drosophila melanogaster*. II. Mutational mosaicism induced without apparent breakage by a monofunctional alkylating agent. *Genetics*, *47*, 561-576

Carlson, E.A., Sederoff, R. & Cogan, M. (1967) Evidence favouring a frame-shift mechanism for ICR-170-induced mutations in *Drosophila melanogaster*. *Genetics*, *55*, 295-313

Carramolino, L., Ruiz-Gomez, M., Carmen de Guerrero, M., Campuzano, S. & Modolell, J. (1982) DNA map of mutations at the *scute* locus of *Drosophila melanogaster*. *EMBO J.*, *1*, 1185-1191

Casida, J.E. (1969) *Insect microsomes and insecticide chemical oxidations*. In: Gilette, J.R., Conney, A.H., Cosmides, G.J. & Estabrook, R.W., eds, *Microsomes and Drug Oxidations*, New York, Academic Press, pp. 517-531

Cavalli-Sforza, L.L. & Bodmer, W.F. (1971) *The Genetics of Human Populations*, San Francisco, W.H. Freeman

Datta, A.R., Randolph, B.W. & Rosner, J.L. (1983) Detection of chemicals that stimulate *Tn9* transposition in *Escherichia coli K12*. *Mol. gen. Genet.*, *189*, 245-250

Donner, M., Hytoenen, S. & Sorsa, M. (1983) Application of the sex-linked recessive lethal test in *Drosophila melanogaster* for monitoring the work environment of a rubber factory. *Hereditas*, *99*, 7-10

Dusenbery, R.L. (1986) Mutations at the *mei-41*, *mus(1)103*, *mus(2)205*, and *mus(3)310* loci of *Drosophila* exhibit differential UDS responses with different DNA damaging agents. *Mutat. Res.* (in press)

Eeken, J.C.J. & Sobels, F.H. (1981) Modification of *MR* mutator activity in repair-deficient strains of *Drosophila melanogaster*. *Mutat. Res.*, *83*, 191-200

Eeken, J.C.J., Sobels, F.H., Hyland, V. & Schalet, A.P. (1985) Distribution of MR-induced sex-linked recessive lethal mutations in *Drosophila melanogaster*. *Mutat. Res.*, *150*, 261-275

Eisen, H.J., Hannah, R.R., Legraverend, C., Okey, A.B. & Nebert, D.W. (1983) *The Ah receptor: controlling factor in the induction of drug-metabolizing enzymes by certain chemical carcinogens and other environmental pollutants*. In: Litwack, G., ed., *Biochemical Actions of Hormones*, New York, Academic Press, pp. 227-258

Engels, W.R. (1979) The estimation of mutation rates when premeiotic events are involved. *Environ. Mutagen.*, *1*, 37-43

Epler, J.L. (1966) Ethyl methanesulfonate-induced lethals in *Drosophila*. Frequency-dose relations and multiple mosaicism. *Genetics*, *54*, 31-36

Fahmy, O.G. & Bird, M.J. (1952) Chromosome breaks among recessive lethals induced by chemical mutagens in *Drosophila melanogaster*. *Heredity, 6*, 149-159

Fahmy, O.G. & Fahmy, M.J. (1961) Cytogenetic analysis of the action of carcinogens and tumor inhibitors in *Drosophila melanogaster*. X. The nature of the mutations induced by the mesyloxy esters in relation to molecular cross-linkage. *Genetics, 46*, 447-458

Falk, R. & Atidia, J. (1975) Mutation affecting taste perception in *Drosophila melanogaster*. *Nature, 254*, 325-326

Fisher, R.A. (1935) The logic of inductive inference. *J.R. stat. Soc. (Series A), 98*, 39-54

Ford, S. & Tomkins, L. (1985) An assay to measure the consumption of attractants in solution. *Drosophila Inform. Serv., 61*, 72

Förster, R.E. & Würgler, F.E. (1984) In-vitro studies on the metabolism of aflatoxin B_1 and aldrin in testes of genetically different strains of *Drosophila melanogaster*. *Arch. Toxicol., 56*, 12-17

Freese, E. (1971) *Molecular mechanism of mutagenesis*. In: Hollaender, A., ed., *Chemical Mutagens, Principles and Methods for their Detection*, Vol. 1, New York, Plenum, pp. 1-56

Fujikawa, K., Inagaki, E., Uchibori, M. & Kondo, S. (1983) Comparative induction of somatic eye-color mutations and sex-linked recessive lethals in *Drosophila melanogaster* by tryptophan pyrolysates. *Mutat. Res., 122*, 315-320

Fujikawa, K., Ryo, H. & Kondo, S. (1985) The *Drosophila* reversion assay using the unstable zeste-white somatic eye color system. *Progr. Mutat. Res., 5*, 310-324

Gateff, E. (1978) *Malignant and benign neoplasms of* Drosophila melanogaster. In: Ashburner, M. & Wright, T.R.F., eds, *The Genetics and Biology of Drosophila*, Vol. 2B, London, Academic Press, pp. 181-275

Gateff, E. (1982) Cancer, genes and development: the Drosophila case. *Adv. Cancer Res., 37*, 33-34

Gatti, M., Pimpinelli, S. & Baker, B.S. (1980) Relationships among chromatid interchanges, sister chromatid exchanges, and meiotic recombination in *Drosophila melanogaster*. *Proc. natl Acad. Sci. USA, 77*, 1575-1579

Gocke, E., Eckhardt, K., King, M.-T. & Wild, D. (1982) Some statistical aspects of spontaneous sex-linked recessive lethal mutations in *Drosophila*. *Mutat. Res., 104*, 239-242

Gollapudi, B.G., Bruce, R.J. & Sinha, A.K. (1985) The role of feeding rejection in *Drosophila* assays. *Mutat. Res., 144*, 13-17

Graf, U., Würgler, F.E., Katz, A.J., Frei, H., Juon, H., Hall, C.B. & Kale, P.G. (1984) Somatic mutation and recombination test in *Drosophila melanogaster*. *Environ. Mutagen., 6*, 153-188

Green, M.M. (1962) Back mutation in *Drosophila melanogaster*. II. Data on additional *yellow* and *white* mutants. *Genetics, 47*, 483-488

Green, M.M. (1967) The genetics of a mutable gene at the white locus of *Drosophila melanogaster*. *Genetics, 56*, 467-482

Green, M.M (1977) Genetic instability in *Drosophila melanogaster*. De novo induction of putative insertion mutations. *Proc. natl Acad. Sci. USA, 74*, 3490-3493

Green, M.M., Todo, T., Ryo, H. & Fujikawa, K. (1986) The genetic-molecular basis for a simple somatic system which detects environmental mutagens in *Drosophila melanogaster*. *Proc. natl Acad. Sci. USA* (in press)

Griesemer, R.A. & Cueto, C., Jr (1980) *Toward a classification scheme for degrees of experimental evidence for the carcinogenicity of chemicals for animals*. In: Montesano, R. Bartsch, H. & Tomatis, L., eds, *Molecular and Cellular Aspects of Carcinogen Screening Tests (IARC Scientific Publications No. 27)*, Lyon, International Agency for Research on Cancer, pp. 259-281

Gubb, D., Roote, J., Gill, S.M.C., Ashburner, M. & Shelton, M. (1985) *A large transposing element in* Drosophila: *interaction between* white+ *and* zeste-1 (Abstract) In: *IXth European Drosophila Research Conference*, 2-7 September 1985, Budapest, Balatonszeplak

Hällström, I. (1986a) *Genetic regulation of the cytochrome P-450 system in* Drosophila melanogaster. In: Ramel, C., Lambert, B. & Magnusson, J., eds, *Genetic Toxicology of Environmental Chemicals*, Part B, *Genetic Effects and Applied Mutagenesis*, New York, Alan R. Liss, pp. 419-425

Hällström, I. (1986b) Genetic variation in cytochrome p-450 dependent demethylation in *Drosophila melanogaster*. *Biochem. Pharmacol.* (in press)

Hällström, I. & Blanck, A. (1985) Genetic regulation of the cytochrome P450 system in *Drosophila melanogaster*. *Chem.-biol. Interact., 56,* 157-171

Hällström, I. & Grafström, R. (1981) The metabolism of drugs and carcinogens in isolated subcellular fractions of *Drosophila melanogaster*. *Chem.-biol. Interact., 34,* 145-159

Hällström, I., Magnusson, J. & Ramel, C. (1982) Relation between the somatic toxicity of dimethylnitrosamine and a genetically determined variation in the level and induction of cytochrome P-450 in *Drosophila melanogaster*. *Mutat. Res., 92,* 161-168

Hällström, I., Blanck, A. & Atuma, S. (1983) Comparison of cytochrome P-450 dependent metabolism in different developmental stages of *Drosophila melanogaster*. *Chem.-biol. Interact., 46,* 39-54

Hällström, I., Blanck, A. & Atuma, S. (1984) Genetic variation in cytochrome P-450 and xenobiotic metabolism in *Drosophila melanogaster*. *Biochem. Pharmacol., 33,* 13-20

Hannah-Alava, A. (1968) Induced crossing-over in the presterile broods of *Drosophila melanogaster* males. *Genetica, 39,* 94-152

IARC (1980) *IARC Monographs on the Evaluation of the Carcinogenic Risk of Chemicals to Humans*, Suppl. 2, *Long-term and Short-term Screening Assays for Carcinogens: A Critical Appraisal*, Lyon, pp. 157-183

Jack, J.W. & Judd, B.H. (1979) Allelic pairing and gene regulation: a model for the *zeste-white* interaction in *Drosophila melanogaster*. *Proc. natl Acad. Sci. USA, 76,* 1368-1372

Jansen, M., Baars, A.J. & Breimer, D.D. (1984) Cytosolic glutathione S-transferase in *Drosophila melanogaster*. *Biochem. Pharmacol., 33,* 3655-3659

Jansen, M., Baars, A.J. & Breimer, D.D. (1986) Microsomal and cytosolic epoxide hydrolase in *Drosophila melanogaster*. *Biochem. Pharmacol., 35,* 2229-2232

Jenkins, J.B. (1967) Mutagenesis at a complex locus in *Drosophila* with the monofunctional alkylating agent, ethyl methanesulfonate. *Genetics, 57,* 783-793

Kastenbaum, M.A. & Bowman, K.O. (1970) Tables for determining the statistical significance of mutation frequencies. *Mutat. Res., 9,* 527-549

Katz, A.J. (1984) Genotoxicity of isopropyl methanesulfonate (iPMS) in somatic cells of *Drosophila* (Abstract). *Genetics, 107,* s55

Katz, A.J. (1985) Genotoxicity of 5-azacytidine in somatic cells of *Drosophila*. *Mutat. Res., 143,* 195-199

Kaufman, T.C. & Suzuki, D.T. (1974) Temperature-sensitive mutations in *Drosophila melanogaster*. XX. Lethality due to translocations. *Can. J. Genet. Cytol., 16,* 579-592

Kidwell, M.G., Kidwell, J.F. & Sved, J.A. (1977) Hybrid dysgenesis in *Drosophila melanogaster*: a syndrome of aberrant traits including mutation, sterility, and male recombination. *Genetics, 86,* 813-833

Kikkawa, H. (1961) Genetical studies on the resistance of parathion in *Drosophila melanogaster*. I. Gene analyses. *Ann. Rep. Sci. Works Fac. Sci. Osaka Univ.*, 9, 1-20

Kilbey, B.J., McDonalds, D.J., Auerbach, C., Sobels, F.H. & Vogel, E.W. (1981) The use of *Drosophila melanogaster* in tests for environmental mutagens. *Mutat. Res.*, 85, 141-146

Kramers, P.G.N. (1977) Mechanism of hycanthone mutagenesis: the induction of temperature-sensitive mutations in *Drosophila melanogaster*. *Mutat. Res.*, 42, 349-356

Kramers, P.G.N. (1982) Studies on the induction of sex-linked recessive lethal mutations in *Drosophila melanogaster* by nitroheterocyclic compounds. *Mutat. Res.*, 101, 209-236

Kramers, P.G.N. (1985) Trenimon-induced sex chromosome loss in *Drosophila*: different dose-response for ring-X loss as compared to rod-X loss and Y-marker loss. *Mutat. Res.*, 142, 29-35

Kramers, P.G.N. & Burm, A.G.L. (1979) Mutagenicity studies with halothane in *Drosophila melanogaster*. *Anesthesiology*, 50, 510-513

Kramers, P.G.N., Bissumbhar, B. & Mout, H.C.A. (1985a) *Studies with gaseous mutagens in* Drosophila melanogaster. In: Waters, M.D., Sandhu, S.S., Lewtas, J. & Claxton, L., eds, *Short-Term Bioassays in the Analysis of Complex Environmental Mixtures*, Vol. IV, New York, Plenum, pp. 65-73

Kramers, P.G.N., Voogd, C.E., Knaap, A.G.A.C. & van der Heijden, C.A. (1985b) Mutagenicity of methyl bromide in a series of short-term tests. *Mutat. Res.*, 155, 41-47

Kulkarni, A.P., Smith, E. & Hodgson, E. (1976) Occurrence and characterization of microsomal cytochrome P-450 in several vertebrate and insect species. *Comp. Biochem. Physiol.*, 54B, 509-513

Lawley, P.D. (1974) Some chemical aspects of dose-response relationships in alkylation mutagenesis. *Mutat. Res.*, 23, 283-295

Lawley, P.D. & Brookes, P. (1963) Further studies on alkylation of nucleic acids and their constituent nucleotides. *Biochem. J.*, 89, 127-138

Lawley, P.D. & Shah, S.A. (1972) Reaction of alkylating mutagens and carcinogens with nucleic acids: detection and estimation of a small extent of methylation at O^6 of guanine in DNA by MMS *in vitro*. *Chem.-biol. Interact.*, 5, 286-288

Lee, W.R. (1976) Chemical mutagenesis. In: Ashburner, M. & Novitski, E., eds, *The Genetics and Biology of Drosophila*, Vol. 1C, London, Academic Press, pp. 1299-1341

Lee, W.R. (1978) *Dosimetry of chemical mutagens in eukaryote germ cells*. In: Hollaender, A. & de Serres, F.J., eds, *Chemical Mutagens*, Vol. 5, New York, Plenum, pp. 177-202

Lee, W.R., Sega, G.A. & Bishop, J.B. (1970) Chemically induced mutations observed as mosaics in *Drosophila melanogaster*. *Mutat. Res.*, 9, 323-336

Lee, A.R., Kirby, C.J. & Debney, C.W. (1976) The relation of germ line mosaicism to somatic mosaicism in *Drosophila*. *Genetics*, 55, 619-634

Lee, W.R., Abrahamson, S., Valencia, R., von Halle, E.S., Würgler, F.E. & Zimmering, S. (1983) The sex-linked recessive lethal test for mutagenesis in *Drosophila melanogaster*. *Mutat. Res.*, 123, 183-279

Lefevre, G. (1981a) A preliminary comparison of the distribution of EMS- and X-ray-induced mutations in *Drosophila melanogaster* (abstract). *Genetics*, 97, s63

Lefevre, G. (1981b) The distribution of randomly recovered X-ray-induced sex-linked genetic effects in *Drosophila melanogaster*. *Genetics*, 99, 461-480

Lifschytz, E. & Falk, R. (1969) Fine structure analysis of a chromosome segment in *Drosophila melanogaster*. *Mutat. Res.*, 8, 147-155

Lim, J.K. & Snyder, L.A. (1974) Cytogenetic and complementation analyses of recessive lethal mutations induced in the X chromosome of Drosophila by three alkylating agents. *Genet. Res.*, 24, 1-10

Lindahl, T. & Ljungquist, E. (1975) *Apurinic and apyrimidinic sites in DNA.* In: Hanawalt, P.C. & Setlow, R.B., eds, *Molecular Mechanisms for Repair of DNA*, New York, Plenum, pp. 31-38

Liu, C.P. & Lim, J.K. (1975) Complementation analysis of methyl methanesulfonate-induced recessive lethal mutations in the zeste-white region of the X-chromosome of Drosophila melanogaster. *Genetics*, 79, 601-611

MacDonald, D.J. & Luker, M.A. (1980) Fluoride: interaction with chemical mutagens in Drosophila. *Mutat. Res.*, 71, 211-218

Mackay, T.F.C. (1984) Jumping genes meet abdominal bristle: hybrid dysgenesis-induced quantitative variation in Drosophila melanogaster. *Genet. Res.*, 44, 231-237

Magnusson, J. & Ramel, C. (1978) Mutagenic effects of vinyl chloride on Drosophila melanogaster with and without pretreatment with sodium phenobarbiturate. *Mutat. Res.*, 57, 307-312

Magnusson, J., Hällström, I. & Ramel, C. (1979) Studies on metabolic activation of vinyl chloride in Drosophila melanogaster after pretreatment with phenobarbital and polychlorinated biphenyls. *Chem.-biol. Interact.*, 24, 287-298

Margolin, B.H., Collings, B.J. & Mason, J.M. (1983) Statistical analysis and sample-size determinations for mutagenicity experiments with binomial responses. *Environ. Mutagen.*, 5, 705-716

Mason, J.M., Strobel, E. & Green, M.M. (1984) *Mu-2*: mutator gene in Drosophila that potentiates the induction of terminal deficiencies. *Proc. natl Acad. Sci. USA*, 81, 6090-6094

Mason, J.M., Valencia, R., Woodruff, R.C. & Zimmering, S. (1985) Genetic drift and seasonal variation in spontaneous mutation frequencies in Drosophila. *Environ. Mutagen.*, 7, 663-676

Mathew, C. (1964) The nature of delayed mutation after treatment with chloroethyl methanesulfonate and other alkylating agents. *Mutat. Res.*, 1, 163-172

McCalla, D.R., Olive, P., Tu, Y. & Fan, M.L. (1975) Nitrofurazone-reducing enzymes in *E. coli* and their role in drug activation *in vivo. Can. J. Microbiol.*, 12, 1484-1491

Merriam, J.R. & Garcia-Bellido, A. (1972) A model for somatic pairing derived from somatic crossing over with the third chromosome rearrangements in Drosophila. *Mol. gen. Genet.*, 115, 302-313

Miglani, G.S. & Mohindra, V. (1985) Detection of chromosomal aberrations in the progenies of EMS-induced recombinants in Drosophila melanogaster. *Drosophila Inform. Serv.*, 122, 122

Molnar, I & Kiss, I. (1980) An easy method for rearing axenic Drosophila lines. *Drosophila Inform. Serv.*, 55, 163-164

Morton, R.A. & Hall, S.C. (1985) Response of dysgenic and non-dysgenic populations to malathion exposure. *Drosophila Inform. Serv.*, 61, 126-128

Muller, H.J.rosophila melanogaster.
Cancer Res., 39, 4224-4227

Obe, G. & Ristow, H. (1977) Acetaldehyde, but not ethanol, induces sister chromatid exchanges in Chinese hamster cells *in vitro. Mutat. Res.*, 56, 211-213

Organization for Economic Cooperation and Development (1984) *Sex-linked Recessive Lethal Test in* Drosophila melanogaster, *Guidelines for Testing of Chemicals (Guideline 477, 2nd Addendum)*, Paris

Oeschger, N.S. & Hartman, P.E. (1970) ICR-induced frameshift mutations in the histidine operon of *Salmonella. J. Bacteriol.*, *101*, 490-504

O'Hare, K. (1985) The mechanism and control of P element transposition in *Drosophila melanogaster. Trends Genet.*, *1*, 125-254

Olsen, O.-.A. & Green, M.M. (1982) The mutagenic effects of diepoxybutane in wild-type and mutagen-sensitive mutants of *Drosophila melanogaster. Mutat. Res.*, *92*, 107-115

Ong, T.M. & de Serres, F.J. (1980) Genetic analysis of *adenine-3* mutants induced by hycanthone, lucanthone and their indazole analogs in *Neurospora crassa. J. environ. Pathol. Toxicol.*, *4*, 1-8

Porsch-Hällström, I. (1984) *Cytochrome P-450 in* Drosophila melanogaster, *Activity, Genetic Variation and Regulation*, University of Stockholm, Doctoral Thesis, ISBN 91-7146-603-7

Purdom, C.E. (1957) Autonomous action of lethal mutations induced in the germ cells of *Drosophila melanogaster* by 2-chloroethyl methanesulfonate. *Nature*, *180*, 81-85

Rasmuson, A. (1985a) Effects of DNA repair deficient mutants on somatic and germ line mutagenesis in the UZ system in *Drosophila melanogaster. Mutat. Res.*, *141*, 29-33

Rasmuson, A. (1985b) Comparative studies of the induction of somatic eye-color mutations in an unstable strain of *Drosophila melanogaster* by MMS and X-rays at different developmental stages. *Mutat. Res.*, *148*, 65-70

Rasmuson, B. & Green, M.M. (1974) Genetic instability in *Drosophila melanogaster. Mol. gen. Genet.*, *133*, 249-260

Rasmuson, B., Westerberg, B.M., Rasmuson, A., Gvozdev, B.A., Blayava, E.S. & Ilyin, Y.V. (1981) Transpositions, mutable genes, and the dispersed gene family *Dm255* in *Drosophila melanogaster. Quant. Biol.*, *45*, 545-551

Rasmuson, B., Rasmuson, A. & Nygren, J. (1984) *Eye pigmentation changes in Drosophila melanogaster as a sensitive test for mutagenicity*. In: Kilbey, B.J., Legator, M., Nichols, W. & Ramel, C., eds, *Handbook of Mutagenicity Test Procedures*, 2nd ed., Amsterdam, Elsevier, pp. 603-613

Reardon, J.T., Liljestrand-Golden, C.A., Dusenbery, R.L. & Smith, P.D. (1986) Molecular analysis of diepoxybutane-induced mutants at the *rosy* locus of *Drosophila melanogaster. Genetics* (in press)

Reddy, J.K., Azarnoff, D.L. & Hignite, C.E. (1980) Hypolipidaemic hepatic peroxisome proliferators form a novel class of chemical carcinogens. *Nature*, *283*, 397-398

Reddy, J.K., Lalwani, N.D., Reddy, M.K. & Qureshi, S.A. (1982) Excessive accumulation of autofluorescent lipofuscin in the liver during hepatocarcinogenesis by methyl clofenapate and other hypolipidemic peroxisome proliferators. *Cancer Res.*, *42*, 259-266

Ryo, H., Ito, K. & Kondo, S. (1981) Increment of recessive lethal mutations and dominant lethals in offspring of *Drosophila melanogaster* on storage of methylmethanesulfonate-treated sperm in females. *Mutat. Res.*, *83*, 179-190

Ryo, H., Yoo, M.A., Fujikawa, K. & Kondo, S. (1985) Comparison of somatic reversions between the ivory allele and transposon-caused mutant alleles at the white locus of *Drosophila melanogaster* after larval treatment with X rays and ethyl methanesulfonate. *Genetics*, *110*, 441-451

Sankaranarayanan, K. & Sobels, F.H. (1976) *Radiation genetics*. In: Ashburner, M. & Novitski, E., eds, *The Genetics and Biology of Drosophila*, Vol. 1C, London, Academic Press, pp. 1089-1250

Schwartz, M. & Sofer, W. (1976) Alcohol dehydrogenase-negative mutants in *Drosophila*: defects at the structural locus. *Genetics*, *83*, 125-136

P.B. & Olson, W.H. (1981) Methods and criteria for deciding whether specific-locus mutation rate data in mice indicate a positive, negative or inconclusive result. *Mutat. Res.*, *83*, 403-418

de Serres, F.J. & Ashby, J., eds (1981) Evaluation of short-term tests for carcinogens. Report of the International Collaborative Program. *Progr. Mutat. Res.*, *1*

Shearman, C.W. & Loeb, L.A. (1977) Depurination decreases fidelity of DNA synthesis *in vitro*. *Nature*, *270*, 537-538

Shelby, M.D. & Stasiewicz, S. (1984) Chemicals showing no evidence of carcinogenicity in long-term, two-species rodent studies: the need for short-term test data. *Environ. Mutagen.*, *6*, 871-878

Shukla, P.T. & Auerbach, C. (1980) Genetic tests for the detection of chemically induced small deletions in Drosophila chromosomes. *Mutat. Res.*, *72*, 231-243

Simmons, M.J. & Karess, R.E. (1985) Molecular and population biology of hybrid dysgenesis. *Drosophila Inform. Serv.*, *61*, 2-7

Simmons, V.F. (1979) In-vitro assay for recombinogenic activity of chemical carcinogens and related compounds with *Saccharomyces cerevisiae D3*. *J. natl Cancer Inst.*, *62*, 901-909

Slizinska, H. (1969) The progressive approximation, with storage, of the spectrum of TEM-induced chromosomal changes in *Drosophila* sperm to that found after irradiation. *Mutat. Res.*, *8*, 165-175

Smith, P.D. & Dusenbery, R.L. (1985) Mutagen sensitivity of *Drosophila melanogaster*. *Mutat. Res.*, *150*, 235-240

Smith, P.D., Reardon, J.T., Liljestrand-Golden, .C.A. & Dusenbery, R.L. (1986) *Genetic and molecular genetic studies of mutation induction in an excision-defective strain of* Drosophila melanogaster. In: Ramel, C., Lambert, B. & Magnusson, J., eds, *Genetic Toxicology of Environmental Chemicals*, Part A, *Basic Principles and Mechanisms of Action*, New York, Alan R. Liss, pp.

Sobels, F.H. & van Stennis, H. (1957) Chemical induction of crossing-over in *Drosophila* males. *Nature*, *179*, 29-31

Sosnowski, S.A., Rajalakshmi, S. & Sarma, D.S.R. (1976) Protection by dimethyl sulfoxide of strand breaks in hepatic DNA induced by dimethylnitrosamine. *Chem.-biol. Interact.*, *15*, 101-104

Spradling, A.C. & Rubin, G.M. (1981) *Drosophila* genome organization: conserved and dynamic aspects. *Ann. Rev. Genet.*, *15*, 219-264

Spradling, A.C. & Rubin, G.M. (1982) Transposition of cloned P elements into *Drosophila* germ line chromosomes. *Science*, *218*, 341-347

Sun, L. & Singer, B. (1975) The specificity of different classes of ethylating agents toward various sites of Hela cell DNA *in vitro* and *in vivo*. *Biochemistry*, *14*, 1795-1802

Surian, A., Kocsis, Z., Pinter, M., Torok, G., Börzsönyi, M. & Szabad, J. (1985) Analysis of the genotoxic activity of four *N*-nitroso compounds by the *Drosophila* mosaic test. *Mutat. Res.*, *144*, 177-181

Suzuki, D.T., Piternick, L.K., Hayashi, S., Tarasoff, M., Baillie, D. & Erasmus, U. (1967) Temperature-sensitive mutations in *Drosophila melanogaster*. I. Relative frequencies among X-ray and chemically induced sex-linked recessive lethals and semi-lethals. *Proc. natl. Acad. Sci. USA*, *57*, 907-912

Suzuki, D.T., Kaufman, T.C., Falk, D. & U.B.C. Drosophila Research Group (1976) *Conditionally expressed mutations in* Drosophila melanogaster. In: Ashburner, M. & Novitski, E., eds, *The Genetics and Biology of Drosophila*, Vol. 1A, London, Academic Press, pp. 208-264

Swaminathan, S. & Lower, G.M. (1978) *Biotransformations and excretion of nitrofurans*. In: Bryan, G.T., ed., *Carcinogenesis, a Comprehensive Survey*, Vol. 4, New York, Raven Press, pp. 59-98

Swenson, D.H. & Lawley, P.D. (1978) Alkylation of DNA by carcinogens dimethyl sulphate, ethyl methanesulfonate, N-ethyl-N-nitrosourea and N-methyl-N-nitrosourea: relative reactivity of the phosphodiester site thymidylyl(3′-5′)thymidine. *Biochem. J.*, *171*, 575-587

Szabad, J. & Bennettova, B. (1985) Analysis of the genotoxic activity of five compounds affecting insect fertility. *Mutat. Res.*, *173*, 197-200

Szabad, J., Soos, I., Polgar, G. & Hejja, G. (1983) Testing the mutagenicity of malondialdehyde and formaldehyde by the *Drosophila* mosaic and the sex-linked recessive lethal tests. *Mutat. Res.*, *113*, 117-133

Tates, A.D. (1971) *Cytodifferentiation During Spermatogenesis in* Drosophila melanogaster, Leiden, University of Leiden

Traut, H. & Scheid, W. (1970) Cytological analysis of partial and total X-chromosome loss induced by X-rays in oocytes of *Drosophila melanogaster*. *Mutat. Res.*, *10*, 583-589

Valencia, R., Abrahamson, S., Lee, W.R., von Halle, E.S., Woodruff, R.C., Würgler, F.E. & Zimmering, S. (1984) Chromosome mutation tests for mutagenesis in *Drosophila melanogaster*. A report of the US Environmental Protection Agency Gene-Tox Program. *Mutat. Res.*, *134*, 61-88

Valencia, R., Mason, J.M., Woodruff, R.C. & Zimmering, S. (1985) Chemical mutagenesis testing in *Drosophila*. III. Results of 48 coded compounds tested for the National Toxicology Program. *Environ. Mutagen.*, *7*, 325-348

Velazquez, A., Creus, A., Xamena, N. & Marcos, R. (1984) Mutagenicity of the insecticide endosulfan in *Drosophila melanogaster*. *Mutat. Res.*, *136*, 115-118

Verburgt, F.G. & Vogel, E. (1977) Vinyl chloride mutagenesis in *Drosophila melanogaster*. *Mutat. Res.*, *48*, 327-336

Vig, B.K. (1977) Genetic toxicology of mitomycin C, actinomycins, daunomycin and adriamycin. *Mutat. Res.*, *49*, 189-238

Vogel, E. (1974) Mutagenic activity of the insecticide oxydemetonmethyl in a resistant strain of *Drosophila melanogaster*. *Experientia*, *30*, 369-397

Vogel, E. (1975) Some aspects of the detection of potential mutagenic agents in *Drosophila*. *Mutat. Res.*, *29*, 241-250

Vogel, E. (1977) *Identification of carcinogens by mutagen testing in* Drosophila: *the relative reliability for the kinds of genetic damage measured*. In: Hiatt, H.H., Watson, J.D. & Winsten, J.A., eds, *Origins of Human Cancer*, Book C, *Human Risk Assessment*, Cold Spring Harbor, NY, CSH Press, pp. 1483-1497

Vogel, E. (1979) Mutagenicity of chloroprene, 1-chloro-1,3-trans-butadiene, 1,4-dichlorobutene-2 and 1,4-dichloro-2,3-epoxybutane in *Drosophila melanogaster*. *Mutat. Res.*, *67*, 377-381

Vogel, E. (1980) Genetical relationship between resistance to insecticides and promutagens in two *Drosophila* populations. *Arch. Toxicol.*, *43*, 201-211

Vogel, E. (1982) *Dependence of mutagenesis in* Drosophila *males on metabolism and germ cell stage*. In: Sugimura, T., Kondo, S. & Takebe, H., eds, *Environmental Mutagens and Carcinogens*, Tokyo, University of Tokyo Press, pp. 183-194

Vogel, E.W. (1984) *A comparison of genotoxic activity in somatic tissue and in germ cells of* Drosophila melanogaster. In: Chu, E.H.Y. & Generoso, W.M., eds, *Mutation, Cancer, and Malformation*, New York, Plenum, pp. 233-255

Vogel, E.W. (1985) The *Drosophila* somatic recombination and mutation assay (SRM) using the white-coral somatic eye color system. *Progr. Mutat. Res.*, 5, 313-317

Vogel, E. & Leigh, B. (1975) Concentration-effect studies with MMS, TEB, 2,4,6-C13-PDMT, and DEN on the induction of dominant recessive lethals, chromosome loss and translocations in *Drosophila* sperm. *Mutat. Res.*, 29, 383-396

Vogel, E. & Lüers, H. (1974) A comparison of adult feeding to injection in *Drosophila melanogaster*. *Drosophila Inform. Serv.*, 51, 113-114

Vogel, E. & Natarajan, A.T. (1979a) The relation between reaction kinetics and mutagenic action of monofunctional alkylating agents in higher eukaryotic systems. I. Recessive lethal mutations and translocations in *Drosophila*. *Mutat. Res.*, 62, 51-100

Vogel, E. & Natarajan, A.T. (1979b) The relation between reaction kinetics and mutagenic action of monofunctional alkylating agents in higher eukaryotic systems. II. Total and partial sex-chromosome loss in *Drosophila*. *Mutat. Res.*, 62, 101-123

Vogel, E.W., Blijleven, W.G.H., Kortselius, M.J.H. & Zijlstra, J.A. (1982) A search for some common characteristics of the effects of chemical mutagens in *Drosophila*. *Mutat. Res.*, 92, 69-87

Vogel, E.W., Zijlstra, J.A. & Blijleven, W.G.H. (1983a) Mutagenic activity of selected aromatic amines and polycyclic hydrocarbons in *Drosophila melanogaster*. *Mutat. Res.*, 107, 53-77

Vogel, E.W., Zijlstra, J.A., Blijleven, W.G.H. & Breimer, D.D. (1983b) *Metabolic activation and mutagenic properties of procarcinogens in* Drosophila. In: Kobler, A.R., ed., *In Vitro Toxicity Testing of Environmental Agents, Current and Future Possibilities*, New York, Plenum, pp. 215-233

Vogel, E.W., Frei, H., Fujikawa, K., Graf, U., Kondo, S., Ryo, H. & Würgler, F.E. (1985) Summary report on the performance of the Drosophila assays. *Progr. Mutat. Res.*, 5, 47-57

Vogel, E.W., Bulyzhenkoy, V.E., Fujikawa, K., Graf, U., Ivanov, V.I., Kilbey, B., Kondo, S. & Laison, K. (1986) Summary report on the performance of the *Drosophila* assays. *Progr. Mutat. Res.*, 6 (in press)

Waters, L.C., Nix, C.E. & Epler, J.L. (1983) Studies on the relationship between dimethylnitrosamine-demethylase activity and dimethylnitrosamine-dependent mutagenesis in *Drosophila melanogaster*. *Chem.-biol. Interact.*, 46, 55-66

Wild, D., King, M.T., Gocke, E. & Eckhardt, K. (1983) Study of artificial flavouring substances for mutagenicity in the *Salmonella*/microsome, Basc and micronucleus test. *Food chem. Toxicol.*, 21, 707-719

Woodruff, R.C. & Brodberg, R.K. (1983) Comparison of mating schemes for screens of chemical induced heritable translocations in *Drosophila melanogaster*. *Mutat. Res.*, 119, 293-297

Woodruff, R.C. & Gander, R.M. (1974) The induction of temperature-sensitive mutations in *Drosophila melanogaster* by the acridine mustard ICR-170. *Mutat. Res.*, 25, 337-345

Woodruff, R.C., Valencia, R., Lyman, R.F., Earle, B.A. & Boyce, J.T. (1980) The mutagenic effect of platinum compounds in *Drosophila melanogaster*. *Environ. Mutagen.*, 2, 133-138

Woodruff, R.C., Phillips, J.P. & Irwin, D. (1983) Pesticide-induced complete and partial chromosome loss in screens with repair-defective females of *Drosophila melanogaster*. *Environ. Mutagen.*, 5, 835-864

Woodruff, R.C., Mason, J.M., Valencia, R. & Zimmering, S. (1984) Chemical mutagenesis testing in *Drosophila*. I. Comparison of positive and negative control data for sex-linked recessive lethal mutations and reciprocal translocations in three laboratories. *Environ. Mutagen.*, 6, 189-202

Woodruff, R.C., Mason, J.M., Valencia, R. & Zimmering (1985) Chemical mutagenesis testing in *Drosophila*. V. Results of 53 coded compounds tested for the National Toxicology Program. *Environ. Mutagen.*, 7, 677-702

Würgler, F.E. (1986) *Mutagenicity assays detecting recombination.* In: Ramel, C., Lambert, B. & Magnusson, J., eds, *Genetic Toxicology of Environmental Chemicals*, Part B, *Genetic Effects and Applied Mutagenesis*, New York, Alan R. Liss, pp. 85-90

Würgler, F.E. & Graf, U. (1985) *Mutagenicity testing with* Drosophila melanogaster. In: Muhammed, A. & von Borstel, R.C., eds, *Basic and Applied Mutagenesis, with Special Reference to Agricultural Chemicals in Developing Countries*, New York, Plenum, pp. 343-372

Würgler, F.E. & Vogel, E.W. (1986) *In-vivo mutagenicity testing using somatic cells of* Drosophila melanogaster. In: de Serres, F.J., ed., *Chemical Mutagens*, Vol. 10, New York, Plenum, pp. 1-72

Würgler, F.E., Graf, U., Frei, H. & Juon, H. (1983) Genotoxic activity of the anti-cancer drug methotrexate in somatic cells of *Drosophila melanogaster. Mutat. Res., 122*, 321-328

Würgler, F.E., Sobels, F.H. & Vogel, E. (1984) Drosophila *as an assay system for detecting genetic changes.* In: Kilbey, B.J., Legator, M., Nichols, W. & Ramel, C., eds, *Handbook of Mutagenicity Tests Procedures*, 2nd ed., Amsterdam, Elsevier, pp. 555-601

Würgler, F.E., Graf, U. & Frei, H. (1985) Somatic mutation and recombination in wings of *Drosophila melanogaster. Progr. Mutat. Res., 5*, 325-340

Yahagi, T., Nagao, M., Seino, Y., Matsushima, T., Sugimura, T. & Okada, M. (1977) Mutagenicities of *N*-nitrosamines on *Salmonella. Mutat. Res., 4*, 121-130

Yang, R.S.H. (1976) *Enzymatic conjugation and insecticide metabolism.* In: Wilkinson, C.F., ed., *Insecticide Biochemistry and Physiology*, New York, Plenum, pp. 177-225

Yanopoulos, G., Pelecanos, M. & Zacharopoulov, A. (1980) Combined action of diethyl sulphate (DES) and of male recombination factor 31.1 MRF in *Drosophila melanogaster. Genetika, 12*, 41-48

Yoo, A.M., Ryo, H., Todo, T. & Kondo, S. (1985) Mutagenic potency of amino acid pyrolysates in the *Drosophila* wing test and its correlation to carcinogenic potency. *Jpn J. Cancer Res. (Gann), 76*, 468-473

Yoon, J.S., Mason, J.M., Valencia, R., Woodruff, R.C. & Zimmering, S. (1985) Chemical mutagenesis testing in *Drosophila.* IV. Results of 45 coded compounds tested for the National Toxicology Program. *Environ. Mutagen., 7*, 349-367

Yoshikawa, I., Ayaki, T. & Oshima, K. (1984a) Comparative studies of dose-response curves for recessive lethal mutations induced by ethyl-nitrosourea in spermatogonia and in spermatozoa of *Drosophila melanogaster. Environ. Mutagen., 6*, 489-496

Yoshikawa, I., Ayaki, T. & Oshima, K. (1984b) *Comparison of dose-response curves for ENU-induced recessive lethal mutations between spermatogonia and spermatozoa in* Drosophila. In: Tazima, Y., Kondo, S. & Kuroda, Y., eds, *Problems of Threshold in Chemical Mutagenesis.* Tokyo, The Environmental Mutagen Society of Japan, pp. 93-97

Younes, M., Siegers, C.-P. & Filser, J.G. (1979) Effect of dithiocarb and dimethyl sulfoxide on irreversible binding of 14C-bromobenzene to rat liver microsomal protein. *Arch. Toxicol., 42*, 289-293

Zachar, Z. & Bingham, P.M. (1982) *Regulation of* white *locus expression: the structure of mutant alleles at the white locus of* Drosophila melanogaster. *Cell, 30*, 529-541

Zeiger, E. (1983) An overview of genetic toxicity testing in the National Toxicology Program. *Ann. N.Y. Acad. Sci., 407*, 387-394

Zeiger, E. & Drake, J.W. (1980) *An environmental mutagenesis test development programme.* In: Montesano, R., Bartsch, H. & Tomatis, L., eds, *Molecular and Cellular Aspects of Carcinogen Screening Tests (IARC Scientific Publications No. 27)*, Lyon, International Agency for Research on Cancer, pp. 303-313

Zijlstra, J.A. & Vogel, E.W. (1984) Mutagenicity of 7,12-dimethylbenz[*a*]anthracene and some aromatic mutagens in *Drosophila melanogaster*. *Mutat. Res.*, *125*, 243-261

Zijlstra, J.A. & Vogel, E.W. (1985) *The possible involvement of monoamine oxidases in the bioactivation and de-activation of mutagens in* Drosophila melanogaster. In: *Fourth International Conference on Environmental Mutagens, Stockholm, 24-28 June 1985, Abstract Book*, p. 106

Zijlstra, J.A., Vogel, E. & Breimer, D.D. (1979) Occurrence and inducibility of cytochrome P-450 and mixed-function oxidase activities in microsomes from *Drosophila melanogaster*. *Mutat. Res.*, *64*, 151-152

Zijlstra, J.A., Vogel, E.W. & Breimer, D.D. (1984) Strain-differences and inducibility of microsomal enzymes in *Drosophila melanogaster* flies. *Chem.-biol. Interact.*, *48*, 317-338

Zimmering, S. & Kammermeyer, K.L. (1982) On the nature of partial losses of the Y chromosome from treatment of ring-X/B^SY_y+ males with diethylnitrosamine (DEN) or procarbazine and mating with repair-deficient *st mus302* females of *Drosophila*. *Mutat. Res.*, *104*, 121-123

Zimmering, S., Mason, J.M., Valencia, R. & Woodruff, R.C. (1985) Chemical mutagenesis testing in *Drosophila*. II. Results of 20 coded compounds tested for the National Toxicology Program. *Environ. Mutagen.*, *7*, 87-100

REPORT 13

ASSAYS FOR ANEUPLOIDY IN *DROSOPHILA MELANOGASTER*

Prepared by:

*F.E. Würgler (Rapporteur), C. Ramel (Chairman),
E. Moustacchi and A. Carere*

1. Introduction

The many experimental advantages of *Drosophila* (see Report 12) have stimulated the development of assays for detecting aneuploidy. There are two groups of tests: (1) those which detect complete (CL) and partial (PL) chromosome loss (also classified as assays for detecting clastogenicity (Valencia et al., 1984)) and (2) those which detect nondisjunction (ND) resulting in chromosome gain as well as chromosome loss (see Table 1).

Table 1. Numbers of compounds in the *Drosophila* Gene-Tox data base evaluated for aneuploidy and correlated with carcinogenicity[a]

Test	Mutagenicity[b]	Carcinogenicity[b]				
		+	-	?	NI	Total
Clastogenesis	+	15	1	0	10	26
	-	4	0	2	7	13
	(+)	1	0	1	1	3
	inc	4	0	4	14	22
	(-)	2	0	1	9	12
	Total	26	1	8	41	76
Nondisjunction	+	4	0	3	8	15
	-	5	0	1	7	13
	(+)	1	0	1	1	3
	inc	2	0	1	6	9
	(-)	0	0	0	4	4
	Total	12	0	6	26	44

[a]From Valencia et al. (1984)

[b]+, positive; -, negative; inc, inconclusive; (+), inconclusive, but looks positive; (-), inconclusive, but looks negative; ?, questionable; NI, not indicated by Griesemer and Cueto (1980)

2. Tests for chromosome loss

2.1 *Genetic principles*

Induced aneuploidy is usually tested by exposing male germ cells to a potential clastogenic agent. The X chromosome in the male may be a rod or a ring chromosome. Chromosomal breaking events may lead to the loss of a rod or of a ring-X chromosome (Leigh, 1976). Ring-X chromosomes are more sensitive to induced loss because, in addition to the chromosomal breakage effects in the ring, sister chromatid exchanges also lead to dicentric chromosomes and consequently to loss (Würgler & Graf, 1980). The X chromosome carries *yellow* (y) as a body-colour marker. The Y chromosome usually employed is the $B^s Yy^+$ chromosome with the dominant B^s ('*Bar of Stone*', a very narrow eye) on the long arm of the chromosome and the dominant wild type allele of *yellow* (y^+) on the short arm. The males are crossed with normal females with two free X chromosomes. The X in the female should be a rod, and it must be an X that is viable in males without a Y chromosome. The inclusion of *white* (or *white-apricot*; w or w^a) in the maternal X chromosome allows detection of aneuploidy arising in either parent. If the maternal X also carries *yellow* (y), the system allows detection of the segregation or loss of the y^+ on the short arm of the Y chromosome. In the cross of $X^{c2}, y/B^s Yy^+$ males with y w females, the usual progeny are *yellow* (y/y w) females and *white Bar* ($B^s Yy^+/y\ w$) males; complete chromosome loss of X or Y leads to exceptional *yellow white* ($O/y\ w$ males), partial loss results in either *yellow white Bar* ($B^s/y\ w$ males or *white* ($y^+/y\ w$) males.

2.2 *Storage effects*

If mutagen-exposed mature sperm are stored in females and used to inseminate an egg only after many days, a time-dependent increase in chromosomal aberrations ('storage effect') may be observed. Storage effects have been found with chromosomal aberrations (heritable 2-3 translocations), ring-X losses and partial chromosomal losses. Different premutational lesions, such as alkylated bases, intra- and interstrand DNA cross-links and DNA-protein cross-links, may be the substrates for the storage effect (for details, see Report 12).

2.3 *Maternal effects*

A large body of experimental evidence supports the concept that mutagens induce premutational lesions in postmeiotic cells, and particularly in the metabolically inert nuclei of mature spermatozoa, which will be repaired or fixed as mutations only after the sperm has entered an egg. To allow the sperm nucleus to restart DNA metabolism, important components of the egg cytoplasm enter the sperm nucleus, e.g., enzymes that have been synthesized on the basis of genetic information from the female that produced the oocyte. If then, for example, defective enzymes functioning in excision repair have been produced by an excision-repair-defective female, the normal level of repair may not be attained and the frequency of mutation fixation may increase. Therefore, if mutagens that induce excisable lesions are tested in mature sperm by crossing the treated males with excision-repair-defective females, an increased frequency of chromosomal loss (and other types of mutations) may be obtained over that seen with a corresponding cross with excision-repair-proficient females. This phenomenon has been studied in several laboratories (for a review, see Würgler et al., 1986). A large number of repair-defective strains exist in Drosophila (for a review, see Boyd et al., 1983). Several have been studied for the occurrence of maternal effects with chemically-treated sperm; these include the excision-repair-defective mei-9 (mei-9^a or mei-9^{L1}) and the postreplication-repair-deficient mus302.

2.4 Data obtained

The phenotypes of F_1 progeny of exposed males mated to test females indicate whole or partial chromosome loss. These flies are counted and their frequency is calculated among total males or among total F_1 flies.

2.5 Test performance

The Gene-Tox report on chromosomal mutation tests (Valencia et al., 1984; Zimmering et al., 1986) included 76 chemicals tested by the aneuploidy test for clastogenesis (Table 1). Of these, 26 were positive and 13 negative; 37 could not be classified definitively as positive or negative, usually owing to insufficient numbers of gametes tested. Among the 26 positive compounds there were 15 carcinogens and one noncarcinogen; among the 13 negatives there were four known and two suspected carcinogens but no noncarcinogen.

A set of 21 compounds tested in the mei-9^a test for chromosome loss was reviewed by Zimmering (1983). Seventeen of the compounds are classified as carcinogens and four are unknown in this regard. At the exposure concentrations reported, 21/21 compounds were positive for complete and partial loss and 21/21 for partial loss in the mei-9^a test. At the same concentration, only 10/20 compounds were positive for complete and partial loss and 2/21 compounds for partial loss with repair-proficient females. Data were compared with results of heritable translocation tests in which the concentration was no lower and the duration of treatment no less than that in the mei-9^a experiments. Of 21 compounds, 14 were judged positive and seven negative in the heritable translocation test; four of the negatives are classified as carcinogens. Most positives required sperm storage.

A final recommendation of the use of the mei-9^a assay cannot be reached on the basis of this limited set of data, in which the majority of compounds are alkylating agents. A larger set of compounds of different chemical classes must be tested. With some compounds, e.g., hexamethylphosphoramide (Vogel et al., 1985), mei-9 females did not show a drastic increase in the recovery of complete or partial losses; the technical aspects of the use of meiotic mutants must be considered carefully.

3. Tests for nondisjunction

The induction and consequences of missegregation (nondisjunction) of chromosomes have been paid increasing attention within the last few years (Zimmermann et al., 1979; Evans, 1985; Dellarco et al., 1986). *Drosophila*, with its well developed genetics, is expected to play an important role in the development of reliable, rapid and inexpensive assay systems for this important type of damage. The present status of aneuploidy testing in *Drosophila* has recently been reviewed (Valencia et al., 1984; Zimmering et al., 1986).

3.1 Background

During meiosis in standard *Drosophila* strains, homologous chromosomes usually pair at prophase I and separate at anaphase I; sister centromeres then separate at anaphase II, and a chromatid is delivered to each of the four meiotic products. In females, meiotic products form a linear array, and the product nearest to the egg axis becomes the functional egg nucleus. In males, all products are functional. Genetic control of cell divisions in males and females by a large number of genes (see, e.g., Sandler et al., 1968) appears to be the same in mitotic cells (gonia) and in meiosis II, but the control of chromosomal behaviour at meiosis I is not (Baker & Carpenter, 1972; for a review, see Baker et al., 1976). The differences are related primarily to the absence of recombination in males (Cooper, 1950). Comparison with mammalian meiosis suggests that meiosis I in male *Drosophila* is aberrant in this respect.

Nondisjunction at the first meiotic division of the sex chromosomes in XX females gives rise to XX and nullo-X eggs. Nondisjunction in XY males produces XY and nullo-XY gametes. At the second meiotic division, XX and YY as well as nullo-X/nullo-Y gametes result from nondisjunction. The recovery frequency of diploid sex chromosome gametes (chromosome gain) is about 0.05%; in contrast, the nullo gametes (chromosome loss) are about three to eight times more frequent. This indicates that other mechanisms in addition to nondisjunction (e.g., chromosomal breakage) contribute to chromosomal loss. The frequency of spontaneous nondisjunction of the autosomes appears to be similar to that of sex chromosomes (for a review of spontaneous nondisjunction in *Drosophila*, see Zimmering, 1976).

Nondisjunction may arise by (a) nonconjunction leading to failure of homologues to separate at anaphase I, (b) failure of sister chromatids to separate at anaphase II following normal segregation of homologues at anaphase I, or (c) precocious separation of centromeres at anaphase I (Davis, 1971; Mason, 1976), Genetic analysis of spontaneously arising exceptional daughters from ordinary XX females suggests that (a) virtually all nondisjunction results from the failure of the reduction division, (b) exceptional females may arise from exchange tetrads, and (c) some 40% of all exceptional females are equational exceptions in that they are homozygous for some part of the chromosome. The frequency of homozygosis for a given gene is proportional to the distance from the centromere (Merriam & Frost, 1964).

Factors affecting meiotic segregation include the presence of an extra chromosome or chromosomal element (Lindsley & Sandler, 1965), inversion heterozygosity (Cooper et al., 1955), meiotic mutants (Baker et al., 1976), temperature (Tokunaga, 1970; Grell, 1971), X-rays (Traut, 1971), mutagens (Valencia et al., 1984; Zimmering et al., 1986) and mutagen-induced chromatid-type heterologous interchanges (Parker, 1970; for review, see Parker & Williamson, 1976).

3.2 Sex chromosome segregation in XX and XY individuals without selection

(a) The conventional scheme

In the conventional assays employing sex chromosome aneuploidy, the segregation of genetically marked sex chromosomes is followed, and exceptional progeny are indicative of nondisjunction. Two basic test protocols have been employed: male treatment and, less frequently, treatment of larval or adult females. In both schemes, the X chromosomes may be marked with y (yellow body colour) and with w or w^a (white or white-apricot eye colour). The Y chromosome may be marked with B^S (dominant Bar of Stone eye) and with y^+ (wild type allele of y). By scoring the F_1 progeny for sex, body colour, eye colour and eye shape, normal progeny may be distinguished from progeny resulting from parental gametes with a loss or a gain of a sex chromosome (for details see Zimmering et al., 1986; Szabad, unpublished data).

It is important to realize that, after exposure of larvae or analysis of progeny from very late broods after adult treatment, mitotic nondisjunction can also be detected in proliferating premeiotic male and female germ-line cells. The study of mitotic nondisjunction should not be neglected, because it will permit determination of differences in response in chiasmatic (female) and achiasmatic (male) meiocytes in mitotic cells. In addition, mitotic nondisjunction in germ line cells will be an important parameter if somatic nondisjunction has to be compared with that occurring in mitotic and meiotic cells of the germ line. Mitotic nondisjunction can play an important role in human carcinogenesis, as shown, for example, with acute leukaemias in Down's syndrome (Rowley, 1981) and with meningiomas of the spinal canal, which almost invariably have a monosomy of chromosome 22 (Zankl & Zang, 1980). The biological significance of chemically induced aneuploidy has recently been reviewed (Oshimura & Barrett, 1986).

Szabad (unpublished data) points out that in the crosses used to detect nondisjunction, newly induced, attached sex chromosomes (XY and XX) can be recovered. In the course of chromosomal pairing, X.Y exchanges between sex chromosomes can take place in the testes (Zimmering, 1976; Lüning, 1982). In females, the production of attached-X chromosomes (XX) may be expected. XX can also result from male germ cells by mutation taking place after insemination of the egg. The presence of a section of the Y chromosome carrying the *bb*-locus facilitates the induction of compound X chromosomes (Leigh, 1972; Graf & Würgler, 1975). Although possible, screening for induced, compound sex chromosomes has rarely been included in studies of chemical mutagens.

(b) The zeste *scheme*

The *zeste* scheme (Zimmering *et al.*, 1986) is a selective free-X protocol involving no lethals and designed to permit the recovery of all exceptional females, i.e., those recovered from (i) no-exchange tetrads, (ii) equational exceptions arising following exchange and nondisjunction, and (iii) meiosis II nondisjunctionals resulting from failure of sister chromatid separation. The scheme takes advantage of the sexual dimorphism of the *zeste* (z) mutation; hemizygous z males are phenotypically wild-type, whereas homozygous z females have yellowish eyes. Homozygous $y\,z$ females are mated with $y^2\,z$ males and $y\,z\,/\,y^2\,z$ females of the F_1 generation mated with attached-XY w males not carrying a free Y chromosome. Normal females and males are phenotypically wild-type, exceptional females have yellowish eyes (z) and yellow body colour (y or y^2). The latter are recognizable products of events (ii) and (iii) described above, and exceptional males have white eyes (w). Accordingly, scoring for exceptionals is rapid and efficient.

3.3 *Semiselective tests*

In these assays, half of the normal progeny die, and the remainder are retained for statistical analysis, so that the effort involved in counting normal progeny is halved. Exceptional progeny resulting from gametes with chromosomal gain or loss are phenotypically distinguishable from the surviving normal progeny. Since these are potentially useful assays, their further development and validation is recommended.

(a) The X lethal scheme

In the scheme of Grell (1981), females carrying normal X chromosomes heterozygous for two closely linked lethals at the far proximal end of the chromosome are mated with appropriately marked males. Specifically, $y\,w^a\,f\,l^1+\,/\,y\,w^a\,f+l^2$ females (which may have been exposed to a test compound throughout the whole larval period) are mated with $y^2\,v\,B\,/\,y^+$ Y males. All normal males die, with the exception of crossovers between the lethals. Normal females are phenotypically yellow and intermediate *Bar*. Maternal nondisjunction leads to $w^a f$ exceptional females and $y^2\,v\,B$ exceptional males. Paternal nondisjunction produces phenotypically *Bar* females. Experimental data obtained with this system have not yet been reported in the open literature.

(b) The free inverted X chromosome (FIX) scheme

A scheme employing heterozygous lethals in heterozygously inverted X chromosomes was designed by Zimmering *et al.* (1986). Females homozygous for y, heterozygous for the complex sc^8 In49 inverted X and heterozygous for two X chromosome lethals (one on each X) are mated with attached-XY w males not carrying a free Y chromosome. All normal males die (except for occasional double crossovers permitting their survival). Normal females are phenotypically wild type, exceptional females *yellow*, and exceptional males *white*. The two exceptional classes represent gain of a maternal X chromosome (female) and loss of maternal X (male).

(c) The compound X (CDX) scheme

The CDX system (Zimmering *et al.*, 1986) is a variation of the lethal-bearing system described above. The Muller C(l)DX compound X chromosome is deficient for *bobbed*. Individuals not carrying a normal allele of *bobbed* die. CDX females of the composition $y\ w\ f\ bb^- / y^+\ Y$ are mated with *y* attached-XY males not carrying a free Y chromosome. Normal females die, normal males are phenotypically *yellow*, exceptional females *white*, and exceptional males wild type.

(d) The killer of prune *scheme*

A semiselective system based on the interaction between *K-pn* (*Killer of prune*) and *pn* (*prune*) has been developed by Foureman (1986a,b). In this system, the female F_1 progeny arising from normal segregational events are eliminated. The surviving individuals include an expected class of normal males arising from normal disjunction in both parents and four major exceptional and phenotypically distinguishable classes arising from either nondisjunction in the first or second meiotic division of the male or from partial recovery of the marked Y in the male resulting from X.Y interchanges or marker loss. A pilot study (Foureman, 1986a,b) using X-rays, heat shock and colchicine indicated that the system is worth further development.

3.4 Selective tests

(a) The translocation XY scheme

The translocation XY system (Craymer, 1974; Foureman, 1979, 1983) monitors chromosome gain only in females. Females homozygous for w^a are mated with males carrying a translocation between the X and the doubly marked Y chromosome $B^S Y y^+$. The translocation breakpoint is in the middle of the X chromosome; the distal segment of the X, X^D, is marked with y^+, and the proximal segment, X^P, with B^S. The chromosomes X^D and X^P behave as a bivalent and segregate at meiosis I in the male. All normal offspring (X/X^D and X/X^P) die. Nondisjunction in the female produces diplo-X and nullo-X eggs. The former are recovered as a hyperploid female (XX/X^D or XX/X^P) and phenotypically w^a or $y\ w^a\ B^S$; the latter are not recoverable. Sample size is based on the number of exceptional females among controls. Positive results were obtained following exposure of adult females to cold shock, X-rays and colchicine, and exposure of larvae and/or pupae to heat shock, colchicine and dimethyl sulfoxide. Negative results were obtained following exposure of female larvae to trifluralin (Foureman, 1983).

(b) *The* maroon-like *scheme*

The scheme of Smith (1983) is designed to detect chromosome gain only. It makes use of the observation that larvae carrying X-linked *mal* (*maroon-like*) mutations are deficient in xanthine dehydrogenase and are therefore killed in the presence of high concentrations of purine. Essentially, females heterozygous for two different complementing *mal* mutations ($mal^1 +$ / $+ mal^2$) are treated as larvae or adults and mated with males carrying the equivalent of $mal^1\ mal^2$. F_1 larvae are treated with high concentrations of purine (6 mM), which kill all larvae exhibiting phenotypically either mal^1 or mal^2 or both (normal females, normal males, and exceptional $mal^1\ mal^2$ males). The only survivors are the XXY females with the complementing constitution $mal^1 +$ / $+ mal^2$. Counts of eggs or of progeny from bottles not treated with purine are used to estimate sample size.

3.5 *Nondisjunction of autosomes*

Bateman (1968) first studied the induction of nondisjunction of chromosome II. Since loss of any of the major autosomes would normally be lethal, the crossing scheme involves a special stock of males, in which chromosome II was present as a compound with the two left arms attached to one centromere and the two right arms to another (compound reverse metacentrics). In spermatogenesis, the two compounds behave as univalents, passing independently to one of the two poles at the first meiotic division, so that aneuploid gametes carrying the left, the right, both or no compounds are believed to be formed in equal proportions. On mating exposed females carrying normal chromosome II to compound-2 males, all normal gametes will produce lethal zygotes when combined with nullisomic or disomic gametes from the compound male. Surviving progeny result from nondisjunction, leading to nullisomic and disomic eggs, and from chromosomal breakage and reunion events producing newly induced compound chromosomes.

With this system, Bateman (1968) and Clark and Sobels (cited in Sankaranarayanan & Sobels, 1976) studied X-ray-induced nondisjunction in females. Compounds that interact with DNA, such as ethyl methanesulfonate, 2,3,5,6-tetraethyleneimino-1,4-benzoquinone and mitomycin C, were negative (Sobels, 1979), but halothane had a weak ability to increase the level of nondisjunction in this system (Clements & Todd, 1981).

Traut (1981, 1984) has combined the conventional scheme for sex chromosome nondisjunction in females with the compound autosome method. Use of this method —the aneuploidy pattern method — showed that spontaneous events, X-rays, bleomycin and colchicine induced different patterns of aneuploid progeny. This technique, used so far only for the study of known aneuploidy-inducing mutagens, might well be developed into a screening method for aneuploidy-inducing chemicals.

3.6 Gender

A panel that reviewed aneuploidy testing in *Drosophila* (Zimmering *et al.*, 1986) expressed the view that the preferred protocol for the detection of induced sex chromosome aneuploidy would be one that employs treatment of females. The principal argument against protocols in which males are treated is the absence of meiotic recombination in *Drosophila* males. The lack of crossing over in spermatocytes vitiates the detection of chemicals which exert their effects by interfering with this process. Since most organisms, including humans, exhibit meiotic crossing over in each sex, the use of *Drosophila* males makes extrapolation to other organisms more precarious.

Szabad (unpublished data) points out that use of adult males should not be recommended for the study of mitotic nondisjunction in germ-line cells for the following reasons: (a) only one-quarter of aneuploid male germ-line cells are capable of sperm production after nondisjunction; (b) of the sperm originating from XYY spermatogonial cells, only one-quarter can be recognized in the subsequent generation (*via* XY-bearing sperm), and (c) it takes a long time to recover progeny derived from XYY spermatogonial cells.

Nonetheless, there is no compelling experimental evidence at this time to justify exclusion of males as a test vehicle (Foureman, 1983; Zimmering *et al.*, 1986).

3.7 Data obtained

Exceptional phenotypes among F_1 progeny allow for the identification of individuals which developed from germ cells with gain or loss of a chromosome. Results are expressed as the frequency of exceptional flies among total flies.

3.8 Test performance

The data on nondisjunction tests, reviewed by Valencia *et al.* (1984) and Zimmering *et al.* (1986), led to the following conclusions regarding test performance. Of the 76 compounds reviewed (Zimmering *et al.*, 1986), judgements on aneuploidy induction were made for 34 compounds. Of these, 17 were judged positive for chromosomal gain (11/34 for chromosomal gain and chromosomal loss, 6/34 for chromosome gain only). Similarly, 17/34 compounds were judged negative for chromosomal gain (11/34 for chromosomal gain and chromosomal loss, 6/34 for chromosomal gain only).

Are any of the compounds found to induce aneuploidy capable of inducing gene mutations and/or chromosome breakage in *Drosophila*, i.e., are any specifically aneuploidy inducers? Table 2 provides information based on the Gene-Tox reports on recessive lethal tests (Lee *et al.*, 1983) and chromosomal mutations (Valencia *et al.*, 1984) and on a review of somatic mutation assays in *Drosophila* (Würgler & Vogel, 1986) for the 17 compounds positive for chromosomal gain. Since ten of the compounds positive for chromosomal gain were positive in one or more other tests, and no reliable data exist regarding results in other tests for the remaining seven compounds, the answer to the question awaits results from further work.

Table 2. Mutagenicity in other *Drosophila* assays of compounds positive in the *Drosophila* aneuploidy assay for chromosomal gain[a]

Compound (CAS Registry No.).	Chromosome[b] gain	Chromosome[b] loss	Recessive lethals[c]	Trans- locations[d]	Clasto- genicity[d]	Somatic assays[e]
Acenaphthene (83-32-9)	+	+	0	0	0	0
Actinomycin D (50-76-0)	+	+	i	0	i	0
Azathioprine (446-86-6)	+	+	0	0	+	0
Bleomycin (11056-06-7)	+	+	0	0	0	+
Caffeine (58-08-2)	+	+	+	0	(+)	+[f]
Clophen A50 (8068-44-8)	+	-	0	0	(-)	0
Colcemid (477-30-5)	+	-	0	0	i	0
p,p'-DDT (50-29-3)	+	+	-	0	+	0
1,2-Dichloroethane (107-06-2)	+	+	+	0	i	+
5-Fluorodeoxyuridine (50-91-9)	+	+	i	0	i	+
Halothane (151-67-7)	+	+	+	0	0	0
Methylmercuric hydroxide (1184-57-2)	+	-	0	0	0	0
N-Methyl-*N'*-nitro-*N*-nitrosoguanidine (70-25-7)	+	-	+	+	+	+
Proflavin (92-62-6)	+	+	0	0	i	+
Satratoxin H (53126-64-0)	+	i	0	0	0	0
Sodium cyclamate (139-05-9)	+	+	i	0	i	0
Vinblastine (865-21-4)	+	-	0	0	0	+

[a]Symbols used: +, positive; -, negative; i, inconclusive; (+), inconclusive, but looks positive (Valencia *et al.*, 1984); (-), inconclusive, but looks negative (Valencia *et al.*, 1984); O, no data available

[b]Zimmering *et al.* (1986)

[c]Lee *et al.* (1983)

[d]Valencia *et al.*, 1984

[e]Würgler & Vogel (1986)

[f]Graf & Wügler (1986)

3.9 *Strengths and weaknesses of the assay systems*

It is premature to discuss the strengths and weaknesses of the aneuploidy test systems in *Drosophila*, since commonly accepted standards for validating such screening assays have not yet been met. These include (a) reproducibility of results within a given laboratory in tests with a large number of compounds, (b) agreement between different laboratories testing a common set of chemicals with the same protocol, and (c) an acceptable level of concordance with test systems in other organisms (Zimmering *et al.*, 1986). Furthermore, it is likely that, in the future, aneuploidy screening will make use of special selective genetic systems, permitting far larger sample sizes to be obtained and providing a firmer data base for judgement on the aneuploidy-inducing ability of chemicals.

4. Acknowledgement

We thank P. Fourman, J. Mason, J. Szabad and S. Zimmering for sending us manuscripts and proofs of unpublished papers and kind permission to include this information in the present report.

5. References

Baker, B.S. & Carpenter, A.T.C. (1972) Genetic analysis of sex chromosomal meiotic mutants in *Drosophila melanogaster*. *Genetics*, *71*, 255-286

Baker, B.S., Boyd, J.B., Carpenter, A.T.C., Green, M.M., Nguyen, T.D. & Smith, P.D. (1976) Genetic control of meiotic recombination and somatic DNA metabolism in *Drosophila melanogaster*. *Proc. natl Acad. Sci. USA*, *73*, 4140-4144

Bateman, A.J. (1968) Non-disjunction and isochromosomes from irradiated chromosome II in *Drosophila*. In: *The Effects of Radiation on Meiotic Systems*, Vienna, International Atomic Energy Agency, pp. 63-70

Boyd, J.B., Harris, P.V., Presley, J.M. & Narachi, M. (1983) Drosophila melanogaster*: a model eukaryote for the study of DNA repair*. In: Friedberg, E.C. & Bridges, B.A., eds, *Cellular Response to DNA Damage*, New York, Alan R. Liss, pp. 107-123

Clements, J. & Todd, N.K. (1981) Halothane and non-disjunction in *Drosophila*. *Mutat. Res.*, *91*, 225-228

Cooper, K.W. (1950) *Normal spermatogenesis in* Drosophila. In: Demerec, M., ed., *Biology of Drosophila*, New York, Hafner, pp. 1-61

Cooper, K., Zimmering, S. & Krivshenko, J. (1955) Interchromosomal effects and segregation. *Proc. natl Acad. Sci. USA*, *41*, 911-914

Craymer, L. (1974) A new genetic testing procedure for potentional mutagens. *Drosophila Inform. Serv.*, *51*,62

Davis, B. (1971) Genetic analysis of a meiotic mutant resulting in precocious sister-centromere separation in *Drosophila melanogaster*. *Mol. gen. Genet.*, *113*, 251-727

Dellarco, V.L., Voytek, P.E. & Hollaender, A. (1985) *Aneuploidy, Etiology and Mechanisms*, New York, Plenum

Evans, H.J. (1985) Etiology and mechanism of aneuploidy. *Trends Genet.*, *1*, 123-124

Foureman, P. (1979) A translocation X;Y system for detecting meiotic nondisjunction and chromosome breakage in males of *Drosophila melanogaster*. *Environ. Health Perspect.*, *31*, 53-58

Foureman, P. (1983) *The Translocation X;Y Test: A Selective System for the Detection of Sex Chromosome Aneuploidy in* Drosophila melanogaster, Thesis, University of Wisconsin, Madison

Foureman, P. (1986a) Use of a semi-selective test for the detection of first and second division nondisjunction in *Drosophila melanogaster*. *Environ. Mutagen.*, *8*, 173-182

Foureman, P.A. (1986b) *Detection of aneuploidy in* Drosophila. In: de Serres, F.J., ed., *Chemical Mutagens, Principles and Methods for Their Detection*, Vol. 10, New York, Plenum, pp. 183-213

Graf, U. & Würgler, F.E. (1975) Compound-X chromosomes from X-rayed mature sperm of *Drosophila melanogaster*. *Arch. Genet.*, *48*, 148-150

Graf, U. & Würgler, F.E. (1986) Investigation of coffee in *Drosophila* genotoxicity tests. *Food chem. Toxicol.* (in press)

Grell, R. (1981) A test for nondisjunction in *Drosophila melanogaster* (Abstract). *Environ. Mutagen.*, *3*, 365

Griesemer, R.A. & Cueto, C. (1980) *Toward a classification scheme for degree of experimental evidence for the carcinogenicity of chemicals from animals*. In: Montesano, R., Bartsch, H. & Tomatis, L., eds, *Molecular and Cellular Aspects of Carcinogen Screening Tests (IARC Scientific Publications No. 27)*, Lyon, International Agency for Research on Cancer, pp. 259-281

Lee, W.R., Abrahamson, S., Valencia, R., von Halle, E.S., Würgler, F.E. & Zimmering, S. (1983) The sex-linked recessive lethal test for mutagenesis in *Drosophila melanogaster*. *Mutat. Res.*, *123*, 183-279

Leigh, B. (1972) Induction of attached-X chromosomes in spermatozoa by X-irradiation. *Drosophila Inform. Serv.*, *48*, 107

Leigh, B. (1976) *Ring chromosome and radiation induced chromosome loss*. In: Ashburner, M. & Novitski, E., eds, *Genetics and Biology of* Drosophila, Vol. 18, London, Academic Press, pp. 505-528

Lindsley, D. & Sandler, L. (1965) Meiotic behavior of grossly deleted X chromosomes in *Drosophila melanogaster*. *Genetics*, *51*, 223-245

Lüning, K.G. (1982) Genetics of inbred *Drosophila melanogaster*. IV. Test for paternal nondisjunction, chromosome loss and X.Y exchange. *Hereditas*, *96*, 97-99

Mason, J. (1976) Orientation disruptor (ord): a recombination-defective and nondisjunction-defective meiotic mutant in *Drosophila melanogaster*. *Genetics*, *84*, 545-572

Merriam, J. & Frost, J. (1964) Exchange and nondisjunction of the X chromosomes in female *Drosophila melanogaster*. *Genetics*, *49*, 109-122

Oshimura, M. & Barrett, J.C. (1986) Chemically induced aneuploidy in mammalian cells: mechanisms and biological significance in cancer. *Environ. Mutagen.*, *8*, 129-159

Parker, D. (1970) Coordinate nondisjunction of Y and forth chromosomes in irradiated compound-X female *Drosophila*. *Mutat. Res.*, *9*, 307-322

Parker, D.R. & Williamson, J.H. (1976) *Aberration induction and segregation in oocytes*. In: Ashburner, M. & Novitski, E., eds, *The Genetics and Biology of Drosophila*, Vol. 1C, London, Academic Press, pp. 1251-1268

Rowley, J.D. (1981) Downs syndrome and acute leukemia: increased risk may be due to trisomy 21. *Lancet*, *ii*, 1020-1022

Sandler, L., Lindsley, D., Nicoletti, B. & Trippa, G. (1968) Mutants affecting meiosis in natural populations of *Drosophila melanogaster*. *Genetics*, *60*, 525-558

Sankaranarayanan, K. & Sobels, F.H. (1976) Radiation genetics. In: Ashburner, M. & Novitski, E., eds, *The Genetics and Biology of Drosophila*, Vol. 1C, London, Academic Press, pp. 1089-1250

Smith, P.D. (1983) A rapid selection technique for detecting meiotic X-chromosomal nondisjunction in *Drosophila melanogaster*. *Mutat. Res.*, *108*, 169-174

Sobels, F.H. (1979) Studies of non-disjunction of the major autosomes in *Drosophila melanogaster*. II. Effects of dose-fractionation, low radiation doses, EMS and ageing. *Mutat. Res.*, *59*, 179-188

Tokunaga, C. (1970) The effect of low temperature and aging on nondisjunction in *Drosophila*. *Genetics*, *65*, 75-94

Traut, H. (1971) The influence of the temporal distribution of the X-ray dose and the induction of the X-chromosome nondisjunction and X-chromosome loss in oocytes of *Drosophila melanogaster*. *Mutat. Res.*, *12*, 321-327

Traut, H. (1981) Aneuploidy patterns in *Drosophila melanogaster*. *Environ. Mutagen.*, *3*, 275-286

Traut, H. (1984) Aneuploidy induced by bleomycin in oocytes of *Drosophila melanogaster*: studies with the aneuploidy pattern method. *Environ. Mutagen.*, *6*, 889-894

Valencia, R., Abrahamson, S., Lee, W.R., von Halle, E.S., Woodruff, F.E., Würgler, F.E. & Zimmering, S. (1984) Chromosome mutation tests for mutagenesis in *Drosophila melanogaster*. A report of the US Environmental Protection Agency Gene-Tox Program. *Mutat. Res.*, *134*, 61-88

Vogel, E.W., van Zeeland, A.A., Raaymakers-Jansen, C.A. & Zijlstra, J.A. (1985) Analysis of hexamethylphosphoramide (HMPA)-induced genetic alterations in relation to DNA damage and DNA repair in *Drosophila melanogaster*. *Mutat. Res.*, *150*, 241-260

Würgler, F.E. & Graf, U. (1980) *Mutation induction in repair deficient strains of* Drosophila. In: Generoso, W.M., Shelby, M.D. & de Serres, F.J., eds, *DNA Repair and Mutagenesis in Eukaryotes*, New York, Plenum, pp. 223-240

Würgler, F.E. & Vogel, E.W. (1986) *In-vivo mutagenicity testing using somatic cells of* Drosophila melanogaster. In: de Serres, F.J., ed., *Chemical Mutagens*, Vol. 10, New York, Plenum, pp. 1-72

Würgler, F.E., Frei, H. & Graf, U. (1986) *Mutagen-sensitive mutants and chemical mutagenesis in* Drosophila. In: de Serres, F.J., ed., *Chemical Mutagens*, Vol. 10, New York, Plenum, pp. 381-425

Zankl, H. & Zang, K.D. (1980) Correlations between clinical and cytological data in 180 human meningiomas. *Cancer Genet. Cytogenet.*, *1*, 351-356

Zimmering, S. (1976) *Genetic and cytogenetic aspects of altered segregation phenomena in* Drosophila. In: Novitsky, E. & Ashburner, M., eds, *Biology of Drosophila*, Vol. 1B, London, Academic Press, pp. 569-613

Zimmering, S. (1983) The *mei-9a* test for chromosome loss in *Drosophila*: a review of assays of 21 chemicals for chromosome breakage. *Environ. Mutagen.*, *5*, 907-921

Zimmering, S., Mason, J.M. & Osgood, C. (1986) Current status of aneuploidy testing in *Drosophila*. *Mutat. Res.*, *167*, 71-87

Zimmermann, F.K., de Serres, F.J. & Shelby, M.D., eds (1979) Systems to Detect the Induction of Aneuploidy by Environmental Mutagens. *Environ. Health Perspect.*, *31*

REPORT 14

SHORT-TERM ASSAYS FOR THE ANALYSIS OF BODY FLUIDS AND EXCRETA

Prepared by:

S. Venitt (Rapporteur), H. Bartsch, G. Becking,
R.P.P. Fuchs, M. Hofnung, C. Malaveille, T. Matsushima,
M. Roberfroid and H.S. Rosenkranz (Chairman)

1. Introduction

Investigating the genotoxic activity of body fluids and excreta using short-term tests can serve three purposes pertinent to the study and possible prevention of human cancer: (i) detection of potentially carcinogenic agents, by assaying *in vitro* the genotoxicity of body fluids and excreta collected from laboratory animals which have been dosed *in vivo* with a test substance; (ii) noninvasive monitoring of human absorption of carcinogens and mutagens in the work-place, by assaying excreta for genotoxic activity; and (iii) study of the etiology of cancer, by assaying body fluids and excreta for genotoxins formed endogenously, or taken into the body as constituents of, for example, food, drinking-water, medicines, drugs, cosmetics, tobacco products and betel quid.

Thus, body-fluid analysis may be performed on material collected from laboratory animals dosed with a test material in order to obtain data on absorption, distribution, metabolism and excretion. It may also be performed on material collected from wild animals, from domestic animals, or from human donors. An assay may be related to the deliberate administration of a substance under controlled conditions (e.g., to a laboratory rodent or to a cancer patient undergoing chemotherapy) or to the investigation of environmental factors (e.g., the ingestion of pesticides by domestic animals, or investigation of diet-related cancer or smoking-related cancer in man).

In this chapter, 'body-fluid analysis' is used as short-hand for the assay of test materials derived from a variety of body fluids and excreta (e.g., saliva, breast aspirates, gastric juice, blood, expectorate, peritoneal fluid, bile, semen, prostatic fluid, vaginal secretions, amniotic fluid, urine, faeces) by well-established in-vitro short-term tests for genotoxicity (e.g., bacterial mutation tests, cytogenetic tests with cultured mammalian cells).

Detailed reviews of body-fluid analysis are available (Legator *et al.*, 1982; Combes *et al.*, 1984). Several topics related to body-fluid analysis are discussed by Berlin *et al.* (1984). The detection of mutagens in human faeces has been reviewed by Venitt (1982) and by Wilkins and Van Tassell (1983); for a short critique of body-fluid analysis in relation to assessment of carcinogen exposure in man, see Garner (1985).

2. Principles and scientific basis of body-fluid analysis

The basic idea of the method is to determine if and to what extent the body fluid or excreta under investigation contains genotoxic activity and to relate the findings to the treatment or exposure received by the donor. The scientific principles of body-fluid analysis fall under three distinct headings: (i) the selection, treatment or exposure of the sample donor; (ii) the collection and treatment of the sample; and (iii) the choice of short-term test and its execution and interpretation, with regard to the nature of the test material. The scientific principles of the short-term tests (e.g., bacterial short-term tests, cytogenetic tests) themselves are dealt with in other reports in this volume.

2.1 *Selection, treatment or exposure of sample donor*

Clearly, the criteria for selecting and treating a sample donor depend on the purpose of the investigation and the nature of the donor. Since the bulk of the published literature deals with just two major subdivisions of the topic, namely, analysis of body fluids from laboratory animals, and analysis of human body fluids and excreta, these are discussed separately.

(a) Laboratory animals

This topic is reviewed in detail by Combes *et al.* (1984).

Short-term tests conducted *in vitro* which employ an exogenous metabolic system do not take account of the processes of absorption, distribution, metabolism and excretion of the test chemical which are characteristic of mammals. Techniques to overcome this deficiency include (i) the host-mediated assay, which is discussed in Report 6; and (ii) the collection of body fluids and excreta from dosed animals and testing them for genotoxicity. In this mode of body-fluid analysis, it is assumed that the test material will be metabolized and excreted in a manner that will allow recovery of genotoxic activity in the chosen analyte. Extraction and fractionation of the body fluid and subsequent assays of fractions for genotoxicity are refinements which add further data on distribution and metabolism. In addition, manipulation of the donor, for example, by treatment with inducers and inhibitors of activating enzymes, may add yet further information.

Scientific judgements are helped by knowledge of the physical and chemical nature of the test material and how this affects its distribution and metabolism. For example, the proportion excreted in the urine, or *via* the bile (with or without enterohepatic circulation) to the faeces, or directly in the faeces, and the ultimate genotoxicity of a given compound depend, *inter alia*, on its molecular weight, its lipophilicity, its polarity, its affinity for plasma proteins, its susceptibility to oxidative metabolism and to conjugation, and the half-life of its metabolites in various biochemical compartments of the body. In addition, knowledge of how the given test material or its metabolites respond to the metabolic activity of the gut flora (e.g., reductive cleavage, further conjugation and deconjugation) may also assist in choosing an animal of the appropriate species, strain, age and sex with which to perform body-fluid analysis. The choice of body fluid or excreta to be analysed, as well as the most appropriate route of administration will also be helped by this knowledge.

Having made these choices, the experimentalist must then choose a further series of conditions for the assay. These will include decisions about the husbandry of the experimental animals (e.g., the nature, quantity and frequency of giving the diet and drinking-water; the numbers of animals per cage; the temperature; freedom from infection; provision of suitable metabolic cages; prevention of contamination of faeces and urine by each other and by food, water or bedding); the number of animals per experimental group; the number of different doses of the test compound; provision of negative and positive controls; and sampling time. Consideration must also be given to the very different social and dietary customs practised by laboratory rodents; for example, coprophagy, virtually unknown in human populations, is common among rats, and may well vitiate an experimental protocol unless steps are taken to prevent it. Many of these latter choices, although influenced by scientific considerations, are more likely to be based on what is practicable and economical in time and money.

Table 1 gives examples of the wide variety of studies that have been performed on body-fluid analysis with experimental animals.

(b) Human donors

Studies using human donors fall into four main groups: (i) methodological, (ii) experimental, (iii) occupational and (iv) etiological. Examples of these different categories are shown in Table 2, which is by no means an exhaustive list — references to other studies are given by Legator *et al.* (1982) and Combes *et al.* (1984). Most of the urine studies listed in Table 2 and in those two reviews employed the XAD-resin-*Salmonella*/microsome test devised by Yamasaki and Ames (1977). This is discussed in section 2.3(*a*)(i).

Body-fluid analysis using human donors is obviously much more limited in scope than studies using experimental animals. Invasive techniques such as endoscopy, intubation or bile-duct cannulation are clearly practicable only in a clinical setting, and material other than urine and faeces (and to a lesser extent, saliva and gastric juice) can be collected only in very small quantities, and only with difficulty. Large-scale studies have therefore been restricted to the use of urine or faeces: this is borne out by inspection of Table 2.

Table 1. Examples of studies of body-fluid analysis in animals (selected references, 1980 to September 1985)

Donor	Test material	Body fluid	Assay system	Reference
Dog, otter sea-gull, cow, horse, sheep, chicken, goose	None	Faeces	Chromosomal aberrations in cultured Chinese hamster CHO cells	Stich et al. (1980)
Rats	Benzidine and other aromatic amines	Urine	Salmonella/microsome assay	Bos et al. (1980)
Rats, mice	Acrylonitrile	Urine	Salmonella/microsome assay	Lambotte-Vandepaer et al. (1980)
Rats	2,4-Diamino-anisole	Urine	Salmonella/microsome assay	Reddy, T.V. et al. (1980)
Rats	Acrylonitrile	Urine	Salmonella/microsome assay	Lambotte-Vandepaer et al. (1981)
Mice	Beef extract	Urine	Salmonella/microsome assay	Dolara et al. (1980)
Mice	2-Amino-fluorene, cyclophosphamide, lucanthone	Urine	L5178Y TK$^{+/-}$ gene mutation assay in cultured mouse cells	Amacher et al. (1981)
Syrian hamsters, rats	Phenacetin	Urine	Salmonella/microsome assay	Camus et al. (1982)
Pigs	High-fat, low-fibre 'western' diet, with and without ethanol	Faeces	Salmonella/microsome assay	Topping et al. (1982)
Rats	Trp-P-1, Trp-P-2	Bile	Salmonella/microsome assay	Dolara et al. (1982)
Fish (bream)	River Rhine water	Bile	Salmonella/microsome assay	van Kreijl et al. (1982)
Mice	Antiamoebic and anthelmintic drugs	Urine	Salmonella/microsome assay	Cortinas de Nava et al. (1983)
Rats	Cisplatin	Urine, plasma	Salmonella/microsome assay	Safirstein et al. (1983)
Baboons	Cigarette smoke	Urine	Salmonella/microsome asay	Marshall et al. (1983)

Table 1. (contd)

Donor	Test material	Body fluid	Assay system	Reference
Rabbits, rats	Benzo[a]pyrene	Bile	*Salmonella*/microsome assay	Chipman et al. (1983)
Rats	Chemically-defined diets, with and without cellulose and pectin	Faecal pellets, luminal gut contents	*Salmonella* fluctuation test	Kuhnlein et al. (1983a)
Mice	Benzo[a]pyrene	Urine	*Salmonella*/microsome assay	Camus et al. (1984)
Rats	1,2-Dimethylhydrazine	Faeces	*Salmonella*/microsome assay	Askew et al. (1984)
Rats	Benzidine	Urine, bile	*Salmonella*/microsome assay	Lynn et al. (1984)
Rats	Beef extract	Stomach contents, bile, urine	*Salmonella*/microsome assay	Munzner & Wever (1984)
Rats	Food colours Brown FK and Red 2G	Urine, faeces	*Salmonella*/microsome assay and fluctuation test	Edwards & Combes (1984)
Rats	Sodium saccharin	Urine	*Salmonella*/microsome assay	Hasegawa et al. (1984)
Rats	Cigarette smoke	Urine	*Salmonella* fluctuation test	Mohtashamipur et al. (1984)
Rats	1,3-Diaminobenzene	Urine	*Salmonella*/microsome assay	Clemmensen & Lam (1984)
Lambs	Sugar beet grown on sludge-amended soil	Blood, urine	*Salmonella*/microsome assay	Telford et al. (1984)
Rats	2-Ethylhexanol-derived plasticizers	Urine	*Salmonella*/microsome assay	DiVincenzo et al. (1985)
Rats	Flunitrazepam	Urine	*Salmonella*/microsome assay	Staiano et al. (1984)
Rats	2-Amino-3-methylimidazo-(4,5-f)quinoline (IQ)	Urine	*Salmonella*/microsome assay	Barnes & Weisburger (1985)

Table 1. (contd)

Donor	Test material	Body fluid	Assay system	Reference
Rats	Azo dyes	Urine	Salmonella/ microsome assay	Joachim et al. (1985)
Syrian hamsters	Schistosoma haematobium	Urine	Salmonella/ microsome assay	Gentile et al. (1985)
Rats	Diesel particulates	Urine	Salmonella/ microsome assay	Belisario et al. (1985)
Rats	Comparison of paired carcinogens and noncarcinogens	Urine	Salmonella/ microsome assay	Malaveille et al. (1986)

The scientific principles underlying the application of body-fluid analysis to human populations are close to those which govern epidemiological studies, because most donors are drawn, not from controlled environments, but from communities. Each donor is genetically unique; each donor creates his or her own *milieu interieur*, both by personal choice (e.g., type and amount of diet; to smoke or not to smoke; to consume or not to consume alcohol) and by imposed conditions, for example, passive smoking, occupation, ill-health or living conditions. Such confounding variables must therefore be taken into account.

(i) *Confounding variables*

To date, the two most serious confounding variables in human body-fluid analysis are smoking and diet. Other possible causes of bias are occupational exposure (e.g., in studies of etiology), drug intake and disease states.

(1) *Smoking*: The first demonstration, by Yamasaki and Ames (1977), that smokers' urine is mutagenic, while nonsmokers' urine is not, has since been confirmed in numerous studies (Table 2). Indeed, in many of these, smoking was the only variable that was consistently associated with urinary mutagenicity. *Thus, urinary mutagenicity studies are worthless unless smoking habits are taken into account.* Even when this is done, experience has shown that the effects of smoking may well overwhelm any rather weak effect of, for example, putative occupational exposure. Simply dividing study groups into smokers and nonsmokers is an inadequate method for controlling for tobacco use, since several studies have revealed a clear correlation between the numbers of cigarettes smoked and the level of urinary mutagenicity (e.g., van Doorn et al., 1979).

Table 2. Examples of body-fluid analysis using human donors

Body fluid or excreta	Purpose of study	Reference
	Methodological	
Urine	Method for assaying mutagenic activity in urine[a]	Yamasaki & Ames (1977)
Faeces	Inhibition of mutagenicity of model mutagens by faecal extracts[a]	Hayatsu et al. (1981)
Urine	Detection of chromosomal anomalies in mammalian cell cultures[b]	Beek et al. (1982)
Urine	Detection of sister-chromatid exchanges in human lymphocytes[b]	Stiller et al. (1982)
Urine	Identification of a urinary mutagen from cigarette smoke[a]	Connor et al. (1983)
Faeces	Extraction-fractionation scheme for non-volatile faecal mutagens[a]	Dion & Bruce (1983)
Faeces	Aerobic and anaerobic methods for detecting faecal mutagens[c]	Venitt & Bosworth (1983)
Urine	Method for increasing sensitivity of bacterial mutation assay[a]	Kado et al. (1983)
Urine	Problems of growth enhancement in fluctuation tests giving rise to spurious data[c]	Gibson et al. (1983)
Gastric juice	Importance of controlling for histidine in bacterial assays[a]	O'Connor et al. (1984)
Urine	Optimal choice of sorbents for preparing urine extracts[a]	De Raat & van Ardenne (1984)
Urine	Effects of storage, sorbents and elution solvent on recovery of mutagenic activity[a]	Ong et al. (1985)
Urine	Use of high-performance liquid chromatography to isolate clastogenic agents from urine[b]	Curtis & Dunn (1985)
Urine	Study of three methods for isolating mutagens from smokers[a]	Mohtashamipur et al. (1985)
	Experimental	
Urine	Detection of mutagenic activity in patients receiving cytotoxic chemotherapy[a]	Minnich et al. (1976)
Urine	Detection of mutagenic activity following drinking of instant coffee[a]	Aeschbacher & Chappuis (1981)
Urine	Detection of mutagenic activity in patients treated with coal-tar and ultraviolet radiation[a]	Wheeler et al. (1981)

Table 2. (contd)

Body fluid or excreta	Purpose of study	Reference
Urine	Detection of mutagens following meal of fried bacon or pork[a]	Baker et al. (1982)
Faeces	Effect of ascorbic acid and α-tocopherol on faecal mutagenicity[a]	Dion et al. (1982)
Urine	Detection of mutagenic activity in patients treated with cisplatin[a]	Safirstein et al. (1983)
Faeces	Effect of dietary modificatons on faecal mutagens[c]	Kuhnlein et al. (1983a,b)
Urine	Detection of mutagenic activity in passive smokers[a]	Bos et al. (1983)
Urine	Study of mutagens in medium-tar- and low-tar-cigarette smokers and non-smokers[c]	Sorsa et al. (1984)
Urine	Time-course study of cigarette smokers[a]	Kobayashi & Hayatsu (1984)
Urine	Time-course study of cigarette smokers[a]	Kado et al. (1985)
Urine	Detection of mutagens following meal of fried beef[a]	Sousa et al. (1985a)
Urine	Detection of mutagens after drinking red wine or grape juice[a]	Sousa et al. (1985b)
Urine	Detection of mutagens following meal of fried beef[a]	Hayatsu et al. (1985a)
Faeces	Detection of mutagens following meal of fried beef[a]	Hayatsu et al. (1985b)
Urine	Effect of diet and smoking on urinary mutagenicity[a]	Sasson et al. (1985)
	Occupational	
Urine	Detection of mutagenic activity in rubber workers[c]	Falck et al. (1980)
Urine	Detection of mutagenic activity in operating-room personnel[a]	Baden et al. (1980)

Table 2. (contd)

Body fluid or excreta	Purpose of study	Reference
Urine	Detection of mutagenic activity in coke-plant workers[a]	Møller & Dybing (1980)
Urine	Detection of mutagenic activity in chemical workers[a]	Dolara et al. (1981)
Urine	Detection of mutagenic activity in pharmacy staff handling cytostatic drugs[a]	Nguyen et al. (1982)
Urine	Detection of mutagenic activity in nurses handling cytostatic drugs[a]	Bos et al. (1982)
Urine	Detection of mutagenic activity in workers exposed to mineral oils and iron oxide particles[a]	Laires et al. (1982)
Urine	Detection of mutagenic activity in carbon-electrode workers[a]	Pasquini et al. (1982)
Urine	Comparison of mercapturic acid levels and mutagenicity in chemical workers[a]	Buffoni et al. (1983)
Urine	Detection of mutagenic activity in chemical and coke-plant workers[a]	Kriebel et al. (1983)
Urine	Detection of mutagenic activity in nurses handling cytostatic drugs; comparison with assays for platinum[a,c]	Venitt et al. (1984)
Urine	Detection of mutagenic activity in nurses handling cytostatic drugs[a]	Barale et al. (1985)
Urine	Detection of mutagenic activity in coal-liquefaction workers[a]	Recio et al. (1984)
Expectorate	Detection of mutagenic activity in sputum/mucus in aluminium workers exposed to polycyclic aromatic hydrocarbons[a]	Krøkje et al. (1985)
Urine	Detection of mutagenic activity in aluminium workers exposed to coal-tar pitch volatiles[a]	Heussner et al. (1985)
Urine	Detection of mutagenic activity in chemotherapy workers[a]	Everson et al. (1985a)
Urine	Detection of mutagenic activity in autopsy workers[a]	Connor et al. (1985)

Table 2. (contd)

Body fluid or excreta	Purpose of study	Reference
	Etiological	
Gastric juice	Detection of mutagenic activity in gastric juice in a high-risk group[a]	Montes et al. (1979)
Faeces	Comparison of mutagenic activity in high- and low-risk bowel-cancer groups[a]	Reddy, B.S. et al. (1980)
Faeces	Comparison of mutagenic activity in vegetarians and nonvegetarians[c]	Kuhnlein et al. (1981)
Saliva	Detection of clastogenic activity in betel-nut and tobacco chewers[b]	Stich & Stich (1982)
Faeces	Comparison of mutagenic activity in high- and low-risk bowel-cancer groups[a]	Mower et al. (1982)
Urine	Detection of mutagenic activity in patients with urinary-tract cancers[a]	Caderni et al. (1982)
Urine	Time-course of mutagens in smokers[a]	Jaffe et al. (1983)
Urine	Detection of mutagenic activity in carcinoma of bilharzial bladder[a]	Everson et al. (1983a)
Urine	Detection of mutagenic activity in liver-cirrhosis patients[a]	Everson et al. (1983b)
Faeces	Detection of mutagens in people consuming a mixed western diet[a]	Reddy et al. (1984)
Faeces	Detection of nuclear aberrations in mouse colon treated *in vivo*	Suzuki & Bruck (1984)
Urine	Detection of mutagenic activity in *bidi* smokers and tobacco chewers[a,d,e]	Menon & Bhide (1984)
Gastric juice	Comparison of mutagenicity in patients with gastric ulcers, carcinoma, resection and drug treatment[a]	Morris et al. (1984)
Urine	Detection of clastogenic agents in coffee drinkers and cigarette smokers[b]	Dunn & Curtis (1985)
Amniotic fluid	Detection of mutagenic activity in amniocentesis samples[a]	Everson et al. (1985b)
Amniotic fluid	Detection of mutagenic activity in amniotic fluid from smokers and nonsmokers[a]	Rivrud et al. (1986)

[a] *Salmonella*/microsome test or modification
[b] Cytogenic test using cultured mammalian cells
[c] Bacterial fluctuation test
[d] Point mutation test using cultured mammalian cells
[e] Micronucleus test (mice, *in vivo*)

Moreover, passive smoking by declared nonsmokers must also be taken into account, since there is evidence that there is increased mutagenicity in the urine of nonsmokers who have been exposed to cigarette smoke merely by sharing a badly ventilated room with heavy smokers (Bos et al., 1983). Independent evidence for the absorption of tobacco smoke constituents during passive smoking is given by the work of Feyerabend et al. (1982), who demonstrated elevated levels of nicotine in saliva and urine in nonsmokers who had each spent a morning sharing an office with smokers. Routine determination of urinary nicotine or cotinine may therefore be a necessary requirement in urinary genotoxicity assays.

The appearance and clearance of mutagens in the urine of smokers also influences the design of urinary mutagenicity studies. Yamasaki and Ames (1977) showed that the urine of smokers is much less mutagenic after 6-8 h of sleep; in a time-course study, Kobayashi and Hayatsu (1984) found that urinary mutagenicity increased rapidly after the start of smoking and decreased to nonsmoking levels in 6-13 h after smoking had ceased. A detailed study of urinary mutagenicity in an occasional smoker (less than one cigarette per week) and of a heavy smoker (20 per day) showed that peak mutagenic activity appeared in the urine 4-5 h after smoking a single cigarette, and that, for the occasional smoker, mutagens were excreted following first-order kinetics (Kado et al., 1985). Potential urine donors (smokers and nonsmokers alike) must therefore be questioned closely about all aspects of their contact with tobacco smoke.

The effect of smoking on the genotoxicity of other body fluids and excreta has not been studied to any great extent; however, the possible confounding effects of smoking should be considered in studies of body fluids other than urine.

(2) *Diet*: A distinction must be drawn between studies of diet *per se*, and diet as a confounding variable in investigations of other topics.

Several independent studies have shown significantly raised levels of urinary mutagenicity within a few hours following meals of fried meat, such as fried bacon or pork (Baker et al., 1982) or fried beef (Hayatsu et al., 1985a; Sousa et al., 1985a). Mutagenic activity in urine was found to decrease within 24 h, although in some cases it remained elevated after this time. Baker et al. (1982) found that urinary mutagenicity levels were not increased in a volunteer who had eaten a meal of bacon or pork cooked in a microwave oven. Thus, the consumption of seared or browned meat is likely to be a major contributor to the appearance of urinary mutagens (and faecal mutagens — see below). Among a variety of potent bacterial mutagens formed during the cooking of food, two (2-amino-3,8-dimethylimidazo[4,5-*f*]quinoxaline (MeIQx) and 2-amino-3-methylimidazo[4,5-*f*]quinoline (IQ)) have been detected in fried beef (Sugimura, 1985). Preliminary evidence (Hayatsu et al., 1985a) suggests that the urinary mutagens detected in their study are metabolites of MeIQx.

Faecal mutagenicity was also increased in subjects consuming fried beef, the faeces remaining mutagenic for three to four days after the meal, even though no further fried meat was consumed in this period. Again, it is likely that as yet unidentified metabolites of MeIQx are responsible for the sharp increase in faecal mutagenicity following a meal of browned meat (Hayatsu et al., 1985b). The consumption of foods that are cooked at high temperatures must therefore be considered another serious confounding variable in body-fluid analysis, especially in studies of urinary or faecal mutagenicity.

The extent to which coffee drinking is a confounding variable in urinary genotoxicity studies cannot be resolved at present, since the two contradictory studies that are available employed very different methods of extraction and assay. Aeschbacher and Chappuis (1981), with a widely used bacterial urine assay for nonpolar organic compounds (see below, and Yamasaki & Ames, 1977), found no evidence that intake of instant coffee raised urinary mutagenicity levels above control values. Significant increases in clastogenic activity in cultured mammalian cells was induced by high-performance liquid chromatography (HPLC) fractions extracted from urine samples donated by coffee drinkers, when compared with appropriate controls (Dunn & Curtis, 1985).

In the only study available, mutagenic activity was not detected in the urine of nonsmokers who had consumed 750 ml of red wine or 1180 ml of grape juice up to 24 h before urine collection (Sousa et al., 1985b).

In a study of the interaction between diet and smoking on urinary mutagenicity, Sasson et al. (1985) found that, when subjects changed from a normal 'western' diet (high fat and meat), to a Vegan diet (no meat, fish, eggs or dairy products) there was a significant increase in urinary mutagenicity in nonsmokers, light smokers and heavy smokers. Urine from several nonsmokers on the Vegan diet was more mutagenic than urine from smokers on the high-meat diet. It was concluded that dietary factors can play a dominant role in the mutagenicity of urine concentrates.

Dietary constituents (e.g., lipids, Hietanen et al., 1982; broiled beef, Pantuck et al., 1976) have been shown to modify the activity of monooxygenases; this could lead to altered excretion of mutagens and represents an indirect effect of diet as a confounding variable in body-fluid analysis.

(3) *Occupation*: In studies of etiology, the occupational history of each donor should be recorded, since it could be a confounding variable.

(4) *Drugs*: Many cytostatic drugs used in cancer chemotherapy and as immunosuppressants are genotoxic, and mutagenic activity has been detected in urine of patients treated with such drugs (see Sorsa et al., 1985). Urine of patients undergoing antibacterial chemotherapy with nitroimidazoles such as metronidazole has been shown to be mutagenic (Legator et al., 1975). In these cases, the confounding variable is the mutagenicity of the drug (or its metabolites) *per se*. However, antibacterial

properties of drugs may also hinder body-fluid analysis if the drug (or its metabolites) is present in the body-fluid extract and if the indicator organisms used in the assay are susceptible to it.

Another possibility is that although a drug might not be mutagenic *per se* it might provoke an indirect effect; for example, inducers of hepatic monooxygenases, such as phenobarbital, might enhance metabolism and lead to excretion of elevated background levels of urinary mutagens.

For all these reasons, a detailed and accurate record of the intake of drugs is therefore an essential requirement for body-fluid analysis using human donors.

(5) *Disease states*: Two types of investigation have been performed: (i) studies of body-fluid genotoxicity associated with a particular disease state, and (ii) studies of etiological agents suspected to be responsible for certain malignancies. The extent to which either of these categories presents a confounding factor in body-fluid analysis is not clear, since much of the data so far adduced has been contradictory.

In the case of urine analysis, Caderni *et al.* (1982) found no difference in urinary mutagenicity between bladder-cancer patients and controls; however, Garner *et al.* (1982) claimed that urine from bladder cancer patients was more mutagenic than urine from controls, although this finding was considered by the authors to require confirmation. Everson *et al.* (1983a) found no convincing evidence that urine of patients with bilharzial-associated bladder cancer was more mutagenic than urine from controls. An association between liver cirrhosis and urinary mutagenicity was claimed by Gelbart and Sontag (1980); however, a similar study by Everson *et al.* (1983b) failed to confirm this association.

Mutagenic activity of gastric juice has been reported in relation to gastric disease (see Table 2). The relevance of genotoxic activity in the gastric juice of different populations to various disease states is still uncertain. The same can be said for studies of faecal genotoxicity, since, although there is a substantial number of studies (see Table 2; Venitt, 1982; Wilkins & Van Tassell, 1983), no coherent picture has yet emerged.

Despite the limitations of current knowledge on the influence of disease states on the appearance of genotoxic activity in body fluids and excreta, certain diseases may predispose individuals to produce genotoxic excreta. Moreover, the genetic background of the donor may influence the excretion of mutagens, as shown in animal studies (Camus *et al.*, 1984).

(6) *Viruses:* Biological fluids and excreta may contain viral agents. Since some viruses are known to be endowed with clastogenic and/or cell transforming abilities, this should be borne in mind when conducting assays with these endpoints.

(ii) *Design of studies using human donors*

Protocols for conducting body-fluid analysis with human donors must take account of several problems:

(1) *Variation between samples from the same donor*: This may include variation in total amount, or in the degree of concentration of the material and its physical properties (e.g., loose stools *versus* hard stools, or concentrated urine *versus* dilute urine). In some urinary mutagenicity studies (e.g., Falck *et al.*, 1980) urinary creatinine levels have been employed in an attempt to account for different rates of glomerular filtration between donors. In studies of faecal mutagenicity, faecal dry weight is often used as a standard measure of dose (reviewed by Venitt, 1982).

(2) *Variation among donors in the same study group*: These will include genetic, environmental and behavioural differences between individuals, which may be difficult to control. Some of the variation may be smoothed out by taking multiple samples and by pooling samples — see below.

(3) *Problems of sample collection*: The timing and frequency of sample collection depend on the type of body fluid or excreta, and the purpose of the study. They also depend on the rates of absorption, distribution and clearance of putative genotoxins; in most cases, collections within one to three days following the exposure under investigation will suffice, unless the study involves continuous exposure and sequential collections.

For the most commonly used materials, namely, urine and faeces, the frequency of collection is determined mainly by the habits and physiology of the donors. It is essential that cross-contamination of faeces and urine be avoided. The use of diuretics or laxatives to induce urination or defaecation in order to standardize collection of samples has not been reported in any of the published studies, and appears to be of extremely dubious scientific merit.

Certain ethnic and religious groups may have deeply held beliefs concerning the handling of excreta, and it may well be difficult or impossible to obtain specimens from such groups.

(4) *Single or multiple samples or pooled samples*: It is virtually impossible for a donor to produce at any one time less than a complete bowel movement or partly to empty the bladder. These quantities represent the lower limits of what is practicable to collect at a given time. Of course, the sample may then be subdivided for a variety of sound reasons. The upper limit of what constitutes a specimen for assay is more problematical, and is associated with the need to reduce variability in sampling and the sensitivity of the chosen short-term test. In the case of urine, a specimen could be the product of a single urination or of urine taken over a stated time and pooled into one large sample; e.g., 24-h urine samples are commonly used in many clinical settings, and

are often used in body-fluid analysis. However, 24-h urine collections are difficult to arrange, being disruptive to the donors and requiring careful management of the donor and of his or her samples. Similar problems arise when collecting stool samples; there is considerable variation in the frequency with which stools are passed and in their bulk. Published studies include those in which products of single bowel movements were assayed, and those in which these were collected over one or more days, pooled and then assayed.

For both urine and faeces, short-term assay of a single sample gives a crude estimate of the *concentration* of genotoxins at one period of time; longer term sequential collections, whether pooled or assayed singly, give a much better estimate of the *amount* of genotoxins excreted by each donor in the study. Sequential collection also indicates temporal changes in genotoxin output which might add useful information on absorption and clearance in relation to the exposure or habit under investigation.

When body fluid samples are small in volume and difficult to collect, or when the number of individual samples is very large, (e.g., nipple aspirates, respiratory expectorate), pooling the samples may be the only method for getting enough material to analyse or for allowing completion of a study. Pooling several samples from the same donor presents no great problem in interpretation; however, pooling samples from several different donors within a single study group, and comparing the genotoxic activity of this pooled sample with that of a pooled sample from a control group (e.g., Krøkje et al., 1985) is much more difficult to interpret. In this method no account can be taken of variation between individual donors within groups. Peto (1983) has discussed the practical and statistical advantages and disadvantages of pooling large numbers of biological samples.

(5) *Samples requiring invasive techniques*: Ethical considerations should be paramount in designing studies which require invasive techniques for collecting samples, and where morbidity is a possible outcome. Gastric juice collection, using intubation or endoscopy, is a routine procedure in gasteroenterology, and assay of gastric juice for genotoxicity does not under these circumstances present ethical problems. However, the collection of gastric juice from healthy volunteers solely for the investigation of its potential genotoxicity (e.g., for obtaining control samples) should be approached more circumspectly, and with due regard to the requirements of ethical committees and prevailing safety codes. The collection of body fluids using invasive techniques may be accompanied by medication (e.g., anaesthetics, anxiolytics, sedatives), which might interfere with the conduct and interpretation of the assay.

(6) *Collection and storage*: Due attention must be paid to correct recording of both the donor and the specimen, to the cleanliness and sterility of the collecting vessel, and to the adequacy of the labelling, bearing in mind that most samples are stored in a deep-freeze. Additives or preservatives should not be used, except when specifically demanded by the nature of the sample (e.g., to prevent post-sampling nitrosation).

(7) *Viruses:* A fact frequently overlooked is the possibility that biological fluids and excreta may contain viral pathogens which are not removed by the chosen method of extraction or fractionation. Addition of such samples to mammalian cell cultures raises the possibility that the cells will become infected and the pathogen multiply. This may represent a safety hazard to laboratory personnel.

(iii) *Use of body-fluid analysis*

Several methods of deploying body-fluid analysis have been used, depending on the purpose of the study. Two main classes are discernible: (1) cross-sectional and (2) time-course studies.

(1) *Cross-sectional studies*: In the cross-sectional method, the 'exposed' group (or groups) is matched with a control 'unexposed' group (or groups), and the appropriate samples of body fluid are collected and assayed for genotoxicity. Within this method, further choices may be made, on the basis of the number of donors and the amount of material obtainable for study. For example, multiple samples may be collected from small numbers of donors, or fewer samples from larger numbers.

(2) *Time-course studies*: In this method, groups of donors are asked to provide samples before, during and after exposing themselves to the subject of the investigation.

Whatever study design is chosen, it is vital to ensure that adequate precautions have been taken to avoid the types of bias discussed above *before* embarking upon the study. The preparation of a comprehensive questionnaire is essential for this task.

2.2 *Selection and treatment of samples*

Most of what follows is equally applicable to material collected from animals and from human donors. The primary sample consists of the raw, undiluted body fluid or excretion product. Whatever genotoxicity assay is employed, the sample must first be collected, recorded, stored, recovered from storage, and if necessary, concentrated, fractionated or extracted. Detailed accounts of these methods are given in the individual papers (e.g., those cited in Tables 1 and 2) and by Legator *et al.* (1982), Venitt (1982) and Combes *et al.* (1984).

(a) *Concentration, extraction, fractionation*

The scientific principles underlying the choice of method for concentration, extraction and fractionation of body fluids and excreta are designed to meet a series of problems peculiar to biological samples. These include the following:

(i) Most short-term tests were developed for assaying single, pure chemicals.

(ii) Body fluids and excreta are complex mixtures containing a variety of substances, some of which may interfere with the chosen short-term assay. Two examples may be cited.

(1) Bacterial mutation tests employing reversion from amino-acid auxotrophy to prototrophy (e.g., the *Salmonella*/microsome test; reversion from histidine-auxotrophy; and tests using tryptophan-requiring strains of *Escherichia coli*) are susceptible to interference by the presence, in the test material, of the required amino acid or its precursor (Yamasaki & Ames, 1977). Fluctuation tests employing these bacteria in liquid suspension are particularly vulnerable, since it is difficult to distinguish enhanced auxotrophic growth from growth of true revertants (Gibson *et al.*, 1983; Venitt & Bosworth, 1983, 1986). The problem of interference by amino acids or their precursors can be avoided by exposing bacteria to the test material for a given time, then centrifuging or filtering the culture in order to separate bacteria from test material. The bacteria can then be resuspended and allowed further growth for fixation and expression of mutation. However, this procedure is considerably more time-consuming than standard methods of assay.

(2) Extracts of human faeces have been shown both to inhibit and to enhance the mutagenicity of reference mutagens in bacterial mutation assays (reviewed by Venitt, 1982).

(iii) Body fluids such as urine and faeces, although available in copious volume, are likely to contain only low concentrations of genotoxic substances.

(iv) Other fluids, such as expectorate, gastric juice, bile, nipple aspirate and sweat, are available in very small quantities, and may contain only low concentrations of genotoxins.

(v) In most cases, the chemical class and identity (and therefore the chemical properties) of putative genotoxins in body fluids are unknown. The value of any study is greatly enhanced by identification of the constituents responsible for the genotoxic activity.

(vi) Putative genotoxins may be conjugated and may be unreactive, or may be very labile in air, light and heat, or normally unreactive constitutents may be activated by air, light or heat. For example, aqueous faecal extracts that are mutagenic when assayed in bacterial fluctuation tests under aerobic conditions were inactive when extracted and assayed aerobically (Venitt & Bosworth, 1983, 1986).

(vii) Individual specimens (e.g., of human urine or faeces) are unique, since each specimen results from the combined effects of all the exposure and behaviour the donor was subject to both in the hours preceding the collection and to longer-term influences such as gut flora and metabolic profile.

(viii) The degree of concentration or dilution of fluid or excreta varies from sample to sample in one individual and between individuals.

(ix) The specimen itself may be metabolically active (e.g., faeces, bacterially-infected urine), and putative mutagens may be formed or destroyed by continued metabolic activity after collection. For example, human faeces become more mutagenic when incubated anaerobically for four days (Wilkins et al., 1980; Reddy et al., 1984).

(x) Constituents of the specimen may be susceptible to physical or chemical reactions that will engender genotoxic substances not normally present in vivo. For example, Miller and Stoltz (1978) found that urine from rats treated with isoniazid became mutagenic only after lyophilization, as did mixtures of isoniazid and urine. Moreover, sterile urine from treated rats became mutagenic after 8-14 days' storage at room temperature.

In practice, methods used for treating body fluids and excreta before assay for genotoxic activity range from no treatment (e.g., testing untreated urine from patients undergoing cytotoxic chemotherapy, Venitt et al., 1984) to complex extraction and fractionation procedures, after employing sorbents, solvent extraction and HPLC (e.g., Curtis & Dunn, 1985; Hayatsu et al., 1985a,b). The chemical identification of novel mutagens from complex biological mixtures is the final aim of body-fluid analysis; see, for example, Hirai et al. (1982), Connor et al. (1983) and Gupta et al. (1983).

The most widely used concentration/extraction method is that devised by Yamasaki and Ames (1977), in which urine is treated with a nonpolar resin (XAD-2), which traps nonpolar compounds and allows the passage of polar constituents (including amino acids, e.g., histidine, which might interfere with reverse mutation assays — see above). The nonpolar fraction is eluted with an organic solvent, which is then evaporated off, leaving a residue, which is then redissolved in a small volume of a solvent compatible with the chosen genotoxicity assay. In order to reveal the mutagenicity of conjugated urinary metabolites, enzymes such as β-glucuronidase or sulfatase are added to the urine concentrate at the time of assay. This method has gained wide acceptance due to its simplicity and speed, since in relatively few operations putative mutagens can be concentrated from large volumes of urine, while the concentration of interfering substances such as histidine is drastically reduced. However, it cannot be regarded as universal since it was designed to trap genotoxins of the classical nonpolar aromatic type, which are by no means representative of all chemical classes known to be genotoxic. For example, polar mutagens such as cisplatin are not retained by XAD-2 (Venitt et al., 1984). Recent work on the improvement of extraction methods includes that of De Raat and van Ardenne (1984), Belisario et al. (1985), Mohtashamipur et al. (1985) and Ong et al., 1985. Another recent innovation is the use of the sorbent 'Blue Cotton' (trisulfo-copper phthalocyanine covalently bound to cotton; Hayatsu et al., 1983), which, when combined with solvent extraction and carboxymethyl cellulose chromatography, has yielded valuable data on mutagenic activity in faeces and urine (Hayatsu et al., 1985a,b). This method is particularly suitable for trapping nonpolar polyaromatic compounds.

2.3 Choice of assay

(a) Assays using bacteria

(i) Point mutation tests

Bacterial point mutation tests which employ reverse mutation from amino acid auxotrophy to prototrophy are the most widely used short-term assays for detecting genotoxic activity in body fluids and excreta. Of this type of assay, the *Salmonella*/-microsome test (Ames *et al.*, 1975; Maron & Ames, 1983) is the most popular, as shown in Tables 1 and 2.

Fluctuation tests, using Ames' *S. typhimurium* strains or tryptophan-requiring strains of *E. coli*, have also been used in many studies. Both methods have been well validated and there is much information on their responses to a wide range of chemical classes under a variety of different conditions of assay and of metabolic activation. It seems likely, therefore, that bacterial reverse-mutation assays will continue to be the assays of first choice for body-fluid analysis. The scientific principles and the advantages and disadvantages of these and other tests involving bacteria are discussed in Report 5.

The only serious practical disadvantage of reverse mutation assays for body-fluid analysis was referred to in section 2.2(*a*)(ii), namely, interference by amino acids (especially histidine or tryptophan) or their precursors present in the test material. When this is suspected, the amino acid content of the test material should be determined and allowance made for the presence of interfering amino acids. This can be done by *o*-phthaldialdehyde derivatization followed by HPLC/flow-fluorimetry (e.g., O'Connor *et al.*, 1984; Venitt & Bosworth, 1986). In addition, bioassays using auxotrophic bacteria allow determination of levels of amino acid precursors (e.g., Venitt & Bosworth, 1983, 1986).

(ii) Forward mutation

Interference by amino acids and their precursors could be avoided by using assays that involve forward mutation for detecting mutations in bacteria. Several systems are available, including 8-azaguanine resistance in *S. typhimurium* (Skopek *et al.*, 1978); forward and reverse mutation at several loci in *E. coli* K-12 343/113 (Mohn & Ellenberger, 1977); arabinose resistance in *S. typhimurium* SV50 (Xu *et al.*, 1984); and the RK test using *E. coli* CHY832 (Hayes *et al.*, 1984). See Report 5 for further details. A survey of the literature revealed no study that employed any of these assay systems for detecting mutagenic activity in body fluids or excreta. There is, however, no reason *a priori* why these or other forward mutation systems should not be used in body-fluid analysis.

(iii) *Induction of DNA damage or repair*

Another method for avoiding the potentially confounding effects of amino acids or precursors in extracts of body fluids is to use a bacterial assay system that employs the induction of DNA repair as an indicator of genotoxicity. There is a variety both of bacterial strains and of assay systems for detecting induction of DNA damage and repair; earlier assays relied on differential killing of repair-proficient and repair-deficient bacteria (IARC, 1980).

A second generation of tests is currently undergoing development and validation, which exploit fusions between operons involved in controlling induced DNA repair and structural genes which code for the synthesis of an easily determined enzyme. Of these assay systems, the SOS chromotest (Quillardet *et al.*, 1982, 1985; Quillardet & Hofnung, 1985) and the *umu*-test (Oda *et al.*, 1985) appear be of particular interest (see Report 5 for further details).

(*b*) *Other assay systems*

Tables 1 and 2 show that very few investigators have chosen to use assay systems other than bacterial mutation tests for body-fluid analysis. There is no underlying scientific reason for this. Cytogenetic tests with cultured mammalian cells seem to offer several advantages in body-fluid analysis; induction of chromosomal anomalies is highly relevant to current theories of carcinogenesis; cytogenetic tests are claimed to be very sensitive to clastogenic agents found, for example, in the urine of coffee drinkers and cigarette smokers (Dunn & Curtis, 1985). However, they are less often used for reasons of time, cost, availability of technical skill and practical feasibility.

(*c*) *Use of metabolic systems*

The background to the scientific principles and use of exogenous metabolic systems in short-term genotoxicity assays is given in Report 15. The use of metabolic systems in body-fluid analysis is widespread, and is in some cases essential in order to demonstrate genotoxic activity. For example, the original observation of the mutagenicity of smokers' urine (Yamasaki & Ames, 1977) required the presence of rat-liver supernatant in the assay. The addition of deconjugating enzymes to the assay system in order to reactivate conjugated putative mutagens, referred to above, may be seen as an adjunct to standard methods of supplying metabolic activity in short-term tests.

3. Relevance of body-fluid analysis to cancer

In the introduction to this Report (Section 1), attention was drawn to three distinct areas of research in which body-fluid analysis could be considered relevant to cancer: (i) detection of potential carcinogens using experimental animals; (ii) biological monitoring of occupational exposure to potential carcinogens; and (iii) identification of etiological agents in human cancer, especially with regard to unknown exogenous and endogenous carcinogens. The relevance to cancer of these three categories depends on the strong qualitative association between genotoxic activity and carcinogenic activity — an association the scientific basis of which has been strengthened by recent findings that certain carcinogens can activate cellular oncogenes.

3.1. *Detection of potential carcinogens in experimental animals*

Body-fluid analysis in experimental animals is one method of extending information gained from in-vitro techniques for detecting the potential carcinogenicity of chemicals, since the in-vivo tests take account of absorption, distribution and metabolism in mammals.

3.2 *Biological monitoring of occupational exposure to carcinogens*

Body-fluid analysis is one of several methods currently under investigation for monitoring individuals or populations occupationally exposed to potential carcinogens. In the present state of knowledge, the occurrence of genotoxic activity in body fluids or excreta is an indicator of short-term exposure and absorption — not of increased risk of cancer either in individuals or in groups. Carefully conducted long-term prospective studies are required to establish the predictive value for cancer risk of body-fluid analysis. Such studies are not yet available.

3.3 *Identification of etiological agents*

The use of body-fluid analysis for studying the etiology of human cancer is based on the premise that genotoxic agents in body fluids and excreta may induce neoplasia in the organ or tissue where exposure occurs, or at distant sites, by circulation of genotoxins in body fluids. Two main methods have been used for deploying body-fluid analysis in the study of etiology: (i) comparison of exposures of groups representing high- and low-risk populations and (ii) the case-control method, in which cancer patients are compared with controls without cancer. Neither approach has yet given decisive evidence that genotoxic activity in body fluids is causally linked to the induction of cancer at sites such as the stomach, urinary bladder and colon.

Since excreta such as urine and faeces are the major vehicles for the elimination of toxic materials from the body, a certain background level of genotoxic activity may well be tolerated through defence mechanisms in those tissues and in organs most intimately in contact with the excreta. If this is the case, the relevance of results gained so far from body-fluid analysis is uncertain. Thorough investigation of base-line levels of genotoxic activity in body fluids is therefore essential.

4. Advantages and disadvantages of body-fluid analysis

4.1 *Advantages*

Body-fluid analysis can be performed with material from animals or people, and, in the case of urine and faeces, is noninvasive. Multiple samples can be collected over time without undue interference with the donor. The effects of habits such as smoking and diet can be investigated. The technique is nonspecific, in that the identity or type of putative genotoxin does not need to be known in advance in order to perform the assay. The same biological endpoint can be used to detect the activity of a variety of different substances. There is a limited amount of evidence that dose-response data can be obtained from body-fluid analysis. Biological activity can be correlated with chemical analysis in the same sample, and a combination of fractionation and short-term assay allows identification of mutagens. The technique is cheap to perform and is relatively sensitive to certain classes of genotoxins.

4.2 *Disadvantages*

Body-fluid analysis using human donors is subject to several important confounding variables, especially in relation to habits (in particular smoking and diet, drug intake and occupation) and to the presence of interfering substances in test samples. Problems of sample collection, storage, extraction, fractionation and methods of assay have yet to be solved. Cumulative exposure cannot be detected, due to the relatively rapid clearance of compounds from and *via* body fluids and excreta. Certain chemical classes (e.g., nitro compounds assayed in bacteria with efficient nitroreductase systems) may be overrepresented, while other classes will be underrepresented. The biological significance of genotoxic activity in body fluids and excreta is not easily interpretable unless the active constituents are identified structurally. For example, urinary mutagenicity assays conducted with rats dosed with pairs of carcinogens and noncarcinogens (hydrocarbons and aromatic amines) did not distinguish the carcinogenic compounds from their noncarcinogenic analogues (Malaveille *et al.*, 1986).

5. Evaluation of results

The scientific value of extensive and well-conducted studies can be seriously compromised by inadequate reporting of full details of all aspects of an investigation. Evaluation of data from body-fluid analysis requires two levels of judgement:

(i) evaluation of the short-term test data themselves, using well-established criteria — reproducibility, statistical significance, dose-responsiveness, use of appropriate negative and positive controls. Raw data (i.e., untransformed data such as colony counts and full tabulations of chromosomal anomalies) should always be published, or be available for independent evaluation. For more information on the performance and evaluation of short-term assays, see, for example, Dean (1983) and Venitt and Parry (1984); and

(ii) evaluation of the data in relation to the attributes of the donor, bearing in mind all the confounding variables related to diet, occupation and drug intake; the nature of the sample and the method of collection, storage and extraction; the possibility of artefacts caused by the presence of interfering substances; the use of positive controls (e.g., samples spiked with reference mutagens); the demonstration of dose-response relationships, or changes in genotoxic activity related to change in exposure or with time.

The statistical significance of such changes should be tested using an appropriate method; in most cases, the nature of the distribution of data from short-term assays of body fluids will be unknown, in which case a nonparametric test is the most appropriate.

6. References

Aeschbacher, H.U. & Chappuis, C. (1981) Non-mutagenicity of urine from coffee drinkers compared with that from cigarette smokers. *Mutat. Res., 89*, 161-177

Amacher, D.E., Turner, G.N. & Ellis, J.H., Jr (1981) Detection of mammalian cell mutagens in urine from carcinogen-dosed mice. *Mutat. Res., 90*, 79-90

Ames, B.N., McCann, J. & Yamasaki, E. (1975) Methods for detecting carcinogens and mutagens with the *Salmonella*/mammalian microsome test. *Mutat. Res., 31*, 347-364

Askew, A.R., Reibelt, L.D. & Visona, A. (1984) Faecal mutagens and intestinal tumours in the dimethylhydrazine-injected rat. *Mutat. Res., 139* 143-147

Baden, J.M., Kelley, M., Cheung, A. & Mortelmans, K. (1980) Lack of mutagens in urines of operating room personnel. *Anesthesiology, 53*, 195-198

Baker, R., Arlauskas, A., Bonin, A. & Angus, D. (1982) Detection of mutagenic activity in human urine following fried pork or bacon meals. *Cancer Lett., 16*, 81-89

Barale, R., Sozzi, G., Toniolo, P., Borghi, O., Reali, D., Loprieno, N. & Della Porta, G. (1985) Sister-chromatid exchanges in lymphocytes and mutagenicity in urine of nurses handling cystostatic drugs. *Mutat. Res., 157*, 235-240

Barnes, W.S. & Weisburger, J.H. (1985) Fate of the food mutagen 2-amino-3-methylimidazo(4,5-f)-quinoline (IQ) in Sprague-Dawley rats. I. Mutagens in the urine. *Mutat. Res., 156*, 83-91

Beek, B., Aranda, T. & Thompson, E. (1982) Induction of sister-chromatid exchanges, cell-cycle delay and chromosomal aberrations by human urine concentrates. *Mutat. Res., 92*, 333-360

Belisario, M.A., Farina, C. & Buonocore, V. (1985) Evaluation of concentration procedures of mutagenic metabolites from urine of diesel particulate-treated rats. *Toxicol. Lett., 25*, 81-88

Berlin, A., Draper, M., Hemminki, K. & Vainio, H., eds (1984) *Monitoring Human Exposure to Carcinogenic and Mutagenic Agents (IARC Scientific Publications No. 59)*, Lyon, International Agency for Research on Cancer

Bos, R.P., Brouns, R.M., van Doorn, R., Theuws, J.L. & Henderson, P.T. (1980) The appearance of mutagens in urine of rats after the administration of benzidine and some other aromatic amines. *Toxicology*, *16*, 113-122

Bos, R.P., Leenaars, A.O., Theuws, J.L. & Henderson, P.T. (1982) Mutagenicity of urine from nurses handling cytostatic drugs, influence of smoking. *Int. Arch. occup. environ. Health*, *50*, 359-369

Bos, R.P., Theuws, J.L.G. & Henderson, P.T. (1983) Excretion of mutagens in human urine after passive smoking. *Cancer Lett.*, *19*, 85-90

Buffoni, F., Santoni, G., Albanese, V. & Dolara, P. (1983) Urinary mercapturic acid in chemical workers and in control subjects. *J. appl. Toxicol.*, *3*, 63-65

Caderni, G., Dolara, P., Constantini, A., Barbagli, G. & Calzolai, A. (1982) Determination of urinary mutagens in patients with urinary tract cancer. *Eur. Urol.*, *8*, 243-246

Camus, A.M., Friesen, M., Croisy, A. & Bartsch, H. (1982) Species-specific activation of phenacetin into bacterial mutagens by hamster liver enzymes and identification of N-hydroxyphenacetin O-flucuronide as a promutagen in the urine. *Cancer Res.*, *42*, 3201-3208

Camus, A.M., Aitio, A., Sabadie, N., Wahrendorf, J. & Bartsch, H. (1984) Metabolism and urinary excretion of mutagenic metabolites of benzo[a]pyrene in C57 and DBA mice strains. *Carcinogenesis*, *5*, 35-39

Chipman, J.K., Millburn, P. & Brooks, T. (1983) Mutagenicity and in-vivo disposition of biliary metabolites of benzo[a]pyrene. *Toxicol. Lett.*, *17*, 233-240

Clemmensen, S. & Lam, H.R. (1984) Mutagens in rat urine after dermal application of 1,3-diaminobenzene. *Mutat. Res.*, *138*, 137-143

Combes, R., Anderson, D., Brooks, T., Neale, S. & Venitt, S. (1984) *The detection of mutagens in urine, faeces and body fluids*. In: Dean, B.J., ed., *Report of the UKEMS Sub-committee on Guidelines for Mutagenicity Testing*, Part II, Swansea, United Kingdom Environmental Mutagen Society, pp. 203-244

Connor, T.H., Ramanujam, V.M.S., Ward, J.B. & Legator, M.S. (1983) The identification and characterisation of a urinary mutagen resulting from cigarette smoke. *Mutat. Res.*, *113*, 161-172

Connor, T.H., Ward, J.B., Jr & Legator, M.S. (1985) Absence of mutagenicity in the urine of autopsy service workers exposed to formaldehyde: factors influencing mutagenicity testing of urine. *Int. Arch. occup. environ. Health*, *56*, 225-237

Cortinas de Nava, C., Espinosa, J., Garcia, L., Zapata, A.M. & Martinez, E. (1983) Mutagenicity of antiamebic and anthelmintic drugs in the *Salmonella typhimurium* microsomal test system. *Mutat. Res.*, *117*, 79-91

Curtis, J.R. & Dunn, B.P. (1985) High-pressure liquid chromatography for isolation of clastogenic agents from urine. *Mutat. Res.*, *147*, 171-177

Dean, B.J., ed. (1983) *Report of the UKEMS Sub-committee on Guidelines for Mutagenicity Testing*, Part I, *Basic Test Battery; Minimal Criteria; Professional Standards; Interpretation; Selection of Supplementary Assays*, Swansea, United Kingdom Environmental Mutagen Society, p. 178

De Raat, W.K. & van Ardenne, R.A. (1984) Sorption of organic compounds from urine in mutagenicity testing: choice of sorbent. *J. Chromatogr.*, *310*, 41-49

Dion, P. & Bruce, W.R. (1983) Mutagenicity of different fractions of extracts of human faeces. *Mutat. Res.*, *119*, 151-160

Dion, P.W., Bright-See, E.B., Smith, C.C. & Bruce, W.R. (1982) The effect of dietary ascorbic acid and alpha-tocopherol on fecal mutagenicity. *Mutat. Res.*, *102*, 27-37

DiVincenzo, G.D., Hamilton, M.L., Mueller, K.R., Donish, W.H. & Barner, E.D. (1985) Bacterial mutagenicity testing of urine from rats dosed with 2-ethylhexanol derived plasticizers. *Toxicology, 34*, 247-259

Dolara, P., Barale, R., Mazzoli, S. & Benetti, D. (1980) Activation of the mutagens of beef extract *in vitro* and *in vivo*. *Mutat. Res., 79*, 213-221

Dolara, P., Mazzoli, S., Rosi, D., Buiatti, E., Baccetti, S., Turchi, A. & Vannucci, V. (1981) Exposure to carcinogenic chemicals and smoking increases urinary excretion of mutagens in humans. *J. Toxicol. environ. Health, 8*, 95-103

Dolara, P., Caderni, G. & Benetti, D. (1982) Activation of Trp-P-1 and Trp-P-2 *in vitro* and *in vivo*. *Nutr. Cancer, 3*, 168-171

van Doorn, R., Bos, R.P., Leijdekkers, C.-M., Wagenaars-Zegers, M.A.P., Theuws, J.L.G. & Henderson, P.T. (1979) Thioether concentration and mutagenicity of urine from cigarette smokers. *Int. Arch. occup. environ. Health, 43*, 159-166

Dunn, B.P. & Curtis, J.R. (1985) Clastogenic agents in the urine of coffee drinkers and cigarette smokers. *Mutat. Res., 147*, 179-188

Edwards, C.N. & Combes, R.D. (1984) Mutagenicity studies of urine and faecal samples from rats treated orally with the food-colourings Brown FK and Red 2G. *Food chem. Toxicol., 22*, 593-597

Everson, R.B., Gad-El-Mawla, N.M., Attia, M.A., Chevlen, E.M., Thorgeirsson, S.S., Alexander, L.A., Flack, P.M., Staiano, N. & Ziegler, J.L. (1983a) Analysis of human urine for mutagens associated with carcinoma of the bilharzial bladder by the Ames *Salmonella* plate assay. *Cancer, 51*, 371-377

Everson, R.B., Flack, P.M. & Sandler, R.S. (1983b) Urinary excretion of mutagens in cirrhosis: limited evidence of an association. *Environ. Res., 32*, 118-126

Everson, R.B., Ratcliffe, J.M., Flack, P.M., Hoffman, D.M. & Watanabe, A.S. (1985a) Detection of low levels of urinary mutagen excretion by chemotherapy workers which was not related to occupational drug exposures. *Cancer Res., 45*, 6487-6497

Everson, R.B., Milne, K.L., Warburton, D., McClamrock, H.D. & Buchanan, P.D. (1985b) Mutagenesis assays of human amniotic fluid. *Environ. Mutagen., 7*, 17-184

Falck, K., Sorsa, M., Vainio, H. & Kilparki, I. (1980) Mutagenicity in urine of workers in rubber industry. *Mutat. Res., 79*, 45-52

Feyerabend, C., Higenbottam, T. & Russell, M.A.H. (1982) Nicotine concentrations in urine and saliva of smokers and non-smokers. *Br. med. J., 284*, 1002-1004

Garner, R.C. (1985) Assessment of carcinogen exposure in man. *Carcinogenesis, 6*, 1071-1078

Garner, R.C., Mould, A.J., Lindsay-Smith, V., Cartwright, R.A. & Richards, B. (1982) Mutagenic urine from bladder cancer patients. *Lancet, ii*, 389

Gelbart, S.M. & Sontag, S.J. (1980) Mutagenic urine in cirrhosis. *Lancet, i*, 894-896

Gentile, J.M., Brown, S., Aardema, M., Clark, D. & Blankespoor, H. (1985) Modified mutagen metabolism in *Schistosoma hematobium*-infested organisms. *Arch. environ. Health, 40*, 5-12

Gibson, J.F., Baxter, P.J., Hedworth-Witty, R.B. & Gompertz, D. (1983) Urinary mutagenicity assays: a problem arising from the presence of histidine associated growth factors in XAD-2 prepared urine concentrates, with particular reference to assays carried out using the bacterial fluctation test. *Carcinogenesis, 4*, 1471-1476

Gupta, I., Baptista, J., Bruce, W.R., Che, C.T., Furrer, R., Gingerich, J.S., Grey, A.A., Marai, L., Yates, P. & Krepinsky, J.J. (1983) Structures of fecapentaenes, the mutagens of bacterial origin isolated from human feces. *Biochemistry*, 22, 241-245

Hasegawa, R., St John, M.K., Cano, M., Issenberg, P., Klein, D.A., Walker, B.A., Jones, J.W., Schnell, R.C., Merrick, B.A. & Davies, M.H. (1984) Bladder freeze ulceration and sodium saccharin feeding in the rat: examination for urinary nitrosamines, mutagens and bacteria, and effects on hepatic microsomal enzymes. *Food chem. Toxicol.*, 22, 935-942

Hayatsu, H., Arimoto, S., Togawa, K. & Makita, M. (1981) Inhibitory effect of ether extract of human faeces on activities of mutagens: Inhibition by oleic and linoleic acids. *Mutat. Res.*, 81, 287-293

Hayatsu, H., Oka, T., Wakata, A., Ohara, Y., Hayatsu, T., Kobayashi, H. & Arimoto, S. (1983) Adsorption of mutagens to cotton bearing covalently bound trisulfo-copper-phthalocyanine. *Mutat. Res.*, 119, 233-238

Hayatsu, H., Hayatsu, T. & Ohara, Y. (1985a) Mutagenicity of human urine caused by ingestion of fried ground beef. *Gann*, 76, 445-448

Hayatsu, H., Hayatsu, T., Wataya, Y. & Mower, H.F. (1985b) Fecal mutagenicity arising from ingestion of fried ground beef in the human. *Mutat. Res.*, 143, 207-211

Hayes, S., Gordon, A., Sadowski, I. & Hayes, C. (1984) RK bacterial test for independently measuring chemical toxicity and mutagenicity: short-term forward selection assay. *Mutat. Res.*, 130, 97-106

Heussner, J.C., Ward, J.B., Jr & Legator, M.S. (1985) Genetic monitoring of aluminium workers exposed to coal pitch tar volatiles. *Mutat. Res.*, 155, 143-155

Hietanen, E., Ahotupa, M., Heikela, A. & Laitinen, M. (1982) *Dietary lipids as modifiers of monooxygenase induction*. In: Hietanen, E., Laitinen, M. & Hanninen, O., eds, *Cytochrome P-450, Biochemistry, Biophysics and Environmental Implications*, Amsterdam, Elsevier, pp. 705-708

Hirai, N., Kingston, D.G.I., Van Tassell, R.L. & Wilkins, T.D. (1982) Structure elucidation of a potent mutagen from human feces. *J. Am. chem. Soc.*, 104, 6149-6150

IARC (1980) *IARC Monographs on the Evaluation of the Carcinogenic Risk of Chemicals to Humans*, Suppl. 2, *Long-term and Short-term Screening Assays for Carcinogens: A Critical Appraisal*, Lyon, pp. 85-106

Jaffe, R.L., Nicholson, W.J. & Garro, A.J. (1983) Urinary mutagen levels in smokers. *Cancer Lett.*, 20, 37-42

Joachim, F., Burrell, A. & Andersen, J. (1985) Mutagenicity of azo dyes in the *Salmonella*/microsome assay using in-vitro and in-vivo activation. *Mutat. Res.*, 156, 131-138

Kado, N.Y., Langley, D. & Eisenstadt, E. (1983) A simple modification of the *Salmonella* liquid-incubation assay. Increased sensitivity for detecting mutagens in human urine. *Mutat. Res.*, 121, 25-32

Kado, N.Y., Manson, C., Eisenstadt, E. & Hsieh, D.P.H. (1985) The kinetics of mutagen excretion in the urine of cigarettAccumulation of mutagenic activity in bile
fluid of river Rhine fish. *Prog. clin. biol. Res.*, 109, 287-296

Kriebel, D., Commoner, B., Bollinger, D., Bronsdon, A., Gold, J. & Henry, J. (1983) Detection of occupational exposure to genotoxic agents with a urinary mutagen assay. *Mutat. Res.*, 108, 67-79

Krøkje, A., Tiltnes, A., Mylius, E. & Gullvag, B. (1985) Testing for mutagens in an aluminium plant. The results of *Salmonella typhimurium* tests on expectorates from exposed workers. *Mutat. Res.*, *156*, 147-152

Kuhnlein, U., Bergstrom, D. & Kuhnlein, J. (1981) Mutagens in feces from vegetarians and non-vegetarians. *Mutat. Res.*, *85*, 1-12

Kuhnlein, U., Gallagher, R. & Freeman, H.J. (1983a) Effects of purified cellulose and pectin fiber diets on mutagenicity of feces and luminal contents of stomach, small and large bowel in rats. *Clin. invest. Med.*, *6*, 253-260

Kuhnlein, H.V., Kuhnlein, U. & Bell, P.A. (1983b) The effect of short-term dietary modification on human fecal mutagenic activity. *Mutat. Res.*, *113*, 1-12

Laires, A., Borba, H., Rueff, J., Gomes, M.I. & Halpern, M. (1982) Urinary mutagenicity in occupational exposure to mineral oils and iron oxide particles. *Carcinogenesis*, *3*, 1077-1079

Lambotte-Vandepaer, M., Duverger van Bogaert, M., de Meester, C., Poncelet, F. & Mercier, M. (1980) Mutagenicity of urine from rats and mice treated with acrylonitrole. *Toxicology*, *16*, 67-71

Lambotte-Vandepaer, M., Duverger van Bogaert, M., de Meester, C., Rollmann, B., Poncelet, F. & Mercier, M. (1981) Identification of two urinary metabolites of rats treated with acrylonitrole; influence of several inhibitors on the mutagenicity of those urines. *Toxicol. Lett.*, *7*, 321-327

Legator, M.S., Connor, T.H. & Stoeckel, M. (1975) Detection of mutagenic activity of metronidazole and niridazole in body fluids of humans and mice. *Science*, *188*, 1118-1119

Legator, M.S., Bueding, E., Batzinger, R., Connor, T.H., Eisenstadt, E., Farrow, M.G., Ficsor, G., Hsie, A., Seed, J. & Stafford, R.S. (1982) An evaluation of the host-mediated assay and body-fluid analysis. A report of the US Environmental Protection Agency Gene-Tox Program. *Mutat. Res.*, *98*, 319-374

Lynn, R.K., Garvie-Could, C.T., Milam, D.F., Scott, K.F., Eastman, C.L., Ilias, A.M. & Rodgers, R.M. (1984) Disposition of the aromatic amine, benzidine, in the rat: characterization of mutagenic urinary and biliary metabolites. *Toxicol. appl. Pharmacol.*, *72*, 1-14

Malaveille, C., Brun, G. & Bartsch, H. (1986) Mutagenicity of urine from rats treated with benzo[*a*]pyrene, pyrene, 2-acetylaminofluorene and 4-acetylaminofluorene in *Salmonella typhimurium* TA100 or TA98 strains. *Progr. Mutat. Res.* (in press)

Maron, D.M. & Ames, B.N. (1983) Revised methods for the *Salmonella* mutagenicity test. *Mutat. Res.*, *113*, 173-215

Marshall, M.V., Noyola, A.J. & Rogers, W.R. (1983) Analysis of urinary mutagens produced by cigarette-smoking baboons. *Mutat. Res.*, *118*, 241-256

Menon, M.M. & Bhide, S.V. (1984) Mutagenicity of urine of bidi and cigarette smokers and tobacco chewers. *Carcinogenesis*, *5*, 1523-1524

Miller, C.T. & Stoltz, D.E. (1978) Mutagenicity induced by lyophilization or storage of urine from isoniazid-treated rats. *Mutat. Res.*, *56*, 289-293

Minnich, V., Smith, M.E., Thompson, D. & Kornfeld, S. (1976) Detection of mutagenic activity in human urine using mutant strains of *Salmonella typhimurium*. *Cancer*, *38*, 1253-1258

Mohn, G.R. & Ellenberger, J. (1977) *The use of* Escherichia coli *K12/343/113(λ) as a multi-purpose indicator strain in various mutagenicity testing procedures*. In: Kilbey, B.J., Legator, M., Nichols, W. & Ramel, C., eds, *Handbook of Mutagenicity Test Procedures,* Amsterdam, Elsevier, pp. 95-118

Mohtashamipur, E., Norpoth, K. & Heger, M. (1984) Urinary excretion of frameshift mutagens in rats caused by passive smoking. *J. Cancer Res. clin. Oncol.*, *108*, 296-301

Mohtashamipur, E., Norpoth, K. & Lieder, F. (1985) Isolation of frameshift mutagens from smokers' urine: experiences with three concentration methods. *Carcinogenesis*, *6*, 783-788

Møller, M. & Dybing, E. (1980) Mutagenicity studies with urine concentrates from coke plant workers. *Scand. J. Work Environ. Health*, *6*, 216-220

Montes, G., Cuello, C., Gordillo, G., Pelon, W., Johnson, W. & Correa, P. (1979) Mutagenic activity of gastric juice. *Cancer Lett.*, *7*, 307-312

Morris, D.L., Youngs, D., Muscroft, T.J., Cooper, J., Rojinski, C., Burdon, D.W. & Keighley, M.R. (1984) Mutagenicity in gastric juice. *Gut*, *25*, 723-727

Mower, H.F., Ichinotsubo, D., Wang, L.W., Mandel, M., Stemmermann, G., Nomura, A., Heilbrun, L., Kamiyama, S. & Shimada, A. (1982) Fecal mutagens in two Japanese populations with different colon cancer risks. *Cancer Res.*, *42*, 1164-1169

Munzner, R. & Wever, J. (1984) Investigations on the detection of mutagenic activity of beef extract in rats after oral administration. *Cancer Lett.*, *23*, 109-114

Nguyen, T.V., Theiss, J.C. & Matney, T.S. (1982) Exposure of pharmacy personnel to mutagenic antineoplastic drugs. *Cancer Res.*, *42*, 4792-4796

O'Connor, H.J., Axon, A.T., Riley, S.E. & Garner, R.C. (1984) Mutagenicity of gastric juice: importance of controlling histidine concentration when using *Salmonella* tester strains. *Carcinogenesis*, *5*, 853-856

Oda, Y., Nakamura, S.-I., Oki, I., Kato, T. & Shinagawa, H. (1985) Evaluation of the new system (*umu*-test) for the detection of environmental mutagens and carcinogens. *Mutat. Res.*, *147*, 219-229

Ong, T.M., Stockhausen, A., Adamo, D. & Whong, W.Z. (1985) The urine mutagenicity assay system. studies relating to recovery, storage and concentration procedures. *Scand. J. Work Environ. Health*, *11*, 45-50

Pantuck, E.J., Hsiao, K.C., Conney, A.H., Garland, W.A., Kappas, A., Anderson, K.E. & Alvares, A.P. (1976). Effect of charcoal-broiled beef on phenacetin metabolism in man. *Science*, *194*, 1055-1057

Pasquini, R., Monarca, S., Sforzolini, G.S., Conti, R. & Fagioli, F. (1982) Mutagens in urine of carbon electrode workers. *Int. Arch. occup. environ. Health*, *50*, 387-395

Peto, R. (1983) The marked differences between carotenoids and retinoids: methodological implications for biochemical epidemiology. *Cancer Surv.*, *2*, 327-340

Quillardet, P. & Hofnung, M. (1985) The SOS Chromotest, a colorimetric assay for genotoxins: procedures. *Mutat. Res.*, *147*, 65-78

Quillardet, P., Huisman, O., D'Ari, R. & Hofnung, M. (1982) SOS Chromotest, a direct assay of induction of an SOS function in *Escherichia coli* K-12 to measure genotoxicity. *Proc. natl Acad. Sci. USA*, *79*, 5971-5975

Quillardet, P., de Bellecombe, C. & Hofnung, M. (1985) The SOS Chromotest, a colorimetric assay for genotoxins: validation study with 83 compounds. *Mutat. Res.*, *147*, 79-95

Recio, L., Enoch, H.G., Hannan, M.A. & Hill, R.H. (1984) Application of urine mutagenicity to monitor coal liquefaction workers. *Mutat. Res.*, *136*, 201-207

Reddy, B.S., Sharma, C., Darby, L., Laakso, K. & Wynder, E.L. (1980) Metabolic epidemiology of large bowel cancer. Fecal mutagens in high- and low-risk population for colon cancer. A preliminary report. *Mutat. Res.*, *72*, 511-522

Reddy, B.S., Sharma, C., Mathews, L. & Engle, A. (1984) Faecal mutagens from subjects consuming a mixed western diet. *Mutat. Res., 135*, 11-19

Reddy, T.V., Benjamin, T., Grantham, P.H., Weisburger, E.K. & Thorgeirsson, S.S. (1980) Mutagenicity of urine from rats after administration of 2,4-diaminoanisole: the effect of microsomal enzymer inducers. *Mutat. Res., 79*, 307-317

Rivrud, G.N., Berg, K., Anderson, D., Blowers, S. & Bjøro, K. (1986) Study of the amniotic fluid from smokers and non-smokers in the Ames test. *Mutat. Res., 169*, 11-16

Safirstein, R., Daye, M. & Guttenplan, J.B. (1983) Mutagenic activity and identification of excreted platinum in human and rat urine and rat plasma after administration of cisplatin. *Cancer Lett., 18*, 329-338

Sasson, I.M., Coleman, D.T., LaVoie, E.J., Hoffman, D. & Wynder, E.L. (1985) Mutagens in human urine: effects of cigarette smoking and diet. *Mutat. Res., 158*, 149-157

Skopek, T.R., Liber, H.L., Krolewski, J.J. & Thilly, W.G. (1978) Quantitative forward mutation assay in *Salmonella typhimurium* using 8-azaguanine resistance as a genetic marker. *Proc. natl Acad. Sci. USA, 75*, 410-414

Sorsa, M., Falck, K., Heinonen, T., Vainio, H., Norppa, H. & Rimpela, M. (1984) Detection of exposure to mutagenic compounds in low-tar and medium-tar cigarette smokers. *Environ. Res., 33*, 312-321

Sorsa, M., Hemminki, K. & Vainio, H. (1985) Occupational exposure to anticancer drugs — potential and real hazards. *Mutat. Res., 154*, 135-149

Sousa, J., Nath, J., Tucker, J.D. & Ong, T.M. (1985a) Dietary factors affecting the urinary mutagenicity assay system. I. Detection of mutagenic activity in human urine following a fried beef meal. *Mutat. Res., 149*, 365-374

Sousa, J., Nath, J. & Ong, T.M. (1985b) Dietary factors affecting the urinary mutagenicity assay system. II. The absence of mutagenic activity in human urine following consumption of red wine or grape juice. *Mutat. Res., 156*, 171-176

Staiano, N., Belisario, M.A., Della Morte, R., Farina, C., Remondelli, P. & Muscettola, G. (1984) Toxic genetic effects of flunitrazepam. *Boll. Soc. Ital. Biol. sper., 60*, 2247-2253

Stich, H.F. & Stich, W. (1982) Chromosome-damaging activity of saliva of betel nut and tobacco chewers. *Cancer Lett., 15*, 193-202

Stich, H.F., Stich, W. & Acton, A.B. (1980) Mutagenicity of fecal extracts from carnivorous and herbivorous animals. *Mutat. Res., 78*, 105-112

Stiller, A., Obe, G., Riedel, L., Riehm, H. & Kappes, C. (1982) Mutagens in human urine: test with human peripheral lymphocytes. *Mutat. Res., 97*, 437-447

Sugimura, T. (1985) Carcinogenicity of mutagenic heterocyclic amines formed during the cooking process. *Mutat. Res., 150*, 33-41

Suzuki, K. & Bruce, W.R. (1984) Human faecal fractions can produce nuclear damage in the colonic epithelial cells of mice. *Mutat. Res., 141*, 35-39

Telford, J.N., Babish, J.G., Dunham, P.B., Hogue, D.E., Miller, K.W., Stoewsand, G.S., Magee, B.H., Stouffer, J.R., Bache, C.A. & Lisk, D.J. (1984) Toxicologic studies with lambs fed sugar beets grown in municipal sludge-amended soil: lowered relative hemoglobin in red blood cells and mutagens in blood and excreta. *Am. J. vet. Res., 45*, 2490-2494

Topping, D.L., Weller, R.A., Nader, C.J., Calvert, G.D. & Illman, R.J. (1982) Adaptive effects of dietary ethanol in the pig: changes in plasma high-density lipoproteins and fecal steroid excretion and mutagenicity. *Am. J. clin. Nutr., 36*, 245-250

Venitt, S. (1982) Mutagens in human faeces: are they relevant to cancer of the large bowel? *Mutat. Res.*, *98*, 265-286

Venitt, S. & Bosworth, D. (1983) The development of anaerobic methods for bacterial mutation assays: aerobic and anaerobic fluctuation tests of human faecal extracts and reference mutagens. *Carcinogenesis*, *4*, 339-345

Venitt, S. & Bosworth, D. (1986) Further studies on the detection of mutagenic and genotoxic activity in human faeces: aerobic and anaerobic fluctuation tests with *S. typhimurium* and *E. coli*, and the SOS Chromotest. *Mutagenesis*, *1*, 49-64

Venitt, S. & Parry, J.M., eds (1984) *Mutagenicity Testing: A Practical Approach*, Oxford, IRL Press, pp. 353

Venitt, S., Crofton-Sleigh, C., Hunt, J., Speechley, V. & Briggs, K. (1984) Monitoring exposure of nursing and pharmacy personnel to cytotoxic drugs: urinary mutation assays and urinary platinum as markers of absorption. *Lancet*, *i*, 74-77

Wheeler, L.A., Saperstein, M.D. & Lower, N.J. (1981) Mutagenicity of urine from psoriatic patients undergoing treatment with coal tar and ultraviolet light. *J. invest. Dermatol.*, *77*, 181-185

Wilkins, T.D. & Van Tassell, R.L. (1983) *Production of intestinal mutagens*. In: Hentges, D.J., ed., *Human and Intestinal Microflora in Health and Disease*, New York, Academic Press, pp. 265-288

Wilkins, T.D., Lederman, M., Van Tassell, R.L., Kingston, D.G.I. & Henion, J. (1980) Characterization of a mutagenic bacterial product in human feces. *Am. J. clin. Nutr.*, *33*, 2513-2520

Xu, J., Whong, W.-Z. & Ong, T.M. (1984) Validation of the *Salmonella* (SV50)/arabinose-resistant forward mutation assay system with 26 compounds. *Mutat. Res.*, *130*, 79-86

Yamasaki, E. & Ames, B.N. (1977) Concentration of mutagens from urine by adsorption with the nonpolar resin XAD2: cigarette smokers have mutagenic urine. *Proc. natl Acad. Sci. USA*, *74*, 3555-3559

REPORT 15
METABOLIC ACTIVATION

Prepared by:

S. Venitt (Rapporteur), H. Bartsch,
G. Becking, R.P.P. Fuchs, M. Hofnung, C. Malaveille,
T. Matsushima, M.L. Mendelsohn, A.E. Pegg,
M. Roberfroid and H.S. Rosenkranz (Chairman)

1. Introduction

The principle underlying the use of all metabolic activation systems is that chemical carcinogens and mutagens may not be carcinogenic or mutagenic *per se* but require conversion by enzymes to reactive intermediates. This concept of metabolic activation was proposed by J. and E. Miller on the basis of their pioneering studies (Miller, 1970; Miller & Miller, 1974). For a description of the enzymes involved and an introduction to the literature, see LaDu *et al.* (1971), Brodie and Gillette (1971), Testa and Jenner (1976), Bartsch *et al.* (1982) and Kalyanaraman and Sivarajah (1984). Some of the principal factors known to influence the amount and activity of these enzymes are summarized in Table 1. Carcinogens and mutagens that require metabolic activation to manifest biological activity are sometimes referred to as 'procarcinogens' and 'promutagens' (see Fig. 1).

Table 1. Some factors affecting enzymes that metabolize xenobiotics

Factor	Reference
Age	Levin & Ryan (1975); Neims *et al.* (1976)
Disease states	Conney *et al.* (1974); Kato (1977)
Diurnal variation	Poley *et al.* (1978)
Hormonal status	Vesell *et al.* (1976)
Housing conditions	Vesell *et al.* (1976)
Immunological factors	Vesell *et al.* (1976)
Inducers	Conney (1967)
Activators and inhibitors	Conney (1967); Cinti (1978)
Nutrition and diet	Cambell & Hayes (1974); Conney *et al.* (1977, 1979)
Sex	Kato (1974); Levin & Ryan (1975)
Species	Williams (1971)
Strains (genetics)	Nebert & Felton (1976); Vesell (1978)
Stress	Vesell *et al.* (1976)

Fig. 1. Metabolic activation of procarcinogens to ultimate carcinogens

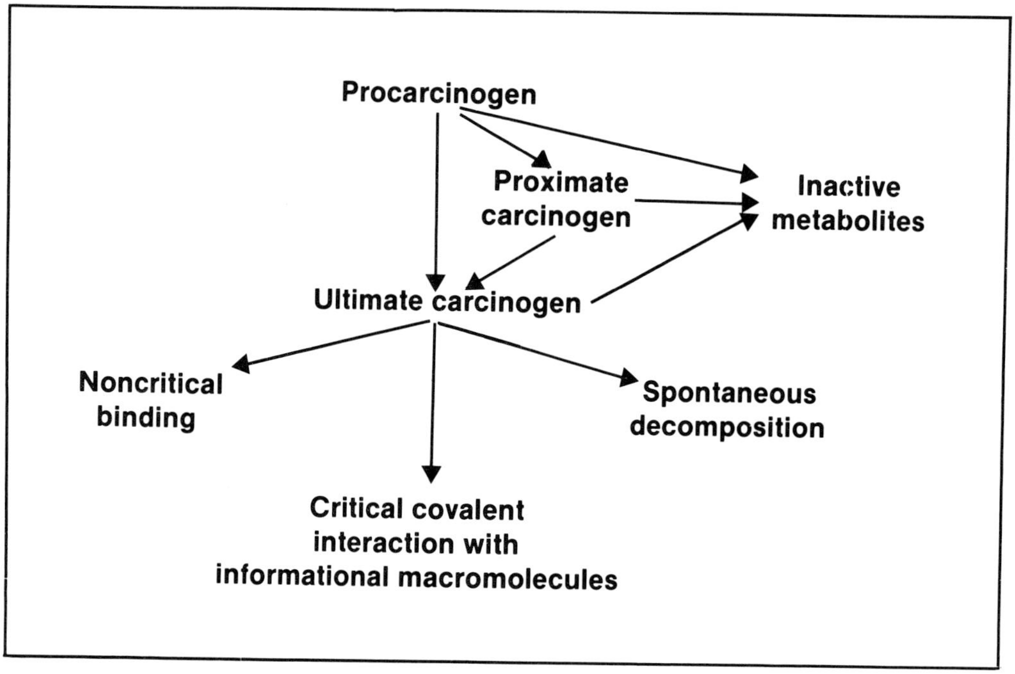

```
                Procarcinogen
                 /    |    \
                /     |     \
            Proximate |   ──→ Inactive
            carcinogen|   ←── metabolites
                \     |     /
                 \    |    /
                Ultimate carcinogen
                /     |     \
               /      |      \
        Noncritical   |    Spontaneous
         binding      |    decomposition
                      ↓
              Critical covalent
               interaction with
          informational macromolecules
```

2. Metabolic activation to electrophilic intermediates

2.1 *Principles*

The metabolism of most xenobiotics involves reactions catalysed by phase-I and phase-II enzymes (Table 2). The primary role of these enzymes is to form hydrophilic metabolites that are easily excreted; however, electrophilic intermediates may also be produced which can react with nucleophilic centres in macromolecules, such as nucleic acids and proteins.

Among the phase-I enzymes, cytochrome(s) P450-dependent mixed-function oxidases have received particular attention, as they are considered to be the major catalysts of the first, very often rate-limiting, step(s) in the pathways leading to the metabolic activation of most chemical carcinogens and mutagens. The importance of mixed-function oxidases is underlined by the large variety of oxygenation reactions that have been described, many of which produce reactive metabolites at carbon, nitrogen or sulfur atoms.

Table 2. Phase-I and phase-II enzymes that catalyse the metabolism of carcinogens

Phase I	Azo-nitro-reductases (M/C)
	Cytochrome P450-dependent monooxygenases (M)
	Cytochrome P450-independent oxidases (M)
	Epoxide hydratase (M/C)
	Hydrolases (M)
	Dehydrogenases (C)
Phase II	Acyl-transferases (C)
	Glucuronyl transferases (M)
	Glutathione S transferases (C)
	Sulfotransferases (C)

[a]M, microsomal enzymes; C, cytosolic enzymes

The mixed-function oxidase system consists of three components (Lu, 1976): (i) reduced nicotinamide adenine dinucleotide phosphate (NADPH) cytochrome-c reductase (NADPH-cytochrome-P450 reductase), (ii) cytochrome P450, and (iii) a phospholipid, phosphatidylcholine. NADPH-cytochrome-P450 reductase uses reducing equivalents supplied by NADPH to reduce cytochrome P450, the terminal oxidase which binds the lipophilic substrate (e.g., drug, carcinogen, steroid) to molecular oxygen. The haemoprotein catalyses incorporation of one of the oxygen atoms into the substrate, and the second atom of oxygen is reduced to water. Cytochrome P450 exists in multiple forms with different but overlapping substrate specificities (Levin et al., 1977; Jacoby, 1980; Lu & West, 1981; Guengerich et al., 1982; Sato & Kato, 1982). Many factors that affect xenobiotic metabolism qualitatively or quantitatively do so by altering the relative amounts or activities of the various forms of cytochrome P450. Of these, induction of the synthesis of increased amounts of enzyme molecules is frequently employed to modify enzyme activities. Phenobarbital, 3-methylcholanthrene and polychlorinated biphenyls are the inducers used most commonly for this purpose.

In many cases, phase-I metabolites (sometimes called 'proximate' carcinogens or mutagens) and certain procarcinogens require metabolism by phase-II enzymes to form 'ultimate' carcinogens or mutagens — highly reactive species which readily form adducts with DNA (see Fig. 1). Many examples of this process have been described (Miller & Miller, 1976; Rannug et al., 1978; Caldwell, 1979; Van Bladeren et al., 1980; Ozawa & Guengerich, 1983; Inskeep & Guengerich et al., 1984).

Phase-I and phase-II enzymes not only play a role in the metabolic activation of various chemicals, but also catalyse detoxification reactions which either favour the excretion of xenobiotics or inactivate their active metabolites. The production of carcinogenic or mutagenic metabolites is thus the result of a balance between activation and detoxification pathways which is often difficult to evaluate (see Fig. 1).

2.2 Cell-free systems used to detect carcinogens and mutagens that require metabolic activation to electrophiles

Cell-free mammalian activation systems have been developed in order to compensate for the limited capabilities of most indicator organisms and cultured cells to metabolize mutagens and carcinogens. Some of the advantages and limitations of the systems currently used in short-term tests for the detection of carcinogens and mutagens are summarized in Table 3. Several other activation systems have been employed, but large data bases are not available for these; they include subcellular fractions from green plants (Plewa & Gentille, 1982), organ explants (Hsu et al., 1978) and isolated perfused liver (Jenssen et al., 1979; Pueyo et al., 1979). Regardless of the source of the exogenous metabolic activating system, it must be remembered that the organism used to detect the mutation may itself be capable of metabolizing the test compound or its metabolites. Eukaryotic organisms are likely to be more metabolically competent than prokaryotes, but the latter also have significant enzyme activities; for example, bacteria contain nitroreductases (Rosenkranz & Speck, 1975), which are capable of efficiently metabolizing foreign compounds, and a number of microorganisms contain oxygenases (Callen, 1982).

The metabolic activation system used most widely in short-term bioassays is the $9000 \times g$ supernatant fraction (S9) of liver from rats pretreated with polychlorinated biphenyls (e.g., Aroclor 1254) (Maron & Ames, 1983). The large data base and general suitability of the activating system described by Ames et al. (1975) for use in conjunction with bacteria as target cells assure its continued use. Except for the additions and modifications outlined below, investigators are strongly urged to follow the procedures described in those publications. When modifications are made, it is most important that the protocols be strictly defined and reported in full detail.

Differences between species in the activity of many of the enzymes associated with xenobiotic metabolism are large (Williams, 1971; Venitt & Forster, 1985). The metabolic activation of a minority of procarcinogens is dependent on the use of S9 from other species, notably the Syrian hamster (e.g., Camus et al., 1982; Haworth et al., 1983).

The basic requirements for the preparation of the S9 were formulated with the Salmonella/microsome assay in mind but are equally applicable to other types of short-term bioassays. Liver S9 preparations have been used successfully to activate a number of different classes of chemicals to products that are mutagenic to cultured mammalian cells (see Report 7). However, the S9 fraction has been observed to be toxic to cultured mammalian cells and can be replaced by a $15\,000 \times g$ (S15) postmitochondrial supernatant fraction (Kuroki et al., 1979).

Table 3. Metabolic activation systems commonly used to metabolize carcinogens and mutagens to ultimately reactive forms

Activation system	Advantages	Disadvantages
Cell-free systems		
Postmitochondrial 9000 × g or 15 000 × g supernatant fraction of rodent liver (S9 or S15)	Easily prepared, contains most of the enzymes associated with xenobiotic metabolism; large data base from previous studies. Good system for routine screening. This system favours activation of carcinogens and mutagens	Large amounts of nucleophiles may bind ultimate carcinogens and mutagens and reduce sensitivity. Difficult to identify particular enzyme(s) involved in activation or detoxification
Microsomal fraction of rodent liver	Contains high levels of cytochrome P450-dependent monooxygenases; recombination with 100 000 × g supernatant permits partial identification of pathways of metabolism	Preparation requires preparative ultracentrifuge and long time. All necessary enzymes may not be present
Purified enzymes, e.g., monooxygenase system, epoxide hydratase glutathione transferase(s)	Biochemically well defined; unequivocal identification of metabolites under same conditions that induce mutations; ratio of various enzymes easily altered	Technical difficulty in preparing enzymes; general knowledge of metabolic fate of each test compound needed. Many important enzymes may be absent. Not suitable for routine screening
Mammalian cells	Cellular integrity and enzyme relationships preserved, endogenous levels of cofactors; intermediate between cell-free and in-vivo activation systems	Technically difficult, and limited sample throughout; more costly than cell-free systems. In some cell lines, important enzymes are present in small amounts or absent
In-vivo host-mediated assay	Compound metabolized and tester organism mutated *in vivo*. Pharmacokinetic parameters of absorption, distribution, metabolism and excretion of compound accounted for	Limited sensitivity throughout for the detection of mutagens and carcinogens. Technically more difficult than in-vitro tests; host reaction against tester organisms. Limited sample throughout compared with in-vitro system

The popularity of Aroclor as an inducer of the cytochrome P450-dependent monooxygenase system used for metabolic activation experiments stems from its ability to induce (in approximately equal amounts) both the major form of cytochrome P450 associated with induction by phenobarbital, and the major form of cytochrome P450 associated with induction by 3-methylcholanthrene (Ryan *et al.*, 1977). Thus, a single preparation of postmitochondrial supernatant prepared from a rat pretreated with a polychlorinated biphenyl mixture has a broad substrate specificity. However, the toxicity and carcinogenicity of polychlorinated biphenyls, coupled with their persistence, has led to concern over their use as routine inducers of the monooxygenase system. Matsushima *et al.* (1976) have recommended combined administration to rats of phenobarbital and 5,6-benzoflavone in order to induce the monooxygenase systems. 5,6-Benzoflavone induces the same haemoprotein as 3-methylcholanthrene but, unlike this and other polycyclic hydrocarbons, is noncarcinogenic. In metabolic activation studies, S9 fraction derived from rats pretreated with phenobarbital and 5,6-benzoflavone metabolized ten promutagens representing a variety of chemical classes to mutagenic products to the same extent as a S9 fraction derived from rats pretreated with a polychlorinated biphenyl mixture. Polycyclic hydrocarbons were activated to mutagens to a slightly lower extent (Matsushima *et al.*, 1976; Gatehouse & Delow, 1979).

(a) Calibration of the 9000 × g supernatant fraction

Variations among different laboratories or in the same laboratory at different times can occur during preparation of S9. The qualitative and quantitative extents to which the various forms of cytochrome P450 are induced vary with different preparations of polychlorinated biphenyls and even among different lots of the same preparation. Thus, the protein content of each preparation should be determined and recorded. If possible, the cytochrome P450 content (Omura & Sato, 1964) of the preparation should also be determined. Comparison of the cytochrome P450 content per mg of protein in untreated and in treated animals will verify the presence and extent of induction of the monooxygenase systems. Additionally, or alternatively, assay of arylhydrocarbon hydroxylase (benzo[*a*]pyrene hydroxylase) (Nebert & Gelboin, 1968) or ethoxyresorufin *O*-deethylase (Lubet *et al.*, 1985) and of aminopyrine *N*-demethylase (Mazel, 1971) or benzophetamine *N*-demethylase (Lu *et al.*, 1972) or pentoxyresorufin *O*-dealkylase (Lubet *et al.*, 1985) will calibrate the preparation with respect to the two major forms of cytochrome P450 induced by polychlorinated biphenyls.

Assay of one or more of the cytochrome P450-independent enzymes that are listed in Table 1 is also recommended; selection of a particular enzyme to assay is dictated in part by its expected involvement in the metabolism of the group of compounds under examination. When enzyme preparations are stored, they should be kept for a period of time not to exceed seven months. Storage temperature should be at or below -80°C. Since the stability of the various enzymes differs during storage, the integrity of the activation system cannot always be assumed from one or two enzyme assays (Hubbard *et al.*, 1985).

(b) Preincubation of test compounds

In the procedure described by Ames *et al.* (1975), metabolic activation of the test compounds takes place in the soft agar overlay. However, certain compounds (e.g., some *N*-nitrosamines) are more efficiently detected as mutagens in liquid suspension. It is recommended that the two assays be combined by preincubating S9, cofactors and test compound with the bacteria at 37°C for 5-20 min before addition of top agar. This method is as simple as the standard plate test, and its utility has been demonstrated (Yahagi *et al.*, 1977; Matsushima *et al.*, 1980).

(c) Dose-response relationship with 9000 × g liver supernatant and with the tested compound

A complete assessment of the presence or absence of mutagenic activity for a compound requires that both the amount of test compound and the amount of enzyme be varied several-fold. Such a protocol assures a wide spectrum of metabolic profiles and the opportunity to define mutagenic potency both in terms of amount of compound and of enzyme or protein needed for activation. A protocol is recommended in which four concentrations of test compound are used in the presence and absence of three different amounts of liver S9. The actual amount of enzyme preparation incubated may vary according to the compound under test. In general, incubations containing 10, 50 and 200 μl S9 from a 25% weight/volume homogenate are suggested (Maron & Ames, 1983; Venitt & Forster, 1985).

Control incubations containing the solvent but not the substrate should be assayed with each amount of enzyme. In addition, as a positive control, mutagenic carcinogens that require metabolic activation should be tested at each enzyme level. Selection of the positive control will be dictated by the sensitivity of the indicator organism and, when possible, by structural similarity with the test compound.

3. Metabolic activation to free-radical intermediates

3.1 *Principles*

Univalent reduction or oxidation of a variety of organic compounds may take place during the metabolism of xenobiotics. Such one-electron exchanges generate free radicals at a carbon, oxygen, nitrogen or sulfur atom (Table 4).

Metabolic pathways for the activation of chemical mutagens and carcinogens have been identified which involve *C*-centred free radicals (for a review, see Cavalieri & Rogan, 1984); *O*-centred free radicals (Bartsch & Hecker, 1971; Bartsch *et al.*, 1972; Floyd *et al.*, 1976a,b; Bachur *et al.*, 1977, 1978; Stier *et al.*, 1982; Cavalieri & Rogan, 1984), an *N*-centred free radical (Mason, 1979; Anderson *et al.*, 1984; Loew & Goldblum, 1985) and an *S*-centred free radical (Watanabe *et al.*, 1980; Mason, 1982; Watanabe *et al.*, 1982). Such free radicals might, like electrophilic intermediates, form covalent adducts with proteins and DNA.

Free radicals might also induce oxidative DNA damage. Since mammals are obligate aerobic organisms, most free radicals (R•) that appear as metabolites of xenobiotics are exposed to and react with molecular oxygen, yielding the superoxide

Table 4. Free-radical metabolites of xenobiotics

Parent compound		Free-radical metabolite	
Haloalkanes	C·	Haloaklyl	(e.g., $Cl_3C·$)
Triarylmethane dyes		Triarylmethyl	$(Ar)_3—C·$
Polycyclic aromatic hydrocarbons		Aryl cation	$Ar^+—C·H$
Alcohols		Hydroxyalkyl	$HO—C·H$ alk
Phenylhydrazine		Phenyl	$C_6H_5·$
Alcohols	O·	Alkoxyl	Alk—O·
Peroxides		Peroxyl	RO—O·
Phenols		Aryloxyl	Ar—O·
Hydroquinones/quinones		Semiquinone	HO—Ar—O·
Nitroxides		Nitroxyl	R_2—NO·
Aromatic amines	N·	Amino	Ar—N·H
Aminophenols		Semiquinoneimine	HO—Ar=N·—R
Phenylhydrazines		Hydrazyl	Ar—NH—N·H
Heterocyclic tertiary amines		Amino cation	$>N·^+—R$
Azo dyes		Azoanion	Ar—N̄—N·—Ar
Tetrazolium cations		Tetrazolinyl	R—C$<^{N·—N—Ar}_{N—N—Ar}$
Nitrite		Nitrogen oxide	N·O_2
Nitroaromatics		Aryl nitro anion	Ar—N·O_2^-
Thiols	S·	Thiyl	R—S·
Sulfite		Sulfuroxyl	S·O_3^-
Organic sulfide		Sulfinium	Ar—S·$^+$—R

anion $O_2^{-·}$. This reduced oxygen species could be a general product of free-radical-mediated metabolic activation of carcinogens and mutagens. The superoxide anion functions both as an oxidant and as a reductant; however, it does not have sufficient reactivity to account for the variety of toxic consequences associated with generation of activated oxygen in biological systems (Fee, 1980). The hydroxyl free radical, HO·, together with H_2O_2 and the singlet oxygen $^1O_2$, seem to be better candidates than the superoxide anion for causing deleterious effects such as DNA damage. (For reviews, see McBrien & Slater, 1982; Thaler-Dao et al., 1984.)

The major biological catalysts of metabolic activation leading to free radicals are cytochrome P450, flavoproteins and peroxidases. Cytochrome P450 and the major flavoproteins (NADPH cytochrome P450 and NADH cytochrome b5 reductases) belong to the microsomal electron transport chain. Their role in mediating the formation of free radicals from xenobiotics is additional to their function as the major catalysts of phase I reactions in drug metabolism (see section 2.1). The situation is quite different with the peroxidases, for which a role as drug metabolizing enzymes was not recognized until recently. In particular, the possible co-oxidation of structurally unrelated chemicals by prostaglandin synthase might be of toxicological significance. Indeed, this complex enzyme system generates the peroxide required to oxidize xenobiotics monovalently. (For recent reviews, see Kalyanaraman & Sivarajah, 1984; Thaler Dao et al., 1984.)

An overview of free radical processes in xenobiotic metabolism is available (Roberfroid et al., 1986). Even though there may be some specific exceptions it appears from that review that: (i) cytochrome P450 is specialized in the production of C-centred free radicals, a process which is not catalysed by peroxidases; (ii) S-centred free radicals are produced both by cytochrome P450 and peroxidases, predominantly prostaglandin endoperoxide synthase; (iii) except for the nitroxyl radical and semiquinones, which have been shown to be products of xenobiotic reactions with cytochrome P450 and flavoproteins, respectively, O-centred free radicals are produced mainly by peroxidases, including the peroxidative component of prostaglandin endoperoxide synthase; (iv) N-centred free radicals are produced predominantly by peroxidases (including prostaglandin endoperoxide synthase), with the exception of those derived from the one-electron reduction of azo- and nitroaromatic compounds, which are formed catalytically by flavoproteins such as NADPH cytochrome P450 reductase.

3.2 Cell-free systems for detecting carcinogens and mutagens metabolically activated to free radicals

Purified horse-radish peroxidase and ram seminal-vesicular microsomes supplemented with arachidonic acid are the most easily available preparations to detect carcinogens and mutagens metabolically activated to free radicals. The procedure for the isolation and storage of ram seminal-vesicular microsomes was described in detail by Guthrie et al. (1982). These preparations have been reported to activate chemicals to active metabolites that bind to DNA and proteins (Mattamal et al., 1981; Guthrie et al., 1982; Anderson et al., 1984); or are mutagenic (Robertson et al. 1983; Battista & Marnett, 1985). However, only a limited number of chemicals have been tested so far.

Other microsomal preparation isolated from various tissues (urinary bladder, kidney, lung, skin, small intestine and liver) of humans, rats, guinea-pigs, rabbits and dogs have also been used to demonstrate the widespread distribution of prostaglandin synthase-dependent metabolic activation of chemicals to molecular species which bind covalently to proteins or nucleic acids (Sivarajah et al., 1981; Nemoto & Takayama, 1984; Wise et al., 1984). All these reports support the concept that prostaglandin synthase-dependent metabolic activation is both tissue- and species-specific and indicate that arachidonic acid-supplemented ram seminal-vesicular microsomes might be used in addition to the classical liver S9 in order to broaden the spectrum of metabolic activation capacities in short-term in-vitro tests. However, comparative studies both of microsomal preparations and of a variety of chemical carcinogens are needed to validate this proposal.

4. Intact cells as activating systems

Intact mammalian cells in culture have been used as a source of enzymes for the activation of carcinogens in short-term bioassays measuring mutations, unscheduled DNA synthesis and transformation (for review, see Langenbach & Oglesby, 1983). The activating cells may themselves serve as the target cells in the assays or they may be cocultured with metabolically-deficient target cells. Primary hepatocytes appear to contain the broadest spectrum of enzymes capable of metabolizing carcinogens, and these cells have been used alone to detect unscheduled DNA synthesis induced by carcinogens and mutagens (Williams, 1976, 1977; Casciano et al., 1978; Michalopoulos et al., 1978; Yager & Miller, 1978) and have been cocultured with other cells which detect mutations (Langenbach et al., 1978a,b; Jones & Huberman, 1980; Amacher & Paillet, 1982; Katoh et al., 1982) or cell transformation (Poiley et al., 1979; Tu et al., 1984). Several studies (Green et al., 1977; Hubbard et al., 1981a,b) have demonstrated the capacity of isolated hepatocytes from mice, rats and hamsters to metabolize various chemicals to reactive species capable of producing mutagenic effects in strains of *Salmonella typhimurium*. Subsequent studies have shown hepatocyte-mediated metabolic activation of: polycyclic aromatic hydrocarbons (San & Williams, 1977; Brouns et al., 1979; Poiley et al., 1980; Glatt et al., 1981; Bos et al., 1983), aromatic amines and amides (Dybing et al., 1979; Brouns et al., 1980; Bos et al., 1982, 1983; Neis et al., 1984), styrene (Belvedere et al., 1984), 1,2-dimethylhydrazine (Kerklaan et al., 1983; Malaveille et al., 1983), procarbazine (Malaveille et al., 1983) and nitrosamines (Bos et al., 1983; Kerklaan et al., 1983). Factors that might affect the mutagenicity of chemicals in the *Salmonella*/hepatocyte assay have been evaluated systematically by Malaveille et al. (1983) and by Williams et al. (1983).

Hepatic epithelial cell lines (Montesano et al., 1973; Tong & Williams, 1978; Mondal et al., 1979) have also been used as a source of activating enzymes. Lethally-irradiated rodent fibroblasts at primary or secondary culture (Huberman & Sachs, 1974) and irradiated BHK-21 cells (Newbold et al., 1977) are examples of nonhepatic cultured cells that have been used for carcinogen activation. Mouse embryo fibroblasts C3H 10T½, which are used in in-vitro cell transformation assays, possess an active prostaglandin synthase which has been shown to co-oxidize benzo[a]pyrene-7,8-dihydrodiol (Boyd et al., 1982) and aflatoxin B_1 (Amstad & Cerutti, 1983) to reactive intermediates. Cultured human bronchial explants have also been used to metabolize polycyclic hydrocarbons to products that are mutagenic to cocultured Chinese hamster V79 cells (Hsu et al., 1978). Experience has shown, however, that the capacity of most cultured cells to metabolize a broad spectrum of potential carcinogens is limited.

Use of cultured cells as an activating system has the following advantages: (i) an intact cellular architecture and arrangement of the various enzyme systems which metabolize xenobiotic compounds and (ii) endogenous levels of cofactors for the various enzymes. However, since the aim of conducting short-term tests *in vitro* is to reveal *any* potential mutagenic activity, it is justifiable to ask whether an in-vitro metabolic system *must* mimic the balance between activation and detoxification characteristic of intact animals. Indeed, it would probably be better to use mixtures of enzymes which mediate all possible modes of activation rather than search for ideal analogues of in-vivo systems.

5. References

Amacher, D.E. & Paillet, S.C. (1982) Hamster hepatocyte-mediated activation of procarcinogens to mutagens in the L5178Y/K mutation assay. *Mutat. Res.*, *106*, 305-316

Ames, B.M., McCann, J. & Yamasaki, E. (1975) Methods for detecting carcinogens and mutagens with the *Salmonella*/mammalian-microsome mutagenicity test. *Mutat. Res.*, *31*, 347-364

Amstad, P. & Cerutti, P. (1983) DNA binding of aflatoxin B_1 by co-oxygenation in mouse embryo fibroblasts C3H/10T1/2. *Biochem. biophys. Res. Commun.*, *112*, 1034-1040

Anderson, B., Larsson, R., Rahimtula, A. & Moldeus, P. (1984) Prostaglandin synthase and horseradish peroxidase catalyzed DNA-binding of p-phenetidine. *Carcinogenesis*, *5*, 161-165

Bachur, N.R., Gordon, S.L. & Gee, M.V. (1977) Anthracycline antibiotic augmentation of microsomal electron transport and free radicals. *Mol. Pharmacol.*, *13*, 901-910

Bachur, N.R., Gordon, S.L. & Gee, M.V. (1978) A general mechanism for microsomal activation of quinone anticancer agents to free radicals. *Cancer Res.*, *38*, 1745-1750

Bartsch, H. & Hecker, E. (1971) On the metabolic activation of the carcinogen N-hydroxy-2 acetylaminofluorene. 3. Oxidation with horse radish peroxidase to yield 2-nitrosofluorene and N-acetoxy-2 acetylaminofluorene. *Biochim. biophys. Acta*, *237*, 567-578

Bartsch, H., Miller, J.A. & Miller, E.C. (1972) N-Acetoxy-N-acetylaminoarenes and nitrosoarenes. One electron non enzymatic and enzymatic oxidation products of various carcinogenic aromatic acetylhydroxamic acids. *Biochim. biophys. Acta*, *273*, 40-51

Bartsch, H., Kuroki, T., Roberfroid, M. & Malaveille, C. (1982) *Metabolic activation systems* in vitro *for carcinogen/mutagen screening tests*. In: de Serres, F.J. & Hollaender, A., eds, *Chemical Mutagens — Principles and Methods for their Detection*, Vol.7, New York, Plenum, pp. 95-161

Battista, J.R. & Marnett, L.J. (1985) Prostaglandin H synthase-dependent epoxidation of aflatoxin B_1. *Carcinogenesis*, *6*, 1227-1229

Belvedere, G., Elovaara, E. & Vainio, H. (1984) Activation of styrene to styrene oxide in hepatocytes and subcellular fractions of rat liver. *Toxicol. Lett.*, *23*, 157-162

Bos, R.P., Van Dosin, R., Yih-Van de Huk, E., Van Gemert, P.S.L. & Henderson, P.T. (1982) Comparison of the mutagenicities of 4-aminobiphenyl and benzidine in the *Salmonella*/microsome, *Salmonella*/hepatocyte and host-mediated assays. *Mutat. Res.*, *93*, 317-325

Bos, R.P., Neis, J.M., Van Gemert, P.S.L. & Henderson, P.T. (1983) Mutagenicity testing with the *Salmonella*/hepatocytes and the *Salmonella*/microsome assays. A comparative study with some known genotoxic compounds. *Mutat. Res.*, *124*, 103-112

Boyd, J.A., Barnett, J.C. & Eling, T.E. (1982) Prostaglandin endoperoxide synthase-dependent cooxidation of (±)trans-7,8-dihydroxy-7,8-dihydrobenzo[*a*]pyrene in C3H/10T½ clone 8 cells. *Cancer Res.*, *42*, 2628-2632

Brodie, B.B. & Gillette, J.R., eds (1971) *Concepts in biochemical pharmacology, Part 2*. In: *Handbook of Experimental Pharmacology,* Vol. 28, Berlin (West), Springer

Brouns, R.E., Bos, R.P., Van Gemert, P.S.L., Yih-Van de Hurk, E.W.M. & Henderson, P.T. (1979) Mutagenic effects of benzo[*a*]pyrene after metabolic activation by hepatic 9000 *g* supernatant or intact hepatocytes. *Mutat. Res., 62*, 19-26

Brouns, R.M.E., Bos, R.P., Van Doorn, R. & Henderson, P.T. (1980) Metabolic activation of 2-acetylaminofluorene by isolated rat liver cells. Involvement of different metabolites causing DNA repair and bacterial mutagenesis. *Arch. Toxicol., 45*, 53-59

Caldwell, J. (1979) The significance of phase II (conjugation) reactions in drug disposition and toxicity. *Life Sci., 24*, 571-578

Callen, D.F. (1982) *Microbial metabolism of environmental chemicals to mutagens and carcinogens.* In: de Serres, F.J. & Hollaender, A., eds, *Chemical Mutagens, Principles and Methods for their Detection*, Vol. 7, New York, Plenum, pp. 163-183

Campbell, C. & Hayes, J.R. (1974) Role of nutrition in the drug metabolizing enzyme system. *Pharmacol. Rev., 26*, 171-197

Camus, A.M., Friesen, M., Croisy, A. & Bartsch, H. (1982) Species-specific activation of phenacetin into bacterial mutagens by hamster liver enzymes and identification of *N*-hydroxyphenacetin o-glucuronide as a pro-mutagen in the urine. *Cancer Res., 42*, 3201-3208

Casciano, D.A., Dan, J.A., Oldham, J.W. & Carie, M.D. (1978) 2-Acetylaminofluorene-induced unscheduled DNA synthesis in hepatocytes isolated from 3-methylcholanthrene treated rats. *Cancer Lett., 5*, 173-178

Cavalieri, E.L. & Rogan, E.G. (1984) *One-electron and two-electron oxidation in aromatic hydrocarbon carcinogenesis.* In: Pryor, W., ed., *Free Radicals in Biology*, Vol. 6, New York, Academic Press, pp. 323-369

Cinti, D.L. (1978) Agents activating the liver microsomal mixed function oxydase system. *Pharmacol. Ther. A*, 727-749

Conney, A.H. (1967) Pharmacological implications of microsomal enzyme induction. *Pharmacol. Rev., 19*, 317-366

Conney, A.H., Craver, B., Kuntzman, R. & Pantuck, E.J. (1974) *Drug metabolism in normal and disease states.* In: Teorell, T., Dedrick, R.L. & Condliffe, P.G., eds, *Pharmacology and Pharmacokinetics*, New York, Plenum, 147-162

Conney, A.H., Pantuck, E.J., Kuntsman, R., Kappas, A. & Alvares, A.P. (1977) Nutrition and chemical biotransformations in man. *Clin. pharmacol. Ther., 22*, 707-719

Conney, A.H., Pantuck, E.J., Pantuck, C.B., Buening, M., Jerina, D.M., Fortner, J.G., Alvares, A.P., Anderson, K.E. & Kappas, A. (1979) *Role of environment and diet in the regulation of human drug metabolism.* In: Estabrok, R.W. & Lindenlarb, E., eds, *The Induction of Drug Metabolism*, Stuttgart, F.K. Schattauer, pp. 583-605

Dybing, E., Soderlund, E., Timm Haug, L. & Thorgirsson, S.S. (1979) Metabolism and activation of 2-acetyl-aminofluorene in isolated rat hepatocytes. *Cancer Res., 39*, 3268-3275

Fee, J.A. (1980) *Is superoxide toxic?* In: Bannister, W.H. & Bannister, J.V., eds, *Biological and Clinical Aspects of Superoxide Dismutase*, New York, Elsevier, pp. 41-48

Floyd, R.A., Soong, L.M. & Culver, P.L. (1976a) Horse radish peroxidase/hydrogen peroxide catalysed oxidation of the carcinogen *N*-hydroxy-*N*-acetyl-2-aminofluorene as effected by cyanide and ascorbate. *Cancer Res., 36*, 1510-1519

Floyd, R.A., Soong, L.M., Walker, R.M. & Stuart, M. (1976b) Lipid hydroperoxide activation of N-hydroxy-N-acetylaminofluorene via a free radical route. *Cancer Res., 36*, 2761-2767

Gatehouse, D.G. & Delow, G.F. (1979) The development of a 'Microtites®' fluctuation test for the detection of indirect mutagens, and its use in the evaluation of mixed enzyme induction of the liver. *Mutat. Res., 60*, 239-252

Glatt, H.R., Billings, R., Platt, K.L. & Oesch, F. (1981) Improvement of the correlation of bacterial mutagenicity with carcinogenicity of benzo[a]pyrene and four of its major metabolites by activation with intact liver cells instead of cell homogenate. *Cancer Res., 41*, 270-277

Green, M.H.L., Bridges, B.A., Rogers, A.M., Horspool, G., Muriel, W.J., Bridges, J.W. & Fry, J.R. (1977) Mutagen screening by a simplified bacterial fluctuation test: use of microsomal preparations and whole liver cells for metabolic activation. *Mutat. Res., 48*, 287-294

Guengerich, F.P., Dannan, G.A., Wright, S.T., & Martin, M.V. (1982) Purification and characterization of microsomal cytochrome P-450s. *Xenobiotica, 12*, 701-716

Guthrie, J., Robertson, I.G.C., Zeiger, E., Boyd, J.A. & Eling, T.E. (1982) Selective activation of some dihydrodiols of several polycyclic aromatic hydrocarbons to mutagenic products by prostaglandin synthetase. *Cancer Res., 42*, 1620-1623

Haworth, S., Lawlor, T., Mortelmans, K., Speck, W. & Zeiger, E. (1983) *Salmonella* mutagenicity test results for 250 chemicals. *Environ. Mutagen., Suppl. 1*, 3-141

Hsu, I.C., Stona, G.D., Autrup, H., Trump, B.F., Selkirk, J.K. & Harris, C.C. (1978) Human bronchus-mediated mutagenesis of mammalian cells by carcinogenic polynuclear aromatic hydrocarbons. *Proc. natl Acad. Sci. USA, 75*, 2003-2007

Hubbard, S.A., Green, M.H.L. & Bridges, J.W. (1981a) *Detection of carcinogens using the fluctuation test with S9 or with hepatocyte activation*. In: Stich, H.F. & San, R.H.C., eds, *Short-term Tests for Chemical Carcinogens*, New York, Springer, pp. 296-305

Hubbard, S.A., Green, M.H.L., Bridges, B.A., Wain, A.J. & Bridges, J.W. (1981b) *Fluctuation tests with S9 and hepatocyte activation*. In: de Serres, F.J. & Ashby, J., eds, *Evaluation of Short-Term Tests for Carcinogens, Report of the International Collaborative Program*, Amsterdam, Elsevier, pp. 361-370

Hubbard, S.A., Brooks, T.M., Gonzalez, L.P. & Bridges, J.W. (1985) *Preparation and characterisation of S9 fractions*. In: Parry, J.M. & Arlett, C.F., eds, *Comparative Genetic Toxicology: The Second UKEMS Study*, London, MacMillan Press, pp. 413-438

Huberman, E. & Sachs, L. (1974) Cell mediated mutagenesis of mammalian cells with chemical carcinogens. *Int. J. Cancer, 13*. 326-333

Inskeep, P.B. & Guengerich, F.P. (1984) Glutathione-mediated binding of dibromoalkanes to DNA: a specificity of rat glutathione-S-transferases and dibromoalkane structure. *Carcinogenesis, 5*, 805

Jacoby, W.B. (1980) *Enzymatic Basis of Detoxication*, New York, Academic Press

Jenssen, D., Beige, B. & Ramel, C. (1979) Mutagenicity testing on Chinese hamster V79 cells treated in the in-vitro liver perfusion system. Comparative investigation of different in-vitro metabolising systems with dimethylnitrosamine and benzo[a]pyrene. *Chem.-biol. Interactions, 27*, 27-39

Jones, C.A. & Huberman, E. (1980) A sensitive hepatocyte-mediated assay for the metabolism of nitrosamines to mutagens for mammalian cells. *Cancer Res., 40*, 406-411

Kalyanaraman, B. & Sivarajah, K. (1984) *The electron spin resonance study of free radicals formed during the arachidonic acid cascade and cooxidation of xenobiotics by prostaglandin synthase*. In: Pryor, W., ed., *Free Radicals in Biology*, Vol. 6, New York, Academic Press, pp. 149-198

Kato, R. (1974) Sex-related differences in drug metabolism. *Drug Metab. Rev., 3*, 1-32

Kato, R. (1977) Drug metabolism under pathological and abnormal physiological states in animals and man. *Xenobiotica, 7*, 25-92

Katoh, Y., Tanaka, M. & Takayama, S. (1982) Higher efficiency of hamster hepatocytes than rat hepatocytes for detecting dimethylnitrosamine and diethylnitrosamine in hepatocyte-mediator Chinese hamster V79 cell mutagenesis assay. *Mutat. Res., 105*, 265-269

Kerklaan, P., Boutier, S. & Mohn, G. (1983) Activation of nitrosamines and other carcinogens by mouse-liver S9, mouse hepatocytes and in the host-mediated assay produces different mutagenic responses in *Salmonella* TA 1535. *Mutat. Res., 110*, 9-22

Kuroki, T., Malaveille, C., Drevon, C., Piccoli, C., Macleod, M. & Selkirk, J.K. (1979) Critical importance of microsome concentration in mutagenesis assay with V79 Chinese hamster cells. *Mutat. Res., 63*, 259-272

LaDu, B.N., Mandel, H.G. & Way, E.L., eds (1971) *Fundamentals of Drug Metabolism and Drug Disposition*, Baltimore, Williams & Wilkins

Langenbach, R. & Oglesby, L. (1983) *The use of intact cellular activation systems in genetic toxicology assays*. In: de Serres, F.J., ed., *Chemical Mutagens. Principles and Methods for their Detection*, Vol. 8, pp. 55-93

Langenbach, R., Freed, H.J. & Huberman, E. (1978a) Liver cell-mediated mutagenesis of mammalian cells by liver carcinogens. *Proc. natl Acad. Sci. USA, 75*, 2864-2867

Langenbach, R., Freed, H.J., Raveh, D. & Huberman, E. (1978b) Cell specificity in metabolic activation of aflatoxin B_1 and benzo[*a*]pyrene to mutagens for mammalian cells. *Nature, 276*, 277-279

Levin, W. & Ryan, D. (1975) *Age and sex differences in the turnover of rat liver cytochrome P-450: the role of neonatal imprinting*. In: Morselli, P.L., Garattini, S. & Sereni, F., eds, *Basic and Therapeutic Aspects of Perinatal Pharmacology*, New York, Raven Press, 265-275

Levin, W., Ryan, D., Huang, M.T., Kawalek, J., Thomas, P.E., West, S.B. & Lu, A.Y.H. (1977) *Characterization of multiple forms of highly purified cytochrome P-450 from the liver microsomes of rats, mice and rabbits*. In: Ullrich, V., Hilldebrandt, A., Roots, I., Estabrook, R. & Conney, A.H., eds, *Microsomes and Drug Oxidations*, Oxford, Pergamon Press, pp. 185-191

Loew, G.H. & Goldblum, A. (1985) Metabolic activation and toxicity of acetaminophen and related analogs. A theoretical study. *Mol. Pharmacol., 27*, 375-386

Lu, A.Y.H. (1976) Liver microsomal drug metabolizing enzyme system: functional components and their properties. *Fed. Proc., 35*, 2460-2463

Lu, A.Y.H. & West, S.B. (1981) *Reconstituted mammalian mixed-function oxidases: requirements, specificities and other properties*. In: Schenkman, J.B. & Kupfer, D., eds, *Hepatic Cytochrome P-450 Monooxygenase System*, Oxford, Pergamon Press, p. 523

Lu, A.Y.H., Somogyi, A., West, S., Kuntzman, R. & Conney, A.H. (1972) Pregnenolone-16α-carbonitrile: a new type of inducer of drug-metabolizing enzymes. *Arch. Biochem. Biophys., 152*, 452-562

Lubet, R.A., Nims, R.W., Mayer, R.T., Cameron, J.W. & Schechtman, L.M. (1985) Measurement of cytochrome P-450 dependent dealkylation of alkoxyphenoxazones in hepatic S9s and hepatocyte homogenates: effects of dicumarol. *Mutat. Res., 142*, 127-131

Malaveille, C., Brun, G. & Bartsch, H. (1983) Studies on the efficiency of the *Salmonella*/rat hepatocyte assay for the detection of carcinogens as mutagens: activation of 1,2-dimethylhydrazine and procarbazine into bacterial mutagens. *Carcinogenesis, 4*, 449-455

Maron, D.M. & Ames, B.N. (1983) Revised methods for the *Salmonella* mutagenicity test. *Mutat. Res.*, *113*, 173-215

Mason, R.P. (1979) *Free radical metabolites of foreign compounds and their toxicological significance.* In: Hodgson, E., Bend, J.R. & Philpot, R.M., eds, *Reviews in Biochemical Toxicology*, Vol. 1, New York, Elsevier, pp. 151-200

Mason, R.P. (1982) *Free radical intermediates in the metabolism of toxic chemicals.* In: Pryor, W., ed., *Free Radicals in Biology*, Vol. 5, New York, Academic Press, pp. 161-222

Matsushima, T., Sawamura, M., Hara, K & Sugimura, T. (1976) *A safe substitute for polychlorinated biphenyls as an inducer of metabolic activation system.* In: de Serres, F.J., Fouts, J.R., Bend, J.R. & Philpot, R.M., eds, *In Vitro Metabolic Activation in Mutagenesis Testing*, Amsterdam, Elsevier, pp. 85-88

Matsushima, T., Sugimura, T., Nagao, M., Yahagi, T., Shirai, A. & Sawamura, M. (1980) *Factors modulating mutagenicity in microbial tests.* In: Norpoth, K.H. & Garner, R.C., eds, *Short-Term Test Systems for Detecting Carcinogens*, Heidelberg, Springer, pp. 273-285

Mattamal, M.B., Zenser, T.V. & Davis, B.B. (1981) Prostaglandin hydroperoxidase-mediated 2-amino-4-(5-nitro-2-furyl)(14C) thiazole metabolism and nucleic acid binding. *Cancer Res.*, *41*, 4961-4966

Mazel, P. (1971) *Experiments illustrating drug metabolism* in vitro. In: LaDu, B.N., Mandel, H.G. & Way, E.L., eds, *Fundamentals of Drug Metabolism and Drug Disposition*, Baltimore, Williams & Wilkins, pp. 546-551

McBrien, D.C.H. & Slater, T.S., eds (1982) *Free Radicals, Lipid Peroxidation and Cancer*, New York, Academic Press

Michalopoulos, G., Sattler, G.L., O'Connor L. & Pitot, H.C. (1978) Unscheduled DNA synthesis induced by procarcinogens in suspensions and primary cultures of hepatocytes on collagen membranes. *Cancer Res.*, *38*, 1866-1871

Miller, E.C. & Miller, J.A. (1974) *Biochemical mechanisms of chemical carcinogenesis.* In: Bush, H., ed., *Biology of Cancer*, New York, Academic Press, pp. 377-402

Miller, E.C. & Miller, J.A. (1976) *The metabolism of chemical carcinogens to reactive electrophiles and their possible mechanisms of action in carcinogenesis.* In: Searle, C.S., ed., *Chemical Carcinogens (ACS Monograph No. 173)*, Washington DC, American Chemical Society, pp. 737-762

Miller, J.A. (1970) Carcinogenesis by chemicals: an overview — G.H.A. Clowes Memorial Lecture. *Cancer Res.*, *30*, 559-576

Mondal, S., Lillehaug, J.R. & Heidelberger, C. (1979) Cell mediated activation of aflatoxin B_1 to transform C3H/10T1/2 cells. *Proc. Am. Assoc. Cancer Res.*, *20*, 62

Montesano, R., Saint Vincent, L. & Tomatis, L. (1973) Malignant transformation *in vitro* of rat liver cells by dimethylnitrosamine and *N*-methyl-*N*'-nitro-N-nitrosoguanidine. *Br. J. Cancer*, *28*, 215-220

Nebert, D.W. & Felton, J.S. (1976) Importance of genetic factors influencing the metabolism of foreign compounds. *Fed. Proc.*, *35*, 1133-1141

Nebert, D.W. & Gelboin, H.V. (1968) Substrate-inducible microsomal aryl hydroxylase in mammalian cell culture. *J. biol. Chem.*, *243*, 6242-6249

Neims, A.H., Warner, M., Loughman, P.M. & Aranda, J.V. (1976) Developmental aspects of the hepatic cytochrome P450 monooxygenase system. *Ann. Rev. Pharmacol. Toxicol.*, *16*, 427-445

Neis, J.M., Van Gemert, P.S.L., Roelofs, H.M.J., Bos, R.P. & Henderson, P.T. (1984) Mutagenicity of benzidine and 4-aminobiphenyl after metabolic activation with isolated hepatocytes and liver 9000 g supernatant from rat, hamster and guinea pig. *Mutat. Res.*, *129*, 13-18

Nemoto, N. & Takayama, S. (1984) Arachidonic acid-dependent activation of benzo[a]pyrene to bind to proteins with cytosolic and microsomal fractions from rat liver and lung. *Carcinogenesis*, *5*, 961-964

Newbold, R.F., Wigley, C.B., Thompson, M.H. & Brookes, P. (1977) Cell-mediated mutagenesis in cultured Chinese hmaster cells by carcinogenic polycyclic hydrocarbons: nature and extent of the associated hydrocarbon-DNA reaction. *Mutat. Res.*, *43*, 101-116

Omura, T. & Sato, R. (1964) The carbon monoxide-binding pigment of liver microsomes. *J. biol. Chem.*, *239*, 2370-2380

Ozawa, N. & Guengerich, F.P. (1983) Evidence for formation of an S-[2-(N^7-guanyl)ethyl]glutathion adduct in glutathion-mediated binding of the carcinogen 1,2-dibromoethane to DNA. *Proc. natl Acad. Sci. USA*, *80*, 5266-5270

Plewa, M.J. & Gentille, J.M. (1982) *The activation of chemicals into mutagens by green plants*. In: de Serres, F.J. & Hollaender, A., eds, *Chemical Mutagens, Principles and Methods for their Detection*, Vol. 7, New York, Plenum, pp. 401-420

Poiley, J.A., Raineri, R. & Pienta, R.J. (1979) Use of hamster hepatocytes to metabolize carcinogens in an in-vitro bioassay. *J. natl Cancer Inst.*, *63*, 519-524

Poiley, J.A., Raineri, R., Andrews, A.W., Cavanaugh, D.M. & Pienta, R.J. (1980) Metabolic activation by hamster and rat hepatocytes in the *Salmonella* mutagenicity assay. *J. natl Cancer Inst.*, *65*, 1293-1298

Poley, G.E., Shively, C.A. & Vesell, E.S. (1978) Diurnal rhythms of aminopyrine metabolism: failure of sleep deprivation to affect them. *Clin. pharmacol. Ther.*, *24*, 726-732

Pueyo, C., Frezza, D. & Smith, B. (1979) Evaluation of three metabolic activation systems by a forward assay in *Salmonella*. *Mutat. Res.*, *64*, 183-194

Rannug, V., Sundrall, A. & Ramel, C. (1978) The mutagenic effect of 1,2-dichloroethane on *Salmonella typhimurium*. I. Activation through conjugation with glutathione *in vitro*. *Chem.-biol. Interactions*, *20*, 1-16

Roberfroid, M.B., Remacle, J. & Viehe, H.G. (1986) *Free radicals and drug design*. In: Testa, B., ed., *Advances in Drug Research*, London, Academic Press (in press)

Robertson, I.G.C., Sivarajah, K., Eling, T.E. & Zeiger, E. (1983) Activation of some aromatic amines to mutagenic products by prostaglandin endoperoxide synthetase. *Cancer Res.*, *43*, 476-480

Rosenkranz, H.S. & Speck, W.T. (1975) Mutagenicity of metronizadole: activation by mammalian liver microsomes. *Biochem. biophys. Res. Commun.*, *66*, 520-525

Ryan, D.E., Thomas, P.E. & Levin, W. (1977) Properties of purified liver microsomal cytochrome P450 from rats treated with the polychlorinated biphenyl mixture Aroclor 1254. *Mol. Pharmacol.*, *13*, 521-532

San, R.H.C. & Williams, G. (1977) Rat hepatocyte primary cell culture-mediated mutagenesis by adult rat liver epithelial cells by procarcinogens. *Proc. Soc. exp. Biol. Med.*, *156*, 534-538

Sato, R. & Kato, R., eds (1982) *Microsomes, Drug Oxidations and Drug Toxicity*, Tokyo, Japan Scientific Societies Press

Sivarajah, K., Lasker, J.M. & Eling, T.E. (1981) Prostaglandin synthase-dependent cooxidation of (±)benzo[a]pyrene-7,8 dihydrodiol by human lung and other mammalian tissues. *Cancer Res.*, *41*, 1834-1839

Stier, A., Clauss, R., Lucke, A. & Reitz, I. (1982) *Radicals in carcinogenesis by aromatic amines*. In: McBrien, D.C.H. & Slater, T.F., eds, *Free Radicals, Lipid Peroxidation and Cancer*, New York, Academic Press, pp. 329-343

Testa, B. & Jenner, P. (1976) *Drug Metabolism, Chemical and Biochemical Aspects*, New York, Marcel Dekker

Thaler-Dao, H., Crastes de Paulet, A. & Paoletti, R., eds (1984) *Isocanoids and Cancer*, New York, Raven Press

Tong, C. & Williams, G.J. (1978) Induction of purine analog-resistant mutants in adult rat liver epithelial lines by metabolic activation-dependent and -independent carcinogens. *Mutat. Res., 58*, 339-352

Tu, A.S., Breen, P.A. & Sivak, A. (1984) Comparison of primary hepatocytes and S9 metabolic activation systems for the CH3-10T½ cell transformation assay. *Carcinogenesis, 5*, 1431-1436

Van Bladeren, P.J., Breimer, D.D., Rotteveel-Smigs, G.M.T. & Mohn, G.R. (1980) Mutagenic activation of dibromoethane and diiodoethane by mammalian microsomes and glutathione-S-transferases. *Mutat. Res., 74*, 341-346

Venitt, S. & Forster, R. (1985) *Bacterial mutagenicity assays: coordinators' report*. In: Parry, J.M. & Arlett, C.F., eds, *Comparative Genetic Toxicology: The Second UKEMS Study*, London, MacMillan Press, pp. 103-144

Vesell, E.S. (1978) Genetic and environmental factors responsible for interindividual variations in drug response. *Adv. Pharmacol. Ther., 6*, 3-12

Vesell, E.S., Lang, C.M., White, W.J., Passananti, G.T., Hill, R.N., Clemens, T.L., Lieu, D.K. & Johnson, W.D. (1976) Environmental and genetic factors affecting the response of laboratory animals to drugs. *Fed. Proc., 35*, 1125-1132

Watanabe, Y., Iyanagi, T. & Oae, S. (1980) Kinetic study on enzymatic S-oxygenation promoted by a reconstituted system with purified cytochrome P-450. *Tetrahedron Lett., 21*, 3685-3688

Watanabe, Y., Iyanagi, T. & Oae, S. (1982) One electron transfer mechanism in the enzymic oxidation of sulfoxide to sulfone promoted by a reconstituted system with purified cytochrome P-450. *Tetrahedron Lett., 23*, 533-536

Williams, G.M. (1976) Carcinogen-induced DNA repair in primary rat liver cell cultures; a possible screen for chemical carcinogens. *Cancer Lett., 1*, 231-236

Williams, G.M. (1977) The detection of chemical carcinogens by unscheduled DNA synthesis in rat liver primary cell cultures. *Cancer Res., 37*, 1845-1851

Williams, K., Inmon, J. & Lewtas, J. (1983) Effect of incubation and activation conditions on the hepatocyte-mediated plate incorporation and preincubation *Salmonella typhimurium*-mutagenesis assays. *Teratog. Carcinog. Mutagen., 3*, 367-376

Williams, R.T. (1971) *Species variations in drug biotransformations*. In: LaDu, B.N., Mandel, H.G. & Way, E.L., eds, *Fundamentals of Drug Metabolism and Drug Disposition*, Baltimore, Williams & Wilkins, pp. 187-205

Wise, R.W., Zenser, T.V., Kadlubar, F.F. & Davis, B.B. (1984) Metabolic activation of carcinogenic aromatic amines by dog bladder and kidney prostaglandin H synthase. *Cancer Res., 44*, 1893-1897

Yager, J.D. & Miller, J.A. (1978) DNA repair in primary cultures of rat hepatocytes. *Cancer Res., 38*, 4385-4395

Yahagi, T., Nagao, M., Seino, Y., Matsushima, T., Sugimure, T. & Okada, M. (1977) Mutagenicities of *N*-nitrosamines on *Salmonella*. *Mutat. Res., 48*, 121-130

REPORT 16

STATISTICAL ANALYSIS OF DATA FROM IN-VITRO ASSAYS OF MUTAGENESIS

Prepared by:

N. Breslow and J. Kaldor

1. Introduction

With the increasing use of short-term screening assays for carcinogenicity and mutagenicity has come the need for appropriate statistical methods for evaluating the data. The two principal end products of such methods are criteria for deciding if an assay result is positive or negative, and a quantitative estimate of the activity of a test agent in an assay. Although a number of statistical procedures have been proposed for the analysis of various short-term tests in recent years, there is as yet no general consensus as to which may be the most appropriate. Earlier writers proposed ad-hoc rules for determination of positivity, such as the requirement of a doubling of the response variable compared with controls (Clive *et al.*, 1979; Chu *et al.*, 1981). Other authors have adopted standard statistical tools, such as t-statistics for comparisons of continuous measurements, and chi-square statistics for comparison of proportions (Weinstein & Lewinson, 1978; Gilbert, 1980; Amphlett & Delow, 1984) or more sophisticated mechanistic modelling of response variables (Margolin *et al.*, 1981; Leong *et al.*, 1985; Kaldor, 1986).

In this paper, we discuss the statistical problems involved in the analysis of data from a broad class of short-term tests, namely those for measuring in-vitro mutagenesis at specific loci. We restrict attention to this group of tests because they are the most widely used, and because, up to now, they have received the most attention from statisticians. However, the discussion should also be relevant to a number of other assays, in particular the in-vitro cell transformation assays, which use very similar experimental procedures but measure a different type of endpoint.

It should be emphasized that we are not advocating any particular statistical procedure over others. Rather, we wish to raise the issues that must be taken into consideration when choosing a statistical method.

2. Experimental background

The goal of in-vitro mutagenesis assays is to determine whether the treatment of a cell culture with a test chemical results in an increase in the fraction of mutants at a specific locus. In order to accomplish this objective, the following experimental steps may be followed:

(a) treatment of cells with the test agent (or with its solvent, in the case of controls);

(b) reculturing of the cells to allow time for the phenotypic expression of the mutations induced during treatment; and

(c) estimation of the number of mutants and total colony-forming units in the culture.

Mutants are counted in a large aliquot from a culture that has been placed under conditions which select for the mutation, while the total cell population is estimated from a very small fraction of the culture placed under standard growth conditions. The counting can be done either by plating the cells onto Petri dishes and counting the number of colonies resulting, or by dividing the culture into microwells and counting the fraction of wells that contain at least one colony. In either case, the result is an estimate of the *mutant fraction (MF)*, which is the ratio of mutant colony-forming units to total colony-forming units in the culture. It is the outcome variable used to compare treated cultures (possibly at several dose levels) with control cultures.

Symbolically, suppose that at the end of expression, a large fraction, a, is sampled from a culture for counting mutant colony-forming units, and that a much smaller fraction, b, is sampled to count total colony-forming units. If Petri dish plating is used, let M and t, respectively, represent the number of colonies counted in each case. We refer to M as the *mutant yield*. The mutant fraction in the culture is estimated by

$$MF = \frac{M/a}{t/b}.$$

If microwells are used, let p_m and p_t represent the fractions of wells with no colony-forming units under selective and normal growth conditions respectively. Then, using the Poisson distribution, we can estimate the corresponding number of colony-forming units in all the wells as $M = -\log p_m$ and $\hat{t} = -\log p_t$, and again estimate

$$MF = \frac{M/a}{\hat{t}/b}.$$

In the treated culture, the mutant fraction is an estimate of the fraction of cells in which the mutation was induced during exposure, provided the following assumptions hold (Lee & Caspary, 1983; Leong et al., 1985; see Report 7, section 2.4):

(a) The mutation has no effect on the viability of the cell after exposure, its growth rate during expression or its ability to form colonies.

(b) No mutant existed on the culture before exposure, nor was any formed spontaneously during expression or selection.

There is no way to correct for the bias induced if the first assumption is not valid. The effect of mutants arising before or after exposure can be taken into account by including the control mutant fraction in the statistical analysis.

The steps outlined above are those of standard mutagenesis assays, such as those with the *hprt* locus in V79 cells, the *hprt* locus in CHO cells and the thymidine kinase locus in mouse lymphoma cells (Amacher et al., 1980; Clive et al., 1979; and Report 7, section 3). However, some simplified protocols have become widely used, most notably those of Ames et al. (1975) for histidine reversion in *Salmonella typhimurium*, and the simplified fluctuation test of Green and Muriel (1976) for tryptophan reversion in *Escherichia coli*. In these protocols, only estimates of mutant yield are available, either directly from colony counts (Ames et al., 1975) or from the number of tubes or wells that become turbid (Green & Muriel, 1976). The price to be paid for experimental simplicity is that no correction can be made for decreasing cell survival with increasing dose, so that at high doses the effects of cell killing predominate over those of mutagenesis.

3. Dose-response functions

Optimal statistical procedures for bioassay design, for significance testing of whether there is a genuine mutagenic response, and for quantitative estimation of potency all require some knowledge of the relationship between the average mutant frequency (or, for assays that provide no data on cell survival, the average mutant yield) and the dose level x. Various functions have been proposed to represent this relationship. Some are simple mathematical expressions, proposed *ad hoc*, that seem to capture the main features of the observed experimental data and at the same time allow the use of relatively elementary statistical techniques. Others are derived from explicit, but speculative, mathematical models for what are believed to be the underlying biological processes. In the following, we outline the most important classes of models that have been proposed.

Haynes and Eckardt (1979) present a comprehensive catalogue of models based on the hit theory, which enjoys wide acceptance in the field of radiobiology. They assume that mutant cells are produced by a 'biological hit' at the cellular level, that toxic death is likewise produced by a 'lethal hit', and, furthermore, that the two processes are stochastically independent. The expected numbers of such hits (H) produced by a dose x of the chemical agent are represented by low order polynomials, namely

$$H_m(x) = m_1 x + m_2 x^2 + ...$$

and (1)

$$H_k(x) = k_1 x + k_2 x^2 + ...$$

for mutant hits (m) and kills (k) (lethal hits), respectively. The absence of a constant term m_o in $H_m(x)$ reflects the fact that the model is for the average *excess* (above background) number of hits produced by the chemical. Lethal hits are assumed not to occur in the absence of exposure to the chemical. For practical purposes, it is sufficient to ignore the higher order terms m_3, k_3, etc. Most important is the coefficient m_1, which represents the initial low-dose slope of the curve of mutant frequency. Statistical tests of the hypothesis that $m_1 = 0$ *versus* alternatives in which $m_1 > 0$ provide an approximate means for evaluating the evidence for a true mutagenic effect, and the point estimate of m_1 provides the most appropriate quantitative measure of mutagenic potency (see section 6 below). The second term, m_2, allows for some upward curvature in the dose-response function at mid-range doses and facilitates tests for a lack of fit of the linear model. The decline in the mutant yield that may accompany a reduction in the number of surviving cells at higher doses is expressed in the parameters k_1 and k_2.

According to the hit-theory model, the expected probabilities of survival (S) and mutation (mutant freqency; MF) following treatment of N_0 cells are

$$S(x) = \exp[-H_k(x)]$$

and (2)

$$MF(x) = 1 - \exp[-H_m(x)] \simeq H_m(x),$$

while the expected mutant yield is

$$M(x) = [1 - \exp[-H_m(x)]] \exp[-H_k(x)] N_0$$

$$\simeq H_m(x) \exp[-H_k(x)] N_0. \quad (3)$$

Markedly different dose-response patterns are produced according to whether the mutation rates $H_m(x)$ are linear (L: $m_1 > 0$, $m_2 = 0$), quadratic (Q: $m_1 = 0$, $m_2 > 0$) or linear/quadratic (LQ: $m_1 > 0$, $m_2 > 0$), and similarly for lethal hits. Figures reproduced here from Haynes and Eckardt (1979) show the expected patterns of $MF(x)$ and $M(x)$

observed under the linear mutagenesis-linear survival (L-L) model when one varies survival (Fig. 1) and mutagenesis (Fig. 2) rates, respectively. Figure 3 graphs the patterns for L, Q and LQ mutagenesis in the presence of LQ survival. Note the shouldered survival curve in this latter pattern, and also the fact that the (purely) quadratic mutagenesis model produces a zero slope in the mutant yield at zero dose.

Fig. 1 Mutation yields $M(x)$, frequencies $MF(x)$ and survivals $S(x)$ plotted over dose in arbitrary units for the purely linear kinetic response pattern (Lk, Lm). For both cases A and B, $m_1 = 10^{-6}$(units)$^{-1}$. In A, $k_1 = 1.23 \times 10^{-1}$(units)$^{-1}$; in B, $k_1 = 3.7 \times 10^{-1}$(units)$^{-1}$, so that the sensitivity of B is three-fold greater than that of A. Note that the frequency (dashed line) is the same for both A and B, although the yield curves are very different. The maxima of both curves occur at the LD_{37} dose. (From Haynes & Eckardt, 1979)

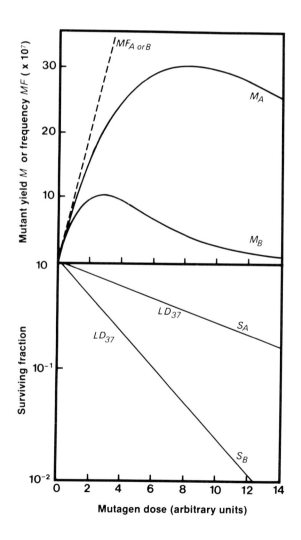

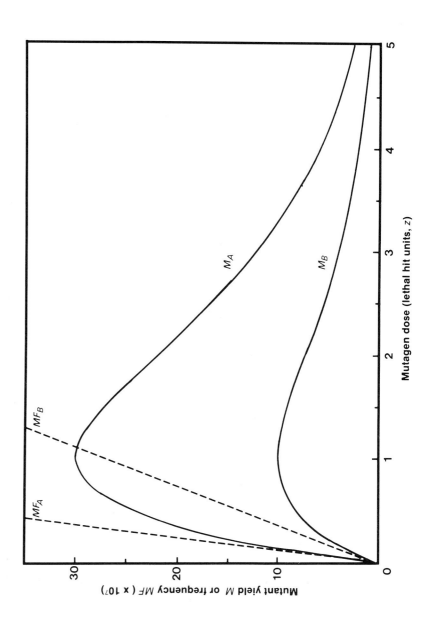

Fig. 2 Mutation yields $M(x)$ and frequencies $MF(x)$ plotted over lethal hits for the purely linear kinetic response pattern (Lk, Lm). The numerical values of m_1 and k_1 are the same as in Figure 1 for both A and B. Survival is not shown, since on a semi-log lethal hit plot, all such curves are merely straight lines passing through the surviving fraction 0.368 for $z = 1$. Note that the frequencies (dashed lines) are given by the initial (dimensionless) slopes of the yield curves (m_1/k_1). In A, $m_1/k_1 = 8.1 \times 10^{-6}$, while in B, $m_1/k_1 = 2.7 \times 10^{-6}$. The maxima of both curves occur at $z = 1$. The dose x in Figure 1 is converted into lethal hit units through the relation $z = k_1 x$. (From Haynes & Eckardt, 1979)

Fig. 3 Calculated yield and survival curves for biphasic, linear-quadratic killing (LQk) and linear (L), quadratic (Q) and biphasic, linear-quadratic (LQ) mutation induction over mutagen dose in arbitrary units. The values of the coefficients of mutation and lethality used were: $m_1 = 10^{-6}$(units)$^{-1}$; $m_2 = 10^{-6}$(units)$^{-2}$; $k_1 = 10^{-1}$(units)$^{-1}$; $k_2 = 10^{-1}$(units)$^{-2}$. Note that the initial slope of the yield curve for kinetic response pattern (LQk, Qm) is zero, while it is non-zero if there is a linear component in mutagenesis. For (LQk, Lm) the yield maximum is at $1/2LD_{14}$; for ($LQkQm$) the maximum shifts to $1/2LD_2$; and for ($LQkLQm$) the maximum falls between these two values. (From Haynes & Eckardt, 1979)

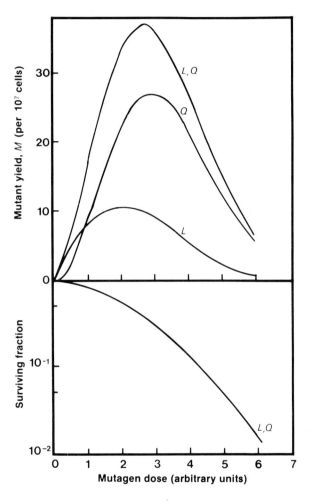

3.1 *Multigeneration exposure*

Although sufficient for most practical purposes, including statistical analysis, the idealized dose-response functions specified by equations (2) and (3) are unlikely to be realized exactly in assays of mutagenic chemicals. Margolin *et al.* (1981) note that some of the radiobiological concepts that underlie the hit-theory formulation are not strictly applicable to microbial reverse mutagenesis assays, in which several generations of cells are exposed to the test chemical before the essential nutrient is exhausted. They derive a more general and complicated model for the mutant yield for the *Salmonella* test (Ames *et al.*, 1975) that depends on (i) the number of generations of auxotrophic growth; (ii) the number of generations during which the chemical retains mutagenic activity; and (iii) the number of generations during which the chemical retains lethal activity. The basic radiobiological equation (3) is obtained as a special case when all three numbers equal one. This agrees with the notion that the effects of a point radiation exposure are limited to a single generation of cells. Margolin *et al.* (1981) note that it is virtually impossible to distinguish dose-response curves for models in which mutagenic activity lasts one generation from those in which it lasts more than one generation in the presence of mild toxicity. Measures of the goodness-of-fit of the model to the data will be nearly identical, even though the slope of the linear dose-response function differs. According to Bernstein *et al.* (1982), this means that the assay 'cannot distinguish between a long-acting weak mutagen and a strong mutagen acting over a short time period'.

3.2 Saturation kinetics

The usual radiobiological explanation for the observation that a dose-response curve is purely quadratic, i.e., has zero slope at $x=0$, is that two hits rather than one are required to produce the mutation. In the case of chemical mutagenesis, such behaviour might also be produced by saturation kinetics that apply to detoxification reactions occurring at the cellular level. Myers *et al.* (1981) note that the strictly increasing form (2) assumed for mutant frequency is likely not to hold for large doses since it will be limited by 'transport across cellular barriers or saturation of bacterial cell with compound'. As an alternative to the one-hit (linear) model, they propose

$$H_m(x) = m_o + V[1-\exp(-m_1 x/V)], \tag{4}$$

where m_o represents the background rate (subtracted out by Haynes and Eckardt) and $m_o + V$ is the upper limit to mutagenicity. This equation may be useful in situations in which saturation phenomena are observed.

3.3 Host susceptibility

The basic hit-theory model may also require modification when there is substantial variability among cells or cell lines in their susceptibility to mutation. Armitage and Spicer (1956) demonstrated that the dose-response function for a heterogeneous collection of cell lines, each of which satisfies the basic linear model

$$H_m(x) = 1-\exp[-m_1 x] \simeq m_1 x \tag{5}$$

but with varying slopes m_1, would be expected to exhibit a downturn from linearity at higher doses. Gart *et al.* (1979) use this formulation to develop a statistical test for the one-hit curve in the cell transformation assay.

3.4 *Semiparametric models*

In view of the inherent uncertainty in the specification of a precise mathematical model for the dose-response function, some research workers prefer to concentrate their attention on the initial portion that is presumed to be linear, without specifying a functional form for the remainder (Bernstein *et al.*, 1982).

4. Sources of variability

Each assay described in Section 2 results in one or more observations on the mutant fraction or the mutant yield at each dose level. For example, a standard *Salmonella* assay may call for three replicate plates to be prepared at a given dose level, each of which is scored for the number of mutant colonies. In the modified fluctuation test of Green and Muriel (1976), the 50 wells or tubes at a given dose level are each scored for the presence or absence of turbidity. Replicate assays conducted at different points in time by the same technicians will yield different results. For example, Table 1 (Simpson & Margolin, 1985) shows the results obtained in three replications of the *Salmonella* assay of Acid Red 114 conducted at different times. There are substantial differences in the counts of revertant colonies between replicate plates within each assay, and even greater differences between replicate assays. Similar variations are observed in cell survival counts for those assays that provide for them. Proper statistical analysis of assay data requires an understanding of the sources and nature of this variation, which are described in more detail below.

4.1 *Sampling variation*

A certain degree of variation in the numbers of mutant or surviving cells following administration of a test chemical results from the fact that the particular cells used in the assay are sampled from a larger population. Due to the inherently stochastic nature of the biological processes involved, some cells will mutate or die when exposed and others will not. The result is that the numbers of mutant and surviving colonies will vary, at a minimum, in accordance with the Poisson sampling distribution for which the variance is equal to the mean. When the final counts of mutant and surviving colonies are of similar magnitude, sampling variation in both the numerator and denominator of the estimate of mutant frequency must be considered.

4.2 *Growth and dilution of cell colonies*

In protocols in which treated cells are allowed to grow for several generations before final plating and counting of mutant and surviving cells, the culture is periodically

Table 1. Revertant colonies after treatment with Acid Red 114 (*S. typhimurium* TA98, hamster liver activation)[a]

Replicate	Dose (µg/l)					
	0	100	333	1000	3333	10 000
1	22	60	98	60	22	23
	23	59	78	82	44	21
	35	54	50	59	33	25
2	19	15	26	39	33	10
	17	25	17	44	26	8
	16	24	31	30	23	
3	23	27	28	41	28	16
	22	23	37	37	21	19
	14	21	35	43	30	13

[a]From Simpson and Margolin (1985)

sampled, diluted and allowed to expand again. Each such dilution increases the variability of the final estimate that will be made of the mutant frequency in the initial cell suspension. First, there is inherent variability in the number of cells in a cell culture that has been allowed to expand for a period of time. Such variation is usually modelled mathematically in terms of a stochastic birth/death process (Cox & Miller, 1965). Second, there is hypergeometric variation associated with the random selection of mutant cells that takes place when the culture is sampled before each dilution. Theoretical expressions for the increase in variance that results from each such step have been derived by Kaldor (1986) and by Leong *et al.* (1985), although the latter ignore the potential variability associated with the birth/death process. (See Report 7, section 2.4 for a more detailed presentation.)

4.3 *Plating and colony formation*

In protocols that call for reculturing of cells, a final sample is taken after sufficient time has elapsed for phenotypic expression, and cells are divided into two subsamples to determine their colony forming ability in the presence and absence of selective conditions. Assuming that each (mutant) cell has an equal chance of being plated and actually forming a colony, the count of the number of colonies formed is well represented as a binomial variable based on sampling from the total number of (mutant) cells in suspension. Colony formation in the *Salmonella* assay for histidine reversion takes place in the same petri dish to which the chemical agent is added, without the intervening steps of cell culture and dilution. In this case, Poisson sampling coupled with binomial variation resulting from random colony formation results in theoretical Poisson variability for the final counts of mutant colonies at each dose level.

4.4 *Additional sources of variability*

In practice, the degree of variability in counts from plate to plate in replicate assays often exceeds the theoretical minimum as specified by the Poisson law for the *Salmonella*-type assays or as determined by the succession of growth and dilution processes for the mammalian cell assays. For example, Table 2 (Margolin et al., 1981) presents results of 20 replicate counts on solvent control plates prepared with *S. typhimurium* TA100 at four laboratories, one of which used two different solvents. There is evidence of substantial 'extraneous' variation in three out of five of the assays. Such variation is often produced by slight procedural errors that may be nearly unavoidable when attempting to follow the experimental protocol. For example, non-homogeneous cell suspensions or titration errors could result in differences in the initial numbers of cells placed on each plate. Likewise, there may be differences in the quantity and purity of the chemical agent, the quantity of histidine or other trace materials, and the quantity of nutrients. Variations in temperature could influence colony formation, and errors and variations in the procedure used to determine the number of colonies on each plate could influence the final count. It is important to try to understand and quantify these extraneous sources of variability, since they may indicate aspects of experimental procedure that need improvement. They also need to be accounted for in the statistical analysis in order to provide a valid variance estimate with which to compare treatment differences and to control the probability of a false positive result.

For tests other than the *Salmonella* assay, no systematic comparison of theoretical and observed variances has been undertaken. Nevertheless, rather than relying on theoretical models for the variability in mutant yields and frequencies, no matter how comprehensive they may seem, it is preferable to conduct simultaneous replications of the entire assay procedure so that the true experimental variation can be measured and used as a basis for scientific inference. Where appropriate, this means making replicate titrations, repeating the entire process of growth and selection of cell suspensions, and making replicate platings for counting of mutant and surviving cells.

4.5 *Variability between laboratories*

Table 2 also shows differences between the mean counts reported by different laboratories, and by the same laboratory under different experimental conditions, that are greater than one can ascribe to within-assay errors. Such systematic differences may be caused by differences in operators, reagents, machines or other aspects of the experimental procedure. This variability is occasionally so great, especially with weaker mutagens, that different laboratories may reach different conclusions about the mutagenicity of the same chemical (Dunkel, 1979; Venitt, 1982; McCann et al., 1984).

Table 2. Twenty replicated solvent control plate counts with *S. typhimurium* TA100 from four laboratories[a]

Laboratory[b]				
1	2	3	4H[c]	4D[c]
81, 86	63, 64	74, 77	168, 171	132, 134
93, 97	65, 68	81, 85	174, 175	144, 145
99, 104	69, 70	87, 88	185, 189	145, 145
105, 110	72, 73	89, 90	190, 191	157, 158
112, 112	75, 80	93, 97	195, 197	158, 161
113, 114	82, 83	98, 99	198, 198	162, 164
115, 117	83, 84	102, 103	203, 205	166, 169
118, 122	84, 85	105, 108	205, 207	174, 177
131, 135	90, 91	108, 110	210, 214	201, 208
155, 183	d d	111, 124	216, 218	208, 219
Mean				
115.10	76.72	96.45	195.45	166.35
Variance				
540.62	81.15	163.10	226.58	631.29
Chi-square test of Poisson variability (20 degrees of freedom)				
89.24[e]	17.98	32.13[f]	22.03	72.10[e]

[a] From Margolin *et al.* (1981)

[b] Revertant colonies per plate listed in ascending order within each laboratory and paired solely for conciseness of presentation

[c] Laboratory 4 conducted this experiment twice, once with water (4H) and once with dimethyl sulfoxide (4D) as the solvent. The same individual did the two tests but on different days. All other laboratories used water.

[d] Contaminated plate, not counted

[e] $p < 0.0001$

[f] $p < 0.05$

Even when the test results reported from several laboratories are all positive, there may be substantial variations in their assessment of mutagenic potency (Venitt & Forster, 1985).

4.6 *Recording errors and outliers*

The data shown in Tables 1 and 2, while demonstrating variability that is clearly in excess of pure sampling error, nevertheless have the virtue that each group of counts for a given assay and dose level are clustered reasonably closely about their mean values. In some experiments, however, aberrant values are recorded that are well outside the 'normal' range. This may result from gross errors of procedure, for example, confusing a high-dose plate with a low-dose one, or from gross errors in the measurement or recording of data, for example, adding or dropping a final digit from the measured count. Careful experimentalists often identify such aberrant data visually and the corresponding datum is either discarded or corrected. However, with the increase in routine testing of large numbers of chemicals and consequent computer analysis of large volumes of data, there is a need for statistical procedures that are reasonably resistant to distortions that may be caused by 'outliers' of this kind.

5. Principles of statistical analysis

Recommended procedures for statistical analysis of data from mutagenesis assays involve the following steps:

(*a*) Recording of the individual counts of numbers of mutant and surviving colonies at each dose level (including control) and use of these original data in the analysis. Measures of mutant frequency at each positive dose level which are corrected for 'background' (Brusick *et al.*, 1980) should generally be avoided because the correlations induced by the correction add further complications to the statistical analysis.

(*b*) Parametric specification of the expected mutant yield or mutant frequency as a function of dose; or semiparametric specification of at least the initial portion of the dose-response curve. In both cases, the model should allow for possible low-dose linearity. Data from previous assays may provide useful information on the functional form of the dose/response relation.

(*c*) Provision for a true experimental variance that changes with dose and may exceed the theoretical variability associated with Poisson or binomial sampling. Ideally, the degree of extraneous variation will be estimated internally from replicate observations recorded at each dose level.

(*d*) Fitting of the model using a robust procedure that corrects for extraneous variability and is resistant to occasionally aberrant data points.

(*e*) Thorough examination of the goodness-of-fit of the model, especially of the critical assumption of linearity in the initial portion of the dose-response function.

(*f*) Testing for mutagenic effect in terms of the statistical significance of the positivity of the initial slope of the dose-response function.

(*g*) Estimation of mutagenic potency in terms of this same slope (see Section 6).

Statisticians have yet to agree on a standard set of methods for statistical analysis of mutagenicity data that meet all the preceding criteria. Further work is needed to develop such methods and to create computer software that will enable their use. The remainder of this section presents a few representative examples of different approaches currently available, with some discussion of their strengths and weaknesses.

5.1 *Least squares regression analysis of mutagenic frequencies*

Snee and Irr have considered the analysis of mutant frequencies measured in assays of CHO (Snee & Irr, 1981) and mouse lymphoma (ML) (Irr & Snee, 1982) cells and *S. typhimurium* (Snee & Irr, 1984). The basic data are derived measurements X_{ijk} of the mutant frequency at the ith dose level of the jth replicate assay of the kth chemical. They transform the data by the power transformation $X \to X^\lambda$, where λ is a value estimated from large amounts of historical data. Its purpose is to provide transformed data which satisfy the assumptions under which standard methods of analysis of variance can be applied. Snee and Irr find that a value of $\lambda = 0.15$ works best for the CHO assays they considered, while either $\lambda = 0.05$ or $\lambda = 0$ ($\log_e$ transform) is used for the ML assays. With λ determined, they conduct an unweighted least squares analysis of variance for each chemical, partitioning the sum of squares for dose into linear, quadratic and higher order terms.

A potential drawback of this approach is that the original data on mutant yields must be transformed before analysis. In some cases, even the dose measures are transformed, which can complicate the estimation of mutagenic potency. Another problem is that there is no intrinsic reason why the same transformation should produce linearity of the dose-response function, equality of variance in the transformed observations at different dose levels *and* approximate normality of residuals, as required for the standard analysis of variance. Moreover, least squares regression procedures are known to be sensitive to the presence of aberrant data (outliers). Nevertheless, the method is easy to implement and seems to yield appropriate tests for the significance of the mutagenic effect (see below).

5.2 *Robust nonlinear regression of mutant yields*

The nonlinear regression approach suggested by Myers *et al.* (1981) for the analysis of data from *Salmonella* assays is designed to overcome the potential problems with simple linear regression. These authors use a hit-theory model for the expected mutant yield as a function of dose — either a linear model with linear lethality,

$$M(x) = (m_0 + m_1 x)\exp(-k_1 x), \tag{6}$$

or else model (4), which incorporates the concept of saturation without toxicity. They conducted an analysis of the variability of replicate counts at different dose levels as a function of the observed mean counts and concluded from their data that the mean-variance relation

$$V(x) = a[M(x)]^b,$$

where $a = 0.11$ and $b = 1.62$, satisfactorily accounted for the observed tendency for the variance to increase with the mean. They propose an iterative regression procedure to fit the model, which involves minimizing the weighted sum of squares of standardized squared residuals $\Sigma Wi\ SR_i^2$. The standardized residual is defined by $SR_i^2 = [Y_i - M(x_i)]/V(x_i)$, where Y_i is the mutant yield at dose x_i, and W_i is a weight chosen to reduce or limit the influence of outlying values of Y_i on the parameter estimates. Specifically, the weight is defined as $W_i = 1$ when $SR_i \leq K$ and $W_i = K/SR_i$ when $SR_i > K$. Here, K is a constant (set equal to 1.5 by Myers *et al.*), the value of which determines the degree to which outliers are discounted. This approach meets nearly all of the previously specified principles of analysis, although the arbitrariness of the choice of K could be questioned. Its application to other assays would require a historical data base in order to estimate parameters a and b in the variance function.

5.3 *Maximum likelihood: binomial or Poisson variation*

Least squares regression analyses, whether linear or nonlinear and whether weighted or not, make no explicit assumptions about the statistical distribution of the experimental errors nor about the magnitude of the error variance. Indeed, one of their major strengths is that the variance is estimated from replicate observations and thus accounts for extraneous sources of variability. A potential weakness, however, is that information could be lost and the analysis made less sensitive than otherwise if, in fact, the experimental variability were determined entirely by random sampling errors. If it appears that such errors predominate, methods based on maximum likelihood theory are available that easily accommodate nonlinear regression functions and are designed specifically to account for the theoretical mean-variance relationship determined by the sampling distribution. However, they can lead to seriously overstated levels of statistical significance if the experimental errors are in excess of those predicted by sampling considerations alone.

The strongest case for the use of maximum likelihood methods can be made for analysis of data on fractions of positive microwells. Provided that reasonable experimental procedures are taken to avoid any 'clustering' of mutated or mutable cells in particular microwells, one can be reasonably confident that the results for different wells are stochastically independent and that the number of positive (turbid) wells at each dose level therefore follows a binomial distribution. Explicit use of the binomial distribution in this setting is essential in order to accommodate the small numbers of positive wells that may be observed at low doses. Collings *et al.* (1981) discuss such analyses, with particular emphasis on tests of the null hypothesis of no mutagenic effect. Specifically, the number of positives among the replicate wells at dose x are assumed to be binomial with probability $P(x) = 1 - \exp[-M(x)]$. Rather than making any parametric assumption about the expected mutant yield $M(x)$, they concentrate on tests of the null hypothesis that the yields are constant *versus* alternatives in which they either increase with increasing dose (mutagenic effect only) or else first increase and then decrease (both mutagenic and toxic effects). However, maximum likelihood estimation and testing for parameters could be used in any of the models proposed for mutant yield in order to estimate mutagenic potency. In fact, the standard trend test for an increase in yield with increasing dose, also discussed by Collings *et al.*, results from the one-hit model, assuming no toxicity.

Stead *et al.* (1981) have suggested that data on counts of mutant colonies should be analysed by maximum likelihood under Poisson theory. They assume that the observed count at dose x is Poisson with mean $M(x)$ given by

$$M(x) = (m_0 + m_1 x^\gamma)\exp(-k_1 x). \tag{7}$$

This is similar to model (6), except for the unknown power of dose γ that is introduced in order to account for possible nonlinearity in the mutant frequency. However, since it leads to a zero ($\gamma > 1$) or infinite ($\gamma < 1$) slope for the dose-response curve at zero dose, it is not as satisfactory as the linear or linear-quadratic models for assessing mutagenic potency. Replicate measurements at each dose level allow testing of the Poisson assumption, and the use of five or more distinct dose levels to estimate the four parameters in (7) allows testing of the adequacy of the dose-response model. Provided that these assumptions seem to be in reasonable accord with the observed data, tests for toxicity ($k_1 = 0$) and mutagenicity ($m_1 = \gamma = 0$) are based on the likelihood ratio criterion. However, when the observed variability between replicate counts exceeds the Poisson minimum, even though not significantly so, it may reflect extraneous variability the presence of which invalidates the assessment of statistical significance.

Gart *et al.* (1979) discuss maximum likelihood estimation from cell transformation data consisting of the numbers of transformed and surviving cell colonies in each replication (dish) of the assay at each dose level. They assume that the total number of transformed colonies at dose x is binomially distributed, with the denominator equal to the number of surviving colonies and transformation probability given by the one-hit curve (5). Transformed colonies are assumed not to occur in the absence of exposure. Tests for the one-hit curve *versus* alternatives of curvature due to host susceptibility have been developed, and these are used to reject from further analysis those data points for higher doses in which the observed proportions of transformed cells are less than those predicted by the one-hit model. Chi-square tests for testing the assumption of binomial variability at each dose level are also described. Even though extraneous variation cannot be precluded at some dose levels in the data they examine, these authors ignore such variation on the grounds that this provides 'a severe test of the model'. They develop a weighted least squares regression procedure for testing the one-hit model *versus* the alternative of host susceptibility in situations in which the binomial denominators are not observed for each replication, but are instead estimated from a more limited number of observations on the surviving colonies at each dose level.

5.4 *Maximum and quasilikelihood procedures that accommodate extraneous variation*

More general theoretical models for the distribution of replicate counts of mutant yield are available that account for the possibility of extraneous variation. Margolin *et al.* (1981) hypothesize that the counts for each plate are indeed Poisson variables, but that the Poisson parameters for different plates at the same dose level are drawn from a 'mixing' distribution. If this mixing distribution is taken to be *gamma* with the usual parametrization, then the observed counts at dose x have a negative binomial distribution. The mean, $M(x)$, is determined by one of the dose-response functions considered earlier, and the variance is

$$M(x)[1 + \sigma^2 M(x)], \tag{8}$$

where $\sigma^2 \geqslant 0$ is an additional parameter that is estimated from the data to account for the extraneous variability. Alternative parametrizations of the mixing *gamma* distribution lead to other mean variance relationships. These authors propose rather complicated dose-response functions, of which the one-hit model with cell kill $M(x) = [1-\exp(-m_0-m_1x)]\exp(-k_1x)$ is a special case. They provide a convincing example in which the extra parameter σ^2 is essential in order to accommodate the extraneous variation and leads to substantial increases in the standard error of the coefficient m_1 of mutagenic effect in comparison with the Poisson sampling model ($\sigma^2 = 0$).

Bernstein *et al.* (1982) consider, in addition to the negative binomial, two other probability models that incorporate an extra parameter to represent the extraneous variation. Both are versions of the generalized Poisson family of Consul and Jain (1973). One has variance $\sigma^2 M(x)$ and the other $[M(x)]^b$, where $b > 0$. The latter model has the same effect on the variance as the Box-Cox transformation (Box & Cox, 1964) used by Snee and Irr (1981), whereby a power transformation of the observed mutant count is assumed to have constant variance. Both Bernstein *et al.* (1982) and Margolin *et al.* (1981) make separate estimates of σ^2 and b from each assay, thus controlling the extraneous variability internally in accordance with the principles of statistical analysis recommended above.

Bernstein *et al.* (1982) propose a 'point rejection' approach, similar in concept to that of Gart *et al.* (1979), for estimation of parameters in a dose-response function that is only partially specified. Starting with the highest dose, they consecutively reject data points until a reasonable fit is obtained to the linear model $M(x) = m_0 + m_1 x$ (no toxicity). The tests are based on likelihood ratios under one of the three generalized Poisson models. Statistical significance of the mutagenic response is evaluated using the statistic $t = \max[m_1/\text{SE}(m_1)]$, where the maximum is taken over the several estimates of m_1 and its standard error obtained at each step of the point rejection process. A

Bonferroni-type correction factor is applied to the resulting p-value to account for the multiplicity of t-statistics computed. The rationale for this empirical approach is two fold. First, these authors believe that the (assumed) initial linear portion of the dose-response curve is the only part of the curve that is relevant for assessment of mutagenic potency: 'We have taken the view that an attempt to model the high dose regions of the curve would result in a net loss in precision, since at high doses effects other than mutagenesis predominate.' Second, they state that the monotonic mean variance relations do not extend to high doses at which the counts are sometimes highly variable even though their averages are falling.

Breslow (1984) proposed a somewhat ad-hoc modification of the method proposed by Margolin *et al.* (1981) that is based on concepts of quasilikelihood (McCullagh & Nelder, 1983). It can be implemented easily using standard statistical programs (Baker & Nelder, 1978). This procedure assumes that the same mean variance equation (8) holds as for the negative binomial, but makes no further assumptions about distributional form. It uses a log-linear approximation to the one-hit dose-response model, given by

$$\log M(x) = m_0 + m_1 \log(x + x_0) - k_1 x, \tag{9}$$

where x_0 is the smallest non-zero dose level. Parameters are first estimated by maximum likelihood under the Poisson model. If the Pearson chi-square goodness-of-fit statistic is no larger than its Poisson expected value of $N-p$, where N is the total number of counts and p is the number of estimated parameters (here, $p = 3$), determination of statistical significance is based on Poisson variation alone. Otherwise, the parameters m_0, m_1, m_2 and σ^2 are determined by jointly solving one equation equating the Pearson statistic to the expected value and three 'quasi-likelihood' equations for m_0, m_1 and m_2 that are determined by the mean variance relation (8).

5.5 *Nonparametric procedures*

Several authors have proposed nonparametric testing procedures for evaluating the statistical significance of mutagenic effect. The null and alternative hypotheses are specified in terms of equality or stochastic ordering of the statistical distributions of the replicate counts at the control (F_0) and ith of I ordered dose levels (F_i). If it can be assumed that these $I+1$ distributions are identical in the absence of a mutagenic effect,

$$H_0: F_0 = F_1 = \ldots = F_I, \tag{10}$$

and satisfy the stochastic ordering relation

$$H_1: F_0 \geq F_1 \geq \ldots \geq F_I \tag{11}$$

in the presence of a genuine mutagenic effect, then standard nonparametric trend tests may be applied. For example, Boyd (1982) suggests using Jonckheere's test, whereby two-sample Mann-Whitney statistics are calculated for each pair of ordered dose levels and then summed. Wahrendorf *et al.* (1985) describe a modification of the procedure of Page (1963) which consists of a weighted sum of the total ranks of the observations in the $I + 1$ dose groups. Typically the weights are given by $I-i$, but other weighting schemes are possible if the alternative hypothesis specifies other than a monotonic increase in the mutant yield. In fact, due to the toxic effects of the agent at higher doses, (10) and (11) often do not adequately describe the null and alternative hypotheses of interest, and this limits the applicability of standard nonparametric procedures.

Simpson and Margolin (1985) have developed new nonparametric procedures that are designed specifically to account for possible toxic effects of the agent on mutant yields. They express the null hypothesis in terms of the stochastic ordering

$$H_0: F_0 \leq F_1 \leq ... \leq F_I (F_0 < F_i \text{ for some } 1 \leq i \leq I).$$

They exclude higher doses at which a significant downturn in yield is observed and then apply a standard nonparametric trend test with the lower, nonexcluded doses. More specifically, they carry out a trend test with the dose level $0, 1, ..., i_m$, where i_m is the maximum index i such that a two-sample test T_i comparing the pooled counts for levels $0, 1, ..., i$ with those for the $i + 1$ dose level exceeds a specified constant c_i. The c_i are chosen to control the overall level of statistical significance. In their examples, Jonckheere's statistic is used to test for trend, while T_i is the two-sample Mann-Whitney statistic.

While such nonparametric procedures offer a convenient solution to the problem of evaluating the statistical significance of the observed mutagenic effect, they do not provide for estimation of mutagenic potency and thus cannot be said to provide a complete statistical analysis of in-vitro test data.

5.6 Comparisons among alternative procedures

Schumacher (1985) has undertaken an extensive comparison of the statistical properties of the parametric and nonparametric procedures discussed above by means of Monte Carlo sampling. Data were generated by computer from nine hypothetical dose-response functions, which are displayed in Table 3. One thousand samples, consisting of three replicate observations from each of the five dose levels, were generated by computer under each of two sampling distributions — the Poisson model and the negative binomial model with $\sigma^2 = 0.02$. Table 4 shows the number of Monte Carlo trials out of 1000 in which each statistical procedure rejected the null hypothesis of no mutagenic effect at the nominal 5% level. The dose-response functions (see Table 3) A1, A2 and A3 all correspond to situations in which there is both mutagenesis and toxicity, B1 and B2 show either no effects or toxic effects only, and C1, C2, C3 and C4 show only mutagenic effects.

Table 3. Hypothetical dose-response functions used in Monte Carlo studies[a]

Profile	Dose				
	0	1	10	100	1000
A1	85	95	105	115	10
A2	85	95	105	100	95
A3	85	85	135	85	85
B1	85	85	85	85	85
B2	85	85	85	75	65
C1	85	92	103	115	125
C2	85	90	100	130	180
C3	85	105	105	105	105
C4	85	85	85	90	180

[a]From Schumacher (1985)

It is apparent that the standard nonparametric trend tests of Page and of Jonckheere, while adequately controlling the nominal significance level under the null hypothesis (B), did very poorly in identifying mutagenic effects in the presence of toxicity (A). The parametric procedures differed substantially in their behaviour under the various hypothetical models, due largely to differences in the dose-response functions that they assumed. Schumacher investigated a large-sample least-squares approximation to the quasilikelihood estimates proposed by Breslow. The version of Myers' method he investigated involved the dose-response function (4) with linear mutagenic effects and exponential cell kill, the parameters being estimated *via* weighted nonlinear least squares regression after preliminary analysis to determine the constants in the mean variance relation. Margolin's procedure was implemented using the one-hit model (6), the parameters being estimated by maximum likelihood under the negative binomial assumption. The Snee and Irr method fitted the simple linear regression model with transformed counts, specifically $(y+1)^{0.15} = m_0 + m_1 x + \epsilon$, where ϵ is a mean-zero error term, using unweighted least squares.

In these numerical studies, Breslow's procedure tended to reject the true null hypothesis (B1) about 30% too often under Poisson sampling (Table 4,B) but achieved the nominal level under negative binomial sampling. The procedures of Myers and of Margolin were too conservative; that of Snee and Irr was just about right. Considering the 14 Monte Carlo studies under alternative hypotheses (A1-A3 and C1-C4 under both Poisson and negative binomial sampling), and counting 0.5 for tied results, Breslow's procedure gave the best result in 9.5, Snee and Irr's in 2.5, Myers in 2 and Margolin's in 0. However, a fairer comparison of these procedures would adjust the critical values of each so that approximately 5% of the B1 samples showed a positive result.

Table 4. Number of rejections of the null hypothesis of no mutagenic activity at the 5% level of significance in 1000 Monte Carlo trials[a]

Test procedure	Poisson sampling				Negative binomial sampling			
	Dose-response function				Dose-response function			
	A1	A2	A3		A1	A2	A3	
Page	0	446	42		0	281	39	
Jonckheere	0	327	37		0	284	30	
Breslow	564	705	196		282	409	192	
Myers	936	187	6		910	84	10	
Margolin	192	120	84		112	76	30	
Snee & Irr	0	32	0		0	53	0	
	B1	B2			B1	B2		
Page	52	0			33	0		
Jonckheere	46	0			32	0		
Breslow	73	5			53	3		
Myers	22	6			26	6		
Margolin	16	40			18	26		
Snee & Irr	54	0			42	0		
	C1	C2	C3	C4	C1	C2	C3	C4
Page	1000	1000	785	813	927	1000	432	759
Jonckheere	1000	1000	792	768	924	998	436	709
Breslow	997	1000	726	1000	852	1000	405	976
Myers	571	920	63	74	314	832	92	72
Margolin	300	850	30	0	176	154	48	16
Snee & Irr	955	1000	154	1000	685	1000	123	1000

[a] From Schumacher (1985)

6. Estimates of mutagenic potency

For many purposes, it is useful to provide a reasonably reliable ranking of chemical agents in terms of mutagenic potency. Precise definition of this quantity is more difficult, however, and several alternative suggestions have been made. One rule of thumb suggested by biologists is simply the dose needed to double the spontaneous rate. Other authors have proposed measures of potency that are based on the initial slope of the dose-response function, assuming that this is linear at zero dose (McCann et al., 1975; Meselson & Russell, 1977; Bartsch et al., 1980). In symbols, if $M(x) \simeq a + bx$ for low doses, the mutagenic potency is defined as the slope m_1, i.e., as the *absolute* increase in mutant yield over background per unit dose. Equivalently, it may be defined as the dose needed to achieve a prescribed absolute increase in the mutant yield, as for example $D_{100} = 100/m_1$, the dose needed to increase the count by 100. Alternatively, if there appears to be an association between spontaneous and induced mutagenesis, it may be preferable to measure the potency in *relative* terms, such as b/a, the fraction by which the spontaneous rate is increased for each unit amount of the chemical agent. Horn *et al.* (1983) demonstrated such an association for some strains of *S. typhimurium* used in the Ames assay (TA98 and TA1537) but found it was absent in others (TA100). However, expressing potency in relative terms did not lead to greater reproducibility when using, as a criterion for comparison, the coefficient of variation of potency estimates that were made in a series of replicate assays. More extensive investigations of this kind could usefully be made.

Most of the parametric methods of statistical analysis discussed earlier model dose-response as a function that is linear at low doses and thus provide a measure of potency in terms of the low-dose slope. Exceptions include one of the procedures discussed by Snee and Irr, in which different power transformations are applied to the mutant fractions *and* to dose before fitting a linear regression relationship. Similarly, the power transform of dose used in the model proposed by Stead *et al.* (1981) to account for possible nonlinearities does not provide a one-parameter low-dose potency estimate. The log-linear formula (13) does describe a linear relationship at low doses.

In general, it is important to establish that the assumption of low-dose linearity is consistent with the data before estimating potency from an experiment. For this purpose, a goodness-of-fit test of the linearity assumption can be used. For example, Bernstein *et al.* (1982) fit a straight line using their point rejection method, *excluding* the data for the negative/solvent control, and then test the fit of the observed control data to the intercept of the fitted line.

Unless the survival curve has a shoulder at the origin, the initial slope of mutant yield *versus* dose confuses mutant and toxic effects. For example, under the one-hit model (6) with background, $b = m_1 - k_1^1 m_0$, whereas mutagenic potency would perhaps be better measured by m_1 alone.

McCann *et al.* (1984) recently calculated mutagenic potencies in terms of the initial slope from a large number of *Salmonella* test results published in the literature. They employed the methods of Bernstein *et al.* (1982) in their calculations. While the results for different chemicals were reasonably well clustered within the million-fold range of potencies observed, they nevertheless revealed marked between-laboratory variability.

One method of reducing the degree of interlaboratory variability of potency estimates in biological assays has been known to biostatisticians for decades (Finney, 1964). It is surprising that it has not seen wider use in the field of mutagenesis. This is to select a standard mutagenic chemical for use as a positive control in a particular assay system for which it is known to give positive results. Different dose levels of the standard are investigated in each assay just as they are for each of the test chemicals examined. Using a type of statistical analysis appropriate to the results observed (e.g., slope ratio, parallel line), the potency of each test chemical is then expressed in terms of the equivalent units of the standard needed to obtain a similar mutagenic response. This standard approach to biological assays could profitably be used in future mutagenesis studies to improve the reliability and comparability of results reported by different laboratories.

7. Acknowledgements

This work was supported in part by USPHS Grant 1-RO1-CA40644.

8. References

Amacher, D.E., Paillet, S.C., Turner, G.N., Ray, V.A. & Salsburg, D.S. (1980) Point mutations at the thymidine kinase locus in L5178Y mouse lymphoma cells. II. Test validation and interpretation. *Mutat. Res.*, 72, 447-474

Ames, B.N., McCann, J. & Yamasaki, E. (1975) Methods for detecting carcinogens and mutagens with the *Salmonella*/mammalian-microsome mutagenicity test. *Mutat. Res.*, 99, 53-74

Amphlett, G.E. & Delow, G.F. (1984) Statistical analysis of the micronucleus test. *Mutat. Res.*, 128, 161-166

Armitage, P. & Spicer, C.C. (1956) The detection of variation in host susceptibility in dilution counting experiments. *J. Hyg.*, 54, 401-404

Baker, R.J. & Nelder, J.A. (1978) *The GLIM System, Release 3*, Oxford, Numerical Algorithms Group

Bartsch, H., Malaveille, C., Camus, A.M., Martel-Planche, G., Brun, G., Hautefeuille, A., Sabadie, N., Barbin, A., Kuroki, K., Drevon, C., Piccoli, C. & Montesano, R. (1980) Validation and comparative studies on 180 chemicals with *S. typhimurium* strains and V79 Chinese hamster cells in the presence of various metabolizing systems. *Mutat. Res.*, 76, 1-50

Bernstein, L., Kaldor, J., McCann, J. & Pike, M.C. (1982) An empirical approach to the statistical analysis of mutagenesis data from the *Salmonella* test. *Mutat. Res.*, 97, 267-281

Box, G.E.P. & Cox, D.R. (1964) An analysis of transformation. *J. R. stat. Soc. B*, 26, 211-252

Boyd, M.N. (1982) Examples of testing against ordered alternatives in the analysis of mutagenicity data. *Mutat. Res.*, 97, 147-153

Breslow, N.E. (1984) Extra-Poisson variation in log-linear models. *Appl. Stat.*, 33, 38-44

Brusick, D.J., Simmon, V.F., Rosenkranz, H.S., Ray, V.A. & Stafford, R.S. (1980) An evaluation of the *Escherichia coli* WP_2 and WP_2 *uvr*A reverse mutation assay. *Mutat. Res.*, 76, 169-190

Chu, K.C., Patel, K.M., Lin, A.H., Tarone, R.E., Linhart, M.S. & Dunkel, V.C. (1981) Evaluating statistical analyses and reproducibility of microbial mutagenicity assays. *Mutat. Res.*, 85, 119-132

Clive, D., Johnson, K.O., Spector, J.F.S., Batson, A.G. & Brown, M.M.M. (1979) Validation and characterization of the L5178Y/$TK^{+/-}$ mouse lymphoma mutagen assay system. *Mutat. Res.*, 59, 61-108

Collings, B.J., Margolin, B.H. & Oehlert, G.W. (1981) Analyses for binomial data, with application to the fluctuation test for mutagenicity. *Biometrics*, 37, 775-794

Consul, P. & Jain, G. (1973) A generalization of the Poisson distribution. *Technometrics, 15*, 791-799

Cox, D.R. & Miller, H.D. (1965) *The Theory of Stochastic Processes*. London, Wiley

Dunkel, V.C. (1979) Collaborative studies on the *Salmonella*/microsome mutagenicity assay. *J. Assoc. off. anal. Chem., 62*, 874-882

Finney, D.J. (1964) *Statistical Methods in Biological Assay*, London, Griffin

Gart, J.J., DiPaolo, J.A. & Donovan, P.J. (1979) Mathematical models and the statistical analyses of cell transformation experiments. *Cancer Res., 39*, 5069-5075

Gilbert, R.I. (1980) The analysis of fluctuation tests. *Mutat. Res., 74*, 283-289

Green, M.H.L. & Muriel, W.J. (1976) Mutagen testing using Trp^+ reversion in *Escherichia coli*. *Mutat. Res., 38*, 3-32

Haynes, R.H. & Eckardt, F. (1979) Analysis of dose-response patterns in mutation research. *Can. J. Genet. Cytol., 21*, 277-302

Horn, L., Kaldor, J. & McCann, J. (1983) A comparison of alternative measures of mutagenic potency in the *Salmonella* (Ames) test. *Mutat. Res., 109*, 131-141

Irr, J.D. & Snee, R.D. (1982) A statistical method for analysis of mouse lymphoma L5178Y cell TK locus forward mutation assay. Comparison of results among three laboratories. *Mutat. Res., 97*, 371-392

Kaldor, J. (1986) Model-based statistical procedures for the analysis of in-vitro mutagenesis assays. *Biom. J., 28*, 469-484

Lee, Y.J. & Caspary, W.J. (1983) Mathematical model of L5178Y mouse lymphoma forward mutation assay. *Mutat. Res., 113*, 417-430

Leong, P.M., Thilly, W.G. & Morgenthaler, S. (1985) Variance estimation in single-cell mutation assays: comparison to experimental observations in human lymphoblasts at 4 gene loci. *Mutat. Res., 150*, 403-410.

Margolin, B.H., Kaplan, N. & Zeiger, E. (1981) Statistical analysis of the Ames *Salmonella*/-microsome test. *Proc. natl Acad. Sci. USA, 78*, 3779-3783

McCann, J., Choi, E., Yamasaki, E. & Ames, B.N. (1975) Detection of carcinogens as mutagens in the *Salmonella*/microsome test: assay of 300 chemicals. *Proc. natl Acad. Sci. USA, 72*, 5135-5139

McCann, J., Horn, L. & Kaldor, J. (1984) An evaluation of *Salmonella* (Ames) test data in the published literature. Application of statistical procedures and analysis of mutagenic potency. *Mutat. Res., 134*, 1-47

McCullagh, P. & Nelder, J.A. (1983) *Generalised Linear Models*, London, Chapman & Hall

Meselson, M. & Russell, K. (1977) *Comparisons of carcinogenic and mutagenic potency*. In: Hiatt, H., Watson, J. & Winsten, J., eds, *Origins of Human Cancer*, Cold Spring Harbor, NY, CSH Press, pp. 1473-1481

Myers, L.E. Sexton, N.H., Southerland, L.I. & Wolff, T.J. (1981) Regression analysis of Ames test data. *Environ. Mutagen., 3*, 575-586

Page, E.B. (1963) Ordered hypothesis for multiple treatments: a significance test for linear ranks. *J. Am. stat. Assoc., 58*, 216-230

Schumacher, M. (1985) *Statistiche Analyse des Ames-Tests*, Dortmund, University of Dortmund

Simpson, D.G. & Margolin, B.H. (1986) Recursive nonparametric testing for dose-response relationships subject to downturns at high doses. *Biometrika, 73*, 589-596

Snee, R.D. & Irr, J.D. (1981) Design of a statistical method for the analysis of mutagenesis at the hypoxanthine-guanine phosphoribosyl transferase of cultured Chinese hamster ovary cells. *Mutat. Res., 85*, 77-93

Snee, R.D. & Irr, J.D. (1984) A procedure for the statistical evaluation of Ames *Salmonella* assay results; comparison of results among four laboratories. *Mutat. Res., 128*, 115-125

Stead, A.G., Hasselblad, V., Creason, J.P. & Claxton, L. (1981) Modeling the Ames test. *Mutat. Res., 85*, 13-27

Venitt, S. (1982) UKEMS Collaborative Genotoxicity Trial. Bacterial mutation tests of 4-chloromethylbiphenyl, 4-hydroxymethylbiphenyl and benzyl chloride. Analysis of data from 17 laboratories. *Mutat. Res., 100*, 91-109

Venitt, S. & Forster, R. (1985) *Bacterial mutagenicity assays: coordinators' report.* In: Parry, J.M. & Arlett, C.F., eds, *Comparative Genetic Toxicology: The Second UKEMS Study*, London, MacMillan Press, pp. 103-144

Wahrendorf, J., Mahon, G.A.T. & Schumacher, M. (1985) A nonparametric approach to the statistical analysis of mutagenicity data. *Mutat. Res., 147*, 5-13

Weinstein, D. & Lewinson, T.M. (1978) A statistical treatment of the Ames mutagenicity assay. *Mutat. Res., 51*, 433-434

REPORT 17

TESTING OF COMPLEX CHEMICAL MIXTURES

Prepared by:

V.J. Feron, R.A. Griesemer, S. Nesnow (Rapporteurs),
F. Anders, P. Bannasch, R. Becker, G. Becking, J.R. Cabral,
G. Della Porta, D. Henschler, N. Ito, R. Kroes, P.N. Magee,
M.L. Mendelsohn, N.P. Napalkov, G.N. Rao and J. Wilbourn

1. Introduction

The testing of complex mixtures is not new. In 1775, Pott reported that scrotal cancer in chimney sweeps was caused by soot, a complex mixture of chemical substances (Potter, 1963). The first demonstration of chemical carcinogenesis in animals was made in 1915 by Yamagiwa and Ichikawa (1915) who showed that painting coal-tar on the ears of rabbits resulted in tumours.

Examples of other widely varying types of complex mixtures are tobacco smoke, fly ash, automotive emissions, contaminated drinking water, cooked and uncooked food and occupational atmospheres. This report deals only with these types of complex mixtures and does not examine simple mixtures of a few pure chemicals. So far as testing is concerned, on the one hand, each complex mixture presents its own specific problems for experimental design and analysis. On the other hand, testing and evaluation of these diverse mixtures share common problems of characterization, instability, changing chemical composition, bioavailability and often a partial lack of correspondence between the material tested and the mixtures to which humans are exposed (Witschi & Munro, 1983).

The use of short-term assays to evaluate the potential health hazards of complex environmental mixtures has been described in a series of publications (Waters *et al.*, 1979, 1981, 1983a, 1985), which contain information on the collection, preparation and chemical characterization of samples of complex mixtures to be used for bioassays. In addition, efforts to integrate chemical and biological data to assess human health hazards are reviewed. Historically, the *Salmonella typhimurium* test has been used primarily, owing to the ease of test performance.

Many of the principles of and limitations to the testing of complex mixtures in short-term assays also apply to the testing of complex mixtures in long-term carcinogenicity assays in animals. The design, conduct and analysis of long-term carcinogenicity studies on complex mixtures are not essentially different from those of long-term tests with individual compounds, but special consideration must be given to the storage and stability of the large quantities of test material required. The complexity and instability of mixtures may restrict the possibilities of testing. Application to the skin of mice has been used widely for semisolid and nonmiscible materials, for example. Other considerations are chemical interactions between compounds in the mixture and toxicological interactions in the animals (National Academy of Sciences, 1980, 1982).

2. Sample selection and collection

2.1 *Representativity of the test material*

In selecting test materials for short-term tests or carcinogenicity assays in experimental animals, the representativity for humans of the sample tested must be considered. Therefore, knowledge of chemical composition is desirable. For instance, chemical composition determines the type of welding fume (Stern, 1983) and the category of coal-tar (IARC, 1984, 1985) and reflects practices occurring in industry. In addition, the composition of a complex mixture to which workers are exposed, even within the same occupational setting, may differ widely in the course of a working day. In such cases, it may be preferable to test an average (representative) exposure mixture. These and additional aspects involved in exposure of complex mixtures in the workplace have been reviewed by Ballantyne (1985).

2.2 *Artefacts produced during the collection and preparation of test samples*

One of the most significant problems in the bioassay of complex mixtures is possible alteration of the sample prior to bioassay, in the form of loss of specific components of the sample or chemical transformation of one or more components.

Raabe (1982) and Jungers and Lewtas (1982) reviewed methods applicable to the collection of airborne particles. In this instance, the biological activity of the sample depends on the collection device employed and the method and solvent used to extract it. Samples may have to be protected from heat, light and humidity to prevent degradation. Fly-ash particles collected at 95°C were mutagenic to *S. typhimurium*, while no mutagenicity could be detected with particles collected at 107°C (Fisher & Chrisp, 1979; Fisher *et al.*, 1979). Presumably, the organic materials that contributed to the mutagenic activity were 'distilled' off the particles at the higher temperature. Similar phenomena have been reported for diesel exhaust emissions (Huisingh *et al.*, 1979). One common problem with particulate samples is that they may not contain the volatile components unless specialized samplers are used. Standard protocols for some short-term tests may not be effective in detecting gaseous and volatile components, but methods that overcome this problem have recently become available (Hatch *et al.*, 1983; Claxton, 1985; Hatch *et al.*, 1986).

The mutagenic activity in *S. typhimurium* of organic residues from waste-water sludges was reported to be dependent on the method of fractionation used, and these data suggested that mutagenic agents were either being created or destroyed by the fractionation procedure (Tabor *et al.*, 1985). The problem of the creation of mutagenic artefacts during the collection of samples is discussed further by Pitts *et al.* (1978). Benzo[*a*]pyrene coated glass-fibre filters that were exposed to atmospheres containing nitrogen dioxide, a contaminant of smog, were subjected to extraction and chemical and biological analysis. The extracts were highly mutagenic to *S. typhimurium* and were found to act directly, unlike the parent benzo[*a*]pyrene. Chemical analysis indicated the formation of nitrated derivatives. Therefore, long sampling times for the collection of particulates on filters can influence the overall activity of the samples by the creation of artefacts through chemical transformation.

2.3 *Methods of extraction and solvent effects*

Methods of extraction, choice of solvents and residual solvent effects can affect the overall activity of a complex mixture (Epler, 1980; Hughes *et al.*, 1980). In general, extraction with hot solvents is to be avoided, as unstable components may be degraded and semivolatiles lost. Soxhlet extraction methods should be performed with care using solvents with low boiling-points. Certain solvents may extract only a portion of the total mixture, and extractions with multiple solvents should be explored (Jungers & Lewtas, 1982). Solvents may introduce a bacteriotoxic residue into a test sample, as reported for toluene extracts of carbon black. The toxic residues were not formed after extraction of the carbon black samples with acetone or benzene (Agurell & Löfroth, 1983).

2.4 *Stability of test samples*

Data on stability during storage and treatment of samples (condensation, concentration, extraction, fractionation) are essential, since changes in chemical composition may influence the results of bioassays (IARC, 1985). Treatments are often necessary because the mixture to which humans are exposed may be incompatible with the test system, due to overall toxicity (Guerin *et al.*, 1979) or because of very low concentrations of the active components (e.g., in drinking-water) (Kopfler, 1981). In the case of drinking-water, organic concentrates and organic fractions are usually tested, with the disadvantage that possible interactive effects may be missed (Loper & Tabor, 1981; Pereira & Bull, 1981; Robinson *et al.*, 1981). The conflicting results of long-term carcinogenicity studies of mutagenic drinking-water concentrates have been ascribed to differences in the extraction, concentration and reconstitution of the samples (Kool *et al.*, 1985).

2.5 *Indicator constituents*

It may also be useful to identify the most abundant or active components in mixtures, making it possible sometimes to draw correlations between chemical composition and specific effects (Dumont et al., 1983). This has led, for example, to the adoption of benzo[a]pyrene as an indicator for environmental air quality. The risk of this approach is that minor components with potent biological activity may be overlooked, such as nitroarenes in diesel emissions and the ambient atmosphere (Rosenkranz et al., 1983). Using an activity-directed fractionation scheme, this group of potent mutagens was detected although present in less than 2% of the mass of the mixture (Löfroth et al., 1980; Rosenkranz et al., 1980).

3. Issues specific to short-term tests

3.1 *Lack of concentration-response*

A difficulty in the interpretation of short-term test data for complex mixtures is the frequent lack of a concentration-response. This may be particularly true of the short-term bioassays that utilize mammalian cells in culture (Curren et al., 1981). Much of the problem can be caused by lack of a soluble sample, especially at the higher concentrations. Complex mixtures composed of organic materials are generally relatively insoluble in aqueous solution and must be delivered to the test organism in organic solvents which are miscible in water. At higher concentrations of test material, the organic mass may precipitate out of solution decreasing the effective concentration. This is not a major problem in *S. typhimurium* bioassays when they are performed with dimethyl sulfoxide as a solvent. Mammalian cells in culture, however, cannot survive high concentrations of dimethyl sulfoxide. A problem inherent in assays of complex mixtures is that the soluble components available to interact with cells may not occur in proportion to their original concentration in the mixture. One possible partial solution to this problem is to use mammalian cells in suspension, as in the L5178Y mouse lymphoma bioassay. Alternatively, for bioassays which use monolayer cultures, it may be advisable to rock the dishes during treatment.

3.2 *Metabolic activation*

Complex mixtures must, of course, be tested with and without an exogenous source of metabolic activation in those short-term bioassay systems that are devoid of or deficient in this activity. However, little attention has been paid to the use of metabolic activation systems other than polychlorinated biphenyl (Aroclor 1254)-induced rat liver. When the effects of uninduced and induced rat and hamster liver fractions in the *S. typhimurium* bioassay were compared using four different complex mixtures, dramatic differences were observed in the metabolic activation capability of the two animal species (Williams & Lewtas, 1985). Epler (1980) examined the effect of inducers on metabolic activation capability and found marked differences in the abilities of phenobarbital, 3-methylcholanthrene, and Aroclor 1254-induced rat livers to activate synfuel fractions to forms mutagenic to *S. typhimurium*. Another variable in metabolic activation is the concentration of microsomal protein added to the assay. Maximal genetic activity may be found with different protein concentrations for different complex mixtures (Hughes *et al.*, 1983). Similar effects have been reported for certain classes of polycyclic aromatic hydrocarbons (Nesnow *et al.*, 1984).

4. Issues common to short-term and long-term tests

4.1 *Use of high concentrations of test material*

Complex mixtures consist of a large number of components; therefore, the effective concentration of any one component is diluted by the accompanying components. To observe detectable effects in short-term tests, it is usually necessary to use high concentrations of test material (> 1.0 mg/ml). For bioassays which use mammalian cells in culture, these high concentrations can alter the pH or the osmolarity of the medium and give spurious and unreproducible results.

In long-term carcinogenicity testing, the issue of high concentrations pertains to new food products, which are often incorporated in experimental diets at levels of 20% and above. This may lead to nutritional imbalances or to nutritional deficiencies due to the presence of components that interfere with micronutrients. It is recommended that preliminary studies be performed to determine the maximum amount of a food product that can be incorporated in the diet without significant impairment of growth and survival. The maximal amount in the diet depends not only on the physical nature and nutritional quality of the test material but also on the presence of natural constituents that exert toxic effects when fed at high dietary levels. Such problems have been encountered in the testing of irradiated onions and mushrooms (Food and Agriculture Organization, 1977).

Reduction of the water content of foods by freeze- or heat-drying may facilitate long-term testing. Long-term studies with freeze-dried irradiated onions did not produce evidence of carcinogenicity; however, removal of constituents with a high vapour pressure during the drying procedure may have altered the toxicological properties of the irradiated products (Food and Agriculture Organization, 1977).

4.2 *Presence of highly toxic constituents*

Prior chemical analysis of mixtures is useful for the detection of highly toxic or irritant constituents, the presence of which is often incompatible with testing for the mutagenic and carcinogenic properties of the complete mixture in sufficiently high concentrations. Any step taken to remove such components or otherwise to make the mixture compatible with the test system, however, entails an alteration in the composition of the test material and therefore reduces the relevance of the results of the bioassay (Guerin *et al.*, 1979).

Fractionation of complex mixtures may aid in separating cytotoxic and genetic and related activities. Hughes *et al.* (1983) reported mutagenic activity in a coal-gasification tar after fractionation into acid, basic and neutral fractions. The majority of the mutagenic activity of that tar sample resided in the basic fraction.

The presence of carbon monoxide, nicotine and strong irritants in cigarette smoke renders it difficult to expose experimental animals continuously to high concentrations of fresh smoke for a long period of time (Binns *et al.*, 1976). In order to diminish the toxicity of smoke, experimental inhalation systems require excessive dilution or intermittent exposures. Intermittent exposure systems have been developed which largely prevent major chemical and physical changes in smoke. Experimental exposure conditions could lead to different results in animals than in humans.

4.3 *Additive, synergistic and antagonistic interactions*

Data on the testing of fractions of complex mixtures in short-term tests may be difficult to interpret since the responses may represent a combination of additive, synergistic and antagonistic interactions. Kaden *et al.* (1979) reported that a methylene chloride extract of kerosene soot contained 18 identified polycyclic aromatic hydrocarbons. Using the forward mutation assay in *S. typhimurium* strain TM677 (8-azaguanine resistance), six of the 18 polycyclic aromatic hydrocarbons were found to be mutagenic. The sum of the mutagenic activities of the individual polycyclic aromatic hydrocarbons at the concentrations present in the methylene chloride extract was twice that of the whole extract. Bioassay of a synthetic mixture of all the polycyclic aromatic hydrocarbons in concentrations simulating their contributions to the original methylene chloride extract gave the same mutagenic activity as the extract, suggesting that antagonistic interactions between the polycyclic aromatic hydrocarbons may have played a role in the overall mutagenic activity. Inhibition of the enzymes that metabolically activate the mutagenic components of the complex mixtures may be responsible for this antagonistic interaction (Haugen & Peak, 1983). Synergistic interactions for mutagenic effects have also been reported for complex mixtures (Li *et al.*, 1983; Schoeny & Warshawsky, 1983) and for synthetic mixtures of environmental agents (Rao *et al.*, 1979).

In long-term testing, similar interactions may occur, as has been shown, for example, for tobacco-smoke condensate (Hoffmann & Wynder, 1971), coal combustion and automobile exhaust (Grimmer *et al.*, 1983), and used engine oil (Grimmer *et al.*, 1982).

4.4 *Screening a series of complex mixtures*

Often, an investigator must identify certain complex mixtures from among a group of related complex mixtures. This selection process is dependent in part on sample availability, the number of samples in the selected group, the cost of performing the bioassay and the biological endpoint used. The tendency in many laboratories is initially to use short-term tests, as they are cheaper and require small amounts of samples. The *S. typhimurium* bioassay is most commonly used for this purpose. Often, the desired result in this initial screen is a qualitative decision on mutagenic activity. Once mutagenic activity has been detected, those samples that are active can be subjected to additional testing in short-term bioassays that use other endpoints and test organisms. As described elsewhere in this volume, short-term bioassays are available that can detect DNA damage, gene mutation, chromosomal damage and morphological transformation. The assemblage of short-term bioassays into tiers, batteries or phased approaches for a more complete evaluation of the genotoxic and carcinogenic potential of complex mixtures has been discussed by Waters *et al.* (1983b).

4.5 *Comparative analysis of samples*

A major use of short-term and long-term bioassays in testing complex mixtures is the quantitative comparison of two or more mixtures. Such comparisons may be performed (a) in evaluations of emissions, effluents or samples from different sources or collected at different times or places, (b) in evaluation of different control or emission technologies, collection devices, fractionation schemes or extraction systems, or (c) in evaluation in respect to a reference standard. Rannug (1983) and Claxton (1983) describe variations in the mutagenic activities in *S. typhimurium* of automotive exhaust samples due to variations in fuels, engines, temperatures of collection and driving cycles. Shimizu *et al.* (1984) reported on the evaluation of the genotoxicity of process streams from a coal gasification system. Using three short-term bioassays — Ames' *S. typhimurium* reverse mutation assay, the Chinese hamster ovary cell/hypoxanthine-guanine phosphoribosyl transferase gene locus mutation assay, and the Chinese hamster lung primary culture/sister chromatid exchange — they concluded that a process being applied to clean up gasifier products was effective in reducing the levels of genotoxic materials. Using morphological cell transformation of Syrian hamster embryo cells as an endpoint, Frazier and Andrews (1983) compared the activity of distillate fractions from synfuel mixtures and found that the heavy distillate fraction contained all of the activity.

In animals, mouse skin tests for tumour initiating activity have been used to compare quantitatively the activity of a wide variety of complex mixtures, including gasoline, diesel, coke oven and roofing tar emissions (Nesnow *et al.*, 1983), while long-term inhalation tests in rats have been performed to compare the carcinogenic activities of shale-oil samples (Holland *et al.*, 1983).

4.6 *Bioassay-directed fractionation*

Bioassay-directed fractionation, in which chemical fractionation schemes are derived on the basis of the locations and types of biological activity, has been a particularly effective tool in characterizing the genotoxic components in complex mixtures from diesel emissions, cigarette smoke condensate, coke oven emissions and roofing-tar emissions using *S. typhimurium* (Austin *et al.*, 1985). Fractionation may also be effective in characterizing components with tumour-promoting activity, as has been investigated for cigarette smoke condensate (Hoffmann & Wynder, 1971; Lazar *et al.*, 1974), and with carcinogenic activity, as has been shown for used engine oil (Grimmer *et al.*, 1982), coal combustion, automobile exhaust (Grimmer *et al.*, 1983) and cigarette-smoke condensate (IARC, 1986). Using a combination of Sephadex LH20 fractionation, high-performance liquid chromatography fractionation, chemical analysis and bioassay, Royer *et al.* (1983) have identified specific fractions and components responsible for the mutagenic activity of coal-gasifier tar.

Bioassay-directed fractionation has been especially useful for testing cigarette smoke (IARC, 1986). Although application of cigarette-smoke condensate to the skin is less relevant than inhalation of whole smoke for estimating risk to smokers, skin painting studies with condensates and certain (e.g., mutagenic) fractions of smoke condensate have produced a vast amount of information on the carcinogenic potential of cigarette smoke and have strongly contributed to the identification of carcinogens, cocarcinogens and tumour promoters in tobacco smoke. In addition, in such studies, a dose-response relationship has been established between the amount of tar applied to the skin and the skin tumour yield. These findings supported epidemiological data relating the amount of cigarette smoke inhaled and the likelihood of cancer of the lungs and oral cavity.

5. Evaluation of data

In principle, evaluation of experimental data on complex mixtures is no different from that of data on single compounds. However, both the qualitative and quantitative aspects are more complicated, and add uncertainties to an area already full of uncertainties (Albert, 1981).

The tumour response obtained with a complex mixture may be the resultant of a mix of initiating, promoting, synergistic, additive, cocarcinogenic, antagonistic and inhibiting activities of the compounds in the mixture. One question is whether, at high concentrations, the contribution of the different activities to the biological effect is proportionally the same as that at low concentrations. Most probably the answer is no, because it is not illogical to assume that a mixture contains cocarcinogenic or antagonistic factors which are less active or inactive at lower exposure levels. Concentration-response and pharmacokinetic studies in animals are helpful but difficult to design without prior knowledge of the types of interactions to be expected.

If one examines the relationship between the potency of a series of complex mixtures in two bioassay systems and finds them to be parallel, it may be assumed that the relationship between the potencies of other complex mixtures will be parallel in the two bioassay systems. This proposed constant relative potency model has been tested with three complex mixtures known to be associated with an increased incidence of lung cancer in exposed populations — emissions from a coke oven, emissions from a roofing-tar pot and cigarette-smoke condensate (Albert *et al.*, 1983; Lewtas *et al.*, 1983). Data on lung cancer from epidemiological studies were expressed in terms of relative potency: coke oven, 1.0; roofing tar, 0.39; cigarette smoke condensate, 0.0024. Samples from the three sources were collected and bioassayed in a variety of short-term in-vitro and in-vivo bioassays. The best correlation between human lung cancer potency data and potency data from short-term tests was found with the mouse skin tumour initiation studies using SENCAR mice, in which the relative potency values obtained were: coke oven, 1.0; roofing-tar, 0.20; cigarette smoke condensate, 0.0011. In another experiment using the mouse skin tumour initation assay, it was found that the relative potency of extracts of diesel exhaust particulates was 0.28. Comparison of this figure with the data on coke oven, roofing tar and cigarette smoke condensate allowed an estimation of the lung cancer response in humans exposed to this type of diesel emission (Albert *et al.*, 1983).

Gibb and Chen (1985) stressed that complex carcinogenic mixtures often contain more than one carcinogen, which increases the likelihood that mixtures act on more than one stage of the carcinogenic process. Using as examples data from animal experiments as well as from epidemiological studies, they showed that the multistage theory of carcinogenesis can be applied to interpret dose-response data on complex mixtures.

6. References

Agurell, E. & Löfroth, G. (1983) *Presence of various types of impurities on carbon black detected by the* Salmonella *assay*. In: Waters, M.D., Sandhu, S.S., Lewtas, J., Claxton, L., Chernoff, N. & Nesnow, S., eds, *Short-Term Bioassays in the Analysis of Complex Environmental Mixtures III*, New York, Plenum, pp. 297-306

Albert, R.E. (1981) *Assessing carcinogenic risk resulting from complex mixtures*. In: Waters, M.D., Sandhu, S.S., Huisingh, J.L., Claxton, L. & Nesnow, S., eds, *Short-term Bioassays in the Analysis of Complex Environmental Mixtures II*, New York, Plenum, pp. 507-512

Albert, R.E., Lewtas, J., Nesnow, S., Thorslund, T.W. & Anderson, E.L. (1983) Comparative potency method for cancer risk assessment: application to diesel particulate emissions. *Risk Anal., 3*, 101-117

Austin, A.C., Claxton, L.D. & Lewtas, J. (1985) Mutagenicity of the fractionated organic emissions from diesel, cigarette smoke condensate, coke oven, and roofing tar in the Ames assay. *Environ. Mutagen., 7*, 471-487

Ballantyne, B. (1985) Evaluation of hazards from mixtures of chemicals in the occupational environment. *J. occup. Med., 27*, 85-94.

Binns, R., Beven, J.L., Wilton, L.V. & Lugton, W.G.D. (1976) Inhalation toxicity studies on cigarette smoke. III. Tobacco smoke inhalation dosimetry study on rats. *Toxicology, 6*, 207-217

Claxton, L. (1983) Characterization of automotive emissions by bacterial mutagenesis bioassay: A review. *Environ. Mutagen.*, 5, 609-631

Claxton, L.D. (1985) Assessment of bacterial mutagenicity methods for volatile and semivolatile compounds and mixtures. *Environ. Int.*, 11, 375-382

Curren, R.D., Kouri, R.E., Kim, C.M. & Schechtman, L.M. (1981) Mutagenic and carcinogenic potency of extracts from diesel related environmental emissions: simultaneous morphological transformation and mutagenesis in Balb/c3T3 cells. *Environ. Int.*, 5, 411-415

Dumont, J.N., Schultz, T.W., Buchanan, M.V. & Kao, G.L. (1983) *Frog embryo teratogenesis assay: Xenopus (fetax) — a short-term assay applicable to complex environmental mixtures.* In: Waters, M.D., Sandhu, S.S., Lewtas, J., Claxton, L., Chernoff, N. & Nesnow, S., eds, *Short-term Bioassays in the Analysis of Complex Environmental Mixtures III*, New York, Plenum, pp. 393-406

Epler, J.L. (1980) *The use of short-term tests in the isolation and identification of chemical mutagens in complex mixtures.* In: de Serres, F.J. & Hollaender, A., eds, *Chemical Mutagens; Principles and Methods for their Detection*, Vol. 6, New York, Plenum, pp. 239-270

Fisher, G.L. & Chrisp, C.E. (1979) *Physical and biological studies in coal fly ash.* In: Waters, M.D., Nesnow, S., Huisingh, J.L., Sandhu, S.S. & Claxton, L., eds, *Application of Short-term Bioassays in the Fractionation and Analysis of Complex Environmental Mixtures*, New York, Plenum, pp. 441-462

Fisher, G.L., Chrisp, C.E. & Raabe, O.G. (1979) Physical factors affecting the mutagenicity of fly ash from a coal-fired power plant. *Science*, 204, 879-881

Food and Agriculture Organization (1977) *Wholesomeness of Irradiated Food, Report of the Joint FAO/IAEA/WHO Expert Committee (WHO Technical Report Series No. 604)*, Rome

Frazier, M.E. & Andrews, T.K., Jr (1983) Transformation of Syrian hamster embryo cells by synfuel mixtures. *J. Toxicol. environ. Health*, 11, 591-606

Gibb, H.J. & Chen, C.W. (1985) *Risk assessment of complex mixtures.* In: Waters, M.D., Sandhu, S.S., Lewtas, J., Claxton L., Strauss, G. & Nesnow, S., eds, *Application of Short-term Bioassays in the Analysis of Complex Environmental Mixtures I*, New York, Plenum, pp. 353-361

Grimmer, G., Dethbarn, G., Brune, H., Deutsch-Wenzel, R. & Misfeld, G. (1982) Quantification of the carcinogenic effect of polycyclic aromatic hydrocarbons in used engine oil by topical application onto the skin of mice. *Int. Arch. occup. environ. Health*, 50, 95-100.

Grimmer, G., Naujack, K.-W., Dethbarn, G., Brune, H., Deutsch-Wenzel, R. & Misfeld, J. (1983) Characterization of polycyclic aromatic hydrocarbons as essential carcinogenic constituents of coal combustion and automobile exhaust using mouse skin-painting as a carcinogen-specific detector. *Toxicol. environ. Chem.*, 6, 97-107

Guerin, M.R., Clark, B.R., Ho, C.-H., Epler, J.L. & Rao, T.K. (1979) *Short-term bioassays of complex organic mixtures: Part I, Chemistry.* In: Waters, M.D., Nesnow, S., Huisingh, J.L, Sandhu, S.S. & Claxton, L., eds, *Application of Short-term Bioassays in the Fractionation and Analysis of Complex Environmental Mixtures*, New York, Plenum, pp. 247-268

Hatch, G.G., Mamay, P.D., Christenson, C.C., Casto, B.C. & Nesnow, S. (1983) Chemical enhancement of viral transformation of Syrian hamster embryo cells by gaseous and volatile chlorinated methanes and ethanes. *Cancer Res.*, 43, 1945-1950

Hatch, G.G., Conklin, P.M., Christenson, C.C., Anderson, T.M., Langenbach, R. & Nesnow, S. (1986) Mutation and enhanced virus transformation of cultured hamster cells by exposure to gaseous ethylene oxide. *Environ. Mutagen.*, 8, 67-76

Haugen, D.A. & Peak, M.J. (1983) Mixtures of polycyclic aromatic compounds inhibit mutagenesis in the *Salmonella*/microsome assay by inhibition of metabolic activation. *Mutat. Res., 116*, 257-269

Hoffmann, D. & Wynder, E.L. (1971) A study of tobacco carcinogenesis. XI. Tumour initiators, tumour accelerators, and tumour promoting activity of condensate fractions. *Cancer, 27*, 848-864

Holland, L.M., Gonzaes, M., Wilson, J.S. & Tilkry, M.I. (1983) *Pulmonary effects of shale dusts in experimental animals.* In: Wagner, W.L., Rom, W.N. & Merchant, J.A., eds, *Health Issues Related to Metal and Non-metallic Mining*, London, Butterworth, pp. 485-496

Hughes, T.J., Pellizzari, E., Little, L., Sparacino, C. & Kolber, A.R. (1980) Ambient air pollutants; collection, chemical characterization and mutagenicity testing. *Mutat. Res., 76*, 51-83

Hughes, T.J., Wolff, T.J., Nichols, D., Sparacino, C. & Kolber, A.R. (1983) *Synergism and antagonism in complex mixtures: unmasking of latent mutagenicity by chemical fractionation.* In: Kolber, A.R., Grant, L.D., DeWoskin, R.S. & Hughes, T.J., eds, *In Vitro Toxicity Testing of Environmental Agents*, New York, Plenum, pp. 115-144

Huisingh, J., Bradow, R., Jungers, R., Claxton, L., Zweidinger, R., Tejada, S., Bumgarner, J., Duffield, F., Waters, M., Simmon, V.F., Hare, C., Rodriguez, C. & Snow, L. (1979) *Application of bioassay to the characterization of diesel particle emissions.* In: Waters, M.D., Nesnow, S., Huisingh, J.L., Sandhu, S.S. & Claxton, L., eds, *Application of Short-term Bioassays in the Fractionation and Analysis of Complex Environmental Mixtures*, New York, Plenum, pp. 381-418

IARC (1984) *IARC Monographs on the Evaluation of the Carcinogenic Risk of Chemicals to Humans*, Vol. 34, *Polynuclear Aromatic Compounds, Part 3, Industrial Exposures in Aluminium Production, Coal Gasification, Coke Production, and Iron and Steel Founding*, Lyon

IARC (1985) *IARC Monographs on the Evaluation of the Carcinogenic Risk of Chemicals to Humans*, Vol. 35, *Polynuclear Aromatic Compounds, Part 4, Bitumens, Coal-tars and Derived Products, Shale-oils and Soots*, Lyon

IARC (1986) *IARC Monographs on the Evaluation of the Carcinogenic Risk of Chemicals to Humans*, Vol. 38, *Tobacco Smoking*, Lyon

Jungers, R.H. & Lewtas, J. (1982) *Airborne particle collection and extraction methods applicable to genetic bioassays.* In: Tice, R.R., Costa, D.L. & Schaich, K.M., eds, *Genotoxic Effects of Airborne Agents*, New York, Plenum, pp. 35-47

Kaden, D.A., Hites, R.A. & Thilly, W.G. (1979) Mutagenicity of soot and associated polycyclic aromatic hydrocarbons to *Salmonella typhimurium*. *Cancer Res., 39*, 4152-4159

Kool, J.H., Kuper, F., van Haeringen, H. & Koeman, J.H. (1985) A carcinogenicity study with mutagenic organic concentrates of drinking water in the Netherlands. *Food. chem. Toxicol., 23*, 79-85

Kopfler, F.C. (1981) *Alternative strategies and methods for concentrating chemicals from water.* In: Waters, M.D., Sandhu, S.S., Huisingh, J.L., Claxton, L. & Nesnow, S., eds, *Short-term Bioassays in the Analysis of Complex Environmental Mixtures II*, New York, Plenum, pp. 141-154.

Lazar, P., Chouroulinkov, I., Izard, C., Moree-Testa, P. & Hemon, D. (1974) Bioassays of carcinogenicity after fractionation of cigarette smoke condensate. *Biomedicine, 20*, 214-222

Lewtas, J., Nesnow, S. & Albert, R. (1983) A comparative potency method for cancer risk assessment: clarification of the rationale, theoretical basis, and application to diesel particulate emissions. *Risk Anal.*, *3*, 133-137

Li, A.P., Brooks, A.L., Clark, C.R., Shimizu, R.W., Hanson, R.L. & Dutcher, J.S. (1983) *Mutagenicity testing of complex environmental mixtures with Chinese hamster ovary cells.* In: Waters, M.D., Sandhu, S.S., Lewtas, J., Claxton, L., Chernoff, N. & Nesnow, S., eds, *Short-term Bioassays in the Analysis of Complex Environmental Mixtures III*, New York, Plenum, pp. 185-198

Löfroth, G., Hefner, E., Alfheim, I. & Moller, M. (1980) Mutagenic activity in photocopies. *Science*, *109*, 1037-1039

Loper, J.C. & Tabor, M.W. (1981) *Detection of organic mutagens in water residues.* In: Waters, M.D., Sandhu, S.S., Huisingh, J.L., Claxton, L. & Nesnow, S., eds, *Short-term Bioassays in the Analysis of Complex Environmental Mixtures II*, New York, Plenum, pp. 155-166

National Academy of Sciences (1980) *Principles of Toxicological Interactions Associated with Multiple Chemical Exposures*, Washington DC, National Academy Press

National Academy of Sciences (1982) *Assessment of Multichemical Contamination. Proceedings of an International Workshop, Milan, Italy, April 28-30, 1981*, Washington DC, National Academy Press

Nesnow, S., Triplett, L.L. & Slaga, T.J. (1983) Mouse skin tumor initiation-promotion and complete carcinogenesis bioassays mechanisms and biological activities of emission samples. *Environ. Health Perspect.*, *47*, 255-268

Nesnow, S., Leavitt, S., Easterling, R., Watts, R., Toney, S.H., Claxton, L., Sangaiah, R., Toney, G.E., Wiley, J., Fraher, P. & Gold, A. (1984) Mutagenicity of cyclopenta-fused isomers of benz[*a*]anthracene in bacterial and rodent cells and identification of the major rat liver microsomal metabolites. *Cancer Res.*, *44*, 4993-5003

Pereira, M.A. & Bull, R.J. (1981) *Short-term methods for assessing in vivo carcinogenic activity of complex mixtures.* In: Waters, M.D., Sandhu, S.S., Huisingh, J.L., Claxton, L. & Nesnow, S., eds, *Short-term Bioassays in the Analysis of Complex Environmental Mixtures II*, New York, Plenum, pp. 167-175

Pitts, J.N., Jr, van Cadwenberghe, K.A., Grosjean, D., Schmid, J.P., Fritz, D.R., Belser, W.L., Jr, Knudson, G.B. & Hynds, G.M. (1978) Atmospheric reactions of polycyclic aromatic hydrocarbons: facile formation of mutagenic nitro derivatives. *Science*, *202*, 515-519

Potter, M. (1963) Percival Pott's contribution to cancer research. *Natl Cancer Inst. Monogr.*, *10*, 1-13

Raabe, O.G. (1982) *Problems associated with assessing the mutagenicity of inhalable particulate mattter.* In: Tice, R.R., Costa, D.L. & Schaich, K.M., eds, *Genotoxic Effects of Airborne Agents*, New York, Plenum, pp. 209-223

Rannug, U. (1983) Data from short-term tests on motor vehicle exhausts. *Environ. Health Perspect.*, *47*, 161-169

Rao, T.K., Ellis, K.B., Tipton, S.C. & Epler, J.L. (1979) Effect of cocarcinogen benzo[*e*]pyrene on microsome-mediated chemical mutagenesis in *Salmonella typhimurium*. *Environ. Mutagen.*, *1*, 105-112

Robinson, M., Glass, J.W., Cmehil, D., Bull, R.J. & Orthöfer, J.G. (1981) *The initiating and promoting activity of chemicals isolated from drinking water in the Sencar mouse: a five-city survey.* In: Waters, M.D., Sandhu, S.S., Huisingh, J.L., Claxton, L. & Nesnow, S., eds, *Short-term Bioassays in the Analysis of Complex Environmental Mixtures II*, New York, Plenum, pp. 177-188

USE OF STRUCTURE-ACTIVITY RELATIONSHIPS IN PREDICTING CARCINOGENESIS

H.S. Rosenkranz, M.R. Frierson & G. Klopman

Case Western Reserve University,
Centre for the Environmental Health Sciences
School of Medicine, Cleveland, OH 44107, USA

1. Introduction

It is essential that computer-aided storage, retrieval and analysis of the tremendous volume of data on carcinogens and potential carcinogens become available to researchers and decision-makers in environmental science. Because of the volume of data being generated in research laboratories throughout the world, there is no longer (if there ever were) single individuals or groups of individuals whose judgements of the carcinogenic and genotoxic effects of new, potentially harmful substances could possibly take into account the whole body of experimental knowledge. Most of the computer programs discussed below are capable, in principle, of doing this. In addition, they do it in a critical fashion, being designed to mimic the analytical aspect of judgement making that is characteristic of the human expert. This is not to say that the computer judgement surpasses or is even equal to that of the human expert, since in many ways artificial intelligence will remain artificial. However, these expert systems have an acceptably high degree of accuracy on several important data bases. There is still room for improvement and therefore continued need for the support and development of these methods.

Several approaches have been applied to the elucidation of structural relationships among carcinogens for predictive purposes. These include the Hansch and SIMCA (Soft, Independent Modelling of Class Analogy) methods, which are based upon extrathermodynamic approches, and the ADAPT (Automatic Data Analysis using Pattern Recognition Techniques), CASE (Computer Automated Structure Evaluation) and Enslein procedures, which are based upon connectivity relationships. Altogether, the classification power of these procedures varies from 75 to 96%, depending upon the data base that is available. Each of these methods, excluding the CASE method, is based upon the questions that are asked by the investigator, and, hence, to a large extent, they are influenced by his intuition, previous knowledge and bias. Thus, depending upon the investigator's acumen, the results may be highly predictive for classes of chemicals for which the basic mechanism of action is understood, for example, for polycyclic aromatic hydrocarbons; predictivity will be enhanced if the investigator includes as one of the descriptors the presence or absence of the bay region. This, of course, calls for prior knowledge. The correlations that are derived by these methods involve time-consuming calculations based upon multivariate regression analyses and are expressed as empirical relationships (see below).

CASE is entirely computer-automated and is not influenced by the investigator's bias, since the descriptors are selected automatically by an artificial intelligence system. The predictions are based upon the presence or absence of readily recognizable contiguous linear or linear/branched fragments embedded in the structure of the parent molecule. The predictive power of CASE is dependent upon the nature of the training set that is used to identify the structural descriptors.

2. Basis of structure-activity procedures

The first attempts to correlate biological activity to molecular properties were made at the turn of the century. Meyer (1899) and Overton (1901) both looked to oil/water partition coefficients to explain the potencies of certain nervous-system drugs (Osman et al., 1979). The first really quantitative approach to the treatment of biological activities and molecular structure was that of Collander (1954), who showed that the rate of transport through cellular material was proportional to the log of P (the partition coefficient between an organic solvent and water). This was extended by Hansch et al. (1963) and Hansch and Fujita (1964), who used the Hammett substituent parameter approach to relate biological response rate constants to the electronic effects (as represented by σ constants) of attached functionalities. The following regression equation has become standard:

$$\log(1/C) = a + (b \times \pi_X) + (c \times \pi_X^2) + (d \times E_s) + (e \times MR_X) + (f \times \sigma_X) \qquad (1)$$

where a, b, c, d, e and f are the regression coefficients. The substituent parameter π_X is the log of the ratio of the partition coefficients of substituent X with respect to a hydrogen substituent. The substituent parameter E_s represents the steric effect of substituent X, MR_X the polarizability of X and σ_X the electronic effect of X, all with respect to a hydrogen substituent. There are a number of other substituent parameters (often called descriptors) with more specialized roles. Since the substituent parameters are defined with respect to hydrogen they can be used only for congeneric data bases. As will be seen later, global properties, such as log P, may be used as a parameter in heterogeneous data bases. For a thorough introduction to the Hansch approach, a number of reviews are available (e.g., Hansch, 1978; Osman et al., 1979).

The Hansch method has some other shortcomings. In many cases of interest (especially for some of the new compounds, for which it is most important to establish potential mutagenicity or carcinogenicity), the physical properties of the compounds and their substituents are unknown, or the data representing the biological activity are not quantitative. Most of the strategies that have arisen to get around these problems do have certain features, if not philosophies, in common. The most important of these is that somehow the actual molecular architecture is ultimately responsible for the observed biological (and chemical) properties.

The Free-Wilson method grew out of a direct response to the lack of physical and biological data on compounds of biological interest (Free & Wilson, 1964). The authors were particularly concerned about filling in the gaps in biological testing data to help establish quantitative structure-activity relationships. It was assumed in the original development of their method that the effect of a substituent on biological activity was additive and independent of the presence or absence of other substituents at other positions. The generalized form of the equation they proposed is:

$$BA = A_{ij} * S_{ij} + k, \tag{2}$$

where A_{ij} is the activity contribution of substituent i at site j and the S_{ij} is an indicator variable (given the value 1 if the substituent i is present at site j and 0 if it is not). The constant k is the average value of the biological activity (BA) and is assigned to a hypothetical structure that has neither substituents nor hydrogen in the sites of interest. The equation may then be solved in a least-squares manner. The most important aspect of this work is that it demonstrated that additivity of substituent contributions may be a valid assumption in some situations. Many of the more useful models in this field incorporate this additivity hypothesis in some way.

The above methods were found to be quite useful in correlating biological activity and structure among small, congeneric data bases such as those typically found in drug design, where they are used extensively. The structure-activity relationship methods most useful in the area of environmental science are, however, designed for much larger (hundreds to thousands of compounds), noncongeneric data bases. These methods may be broadly categorized as pattern recognition methods. Pattern recognition methods can be used to identify similarities (patterns) among many compounds of each activity class. The methods basically differ in the way in which they define this similarity, which may be in terms of physical properties, molecular substructures or functionalities, or even the number and type of atoms. In addition, the degree of knowledge called for on the part of the researcher about the systems he is studying may differ greatly. Each of the approaches has had some success in the domains in which they have been applied.

2.1 ADAPT (Automated Data Analysis using Pattern Recognition Techniques)

(a) Description, including pattern recognition statistics

The ADAPT (Automated Data Analysis using Pattern Recognition Techniques) program (Stuper et al., 1977) was one of the first 'no-model' approaches to structure-activity relationships, which is based on the similarity of a test compound to members of a training set. Similarity is defined by a number of descriptors, including such varied quantities as atom type counts, bond type counts, molecular weights and the number of rings and number of ring atoms, which are supplied at the discretion of the researcher and evaluated for usefulness by the statistical procedures described below. Also included are substructure and environment descriptors based on molecular connectivities, partial atomic charges, partition coefficients and geometric descriptors (such as principal (mass) moments). The ADAPT program was one of the first methods that could be applied to relatively large, noncongeneric data bases. Several publications review the method in varying detail (Stuper et al., 1977; Jurs et al., 1978; Stuper et al., 1979).

(b) Statistical analyses

Once descriptor generation is finished, the user of ADAPT must choose those descriptors from the rather extensive list above that will be submitted for statistical analysis. (In one example cited, Jurs pointed out that over 50 parameters were developed for each of 200 compounds in a data-base analysed.) An equation defining the compound as a point in the chosen parameter space may be written

$$X_a = x_1 i + x_2 j \ldots + x_n k, \tag{3}$$

where X_a is the vector of compound a in the parameter space represented by the n x components (descriptors).

The ADAPT program has two kinds of statistical packages, described by Stuper et al. (1977) as parametric and nonparametric. The parametric routines are based on mean vectors, covariance matrices or other statistical measures of the distributions (classes) being separated. The nonparametric routines are based on other measures of similarity, such as the Euclidean distance between compounds in the parameter space. Most of these methods are used to obtain an optimal discriminant function (or 'decision surface', a more colourful term) between two classes of objects (in this case, compounds). In conjunction with these programs is an updating routine which allows the discriminant function to be updated in the light of new data; thus, the program is heuristic. This is an especially important feature in the environmental field. The actual statistical and pattern recognition routines used were obtained from Dixon (1977).

The last step in ADAPT analysis of a data base is feature selection. This is done interactively through several routines using variance selection techniques (Zander et al., 1975).

2.2 The Enslein approach

The Enslein group has developed an approach based on substructure fragments (Enslein & Craig, 1978, 1982; Enslein et al., 1983a,b). The substructure fragments have the same relationship to molecules as words to a sentence, and can be strung together to form molecules.

The 'dictionary' of substructural keys or fragments is obtained from the Wiswesser line notation, as found in the CROSSBOW program (Eakin et al., 1974), with some modifications. The original CROSSBOW set of keys consisted of only 148 substructures; however, Enslein has added others to improve representation and to include certain types of fragments suspected of being related to specific biological activities such as carcinogenesis (Enslein et al., 1983a). The total number of substructural keys at last count was 350.

The model building consists of, as a first step, the cluster analysis procedure of Hall and Khanna (1977). This separates the data base into structured clusters or classes and may indicate outliers. The next step is the choice of substructural keys and other descriptors (which may include physical properties and molecular weight in order of priority), which compels the discriminant analysis routine to accept higher priority features first, such as those which are suspected of being more relevant to activity (such as bay-region). This apparently results in a better quality discriminant equation. Next begins the stepwise discriminant analysis procedure which is the heart of the program, identifying which of the substructure and other descriptors are to be considered important for activity or lack of activity. This set of descriptors, coming out of the discriminant analysis, is then scrutinized, and those descriptors contributing least to the separation of active from inactive compounds are removed one by one until the remaining descriptors meet a predetermined inclusion criterion. The next step is a ridge regression procedure (Marquardt & Snee, 1975), which eliminates descriptors that may cause instabilities in the discriminant equation. As a last step, all the surviving descriptors are submitted to another discriminant analysis. This results in a final set of descriptors and their associated coefficients.

There are two sets of coefficients: one for activating features and one associated with inactivating features. One obtains a projected probability P of activity for a given compound by combining cumulative discriminant scores for the activating features and the discriminant scores for the inactivating features in the following equation:

$$P = \frac{\exp(\text{activating score})}{\exp(\text{inactivating score}) + \exp(\text{activating score})} \quad (4)$$

Confidence levels may be assigned to this estimate on the basis of how many features of the compound are found to be represented in the original data base (training set).

The accuracy of the discriminant analysis may be evaluated in a number of ways. The most formal of the procedures is the two-sample Kolmogorov-Smirnov test (Siegel, 1956). This test is based on the idea that the distribution of the misclassified compounds must conform to a binomial distribution centred on a probability of 0.50. In addition, test compounds of known activity may be submitted to prediction which were not included in the training set.

2.3 The SIMCA method

The SIMCA (one meaning of the acronym is Soft, Independent Modelling of Class Analogy) method of Wold and Sjostrom (1977) is one of a variety of methods employing Hansch-type variables and the pattern recognition method of mathematical analysis. As in all of the methods described, one part of the analysis consists of defining the compounds of a training set as points in a multidimensional space. The vectors representing the points consist of variables thought to be related to the activity of the compounds, such as the Hansch physicochemical variables described above. The mathematical treatment that the SIMCA method uses is the principal component method. The concept of principal component methods is illustrated in Figure 1. In this simple case, we have two classes of 'compounds', with different biological activities. On each axis of the graph, we have one kind of physicochemical property. As in any successful resolution by this method, the two classes of compounds have ended up rather well separated as to class. Within each class, an equation may be developed which fits the activity using the property variables

$$Y_{ik}^q = m_i^q + \sum^A b_{ia}^q * u_{ak}^q + e_{ik}^q, \tag{5}$$

where the Y_{ik}^q are the measured activities for object k in class q, m_i^q and b_{ia}^q define the position and direction of the class (in terms of the property variables), u_{ak}^q defines the position and direction of the particular object k within the class (again, in terms of the property variables) and the e_{ik}^q are the error quantities. The quantity A is the dimensionality needed to locate the objects in the property space (a dimension of two is indicated in Figure 1). This fact, that one obtains an equation relating activity to the parameters that have been put in, makes this method somewhat different from other pattern recognition methods, which are basically classification schemes. SIMCA is also related to factor analysis, using an equation similar to (5), but this does no preliminary classification and works only on the training set as a whole. To summarize, SIMCA first classifies similar compounds on the basis of activity class, then derives an equation for each class describing the data structure within that class.

2.4 The CASE (Computer Automated Structure Evaluation) method

(a) General description

Most of the patterns discussed require a great deal of input from the researcher and hence (from one perspective) a great deal of bias. This is particularly so with regard to

Fig. 1. Illustration of the principal component method of structure-activity relationship analysis as used in conjunction with SIMCA

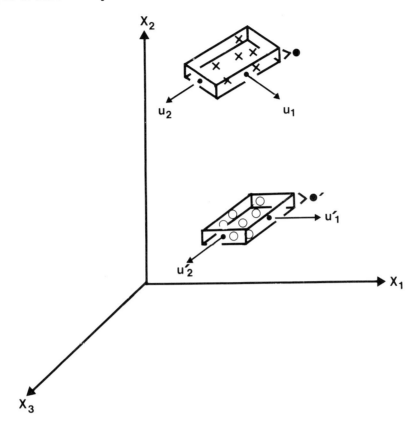

the choice of descriptors. The Computer Automated Structure Evaluator (CASE; Klopman, 1984) is a method that makes only the most basic assumptions about the relationship between molecular structure and biological function. The program was designed to be as automatic as possible and to introduce minimal bias.

(b) Descriptors

The descriptors are based on the substructures (fragments) of the molecules of the training set. The fragments are generated by decomposing each molecule of the data base into all possible pieces, consisting of three to ten contiguously connected heavy atoms together with their attached hydrogens. (Clearly, this generates a number of redundancies, which are dealt with at a later stage of the analysis.) Each fragment also possesses three labels, one defining bond multiplicities, one indicating the presence and position of attached functionalities, such as $-NH_2$, $-COOH$, and the last specifying whether the fragment originated from an active or inactive molecule. In its use of fragment descriptors, CASE is similar to the Enslein approach; however, the Enslein approach possesses a dictionary of fragments, and the CASE method an 'open-ended' set of substructure fragments. By 'open-ended' is meant that no restriction is placed on the choice of fragments representing the activity/inactivity of a compound; this may be *any* fragment in the molecule, as long as it possesses at least three and no more than ten heavy atoms. In the Enslein procedure, the activating and inactivating fragments must come from the dictionary of available fragments, and, while this may be an extensive list, the substructure one needs to predict activity or inactivity may not be included. The Enslein procedure also differs from CASE in the set of statistical routines used to evaluate the importance of any given substructural key and in the representation of those keys. In addition, with Enslein's method, some keys may be marked or given priority to encourage the statistical packages to include certain substructures.

(c) Analysis

Before any further analysis, the redundant fragments are removed. A statistical analysis of the remaining fragment distribution in each of the two pools, inactive and active, is made and a comparison done. Assuming a binomial distribution, any fragment is considered irrelevant to the biological activity of interest if its distribution among the active and inactive fragments is similar to its distribution among the total sample of molecules. A fragment is considered relevant to the biological activity if the probability of its occurrence in the distribution is less than 5% on the basis of a chance (binomial) distribution. Such a fragment retains the activity label of its origin.

(d) Testing new molecules

After the program has been trained for a particular data base, new compounds of unknown activity may be tested. Submitting a structure to analysis (with a key indicating that the activity is unknown) results in comparison of its fragments with the relevant fragments obtained from the training set. If overlaps are found, a probability that the new compound possesses the activity may be calculated on the basis of the probability that the overlapping fragments were relevant to activity in the training set. If the actual activity of this (or another compound) becomes known, the training analysis is performed again with the new data, thus updating and correcting the ability of the program to project. It therefore has the characteristics of a learning machine.

(e) Quantitative aspects

The program can also be used to estimate potencies on the basis of fragments, log P's and log P^2's. Additivity of the contributions of fragments to the activity is assumed:

$$A_m = a_i f_i + \log P_m + \log P_m^2 + k, \tag{6}$$

where A_m is the calculated activity for molecule m, a_i is the activity contribution of fragment i, and f_i is either an indicator variable or the number of occurrences of fragment i in molecule m (the choice of which is a program option). A linear multiple regression analysis is performed with the above equation, using the partial and full F-tests to determine whether a variable should be included or not (Bevington, 1969). Critics of multiple linear regression analysis have pointed out that many of the standard tests of significance do not properly reflect the probability of obtaining chance correlations (especially when 'large' numbers of variables are screened) (Topliss & Edwards, 1979; Wold & Dunn, 1983). The CASE method is especially sensitive to this criticism since it creates and screens independent variables (fragments) that comprise 50-100% of the total number of dependent variables (activities of compounds).

It was found that, as a general rule, the safeguards that are built into the CASE program, i.e., eliminating redundant fragments (colinearity, which is one facet of the problem), accepting into the regression analysis only those fragments highly probably associated with activity, and including partial and full F-tests (Bevington, 1969; Weissberg, 1980), are generally sufficient to prevent chance correlations (Klopman & Kalos, 1985).

3. Relative costs and equipment needed for each procedure

The programs discussed are largely comparable in cost and equipment requirements. The ADAPT program was originally designed to run on the MODCOMP/II/25 16 bit computer but is now available from the scientific software company Molecular Design, Ltd, whose programs can run on several mainframe and minicomputer systems. The CASE program currently runs on the VAX family of Digital Equipment Corporation computers. Both ADAPT and CASE are integrated with programs that allow graphical input of chemical structures. The terminals required (Tektronix brand or Tektronix emulators) for the graphical input of the CASE program are not expensive — about US$ 1000 for the simplest emulator; however, the graphical input, while convenient, is not required for CASE.

4. Applications to carcinogenicity testing

4.1 *Mutagenicity and carcinogenicity studies using ADAPT*

Jurs' group has been one of the most active in studying structure-activity relationships of mutagens and carcinogens. One of their earliest studies (Chou & Jurs, 1979) was done on 153 compounds, 118 carcinogens and 35 noncarcinogens containing an N-nitroso group, using the ADAPT program. In this pattern recognition method, the compounds are defined in a variable space, which is searched for a discriminant separating carcinogens from noncarcinogens. The variables belong to several different classes: topological, geometrical, physiological and electronic. Fifteen descriptors were found useful in discriminating between carcinogens and noncarcinogens in this data base, including six fragment descriptors (numbers of C atoms, N atoms, bonds, aromatic rings, double bonds and basis rings), five molecular connectivity variables, one geometric variable, one environment and two σ charge descriptors. The ADAPT program has a number of options for statistical pattern searching, including quadratic Bayes, linear Bayes, K-nearest neighbor ($K = 3$), iterative least squares linear discriminants, simplex optimization and the linear learning machine. In this instance, the linear learning machine and the iterative least squares routine produced the best separations between carcinogens and noncarcinogens. The overall predictive ability for the linear learning machine was evaluated at 91% (by removal of ten compounds from the 153, analysing the remaining 143, predicting the placement of the excluded ten and repeating this process ten to 15 times to obtain an average percentage correctly predicted). Jurs demurs to generalize his descriptors, saying that they are useful only in context; however, he indicates that the presence of the two electronic descriptors may indicate the importance of α-hydroxylation to the carcinogenic process. This had been implied, for instance, by the experimental work of Camus et al. (1978). In addition, the ratio of large to intermediate principal (mass) moments was retained, suggesting that molecular shape is crucial for carcinogenicity.

Another data base treated by the ADAPT program consisted of 157 aromatic amines (Yuta & Jurs, 1981) tested in rats by oral administration and by injection. Tumours occurred at various organ sites, including breast, ear duct and liver. Less specific tumour sites were combined into one category. The analysis was done according to tumour site and administration protocol; in addition, a mixed data base was generated by combining all results and analysed by ADAPT. The same classes of descriptors and the same statistical evaluations were used as before. The K-nearest neighbour was the least useful discriminant generator, being unable to classify 30-40% of the compounds. The linear discriminant generated by least squares produced the best separation, with 90% correct. Some differences were seen in the quality of the separations, depending on which subset of the data base was being examined. In general, the subsets of mixed biological endpoints were more difficult to separate, and the subset consisting of mixed tumour sites and mixed administration routes was the most difficult to fit. The overall predictive ability was 85-90% for the mixed endpoint subset of the aromatic amine data base.

Some of the more important descriptors in each subset of the data base were the number of basis rings, number of oxygen atoms and molecular connectivity. Jurs considers that the importance of the basis ring suggests that molecular volume may be crucial.

ADAPT was also used to evaluate a chemically diverse data base of 130 carcinogens and 79 noncarcinogens (Jurs et al., 1979). The learning machine was the best classifier, correctly placing approximately 90% of the original total data base. When 17 misclassified compounds were removed, the learning machine required 26 descriptors to correctly place the remaining 192 compounds. The predictive ability for this reduced data base was 85%; the predictive ability among carcinogens was 90%, and that for noncarcinogens, 78%.

A portion of the data base collated by Dipple (1976), consisting of 200 polycyclic aromatic hydrocarbons, was also analysed with ADAPT (Yuan & Jurs, 1980). The descriptors, from each of the descriptor categories, included the substructure representing the bay-region (it should be recalled that the substructures are entered by the researcher before the analysis begins), log P, descriptors representing the area of the molecules and their volume and shape, as well as the number of rings in the structure. Electronic descriptors included that of the average σ charge of the substructure representing the bay-region.

Yuan and Jurs were able to obtain a decision surface (discriminant) of 28 terms, which correctly separated 191 (96%) of the 200 compounds into active and inactive classes. In predictive trials, the same decision surface correctly identified approximately 90% of the compounds as carcinogens or noncarcinogens.

An extended version of the N-nitroso data base previously reported by the Jurs group was recently evaluated (Rose & Jurs, 1982), consisting of 150 compounds (112 carcinogens and 38 noncarcinogens) from the literature. The usual descriptors were used, with the addition of one new additional descriptor: a symmetry parameter. The symmetry parameter is defined as the ratio of the number of uniquely connected atoms over the total number of connected atoms. Six substructures were chosen, with the expectation that positions α and β to the N-nitroso group might be especially important in biological α-hydroxylation of N-nitroso compounds (Rose & Jurs, 1982).

With 15 descriptors, 93% (140/150) of the compounds were correctly placed as carcinogens or noncarcinogens. The percentage correctly expected on the basis of guessing may be calculated for comparison:

$$P = [(a/T)^2 + (b/T)^2], \tag{7}$$

where a is the number of compounds in one category, b is the number of compounds in the other category and T is the total number of compounds. The value of P for this data base distribution is 62%. The overall predictive ability has been evaluated by Rose and Jurs to be about 86%, and that for carcinogens only, slightly higher, at 90%.

For this data base, Rose and Jurs also explained which descriptors accounted for how many of the correctly placed compounds. The most positive charge of certain substructures accounted for approximately 13% of the observed separations. This descriptor was reported to be important not only as an indicator of reactivity but also as a determinant of transport properties. The molecular size and shape descriptors (basis rings, principal moments and their ratios) accounted for nearly 20% of the separability. The four substructures representing the α and β positions to the N-nitroso group were able to separate 97% of the data, again supporting the importance of these positions in activation to ultimate carcinogenic species.

4.2 Carcinogenicity studies utilizing SIMCA

Only a few studies have utilized SIMCA in the analysis of carcinogenic data bases. Dunn and Wold (1981) looked at the carcinogenic activity of 61 substituted N-nitroso compounds, comprising 50 carcinogens and 11 noncarcinogens. The variables used in the initial principal component analysis were Rekker's f value, MR, σ^*, E_s and the steric parameters L (length of the substituent) and B_4 (width of substituent perpendicular to L and at the widest point) of Verloop et al. (1976). Log P of the molecules was also used. This step failed to indicate any structure for the group as a whole, indicating that the set does not represent a single class. The next step was to produce a class or classes independently showing some structure. The inactive compounds were assumed to have no structure and were no longer treated. The active compounds were divided into three somewhat arbitrary classes. The first division was accomplished by considering which of the active compounds required metabolic activation to express their carcinogenic properties and those which did not. The set requiring activation was further broken down into two more sets: one that would be expected to yield a cation on activation and the other a diazoalkane. These groups did exhibit some structure, that is, their activity could be described by an equation like (5) within their class. In fact, for $-2 < u_1 < 0$ and $-2 < u_2 < 0$, there were two separated clusters of compounds, which had some biological significance: one cluster represented compounds inducing tumours in the oesophagus, liver and nasal turbinates, and the other cluster consisted of compounds causing tumours only in the liver. (All compounds were administered orally.) In the case of compounds causing oesophageal cancer, σ^* was an important physicochemical factor, electron withdrawal or electron neutral substituents having greater activity.

Dunn and Wold (1978) also evaluated the carcinogenic potential of the 4-nitroquinoline-1-oxides. Again, Hansch-like variables were used. The noncarcinogens had no internal structure as a whole. Four components ($A = 4$ in equation (5)) were used to describe the variation in the active class.

The SIMCA method was also used to analyse class structure among 32 polycyclic aromatic hydrocarbons (Norden et al., 1978). A mixture of theoretical and experimentally measured variables was used in the analysis. The theoretical variables included the Pullman indices for bond localization and carbon localization energies for the K and L regions (Pullman & Pullman, 1955).

4.3 Application of the Enslein method to carcinogenicity and mutagenicity studies

Enslein and Craig (1982) used their methods on a data base collected from the *IARC Monographs on the Evaluation of the Carcinogenic Risk of Chemicals to Humans*, Volumes 1-17 (IARC, 1972-1978). They used 343 of the compounds available from this data base: 120 noncarcinogens and 223 carcinogens. (Although IARC has defined six categories of degree of evidence for carcinogenicity, Enslein and Craig combined noncarcinogens with indefinite human carcinogens and combined definite human and animal carcinogens.) The data base was chemically quite diverse. The descriptors were substructural fragments represented by WLN codes, molecular weight and the molar refractivity. Some additional fragment descriptors (keys) were employed that were derived from known carcinogens or were required to code special functions.

The discriminant equation, utilizing 79 descriptors (all but one, the molecular weight, of a substructural type), correctly classified 87-91% of the carcinogens and 78-80% of the noncarcinogens. The indeterminancy comes about because the developed discriminant equation defines only a probability of a compound being a carcinogen, resulting in regions around 50% that are undefinable in terms of whether a compound is a carcinogen or not. Enslein and Craig did not attempt in this study to evaluate a test set of compounds that were not in their original data base; however, they pointed out that compounds that do not possess features of the training set are likely to be misclassified. For this reason, eight compounds were 'strongly misclassified', meaning they were given high probabilities of belonging to a class by the discriminant equation, but that this predicted class was incorrect. One, 2,4-diaminotoluene, was misclassified in every model that the Enslein group tried. Enslein and Craig suggest that this compound be re-examined experimentally.

The many relatively unranked descriptors utilized in this study do not lend themselves to a mechanistic interpretation. Indeed, Enslein and Craig do not recommend trying to interpret their results in this way. The discriminant equation derived is for purposes of classification only and only within the chemical space represented by the compounds of the data base.

The Enslein group also looked at a data base of 532 mutagens and nonmutagens (Enslein *et al.*, 1983a). The data came from two sources: the Environmental Mutagenesis Information Center (EMIC), Oak Ridge, Tennessee, USA and the US National Toxicology Program (NTP). Since these sources report mutagenic activity in a number of different tests, some criteria had to be developed to define active and inactive classes for use in the discriminant analysis. Mutagens were defined as those compounds that gave at least one positive response (with or without S9 activation) in the *Salmonella typhimurium* strains tested (TA98, TA100, TA1535, TA1537 and TA1538). A compound was defined as a nonmutagen when it gave negative results in the three strains TA1535, TA1537 and TA1538. This results in 301 mutagens and 231 nonmutagens. Of these, 264 of the mutagens and 208 of the nonmutagens were used in developing the discriminant equation, and 37 mutagens and 23 nonmutagens were held back for evaluating the model.

The substructural fragments were the CROSSBOW WLN-based keys discussed above (Eakin *et al.*, 1974) plus the additional keys developed by Enslein to better represent his data bases and to provide substructural keys that were thought to be related to carcinogenic activity, which, however, were not further identified (Enslein *et al.*, 1983a). Molecular weights were also included in this data base as descriptors.

The model building routine was as previously described. The discriminant equation ultimately derived was able correctly to classify 86% of the mutagens in the data base. Classification of the set of test mutagens was about 80% correct. Although 20 descriptors were described as being most important, Enslein again cautioned against mechanistic interpretation, since apparently their ranking could change greatly with changes in the data base, and the fluctuations in descriptor importance could be damped to near zero by the inclusion of approximately 1500 compounds in the data base. This number was estimated from results in a rat oral LD_{50} data base treated by this method (Enslein *et al.*, 1983b).

One systematic misclassification was the class of mutagens known as intercalators. This is easily explained by consideration of the types of descriptors used in the Enslein method. This particular class of mutagens is probably strongly dependent on topological features and possibly on the ability to bind hydrogen (Enslein *et al.*, 1983a). The coded substructures are expressed in two dimensions and therefore would not be expected to provide much information of a topological nature.

4.4 *CASE analyses related to mutagenicity and carcinogenicity*

The CASE program has shown itself useful in the analysis of both carcinogenicity and mutagenicity data. One of the first CASE analyses of a mutagenicity data base looked at 53 nitroarenes tested in *S. typhimurium* TA98 in the absence of S9 activation (Rosenkranz & Mermelstein, 1983). There were 26 compounds classified as mutagens, 22 as nonmutagens and five as marginal. The CASE analysis was able correctly to classify 89% of these compounds on the basis of structural information alone. Structural fragments known to be involved in mutagenesis as well as some interesting new ones were identified (Klopman & Rosenkranz, 1983). Some of the fragments found to be associated with mutagenicity in that study are shown in Figure 2. Of particular interest are fragments 1 and 3. Fragment 1 indicates that there must be continuation of aromaticity *para* to the nitro substituent; however, if this position is unsubstituted, i.e., a hydrogen is present, the resulting 'biophobe' is actually deactivating.

The compounds in this data base that were not correctly classified were mainly those containing more than five rings. The implication is that, although these compounds may contain some of the same activating features as the smaller ring structures, their bulk may prevent them from crossing cell membranes.

Another study looked at 56 polycyclic aromatic hydrocarbons: 17 carcinogenic, 33 noncarcinogenic and six only marginally carcinogenic or indeterminate (Klopman *et al.*, 1985a), from the literature (Arcos & Argus, 1968). The analysis identified three fragments that had >99% chance of being associated with the carcinogenicity (or lack of) for the training set (see Fig. 3). In spite of the fact that the program received no information other than the structures and activities of the compounds, these fragments

Fig. 2. Fragments identified as especially important in determining mutagenic properties of 53 nitroarenes tested against *Salmonella typhimurium* strain TA98 without S9 activation. Fragments 1 and 2 activate mutagenicity, while 3 and 4 are inactivating.

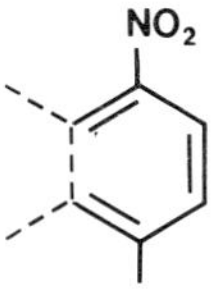

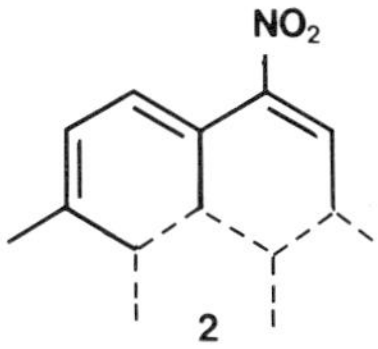

represent essential features of the bay region, 'L' region (Pullman & Pullman, 1955; Pullman, 1964) and pseudo-bay region (Perin-Roussel *et al.*, 1984). Of these compounds, 86% were correctly placed in their respective activity categories. Essentially the same success rate was obtained for another polycyclic aromatic hydrocarbon data base (Klopman, 1984).

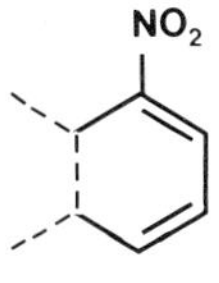

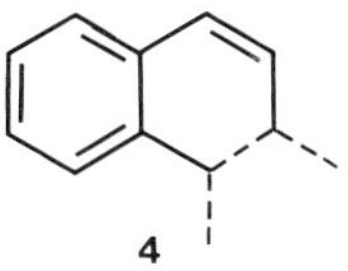

Fig. 3. Fragments identified as being important to carcinogenicity in a polycyclic aromatic hydrocarbon data base of 56 compounds and CASE analysis. Fragment 1 activates carcinogenicity and is representative of the 'bay-region'; fragment 2 is also activating and is representative of the 'pseudo-bay-region'; while fragment 3 is inactivating and is representative of the L-region.

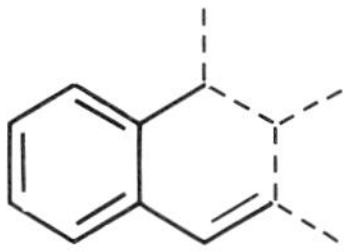

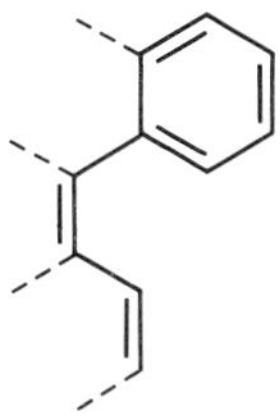

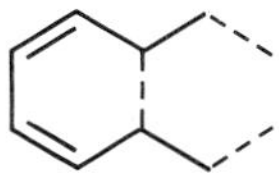

Mitchell et al. (1986) evaluated substituted polycyclic aromatic hydrocarbons that had been tested for both carcinogenic and mutagenic activity with the CASE program. Upon comparison of the results, few overlapping fragments were seen, confirming previous observations that, for instance, the bay region is important in carcinogenesis but much less so in mutagenesis.

Aromatic amines tested in *S. typhimurium* strains TA98 and TA100 with S9 activation have also been analysed by CASE (Klopman et al., 1985b). The compounds studied were monocyclic and polycyclic aromatic amines; 80 were tested in TA98 and 107 in TA100. CASE fragment analysis showed that (1) substitution of the nitrogen functionality by alkyl groups tended to be a deactivating feature in both the TA98 and TA100 data bases, and (2) the deactivating fragment $-CH_2CH_2OH$ decreased activity whether attached to a nitrogen or not. Within the broad classifications of mutagen or nonmutagen, CASE placed 88% of the compounds of the TA98 data base and 84% of the compounds in the TA100 data base. However, since the data were entered in a semiquantitative form, a fragment-based QSAR was possible. The results of this analysis are shown in the 'truth matrix' summary of Table 1. It is clear that, ideally, all of the off-diagonal elements should be zero; most are fairly small.

Table 1. Summary of results calculated by CASE and of experimental results for amines tested in *S. typhimurium* with S9

Experimental	Calculated					
	-	+	++	+++	++++	Total
S. typhimurium TA98						
-	28	6	0	0	0	34
+	5	2	1	0	0	8
++	0	9	2	2	1	14
+++	1	5	1	4	1	12
++++	0	0	0	3	14	17
Total	34	22	4	9	16	85
S. typhimurium TA100						
-	39	1	0	2	1	43
+	6	0	0	2	0	8
++	5	0	0	2	3	10
+++	0	0	0	5	6	11
++++	2	0	0	5	28	35
Total	52	1	0	16	38	107

Finally, for a set of 19 compounds tested in the TA98 strain, 84% were placed correctly within the general categories of mutagens and nonmutagens.

A pesticide data base has also been analysed by CASE (Klopman et al., 1985c). The 54 compounds examined had been tested in five short-term assays: the *S. typhimurium* mutagenicity assay (SAL), a specific gene mutation assay using mouse lymphoma L5178Y cells (L5T), prokaryotic differential survival assays using, respectively, pairs of *Escherichia coli* $polA^+$ and $polA_1^-$ (REP) and *Bacillus subtilis* rec^+ and rec^- (REW) and mutagenicity in the yeast *Saccharomyces cerevisiae* (YE3). The analysis of the five tests is depicted in Figure 4. Fragments 1, 2 and 4 were found to be activating in compounds tested in SAL, YE3 and L5T; fragments 5 and 6 were inactivating in compounds in the same tests. These findings indicate the possibility of a mechanistic relationship between mutagenicity in a prokaryote (*S. typhimurium*), recombinogenicity (i.e., DNA damage) in a eukaryote *(S. cerevisiae)* and mutagenicity in mammalian cells (L5T). There was not much overlap between fragments on the analysis and comparison of the REP and REW data bases. Fragments 2, 7 and 8 were activating and fragment 9 was inactivating in the REW data base. In the REP data base, there were three activating fragments: 2, 10 and 11. The differences between the results of REP and REW were imputed to differences in cell-wall structure between the two organisms (*E. coli* being gram-negative and *B. subtilis* gram-positive). Gram-negative organisms have a cell wall that facilitates passage of lipophilic molecules, while gram-positive organisms have a cell wall that facilitates passage of hydrophilic species.

A comparison across the battery of five tests showed fragment 2 as activating in all the tests. There was no other overlap between REP and REW and the other three tests, consistent with a classification scheme of Waters et al. (1982).

The Jurs data base of carcinogens (Yuan & Jurs, 1980) was submitted to CASE analysis, since it was thought that a direct comparison of these two methods might be informative. Jurs was able correctly to classify 95% of the original 200 polycyclic aromatic hydrocarbons using 28 descriptors. The latest version of CASE treating the same 200 compounds in a semiquantitative fashion correctly classified 89% (Klopman, G. & Frierson, M.R., unpublished data), using less than half as many descriptors as in the ADAPT program. The discrepancy may be due to the fact that investigation of the carcinogenicity of polycyclic aromatic hydrocarbons is a very well-developed area of studies of structure-activity relationships. The ADAPT program takes advantage of this knowledge in its design and can accept almost any descriptor one may wish to include. In this particular case, information about the bay-region and charge densities around the periphery of the compounds, both of which have been correlated to carcinogenicity, were included (Yuan & Jurs, 1980). The CASE program designs its classification scheme without any input or knowledge from the researcher apart from the structure of the molecule and its biological activity. Clearly, this may be a disadvantage in cases that have been well studied for structure-activity relationships, but it may be a distinct advantage in those cases where little is known about what makes a structure biologically active.

Fig. 4. Fragments identified as being important in the CASE analysis of data bases consisting of experimental testing results in the following five testing procedures: the *Salmonella typhimurium* mutagenicity assay (**SAL**), a specific gene mutation assay using mouse lymphoma L5178Y cells (**L5T**), prokaryotic DNA repair assays using, respectively, pairs of *Escherichia coli polA*$^+$ and $polA_1-$ (**REP**) and *Bacillus subtilis rec*$^+$ and *rec*$^-$ (**REW**) and mutagenicity in the yeast *Saccharomyces cervisiae* (**YE3**).

Details of the relationships between tests and fragments are given in the text.

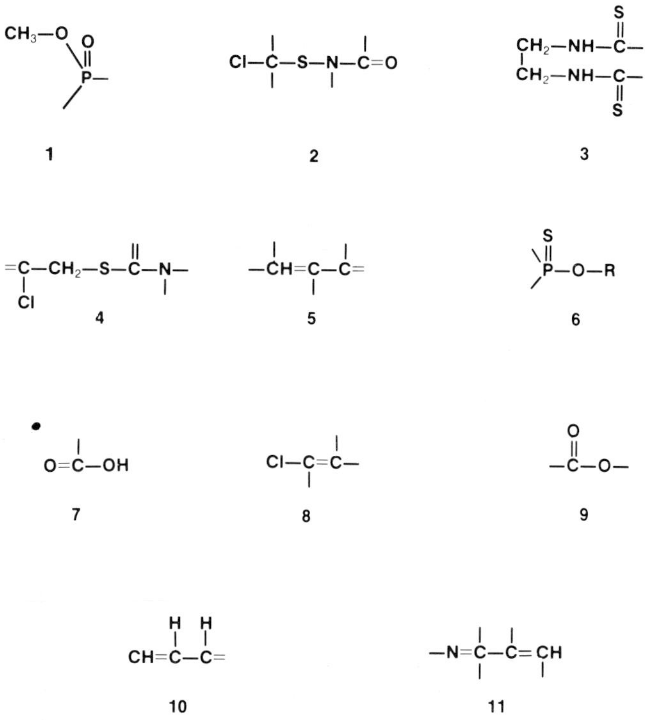

5. Future developments and summary

Possibly, one of the greatest shortcomings of all of the present methods for relating chemical structure quantitatively to biological activity is that the tools for dealing with the three-dimensional nature of biologically active molecules are still very crude. This is important, since biological specificity is predicated on the shape and points of interaction of molecules, including the handedness of the molecules and is a particularly important facet in evaluating potentially harmful effects of chemicals that interact with DNA — for example, the powerful intercalating agents. Many laboratories are hard at work to improve this situation.

While the relative merits of the programs described above have not been fully evaluated, the CASE program has, in principle, certain attractive features. The most important is its ability to develop descriptors automatically on the basis of biological testing results and chemical structure, in the absence of any other specialized knowledge on the part of the researcher. The fact that the resulting descriptors are based largely on molecular fragments makes the results entirely sensible to scientists of many disciplines — important in an interdisciplinary field like environmental health sciences.

Irrespective of the choice of an expert system in this area (if that is indeed important) is the continued growth and health of this newly developing theoretical side to the environmental health sciences.

6. References

Arcos, J.C. & Argus, M.F. (1968) Molecular geometry and carcinogenic activity of aromatic compounds: new perspectives. *Adv. Cancer Res.*, *11*, 305-471

Bevington, P.R. (1969) *Data Reduction and Error Analysis for the Physical Sciences*, New York, McGraw-Hill, pp. 164-203

Camus, A.M., Wiessler, M., Malaveille, C. & Bartsch, H. (1978) High mutagenicity of N-(α-aryloxy)alkyl-N-alkyl-N-alkylnitrosamines in *S. typhimurium*: model compounds for metabolically activated N,N-dialkylnitrosoamines. *Mutat. Res.*, *49*, 187-194

Chou, J.T. & Jurs, P.C. (1979) Computer-assisted structure-activity studies of chemical carcinogens. An N-nitroso compound data set. *J. med. Chem.*, *22*, 792-797

Collander R. (1954) The permeability of Nitella cells to nonelectrolytes. *Physiol. plant.*, *7*, 420-445

Dipple, A. (1976) *Polynuclear aromatic carcinogens*. In: Searle, C.E. ed., *Chemical Carcinogens* (*ACS Monograph 173*), Washington DC, American Chemical Society, pp. 245-314

Dixon, W.J., ed. (1977) *BMD Biomedical Computer Programs* (*BMDP7M, Stepwise Discriminant Analysis*), Berkeley, University of California Press, p. 233

Dunn, W.J., III & Wold, S. (1978) A structure-carcinogenicity study of 4-nitroquinoline-1-oxides using the SIMCA method of pattern recognition. *J. med. Chem.*, *21*, 1001-1007

Dunn, W.J., III & Wold, S. (1981) The carcinogenicity of N-nitroso compounds: a SIMCA pattern-recognition study. *Biorg. Chem.*, *10*, 29-45

Eakin, D.R., Hyde, E. & Parker, G. (1974) Use of computers with chemical structural information —ICI (Imperial Chemical Industries, Ltd) CROSSBOW system. *Pestic. Sci.*, *5*, 319-326

Enslein, K. & Craig, P.N. (1978) A toxicity estimation model. *J. environ. Pathol. Toxicol.*, *2*, 115-132

Enslein, K. & Craig, P.M. (1982) Carcinogenesis: a predictive structure-activity model. *J. Toxicol. environ. Health*, *10*, 521-530

Enslein, K., Lander, T.R. & Strange, J.R. (1983a) Teratogenesis: a statistical structure-activity model. *Teratog. Carcinog. Mutagen.*, *3*, 429-438

Enslein, K., Lander, T.R., Tomb, M.E. & Landis, W.G. (1983b) Mutagenicity (Ames): structure-activity model. *Teratog. Carcinog. Mutagen.*, *3*, 503-513

Free, S.M. & Wilson, J.W. (1964) A mathematical contribution to structure-activity studies. *J. med. Chem.*, 7, 395-399

Hall, D.J. & Khanna (1977) *The ISODATA method*. In: Enslein, K., Wilf, H.S. & Ralston, A., eds, *Statistical Methods for Digital Computers*, New York, Wiley

Hansch, C. (1978) *Recent advances in biochemical QSAR*. In: Chapman, N.B., & Shorter, J., eds, *Correlation Analysis in Chemistry*, New York, Plenum

Hansch, C. & Fujita, T. (1964) ρ - σ - π analysis. A method for the correlation of biological activity and chemical structure. *J. Am. chem. Soc.*, 86, 1616-1626

Hansch, C., Muir, R.M., Fujita, T., Maloney, P.O., Geiger, C.F. & Streich, M.J. (1963) The correlation of biological activity of plant growth regulators and chloromycetin derivatives with Hammett constants and partition coefficients. *J. Am. chem. Soc.*, 85, 2817-2824

IARC (1972-1978) *IARC Monographs on the Evaluation of the Carcinogenic Risk of Chemicals to Humans*, Vols 1-17, Lyon

Jurs, P.C., Chou, J.T. & Yuan, M. (1978) *Studies of chemical structure-biological activity relations using pattern recognition*. In: Olson, E.C. & Christoffersen, R.E., eds, *Computer Assisted Drug Design (ACS Symposium Series 112)*, Washington DC, American Chemical Society, pp. 103-129

Jurs, P.C., Chou, J.R. & Yuan, M. (1979) Computer-assisted structure-activity studies of chemical carcinogens. A heterogeneous data set. *J. med. Chem.*, 22, 476-483

Klopman, G. (1984) Artificial intelligence approach to structure-activity studies. Computer automated structure evaluation of biological activity of organic molecules. *J. Am. chem. Soc.*, 106, 7315-7321

Klopman, G. & Kalos, A. (1985) Causality in structure-activity studies. *J. Comput. Chem.*, 6, 492-506

Klopman, G. & Rosenkranz, H.S. (1983) Structural requirements for the mutagenicity of environmental nitroarenes. *Mutat. Res.*, 126, 227-238

Klopman, G., Namboodiri, K. & Kalos, A. (1985a) *Computer automated evaluation and prediction of the Iball index of carcinogenicity of polycyclic aromatic hydrocarbons*. In: Rein, R., ed., *Molecular Basis of Cancer*, Part A, *Macromolecular Structure, Carcinogens and Oncogenes*, New York, Alan R. Liss, pp. 287-298

Klopman, G., Contreras, R., Rosenkranz, H.S. & Waters, M.D. (1985b) Structure-genotoxic activity relationship of pesticides: comparison between the results of several short-term assays. *Mutat. Res.*, 147, 343-356

Klopman, G., Frierson, M.R. & Rosenkranz, H.S. (1985c) Computer analysis of toxicological databases: mutagenicity of aromatic amines in *Salmonella* tester strains. *Environ. Mutagen.*, 7, 625-653

Marquardt, D.W. & Snee, R.D. (1975) Ridge regression in practice. *Am. Stat.*, 29, 3-20

Meyer, H. (1899) On the theory of alcohol-induced narcosis. I. Which properties of anaesthetics cause narcotic effects? [in German]. *Arch. exp. Pathol. Pharmakol.*, 42, 110-118

Mitchell, C.S., Klopman, G. & Rosenkranz, H.S. (1986) *Computer automated evaluation of mutagenicity and carcinogenicity of selected polycylic aromatic hydrocarbons*. In: *Polycyclic Aromatic Hydrocarbons (9th International Symposium)*, Columbus, OH, Battelle Press (in press)

Norden, B., Edlund, U. & Wold, S. (1978) Carcinogenicity of polycyclic aromatic hydrocarbons, studied by SIMCA pattern recognition. *Acta chem. Scand., Ser. B.*, *32*, 602-608

Osman, R., Weinstien, H. & Green, J.P. (1979) *Parameters and methods in quantitative structure activity relationships.* In: Olson, E.C. & Christoffersen, R.E., eds, *Computer Assisted Drug Design (ACS Symposium Series 112)*, Washington DC, American Chemical Society, pp. 21-77

Overton, E. (1901) *Studies on Narcosis* [in German], Jena, Fischer

Perin-Roussel, O., Ekert, B., Barat, N. & Zajdela, F. (1984) DNA-protein crosslinks induced by exposure of cultured mouse fibroblasts to dibenzo[*a,e,*]fluoranthene and its bay and pseudo-bay region dihydrobiols. *Carcinogenesis*, *5*, 379

Pullman, A. & Pullman, B. (1955) Electronic structure and carcinogenic activity of some aromatic molecules. *Adv. Cancer Res.*, *3*, 117-169

Pullman, B. (1964) Electronic aspects of the interactions between the carcinogens and possible cellular sites of their activity. *J. cell. comp. Physiol.*, *64* (Suppl. 1), 109

Rose, S.L. & Jurs, P.C. (1982) Computer-assisted studies of structure-activity relationships of *N*-nitroso compounds using pattern recognition. *J. med. Chem.*, *25*, 769-776

Rosenkranz, H.S. & Mermelstein, R. (1983) Mutagenicity and genotoxicity of nitroarenes: all nitro-containing chemicals were not created equal. *Mutat. Res.*, *114*, 217-267

Siegel, S. (1956) *Non-parametric Statistics for the Behavioral Sciences*, New York, McGraw-Hill, pp. 127-136

Stuper, A.J., Brugger, W.E. & Jurs, P.C. (1977) *A computer system for structure-activity studies using chemical structure information handling and pattern recognition techniques.* In: Kowalski, B.R., ed., *Chemometrics: Theory and Application (ACS Symposium Series 52)*, Washington DC, American Chemical Society, pp. 165-191

Stuper, A.J., Brugger, W.E. & Jurs, P.C., eds (1979) *Computer-assisted Studies of Chemical Structure and Biological Function*, New York, Wiley

Topliss, J.G. & Edwards, R.P. (1979) Chance factors in studies of quantitative-structure activity relationships. *J. med. Chem.*, *22*, 1238-1244

Verloop, A., Hoogenstraaten, W. & Tipker, J. (1976) *Development and application of new steric parameters in drug design.* In: Ariens, E.J., ed., *Drug Design*, Vol. 7, New York, Academic Press, pp. 165-207

Waters, M.D., Sandhu, S.S., Simmon V.F., Mortelmans K.E., Mitchell A., Jorgenson, T.A., Jones, D.C.L., Valencia, R. & Garrett, N.E. (1982) *Study of pesticide genotoxicity.* In: Fleck, R.A. & Hollaender, A., eds, *Genetic Toxicology*, New York, Plenum, pp. 275-326

Weissberg, S. (1980) *Applied Linear Regression*, New York, Wiley

Wold, S. & Dunn, W.J. (1983) Multivariate quantitative structure-activity relationships (QSAR): conditions for their applicability. *J. chem. Inf. comp. Sci.*, *23*, 6-13

Wold, S. & Sjostrom, M. (1977) *SIMCA: A method for analyzing chemical data in terms of similarity and analogy.* In: Kowalski, B.R., ed., *Chemometrics: Theory and Application (ACS Symposium Series 52)*, Washington DC, American Chemical Society, pp. 243-282

Yuan, M. & Jurs, C. (1980) Computer-assisted structure-activity studies of chemical carcinogens: a polycyclic aromatic hydrocarbon data set. *Toxicol. appl. Pharmacol.*, *52*, 294-312

Yuta, K. & Jurs, P.C. (1981) Computer-assisted structure-activity studies of chemical carcinogens. Aromatic amines. *J. med. Chem.*, *24*, 241-251

Zander, G.S., Stuper, A.J. & Jurs, P.C. (1975) Non-parametric features selection in pattern recognition applied to chemical problems. *Anal. Chem.*, *47*, 1085-1093

GENETIC TOXICOLOGY AT THE CROSSROADS: A PERSONAL OVERVIEW OF THE DEPLOYMENT OF SHORT-TERM TESTS

B.A. Bridges

*MRC Cell Mutation Unit, University of Sussex,
Falmer, Brighton, Sussex BN1 9RR, UK*

The contents of this book are an impressive testimony to the massive investment in genetic toxicology over the last 15 years. During this time, short-term tests, usually based upon mutagenicity or some other response to DNA damage, have proved of considerable value in the detection of possible carcinogens, particularly in complex mixtures such as dietary components and vehicle exhausts. They are currently being integrated into epidemiological studies with a view to identifying the causes of known human hazards (e.g., Friesen *et al.*, 1985; Nair *et al.*, 1985) and are proving valuable in exploring the organ specificity of chemical carcinogens. Yet, for all the progress that has been made, the deployment of these tests to predict the carcinogenic potential of individual chemicals in a regulatory context is still confused. It seems important that the profession should get to grips with the problems and resolve them before confusion becomes widespread.

Short-term tests can also be used to detect potential germ cell mutagens, and the problems arising in that connection have been discussed elsewhere (Bridges & Mendelsohn, 1986). Here, however, we are concerned with the objective of predicting whether or not a substance has the ability to cause cancer in mammals — in particular, humans. In the early days, it seemed that prediction could be accurate 90% of the time; but, since then, accuracy has declined, and the most recent study indicated that in the Ames *Salmonella* assay, the predictivity of a mutagenic response for carcinogenicity was 75% and that of a nonmutagenic response for noncarcinogenicity, 48% (Zeiger & Tennant, 1986). In this study, results for 210 compounds tested according to standard protocols in the *Salmonella* assay and in two mammalian cell assays were compared with long-term bioassay results from programmes of the US National Cancer Institute and US National Toxicology Program. If the performance of short-term tests is really so poor, it could be argued on the one hand that they have little practical value. On the other hand, short-term tests have already proved their worth in many specific instances and have gained the confidence of many scientists who understand their potential and their limitations.

Most of the problems started to arise when we began to think of these studies as routine tests to be conducted under defined protocols, rather than as experiments designed to test specific hypotheses. Once we had jumped this barrier, we were no longer obliged to think about what we were doing in a scientific manner and could focus our attention solely on details of protocol and statistical analysis. Although I do not dispute the importance of these factors, their overemphasis has sometimes led to a neglect of the importance of mechanistic understanding. This trend must be reversed if the proper place of short-term tests is to be realised. It is now abundantly clear that an evaluation of the potential carcinogenicity of an agent requires an integrated assessment of all available information, starting from first principles. A checklist approach (i.e., an approach which involves the accumulation of a certain number of positive results, regardless of what they mean in a specific instance) is a recipe for disaster. In the same way, the choice of short-term tests in a regulatory context must be founded on basic scientific principles and may not necessarily be the same for each agent. Let us consider some of the factors that must be taken into account.

What is a carcinogen?

If carcinogenicity is to be successfully predicted by short-term tests, one must first ask how well the carcinogenicity of a substance can be determined for the purpose of a validation standard. Here we come up against the problem that there are clear species differences in carcinogenicity; in other words, 'carcinogens' do not appear to express their activity in every species.

What this probably means in most cases is that potency varies from one species to another and that in some cases it falls below the level of detectability in a standard bioassay. An excellent example is isoniazid, which is reproducibly carcinogenic in mice but not in hamsters and probably not in rats. The likely explanation seems to be that, in mice, the metabolism is such as to produce a high level of hydrazine, which is a known mutagen. In other species, the level is lower and the carcinogenic effect is at or below the level at which it can be measured in a bioassay (for discussion, see Jansen *et al.*, 1980). In instances such as this, it is usual to find that metabolic differences between species are quantitative rather than qualitative, although their magnitude is enough to affect the probability of detection in a standard carcinogenicity bioassay.

In the data base reviewed by Zeiger and Tennant (1986), of those substances carcinogenic to mice or rats only 50% were carcinogenic to both. The predictivity of mice for a substance carcinogenic to rats was 64%, and that of rats for a substance carcinogenic to mice, 69%. This result has several profound implications for the field of carcinogenesis and risk assessment, but one in particular is relevant to short-term testing: assuming that predictability between rodents and nonrodents is similar to that between mice and rats, a significant number of real carcinogens is likely to be missed by using only mice and rats. A significant number of apparent 'false'-positive results in short-term tests would therefore probably be found to be true if bioassays were carried out in more animal species.

In view of the importance of this point, it seems essential to subject it to experimental investigation. This could be done by identifying, say, a dozen substances for which there is sufficient evidence of mutagenicity *in vivo* in rats or mice and subjecting them to long-term bioassays in one or more additional species. Work should also be carried out to determine why the mutagenic effect in rats or mice was not sufficient to permit detection of carcinogenic activity in those species. Mendelsohn (1986) has recently discussed the problem of prediction of carcinogenicity by short-term tests and has itemized numerous possible explanations for the discrepancies.

The fact that many carcinogens appear not to express their activity in all species also has profound implications for the extrapolation of effects from laboratory animals to humans, as must be done in the process of risk assessment. To dwell further on this aspect is outside the present context, but it must not be overlooked in the overall problem of prediction of human carcinogenic risk.

There are different types of carcinogen

It is unwise to demand more of any test than it is capable of achieving. Short-term tests that are based upon mutagenicity and DNA damage cannot detect (except by chance) substances that are carcinogenic by virtue of other modes of action. It is also clear that other modes of action besides mutation are able to increase tumour incidence in bioassays. Tumour promotion is one (see Report 10); although promoters are in principle noncarcinogenic by definition, if the investigator works hard enough individual promoters can often be shown to increase tumour incidence, presumably because they act on rare, spontaneously initiated cells. Other types of carcinogenesis have been termed 'epigenetic' and 'hormonal'; agents thought to act in these ways are generally inactive in mutagenicity assays. A fourth type of carcinogen is claimed to be active only at near-toxic doses, at which the homeostasis of the animals is disturbed. The involvement of genotoxicity has to be rather precisely defined where these particular substances are concerned. Damage to genetic material may indeed occur at near-toxic doses and may even be involved in carcinogenesis under such conditions. Genotoxicity may also be detectable at high doses in mammalian cells and other systems, presumably for related reasons (see, for example, Cole *et al.*, 1986). It is important to note, however, that such agents may not show stochastic mutagenic activity typical of classical genotoxic agents and might therefore be effectively noncarcinogenic at subtoxic dose levels. (For further discussion of nongenotoxic carcinogens, see IARC, 1983; Upton *et al.*, 1984; Ashby, 1986.) There is a real need for further work on substances in these categories, and it should not be assumed that they can be as easily identified as the foregoing paragraph may suggest.

The various modes of carcinogenicity are of course not mutually exclusive, and it may well emerge that substances able to act in more than one mode thereby have greatly enhanced potency. From the point of view of the validation of short-term tests, however, it may be seen that a collection of animal bioassay results identifies a mixed group of carcinogens and does not constitute a valid data base against which to compare the results of short-term tests that are inherently specific for genotoxic carcinogens. The validation of short-term genotoxicity tests can, however, now be approached in another way.

The validity of short-term mutagenicity tests in whole mammals

In the last 15 years, ample justification has been provided for the proposition that mutational events (in a broad sense) are involved in the multistep processes we know as carcinogenesis. In particular, this has been confirmed in the last few years in specific detail for the expression of various *onc* genes, well documented by Ramel (1986 and this volume). Given the wide range of mutational changes that have been associated with *onc* gene activation and with cancer, and the fact that most mutagens can give rise to a wide variety of different types of mutation, it follows that any substance for which there is sufficient evidence of mutagenesis in rats or mice ought to be regarded as a potential carcinogen. Even in those rare cases in which good positive data on mammalian mutagenesis are accompanied by apparently innocent carcinogenicity data, it would be rash to assume without further work that the substance would be similarly non-carcinogenic to humans. In this context, it is obviously relevant that carcinogenesis is a multistep process. An agent that does not induce all the steps may not be easily detected in a clean experimental cancer bioassay, yet may still be hazardous in the dirty real world inhabited by humans.

It would be foolish to deny that there are factors operating to affect carcinogenic potency over and above those which affect mutagenic potency. Thus, although in-vivo mutagenicity can be held to indicate carcinogenic *potential*, it has little to tell us about carcinogenic *potency*, except, perhaps, for some specific groups of closely related substances.

Notwithstanding, it can be argued that the time has come to regard mutagenicity in whole mammals as having a definitive status with regard to the genotoxic component of carcinogenesis. This assertion is based primarily on mechanistic understanding of *onc* gene activation and cannot be tested by a comparative exercise of validation; since genotoxicity tests are needed to define genotoxic carcinogens, the argument is circular. If, however, the validity of in-vivo short-term mutagenicity tests is accepted on mechanistic grounds, then the only validation that needs to be considered concerns the quality of the experiments and of the data generated, the statistical analysis of the results, and closeness of the endpoints actually measured to those known to be involved in carcinogenesis.

In the fullness of time, we may hope for practical assays that measure *onc* gene activation specifically both *in vitro* and *in vivo*, but that time has not yet come. Moreover, the number of *onc* genes likely to exist and the variety of ways in which activation may occur mean that such assays are unlikely to be general enough to screen for genotoxic carcinogens. Their role is likely to be in providing a more detailed mechanistic understanding of how individual substances act, and perhaps also to detect types of activation that do not involve mutations.

The in-vivo battery

If in-vivo (whole mammal) mutagenicity is accepted as a surrogate for the genotoxic component of carcinogenicity, then consideration can be given to the tests that should belong in such a battery. A number of tests can be used to detect substances distributed in the blood; some of these may also be expected to pick up substances of short lifetime generated by activation in the bone marrow or testis (see Reports 7 and 8). It is also important that substances produced by activation in the gut and the liver be detected. Ashby (1986) has proposed the unscheduled DNA synthesis assay in liver (Mirsalis & Butterworth, 1980) plus the micronucleus test as the optimal in-vivo battery. One aspect of this question concerns the extent to which one should try to detect substances that are activated to short-lived species in specific organs. The limitations are analogous to those restricting animal carcinogenicity bioassays, in which not all sites can be used in practice to show a significant increase, for purely statistical reasons, and in which more than two species should really be examined in order to have a 99% chance of detecting a carcinogen.

If maximal sensitivity for detection of genotoxic activity is desired, there are numerous in-vivo endpoints that may be employed that are not strictly mutational, for example, unscheduled DNA synthesis, sister chromatid exchange, DNA adducts and DNA strand breakage. Since none of these in themselves can involve *onc* gene activation, they would need to be validated against results with genuine mutation assays. If one wishes to lay down criteria for in-vivo mutagenicity tests in which a positive result would command confidence, then assays that involve gene mutation or chromosomal aberrations (numerical or structural) are essential.

While I believe that it is possible to select a small battery of short-term tests in whole animals, it must be accepted that there is still scope for improvement in assay design in terms of specificity, economy of animals and animal welfare.

The in-vitro battery

If short-term mutagenicity tests in whole mammals are regarded as indicating the potential of a substance for genotoxic carcinogenicity, then in-vitro tests can be regarded as suggesting the possibility of in-vivo mutagenicity. There is, nevertheless, an important difference between the bases of these two relations which affects the reliance that may be put upon them. In the relation between in-vivo mutagenicity and genotoxic carcinogenicity, absorption, metabolism and distribution are common factors and the molecular endpoints are similar, so that prediction can be made with some confidence. In the relation between in-vitro and in-vivo mutagenicity, not only may the endpoint and the structure of the genome be very different, but the metabolism is likely to be different, and absorption and distribution dramatically so. Thus, one may wonder that the prediction of in-vivo mutagenicity by in-vitro tests is as good as it is.

It may not matter overmuch if an in-vitro test is inclined to be too sensitive, since no one should consider a substance as a potential carcinogen solely on the basis of positive in-vitro test results (see ICPEMC Committee 2, 1982). A proper consideration of short-term in-vivo data, together with structural metabolic and chemical information, can enable the identification of those substances that present no significant in-vivo risk.

A possible problem in using short-term in-vitro tests is that a substance with an unusual activation pathway may be missed. Modifications, particularly of the *Salmonella* assay, go some way towards minimizing this drawback (see Report 5). Experience has shown, however, that there are remarkably few genotoxic carcinogens that show this sort of genotoxicity. Most are positive in a range of short-term tests. Increasing the number of different assays, moreover, in an attempt to avoid missing such substances leads to a problem that remains a severe limitation of short-term tests and which became apparent in the first collaborative trial of the International Programme of Chemical Safety. In the results of 13 laboratories on the *Salmonella* assay (mostly with fairly similar or identical protocols), the proportion of false results compared with the consensus assignment was about 12% (Bridges *et al.*, 1981). By insisting on repeat experiments and by improving protocols, one might expect this figure to decline to around 5%, most of which is probably attributable to imperfect technical practice or statistical anomaly. Incidentally, the scope for both of these sources of misinformation is just as great with long-term carcinogenicity bioassays, and I am sure that the false result percentages would be comparable if they could ever be known.

The problem of isolated false results is magnified by the fact that the more systems one employs in order to detect substances acting by obscure pathways the more likely one is to encounter isolated positive results. Some of these may be genuine, but at present we cannot readily distinguish them from positives arising by pathways that are irrelevant to the situation in the whole animal, or arising from the action of impurities which become significant at the massive concentrations often employed.

The message seems to be clear that a relatively small number of in-vitro tests must suffice to detect most of those agents that might be mutagenic *in vivo*. If one or more of these tests is convincingly positive, there is nothing to be gained by further in-vitro testing — the focus of attention must move *in vivo*. If the in-vitro tests are negative, additional testing is likely to produce only isolated positive results, the significance of which is imponderable. The number of in-vitro assays needed to detect generally genotoxic agents is thus small; Ashby (1986) suggests two. Additional assays would be needed only if borderline or suspect results are obtained. If, together with these mutagenicity tests, one could add a valid assay for abnormal chromosomal segregation (see Report 7) and another predictive for tumour promotion (see Report 10), one would probably have gone as far as is likely to be attainable in the immediate future. It cannot be emphasized too strongly that the design of a battery of in-vitro tests is a scientific business and needs to be undertaken for each substance with full regard for all available information, including chemical structure and chemical and physical properties. The *Salmonella* assay, for example, is clearly of little value for certain chemical classes, including phthallates and chlorinated organics (Zeiger & Tennant, 1986). This being so, the practice of 'blind' testing, without all this information being taken into account, is likely to lead to problems.

Promotion and abnormal chromosomal segregation are ripe for development both at the testing level and in terms of their mechanistic involvement in carcinogenesis. A third area in which future hopes must lie is in-vitro transformation (see Report 9). An understanding of its mechanistic basis should not be too long delayed. These three areas may in time come to complement the orthodox tests based on DNA damage and mutagenicity and enable prediction to be made of many of the nongenotoxic categories of carcinogen. I am convinced, however, that this will gain acceptance only with a mechanistic understanding and not by virtue of statistical correlations.

Conclusions

This paper may be summarized as follows:

1. Two main difficulties rule out the validation of short-term tests for predicting the carcinogenicity of various substances by comparing test results with long-term bioassay results. Firstly, carcinogenicity is not expressed in every species to a detectable extent, so that long-term bioassay results in two rodent species underestimate the proportion of chemicals with carcinogenic potential. Secondly, genotoxicity tests can be expected to detect only those compounds that act through a genotoxic mechanism. Ideally, one would like to validate the tests only against known genotoxic carcinogens, but because these can be identified only by genotoxicity tests this approach involves a circular argument and is therefore invalid.

2. It can be argued that enough is now known about the involvement of gene, chromosomal and aneuploidy mutations in the activation and expression of *onc* genes that the ability to cause such mutations in whole animals can be taken on mechanistic grounds to indicate a potential for genotoxic carcinogenicity, without need of further validation by comparative studies.

3. Mutagenesis studies in whole mammals thereby acquire a definitive status in their own right, as indicating carcinogenic *potential* of a substance. Discussion should now focus on the identification of the best whole mammal short-term tests currently available and on the future development of such tests.

4. In-vitro tests indicate the possibility of mutagenic activity in mammals but have no definitive status, since there are numerous reasons why the activity may not be expressed to a significant extent in whole animals. The in-vitro test battery should comprise as few tests as possible and these tests should have high, broad spectrum sensitivity. Enlargement of the battery with a view to detecting substances acting by obscure or minor pathways will inevitably result in occasional positive results. Their significance, however, cannot be appreciated without a great deal of work, which can be contemplated only in exceptional circumstances.

5. Agents that cause cancer by nongenotoxic mechanisms will not be detected by these tests, except by chance.

Epilogue

These comments are, of course, my own, and I should not be assumed to be wearing the hat of any organization with which I may be associated.

It is some years since I proposed a system in which in-vivo (whole mammal) and in-vitro tests were ascribed different statuses and weights in a three-tier evaluation protocol (Bridges, 1973, 1976). The information and arguments presented here were not available at that time. Although some minor modification of the scheme now seems justified, the case for such an approach seems to have been strengthened.

References

Ashby, J. (1986) The prospects for a simplified and internationally harmonized approach to the detection of possible human carcinogens and mutagens. *Mutagenesis, 1*, 3-16

Bridges, B.A. (1973) Some general principles of mutagenicity screening and a possible framework for testing procedures. *Environ. Health Perspect., 6*, 221-227

Bridges, B.A. (1976) *Use of a three-tier protocol for evaluation of long-term toxic hazards particularly mutagenicity and carcinogenicity.* In: Montesano, R., Bartsch, H. & Tomatis, L., eds, *Screening Tests in Chemical Carcinogenesis (IARC Scientific Publication No. 12)*, Lyon, International Agency for Research on Cancer, pp. 549-559

Bridges, B.A. & Mendelsohn, M.L. (1986) *Recommendations for screening for potential human germ cell mutagens: an ICPEMC working paper.* In: Ramel, C., Lambert, B. & Magnusson, J., eds, *Genetic Toxicology of Environmental Chemicals*, Part B, *Genetic Effects and Applied Mutagenesis*, New York, Alan R. Liss, pp. 51-65

Bridges, B.A., McGregor, D., Zeiger, E., Bonin, A., Dean, B.J., Lorenzo, F., Garner, R.C., Gatehouse, D., Hubbard, S., Ichinotsubo, D., MacDonald, D., Martire, G., Matsushima, T., Mohr, G., Nagao, M., Richold, M., Rowland, I., Simmon, V., Skopek, T., Truman, R. & Venitt, S. (1981) *Summary report on the performance of bacterial mutation assays.* In: de Serres, F.J. & Ashby, J., eds, *Evaluation of Short-Term Tests for Carcinogens*, Amsterdam, Elsevier, pp. 49-67

Cole, J., Muriel, W.J. & Bridges, B.A. (1986) The mutagenicity of sodium fluoride to L5178Y [wild-type and TK$^{+/-}$ (3.7.2.c)] mouse lymphoma cells. *Mutagenesis, 1*, 157-167

Friesen, M., O'Neill, I.K., Malaveille, C., Garren, L., Hautefeuille, A., Cabral, J.R.P., Galendo, D., Lasne, C., Sala, M., Chouroulinkov, I., Mohr, U., Turusov, V., Day, N.E. & Bartsch, H. (1985) Characterization and identification of six mutagens in opium pyrolysates implicated in oesophageal cancer in Iran. *Mutat. Res., 150*, 177-191

IARC (1983) *Approaches to Classifying Chemical Carcinogens According to Mechanisms of Action (IARC Internal Technical Report No. 83/001)*, Lyon

ICPEMC Committee 2 (1982) Final report. Mutagenesis testing as an approach to carcinogenesis. *Mutat. Res., 99*, 73-91

Jansen, J.D., Clemmeson, J. & Sundaram, K. (1980) Isoniazid — an attempt at retrospective prediction. *Mutat. Res., 76*, 85-112

Mendelsohn, M.L. (1986) Can chemical carcinogenicity be predicted by short-term tests? *Ann. N.Y. Acad. Sci.* (in press)

Mirsalis, J.C. & Butterworth, B.E. (1980) Detection of unscheduled DNA synthesis in hepatocytes isolated from rats treated with genotoxic agents. *Carcinogenesis, 1*, 621-625

Nair, J., Ohshima, H., Friesen, M., Croisy, A., Bhide, S.F. & Bartsch H. (1985) Tobacco-specific and betel nut-specific compounds: occurrence in saliva and urine of betel quid chewers and formation *in vitro* by nitrosation of betel quid. *Carcinogenesis, 6*, 295-303

Ramel, C. (1986) Deployment of short-term assays for the detection of carcinogens — genetic and molecular considerations. *Mutat. Res.* (in press)

Upton, A.C., Clayson, D.B., Jansen, J.D., Rosenkranz, H.S. & Williams, G.M. (1984) Report of ICPEMC Task Group 5 on the differentiation between genotoxic and non-genotoxic carcinogens. *Mutat. Res., 133*, 1-49

Zeiger, E. & Tennant, R.W. (1986) *Mutagenesis, clastogenesis, carcinogenesis: expectations, correlations and relations*. In: Ramel, C., Lambert, B. & Magnusson, J., eds, *Genetic Toxicology of Environmental Chemicals*, Part B, *Genetic Effects and Applied Mutagenesis*, New York, Alan R. Liss, pp. 75-84

DEPLOYMENT OF SHORT-TERM ASSAYS FOR THE DETECTION OF CARCINOGENS: GENETIC AND MOLECULAR CONSIDERATIONS[1]

C. Ramel

Department of Genetic and Cellular Toxicology,
Wallenberg Laboratory, University of Stockholm,
S-106 91 Stockholm, Sweden

Summary

The deployment of short-term assays for the detection of carcinogens must be based inevitably on the genetic alterations actually involved in carcinogenesis. This paper gives an overview of oncogene activation and other mutagenic events associated with cancer induction. It is emphasized that there are indications of DNA alterations in carcinogenicity, which are not in accordance with 'conventional' mutations and mutation frequencies, as measured by short-term assays of point mutations, chromosomal aberrations and numerical chromosomal changes. Discrepancies between DNA alterations in carcinogenicity and the endpoints of short-term assays in current use include transpositions, insertion mutations, polygene mutations, gene amplifications and DNA methylations. Furthermore tumorigenicity may imply induction of genetic instability, followed by a cascade of genetic alterations.

The evaluation of short-term assays for carcinogenesis involves mainly two correlations, that is, between mutation and animal cancer data on the one hand and between animal cancer data and human carcinogenicity on the other. It should be stressed that animal bioassays for cancer generally imply tests specifically for chemicals that function as complete carcinogens, which may be a rather poor reflection of the actual situation in human populations.

The primary aim of short-term mutagenicity assays is to provide evidence of whether a compound can be expected to cause *mutations* in humans, and such evidence has to be considered seriously, even against a background of negative cancer data. For an evaluation of data from short-term assays, the massive amount of empirical data from different assays should be used, and new computer systems can be expected to provide improved predictions of carcinogenicity.

---

[1] A version of this paper also appears in *Mutation Research* (in press)

Introduction

The correlation between carcinogenicity and mutagenicity has been the foundation on which the use of mutagenicity assays to detect carcinogenic chemicals has been based. Previously, this correlation was largely operational, although there have been several indications of a causal relationship between carcinogenicity and mutagenicity. Thus, the carcinogenic property of cells is transmitted from cell generation to cell generation; tumours are usually of monoclonal origin, specific mutational changes are observed in certain specific neoplasms; hereditary defects in DNA repair lead to an increased risk for cancer; mutagenic and carcinogenic chemicals share electrophilic properties; and carcinogenic chemicals are mostly also mutagenic in experimental systems. All these facts are primarily compatible with the involvement of mutational events in carcinogenesis.

During the last few years, our knowledge of the sequence of events leading to cancer has undergone a dramatic change. The recognition of specific viral and cellular oncogenes that are involved in the transformation of normal cells to malignant cells has opened up an entirely new approach to the study of the mechanisms of cancer. Tumour formation has been linked with the activation of oncogenes, often through specific, definable mutagenic events. Studies of the gene products of oncogenes have revealed connections with growth factors and their receptors and indications of their mode of action through kinase activity by specific amino acid phosphorylation. These data seem to provide an inescapable causal correlation between carcinogenicity and mutagenicity, although it is not possible to claim universal validity of this correlation for all forms of cancer.

Oncogene activation

The phylogenetic conservation of cellular oncogenes points to the fact that they fulfil essential biological functions. Investigations of oncogene products and their properties indicate that they play an important role in differentiation and cell proliferation. Obvious links in that respect are furnished, for instance, by the *sis* oncogene, which produces its own growth factor, closely related to the platelet-derived growth factor, and *erb*-B, whose gene product is related to the receptor of the epidermal growth factor. Several oncogenes function as kinases, specifically phosphorylating tyrosine, which process has been implicated in the cellular response to mitogenic growth factors. Activation of oncogenes implies changes of the genetic regulation of these processes. Some of these changes have been shown to be brought about by different mutagenic events, spontaneous as well as induced. The correlation between mutagenesis and carcinogenesis therefore can be assumed to reflect not only an empirical association, but a real causal relationship. However, the multistage process of tumorigenesis complicates this relationship between mutations and cancer. As a background for discussions of the use of short-term tests for the evaluation of carcinogenic activity, it is important to examine what is known about the actual relationship between mutations, oncogene activation and tumour formation.

The activation of oncogenes has been shown to be associated with different mutagenic changes, which can occur spontaneously, by viral insertion mechanisms and by chemical or physical mutagens. There are principally two mechanisms for oncogene activation through these mutational changes: one implies an altered gene product, the other an alteration of the expression of oncogenes. The mutations involved can be illustrated by the two most intensely studied families of oncogenes, *ras* and *myc*. *Ras* represents activation through an altered gene product, and *myc*, activation through a change in level of transcription and expression. These two oncogenes also illustrate other classifications of oncogenes: *ras* genes are transforming oncogenes, causing neoplastic transformation in established immortal cell lines, while *myc* is an immortalizing oncogene.

The gene product of *ras*, p21, is a protein localized at the inner cellular membrane, with the characteristic of binding guanine triphosphate. The protein coded by *myc* is a nuclear protein.

Restriction of DNA, DNA sequencing and particularly the development of transfection assays (Shih *et al.*, 1979) have made it possible to study the mutational changes involved in oncogene activation in great detail. This work has been particularly fruitful with respect to the activation of *ras* oncogenes, which also provided evidence of the oncogenic effect of single base substitutions. By transfection assays with DNA from human bladder tumour cell lines and the established fibroblast cell line NIH 3T3, it was shown independently in 1982 by three research groups (Reddy *et al.*, 1982; Tabin *et al.*, 1982; Taparowsky *et al.*, 1982) that DNA from cancer cells could transform recipient cells and that this transformation was linked to a single base substitution of codon 12 in the cellular Harvey sarcoma oncogene (c-Ha-*ras*). This base substitution, a transversion from G to T, caused a corresponding change of the amino acid glycine to valine in the gene product p21. Using DNA transfection assays, it has been shown that 10-30% of human tumours contain altered forms of one of the three *ras* oncogenes, Ha-*ras*, Ki-*ras* and N-*ras* (Bos *et al.*, 1985). These alterations involve amino acids 12, 13 and 61 through a single base-pair mutation, implying the substitution of glycine, glycine and glutamine, respectively, in the gene product p21. In-vitro mutagenesis experiments have furthermore shown that mutations at amino acids 59 and 63 can lead to transforming activity (Fasano *et al.*, 1984). It should be emphasized that the altered *ras* oncogene has been detected only in tumour cells and not in normal cells of the same individual (Feig *et al.*, 1984; Santos *et al.*, 1984; Bos *et al.*, 1985).

In experimental systems, activation of *ras* oncogenes has been observed after various treatments with mutagenic agents. Thus, a transforming *ras* oncogene was obtained in guinea-pig cell lines after treatment with 3-methylcholanthrene, benzo[*a*]-pyrene, *N*-methyl-*N'*-nitro-*N*-nitrosoguanidine and *N*-nitrosodiethylamine (Sukumar *et al.*, 1984). Marshall *et al.* (1984) treated c-Ha-*ras* protooncogenes in two plasmids *in vitro* with benzo[*a*]pyrene dihydrodiol epoxide and demonstrated by transfection of NIH 3T3 cells a transforming property as a result of the treatment. Activation of the oncogene occurred in DNA and not in precursor nucleotides and independently of the form in which the plasmid DNA was treated — supercoiled, relaxed circular or linear DNA.

Induced activation of the *ras* oncogenes involves characteristic, puzzling specificities. Zarbl *et al.* (1985) studied induction of mammary tumours in rats with *N*-methyl-*N*-nitrosourea and 7,12-dimethylbenz[*a*]anthracene. By restriction analysis, it was shown that *N*-methyl-*N*-nitrosourea caused a change of G in codon 12 in 31 out of 38 tumours tested in one experimental set. In 36 of 48 tumours showing this pattern of restriction analysis, the actual change of codon 12 was from G to A at position 35. Three tumours induced by 7,12-dimethylbenz[*a*]anthracene, however, were all normal with respect to codon 12 but had an alteration of codon 61.

Guerrero *et al.* (1984a,b) showed that in mouse lymphomas induced by *N*-methyl-*N*-nitrosourea and γ radiation, different *ras* oncogenes were activated — N-*ras* in chemically-induced tumours and Ki-*ras* by radiation. Sequence analysis of Ki-*ras* induced by γ radiation again revealed the specific base substitution of the second base in codon 12 from G to A, the same base pair involved in so many tumours both in animals and humans.

Although *ras* activation has an effect on a wide spectrum of tumours, there also seems to be specificity in the activation of the *ras* oncogenes. Thus, Bos *et al.* (1985) recently showed that five out of six acute myeloid leukaemic tumours in humans contained an activated N-*ras* oncogene which, contrary to all previously investigated tumours, had a base substitution not in codon 12 or 61 but in codon 13, in two cases resulting in a change from glycine to aspartic acid and in three cases from glycine to valine. In nine other cases of acute myeloid leukaemic cells, activation of N-*ras* was also found. This activation of N-*ras* is probably not the only cause of the transfection, and other genetic changes are also involved.

On the basis of what we know of the process of these oncogene activations, it can be assumed that they operate in heterozygous form. This has been born out by experimental evidence. Thus, Zarbl *et al.* (1985) found in *N*-methyl-*N*-nitrosourea-induced tumours in rats both the restriction pattern of the protooncogene and the activated Ha-*ras*-1. Santos *et al.* (1984), however, have also reported homozygosity for an activated form of Ki-*ras*. Capon *et al.* (1983) found a functional homozygosity in a cell line for Ki-*ras*-2.

The homozygosity of recessive genes involved in certain cancer forms has been postulated for a long time (Knudsen, 1971). This hypothesis has been strongly supported in recent years for hereditary retinoblastoma and Wilms' tumour (Murphree & Benedict, 1984). The retinoblastoma gene represents a class of cancer genes which, in contrast to oncogenes, act as suppressors in the regulation of differentiation and cell proliferation. Inactivation of that function results in tumour development. One of the retinoblastoma alleles is apparently inactivated in the hereditary form, and tumour induction is caused by a subsequent repression or inactivation of the other allele. This functional homozygosity of mutant alleles can occur by inactivation or loss of the normal allele, by chromosomal loss, deletion, nondisjunction or reduplication, mitotic recombination, gene mutation or gene inactivation, for instance through translocation to the X chromosome.

In some cases it can be assumed that activation of oncogenes occurs at an early stage of tumour development corresponding to cancer initiation. Thus, the activation of Ha-*ras*-1 in rat mammary carcinoma by treatment with *N*-methyl-*N*-nitrosourea, studied by Zarbl *et al.* (1985), must have occurred early in the process of cancer development, due to the short life of the mutagenic compound. Activation of c-Ha-*ras* in skin tumours occurs both in benign papillomas and in carcinomas developing from papillomas, again indicating an early stage of oncogene activation (Balmain *et al.*, 1984). Experimental data also indicate that activated *ras* may be involved at a late stage of carcinogenesis. Albino *et al.* (1984) showed by transfection assay that only one of five melanoma cell lines from independent metastatic foci in one patient contained activated *ras* oncogene. Tainsky *et al.* (1984), studying tumour induction in human PA1 teratocarcinoma cells, found that, although early-passage cells caused tumour formation in nude mice only, later passages showed transforming activity with base substitution in codon 12.

The activation of *ras* oncogenes is essentially associated with a qualitative mutational change of the gene product. The other mechanism of oncogene activation implies a quantitative change of gene expression, which has been shown to occur by means of at least four different mechanisms — first, the integration of retroviral transcriptional promoter which replaces the normal promotor, second, amplification of c-oncogenes, third, increase in oncogene transcription by means of enhancers, which can act both upstream and downstream of an oncogene and, fourth, chromosomal rearrangements that place an oncogene in the vicinity of immunoglobulins, the promotor of which increases transcription of the oncogene (for reviews, see Land *et al.*, 1983a,b; Klein & Klein, 1984).

C-*myc* constitutes the best studied example of retroviral oncogenes that presumably act by means of an alteration of gene expression by translocation and amplification, as in Burkitt's lymphoma and mouse plasmacytomas. Gene amplications are often coupled to translocations of *myc* oncogene. This applies to c-*myc* in leukaemia, lung carcinoma and neuroendocrine tumours and particularly for the related N-*myc* in neuroblastoma. An increased expression of *myc* is, however, not likely to be a general explanation of its oncogenic effect, as the breakpoints in the translocations in Burkitt's lymphoma can occur both upstream and downstream of c-*myc*. Accordingly, there is not always increased expression of *myc* in translocations, but there may be wrong timing of *myc* expression. Davis *et al.* (1984) have reported that qualitative changes of the nucleotides in the first exon of translocated c-*myc* occur, indicating a loss of transcriptional control by preventing the binding of repressor protein to exon 1. Mutational changes in the second exon of *myc* have also been reported, which probably occur as a result of the translocation to immunoglobulin loci (Rabbitts *et al.*, 1983).

Myc evidently acts at another level of tumour induction than *ras* — *myc* primarily causing immortalization of the cells, *ras* an abnormal growth control — and these functions are genetically separate. The borderline between *ras* and *myc* types of oncogene activation is not sharp, however, as indicated by the fact (mentioned above) that activation of *myc* is often associated with nucleotide changes. Spandidos and Wilkie (1984) showed, however, that a complete transforming property involving both the *ras* and *myc* functions can be acquired with activated *ras* only if it is coupled in a plasmid to a transcriptional enhancer from simian virus 40 or the Moloney murine leukaemia virus long-terminal repeat.

In the present context, it is important to establish that oncogene activation involves a variety of mutational changes — base substitutions, translocations, deletions, insertions of viral DNA and gene amplifications. The extent to which these mutations can be brought about by exogenous mutagens is not clear, but experimental activation of *ras* oncogenes by mutagenic agents has been performed by many research groups.

To the list of genetic alterations that possibly act at early stages of carcinogenesis can be added mitotic recombination and nondisjunction, presumably involved in the etiology at least of retinoblastoma, as mentioned above.

Other genetic mechanisms for DNA alteration may operate in the course of tumour formation. It is known from studies with *Drosophila melanogaster* that a large proportion of spontaneous mutations are in fact insertion mutations caused by a series of endogenous transposable elements similar to retrovirus, and it is unlikely that this mechanism is absent in mammals, although clear proof is lacking. Brickell *et al.* (1983) have reported, however, an increased rate of transcription of a histocompatibility antigen as a general trait in tumours of mice. The corresponding gene contains a widely distributed repetitive element in the mouse genome, indicating a transposable element. Insertion mutations after infection with Moloney murine leukaemia virus have been reported for the hypoxanthineguanine phosphoribosyl transferase (*hprt*) locus in mammalian cells (King *et al.*, 1985).

Tumorigenicity is closely linked to regulation of transcription and differentiation, and there are several indications that methylation of DNA is involved in this process. Unfortunately, very little seems to be known of the possible role of methylation in neoplastic transformation, but hypomethylation of some DNA sequences has been reported in some humans cancers (Feinberg & Vogelstein, 1983). Holliday and Pugh (1975) presented a model in which methylation of DNA adjacent to structural genes determines transcription and differentiation, including a method for 'counting' the number of cell divisions in different tissues. Holliday (1979) extended this model to include the induction of cancer by means of a loss of methylation in connection with repair of DNA. Although methylation of DNA is clearly not the only mechanism for turning genes on and off during differentiation, its occurrence, at least in vertebrates, is sufficiently well established (for a review, see Kolata, 1985) that it must be taken into consideration also in connection with carcinogenesis.

Mutational changes may be involved not only at early stages of tumorigenicity. Although tumour promotion has often been considered an epigenetic event, the work of Cerutti (for a review, see Cerutti, 1985) indicates that promotors such as phorbol esters act by causing an oxidative burst and the production of superoxide radicals and hydrogen peroxide as well as a clastogenic factor. Phorbol esters have also been shown to cause aneuploidy (Parry *et al.*, 1981), and an increase in the frequency of sister chromatid exchange has been reported by one research group (Kinsella & Radman, 1978). It has also been shown that conversion of benign papillomas to malignant carcinomas, which is a late event in the mouse skin tumour assay, is increased by the mutagens urethane, N-methyl-N'-nitro-N-nitrosoguanidine and 4-nitroquinoline-N-oxide but not by the promotor 12-O-tetradecanoylphorbol 13-acetate (Hennings *et al.*, 1983).

Progression of tumours involves various chromosomal and ploidy changes, which are important for the development of malignancy (Klein & Klein, 1985).

Almost all types of mutagenic event have been linked to the process of tumour formation at one stage or another. Conversely — and of special importance in the present context — there is probably no genetic lesion that can be considered unimportant for carcinogenicity. This obviously must be taken into consideration when applying test methods to reveal carcinogenic potential of chemicals.

Multistage carcinogenicity

A wealth of data strongly indicates that carcinogenicity is a multistep process and is usually divided into three stages — initiation, promotion and progression. Further division of stages occurs for instance in promotion (Slaga, 1980), and the same is obviously true for progression (Klein & Klein, 1985). However, recognition of defined stages in carcinogenicity is often not possible, as many analyses have indicated progressive changes towards increased malignancy. Certain clones of adenovirus-transformed rat embryo cells thus undergo a sequence of progressive changes with serial passages (Fisher *et al.*, 1979). Chou (1980) analysed animal cancer data mathematically by the median-effect principle. The analysis suggested that chemical carcinogens exert their effects according to the mass-action law and that the data were compatible with either multiple hits or a slow transition of the identities of genetic materials. The data used for this analysis (Peto *et al.*, 1975) indicate that cancer incidence after repeated application is related to the length of time of exposure. Druckrey (1972) reported a drastic difference in dose-response to N-nitrosodiethylamine for different exposure regimes in rats. Prolonged low-dose application decreased the total dose required to induce carcinomas by 15 times as compared to a high-dose regime. These data are compatible with a progressive process in tumour formation, composed of a series of stochastic changes followed by selection. If some of the changes imply mutagenic events, continuous application of mutagenic agents would greatly increase the chance of carcinogenicity. Farber (1980) described the process of chemical carcinogenesis as being made up of a number of discrete, discontinuous changes, like a mutation, each of which is followed by the selection of one or more types of altered cells.

At a molecular level, chemical induction of tumours is indeed not easy to understand, considering the multistage process of carcinogenicity. The original observation of the transforming property of activated *ras* oncogene in transfection assays was made in the established cell culture NIH 3T3, but the same transfection system was inefficient with normal diploid cells. It turned out that the *ras* oncogene only provided the cell transformation property to recipient cells, but the immortalization property had also to be supplied in order to cause a neoplastic transformation. This was acquired by combined transfection with activated *ras* oncogene and an activated *myc* oncogene, the latter providing immortalization (Land *et al.*, 1983a,b).

Transformation of diploid cells by transfection could also be brought about by combining activated *ras* with large T antigen from DNA virus, which causes immortalization of the cells (Ruley, 1983). Finally, the immortalization step could be provided by treatment with chemical carcinogens (Newbold & Overell, 1983). It can be assumed that these DNA alterations corresponded to an initiation stage of carcinogenesis, indicating that already this early part of cancer induction requires more than one mutational event. Against this background, the induction of cancer by chemicals and radiation becomes difficult to reconcile with normal mutation frequencies. The experiments of Guerrero *et al.* (1984a,b), on mouse lymphomas induced by N-methyl-N-nitrosourea and γ radiation, referred to above, are of interest in this context: 600 rads γ radiation activated Ki-*ras* and in all three tested tumours the activation implied the same base substitution — from G to A in codon 12. With the dose applied, such specific base substitution is statistically possible among the 10^7—10^8 lymphocytes available, but there is hardly any reasonable statistical chance, in view of known mutation frequencies, for another genetic mutation in the same cell, if transformation requires at least two mutagenic events.

Cellular transformations sometimes occur at such high frequencies that the process is difficult to explain by anything like normal mutation rates. It has been shown that carcinogen-induced transformation of mammalian cells involves poly(ADP-ribose) transferase (Kun *et al.*, 1983; Borek *et al.*, 1984). Transformation is suppressed by inhibitors of this enzyme, such as benzamides, which usually has an opposite effect on mutations, increasing the frequency. This led Borek *et al.* (1984) to suggest that the two processes involve different regulatory mechanisms: transformation may be mediated by ADP-ribosylation, while mutations are mediated through DNA repair.

How the sequence of events after initiation, leading to malignancy, is triggered by the initial carcinogen treatment is also difficult to visualize. As pointed out by Farber (1980) the episodes after the first rare event and the first period of selection are no longer dependent upon the operation of any recognizable environmental influence and appear to be self generating.

Although it is far from clear how the combined activation of oncogenes occurs, in some cases the activation of oncogenes is not independent. As mentioned above, the *sis* oncogene produces a growth factor, closely related to the platelet-derived growth factor, and this in its turn increases *myc* expression 40-fold. Oncogenes may therefore fit into a cascading hierarchy, in which the action of one controls the others (see Marx, 1983 for a review).

Cancer testing — shortcomings and problems

The application of mutagenicity testing to carcinogenicity is usually considered as a substitute for conventional animal cancer tests, necessitated above all by the large expenditure of money and time associated with cancer tests. Although this is essentially true, there are also other problems to be considered. The use of short-term mutagenicity tests to detect human carcinogens rests in practice on two correlations — that between mutagenicity test results and animal cancer data and that between animal cancer data and human carcinogenicity. The strength of the latter correlation also has to be taken into account when interpreting mutagenicity test data. The relationship between animal cancer data and the situation in humans is problematic. Apart from the obvious possibility of metabolic and other differences in response between test animals and humans, which applies to all toxicological evaluations, the multistage carcinogenicity process implies an additional problem. With few exceptions, animal cancer tests in fact evaluate only the property of individual chemicals to act as complete carcinogens. This may be a poor reflection of the actual exposure situation in humans, in which combinations of initiators and promoters can often be suspected to occur. For instance, chemicals that function as strong initiators and poor promoters are poor carcinogens in animal cancer tests. In combination with an appropriate promoter in human exposure — for instance, tobacco smoking — the situation may be drastically different. An example of the requirement for both promoter and initiator in an experimental system is provided by *trans*-4-acetylaminostilbene (Hilpert *et al.*, 1983). This compound produces DNA lesions in rat liver and yet is an incomplete carcinogen for that tissue; only in combination with a promoter were neoplastic lesions produced. Mammary tumours were induced only by combining treatment with *trans*-4-acetylaminostilbene with treatment with a a promoter; as such, diethylstilboestrol was particularly efficient, indicating a hormonal influence on promotion.

Another obvious means of missing the carcinogenic property of a test chemical in animal assays is the low sensitivity of the system, due to the limited number of animals that can be used. This is, of course, particularly evident for toxic chemicals, which cannot be tested in higher doses. However, in this case, the chance of missing an effect can at least be calculated statistically as type 2 error.

When evaluating the net relationship between mutagenicity and carcinogenicity from test data, it must be stressed that 'false'-positive mutagenicity data imply a very different and far less reliable parameter for the accuracy of mutagenicity testing performance than false-negative mutagenicity data.

Mutagenicity tests — shortcomings and problems

There are several reasons why a complete correlation between mutagenicity and carcinogenicity tests cannot be expected. In-vitro systems do not truly reflect metabolism, repair, cell proliferation and chromosomal structure *in vivo*, but even in-vivo mutagenicity tests take these properties into account only in specific tissues. It is also evident that mutational changes constitute only a part of the complex sequence of events leading to malignancy. The fact that such high correlations between mutagenicity and carcinogenicity test data have in fact been found nevertheless speaks in favour of mutagenic events being critical in carcinogenicity.

Over 100 test systems are in more or less regular use (Hollstein *et al.*, 1979), covering different types of mutagenic lesions. However impressive this list of test systems may be, they are designed to test more or less 'conventional' structural mutational changes, which may not be the only ones of importance in carcinogenicity, as touched upon before. The 'cascade' of effects during the process of tumour induction involves, at least to some extent, alterations at the DNA level. As pointed out above, it is difficult to visualize simple mutational events in this context. It cannot be excluded that more 'unconventional' mechanisms operate, in view of the instability of the genetic material. As pointed out by Haynes (1985), it is possible that recombinational processes associated with rearrangement of DNA sequences are a normal feature in cell differentiation and gene expression not only in cells involved in the immune system but also in other tissues. If this is the case, one might expect mistakes to be made during the various processes of DNA alignment, breakage, synthesis and splicing which presumably are involved in such phenomena. It may be of interest in this connection to recall the possibility outlined above that cellular transformation does not depend on regular mutagenic events but on poly(ADP) ribosylation. It is conceivable that an instability in such processes can be brought about in carcinogenicity. McClintock's discovery of transposing elements was preceded by her observations of genetic imbalance from 'genomic stress', which caused a cascade of genetic alterations (for a review, see

McClintock, 1984). Reverse transcription has turned out to be rele[vant?] but evidence is rapidly accumulating that it applies also to higher or[ganisms?] humans (for a review, see Baltimore, 1985). Reverse transcription [and tran-?] sitions, which may operate or affect the genetic regulation of growth and, in turn, be involved in tumorigenicity. In *Drosophila*, tra[nspositions are] induced by a specific factor, the *p*-factor, which occurs in many natural populatio[ns]. The *p*-factor presumably contains a specific enzyme for this insertion of DNA sequences, a 'transposase' (see Rubin & Spradling, 1982; Spradling & Rubin, 1982). Although there is no comparable factor known in mammals, the occurrence of a similar mechanism cannot be ruled out, and such a situation would cause abnormally high mutation frequencies.

Another DNA alteration not related to ordinary mutagenesis is methylation of DNA, which has been implicated in both differentiation and carcinogenesis, as mentioned above. At present, there seems to be no compelling evidence for the involvement of methylation in carcinogenesis, however.

The induction of mutations involves not only straightforward mutagenic interactions between chemicals and DNA or chromosomes; in many cases, various types of indirect effects and interactions operate, and such processes are not always revealed in mutagenicity test systems. Some such possibilities may be mentioned.

Chemicals may act by causing an imbalance of the nucleotide pool, and such imbalances can have profound effects on the genome. Haynes (1985) points out that essentially all known forms of genetic alterations, from point mutations to oncogenic transformation and teratogenesis, have been observed subsequent to induced deoxyribonucleotide triphosphate pool disturbances in appropriate in-vivo assay systems. While some of these effects could be selection artefacts, it is clear that not all are.

The generation of oxygen radicals by chemicals constitutes another indirect mechanism for mutagenicity, which has attracted much attention lately. Ames (1983) pointed out that many of the known naturally occurring mutagens and carcinogens have properties that could result in the formation of radicals, rather than any direct interaction with DNA. Cerutti (for a review, see Cerutti, 1985) has found that lipid peroxidation and radical formation is a common property of promoters. Radical formation has also been implicated in tumour formation *in vivo* (Emanuel, 1973). The formation of oxygen radicals is a continuous process in all aerobic organisms, and presumably the net effect of oxygen radicals depends on a balance between the formation of radicals and the efficiency of the defense mechanisms against radicals, such as catalase, superoxide dismutase, glutathione peroxidase and vitamins A, C and E. The effect of radicals is therefore also highly dependent on the state of the defense mechanism. A binding or deficiency of radical-scavenging enzymes or vitamins will evidently be of importance. As an example, some thiurams are potent mutagens in

Salmonella without showing any corresponding binding to DNA. The mutagenicity is strongly dependent on the oxygen tension, and a pronounced synergism occurs with the radical-generating vitamin K (menadione). The background of the mutagenic effect of thiurams seems to be inactivation of defense enzymes against oxygen radicals, especially superoxide dismutase (Rannug & Rannug, 1984).

Other experimental results are difficult to explain on the basis of ordinary mutational events. This applies to the increased incidence of cancer in offspring and further generations after treatment of male and female germ cells with carcinogenic agents (Tomatis *et al.*, 1975; Nomura, 1982, 1986); Nomura (1982, 1986) reported from large-scale experiments that treatment of male mice with X-rays gave increased cancer incidence in F_2 and F_3 offspring of such high frequency that it can be explained only by one of three possibilities: involvement of a very large number of genetic sites, a remarkably high mutation frequency or a mechanism other than conventional mutations. An unexpectedly high frequency of polygenic mutations has been recorded at least in *Drosophila* (for a review, see Ramel, 1983). Mutations in polygenes that regulate quantitative characteristics such as viability, fecundity, intelligence and reaction speed may constitute a far more important genetic problem in human populations than major 'conventional' mutations. If mutagenic events involving polygenic systems are also of importance in carcinogenicity, our present battery of standard mutagenicity test systems is certainly not equipped to detect them, other than through extrapolation from major mutations.

Technical development and short-term testing

Since the IARC meeting on long-term and short-term screening assays for carcinogenicity (IARC, 1980), basic knowledge about the processes of mutagenicity and carcinogenicity has gone through a rapid and spectacular development. This applies in particular to analyses of viral and cellular oncogenes, their activation, expression and biological functions, as summarized above. Behind this development lies the equally dramatic process in molecular biology of knowledge about the genetic material. Recombinant DNA techniques with the use of restriction enzymes and restriction pattern analyses, cloning, DNA sequencing and blotting techniques have produced a revolution in biology. Through these techniques, it has been possible to analyse gene structure, gene regulation, DNA repair, mutations and other central issues at a molecular level in an unprecedented way (for a review, see Mekler *et al.*, 1985). This new development has provided essential background information for a better understanding of processes of importance in mutagenicity and carcinogenicity and therefore also of importance for practical screening to detect carcinogenic properties. At the applied and practical levels of testing, these techniques have not yet received any widespread use; however, they have an obvious potential in the near future for more applied purposes. Several DNA methods for possible practical use in monitoring human populations for mutation rates were identified at a recent meeting sponsored by ICPEMC and the US Department of Energy. These methods include restriction fragment length polymorphism, direct sequencing of DNA, gradient denaturation gels, heteroduplex DNA, subtractive hybridization with synthetic oligonucleotides and the use of RNAase A to cleave C:A mismatches in RNA:DNA heteroduplex (Mendelsohn, 1986).

Recombinant DNA techniques have been used to study mutations in a variety of organisms, but the data on mammalian genes is so far limited. Cloning of genes for *hprt* in rodents and humans (Konecki *et al.*, 1982) has made possible molecular analyses of mutations in this classical locus. It has been found for instance, that in-vivo *hprt* mutations in human lymphocytes include a remarkably high frequency (57%) of deletions, exon amplications and other gene alterations that are not revealed cytogenetically (Turner *et al.*, 1985). It is clear that this technique will open new possibilities for analysing the molecular mechanism of mutations by chemical mutagens and carcinogens, which in the future may be applied on a sufficient scale to be used practically.

The relationship between exposure to a genotoxic agent and the final effect on the organism is a central issue in risk estimation. This includes several problems — measurement of primary interactions with DNA, induction of DNA lesions and processing of these lesions to manifest mutations. These problems can be attacked by molecular methods. Although the use of monoclonal antibodies to detect defined lesions in DNA requires lengthy and cumbersome construction of specific antibodies, the technique has improved considerably and can be expected to be suitable also for more general practical applications in future. Measurement of DNA alkylation is another method that can give information on doses received at the DNA level from exposure both in experimental organisms and humans. Another measure of biological doses makes use of alkylation of haemoglobin, from which DNA alkylation can be calculated and the risk be estimated (Ehrenberg & Osterman-Golkar, 1980).

The processing of DNA lesions to mutations involves sieving of DNA repair, which has been shown in recent investigations to be extremely versatile with specific enzymes and proteins for different lesions. The interplay between exposure to mutagenic agents and repair is crucial for dose-response relationships and for differences in response to carcinogens in different tissues. The measurement of single-strand breaks by alkaline elution, DNA upspinning and centrifugation techniques have been particularly useful for measuring mutagenic exposure (for a review, see Friedberg *et al.*, 1981).

New emphasis in cancer testing

Extensive epidemiological analyses of cancer incidence have caused a certain shift in emphasis concerning the causes of cancer in human populations. The largest risk for cancer has been attributed to diet, while occupational exposure seems to be responsible for only a few percent of cancer cases (Doll & Peto, 1981). There are strong indications that the high contribution of the diet is due to naturally occurring compounds rather than to pesticide residues or artificial food additives. A steadily increasing number of natural mutagenic and carcinogenic compounds is being recognized in ordinary food. Of particular importance in this respect are the pyrolysis products of amino acids, which have been identified, synthesized and studied with respect to mutagenicity and carcinogenicity by Sugimura and his collaborators (Sugimura & Sato, 1983; Sugimura, 1985).

As mentioned above, Ames (1983) emphasized that many dietary carcinogens can be assumed to act by the production of oxygen radicals. Well-known carcinogens, such as benzo[*a*]pyrene, which hitherto have been considered mutagenic solely because they have reactive metabolites that bind directly to DNA, may in fact also act by the generation of oxygen radicals (Chesis *et al.*, 1984). Cerutti (1985) pointed out that known promoters have the common denominator of being 'membrane active', causing lipid peroxidation and generation of oxygen radicals. These findings have directed much attention to oxygen stress and the reaction of cells to such stress. The response to this stress is very similar to the response to heat shocks; and similar signal substances, polyadenylated nucleotides, seem to be involved (Lee *et al.*, 1983). At a more practical level, the mutagenic effect of oxygen radicals is at least predominantly on A—T base pairs rather than on C—G, as is the case for most other chemical mutagens. Ames' group (Maron & Ames, 1983) has therefore used peroxide-sensitive *Salmonella* strains to construct new tester strains, based on A—T base-pair substitution and therefore suitable for the detection of lesions by radical-generating mechanisms.

The induction of mutations and cancer by chemicals in animals and humans is dependent not only on the genotoxic potency of the chemicals but also on the occurrence and status of various endogenous and exogenous defence mechanisms. In fact, the net result of exposure to a carcinogen is to a great extent governed by a balance between opposite forces — activation and deactivation of procarcinogens, promotion and antipromotion, formation of reactive radicals and scavenging of such radicals. In order to prevent human cancer, recognition of factors that modify the effect of carcinogenic agents may be just as essential as identification of carcinogens themselves (Ramel, 1984). The large variation in cancer incidence among different human populations cannot be attributed only to exposure to genotoxic chemicals but also to the variation in efficiency of appropriate defence mechanisms. This is particularly important in connection with oxygen radicals, which are generated continuously in aerobic organisms and against which the cells have developed elaborate defense mechanisms. The extent to which the balance between radical formation and radical scavenging can be upset *in vivo* in higher organisms is a question of primary importance for evaluating the risk of many chemicals.

Viewpoints on the strategy of testing

The viewpoints expressed at the previous IARC meeting (IARC, 1980) by the working group on the rationale for deployment of short-term assays for evidence on carcinogenicity, with Drake and de Serres as rapporteurs, are still valid, but recent developments in basic and applied research within this area justify some shift in emphasis in testing and some general comments on the strategy of testing.

The aim of short-term mutagenicity testing in the present context is the detection of carcinogenic activity of chemicals. With reference to what has been said above, negative experimental and epidemiological cancer data cannot indicate definitely the carcinogenic effect of a chemical, and in particular not its cocarcinogenic potency. Conversely, a chemical classified falsely as positive from extrapolation from mutagenicity data may turn out to be a true positive after further investigation. Therefore, a primary aim of mutagenicity testing is to provide sufficient evidence of whether a compound can be expected to be *mutagenic* in humans. Even against a background of negative cancer data, observation of a mutagenic effect in several test systems must also be taken seriously from the point of view of carcinogenicity, unless sufficient evidence of the mechanism of action can furnish information to the contrary. Such information is difficult, but not impossible, to obtain. Dichlorvos is an example of a compound the carcinogenicity of which appears to be at least very low in spite of positive in-vitro mutagenicity test data. This evaluation can be made on the basis of its mechanism of action in higher organisms (see Ramel *et al.*, 1980).

For practical screening of mutagens, the new techniques mentioned above can be assimilated only gradually into testing protocols. At present, sufficiently validated standard test protocols must be used for more routine purposes. However, in view of our lack of knowledge about the actual genetic mechanisms behind neoplastic transformation and the possible involvement of 'unconventional' mutagenic processes, as touched upon above, testing requirements may change rapidly with new insights into the genetic mechanisms of carcinogenesis. The endpoint in short-term assays that theoretically contributes the closest link to carcinogenicity is cellular transformation, although the practical experimental procedure involves many problems (see Brookes, 1981).

As mentioned above, an important conclusion from recent knowledge of oncogene activation and genetic regulation of neoplastic transformation is the fact that essentially all kinds of mutational endpoints may be involved in one way or another in carcinogenesis. Therefore, complete testing for carcinogenicity must cover a very wide spectrum of genetic endpoints. The predictive value and accuracy of short-term tests for different endpoints varies. One endpoint that may very well be of great importance to tumour formation is nondisjunction (Klein & Klein, 1984; Murphree & Benedict, 1984). It is unfortunate that reliable short-term test methods for detecting nondisjunction are largely lacking or at least insufficiently validated. There is a great need for development in this area.

The actual strategy of testing cannot be decided upon according to a rigid schedule but must be adapted to circumstances, such as the extent and mode of exposure, the nature and properties of the chemicals and pharmacokinetic data. Test systems recommended for premarketing evaluations by the Organization for Economic Cooperation and Development and the European Economic Community constitute minimal criteria, which are often not sufficient for reliable identification of mutagenic and carcinogenic activity. The requirement of a large battery of tests, however, inevitably results in an accumulation of data pointing in both positive and negative directions. The large international collaborative test programmes, such as that initiated by the UK Medical Research Council (de Serres & Ashby, 1981) and the in-vitro (Ashby et al., 1985) and in-vivo (Ashby et al., 1986) programmes organized by the International Programme on Chemical Safety (IPCS), the World Health Organization (WHO), the International Labour Organization (ILO) and the United Nations Environmental Programme (UNEP), have furnished invaluable model material for handling such data. The large data base of published material from different test systems in the US Gene-Tox programme, reported in a series of articles in *Mutation Research*, also constitute an invaluable source of basic material for investigations and for the development of improved test strategies.

The handling of these massive amounts of data opens up new possibilities for a rational use of accumulated information for safer interpretation of data and for formulating more rational test strategies. At the same time, this undertaking requires new and more sophisticated computerized methods. As an example of such approaches, Rosenkranz and coworkers have developed two computer systems, the CPBS (Carcinogenicity Prediction and Battery Selection) program and the CASE (Computer Automated Structure Evaluation) program for computer analysis of data (see Rosenkranz et al., 1985 and this volume). The CPBS procedure can be used to predict the carcinogenicity of chemicals on the basis of short-term test results. Different tests are given different weights according to their known performance characteristics, and a prediction of carcinogenicity can be obtained even with mixtures of positive and negative test results. The CASE system is based on the substructures of chemicals, to predict carcinogenicity both qualitatively and quantitatively. The programme constitutes an artificial intelligence, which can be fed new pieces of information continously to increase the accuracy of predictions based on chemical structures. Other computerized structure-activity correlation programs have been developed by Tinker (1981a,b). Cluster and multivariate analyses are very useful for the analysis of complex data (Wold et al., 1984; Benigni & Giuliani, 1985), and these methods can be expected to be important in the near future for the analysis of mutagenicity data in general.

The use of these computer programs, and in particular the CPBS method of Rosenkranz et al. (1985), makes it possible to recognize assays that overlap and are not independent of each other, as they measure the same genetic mechanism or endpoint.

Examples of pairs of such overlapping assays are *Salmonella* and unscheduled DNA synthesis, *Drosophila* recessive lethals and unscheduled DNA synthesis, *Salmonella* and *Drosophila* recessive lethals (Pet-Edwards et al., 1985). Such associated assays should not be included in the same test battery, as they provide no more information and add to the cost (see Rosenkranz et al., 1984a).

The presentation of test data from short-term tests in a surveyable manner is important and difficult. A graphical method with an evaluation of the data has been suggested by Garrett et al. (1984). This procedure was discussed for possible adaptation to the *IARC Monographs on the Evaluation of the Carcinogenic Risk of Chemicals to Humans* at the Fourth International Conference on Environmental Mutagens in Stockholm in 1985.

It is not the intention of this paper to elaborate on the actual composition of test batteries for various purposes, but some general remarks may be appropriate.

First, the assays to be used should be adapted as far as possible to the nature of the test substances. It is known that certain classes of chemicals perform better in one test system than another. Assays with microorganisms and in particular Ames' test on *Salmonella* will constitute a core in practically all testing strategies, because of their sensitivity. The *Salmonella* assay is an indispensible tool not only for the testing of individual chemicals, but also for the recognition of mutagenic impurities and for the identification of mutagenic compounds in complex mixtures, that is, in the form that exposures generally occur in real life. It is also evident, however, that bacterial tests do not pick up all classes of carcinogens with similar efficiency (see Rinkus & Legator, 1975). Thus, chlorinated compounds are much less efficiently identified than polyaromatic hydrocarbons. As emphasized by Ashby (1983, 1986), it is important that the different predictive powers of in-vitro and in-vivo short-term tests for carcinogenicity be taken into consideration. The response of an intact organism will inevitably differ from that of in-vitro systems at several levels — uptake, distribution, activation, detoxification and so on. As was evident in the large IPCS collaborative test programme on in-vivo tests, mentioned above (Ashby et al., 1986), an impressive number of in-vivo tests exists in rodents, although many of them require further validation. A promising area is measurement of unscheduled DNA synthesis *in vivo* in liver (Mirsalis & Butterworth, 1980). Lutz et al. (1984), studying the effect of carcinogens on rat liver, showed that carcinogens could be separated into roughly two groups — those that bind covalently to DNA but do not affect cell division and those that do not bind to DNA but induce cell division. The former group seems to correspond to initiating carcinogens, and the second to promoters. In the IPCS collaborative studies, *in-vivo* measurements of unscheduled DNA synthesis in rat liver allowed the recognition of chemicals that perform unscheduled DNA synthesis and those that induce cell division at the same time (Ashby et al., 1986). The possibility of using this method for simultaneous recognition of initiating and promoting agents is presently being explored (Beije, personal communication).

The unscheduled DNA synthesis technique is not limited to investigations of the liver but can also be applied to other tissues, such as the stomach (Furihata & Matsushima, 1982), tracheal epithelial cells (Doolittle & Butterworth, 1984), bronchial epithelial cells (Doolittle *et al.*, 1985), pancreatic cells (Steinmetz & Mirsalis, 1984), kidney cells (Tyson & Mirsalis, 1986) and spermatocytes (Working & Butterworth, 1984).

The fruitfly, *Drosophila malanogaster*, constitutes a useful in-vivo system, as many of its physiological and enzymatic systems are the same as or very closely related to those of mammals. The sex-linked recessive lethal system has shown high sensitivity as well as specificity (Rosenkranz *et al.*, 1984b). Additional properties of importance relate to the fact that this organism is the genetically best known higher organism, and a wide spectrum of test systems are available.

References

Albino, A.P., Lestrange, R., Oliff, A.I., Furth, M. & Old, L.J. (1984) Transforming *ras* genes from human melanoma. A manifestation of tumor heterogeneity. *Nature, 308*, 69-72

Ames, B.N. (1983) Dietary carcinogens and anticarcinogens. *Science, 221*, 1256-1263

Ashby, J. (1983) The unique role of rodents in the detection of possible human carcinogens and mutagens. *Mutat. Res., 115*, 177-213

Ashby, J. (1986) The prospects for a simplified and internationally harmonized approach to the detection of possible human carcinogens and mutagens. *Mutagenesis, 1*, 3-16

Ashby, J., de Serres, F., Draper, M., Ishidate, M., Jr, Margolin, B.H., Matter, B.E. & Shelby, M.D. (1985) *Evaluation of Short-term Tests for Carcinogenicity. Report of the International Programme on Chemical Safety's Collaborative Study on In Vitro Assays*, Amsterdam, Elsevier

Ashby, J., de Serres, F., Draper, M., Ishidate, M., Jr, Margolin, B.H., Matter, B.E. & Shelby, M.D. (1986) *Evaluation of Short-term Tests for Carcinogenicity. Report of the International Programme on Chemical Safety's Collaborative Study on In Vivo Assays*, Amsterdam, Elsevier (in press)

Balmain, A., Ramsden, M., Bowden, G.T. & Smith, J. (1984) Activation of the mouse cellular Harvey-*ras* gene in chemically induced benign skin papillomas. *Nature, 307*, 658-660

Baltimore, D. (1985) Retroviruses and retrotransposons: the role of reverse transcription in shaping the eukaryotic genome. *Cell, 40*, 481-482

Benigni, R. & Giuliani, A. (1985) Cluster analysis of short-term tests: a new methodological approach. *Mutat. Res., 147*, 139-151

Borek, C., Morgan, W.F., Ong, A. & Cleaver, J.E. (1984) Inhibition of malignant transformation *in vitro* by inhibitors of poly (ADP-ribose) synthesis. *Proc. natl Acad. Sci. USA, 81*, 243-247

Bos, J.L., Toksoz, D., Marshall, C.J., Verlaan-de Vries, M., Veeneman, G.H., van der Eb, A.J., van Boom, J.H., Janssen, J.W.G. & Steerorden, A.C.M. (1985) Amino-acid substitutions at codon 13 of the N-*ras* oncogene in human acute myeloid leukaemia. *Nature, 315*, 726-730

Brickell, P.M., Latchman, D.S., Murphy, D., Willison, K. & Rigby, P.W. (1983) Activation of a Qa/Tla class I major histocompatibility antigene gene is a general feature of oncogenesis in the mouse. *Nature, 306*, 756-760

Brookes, P. (1981) Critical assessment of the value of in-vitro cell transformation for predicting in-vivo carcinogenicity of chemicals. *Mutat. Res.*, *86*, 233-242

Capon, D.J., Seeburg, P.H., McGrath, J.P., Hayflick, J.S., Edman, U., Levinson, A.D. & Goeddel, D.V. (1983) Activation of Ki-*ras* 2 gene in human colon and lung carcinomas by two different point mutations. *Nature*, *304*, 507-513

Cerutti, P.A. (1985) Prooxidant states and tumor promotion. *Science*, *227*, 375-381

Chesis, P.L., Levin, D.E., Smith, M.T., Ernster, L. & Ames, B.N. (1984) Mutagenicity of quinones: pathways of metabolic activation and detoxification. *Proc. natl Acad. Sci. USA*, *81*, 1696-1700

Chou, T.-C. (1980) Comparison of dose-effect relationships of carcinogens following low-dose chromic exposure and high-dose single injection: an analysis by the median-effect principle. *Carcinogenesis*, *1*, 203-213

Davis, M., Malcolm, S. & Rabbitts, T.M. (1984) Chromosome translocation can occur on either side of the c-myc oncogene in Burkitt lymphoma cells. *Nature*, *308*, 286-288

Doll, R. & Peto, R. (1981) The causes of cancer: quantitative estimates of avoidable risks of cancer in the United States today. *J. natl Cancer Inst.*, *66*, 1191-1308

Doolittle, D.J. & Butterworth, B.E. (1984) Assessment of chemically induced DNA repair in rat tracheal epithelial cells. *Carcinogenesis*, *5*, 773-779

Doolittle, D.J., Furtong, J.W. & Butterworth, B.E. (1985) Assessment of chemically induced DNA repair in primary cultures of human bronchial epithelial cells. *Toxicol. appl. Pharmacol.*, *79*, 28-38

Druckrey, H. (1972) Present status of cancer research [in German]. *Arzliche Prax.*, *24*, 1-18

Ehrenberg, L. & Osterman-Golkar, S. (1980) Alkylation of macromolecules for detecting mutagenic agents. *Teratog. Carcinog. Mutagen.*, *1*, 105-127

Emanuel, N.M. (1973) Kinetics and the free-radical mechanisms of tumor growth. *Ann. N.Y. Acad. Sci.*, *222*, 1010-1030

Farber, E. (1980) *Reversible and irreversible lesions in processes of cancer development.* In: Montesano, R., Bartsch, H. & Tomatis, L., eds, *Molecular and Cellular Aspects of Carcinogen Screening Tests (IARC Scientific Publications No. 27)*, Lyon, International Agency for Research on Cancer, pp. 143-151

Fasano, O., Aldrich, T., Tamanoi, F., Taparowsky, E., Furth, M. & Wigler, M. (1984) Analysis of the transforming potential of the human H-*ras* gene by random mutagenesis. *Proc. natl Acad. Sci. USA*, *81*, 4008-4012

Feig, L.A., Bast, R.C., Jr, Knapp, R.C. & Cooper, C.M. (1984) Somatic activation of *ras* gene in a human ovarian sarcoma. *Science*, *223*, 698-701

Feinberg, A.P. & Vogelstein, B. (1983) Hypomethylation distinguishes genes of some human cancers from their normal counterparts. *Nature*, *301*, 89-92

Fisher, P.B., Goldstein, N.I. & Weinstein, I.B. (1979) Phenotypic properties and tumor promotor-induced alterations in rat embryo cells transformed by adenovirus. *Cancer Res.*, *39*, 3051-3057

Friedberg, E.C. & Hanawalt, P.C. (1981) *DNA Repair. A Laboratory Manual of Research Procedure*, New York, Marcel Dekker

Furihata, C. & Matsushima, T. (1982) *Unscheduled DNA synthesis in rat stomach. Short-term assay of potential stomach carcinogens.* In: Bridges, B., Butterworth, B.E. & Weinstein, I.B., eds, *Indicators of Genotoxic Exposure (Banbury Report 13)*, Cold Spring Harbor, NY, CSH Press, pp. 123-135

Garrett, N.E., Stack, H.F., Gross, M.R. & Waters, M.D. (1984) An analysis of the spectra of genetic activity produced by known or suspected human carcinogens. *Mutat. Res., 134*, 89-111

Guerrero, I.A., Villasante, A., Corces, V. & Pellicer, A. (1984a) Activation of a c-K-*ras* oncogene by somatic mutation in mouse lymphomas induced by gamma radiation. *Science, 225*, 1159-1162

Guerrero, I.A., Calzada, P., Mayer, A. & Pellicer, A. (1984b) A molecular approach to leukemogenesis: mouse lymphomas contain an activated c-*ras* oncogene. *Proc. natl Acad. Sci. USA, 81*, 202-205

Haynes, R.H. (1985) *Molecular mechanisms in genetic stability and change: the role of deoxyribonucleotide pool balance.* In: de Serres, F.J., ed., *Genetic Consequences of Nucleotide Pool Imbalance*, New York, Plenum, pp. 1-23

Hennings, H., Shores, R., Wenk, M.L., Spangler, E.F., Tarone, R. & Yuspa, S.H. (1983) Malignant conversion of mouse skin tumours is increased by human initiators and unaffected by tumour promoters. *Nature, 304*, 67-69

Hilpert, D., Romen, W. & Neumann, H.-G. (1983) The role of partial hepatectomy and of promoters in the formation of tumours in non-target tissues of *trans*-4-acetylaminostilbene in rats. *Carcinogenesis, 4*, 1519-1525

Holliday, R. (1979) A new theory of carcinogenesis. *Br. J. Cancer, 40*, 513-522

Holliday, R. & Pugh, J.E. (1975) DNA modification mechanisms and gene activity during the development. *Science, 187*, 226-232

Hollstein, M.J., McCann, J., Angelosanto, F.A. & Nichols, W.W. (1979) Short-term tests for carcinogens and mutagens. *Mutat. Res., 65*, 173-226

IARC (1980) *IARC Monographs on the Evaluation of the Carcinogenic Risk of Chemicals to Humans*, Suppl. 2, *Long-term and Short-term Screening Assays for Carcinogens: A Critical Appraisal*, Lyon

King, W., Patel, M.D., Lobel, L.I., Goff, S.P. & Nguyen-Huu, M.C. (1985) Insertion mutagenesis of embryonal carcinoma cells by retrovirus. *Science, 228*, 554-558

Kinsella, A.R. & Radman, M. (1978) Tumor promoter induces sister chromatid exchanges: relevance to mechanism of carcinogenesis. *Proc. natl Acad. Sci. USA, 75*, 6149-6153

Klein, G. & Klein, E. (1984) Oncogene activation and tumor progression. *Carcinogenesis, 5*, 429-435

Klein, G. & Klein, E. (1985) Evolution of tumours and the impact of molecular oncology. *Nature, 315*, 190-195

Knudsen, A.G. (1971) Mutation and cancer: statistical study of retinoblastoma. *Proc. natl Acad. Sci. USA, 68*, 820

Kolata, G. (1985) Fitting methylation into development. *Science, 228*, 1183-1184

Konecki, D.S., Brennand, J., Fusco, J.C., Caskey, C.T. & Craigh Chinault, A. (1982) Hypoxanthine-guanine phosphoribosyl-transferase genes of mouse and Chinese hamster. Construction and sequence analysis of cDNA recombinants. *Nucleic Acids Res., 10*, 6763-6775

Kun, E., Kirsten, E., Milo, G.E., Kurian, P. & Kumari, H.L. (1983) Cell cycle dependent intervention by benzamide of carcinogen-induced neoplastic transformation and in-vitro poly (ADP ribosylation) of nuclear proteins in human fibroblasts. *Proc. natl Acad. Sci. USA, 80*, 7219-7223

Land, H., Parada, L.F. & Weinberg, R.A. (1983a) Cellular oncogenes and multistep carcinogenesis. *Science, 222*, 771-778

Land, M., Parada, L.F. & Weinberg, R.A. (1983b) Tumorigenic conversion of primary embryo fibroblasts requires at least two cooperating oncogenes. *Nature, 304*, 596-602

Lee, P.C., Bochner, B.R. & Ames, B.N. (1985) ApppA, heat-shock stress, and cell oxidation. *Proc. natl Acad. Sci. USA, 80*, 7496-7500

Lutz, W.K., Büsser, M.-T. & Sagelsdorff, P. (1984) Potency of carcinogens derived from covalent DNA binding and stimulation of DNA synthesis in rat liver. *Toxicol. Pathol., 12*, 106-111

Maron, D.M. & Ames, B.N. (1983) Revised methods for the *Salmonella* mutagenicity test. *Mutat. Res., 113*, 173-215

Marshall, C.J., Vousden, K.M. & Phillips, D.M. (1984) Activation of c-Ha-*ras*-1 proto-oncogene by in-vitro modification with a chemical carcinogen, benzo[*a*]pyrene diol-epoxide. *Nature, 310*, 586-589

Marx, J.L. (1983) Cooperation between oncogenes. *Science, 222*, 602-603

McClintock, T. (1984) The significance of response of the genome to challenge. *Science, 226*, 792-801

Mekler, P., Delehanty, J.T., Lohman, P.H.M., Browner, J., Putte, P.V.D., Pearson, P., Pouwels, P.H. & Ramel, C. (1985) The use of recombinant DNA technology to study gene alteration. ICPEMC Publication No. 11. *Mutat. Res., 153*, 13-55

Mendelsohn, M.L. (1986) *Prospects for DNA methods to measure human heritable mutation rates.* In: Ramel, C., Lambert, B. & Magnusson, J., eds, *Genetic Toxicology of Environmental Chemicals*, Part B, *Genetic Effects and Applied Mutagenesis,* New York, Alan R. Liss, pp. 337-344

Mirsalis, J.C. & Butterworth, B.E. (1980) Detection of unscheduled DNA synthesis in hepatocytes isolated from rats treated with genotoxic agents: an in vivo-in vitro assay for potential carcinogens and mutagens. *Carcinogenesis, 1*, 621-625

Murphree, A.L. & Benedict, W.F. (1984) Retinoblastoma: clues to human oncogenesis. *Science, 223*, 1028-1033

Newbold, R.F. & Overell, R.W. (1983) Fibroblast immortality is a prerequisite for transformation by EJ c-Ha-Ras oncogene. *Nature, 304*, 648-651

Nomura, T. (1982) Parental exposure to X rays and chemicals induces heritable tumours and anomalies in mice. *Nature, 296*, 575-577

Nomura, T. (1986) *Further studies on X-ray and chemically induced germ-line alterations causing tumors and malformations in mice.* In: Ramel, C., Lambert, B. & Magnusson, J., eds, *Genetic Toxicology of Environmental Chemicals*, Part B, *Genetic Effects and Applied Mutagenesis*, New York, Alan R. Liss, pp. 13-20

Parry, J.M., Parry, E.M. & Barrett, J.C. (1981) Tumour promotors induce mitotic aneuploidy in yeast. *Nature, 294*, 363-365

Pet-Edwards, J., Chankong, V., Rosenkranz, H.S. & Haimes, Y.Y. (1985) Application of the carcinogenicity prediction battery selection (CPBS) method to the Gene-Tox data base. *Mutat. Res., 153*, 187-200

Peto, R., Roe, F.J.C., Lee, P.N., Levy, L. & Clack, J. (1975) Cancer and ageing in mice and men. *Br. J. Cancer, 32*, 411-426

Rabbitts, T.M., Hamlyn, P.H. & Baer, R. (1983) Altered nucleotide sequence of a translocated c-*myc* gene in Burkitt lymphoma. *Nature, 306*, 760-765

Ramel, C. (1983) Polygenic effects and genetic changes affecting quantitative traits. *Mutat. Res., 114*, 107-116

Ramel, C. (1984) General environmental modifiers of carcinogenesis. *Acta pharmacol. toxicol., 5* (Suppl. 11), 181-196

Ramel, C., Drake, J. & Sugimura, T. (1980) An evaluation of the genetic toxicity of dichlorovos. *Mutat. Res., 76*, 297-309

Rannug, A. & Rannug, U. (1984) Enzyme inhibition as a possible mechanism of the mutagenicity of dithiocarbamic acid derivatives in *Salmonella typhimurium. Chem.-biol. Interact.*, *49*, 329-340

Reddy, E.P., Reynolds, R.K., Santos, E. & Barbacid, M. (1982) A point mutation is responsible for the acquisition of transforming properties by the T24 human bladder carcinoma oncogenes. *Nature*, *300*, 149-152

Rinkus, S.J. & Legator, M.S. (1975) Chemical characterization of 465 known or suspected carcinogens and their correlation with mutagenic activity in the *Salmonella. Cancer Res.*, *39*, 3289-3318

Rosenkranz, H.S., Klopman, G., Chankong, V., Pet-Edwards, J. & Haimes, Y.Y. (1984a) Prediction of environmental carcinogens: a strategy for the mid-1980s. *Environ. Mutagen.*, *6*, 231-258

Rosenkranz, H.S., Pet-Edwards, J., Chankong, V. & Haimes, Y.Y. (1984b) Assembling a battery of assays to predict carcinogenicity: a case study. *Mutat. Res.*, *14*, 65-68

Rosenkranz, H.S., Mitchell, C.S. & Klopman, G. (1985) Artificial intelligence and Bayesian decision theory in the prediction of chemical carcinogens. *Mutat. Res.*, *150*, 1-11

Rubin, G.M. & Spradling, A.C. (1982) Genetic transformation of *Drosophila* with transposable element vectors. *Science*, *218*, 348-353

Ruley, H.E. (1983) Adenovirus early region 1A enables viral and cellular transforming genes to transform primary cells in culture. *Nature*, *304*, 602-606

Santos, E., Martin-Zanca, D., Reddy, E.P., Pierotti, M.A., Della Porta, G. & Barbacid, M. (1984) Malignant activation of a K-*ras* oncogene in lung carcinoma but not in normal tissue of the same patient. *Science*, *223*, 661-664

de Serres, F.J. & Ashby, J., eds (1981) *Evaluation of Short-term Tests for Carcinogens. Report of the International Collaborative Program*, Amsterdam, Elsevier

Shih, C., Shilo, B.-Z., Goldfarb, M.P., Dannenberg, A. & Weinberg, R.A. (1979) Passage of phenotypes of chemically transformed cells *via* transfection of DNA and chromatin. *Proc. natl Acad. Sci. USA*, *76*, 5714-5718

Slaga, T.J. (1980) *Cancer: etiology mechanisms and prevention — a summary*. In: Slaga, T.J., ed., *Modifiers of Chemical Carcinogenesis*, New York, Raven Press, pp. 243-262

Spandidos, D.A. & Wilkie, N.M. (1984) Malignant transformation of early passage rodent cells by a single mutated human oncogene. *Nature*, *310*, 469-475

Spradling, A.C. & Rubin, G.M. (1982) Transposition of cloned P elements into *Drosophila* germ line chromosomes. *Science*, *218*, 341-347

Steinmetz, K.L. & Mirsalis, J.C. (1984) Induction of unscheduled DNA synthesis in primary cultures of rat pancreatic cells following in-vivo and in-vitro treatment with genotoxic agents. *Environ. Mutagen.*, *6*, 321-330

Sugimura, T. (1985) Carcinogenicity of mutagenic heterocyclic amines formed during the cooking process. *Mutat. Res.*, *150*, 33-41

Sugimura, T. & Sato, S. (1983) Carcinogenicity of mutagenic heterocyclic amines formed during the cooking process. *Cancer Res.*, *43*, 2415-2421

Sukumar, S., Pulciani, S., Doniger, J., di Paolo, J.A., Evans, C.M., Zbar, B. & Barbacid, M. (1984) A transforming *ras* gene in tumorigenic guinea pig cell lines initiated by diverse chemical carcinogens. *Science*, *223*, 1197-1199

Tabin, C.J., Bradley, S.M., Bargmann, C.I., Weinberg, R.A., Papageorge, A.G., Skolnick, E.M., Dahr, R., Lowy, D.R. & Chang, E.H. (1982) Mechanism of activation of a human oncogene. *Nature*, *300*, 143-149

Tainsky, M.A., Cooper, C.S., Gioranella, B.C. & Van de Woude, G.F. (1984) An activated *ras* gene; detected in late but not early passage human PA1 teratocarcinoma cells. *Science, 225*, 643-645

Taparowsky, E., Suard, Y., Fasano, O., Schimiza, K., Goldfarb, M. & Wigler, M. (1982) Activation of the T24 bladder carcinoma transforming gene is linked to a single amino acid change. *Nature, 300*, 762-765

Tinker, J.F. (1981a) A computerized structure-activity correlation. Program for relating bacterial mutagenesis activity to chemical structure. *J. comput. Chem., 2*, 231-243

Tinker, J.F. (1981b) Relating mutagenicity to chemical structure. *J. chem. Inf. Comput. Sci., 21*, 3-7

Tomatis, L., Hilfrich, J. & Turusov, V. (1975) Occurrence of tumors in F_1, F_2 and F_3 descendants of BD rats exposed to *N*-nitrosomethylurea during pregnancy. *Int. J. Cancer, 15*, 385-390

Turner, D.R., Morley, A.A., Haliandros, M., Kutlace, R. & Sanderson, B.J. (1985) In-vivo somatic mutations in human lymphocytes frequently result from major gene alterations. *Nature, 315*, 343-345

Tyson, C.K. & Mirsalis, J.C. (1986) Measurement of unscheduled DNA synthesis in rat kidney cells following in-vivo treatment with genotoxic agents. *Environ. Mutagen.* (in press)

Wold, S., Albano, C., Dunn, W.J., III, Ebersen, K., Hellberg, S., Johansson, E., Lindberg, W. & Sjöström, M. (1984) Modelling data tables by principle components and PLS: class patterns and quantitative predictive relations. *Analusis, 12*, 477-485

Working, P.K. & Butterworth, B.E. (1984) An assay to detect chemically induced DNA repair in rat spermatocytes. *Environ. Mutagen., 6*, 273-286

Zarbl, H., Sukumar, S., Arthur, A.V., Martin-Zanca, D. & Barbacid, M. (1985) Direct mutagenesis of Ha-*ras*-1 oncogenes by *N*-nitroso-*N*-methyl-urea during initiation of mammary carcinogenesis in rats. *Nature, 315*, 382-385

SUBJECT INDEX

A

Absorption, of chemicals, 22, 26, 27, 163, 409, 422, 429, 524
2-Acetylaminofluorene (AAF)
  in assay for liver initiation-promotion, 107-110
  in assay for thyroid initiation-promotion, 113
  in *Drosophila*, 370, 377
  in host-mediated assay, 164
  in human lymphoblast mutation assay, 188
  in tests for aneuploidy in fungi, 324
Activation, metabolic, 129, 130, **439-455**, 508, 545
  exogenous systems
    by hepatocytes, 137
    by mammalian cells, 448-449
    for complex mixtures, 486-487, 488
    in assays in fungi, 303, 307, 309, 310, 311, 313, 332, 338
    in bacterial tests, 143, 146, 151, 156, 157
    in cell transformation assays, 268, 271, 274, 276
    in intercellular communication assays, 296
    in mammalian cells in culture, 197, 211-212, 213, 221, 442
    in body-fluid analysis, 428
    in *Drosophila*, 356-361
Administration, route of
  in mammals, 23, 26-29, 31, 71, 112, 113, 164, 201, 212, 411
  in *Drosophila*, 373, 375-376, 380
Aerosol, testing of, 27
Aflatoxin
  activation of, 448
  adducts, 133
  in *Drosophila*, 356, 359, 369
  in human fibroblast mutation assay, 183
  in human lymphoblast mutation assay, 188
  in skin tumour assay, 105
Age
  and disease, 61-62
  and mutation, 192
  effect on DNA repair, 33, 87
  effect on metabolism, 33, 87, 439
  of experimental animals, 25, 32, 36, 411
  -specific incidence rate, 68, 69
Ah locus, 357-359
Alkaline elution, 133-134, 135, 137
Alkaline phosphatase
  assay, 147
  deficiency, 93, 111
  synthesis, 156
Alkaline sucrose gradient, 133-134, 135
Alkylating agent
  as inducers of sister chromatid exchange, 207
  covalent binding of, 129
  in cell transformation assays, 271, 276, 277
  in *Drosophila*, 365, 369, 374

Ames test (see also *Salmonella typhimurium*), 147, 151-154, 478, 519, 545
Analysis
  of tissues and physiological fluids, 31, **409-438**
  statistical (*see* Statistics)
Anchorage-independent growth (*see also* Growth), 272, 278, 279, 280, 291
Aneuploidy, 529, 535
  accuracy of tests for, 222-223
  in *Drosophila*, 219, 381, **395-407**
  in fungi, 303, 306, **313-339**
  in human cells, 167
  in mammalian cells, 8, 171, 200, 204, **218-226**
  in mammals *in vivo*, 523, 525
  in yeast, 287, 288
  relationship to reproductive pathology, 218-219
  relationship to tumours, 219-220, 223
Animal, experimental
  husbandry, 18, 23, 24, 37, 38, 41-42, 49, 411
  reception of, 48-50
Aspergillus nidulans, tests using, 303, 306-307, 309, 313, 317, 322, 327-339
Ataxia telangiectasia, 205, 217
5-Azacytidine, 194, 356
8-Azaguanine resistance, 154, 180-181, 183, 193, 196, 306, 427, 488

B

Bacillus subtilis, tests using, 147, 156, 513-514
Bacteria (*see also individual species*)
  differential survival of, 156-157, 163, 164
  in experimental animals, 60-61
  in host-mediated assay, 163-164
  in metabolic activation in *Drosophila*, 360
  short-term assays using, **143-161**, 202-203, 412-414, 427-428
Bacteriophage (*see also* Phage), 143, 163
BALBc/3T3 cells, transformation assay, 269, **272-275**, 289, 290-291, 292
Base, DNA (*see* DNA, base)
Battery, of short-term tests, 9, 151-153, 489, 513, 523-525, 526, 540, 544, 545
Bay region, 498, 501, 507, 511-512, 513
B-cell, human mutation assay, 177
Bedding, of experimental animals, 50-51, 64
Benzo[*a*]pyrene
  adducts, 133
  as indicator of air quality, 486
  as oncogene activator, 531
  as promoter, 106
  diol epoxide, 131, 532
  hydroxylase, 444

—553—

SUBJECT INDEX

Benzo[a]pyrene (contd)
    in body-fluid analysis, 413
    in *Drosophila*, 357, 358, 370, 377, 381
    in human fibroblast mutation assay, 182
    in human lymphoblast mutation assay, 185
    in mouse spot test, 203
    in production of hyperplasia, 92
    in specific-locus test, 246
    in Syrian hamster embryo-cell transformation assay, 289
    in tests for aneuploidy, 224, 324
    mode of action, 542
    mutagenicity of, 485
Bias
    in analysis of long-term studies in animals, 20, 68
    in analysis of mutagenesis assays, 459
    in analysis of structure-activity relationships, 502-503
    in body-fluid analysis, 424
    in gene mutation assays, 169, 177-178
Binding
    covalent, to DNA, 129, 130, 440, 545
    to nucleic acids, 447
    to nucleophiles, 443
    to proteins, 447
Binomial distribution, 378, 466, 471-473, 504
Biostatistics (*see* Statistics)
Bleomycin
    in chromosomal damage, 171-172, 208
    in nondisjunction in *Drosophila*, 402, 404
    in nondisjunction in fungi, 335
    in sex-linked recessive lethal test, 371-372
    in sister-chromatid exchange induction, 208
    in T-cell mutations, 191
    in yeast assays, 317
Bloom's syndrome, 205, 207, 217
Body fluids, testing of, 146, **409-438**
    relevance to cancer, 428-429
Body weight, of experimental animals, 20, 25, 46
    gain, 35, 45, 54, 65-66
    observation of, 39, 56, 57
    randomization by, 38, 52
Bone marrow cells, 209, 210, 212, 220-221, 376, 418
5-Bromodeoxyuridine (BUdR), 136, 187, 206-207

C

Cages, for experimental animals, 20, 27, 28, 42, 46, 49, 411
    laminar flow, 50
    randomization of, 38
    selection of, 50-51
    solid-bottom, 50, 56
    wire-bottom, 51, 56, 60
Carcinogen
    complete, 104, 105, 529, 537
    mutagenic, 115-116
    naturally occurring, 356
    organ-specificity of, 519
    potential, 7, 8, 143, 203, 409, 428, 429, 497
    proximate, 440, 441
    structure-activity relationships of, 505-514
    ultimate, 129, 439, 440, 441
Carcinogenesis (*see also* Carcinogenicity)
    complete, 104, 105, 106
    definition of, 520-521
    DNA damage in, 143
    mechanisms of, 7, 9, 33, 148, 268, 272, 522, 540
    prediction of, 497-517
    relation to body-fluid analysis, 418, 429
    relation to endpoints of bacterial tests, 148
    stages of, 8, **103-126**, 381, 491
Carcinogenicity
    and aneuploidy, 318-319, 333-336, 338
    and chromosomal damage, 172
    and mutagenicity in *Drosophila*, 370, 372-373, 375-378, 380, 395, 397
    and sex-linked recessive lethal test, 363
    assays, long-term in experimental animals, 7, **13-126**, 483, 524, 537-538
    of complex mixtures, 483-486, 487-491
    mechanisms of, 535-537
    modes of, 521-522
    potential
        with alkaline elution assay, 134
        with bacterial tests, 146
        with mammalian cell assays, 199, 220, 221, 288
        with mammals *in vivo*, 526
    prediction of, 170-171, 213, 214, 303, 499, 519, 522, 529-551
Cell line (*see also individual lines*, Mammalian cells)
    as source of activation, 448
    human, 169, 177, 178, 179-191, 221, 224-225, 226
    rodent, 194, 195-196, 221, 224-225, 226, 269
Cell-to-cell communication (*see* Intercellular communication)
Cell transformation (*see* Transformation, neoplastic)
C3H 10T½ cells, 269, 288
    as activating system, 448
    transformation assay, **275-277**, 290-291
Chromatid (*see also* Sister chromatid exchange), 205, 206, 398
    break, 208, 215
    deletion, 257
    gap, 215
    interchange, 257
    structure, 311
Chromosome
    aberrations, 163, 168, 170, 209, 303, 331, 523, 529
        in Chinese hamster ovary cells, 412
        in *Drosophila*, 352, 354, 366, 367, 370, 371, 374, 396
        spontaneous, 205
    assays, 169-170, 171
    breaks, 205, 213, 260, 332, 398, 402, 403
    damage, 171-172, 200, 415, 489
    deletion
        in fungi, 303, 311, 331, 332
        in mammalian cells, 8, 167, 204, 205, 216, 533

SUBJECT INDEX

Chromosome (contd)
  fragments, 205
  gain, 395, 400, 401, 403, 404
  gaps, 213
  hybridization, 115
  loss, 395, 396-397, 398, 400, 403, 404, 533
  nondisjunction (see Nondisjunction)
  numerical changes (see Aneuploidy)
  rearrangement, 204, 205, 209, 311, 312, 313
  relevance of endpoints to cancer, 216-218, 428
  segregation, 525
  structural changes in, 203-206, 210-218
  translocation
    in human cancer, 216
    in other human cells, 216-217
    in mammalian cells, 8, 167, 205
Cigarette smoking (see Tobacco)
Cisplatin
  in body-fluid analysis, 412, 426
  in DNA adducts, 131
  in host-mediated assay, 164
  in storage effects in *Drosophila*, 375
  patients treated with, 416
Clastogen, 205-206, 209, 217, 396
Cofactors, in metabolic system, 146, 271, 443, 445, 448
Colony-forming unit, 458, 466
Complex mixture
  and chromosomal aberrations, 206
  body fluids and excreta as, 424-425
  identification of components in, 9, 146, 426, 519, 545
  testing of, **483-494**
Contamination
  in diet, 44
  of laboratory, 26, 29
Correlation between genetic activity and
  carcinogenicity (see also Carcinogenicity,
  Validation), 9, 148-149, 312, 319, 338, 363-364,
  367, 372-373, 375-378, 380, 395, 397, 418, 429,
  530, 537-538
Covalent binding index, 130
Cyclophosphamide
  in body-fluid analysis, 412
  in chromosomal damage, 172
  in preneoplastic lesions of the urinary bladder, 93
  in specific-locus test, 247
  in tests for aneuploidy in fungi, 412
Cytochrome
  b_5 in *Drosophila*, 357
  in metabolic activation systems, 440-451
  P448 in cultured mammalian cells, 277
  P450 as indicators of carcinogen-induced
    liver foci, 87
  P450 in *Drosophila*, 352, 356-358, 360, 370, 380
  P450 in human cells, 173, 179
  P450 in yeast, 310-311

D

Deletion mutation, 351, 353, 354, 362, 368, 370, 371, 380
Density, cell, 175, 197
  artefacts, 178, 192
  inhibition of growth, 274, 275

Design, of bioassays, 8, 17, 18-46, 459, 483
Detoxification, metabolic, 30, 32, 35, 211, 309, 359,
  441-442, 443, 449, 464, 545
Diet
  administration of test chemicals in, 19, 26, 64, 487
  and human cancer, 7, 542
  as confounder in body-fluid analysis, 419, 431
  choice of, 42-45, 411
  effect on metabolism, 439
  effect on pharmacokinetics, 32, 33, 37
  storage of, 44, 49
  testing of, 412, 413, 415, 416, 418, 419-420, 430, 519
Differentiation, cell, 115, 116, 117, 288, 530, 538
  inhibition of, 287, 533
7,12-Dimethylbenz[*a*]anthracene (DMBA)
  as initiator, 104, 105, 106, 107
  as positive control, 72
  in aneuploidy test, 225
  in *Drosophila*, 370, 371-372, 374
  in induction of lung preneoplasia, 92
  in mammary tumour induction, 532
  in specific-locus test, 247
  model for pancreatic foci, 91
Dimethyl sulfoxide (DMSO)
  in Ames test, 468, 486
  in *Drosophila*, 374
  in fish, 117
  in fungi, 326
Diphtheria toxin resistance, 182-183
Diploid fibroblasts, human, 137, 221-222, 226
DNA
  adduct, 129-133, 135, 523
    analysis of, 130-133
    formation, 441, 445
    in humans, 169
    level in relation to dose, 30, 32
    mitochondrial, 312
    relationship to chromosome damage, 204, 214
    relationship to tumour response, 33
  amplification, 208
  base
    deletion, 148, 168, 173, 248, 253, 365
    disturbance, 204
    insertion, 148, 168, 173, 248, 253
    loss, 135
    -pair substitution, 172, 173, 253, 353, 371
    substitution, 148, 168, 248, 303
    translocation, 148
  covalent binding to, 129, 130, 440
  crosslink, 375, 396
  damage, **129-142**, 200
    detection of, 130-134, 428, 489
    in *Drosophila*, 360-361, 374, 397
    in human tissues, 8, 33, 193
    in yeast, 513
    response of bacteria to, 153-157
  interstrand crosslinks, 207, 214, 375
  lesions
    chemical nature of, 144
    progression of, 197
    reversion of, 144

SUBJECT INDEX

DNA (contd)
    methylation, 194-195, 217, 529, 534-535, 539
    miscoding, 144
    -protein crosslinks, 207
    repair, 32, 33, **129-142**, 200, 258, 428
        -deficient bacteria, 143, 156-157, 163
        detection of, 134-138
        excision, 129, 133, 134, 135, 144, 151, 156, 198, 375, 396, 397, 521
        incision, 130, 134, 135, 156
        ligation, 130, 135, 137
        recombination, 129, 134, 163, 303, 304, 307, 309, 322, 339, 351, 355, 367-373, 380, 398
        resynthesis, 130, 135
        synthesis, replicative, 136
        synthesis, unscheduled, 130, 135-138, 448, 523, 545, 546
    single-stranded breaks, 133, 134, 137, 163, 204, 288, 310, 365, 374, 523, 541
Dominant-cataract mutation test, **250-253**, 255, 256, 257, 261-262
Dominant-lethal mutation, 250, 374
    assay, **257-259**
Dose, animal experiments in
    effect on pharmacokinetics, 33
    in proportion to level administered, 30, 541
    level, 26, 34, 36, 70, 521, 536
    maximum tolerated, 25, 31, 32, 35, 66
    selection of, 31, 34-36, 53-54, 65
    single, 32
Dose-response relationship, 27, 34, 36
    and exposure time, 536, 541
    for metabolic activation, 445
    in bacterial tests, 148, 156
    in body-fluid analysis, 430, 431
    in cell transformation assays, 272, 274, 276, 277, 279
    in *Drosophila*, 366, 371
    in fungi, 304, 307, 309, 320, 323
    in mammalian cell assays, 199, 222, 226
    in mutagenesis assays, 459-465, 469, 470, 472, 473, 474
    in skin tumour assays, 490
    model, 70, 459-465, 469, 470, 472, 473, 474, 475-477, 478
    non-linear, 30
    of complex mixtures, 486, 491
Drinking-water
    administration of test compound in, 26, 30, 56, 485
    contamination of, 19, 483
    for experimental animals, 411
Drosophila melanogaster
    aneuploidy in (*see* Aneuploidy)
    genetic activity in, **351-393**, 546
    metabolic activation in, 360
    mobile elements in, 354-355
    mutations in, 254, 534, 539, 540, 545
Drug
    administration of, 29
    as confounder in body-fluid analysis, 430, 431
    cytostatic, mutagenicity of urine of patients, 420-421
    metabolism (*see* Enzyme, Metabolism)

E

Electrophile, 197, 204, 207
    metabolism to, 40, 143, 356, 440-445
Endonuclease, 145, 156
    micrococcal, 132
    uvr, 144
Enzyme (*see also individual enzymes*, Metabolism)
    changes in hyperplastic bladder, 111
    changes in preneoplastic lesions, 86-87, 89, 92, 93
    drug-metabolizing, 43, 173
    for metabolic activation, 439, 444
    in body-fluid analysis, 426, 428
    inducers, 410, 439
    inhibition of, 410, 439, 488
    phase I and phase II, 440-442, 446
    xenobiotic-metabolizing, 352, 356-360, 369
Epithelial cells
    in culture, 268, 278
    transformation assays, 280
    unscheduled DNA synthesis in, 546
Epoxide hydrolase, 86, 357, 441, 443
Escherichia coli
    genetic network of, 144
    in human cell mutation assays, 177
    K12/343/113, 147, 154, 157, 164, 427
    tests using, 147, 154, 156, 425, 427, 459, 513-514
    transposition in, 355
    WP2, 147, 154
Ethanol
    as solvent, 374
    in body-fluid analysis, 412
    in tests for aneuploidy in fungi, 334, 337, 338
Ethylene oxide, 73, 172, 249, 260
Ethylmethane sulfonate
    in dominant lethal test in *Drosophila*, 374
    in feeding solutions in *Drosophila*, 373
    in heritable translocation assay, 261
    in human lymphoblasts, 188
    in mitotic recombination test in *Drosophila*, 355-356
    in nondisjunction in *Drosophila*, 402
    in sex-linked recessive lethal test in *Drosophila*, 352, 353, 364-365, 366
    in specific-locus mutation assay, 247, 254
    in tests for aneuploidy in fungi, 318, 319, 326
N-Ethyl-*N*-nitrosourea
    in biochemical mutation test, 254-256
    in dominant-cataract mutation test, 251-253
    in *Drosophila*, 365, 366, 371-372
    in fish assay, 116, 117
    in human fibroblast mutation assay, 182
    in mouse lymphocytes, 201
    in specific-locus test, 247, 249
Excreta, analysis of (*see also* Faeces, Urine), **409-438**
Excretion, of test chemicals, 22, 26, 163, 409

SUBJECT INDEX

Exposure
  duration of, 36
  levels, 22, 541
  monitoring, 146, 171, 203, 213, 220, 378, 409, **410-424**, 428, 429
  occupational (*see* Occupation)
Extrapolation
  of bacterial test results to carcinogenesis, 148
  of high to low doses, 309
  of long-term carcinogenicity test results to humans, 30, 33-34
  of results in *Drosophila* to mammals, 352

F

Faeces
  analysis of, 409, 411-413, 415-417, 418, 425, 426, 430
  collection of, 422-423
False-negative result
  in long-term carcinogenicity tests, 25, 37, 66, 338
  in Strain A mouse lung tumour assay, 72
False-positive result
  in bacterial tests, 157, 467
  in short-term tests, 521, 524, 538, 543
  in Strain A mouse lung tumour assay, 72
Fanconi's anaemia, 205, 207, 217
Feed
  administration
    to *Drosophila*, 369, 373, 380
    to mammals, 44-45
  consumption, 20, 44, 46, 56, 57
Fibroblast
  human, 179, 198, 221, 531
    mutation assay, 180-183
    transformation of, 278, 294
  rodent, 195, 196, 221, 268, 448
  virus-infected, assay with, 277-278
Fish, use of
  in body-fluid analysis, 412
  in carcinogenicity testing, 23, **114-116**
Fluctuation test, 413, 415, 418, 425, 427, 459, 465
Focus
  preneoplastic in rat liver, 85-90, 108
  transformed
    in cell transformation assays, 267, 270, 272, 273, 275, 278, 290, 291
    scoring, 273, 274, 275-276
Formaldehyde, 189, 290, 325, 334, 338, 353, 355
Free radical, 133, 206, 207, 288, 337, 338, 445-447
Fungi (*see also individual fungi*, Yeast)
  aneuploidy in (*see* Aneuploidy)
  assays for genetic changes in, **303-349**
  in host-mediated assay, 163

G

β-Galactosidase assay, 147, 155
Gap junction, 291-292
Gastric juice, 409, 415, 418, 425
  collection of, 423

Gene
  amplification, 8, 167, 201, 216, 288, 529
  conversion
    and carcinogenesis, 204
    in *Drosophila*, 356, 367
    in fungi, 304, 307, 309, 317, 319, 339, 370, 372
    spontaneous, 316
  hybridization, 115
  mutation
    in *Drosophila*, 351, 356, 357, 362, 368, 370, 371, 380, 381, 403
    in fungi, 303-312, 315, 319, 337, 338, 339
    in human cells, 172-193, 262
    in mammalian cells, 8, 167, 169, 171-203, 207, 208, 216, 412, 513-514
    in vivo, 523, 525
    relationship to cancer, 172
    relationship to mutations in cancer, 173
  regulatory, 115
Germ-cell mutation, assays for, **245-265**, 519, 540
γ-Glutamyl transpeptidase
  as marker for carcinogen-induced foci, 86, 87, 93, 108, 110
  in hyperplastic bladder, 111
Glutathione, 87
  deficiency, 154
  depletion, 30, 32
  peroxidase, 539
  S-transferase, 87, 108, 357, 441, 443
Glycogen storage, 86, 88, 92, 93
Good laboratory practice, 18, 41
Goodness-of-fit, 464, 469, 474, 478
Growth
  altered, potential, 267
  in agar (*see also* Anchorage-independent growth), 267, 269, 278-279
  factor, 274

H

Hamster
  baby, kidney cells (BHK), 196, 224, 280, 448
  Chinese, cells, (CHO, V79), 170, 194, 196, 197, 198, 199, 208, 210, 224-225, 292-296, 412, 448, 459, 470, 489
  hepatocytes, 448
  in host-mediated assay, 163
  in long-term carcinogenicity tests, 43, 56, 73
  in skin tumour assays, 104-105
  liver in activation systems, 466, 486
  preneoplastic lesions in, 92
  spontaneous tumours in, 61, 62
  Syrian golden, use of
    embryo (SHE) cell transformation assay, **269-272**, 289-290, 292, 489
    in body-fluid analysis, 412, 414
    in long-term carcinogenicity tests, 23, 61, 62
    in metabolic activation, 442

SUBJECT INDEX

Heat
    as mutagen, 180, 184
    shock, 144, 401, 542
Hepatocytes, 137, 448
Heritable translocation test
    in *Drosophila*, 351, 354, **366-367**, 375, 379, 396, 397
    in mammals, 257, **259-261**, 262
Heterozygote
    fish, 116
    rodents, 196, 202, 247
Histidine (*his*)
    controlling for in bacterial assays, 415, 426, 427, 467
    reversion, 147, 151, 425, 459, 466
    strains, 152-153
'Hit' theory, 459-465, 470, 472, 473, 476, 478
Host-mediated assay, 146, **163-165**, 197, 410, 443
Hypoxanthine phosphoribosyltransferase (HPRT) gene (*hprt*), 177, 194, 200, 534
    mutation assay, 179, 197, 459, 489, 541
    loss of activity, 177, 196
    mutants, 192, 193, 201, 292-295

I

'Immortalization', 267, 268-269, 272, 277, 279-280, 531, 534, 536
Immunochemistry, 8, 131-132, 439
Inductest, 147
Inhalation, exposure by
    chambers for, 27-29
    of *Drosophila*, 373, 380
    of mammals, 19, 20, 21, 27-29, 30, 35, 39, 40, 42, 46, 51, 58, 488, 489, 490
    particle size, 21, 28-29
Initiation
    assays for, **103-126**, 489, 491
    effects of carcinogens, 90
    mechanisms of, 200, 311, 336, 545
    mutations in, 381
    spontaneous, 521
Intercellular communication, block of, 201, 288, **291-296**
Interim sacrifice, 37, 46, 69

K

Kidney
    preneoplastic lesions in, 92-93
    promotion of tumours in, 113
    unscheduled DNA synthesis in, 546
Kinetics (*see* Pharmacokinetics)

L

Least squares, 470, 471, 472, 476, 499, 506
Lipid peroxidation, 87, 539, 542
Liquid preincubation assay, 146, 444-445
Liver
    assays for initiation-promotion in, 107-110
    in metabolic activation systems, 442-445, 486
    preneoplastic lesions in, 85-91
    unscheduled DNA synthesis in, 545

Log-linear approximation, 474, 478
Low-dose linearity, 199, 469, 478
Lung
    cancer in humans, 490-491
    initiation-promotion assay, 112-113
    preneoplastic lesions in, 92
    tumours in Strain A mice, 71-72
Lymphoblasts, human, 175, 177, 179, 184-189
Lymphocytes
    human, 175, 179, 192, 208, 209, 210, 211, 220, 221, 415, 541
    mouse, 201

M

Mammalian cells (*see also individual cells*, Hamster, Mouse, Rat)
    aneuploidy in (*see* Aneuploidy)
    assays for genetic changes in, **167-243**, 513, 519
    for complex mixtures, 486, 487
    statistical analysis of, 467
    in host-mediated assay, 163
    in metabolic activation, 442, 448-449
Mammary-gland tumour, 43, 50, 61, 532, 533, 537
    induction in Sprague-Dawley rat, 72-73
Marker, genetic (*see also specific markers*), 179, 180-191, 196-198, 204, 247, 313, 319, 327, 328, 354, 367
Maximum likelihood, 471-474, 476
Melanoma in fish, 115-116
Metabolic cooperation (*see also* Intercellular communication), 291, 292-295
Metabolism (*see also* Activation, metabolic, Cytochrome, Enzyme)
    in *Drosophila*, 352, 381
    information on, 22, 33, 212, 409, 410
    in vivo, 429, 524
    kinetics of, 27
    of foreign compounds, 43, 163, 169
    oxidative, 411
    species differences in, 173, 313, 520
Metabolizing system (*see* Activation, metabolic)
3-Methylcholanthrene
    as enzyme inducer, 441, 443, 486
    as initiator, 112
    as oncogene activator, 531
    in cell transformation assay, 289
Methylmethane sulfonate
    in aneuploidy tests in fungi, 317, 318, 319, 322, 334, 337, 338, 339
    in dominant-lethal mutation assay, 247
    in heritable translocation assay, 261
    in host-mediated assay, 164
    in human lymphoblasts, 189
    in sex-linked recessive lethal assay, 352, 365, 371-372, 374
    in specific-locus mutation assay, 247
N-Methyl-N'-nitro-N-nitrosoguanidine (MNNG)
    as initiator, 290
    as oncogene activator, 531
    in aneuploidy test in mammalian cells, 224, 225

N-Methyl-N'-nitro-N-nitrosoguanidine (MNNG) (contd)
  in aneuploidy tests in fungi, 322, 326, 334, 337, 338
  in cell transformation assays, 276, 289
  in *Drosophila*, 404
  in human fibroblast mutation assay, 183
  in human lymphoblast mutation assay, 187-188
  in mouse spot test, 203
  in skin tumour assay, 105, 535
  in specific-locus test, 247
N-Methyl-N-nitrosourea (MNU)
  in *Drosophila*, 372
  in fish assay, 116, 117
  in host-mediated assay, 164
  in human fibroblast mutation assay, 182
  in human lymphocyte mutation assay, 184-185
  in lymphoma induction, 532, 536
  in mammary tumour induction, 532, 533
  in skin tumour assay, 105
  in thyroid tumour assay, 113
  in urinary bladder tumour assay, 111
Micronucleus formation, assays for, 203-204, 209-218, 376, 418, 523
Mitomycin C
  bacteria sensitive to, 8, 152
  in aneuploidy tests in fungi, 326
  in aneuploidy tests in mammalian cells, 225
  in *Drosophila*, 353
  in heritable translocation test, 261
  in host-mediated assay, 164
  in human T-cell mutation assay, 191
  in nondisjunction assay in *Drosophila*, 402
  in specific-locus test, 247
Mixed-function oxidase (*see also* Cytochrome), 197, 360, 440
Model
  building, 510
  for carcinogenic activity, 71-73, 103-126
  for distribution of replicate counts, 473-474
  for extrapolation, 148
  for induction of skin tumours, 27, **103-126**
  for preneoplastic lesions, 85, 89, 91
  for response variables, 457, 459-465, 469, 470
  for statistical analysis of long-term carcinogenicity tests, 68-69, 70
Monoclonal antibody
  against BUdR-containing DNA, 207
  in detection of DNA adducts, 131-132, 541
Mono-oxygenase, hepatic (*see also* Cytochrome), 420, 421, 441, 443, 444
Mouse
  bone marrow, micronuclei in, 209, 210, 212, 418
  cell lines, 172
  hepatocytes, 448
  in body-fluid analysis, 412-413
  in dominant-lethal mutation assay, 257-259, 262
  in host-mediated assay, 163-164
  liver tumour assay, 107
  lung tumour assay, 112-113
  lymphocytes, 201
  lymphoma, 532, 536
  lymphoma cells (L5178Y), 196, 197, 199, 412, 459, 470, 486, 513-514
  plasmacytoma, 533
  preneoplastic lesions in, 85, 86, 87, 88, 89, 93
  skin tumour assay, 104-107, 289, 484, 489, 491, 535
  specific-locus test, 246-250, 262
  spontaneous tumours in, 61
  spot test, 201-203, 372
  strain A, lung tumours in, 71-72
  use of in carcinogenicity testing, 23, 27, 36, 50, 51, 56, 61, 73, 170, 520
Multivariate regression analysis, 498, 505, 544
Mutagenesis (*see also* Mutation)
  and DNA damage, 129-130
  assays for in bacteria, 151-154
  assays for in mammalian cells, 167-203
  mechanisms of, 9, 144, 146, 312, 540
  statistical analysis of tests for, 457-481
  structure-activity relationships, 505-514
Mutant
  background, 176
  drug-resistant, 196
  fraction, 173-193, 458-459, 465, 478
  frequency, 307-308, 378, 459, 460, 464, 466, 467, 469, 470, 472
  Mendelian, 246-257
  recessive-lethal, 361
  selection of, 197
  spontaneous, 176, 199
  temperature-sensitive, 219, 371
  yield, 307-308, 458, 459, 460, 461, 463, 464, 465, 467, 469, 470-472, 475, 478
Mutation (*see also* Deletion mutation, Dominant-cataract mutation test, Dominant-lethal mutation assay, Gene mutation, Germ-cell mutation, Sex-linked recessive lethal mutation, Specific-locus test)
  biochemical, 253-257, 261-262
  chromosomal, 200, 245-246, 257-261, 262, 403
  definition of, 148
  delayed, 363-365
  dominant, 196, 249, 250-253
  forward
    in bacteria, 130, 149, 153-154, 427, 488
    in fungi, 303, 304, 306, 309
    in human cells, 179
    in rodent cells, 196
  frameshift, 148, 152, 153, 173
    in *Drosophila*, 353
    in fungi, 303, 304
  frequency, 460, 461, 529
  heritable, 171, 245, 351
  in human cancer, 173
  insertion, 534
  Mendelian, 246-257
  'null' (nonsense), 248, 253-254
  petite, 316, 339

SUBJECT INDEX

Mutation (contd)
  point, 163, 200, 219, 303, 304, 333-336, 352, 418, 427, 529
  protein-charge, 252, 253-257
  rates, 256-257, 262, 379, 477, 541
  recessive, 196, 250-251, 351, 381
  reverse, 151, 196, 303, 304, 306, 309, 464, 489
  spontaneous, 196, 197-198, 199, 245-246, 260, 354-355, 477-478, 534
  X-linked, 196

N

Neurospora crassa, tests in, 303, 306, 313, 319-323, 332, 353
4-Nitroquinoline-*N*-oxide
  in cell transformation assay, 289
  in fungi, 308, 325, 334, 338
  in host-mediated assay, 164
  in human fibroblast mutation assay, 182
  in human lymphoblast mutation assay, 188-189
  in lung tumour assay, 113
  in skin tumour assay, 105, 535
  structure-activity relationships, 508
Nitroreductase
  deficiency, 154
  presence of, 441, 442
N-Nitrosamine (*see also individual compounds*)
  activation of, 448
  contamination with, 19, 44
  detection of, 444
  in aneuploidy tests in fungi, 332
  in cell transformation assays, 271
  in DNA damage, 131
  in *Drosophila*, 356, 359, 365
  in induction of liver foci, 86, 88
  in induction of lung preneoplasia, 92
  in induction of pancreatic foci, 91
  in induction of renal preneoplasia, 92
  in induction of urinary bladder preneoplasia, 93
  structure-activity relationships, 505-506, 507-508
N-Nitrosodiethylamine
  as initiator, 108-110, 117
  as oncogene activator, 531
  dose-response to, 536
  in cell transformation assays, 271
  in heritable translocation test, 354
  in inducing chromosomal mutations in *Drosophila*, 366, 367
  in inducing delayed mutations in *Drosophila*, 365
  in mouse spot test, 372
  in progression of altered foci, 88
  in somatic mutation assays in *Drosophila*, 369, 372
  in specific-locus mutation test, 247
  metabolic activation of, 372
N-Nitrosodimethylamine
  activity in *Drosophila*, 358
  as initiator, 112
  in host-mediated assay, 164
  in mouse spot test, 372
  in somatic mutation assays in *Drosophila*, 369, 372

Nondisjunction (*see also* Aneuploidy)
  detection of, 221, 351
  in aneuploidy, 320, 323, 395
  in *Drosophila*, 365, 380, 395, 398-405
  in study of chromosomal mutation, 257
  in yeast systems, 204, 303
  relation to tumour formation, 219, 533, 534, 543
  spontaneous, 398
Non-linear regression, 470-471
Non-parametric testing, 431, 474-477, 500

O

Occupation
  and cancer, 7
  as confounding variable in body-fluid analysis, 414, 420, 431
  atmosphere of, 483
  exposures in, 28, 409, 411, 414, 416-417, 428, 429, 430
Oncogene
  activation of, 149, 216, 218, 429, 522, 523, 525, 529-537, 540, 543
  c-erb, 114, 530
  c-src, 114
  c-yes, 115
  effect of mutations on, 149
  E1a, 279
  induction of rodent cells, 269
  in human tumours, 167, 173, 216, 218, 522, 530-533
  in yeast, 311
  myc, 167, 272, 531, 533-534, 536, 537
  ras, 167, 172, 269, 272, 531-534, 536
  sis, 530, 537
  tu, 114-115
Ouabain resistance, 196, 197
Oxidation, 535
  bacteria sensitive to, 8, 152, 542
  damage by, 370
Oxygen
  radical, 291, 539-540, 542
  stress, 144, 542

P

Pancreas
  preneoplastic changes in, 91-92
  unscheduled DNA synthesis in, 546
Papilloma, 24, 93, 104-105, 107, 110, 533, 535
Partial hepatectomy, 108-110, 291
Pattern recognition, 499-500, 502, 505
Peroxide
  as tumour promoter, 106
  formation in diet, 44
  in metabolic activation, 446-447
Personnel, laboratory
  exposure of, 26, 45-46
  for long-term carcinogenicity tests, 18-20, 47, 48, 49, 54-55, 63-64
Pesticide
  contamination with, 19, 22, 44, 542
  structure-activity relationships, 513

Pesticide
    testing of, 43, 376-377
    use of, 42, 409
pH
    effect on intercellular communication, 291, 294
    effect on mammalian cell assays, 487
    effect on yeast assays, 311
    urinary, in experimental animals, 59
Phage (*see also* Bacteriophage)
    induction of, 144, 147, 154, 155
    λ, 145, 155
    resistance, 177
Pharmacokinetics
    and alkaline elution, 134
    in design of experiments, 8, 17, 18, 22, 23, 26, 27, 30-34, 35, 36, 212, 544
    in *Drosophila* assays, 380, 381
    in evaluation of results, 8, 30, 490
    in experimental animals, 72
    in metabolic activation systems, 443
    of cell proliferation, 88
    of mutagens, 419
Phenobarbital
    as enzyme inducer, 441, 443, 444, 486
    as promoter, 90, 107, 109, 110, 114, 294
    in aneuploidy tests in fungi, 336
    in body-fluid analysis, 421
    in *Drosophila*, 359, 377
    in fish assay, 117
Phenotype
    change in, 148, 267
    expression of, 130, 197, 367, 381, 458, 466
    markers, 278
    mutant, 168, 251
    neoplastic, 287
    of dominant mutations, 250-251
    recessive, 248
    selection of, 169, 170, 177
    stability of, 87-88, 194, 195, 270
    transformed, 217
Phorbol ester, 288-292, 295, 535
Plasmid, 532, 534
    pkM101, 151, 152, 154
Plate incorporation assay, 146
Point
    estimate, 460
    rejection, 473-474, 478
Poisson distribution, 379, 458, 465, 466-467, 468, 469, 471-474, 475-477
polA gene (*see also* Polymerase, DNA), 147, 156, 157, 513-514
Polychlorinated biphenyl, as enzyme inducer, 441-444, 486
Polycyclic aromatic hydrocarbon (PAH)
    activation of, 444, 448, 487
    adducts, 132, 133
    antibodies to, 131
    as initiator, 290
    covalent binding of, 130
    exposure to, 417
    in bacterial tests, 545
    in body-fluid analysis, 426
    in carcinogenicity model, 72
    in cell transformation assay, 271, 273, 276, 277
    in complex mixtures, 488
    in *Drosophila*, 357-359, 370, 374
    in human lymphoblast mutation assay, 185-186
    in liver tumour assay, 108
    in skin tumour assay, 105
    prediction of carcinogenicity of, 498, 507, 508, 510-513
Polymerase, DNA, 145, 156
^{32}P-Post-labelling, 129, 132
Potency
    carcinogenic, 504-505, 520, 522
    mutagenic, 445, 459, 460, 468, 469, 470, 472, 474, 475, 477-478, 522, 542
    of complex mixtures, 490-491
Preneoplastic lesion, 67, **85-101**
    genetic change as surrogate for, 168
    in biliary system, 91
    in kidney, 92-93
    in liver, 85-91, 108
    in lung, 92
    in pancreas, 91-92
    in urinary bladder, 93, 111
    role in carcinogenesis, 8
    spontaneous, 87
Progression, tumour, 33, 89, 217, 272, 313, 535
Promotion, tumour, 535
    activity, 312, 521
    assays for, **103-126**, 268, **287-302**, 525
    in liver foci bioassay, 90
    mechanisms of, 33, 200, 287, 535, 537, 545
β-Propiolactone
    in *Drosophila*, 369
    in human lymphoblast mutation assay, 189
    in skin tumour assay, 105
Protein
    adducts, 445
    binding to, 130
    -charge mutation, 252, 253-257
    content of S9, 444, 445, 487
    deficiency, 43
    -DNA crosslinks, 207, 375, 396
    in experimental diet, 43
Purity, of test substances, 21, 22

Q

Quasilikelihood, 473-474, 476

R

Rabbit, use of
    in body-fluid analysis, 413
    in long-term carcinogenicity tests, 73
    in skin tumour assays, 104

SUBJECT INDEX

Radiation
  ionizing
    as initiator, 113
    γ, 532, 536
    in biochemical mutations, 254
    in dominant-mutation tests, 251-252
    in gene mutation in fungi, 307, 310
    in human lymphoblast mutation test, 179
    in germ-cell mutation assay, 248
    in mitotic recombination in *Drosophila*, 356
    in mutagenesis assays, 464
    in sister chromatid exchange, 208
    in structural chromosomal aberrations, 205, 206, 208
  ultraviolet
    as initiator, 290, 291
    bacteria sensitive to, 152
    in fungi, 308, 310, 314
    in human fibroblast mutation assay, 180-181, 198
    in human lymphoblast mutation assay, 189
    in human T-cell mutation assay, 190-191
    in mammalian cell cultures, 415
Randomization of experimental animals, 37, 38-39, 52
Rat
  breast cancer induction in female Sprague-Dawley, 72-73
  hepatocytes, 448
  liver, epithelial cells, 196, 197
  liver in metabolic systems, 151, 442-445, 486
  preneoplastic lesions in, 85-87, 90, 91-93
  spontaneous tumours in, 61, 62
  use of in body-fluid analysis, 412-414
  use of in long-term carcinogenicity tests, 23, 27, 36, 43, 56, 61, 73, 170, 489, 506, 520
recA gene, 145
  in assays, 147, 156, 157, 513-514
  independence, 145
Recessive lethal (*see* Sex-linked recessive lethal mutation)
Restriction-fragment-length polymorphism, 204, 541
Reversion (*see also* Mutation), 149, 151-152, 154
Risk, human
  carcinogenic, 34, 173, 214, 246, 520-521, 541-542
  genetic, 252-253, 353, 376, 381

S

S9 (9000 × g supernatant fraction of rat liver; *see also* Activation, metabolic), 442, 509, 510, 512
Saccharin
  as promoter, 111, 112, 117, 290, 294, 295
  in aneuploidy tests in fungi, 326
  in body-fluid analysis, 413
Saccharomyces cerevisiae, tests using, 303, 304-306, 310, 313, 314-319, 322, 339, 370, 513-514
Salmonella typhimurium
  genetic network of, 144
  -/hepatocyte assay, 448
  -/microsome assay (*see also* Ames test), 147, 151-154, 171, 202, 309, 376, 442, 459, 468, 478, 509, 511-514, 519, 524, 525, 540, 545
  in body-fluid analysis, 411, 412-414, 418, 425, 427
  in testing complex mixtures, 483-486, 488-490
  statistical analysis of, 464, 465-466, 467, 470, 478
Schizosaccharomyces pombe, tests using, 303, 306
Sensitivity
  of aneuploidy tests, 223
  of bacterial tests, 146, 147, 149-150, 155, 157
  of BALBc/3T3 cell transformation assay, 273
  of fungal strains, 310
  of host-mediated assay, 164
Serum, fetal bovine, in cell transformation assays, 268, 269, 272, 273, 274, 275, 291
Sex-linked recessive lethal mutation, 163, 306, 352-354, 359, 365, 545
  assay, 351, 356, **362-366**, 367, 371-372, 375, 376, 378, 379, 381, 403, 404, 546
  temperature-sensitive, 353
Significance, statistical, 67, 223, 294, 378-379, 430, 459, 470, 471, 472, 473, 474, 475, 505
Sister chromatid exchange (SCE), 170, 203-205, 206-208, 535
  assays in mammalian cells, 210-218, 287, 415, 489
  in *Drosophila*, 352, 375, 396
  in mammals *in vivo*, 523
  relationship to carcinogenesis, 217-218
  relationship to chromosomal aberrations, 208
  relationship to gene mutations, 208
  spontaneous, 207
Skin
  exposure by, 27
  initiation-promotion model, 104-106, 535
  tumours, 50, 533
Smoking (*see* Tobacco)
Somatic mutation and recombination tests (SMART), 367-373, 380
Sordaria brevicollis, tests using, 321, 322, 323, 327, 339
SOS system
  chromotest, 156, 428
  genes, 143, 151
  molecular mechanisms of, 144-145
  mutation of, 144
  tests, 147, 155-156
Species
  choice of for experiments, 23, 32, 411
  differences, 173, 198, 199, 313, 439, 442, 520-521, 525
Specific-locus test, 200, 203, **246-250**, 251, 255, 256, 260, 261-262, 366, 458
Sperm
  of *Drosophila*, 374
  of humans exposed to mutagens, 245-246
Spermatid, 261, 359
Spermatocyte, 359, 546
Spermatogonia, 249, 365-366
Spermatozoa
  of *Drosophila*, 359, 365
  of humans exposed to X-rays, 246

SUBJECT INDEX

Spindle fibre, mitotic, 200, 204, 209, 214, 219, 220, 222-225, 313, 322, 338
    poison, 319, 337
Spot test, mouse, 201-203, 372
Stability of test compounds, 21, 22, 26, 45, 484, 485
Statistics, for analysis of results
    of assays of genetic changes in mammalian cells, 170, 199, 215, 223, 523
    of body-fluid analyses, 431
    of dominant-lethal assay, 259
    of *Drosophila* assays, 378-379
    of initiation-promotion assays, 107
    of in-vitro assays of mutagenesis, **457-481**, 520
    of long-term carcinogenicity tests, 17, 20, 25, 37, 38, 52, 63, **67-71**
    of metabolic cooperation assays, 294
    of structure-activity relationships, 500, 504, 506
Storage
    effect, in *Drosophila*, 374-375, 396
    of test substances, 21, 22, 45, 484, 485
Strain
    animal
        choice of, 24-26
        differences in, 87, 106
        information on, 37, 411
    bacterial
        battery of, 149, 151, 152
        for DNA damage and repair, 428
        number needed, 148
        S. typhimurium, 151-153, 478, 509, 512, 542
    Drosophila, 352, 357, 359
    effect on metabolism, 439
    fish, 115-116
    fungal, 314-319, 327-331, 337, 339
Structure-activity relationship
    in predicting carcinogenicity, 9, **497-517**, 544
    in tumour promotion, 105-106
Superoxide
    anion, 445-446, 535
    dismutase, 539-540
Susceptibility
    host, 464-465, 472
    to tumour induction, 23, 24

T

T-cells, human, 179, 190-192
Teratogenicity, 202, 292
12-*O*-Tetradecanoylphorbol 13-acetate (TPA)
    in aneuploidy tests in fungi, 318, 319, 339
    in aneuploidy tests in mammalian cells, 224, 225
    in fish assay, 117
    in skin tumour assay, 104, 106, 107, 535
    in tumour promoter assay, 287, 289-291, 295
6-Thioguanine
    in metabolic cooperation assay, 292-293, 294-295
    resistance in human cell mutation assay, 177, 179, 180-193
    resistance in rodent cell mutation assay, 196, 197, 201

^{3}H-Thymidine
    as mutagen, 184
    incorporation, 135, 136-137, 192
    in induction of sister chromatid exchange, 206-207
Thymidine dimer, 131, 207
Thymidine kinase (TK), 179, 194, 196, 459
Time to tumour, 52, 107
Tobacco
    chewing, 7, 418
    in chromosomal damage, 172
    interaction with diet, 420
    mutagenicity in urine of chewers, 418
    mutagenicity in urine of smokers, 416, 418, 428
    smoke, 29, 290, 291, 483
        condensate, 488, 490, 491
        testing of, 488, 490
    smoking, 7, 430, 537
        as confounder in body-fluid analysis, 414, 419
Toxicity
    in bacteria, 155, 156, 472, 475-476, 478, 485
    in experimental animals, 19, 22, 35, 37, 43, 66, 134, 487, 488, 521
    in fungi, 310
    in mammalian cells, 200-201, 210, 212, 219, 222, 226, 275
Transformation, neoplastic
    clonal, 270, 272
    enhancement of, 287, 288-291, 292
    frequency, 270, 273, 274, 276, 279, 280
    in fish, 116
    in mammalian cells, 85, 200, 208, **267-286**, 448, 489, 525, 531, 536-537, 543
    mechanisms of, 313, 530
    morphological, 267, 271
    of human cells, 278-280
    scoring, 268, 270, 272, 273, 275
    spontaneous, 270, 274-275, 277, 280
    statistical analysis of, 457, 465, 472
Translocation, gene, 533-534
Trifluorothymidine resistance, 184-189, 199
Tumour (*see also individual organs*)
    background, 25, 36
    benign, 67, 85, 106
    categories of, 66-67
    fatal, 37, 63, 69-70
    genetic changes in, 167
    incidental, 37, 63, 69-70
    in *Drosophila*, 361
    observation of, 39, 56, 63
    production by transformed cells, 267, 272, 274, 275-276
    spontaneous, 23, 24, 61, 62
    transplantable, 220
Tumour-promoting agent (*see also individual agents, Promotion*)
    assays that may be predictive of, **287-302**, 490
    in cultured cells, 201, 272, 287, 521
Two-generation study, 35, 36, 53-54

U

umu test, 145, 156, 428
Urethane
    as initiator, 112
    as positive control, 72
    in fungi, 326
    in skin tumour assay, 105, 535
Urinary bladder
    preneoplastic lesions in, 92
    two-stage models in, 111-112
Urine
    adducts in, 133
    analysis of, 409, 411-414, 415-423, 425-426, 428, 430
    collection of, 422-423
    concentration of, 426
    examination of, 59, 111
uvr gene, 151, 152, 154, 155, 157

V

Validation, for predicting carcinogenicity
    of adduct formation, 132
    of assays in *Drosophila*, 351, 363-364, 369, 370-371, 375-378, 380, 395, 397, 403-404
    of bacterial tests, 149-151, 156
    of cell transformation assays, 267-268, 271, 273, 276
    of chromosomal assays, 170-171
    of clastogenicity tests, 217-218, 223, 225
    of fungus assays, 303, 308-309, 317-319, 322-326, 332-339
    of host-mediated assay, 164-165
    of mouse spot test, 202-203
    of rat-liver foci assay, 90, 108-109
    of short-term assays, 8, 170, 375, 520, 522, 525, 529
    of subcutaneous injection model, 73
    of tests in whole mammals, 522-523

Variation
    between laboratories, 338, 405, 444, 467-468, 478
    experimental, 469, 471, 472
    extraneous, 471, 472-474
    intralaboratory, 467
    sources of in mutagenesis assays, 465-469
Vehicle, 22, 37, 72, 73
Virus
    as clastogens or cell transformants, 421
    in body-fluid analysis, 424
    in experimental animals, 20, 48, 59, 60, 61
    oncogenic, 273, 530-536, 540
    producing benign tumours, 67

X

Xeroderma pigmentosum, 181-183, 198, 207, 217
X-rays
    as initiator, 113, 290, 291
    bacteria sensitive to, 8, 152,
    effect on offspring, 540
    effect on spermatogonia, 252, 255
    effect on spermatozoa, 246
    in *Drosophila*, 352, 355, 399, 401, 402
    in fish assay, 117
    in fungi, 314
    in human fibroblast mutation assay, 180
    in human lymphoblast mutation assay, 184
    in human T-cell mutation assay, 190
    in rodent cell mutation assay, 197

Y

Yeast (*see also individual yeasts*, Fungi), 204, 287, 288, 303, 313, 314-319, 372

PUBLICATIONS OF THE INTERNATIONAL AGENCY FOR RESEARCH ON CANCER

SCIENTIFIC PUBLICATIONS SERIES

(Available from Oxford University Press)

No. 1 LIVER CANCER (1971)
176 pages; out of print

No. 2 ONCOGENESIS AND HERPES VIRUSES (1972)
Edited by P.M. Biggs, G. de-Thé & L.N. Payne
515 pages; out of print

No. 3 N-NITROSO COMPOUNDS - ANALYSIS AND FORMATION (1972)
Edited by P. Bogovski, R. Preussmann & E.A. Walker
140 pages

No. 4 TRANSPLACENTAL CARCINOGENESIS (1973)
Edited by L. Tomatis & U. Mohr,
181 pages; out of print

No. 5 PATHOLOGY OF TUMOURS IN LABORATORY ANIMALS. VOLUME 1. TUMOURS OF THE RAT. PART 1 (1973)
Editor-in-Chief V.S. Turusov
214 pages

No. 6 PATHOLOGY OF TUMOURS IN LABORATORY ANIMALS. VOLUME 1. TUMOURS OF THE RAT. PART 2 (1976)
Editor-in-Chief V.S. Turusov
319 pages

No. 7 HOST ENVIRONMENT INTERACTIONS IN THE ETIOLOGY OF CANCER IN MAN (1973)
Edited by R. Doll & I. Vodopija,
464 pages

No. 8 BIOLOGICAL EFFECTS OF ASBESTOS (1973)
Edited by P. Bogovski, J.C. Gilson, V. Timbrell & J.C. Wagner,
346 pages; out of print

No. 9 N-NITROSO COMPOUNDS IN THE ENVIRONMENT (1974)
Edited by P. Bogovski & E.A. Walker
243 pages

No. 10 CHEMICAL CARCINOGENESIS ESSAYS (1974)
Edited by R. Montesano & L. Tomatis,
230 pages

No. 11 ONCOGENESIS AND HERPESVIRUSES II (1975)
Edited by G. de-Thé, M.A. Epstein & H. zur Hausen
Part 1, 511 pages
Part 2, 403 pages

No. 12 SCREENING TESTS IN CHEMICAL CARCINOGENESIS (1976)
Edited by R. Montesano, H. Bartsch & L. Tomatis
666 pages

No. 13 ENVIRONMENTAL POLLUTION AND CARCINOGENIC RISKS (1976)
Edited by C. Rosenfeld & W. Davis
454 pages; out of print

No. 14 ENVIRONMENTAL N-NITROSO COMPOUNDS — ANALYSIS AND FORMATION (1976)
Edited by E.A. Walker, P. Bogovski & L. Griciute
512 pages

No. 15 CANCER INCIDENCE IN FIVE CONTINENTS. VOL. III (1976)
Edited by J. Waterhouse, C.S. Muir, P. Correa & J. Powell
584 pages

No. 16 AIR POLLUTION AND CANCER IN MAN (1977)
Edited by U. Mohr, D. Schmahl & L. Tomatis
331 pages; out of print

No. 17 DIRECTORY OF ON-GOING RESEARCH IN CANCER EPIDEMIOLOGY 1977 (1977)
Edited by C.S. Muir & G. Wagner,
599 pages; out of print

No. 18 ENVIRONMENTAL CARCINOGENS. SELECTED METHODS OF ANALYSIS
Editor-in-Chief H. Egan
Vol. 1. ANALYSIS OF VOLATILE NITROSAMINES IN FOOD (1978)
Edited by R. Preussmann, M. Castegnaro, E.A. Walker & A.E. Wassermann
212 pages; out of print

SCIENTIFIC PUBLICATIONS SERIES

No. 19 ENVIRONMENTAL ASPECTS
OF N-NITROSO COMPOUNDS (1978)
Edited by E.A. Walker, M. Castegnaro,
L. Griciute & R.E. Lyle
566 pages

No. 20 NASOPHARYNGEAL
CARCINOMA: ETIOLOGY AND
CONTROL (1978)
Edited by G. de-Thé & Y. Ito,
610 pages; out of print

No. 21 CANCER REGISTRATION
AND ITS TECHNIQUES (1978)
Edited by R. MacLennan, C.S. Muir,
R. Steinitz & A. Winkler
235 pages

No. 22 ENVIRONMENTAL CARCINOGENS.
SELECTED METHODS OF ANALYSIS
Editor-in-Chief H. Egan
Vol. 2. METHODS FOR THE MEASURE-
MENT OF VINYL CHLORIDE IN
POLY(VINYL CHLORIDE), AIR, WATER
AND FOODSTUFFS (1978)
Edited by D.C.M. Squirrell & W. Thain,
142 pages; out of print

No. 23 PATHOLOGY OF TUMOURS IN
LABORATORY ANIMALS. VOLUME II.
TUMOURS OF THE MOUSE (1979)
Editor-in-Chief V.S. Turusov
669 pages

No. 24 ONCOGENESIS AND HERPES-
VIRUSES III (1978)
Edited by G. de-Thé, W. Henle & F. Rapp
Part 1, 580 pages
Part 2, 522 pages; out of print

No. 25 CARCINOGENIC RISKS -
STRATEGIES FOR INTERVENTION
(1979)
Edited by W. Davis & C. Rosenfeld,
283 pages; out of print

No. 26 DIRECTORY OF ON-GOING
RESEARCH IN CANCER EPI-
DEMIOLOGY 1978 (1978)
Edited by C.S. Muir & G. Wagner,
550 pages; out of print

No. 27 MOLECULAR AND CELLULAR
ASPECTS OF CARCINOGEN
SCREENING TESTS (1980)
Edited by R. Montesano, H. Bartsch & L. Tomatis
371 pages

No. 28 DIRECTORY OF ON-GOING
RESEARCH IN CANCER EPI-
DEMIOLOGY 1979 (1979)
Edited by C.S. Muir & G. Wagner,
672 pages; out of print

No. 29 ENVIRONMENTAL CARCINOGENS.
SELECTED METHODS OF ANALYSIS
Editor-in-Chief H. Egan
Vol. 3. ANALYSIS OF POLYCYCLIC
AROMATIC HYDROCARBONS IN
ENVIRONMENTAL SAMPLES (1979)
Edited by M. Castegnaro, P. Bogovski,
H. Kunte & E.A. Walker
240 pages; out of print

No. 30 BIOLOGICAL EFFECTS OF
MINERAL FIBRES (1980)
Editor-in-Chief J.C. Wagner
Volume 1, 494 pages
Volume 2, 513 pages

No. 31 N-NITROSO COMPOUNDS:
ANALYSIS, FORMATION AND
OCCURRENCE (1980)
Edited by E.A. Walker, M. Castegnaro,
L. Griciute & M. Börzsönyi
841 pages; out of print

No. 32 STATISTICAL METHODS IN
CANCER RESEARCH
Vol. 1. THE ANALYSIS OF CASE-
CONTROL STUDIES (1980)
By N.E. Breslow & N.E. Day
338 pages

No. 33 HANDLING CHEMICAL
CARCINOGENS IN THE LABORATORY -
PROBLEMS OF SAFETY (1979)
Edited by R. Montesano, H. Bartsch,
E. Boyland, G. Della Porta, L. Fishbein,
R.A. Griesemer, A.B. Swan & L. Tomatis,
32 pages

No. 34 PATHOLOGY OF TUMOURS
IN LABORATORY ANIMALS. VOLUME
III. TUMOURS OF THE HAMSTER
(1982)
Editor-in-Chief V.S. Turusov,
461 pages

No. 35 DIRECTORY OF ON-GOING
RESEARCH IN CANCER EPIDEMIOLOGY
1980 (1980)
Edited by C.S. Muir & G. Wagner,
660 pages; out of print

SCIENTIFIC PUBLICATIONS SERIES

No. 36 CANCER MORTALITY BY OCCUPATION AND SOCIAL CLASS 1851-1971 (1982)
By W.P.D. Logan
253 pages

No. 37 LABORATORY DECONTAMINATION AND DESTRUCTION OF AFLATOXINS B_1, B_2, G_1, G_2 IN LABORATORY WASTES (1980)
Edited by M. Castegnaro, D.C. Hunt, E.B. Sansone, P.L. Schuller, M.G. Siriwardana, G.M. Telling, H.P. Van Egmond & E.A. Walker,
59 pages

No. 38 DIRECTORY OF ON-GOING RESEARCH IN CANCER EPIDEMIOLOGY 1981 (1981)
Edited by C.S. Muir & G. Wagner,
696 pages; out of print

No. 39 HOST FACTORS IN HUMAN CARCINOGENESIS (1982)
Edited by H. Bartsch & B. Armstrong
583 pages

No. 40 ENVIRONMENTAL CARCINOGENS. SELECTED METHODS OF ANALYSIS
Editor-in-Chief H. Egan
Vol. 4. SOME AROMATIC AMINES AND AZO DYES IN THE GENERAL AND INDUSTRIAL ENVIRONMENT (1981)
Edited by L. Fishbein, M. Castegnaro, I.K. O'Neill & H. Bartsch
347 pages

No. 41 *N*-NITROSO COMPOUNDS: OCCURRENCE AND BIOLOGICAL EFFECTS (1982)
Edited by H. Bartsch, I.K. O'Neill, M. Castegnaro & M. Okada,
755 pages

No. 42 CANCER INCIDENCE IN FIVE CONTINENTS. VOLUME IV (1982)
Edited by J. Waterhouse, C. Muir, K. Shanmugaratnam & J. Powell,
811 pages

No. 43 LABORATORY DECONTAMINATION AND DESTRUCTION OF CARCINOGENS IN LABORATORY WASTES: SOME *N*-NITROSAMINES (1982) Edited by M. Castegnaro, G. Eisenbrand, G. Ellen, L. Keefer, D. Klein, E.B. Sansone, D. Spincer, G. Telling & K. Webb
73 pages

No. 44 ENVIRONMENTAL CARCINOGENS. SELECTED METHODS OF ANALYSIS
Editor-in-Chief H. Egan
Vol. 5. SOME MYCOTOXINS (1983)
Edited by L. Stoloff, M. Castegnaro, P. Scott, I.K. O'Neill & H. Bartsch,
455 pages

No. 45 ENVIRONMENTAL CARCINOGENS. SELECTED METHODS OF ANALYSIS
Editor-in-Chief H. Egan
Vol. 6: *N*-NITROSO COMPOUNDS (1983)
Edited by R. Preussmann, I.K. O'Neill, G. Eisenbrand, B. Spiegelhalder & H. Bartsch
508 pages

No. 46 DIRECTORY OF ON-GOING RESEARCH IN CANCER EPIDEMIOLOGY 1982 (1982)
Edited by C.S. Muir & G. Wagner,
722 pages; out of print

No. 47 CANCER INCIDENCE IN SINGAPORE (1982)
Edited by K. Shanmugaratnam, H.P. Lee & N.E. Day
174 pages; out of print

No. 48 CANCER INCIDENCE IN THE USSR Second Revised Edition (1983)
Edited by N.P. Napalkov, G.F. Tserkovny, V.M. Merabishvili, D.M. Parkin, M. Smans & C.S. Muir,
75 pages

No. 49 LABORATORY DECONTAMINATION AND DESTRUCTION OF CARCINOGENS IN LABORATORY WASTES: SOME POLYCYCLIC AROMATIC HYDROCARBONS (1983)
Edited by M. Castegnaro, G. Grimmer, O. Hutzinger, W. Karcher, H. Kunte, M. Lafontaine, E.B. Sansone, G. Telling & S.P. Tucker
81 pages

No. 50 DIRECTORY OF ON-GOING RESEARCH IN CANCER EPIDEMIOLOGY 1983 (1983)
Edited by C.S. Muir & G. Wagner,
740 pages; out of print

SCIENTIFIC PUBLICATIONS SERIES

No. 51 MODULATORS OF EXPERIMENTAL CARCINOGENESIS (1983)
Edited by V. Turusov & R. Montesano
307 pages

No. 52 SECOND CANCER IN RELATION TO RADIATION TREATMENT FOR CERVICAL CANCER: RESULTS OF A CANCER REGISTRY COLLABORATION (1984)
Edited by N.E. Day & J.C. Boice, Jr,
207 pages

No. 53 NICKEL IN THE HUMAN ENVIRONMENT (1984)
Editor-in-Chief, F.W. Sunderman, Jr,
529 pages

No. 54 LABORATORY DECONTAMINATION AND DESTRUCTION OF CARCINOGENS IN LABORATORY WASTES: SOME HYDRAZINES (1983)
Edited by M. Castegnaro, G. Ellen, M. Lafontaine, H.C. van der Plas, E.B. Sansone & S.P. Tucker,
87 pages

No. 55 LABORATORY DECONTAMINATION AND DESTRUCTION OF CARCINOGENS IN LABORATORY WASTES: SOME N-NITROSAMIDES (1984)
Edited by M. Castegnaro,
M. Benard, L.W. van Broekhoven,
D. Fine, R. Massey, E.B. Sansone,
P.L.R. Smith, B. Spiegelhalder,
A. Stacchini, G. Telling & J.J. Vallon,
65 pages

No. 56 MODELS, MECHANISMS AND ETIOLOGY OF TUMOUR PROMOTION (1984)
Edited by M. Börzsönyi, N.E. Day, K. Lapis & H. Yamasaki
532 pages

No. 57 N-NITROSO COMPOUNDS: OCCURRENCE, BIOLOGICAL EFFECTS AND RELEVANCE TO HUMAN CANCER (1984)
Edited by I.K. O'Neill, R.C. von Borstel, C.T. Miller, J. Long & H. Bartsch,
1013 pages

No. 58 AGE-RELATED FACTORS IN CARCINOGENESIS (1985)
Edited by A. Likhachev, V. Anisimov & R. Montesano
288 pages

No. 59 MONITORING HUMAN EXPOSURE TO CARCINOGENIC AND MUTAGENIC AGENTS (1984)
Edited by A. Berlin, M. Draper, K. Hemminki & H. Vainio
457 pages

No. 60 BURKITT'S LYMPHOMA: A HUMAN CANCER MODEL (1985)
Edited by G. Lenoir, G. O'Conor & C.L.M. Olweny
484 pages

No. 61 LABORATORY DECONTAMINATION AND DESTRUCTION OF CARCINOGENS IN LABORATORY WASTES: SOME HALOETHERS (1984)
Edited by M. Castegnaro, M. Alvarez, M. Iovu, E.B. Sansone, G.M. Telling & D.T. Williams
55 pages

No. 62 DIRECTORY OF ON-GOING RESEARCH IN CANCER EPIDEMIOLOGY 1984 (1984)
Edited by C.S. Muir & G.Wagner 728 pages

No. 63 VIRUS-ASSOCIATED CANCERS IN AFRICA (1984)
Edited by A.O. Williams, G.T. O'Conor, G.B. de-Thé & C.A. Johnson,
773 pages

No. 64 LABORATORY DECONTAMINATION AND DESTRUCTION OF CARCINOGENS IN LABORATORY WASTES: SOME AROMATIC AMINES AND 4-NITROBIPHENYL (1985)
Edited by M. Castegnaro, J. Barek, J. Dennis, G. Ellen, M. Klibanov, M. Lafontaine, R. Mitchum, P. Van Roosmalen, E.B. Sansone, L.A. Sternson & M. Vahl
85 pages

No. 65 INTERPRETATION OF NEGATIVE EPIDEMIOLOGICAL EVIDENCE FOR CARCINOGENICITY (1985)
Edited by N.J. Wald & R. Doll
232 pages

No. 66 THE ROLE OF THE REGISTRY IN CANCER CONTROL (1985)
Edited by D.M. Parkin, G. Wagner & C.S. Muir
155 pages

SCIENTIFIC PUBLICATIONS SERIES

No. 67 TRANSFORMATION ASSAY OF
ESTABLISHED CELL LINES:
MECHANISMS AND APPLICATIONS (1985)
Edited by T. Kakunaga & H. Yamasaki
225 pages

No. 68 ENVIRONMENTAL CARCINOGENS.
SELECTED METHODS OF ANALYSIS
VOL. 7. SOME VOLATILE HALOGENATED
HYDROCARBONS (1985)
Edited by L. Fishbein & I.K. O'Neill
479 pages

No. 69 DIRECTORY OF ON-GOING
RESEARCH IN CANCER EPI-
DEMIOLOGY 1985 (1985)
Edited by C.S. Muir & G. Wagner
756 pages

No. 70 THE ROLE OF CYCLIC NUCLEIC
ACID ADDUCTS IN CARCINOGENESIS AND
MUTAGENESIS (1986)
Edited by B. Singer & H. Bartsch
467 pages

No. 71 ENVIRONMENTAL CARCINOGENS.
SELECTED METHODS OF ANALYSIS
VOL. 8. SOME METALS: As, Be, Cd, Cr, Ni,
Pb, Se, Zn (1986)
Edited by I.K. O'Neill, P. Schuller & L. Fishbein
485 pages

No. 72 ATLAS OF CANCER IN SCOTLAND
1975-1980: INCIDENCE AND EPI-
DEMIOLOGICAL PERSPECTIVE (1985)
Edited by I. Kemp, P. Boyle, M. Smans & C. Muir
282 pages

No. 73 LABORATORY DECONTAMI-
NATION AND DESTRUCTION OF
CARCINOGENS IN LABORATORY
WASTES: SOME ANTINEOPLASTIC
AGENTS (1985)
Edited by M. Castegnaro, J. Adams,
M. Armour, J. Barek, J. Benvenuto,
C. Confalonieri, U. Goff, S. Ludeman,
D. Reed, E.B. Sansone & G. Telling
163 pages

No. 74 TOBACCO: A MAJOR INTER-
NATIONAL HEALTH HAZARD (1986)
Edited by D. Zarkdze & R. Peto
324 pages

No. 75 CANCER OCCURRENCE IN
DEVELOPING COUNTRIES (1986)
Edited by D.M. Parkin
339 pages

No. 76 SCREENING FOR CANCER OF THE
UTERINE CERVIX (1986)
Edited by M. Hakama, A.B. Miller & N.E. Day
311 pages

No. 77 HEXACHLOROBENZENE:
PROCEEDINGS OF AN INTERNATIONAL
SYMPOSIUM (1986)
Edited by C.R. Morris & J.R.P. Cabral
668 pages

No. 78 CARCINOGENICITY OF CYTOSTATIC
DRUGS (1986)
Edited by D. Schmähl & J. Kaldor
330 pages

No. 79 STATISTICAL METHODS OF CANCER
RESEARCH, VOL. 3, THE DESIGN AND
ANALYSIS OF LONG-TERM ANIMAL
EXPERIMENTS (1986)
By J.J. Gart, D. Krewski, P.N. Lee,
R.E. Tarone & J. Wahrendorf
219 pages

No. 80 DIRECTORY OF ON-GOING
RESEARCH IN CANCER EPIDEMIOLOGY
1986 (1986)
Edited by G. Wagner & C. Muir
805 pages

No. 81 ENVIRONMENTAL CARCINOGENS.
METHODS OF ANALYSIS AND EXPOSURE
MEASUREMENT VOL. 9. PASSIVE
SMOKING
Edited by I.K. O'Neill, K.D. Brunnemann,
B. Dodet & D. Hoffmann
(in press)

No. 82 STATISTICAL METHODS IN
CANCER RESEARCH, VOL. 2
THE DESIGN AND ANALYSIS OF
COHORT STUDIES
By N.E. Breslow & N.E. Day
(in press)

No. 83 LONG-TERM AND SHORT-TERM
ASSAYS FOR CARCINOGENS:
A CRITICAL APPRAISAL (1986)
Edited by R. Montesano, H. Bartsch,
H. Vainio, J. Wilbourn & H. Yamasaki
564 pages

No. 84 THE RELEVANCE OF N-NITROSO
COMPOUNDS TO HUMAN CANCER:
EXPOSURES AND MECHANISMS
Edited by H. Bartsch, I.K. O'Neill &
R. Schulte-Hermann
(in press)

NON-SERIAL PUBLICATIONS

(Available from IARC)

ALCOOL ET CANCER (1978)
By A.J. Tuyns (in French only)
42 pages

CANCER MORBIDITY AND CAUSES OF
DEATH AMONG DANISH BREWERY
WORKERS (1980)
By O.M. Jensen
145 pages

DIRECTORY OF COMPUTER SYSTEMS
USED IN CANCER REGISTRIES (1986)
By H.R. Menck & D.M. Parkin
236 pages

IARC MONOGRAPHS ON THE EVALUATION OF THE CARCINOGENIC RISK OF CHEMICALS TO HUMANS
(English editions only)

(Available from WHO Sales Agents)

Volume 1
Some inorganic substances, chlorinated hydrocarbons, aromatic amines, N-nitroso compounds, and natural products (1972)
184 pp.; out of print

Volume 2
Some inorganic and organometallic compounds (1973)
181 pp.; out of print

Volume 3
Certain polycyclic aromatic hydrocarbons and heterocyclic compounds (1973)
271 pp.; out of print

Volume 4
Some aromatic amines, hydrazine and related substances, N-nitroso compounds and miscellaneous alkylating agents (1974)
286 pp.

Volume 5
Some organochlorine pesticides (1974)
241 pp.; out of print

Volume 6
Sex hormones (1974)
243 pp.

Volume 7
Some anti-thyroid and related substances, nitrofurans and industrial chemicals (1974)
326 pp.; out of print

Volume 8
Some aromatic azo compounds (1975)
357 pp.

Volume 9
Some aziridines, N-, S- and O-mustards and selenium (1975)
268 pp.

Volume 10
Some naturally occurring substances (1976)
353 pp.; out of print

Volume 11
Cadmium, nickel, some epoxides, miscellaneous industrial chemicals and general considerations on volatile anaesthetics (1976)
306 pp.

Volume 12
Some carbamates, thiocarbamates and carbazides (1976)
282 pp.

Volume 13
Some miscellaneous pharmaceutical substances (1977)
255 pp.

Volume 14
Asbestos (1977)
106 pp.

Volume 15
Some fumigants, the herbicides 2,4-D and 2,4,5-T, chlorinated dibenzodioxins and miscellaneous industrial chemicals (1977)
354 pp.

Volume 16
Some aromatic amines and related nitro compounds - hair dyes, colouring agents and miscellaneous industrial chemicals (1978)
400 pp.

Volume 17
Some N-nitroso compounds (1978)
365 pp.

Volume 18
Polychlorinated biphenyls and polybrominated biphenyls (1978)
140 pp.

Volume 19
Some monomers, plastics and synthetic elastomers, and acrolein (1979)
513 pp.

Volume 20
Some halogenated hydrocarbons (1979)
609 pp.

Volume 21
Sex hormones (II) (1979)
583 pp.

Volume 22
Some non-nutritive sweetening agents (1980)
208 pp.

Volume 23
Some metals and metallic compounds (1980)
438 pp.

IARC MONOGRAPHS SERIES

Volume 24
Some pharmaceutical drugs (1980)
337 pp.

Volume 25
Wood, leather and some associated industries (1981)
412 pp.

Volume 26
Some antineoplastic and immuno-suppressive agents (1981)
411 pp.

Volume 27
Some aromatic amines, anthraquinones and nitroso compounds, and inorganic fluorides used in drinking-water and dental preparations (1982)
341 pp.

Volume 28
The rubber industry (1982)
486 pp.

Volume 29
Some industrial chemicals and dyestuffs (1982)
416 pp.

Volume 30
Miscellaneous pesticides (1983)
424 pp.

Volume 31
Some food additives, feed additives and naturally occurring substances (1983)
314 pp.

Volume 32
Polynuclear aromatic compounds, Part 1, Environmental and experimental data (1984)
477 pp.

Volume 33
Polynuclear aromatic compounds, Part 2, Carbon blacks, mineral oils and some nitroarene compounds (1984)
245 pp.

Volume 34
Polynuclear aromatic compounds, Part 3, Industrial exposures in aluminium production, coal gasification, coke production, and iron and steel founding (1984)
219 pp.

Volume 35
Polynuclear aromatic compounds, Part 4, Bitumens, coal-tar and derived products, shale-oils and soots (1985)
271 pp.

Volume 36
Allyl Compounds, aldehydes, epoxides and peroxides (1985)
369 pp.

Volume 37
Tobacco habits other than smoking; betel-quid and areca-nut chewing; and some related nitrosamines (1985)
291 pp.

Volume 38
Tobacco smoking (1986)
421 pp.

Volume 39
Some chemicals used in plastics and elastomers (1986)
403 pp.

Volume 40
Some naturally occurring and synthetic food components, furocoumarins and ultra-violet radiation (1986)
444 pp.

Volume 41
Some halogenated hydrocarbons and pesticide exposures (1986)
434 pp.

Supplement No. 1
Chemicals and industrial processes associated with cancer in humans (IARC Monographs, Volumes 1 to 20) (1979)
71 pp.; out of print

Supplement No. 2
Long-term and short-term screening assays for carcinogens: a critical appraisal (1980)
426 pp.

Supplement No. 3
Cross index of synonyms and trade names in Volumes 1 to 26 (1982)
199 pp.

Supplement No. 4
Chemicals, industrial processes and industries associated with cancer in humans (IARC Monographs, Volumes 1 to 29) (1982)
292 pp.

Supplement No. 5
Cross Index of Synonyms and trade names in Volumes 1 to 36 (1985)
259 pp.

INFORMATION BULLETINS ON THE SURVEY OF CHEMICALS BEING TESTED FOR CARCINOGENICITY

(Available from IARC)

No. 8 (1979)
Edited by M.-J. Ghess, H. Bartsch
& L. Tomatis
604 pp.

No. 9 (1981)
Edited by M.-J. Ghess, J.D. Wilbourn,
H. Bartsch & L. Tomatis
294 pp.

No. 10 (1982)
Edited by M.-J. Ghess, J.D. Wilbourn &
H. Bartsch
326 pp.

No. 11 (1984)
Edited by M.-J. Ghess, J.D. Wilbourn,
H. Vainio & H. Bartsch
336 pp.

No. 12 (1986)
Edited by M.-J. Ghess, J.D. Wilbourn,
A. Tossavainen & H. Vainio
389 pp.

THE LIBRARY
UNIVERSITY OF CALIFORNIA
San Francisco
(415) 476-2335